Niessner

Recycling of Plastics

Norbert Niessner

Recycling of Plastics

HANSER
Hanser Publishers, Munich

FSC
www.fsc.org
MIX
Paper from responsible sources
FSC® C083411

Distributed by:
Carl Hanser Verlag
Postfach 86 04 20, 81631 Munich, Germany
Fax: +49 (89) 98 48 09
www.hanserpublications.com
www.hanser-fachbuch.de

Editor: Mark Smith
Production Management: Der Buchmacher, Arthur Lenner, Windach
Cover concept: Marc Müller-Bremer, www.rebranding.de, Munich
Cover picture: © stock.adobe.com/SMAK_Photo and also © shutterstock.com/wittaya photo und CC7
Cover design: Max Kostopoulos
Typesetting: Eberl & Koesel Studio, Altusried-Krugzell, Germany
Printed and bound by CPI books GmbH, Leck
Printed in Germany

ISBN: 978-1-56990-856-3
E-Book ISBN: 978-1-56990-857-0
E-Pub ISBN: 978-1-56990-858-7

Preface

Plastics have offered and still offer an unprecedented step-up in quality of life. Their beneficial effects range from reduction of food waste by extension of shelf life, affordable healthcare measures through mass production of diagnostics or infusion kits, significant carbon dioxide savings through lightweight applications in automotive and aircraft, up to a wide range of electronic gadgets in lightweight yet stiff and tough protective housings. A replacement or even no-plastics strategy would not only affect lifestyle, but also increase the emission of greenhouse gases such as carbon dioxide significantly.

At the same time it is clear that environmental pollution by plastic parts must not continue as it has done until now. While recovery of precious metals – even from landfills – is already in place in many parts of the world, it is about time to treat plastic items at the end of their normal lifecycle not as waste anymore, but as a source of precious raw materials. Plastics are not the problem, but rather insufficient collection schemes and recycling initiatives.

This book intends to show the state of the art in plastics recycling, the value chains, the challenges, and the solutions. The following paragraphs give a very condensed overview of the contributions and their authors (those who (co)wrote more than one contribution are mentioned just once here – see the table of contents or "The Authors" for a more complete survey).

We introduce structure-property relationships in plastics with a contribution by leading polymer scientists from Bayreuth University (Prof. Hans-Werner Schmidt Tristan Kolb, and Andreas Schedl), will touch the economics and deep-dive into the important topics of legislation and political/legal boundaries, contributed by two major stakeholders in this field (Accenture with Michael Ulbrich, Marvin Stiefermann, and Sören Hörnicke, and BASF SE with Klaus Wittstock and Victoria Wessolowski). By this dual approach we emphasize the overarching importance of politics and legislation for the successful, global implementation of plastics recycling.

LCA contributions by BASF SE (Prof. Peter Saling), NMB (Thomas Neumeyer, Regino Weber), and TU Braunschweig (Prof. Stephan Scholl, Mandy Paschetag, and Hannes Schneider) complete the first part (Part A).

The recycling chapters will be introduced by Prof. Christian Hopmann, Prof. Rainer Dahlmann, and Martin Facklam (IKV Aachen), followed by a split into mechanical and advanced recycling (by chemistry-based measures), including contributions by Prof. Achim Schmiemann (IFR); Marco Amici (INEOS O&P); Prof. Thomas Schröder (ikd Darmstadt); Hermann van Roost (SCS Brussels); Eike Jahnke and Bianca Wilhelmus (INEOS Styrolution); Hannah Mangold and Caroline Beyer (BASF); Jason Leadbitter (INEOS INOVYN); Prof. Dieter Stapf (KIT); Prof. Thorsten Gerdes, Achim Schmidt-Rodenkirchen, Stephan Aschauer, and Klaus Hintzer (InVerTec); Adrian Griffiths, Stephanie Loo, and Alejandro Sánchez (Recycling Technologies); and Carsten Eichert and Vitalij Salikov (Rittec). A section on the dissolution technology is provided by the Fraunhofer Freising team around Andreas Mäurer and Martin Schlummer.

Contributions by two key representatives from the recycling and sorting industry – by Jürgen Ephan and Arne Köhne (REMONDIS) and Jürgen Priesters (TOMRA) – complement the view on the value chain, introduced by Raphael Kiesel.

We conclude our book with contributions on reuse examples, design for recycling, and future developments from Prof. Volker Altstädt, Prof. Holger Ruckdäschel, Tobias Standau, Matthias Mühlbacher, Josefa Nüßlein, Robin Fachtan (NMB), and Eric Homey.

Our book is written for the decision makers in companies, public institutions, and government/administration, for the student, as well as for the experts from academia and industry. We hope that all stakeholders will receive a comprehensive overview, allowing the identification of the best solution for each recycling challenge.

Norbert Niessner, Editor August 2022

Foreword

Recycling of plastics is a very important topic for society, policymakers, and industry.

Plastic is a material we use every day. It helps feed the world sustainably, keeps us warm, allows us to get from A to B, and even saves lives. We cannot achieve our climate ambitions without plastics.

But we also know that plastics, plastics production and plastics waste management must be sustainable.

To transition to the EU's net zero emissions goals by 2050 requires the European plastics system to realize drastic emissions reductions across the entirety of the value chain, from raw materials to end-of-life. The stakes are high as we know societal sustainability goals, triggered by new EU legislation and consumer/citizen demand, are not just confined to Europe but impact global trade systems too.

Circularity is absolutely fundamental to our industry's transition. It is the most important medium-term lever and recycling is an essential component of circularity. "Reshaping Plastics", an independent report commissioned by Plastics Europe in 2022, concluded that by 2050 the plastics system could achieve 78% circularity with 48% being recycled.

Plastics Europe and our members recognize systemic change is essential to achieve this and needs to be accelerated. European plastics manufacturers are already undertaking huge investments and a far-reaching reorganization of their production and technology base to increase circularity and support recycling.

But we are under no illusions about the scale and complexity of this transition. The European plastics system is extremely diverse and complex. We are talking about multiple supply chains, delivering thousands of products and a plethora of industrial and consumer applications. There is no silver bullet solution.

Whilst plastics manufacturers have a very important role to play in supporting the transition to net zero, we represent only one part of the European plastics system. We do not have all the answers or levers to the recycling challenges we and the plastics system face and cannot do it alone.

Policymakers also have a vital role to play. Accelerating the transition requires an enabling policy framework that incentivizes and fosters a climate of creative collaboration and competition. One that helps to maximize investment and innovation from both private and public actors.

Deeper collaboration between the plastics industry, value chain, policy makers, increased investment and innovation – all of this is vital and it has to be based on a deeper and science-based understanding of the challenges and opportunities that we face in relation to recycling. This is why I welcome this book.

Packed with vital insights and detailed analysis, "Recycling of Plastics" is an invaluable resource for all of us as we rise to the collective challenge and opportunity of creating a circular plastics economy.

Happy reading!

Virginia Janssens

Plastics Europe – Managing Director

Foreword 2

Plastics are ubiquitous in our everyday lives. Economically, they are one of the most prolific materials, with above average global growth rates surpassing other key materials such as paper, glass, or aluminum.

This is not least due to the remarkable properties of plastics, which bring many benefits to the individual consumer as well as society as a whole. For example, plastics are used in packaging to protect goods, also helping to reduce food waste, and to build lighter and more fuel-efficient vehicles, helping to reduce greenhouse gas emissions.

However, more recently, the focus of the public perception of plastics has shifted towards their end of life. Globally, only a small portion of plastic waste is collected, sorted, and recycled, while too much of it is still incinerated, landfilled, or, even worse, leaked into the environment. Policymakers, NGOs, plastic users and end consumers, as well as investors increasingly demand that the plastics industry accepts its share of responsibility and takes concrete action against plastic pollution and greenhouse gas emissions by transitioning from a linear to a circular economy.

Indeed, circularity is an integral aspect of sustainability. Implementing "reduce, reuse, and recycle" (3R) is one key approach to addressing the urgent need to mitigate climate change, in line with the Paris Agreement at a global level, as well as, more specifically, the European Green Deal and the Circular Economy Action Plan (CEAP) at the European Union (EU) level. Our common goal must be a circular, net-zero carbon plastics economy. In the end, further increasing collection, sorting, and recycling of plastic waste will be instrumental to achieve this, in addition to reuse models and renewable materials. This includes standardizing and designing plastic products to increase their ability to be recycled and incorporate recycled content, with closed-loop systems as the ultimate goal, where they are the most sustainable solution.

Also in support of fact-based policy and other decision making, industry will continue to contribute to meeting expectations regarding transparency. This includes aspects such as data collection, traceability, monitoring, and accountability, as well as providing comprehensive scientific evidence on the life cycle impacts of its products.

The global transition to a sustainable, circular economy cannot be achieved with isolated solutions. Within an enabling policy framework, cooperation along the value chains is key to achieve these objectives. In the EU, the Circular Plastics Alliance (CPA) and polymer value chain associations such as PETCORE (PET), PCEP (PE, PP), Vinylplus (PVC), and SCS (Styrenics) are visible examples of such cooperation.

This book provides a strong fact base for all stakeholders involved or interested in the circular economy for plastic. To help to better understand feasibilities, limitations, and opportunities, it lays out the potential of existing mid- and long-term solutions, enabling conditions, as well as trends and future developments. This includes standardization and the huge but not yet fully tapped potential of design for recycling.

Against the backdrop of the above-mentioned as well as more recent developments, the release of this book is timely, as it presents a technology roadmap and possible choices regarding circular feedstocks, which will also help to meet the urgent strategic needs to save fossil resources such as crude oil and natural gas.

I hope that with its diverse content and contributing authors, this book may inspire all its readers to closely communicate and collaborate with each other. With commitment, concrete action, and solutions, together we can deliver an answer to the dual challenge of climate change and plastics pollution by realizing the circular economy for plastics.

Jens Kathmann

Secretary General, Styrenics Circular Solutions (SCS)

The Authors

The Editor

Norbert Niessner

Dr. Norbert Niessner is Innovation Director at INEOS Styrolution. He is a Chemistry graduate from Marburg University, Germany, at which he also obtained his PhD in Polymer Chemistry. From 1989 to 1990, he worked as a post-doctoral fellow at IBM Almaden Research Labs in San Jose, California. After joining BASF in 1990, he held several positions in R&D and marketing & sales in Europe, as well as in Asia. In 2011, he joined the newly created BASF/INEOS joint venture "Styrolution", now a 100% INEOS company.

During his career, Dr. Niessner gained experience in building and leading global R&D and business management teams in the polymers industry worldwide, with focus Europe and Asia. In terms of patents/applications, he is one of the leading inventors in styrenic polymers. His publication/patent list comprises more than 300 tech/scientific papers, book chapters, and original inventions. Special fields of interest are, amongst others, liquid crystalline polymers, development of world-scale polymer manufacturing technologies and product lines, polymers for 3D printing, and advanced recycling technologies.

The Coauthors

Chapter/Section	Title	Author*
2	Introduction to Major Thermoplastic Polymers in the Context of Recycling	▪ Dr. Tristan Kolb, Neue Materialien Bayreuth GmbH, Bayreuth ▪ Andreas Schedl, Neue Materialien Bayreuth GmbH, Bayreuth ▪ Prof. Hans-Werner Schmidt, University of Bayreuth
3	Legislation	▪ Michael Ulbrich ▪ Sören Hörnicke Accenture GmbH, Kronberg im Taunus
4	Economics	▪ Michael Ulbrich ▪ Dr. Marvin Stiefermann Accenture GmbH, Kronberg im Taunus
5	Political Boundary Conditions – Plastics Waste Management	▪ Dr. Klaus Wittstock ▪ Victoria Wessolowski BASF SE, Ludwigshafen am Rhein
6.1	Introduction: What Is Life Cycle Assessment?	▪ Prof. Dr. Peter Saling, BASF SE, Ludwigshafen am Rhein
6.2	Life Cycle Assessment for Different Recycling Routes of Polystyrene	▪ Regino Weber ▪ Dr.-Ing. Thomas Neumeyer Neue Materialien Bayreuth GmbH, Bayreuth
6.3	Life Cycle Assessment of a New Monomer Recycling Process for PET	▪ Dr.-Ing. Mandy Paschetag ▪ Hannes Schneider ▪ Prof. Dr.-Ing. Stephan Scholl Technische Universität Braunschweig
7	Overview of Major Recycling Technologies	▪ Prof. Dr.-Ing. Christian Hopmann ▪ Prof. Dr. rer. nat. Rainer Dahlmann ▪ Dr.-Ing Martin Facklam Institute for Plastics Processing (IKV) in Industry and Craft at RWTH Aachen University
8.1	Basic Principles	▪ Prof. Dr.-Ing. Achim Schmiemann, Ostfalia University of Applied Sciences, Wolfsburg
8.2	Mechanical Recycling of Polyolefins	▪ Dr. Marco Amici, INEOS Olefins and Polymers Europe, Rolle
8.3	Mechanical Recycling of PET	▪ Prof. Dr.-Ing. Thomas Schröder, Darmstadt University of Applied Sciences
8.4.1	Mechanical Recycling of Polystyrene	▪ Herman Van Roost, Styrenics Circular Solutions, Brussels
8.4.2	Mechanical Recycling of Styrene Copolymers	▪ Dr. Eike Jahnke, INEOS Styrolution Europe GmbH, Frankfurt am Main
8.5	Mechanical Recycling of Engineering Thermoplastics	▪ Dr. Hannah Mangold ▪ Dr. Caroline Beyer BASF SE, Ludwigshafen am Rhein

Chapter/Section	Title	Author*
8.6	Mechanical Recycling of PVC	▪ Dr. Jason Leadbitter, INEOS INOVYN, Newton Aycliffe
8.7	Compatibilization of Blends for Mechanical Recycling	▪ Dr. Norbert Nießner ▪ Dr. Bianca Wilhelmus INEOS Styrolution Group GmbH, Frankfurt am Main
9.1.1	An Overview of Major Recycling Technologies	▪ Dr. Bianca Wilhelmus, INEOS Styrolution Group GmbH, Frankfurt am Main
9.1.2	Petrochemical Feedstock Recycling	▪ Prof. Dr.-Ing. Dieter Stapf, Karlsruhe Institute of Technology
9.1.3	Plastic Recycling by Depolymerization	▪ Achim Schmidt-Rodenkirchen, InVerTec e.V., Bayreuth ▪ Dr. Stephan Aschauer, Recenso GmbH, Remscheid ▪ Prof. Dr.-Ing. Thorsten Gerdes, InVerTec e.V., Bayreuth ▪ Dr. Norbert Nießner, INEOS Styrolution Group GmbH, Frankfurt am Main
9.2	Chemical Recycling of Polyolefins	▪ Adrian Griffiths ▪ Dr. Alejandro Sánchez ▪ Stephanie Loo Recycling Technologies Ltd, Swindon
9.3	Chemical Recycling of PET	▪ Carsten Eichert, Rittec Umwelttechnik GmbH, Lüneburg ▪ Dr.-Ing. Vitalij Salikov, Rittec Umwelttechnik GmbH, Lüneburg ▪ Prof. Dr.-Ing. Stephan Scholl, Technische Universität Braunschweig
9.4	Chemical Recycling of Polystyrene	▪ Dr. Norbert Nießner, INEOS Styrolution Group GmbH, Frankfurt am Main ▪ Achim Schmidt-Rodenkirchen, InVerTec e.V., Bayreuth
9.5	Chemical Recycling of Engineering Thermoplastics	▪ Dr. Hannah Mangold ▪ Dr. Caroline Beyer BASF SE, Ludwigshafen am Rhein
9.6	Chemical Recycling of PTFE (as a Model for Other Polymers)	▪ Achim Schmidt-Rodenkirchen ▪ Dr. Klaus Hintzer ▪ Prof. Dr.-Ing. Thorsten Gerdes InVerTec e.V., Bayreuth
10	Dissolution	▪ Dr. Andreas Mäurer ▪ Dr. Martin Schlummer Fraunhofer Institute for Process Engineering and Packaging IVV, Freising
11	Thermochemical Processes	▪ Prof. Dr.-Ing. Dieter Stapf, Karlsruhe Institute of Technology

Chapter/Section	Title	Author*
12	Recycling of Composites	▪ Mathias Mühlbacher ▪ Prof. Dr.-Ing. Volker Altstädt Neue Materialien Bayreuth GmbH, Bayreuth
13	Comparison of Global Recycling Processes	▪ Raphael Kiesel, Laboratory for Machine Tools and Production Engineering (WZL) of RWTH Aachen
14	Value Chain EMEA – Collecting and Sorting	▪ Jürgen Ephan ▪ Arne Köhne REMONDIS Recycling GmbH & Co. KG, Lünen
15	Recycling Value Chain and Detection/Purification Technologies	▪ Jürgen Priesters, TOMRA Sorting GmbH, Mülheim-Kärlich
16	Reuse	▪ Josefa Nüßlein ▪ Dr.-Ing. Thomas Neumeyer Neue Materialien Bayreuth GmbH, Bayreuth
17	Design for Recycling	▪ Robin Fachtan ▪ Prof. Dr.-Ing. Volker Altstädt Neue Materialien Bayreuth GmbH, Bayreuth
18	Future Trends	▪ Prof. Dr.-Ing. Holger Ruckdäschel, Neue Materialien Bayreuth GmbH, Bayreuth ▪ Tobias Standau, University of Bayreuth
19	Future Developments	▪ Prof. Dr.-Ing. Achim Schmiemann ▪ Eric Homey Ostfalia University of Applied Sciences, Wolfsburg

* Order according to chapter/section number.

Glossary

AA	Acrylic acid
ABS	Acrylonitrile–butadiene–styrene copolymer
AN	Acrylonitrile
API	A measure of how heavy or light a petroleum liquid is compared to water: if API > 10, then the liquid is lighter and floats on water; if API < 10, it is heavier and sinks
APC	Aliphatic polycarbonates
ASR	Automotive shredder residue
ASTM	American Society for Testing and Materials; international standardization organization
BGL	British Gas Lurgi Process
BHET	Bis(2-hydroxyethyl) terephthalate
BHETA	Bis(2-hydroxyethylene) terephthalamide
BMBF	German Ministry of Education and Research
BMC	Bulk molding compound
BPA	Bisphenol-A (monomer for polycarbonate)
BTX	Benzene, toluene, xylenes fraction
CAPEX	Capital expenditure
CEAP	EU Circular Economy Action Plan
CEFIC	European Chemical Industry Council
CFC	Carbon fiber composites
CFF	Circular footprint formula
Citeo	Non-profit company founded from the merger of Eco-Emballages and Ecofolio
CO_2e	Carbon dioxide equivalents
CPA	Circular Plastics Alliance of EU Commission
CPE	Chlorinated Polyethylene
CPO	Circular Polyolefins Project
CSS	EU Chemicals Strategy for Sustainability
CW	Commercial waste / consumer waste[1]

[1] Which definition applies is given in the relevant chapter.

DEHP	Diethyl hexyl phthalate (plasticizer)
DfR	Design for recycling
DfS	Design for sustainability
DMT	Dimethyl terephthalate
DSC	Differential scanning calorimetry
DSD	Duales System Deutschland – German collection system
DSP	Selective dissolution and precipitation
ECHA	European Chemicals Agency
ED	Endocrine disruptor
EEA	Eco efficiency analysis
EfW	Energy from Waste
EG	Ethylene glycol
EGMA	Ethylene–glycidyl methacrylate
EHS	Environment, health, safety
EHS&S	Environment, health, safety, and sustainability
ElektroG	German Electrical and Electronic Equipment Act
ELV	End-of-life vehicle
EPA	U.S. Environmental Protection Agency
EPC	Engineering, procurement, construction
EPDM	Ethylene–propylene–diene terpolymer
EPM	Poly(ethylene-co-propylene)
EPR	Extended producer responsibility
EPS	Expanded polystyrene
ETICS	External Thermal Insulation Composite System
EVA	Ethylene–vinyl acetal copolymer
EVOH	Ethylene–vinyl alcohol copolymer
FBP	Final boiling point of a distillation fraction
FBR	Fluid-bed or fluidized-bed reactor
FDA	US Food and Drug Administration
FEP	Freshwater eutrophication potential
FFF	Fused filament fabrication, a 3D printing technology
GewAbfV	German industrial waste ordinance
GHG	Greenhouse gas
GMA	Glycidyl methacrylate
GWP	Global warming potential
GPPS	General purpose polystyrene
Green Dot	German dual system of waste collection ("Grüner Punkt")
HALS	Hindered amine light stabilizer (UV stabilization additives)
HBCDD	Hexabromocyclododecane (flame retardant)
HCFC	Hydrochlorofluorocarbon
HDPE	High-density polyethylene

HEMA	Hydroxyethyl methacrylate
HIPS	High-impact polystyrene
HSP	Hansen Solubility Parameter
HTW	High-temperature Winkler process
IBP	Initial boiling point for a distillation fraction
ISO	International Organization for Standardization
KrWG	Kreislaufwirtschaftsgesetz (German circular economy law)
LB	Liquid boards
LCA	Life cycle analysis
LCI	Life cycle inventory
LCIA	Life cycle impact analysis
LDPE	Low-density polyethylene
LLDPE	Linear low-density polyethylene
LWP	Lightweight packaging waste
LWP-SR	Lightweight packaging waste sorting residue
MA	Maleic anhydride
MERF	Material recovery facility or multi reuse facility
MFI	Melt flow index
MMA	Methyl methacrylate
MP	Mixed plastics
MRF	Multi reuse facility
MSW	Municipal solid waste
MSWI	Municipal solid waste incinerator
NIR	Near infrared technology (used for e.g. plastics sorting)
NGO	Non-governmental organization
ODP	Ozone depletion potential
OECD	Organization for Economic Co-operation and Development
OEFSR	Organization Environmental Footprint Sector Rules
OEM	Original equipment manufacturer
OPEX	Operational expenditure
PA	Polyamide
PAr	Polyarylate
PBAT	Poly(butylene adipate)-co-terephthalate
PBS	Polybutylene succinate
PBT	Persistent, bioaccumulative and toxic substances
PBT	Polybutylene terephthalate
PC	Polycarbonate
PCB	Paper and cardboard
PCB	Printed circuit boards
PCB	Polychlorinated biphenyl
PCR	Postconsumer recycling

PE	Polyethylene
PEF	Polyethylene furanoate
PEFCR	Product Environment Footprint Category Rules
PET	Polyethylene terephthalate
PHA	Polyhydroxyalkanoate
PHB	Polyhydroxybutyrate
PIC	Prior informed consent
PIR	Postindustrial recycling
PLA	Polylactic acid
PMMA	Poly(methyl methacrylate)
PO	Polyolefin
POM	Polyoxymethylene
POP	Persistent organic pollutant
PP	Polypropylene
PPE	Poly(2,6-dimethyl-1,4-phenylene ether)
PPO	Polyphenylene oxide, often synonymous with PPE
PPP	Polluter pays principle
PPWD	EU Packaging & Packaging Waste Directive
PRO	Producer responsibility organization
PS	Polystyrene
PVC	Polyvinyl chloride
PTFE	Polytetrafluoroethylene
PTT	Polytrimethylene terephthalate
PUR	Polyurethane
RDF	Refuse derived fuel (waste fraction)
R&D	Research and development
REACH	EU Registration, Evaluation, Authorization, Restriction of Chemicals
RED	Relative energy difference
RHW	Residual household waste
RPW	Residual plastics waste
RVM	Reversed vending machine
SAN	Styrene-acrylonitrile copolymer
SBC or SBS	Styrene-butadiene block copolymer
SCS	Styrenics Circular Economy
SDGs	UN Sustainable Development Goals
SEBS	Hydrogenated styrene-butadiene block copolymer
SF	Substitute fuels
SHVC	Substances of very high concern
SMA	Styrene-maleic anhydride copolymer
SMC	Sheet molding compound
SMMA	Styrene-methyl methacrylate copolymer

SPE	Society of Plastics Engineers
Sphera	Provider of Environmental, Social and Governance (ESG) performance and risk management software, with a focus on EHS&S
SPI	Sustainable Products Initiative
SR	Shredder residues
SRF	Solid recovered fuel
SSP	Solid state polycondensation
SVZ	Sekundärverwertungszentrum (secondary reutilization center)
TBS	Tracer based sorting
TEM	Transmission electron microscopy
TPA	Terephthalic acid
TPE	Thermoplastic elastomer
TRL	Technology readiness level
TÜV	Technischer Überwachungs-Verein (German Technical Inspection Association)
UVP	Unique value proposition
VerpackG	Verpackungsgesetz (German packaging act)
vPvB	Very persistent, very bioaccumulative substances
WEEE	Waste from electrical and electronic equipment
WFD	EU Waste Framework Directive
WTE	Waste-to-energy (incineration with energy recovery)
XPS	Extruded polystyrene
ZSVR	German Central Packaging Register

Contents

PART A

Fundamentals; Legislative, Economic, and Political Considerations; LCA

1 Introduction

Norbert Nießner

The plastics story is a success story. Already the alchemists dreamt of creating new matter having special, exceptional properties. The area of synthetic polymers started with a secret recipe, developed by businessman and alchemist Bartholomäus Schobinger and disclosed to Benedictine monk Wolfgang Seidel in 1530 [1]. Schobinger treated cheese several times with boiling water, while he decanted the milky part and collected the thick slurry. The slurry was separated and could subsequently be cast into any shape. The new material was a substitute for natural materials like cattle horn.

In the 19th century, breakthrough inventions paved the way towards plastics. Some highlights:

- 1823: Charles Mackintosh fabricated the first waterproof raincoat based on natural rubber.
- 1835: Eduard Simon extracted styrene from the *Styrax* tree; it would become the building block for polystyrene in the late 1920s.
- 1839: Charles Goodyear invented the "vulcanization" (chemical cross-linking) of natural rubber with sulfur and founded the basis for the tire industry.
- 1846: Christian Friedrich Schönbein developed nitrocellulose by treating celluloses with sulfuric and nitric acid. A weakly nitrated cellulose recipe became the basis for *collodion.*
- 1856: Alexander Parkes invented cellulose polymers.
- 1870: John Wesley Hyatt invented *Celluloid*, based on nitrocellulose. Billiard balls made of this material could explode after collision.
- 1872: Johann Friedrich Wilhelm Adolf von Baeyer produced the first phenol-formaldehyde resin.

The first half of the 20th century paved the way for synthetic polymers becoming "mass polymers":

- 1907: Leo Hendrik Baekeland optimized the process conditions of the phenol-formaldehyde condensation and created the first fully synthetic resin, *Bakelite.*
- 1911: Chemist Richard Escales created the German word *Kunststoffe* for synthetic matter.

- 1920: Hermann Staudinger founded the modern polymer science, being awarded with the first Nobel Prize for macromolecular chemistry in 1953.
- 1929: Hermann Mark received a patent on synthetic styrene, as basis for polystyrene (patented 1930).
- 1931: Otto Röhm developed synthetic glass from methyl methacrylate (Plexiglas®, first industrial production 1935).
- 1935: Wallace Hume Carothers invented polyamide-66 (nylon).
- 1936: First industrial polystyrene production at BASF, Germany.
- 1937: First polyethylene pilot plant at ICI, UK; Otto Bayer invented polyurethane; first industrial PVC production in Germany.
- 1947: DuPont developed the engineering plastic polyformaldehyde.
- 1949: Fritz Stastny invented expandable polystyrene (Styropor®).
- 1953: Karl Ziegler polymerized ethylene with his catalysts at ambient temperature and pressure.
- 1956: Giulio Natta polymerized propylene, based on Ziegler catalysts.

1.1 Industrial Aspects

Since then, synthetic polymers showed an unprecedented growth, as can be seen in Figure 1.1.

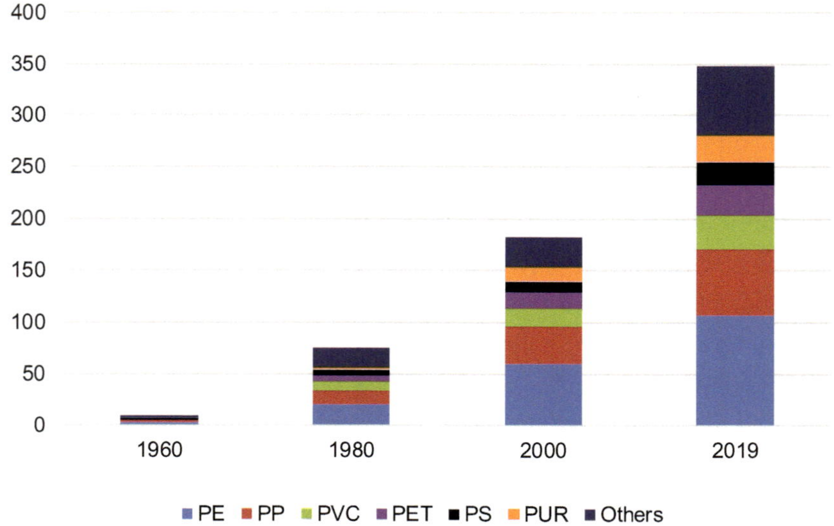

Figure 1.1 Global annual primary plastics production in million t/a [2–4]

The driving force for these lightweight, artificial "plastics" was not only the substitution of heavier, traditional materials like wood, glass, or metal. The main motivation was the myriad of application developments that allow today's modern lifestyle. Examples are given below, with discussion of the possibilities for recycling these products.

- Housings for household goods and electronic equipment: typically, several plastics types are combined in one application. As with PET bottle recycling it is possible to sort out large parts made of a single material (such as refrigeration in-liner, made of polystyrene; or vacuum cleaner housing made of ABS or polypropylene, as shown in Figure 1.2) and to selectively recycle such materials. Mechanical recycling - potentially in combination with an "impact booster" to compensate for loss in impact resistance during thermal treatment in recycling processes - is the choice here.

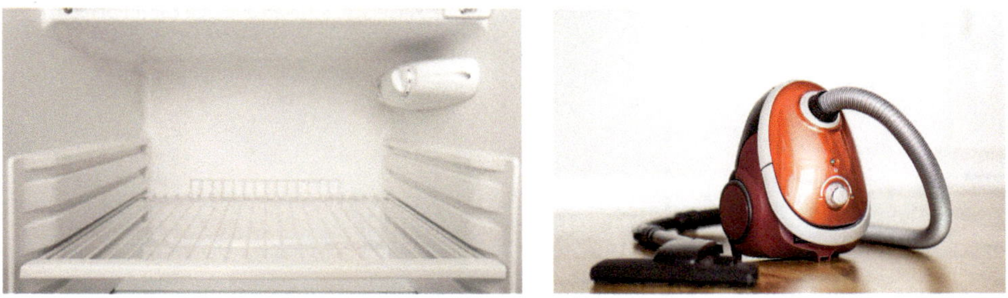

Figure 1.2 Refrigerator in-liner (left) and vacuum cleaner housing (right) as examples for plastics in household goods [3] (Picture source: INEOS Styrolution (left) and Obak/Shutterstock.com (right))

- Medical applications, from infusion sets to therapeutic devices: such products are often incinerated, for obvious reasons. However, a potential recycling pathway could be thermal depolymerization, which works especially well with polymers having a low ceiling temperature T_c - such as PMMA or polystyrene, see Figure 1.3. As many parts of infusion sets (drip chamber, clamp) are made of polystyrene or related materials, depolymerization could become a promising recycling pathway, once fully commercially established.

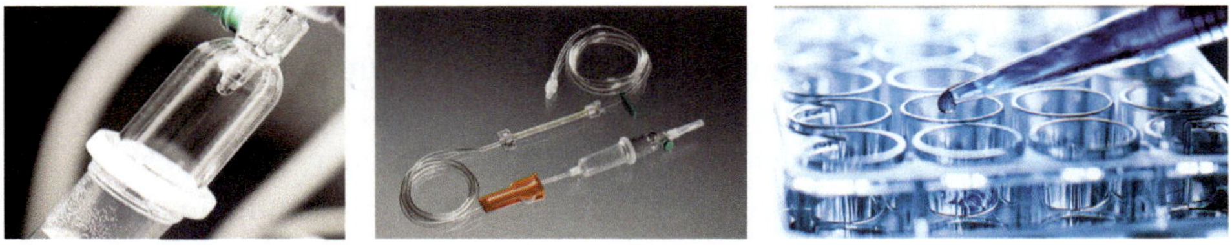

Figure 1.3 Medical applications: drip chamber from infusion sets (left), transparent housings and connectors (middle), and lab ware for diagnostics (right) [3] (Picture source: INEOS Styrolution (left and middle) and Shutterstock.com (148514897, right))

- High-strength fiber composites for lightweight aircraft and automotive applications, e.g. wind shields for trucks or carbon-reinforced automotive body panels, see Figure 1.4: While sorting and recycling of automotive parts has a long and successful history, products based on glass fiber or carbon fiber reinforced composites can in many cases only be downcycled. Especially in the case of non-melting, thermoset (duroplastic) matrix polymers, the parts are crushed, and long fibers are chopped into short fibers and hence lose their value in mechanical reinforcement. The emerging "organosheets" (thermoplastic based fiber composites) might have an advantage here, because at elevated temperatures, the polymer matrix melts, allowing for a separation of matrix and fibers. For thermoset parts, solvolysis pathways are under development, cleaving the otherwise unprocessable matrix back into its monomers, which can be recovered, purified, and reused.

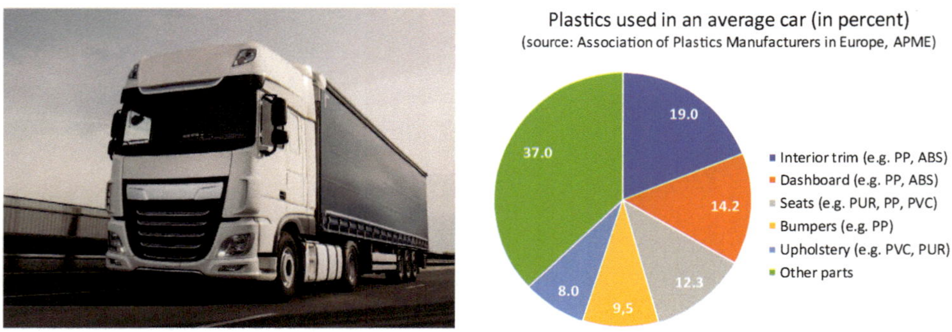

Figure 1.4 Truck plastics parts are partly made from thermoplastic polymers and partly from thermoset (often fiber reinforced) plastics to guarantee high strength at lowest possible weight. Approx. 7% of the global plastics consumption is used for the automotive and two-wheeler industry. A large portion are exterior trim applications [3] (Picture source: BoJack/Shutterstock.com (left))

- Synthetic fibers, securing supply of textiles way beyond the availabilities of natural fibers
- Synthetic rubber and artificial leather
- Toy building bricks
- Lightweight, transparent, and safe plastic bottles: PET bottles are a great example of a commercially and ecologically feasible substitution of traditional materials (here: glass, see Figure 1.5). Due to its excellent combination of impact resistance, thermal and mechanical stability, transparency, low weight, and organoleptically neutral behavior, PET is the choice for packaging mineral water, soft drinks, etc., safely and economically, while providing several established as well as new ways for recycling. The classical recycling path comprises the introduction of cleaned, shredded bottle grade flakes into new PET, prior to manufac-

turing of the bottles in a stretch-blow process. The fact that PET – as an important member of the class of polycondensates – is prone to hydrolytic degradation is addressed by subjecting post-consumer PET bottle granules to a post-condensation process to increase the molecular weight. Finally, polycondensates like PET can be cleaved into their building blocks and hence be subjected to depolymerization recycling.

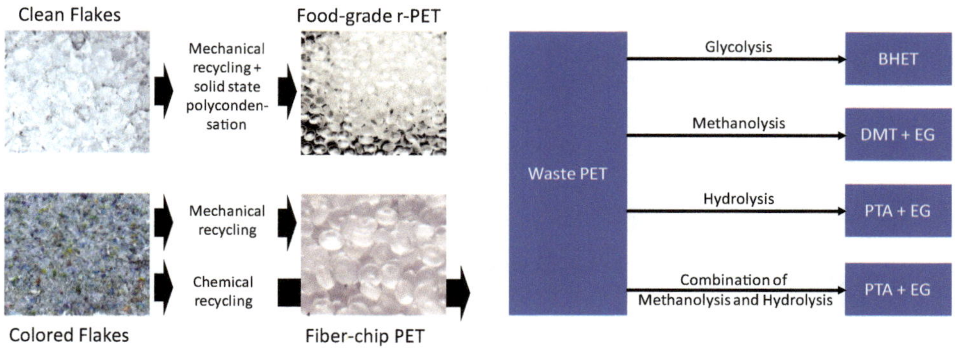

Figure 1.5 Schematic representation of PET bottle recycling: mechanical recycling (left), depolymerization (solvolysis) (right) [3] (Picture source: Nutthpol Kandaj/StanislauV/RecycleMan/Afanasiev Andrii/Shutterstock.com (left) and INEOS (right))

- Multilayer packaging: transparent and tough, barrier against oxygen and water, thermoformable, and lightweight. Multilayer packaging allows safe storage of food (see Figure 1.6), and represents the best possible combination of weight and packaging efficiency (shelf life). However, it presents a recycling challenge: a delamination of layers, followed by mechanical recycling by product class is not possible. However, feedstock recycling (thermal decomposition and feed-back of small molecules into existing chemistry infrastructure) is an option.

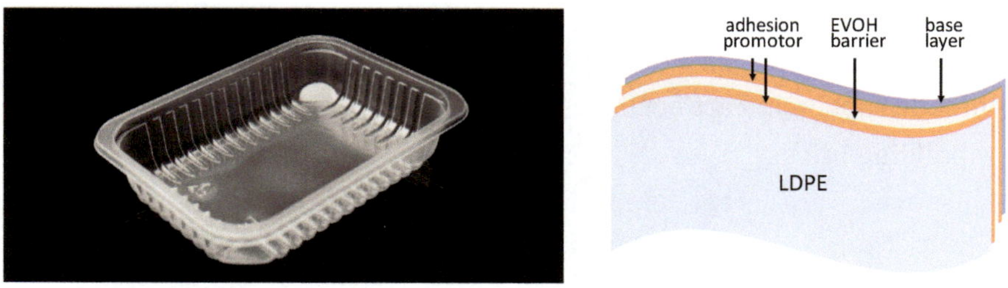

Figure 1.6 Multilayer packaging: transparent and tough, barrier against oxygen and water, thermoformable, and lightweight. Allows safe storage of food [5] (Picture source: nikkytok/Shutterstock-com (left) and INEOS Styrolution (right))

In line with those application developments went the development of new and/or improved processing technologies, such as the following:

- Blown film and sheet extrusion (Figure 1.7)

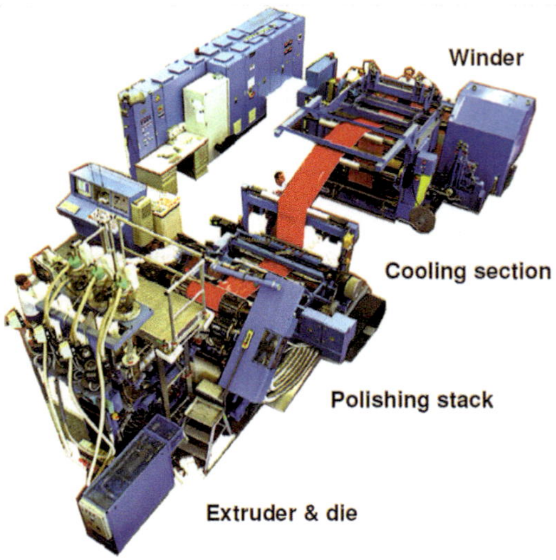

Figure 1.7
Sheet extrusion of polymers [3] (courtesy of Neue Materialien Bayreuth GmbH)

- Blow molding of hollow plastic parts
- Fiber composite manufacturing for large, lightweight automotive parts
- Advanced injection molding technologies, allowing addition of gases and co-injection of different thermoplastics (Figure 1.8)

Figure 1.8 Advanced injection molding machine, allowing addition of gases and co-injection of different thermoplastics (courtesy of Neue Materialien Bayreuth GmbH)

Hence, the development of new plastics created a flourishing industry continuously striving for improvement, efficiency, and high-quality in the processing of plastics, which in itself spurred the development of performance-optimized materials.

To round up the plastics story, we have to take into consideration that even modern electronics would not have been possible without polymer-based photolithography, and protection of ships, bridges, and other steel construction would not be possible without high-performance coatings, based on polymer formulations. The construction of our tallest buildings worldwide would not be possible without polymer-based additives allowing high flow and pumping of concrete over several hundreds of meters in height.

Taking all this together, polymers and plastics are those materials that determine more than anything else our modern lifestyle with increasing health, wealth, and wellbeing even for the poor in this world. *"Plastics have made it possible for us to push the limits and go further, faster and safer than we have dared to go before"* [6].

1.2 Plastics Pollution

Parallel to the unique success story of plastics, focus was laid largely on "linear economics". But since the 1970s, a major revolution, thinking in terms of "circular economy" and reuse of plastics, can be observed. Unfortunately, globally still only a limited amount of end-of-life plastic products are recovered and recycled or used for thermal recovery. The remaining part of the plastic waste ends up in landfills, where it can be carried away by winds, or even in rivers or the ocean. Together with carelessly discharged plastic (e.g., packaging and construction materials), plastic waste enters the environment from year to year. Since plastic degrades slowly, its accumulation in marine, freshwater, and terrestrial ecosystems is of increasing scientific and public concern.

Figure 1.9 Accumulation of (mainly) plastic garbage on the ocean (Picture source: Iurii Stepanov/Shutterstock.com)

Today, between 4 and 12 Mt/a of plastics enters the ocean (Figure 1.9 and Figure 1.10).

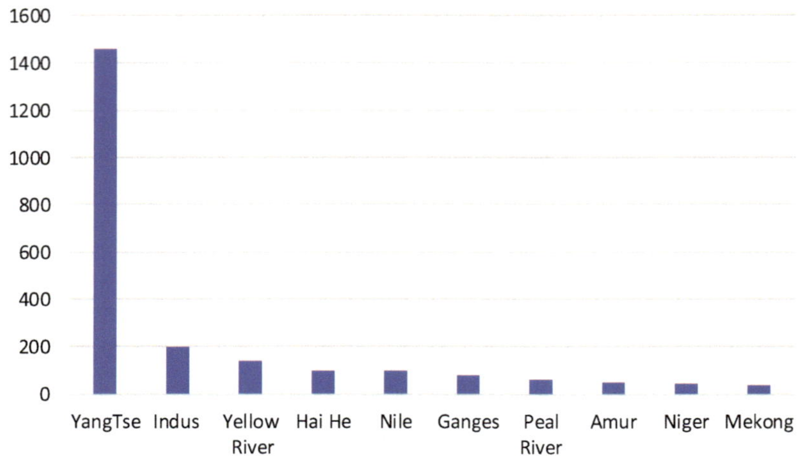

Figure 1.10 Amount of plastics waste (in kt/a) entering the ocean from major rivers [7]

The global plastic pollution and ubiquitous distribution came along with the observation that plastic in the environment does in general not tend to vanish by UV or other types of degradation. Rather than degrading, plastic mostly disintegrates into smaller particles.

More specifically, a combination of mechanical abrasion, UV radiation, and (micro) biological degradation in the environment causes the formation of tiny plastic fragments/particles, which are termed "secondary microplastic". Fibers from synthetic clothes, which are released by washing and enter aquatic ecosystems via sewage treatment plants, can also be assigned to "secondary microplastic". Apart from these, the so-called "primary microplastic" is designed and produced on purpose: such microparticles are broadly applied, for example in industrial cleaners and personal care products (toothpaste, facial and body scrubs, etc.). Virgin plastic pellets, which serve as precursor for manufactured plastic products, are also classed as "primary microplastic" [8]. Such particles can accumulate in organisms and hence become part of the food chain.

As a consequence, do we need to phase out and avoid plastics in future? And what are the materials that (re)substitute plastics? Are bio-plastics the solution to the waste problem? Prof. Gerrit Luinstra, chair of technical polymers at Hamburg University, recently mentioned in an interview: *"Unfortunately it is often much easier to change materials, rather than human behavior. Plastics are basically great materials. We should treat them as if they were gold"* [9].

In a recent "TED Talk", Prof. Kim Ragaert explained it like this: *"Imagine, someone parked a car in the middle of a highway. Who is to blame: the car or the idiot who left the car there?"* [10].

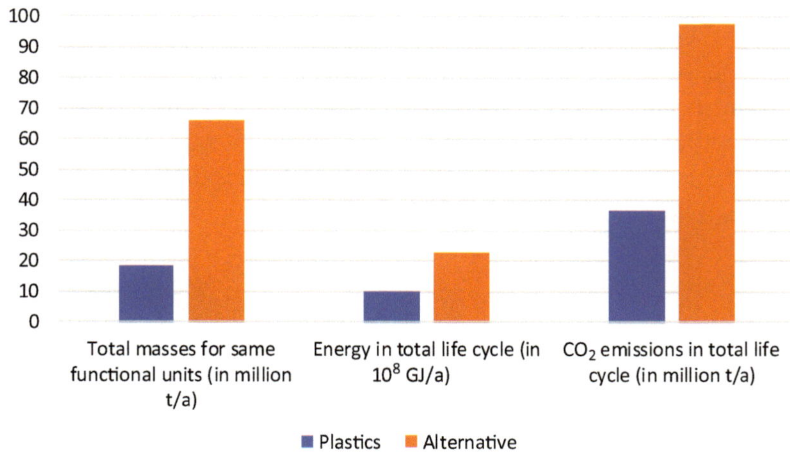

Figure 1.11 What is the environmental impact of switching parts of the plastic production to alternative materials [10]

In the same event, Prof. Ragaert (Maastricht University – Chair of Circular Plastics) outlined the consequence of re-substitution of plastics by traditional materials – knowing however that many of today's applications exist only because polymers as unique building blocks make them possible. The result is overwhelming (Figure 1.11): a multiple of energy will be required, and triple the CO_2 emissions would have to be accepted if we substitute all plastics with metal, glass, wood, etc. This scenario impressively shows what we would have to expect with a plastics ban. Therefore – and especially in the light of finding low CO_2 solutions for our planet – we need to make use of the energy and CO_2 savings potential offered by modern plastics. And – at the same time – we need to bundle all our forces to treat plastic parts at the end of their lives not as waste, but as precious resources for the next cycle. Only then we will master the challenge of reducing greenhouse gas emissions while not compromising our modern lifestyle.

1.3 Plastics Pollution: Bio Solution?

At this stage we briefly want to enter the discussion on bio-based or bio-degradable plastics. While in some people's minds "bio-plastics" will overcome the major challenges created by plastics, others believe that their CO_2 footprint is not always as favorable as it seems at first glance, and, except for the routes to produce such plastics from second- and third-generation feedstock [11], there might as well be competition with food in growing plants as the source for such plastics. Figure 1.12 and Figure 1.13 show two important types of bioplastics and outline their formation, properties, and major applications.

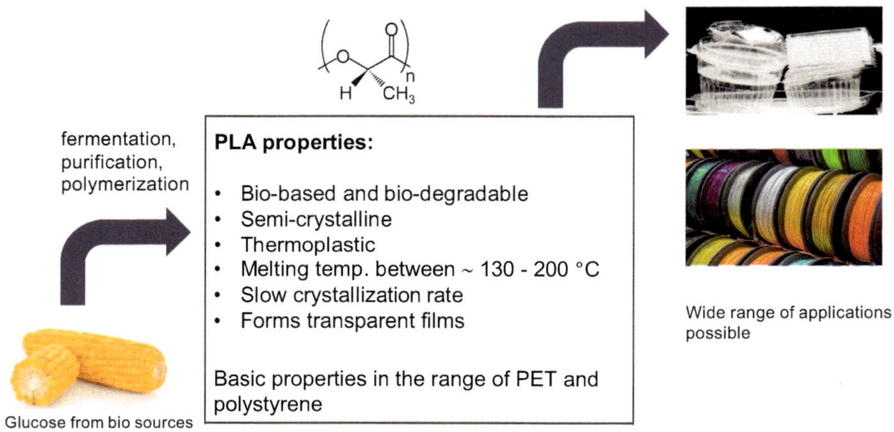

Figure 1.12 Formation, properties, and applications of PLA (polylactic acid) [3] (Picture source: rakasina/Shutterstock.com (left), Konektus Photo/Shutterstock.com (top right), MarinaGrigorivna/Shutterstock.com (mid right))

Bio-degradable plastics are a special case and qualify for those applications with a more or less controlled shelf life, temperature, and/or residence time in the environment. Such bio-degradable plastics typically require a controlled, humid, and oxygen-enriched environment at elevated temperatures over several days to degrade [12]. These conditions, however, can often not be realized in today's day-to-day infrastructure. For this reason, although bio-degradable plastics are an important part of today's plastic product portfolio, they cannot be regarded as the universal solution against plastic pollution.

Degradation of conventional plastics (those with only C–C bonds in the main chain) has been reported [13]. This field of research is comparably young, and gives hope for new solutions to today's waste problems. However, a holistic insight into all major degradation products, their accumulation in the environment, and the effect, for example, on microorganisms is still pending.

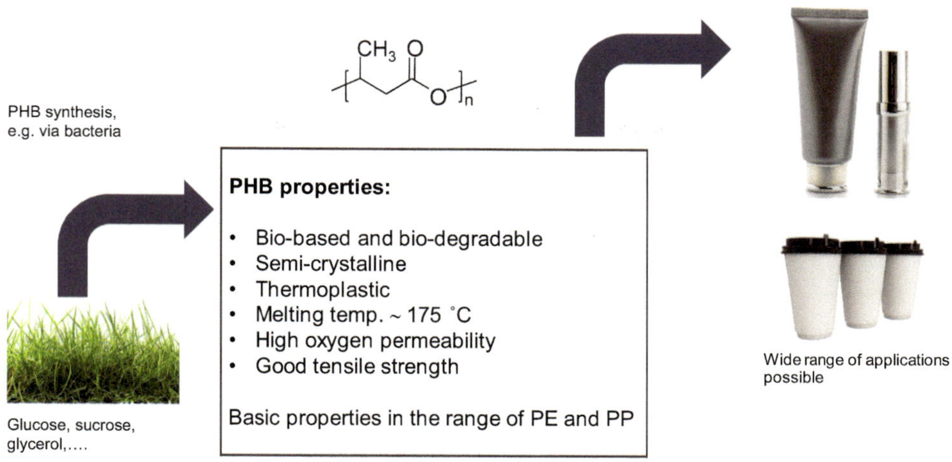

Figure 1.13 Formation, properties, and applications of PHB (polyhydroxybutyrate) [3] (Picture source: Pornsawan Baipakdee/Shutterstock.com (left), Lifestyle Travel Photo/Shutterstock.com (top right), CG_dmitriy/Shutterstock.com (mid right))

Another strong trend is the use the existing petrochemical infrastructure with refineries and steam crackers to produce monomers. By the example of "styrenic polymers", the process is explained as follows (Figure 1.14):

- Everything starts with the biomass source. Food crops and/or products from food crops like vegetable oil are obvious sources. However, the cultivation of such crops for biopolymer production competes with food production. Therefore research groups are developing concepts to start with non-food bio sources (second generation) and even crops from specially engineered plants, like algae reactors (third generation).
- Classical monomers like ethylene, propylene, styrene, acrylonitrile, or butadiene can be produced by using plant biomass as starting material and eliminating oxygen during chemical conversion to the desired monomers. Such multi-step conversions often consume high amounts of energy and/or result in rather moderate yields.
- Alternatively, bio-based monomers can be produced using existing chemistry infrastructure, requiring to convert bio-mass (e.g., palm oil) into hydrocarbons by chemical treatment (e.g., hydrogenation).
- Due to the conversion of carbohydrates (general formula $C_nH_{2n}O_n$) to mostly hydrocarbons, a large weight-% of the bio-based mass is lost.
- Such processes are currently typically available in a pilot scale or small commercial scale only.

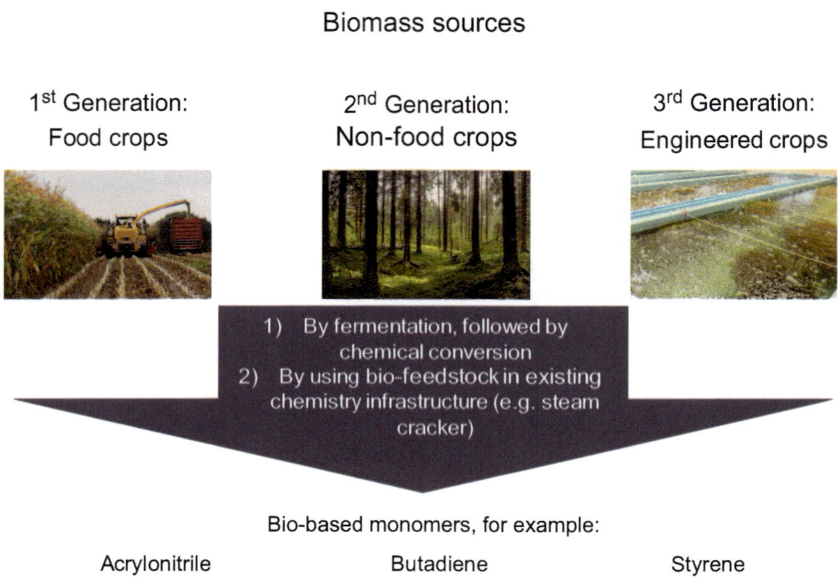

Figure 1.14 Current status of bio-based monomers for the example of ABS (acrylonitrile-butadiene-styrene) [3] (Picture source: Claudia Harms-Warlies/Shutterstock.com (left), Lillac/Shutterstock.com (mid), Pataporn Kuanui/Shutterstock.com (right))

As bio-based polymers are made of fresh biomass, common understanding is that such polymers show the same high ^{14}C isotope content as the biomass source. Hence, some polymer producers demonstrate the bio content of their products via ^{14}C measurement. For bio-polymers produced "on purpose", such as polylactic acid (PLA), this concept works. However for the sake of fast implementation of biomass-based feedstock in existing commodity plastics, it is necessary to co-feed such bio-based feedstock, together with fossil based feedstock, using the large, existing chemical infrastructure. As the supply of bio-based feedstock is still not sufficient to cover the demand of traditional polymers, it is economically not yet always feasible to run such large scale polymerization reactions in existing plants based on 100% bio-feedstock.

A probably more feasible and smarter alternative to the direct polymerization of bio monomers is the "mass balance concept" (Figure 1.15): Here, the large industrial equipment is used for making "bio-attributed" polymers. Both the fossil-based feedstock and the bio-based feedstock (usually available in smaller amounts) are fed into a steam cracker, producing naphtha, which is subsequently converted into monomers. By a certified process, the bio-molecules entering the steam cracker are – by an accounting process – attributed to an equivalent amount of downstream monomers, and later on to the respective polymers. It is not guaranteed that you can find those bio-molecules in your polymer by ^{14}C isotope counting. But it is certified that an equivalent amount of bio-based monomer/polymer has been produced.

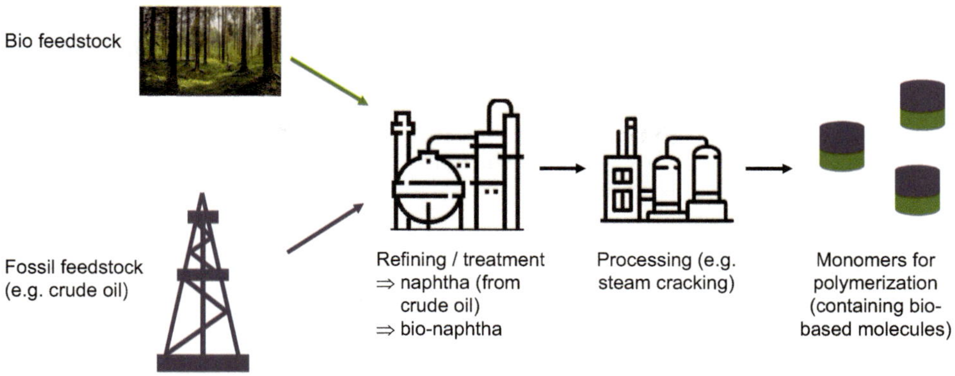

Figure 1.15 Schematic description of bio-based monomers for existing polymers: defined and controlled content of bio-based product in final product [3] (Picture source: Lillac/Shutterstock.com (top left))

The benefits of such a process are:

- Use of existing infrastructure
- No "green-washing": the amount of monomer/polymer that has been produced based on bio-feedstock input to the steam cracker is certified.
- Especially in these times of establishing a circular economy and moving away from a linear economy, the availability of adequate bio-feedstock does not yet reach the availability of crude oil based feedstock. Nevertheless, we can in principle use any bio-based feedstock quantity to produce bio-attributed polymer in the corresponding quantity.

A schematic depiction of the process is given in Figure 1.16.

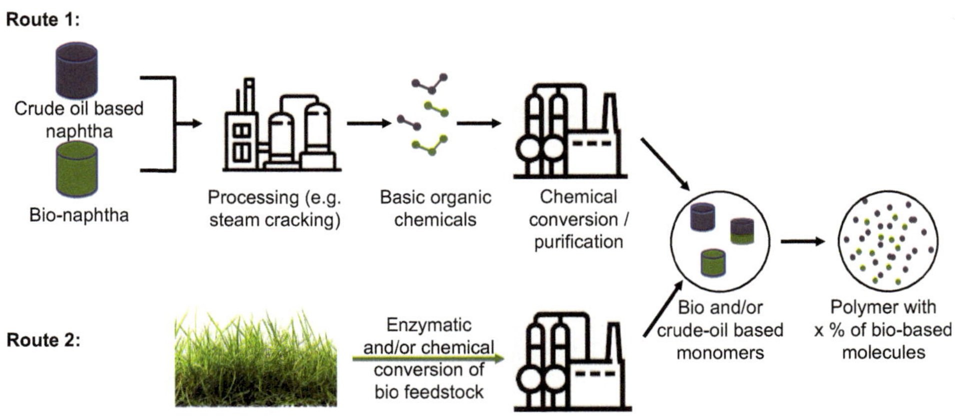

Figure 1.16 Schematic description of bio-mass balance: fluctuating and uncertain content of bio-based monomers in final product; therefore, allocation of bio-molecules per certificate [3] (Picture source: Pornsawan Baipakdee/Shutterstock.com (bottom left))

Bringing all aspects together, we have a set of measures that can help us to reduce the environmental impact of plastics. However, along with strengths and opportunities, every measure also shows certain weaknesses and threats, as outlined in Table 1.1 to Table 1.5 (distinguishing between "weaknesses" and "threats" is difficult, and therefore both are placed in the same column).

Table 1.1 Direct Substitution of Fossil-Based Raw Materials by Bio-Based Building Blocks, Examples: Butadiene from Bioethanol, Acrylonitrile by Fermentation of Biomass (via Glycerol)

Strengths	Weaknesses/Threats	Opportunities
▪ Drop-in solution, same properties of the resulting polymers ▪ No need to change polymerization technology	▪ Often many reaction steps, sometimes low yield ▪ Processes mostly still under development/ at lab or pilot scale ▪ High cost compared to established fossil based solutions ▪ It is questionable whether it is economically and ecologically meaningful to eliminate approx. 60–70 wt% of a bio-based molecule in order to convert it into a "classical" monomer with C=C bond	▪ Independence of fossil feedstock ▪ "Green" source

Table 1.2 Bio Attribution: Bio-Based Feedstock Is Used in an Upstream Process Part (e. g., in the Production of Naphtha), Resulting in Monomers and Polymers Containing a Defined Amount of Bio-Based Carbon. The Mass-Balance of this Procedure is Certified by Independent Institutes

Strengths	Weaknesses/Threats	Opportunities
▪ Drop-in solution, same properties of the resulting polymers ▪ No need to change polymerization technology	▪ Might require special cracker set-up ▪ Requires bio-feedstock treatment to meet cracker-feed specifications" ▪ Possibly feedstock competition to food or energy crops	▪ Independence of fossil feedstock ▪ "Green" solution

Table 1.3 Dedicated Bio-Based Polymers

Polymer	Strengths	Weaknesses/ Threats	Opportunities
PHA (polyhydroxyalkanoate)	▪ Broad feedstock base ▪ Low oxygen and water permeability	▪ High investment for commercial plants	▪ Biodegradable under controlled conditions
PHB (polyhydroxybutyrate)	▪ Feedstock can be biomass waste ▪ Good mechanical & thermal properties	▪ High investment for commercial plants	▪ Biodegradable under controlled conditions

Polymer	Strengths	Weaknesses/ Threats	Opportunities
PLA (polylactic acid)	▪ Already established	▪ High investment for commercial plants ▪ Partly high injection molding cycle times	▪ Biodegradable under controlled conditions
PBAT (poly(butylene adipate)-co-terephthalate)	▪ High toughness		▪ Biodegradable under controlled conditions
PEF (polyethylene furanoate)	▪ Good barrier, thermal & mechanical properties ▪ Made by chemical reaction, not microbiotic	▪ High investment for commercial plants, multi-step process	▪ Partly improved properties compared to PET
PET (polyethylene terephthalate)	▪ Ethylene glycol can be made from bio sources	▪ TPA unit based on bio sources still challenging	▪ "Drop-in" for existing PET applications
APC (aliphatic polycarbonates)	▪ Very good optical and mechanical properties	▪ Currently only partially bio-based	▪ Already used in automotive applications
PTT (polytrimethylene terephthalate)	▪ Better tensile strength, flexural strength, and stiffness than PBT	▪ TPA unit based on bio sources still challenging	
PA	▪ Sustainable "bio" PA grades often based on castor or palm oil	▪ Multi-step process	▪ Drop-in solution for many applications
PBS (polybutylene succinate)	▪ Mechanical properties between PE and PP and can be improved by blending	▪ High investment for commercial plants	▪ Biodegradable under controlled conditions

Table 1.4 Carbon Capture and Utilization: CO_2 is Collected at Stationary Sources (such as Power Plants, Municipal Waste Incineration or as By-Product in Chemical Reactions) and/or Converted into Platform Chemicals

Strengths	Weaknesses/Threats	Opportunities
▪ Removal of CO_2, positive effect on global warming	▪ High energy demand; energy should come from renewable sources ▪ Chemical conversion: "green" hydrogen required, produced with renewable energy ▪ Processes and technology still under development ▪ Uncontrolled release of CO_2 needs to be strictly avoided	▪ No competition to food; uses a "waste" gas ▪ Positive examples exist: e.g. "Carbon4PUR", Covestro ▪ Positive examples: lab production of industrially relevant intermediates via electrochemistry of CO_2

Table 1.5 Lignin-to-Chemicals: Conversion of the Phenolic Natural Polymer Lignin, Which is the Second Most Unused Biomass Waste, into Aromatic Chemicals

Strengths	Weaknesses/Threats	Opportunities
▪ Sustainable solution (reducing GHG emissions) ▪ Approx. 9 million t lignin waste from industry available per year	▪ Pretreatment necessary ▪ Pyrolysis can lead to high coke/char amount ▪ Low yield of aromatic oil compared to other oils ▪ Contamination from salts, carbohydrates, particulates, volatiles possible	▪ Practically unlimited bio resources

The mentioned options to integrate bio-based content or CO_2 into plastics are versatile and quite different in their characteristics. This makes it difficult to compare the specific energy consumption, CO_2 savings, and other environmental impacts for all options.

A generally accepted tool to compare environmental impacts is the "life cycle analysis" ("LCA", see also Chapter 6 of this book). While LCA comparison of related processes (e.g., various recycling methods) are already complex enough, we would therefore suggest to assess the options to integrate bio-based content or CO_2 qualitatively:

- Direct substitution of fossil-based raw materials by bio-based building blocks (Table 1.1) seems to be the easiest way to save CO_2 emissions. However, the direct routes often require a multistep synthesis and yield/energy requirements counterbalance the positive effect of CO_2 reduction. Examples for bio-based routes to formerly fossil-based raw materials:
  - *Ethylene* can be obtained by fermentation of e.g. sugarcane and separation of the resulting bio ethanol, followed by dehydration [14].
  - *Terephthalic acid* (used as comonomer in PET) can be produced e.g. from the biomass-derived furfural. Furfural can be oxidized and dehydrated into maleic anhydride, which reacts in a Diels-Alder reaction with furan. The dehydration of that Diels-Alder (DA) adduct results in the formation of terephthalic acid [15].
  - *Butadiene* can be produced by dehydrogenation of ethanol to acetaldehyde, followed by aldol condensation and dehydration in a one-pot process over a MgO-SiO_2 catalyst (Lebedew process) with up to 70% yield [16]. A different pathway leads from ethanol over butanediol to butadiene via dehydration [17].
  - *Styrene* can be obtained via pyrolysis of biomass to a mixture of benzene, toluene, and xylene (BTX) and subsequent reaction of benzene with ethylene to form styrene [17]. One biotechnological route is the deamination of L-phenylalanine to trans-cinnamic acid by Phenylalanine Ammonia Lyase (PAL), an enzyme present in yeast *Rhodotorula glutinis* and in *E. coli* [18].

- *Acrylonitrile* (e.g., used to produce SAN and ABS) can be produced via several routes, for example: 1) by oxidative decarboxylation of glutamic acid (chemical or biotechnological route), followed by a decarbonylation-elimination step, 2) by oxidation of a mixture of glycerol, ammonia, and air, and further conversion into acrylonitrile in a multistep reaction [19].

Except the ethylene formation from bio-ethanol, most of these reactions have not yet reached fully commercial (world) scale, and hence bio-based "classical" monomers are still under broad investigation.

- The use of bio-naphtha (Table 1.2) in a bio-attribution process seems to be a most straightforward method to save CO_2 emissions. Such a process is mainly based on the existing (modified) petrochemical infrastructure. Together with best suitable recycling methods, this combination of measures would be an efficient start of a true circular economy. Note: today's petrochemical infrastructure is so large that it cannot be fully operated based on bio-feedstock. Hence the products (e.g., monomers for plastics) would be based on a mix of fossil and bio-based feedstock, and such bio-based molecules can then only be allocated to the corresponding amount of products in a certified process [20].
- The use of bio-based plastics as substitute for existing plastics is an alternative that is most easily understood by consumers as a sustainability measure (Table 1.3). The strength of this route is the direct allocation of bio-molecules to plastics [21] and the possibility to measure the bio-based content via the radio carbon (^{14}C isotope) method. Nevertheless this approach often is in competition with food, at least as long as the feedstock materials are the commonly used crops, sugarcane, etc. It is a frequent misperception that bio-based plastics are always bio-degradable too. And if they are, the degradation process often requires quite optimal conditions to be completed within weeks or months, such as elevated temperature and air supply, available only at selected industrial facilities but not in the typical landfill site or even residential gardens. Last but not least, also bio-plastics should be subject to recycling, because with degradation, the material will be lost for the production of new polymers.
- Carbon capture and utilization (CCU, Table 1.4) measures deal with physical "lock-down" or chemical conversion of CO_2 in order to reduce the man-made climate change. Examples are storage of large amounts of CO_2 in subterranean caverns, and conversion of CO_2 into insoluble carbonates ("CarbFix", [22]) or other chemicals. While a safe storage of large amounts of CO_2 poses a specific challenge, the conversion to chemicals is a proven path forward (examples given in [23]). As CO_2 is a "low energy" molecule, it requires high energy reaction partners (plus, potentially, catalysts) to convert this greenhouse gas into chemicals otherwise being produced based on fossil feedstock. Recently, the electrocatalytic reduction of carbon dioxide, powered by renewable electricity, to produce valuable feedstocks has been reported [24]. Such feedstocks could act as a CO_2 sink in converting them into plastics.

- By converting CO_2 into plant biomass and O_2, nature has successfully done this job already for millions of years. Therefore, forestation projects are a meaningful way to capture CO_2 sustainably, but as a self-regulating forest takes many decades to grow, it is a rather slow process.
- The "lignin-to-chemicals" route is highly interesting, as the starting material is readily available (Table 1.5). Whether or not the many different possible reaction products of lignin in a "bio refinery" will establish themselves as alternative to the traditional, crude oil based petrochemicals remains to be seen [25].

1.4 Recycling Methods: Closing the Loop

A circular plastics economy is not possible without recycling. Mechanical recycling of post-industrial and also post-consumer plastics has been practiced for many years now. While post-industrial thermoplastics are mostly reprocessed without property losses, mechanical recycling of post-consumer plastic waste often ended in "downcycling". Mixed waste or contamination of the input resulted in properties inferior to virgin product properties. Examples:

- Mixed plastics to flower pots
- Bottle PET to fiber PET for clothing
- Materials formerly used for food packaging are reused in non-food applications like electronics housings

While there is nothing wrong with reuse of material in less demanding applications, a true circular solution however requires serving the high quality markets (such as food packaging or healthcare applications) with recycled material too. Especially in the light of demanding applications, advanced chemical recycling technologies were developed, such as the de-polymerization of PET, polystyrene, or PMMA, aiming to recover the respective monomer, which is re-polymerized to plastics with the same purity range as the original polymer, allowing them to be used for the same applications.

Recently, mechanical recycling methods experienced significant progress to achieving high purity plastics as well. The "super sorting" of post-consumer polystyrene waste ("supercleaning" technology of machine manufacturer Gneuss), followed by a thorough degassing step, is able to provide very clean polystyrene, suitable for food packaging applications. In cooperation with the Fraunhofer-Institut für Verfahrenstechnik und Verpackung IVV, SCS (Styrenic Circular Solutions) performed so-called challenge tests, which confirmed the ability of this new mechanical recycling technology to remove waste-stream induced impurities from polystyrene. The challenge test revealed the very good cleaning efficiency of the technology used, leading to excellent purity levels of the r-PS. This result is supported by the intrinsic properties of polystyrene being a low diffusion polymer, which prevents waste impurities entering into or migrate through the polymer matrix. The resulting quality of the r-PS provides producers of packaging for the food industry with utmost flexibility, as it can be used in their existing industrial production processes such as extrusion and thermoforming. Members of SCS delivered the proof of concept that mechanically recycled polystyrene can serve as a drop-in solution in form, fill, and seal (FFS) production lines. Multiple variants for packaging are possible: single layer, co-extrusion of the r-PS with virgin polystyrene, or use of the recyclate as middle layer between virgin polystyrene (A-B-A) [26]. The test results enable the first application for an opinion of the European Food Safety Authority (EFSA) on the use of recycled polystyrene (r-PS) as food contact material [26].

Additionally, dissolution technologies have developed during the past 30 years into a high quality and moderate energy consuming alternative to mechanical recycling. Several examples are known, like Polystyvert in Canada [27], Foam Cycle in the USA [28] or Fraunhofer in Germany [29], with the CreaSolv® process being one of the most prominent ones. This process, developed by Andreas Mäurer and Martin Schlummer of Fraunhofer-Institut Verfahrenstechnik und Verpackung IVV in Freising, Germany, uses the specific solubility of polymers in solvents and solvent combinations in order to separate dedicated plastic materials from a "mono" plastic application, from blends, or from coextrusion layer food packaging. Re-precipitation and work-up results in a thoroughly purified polymer. Dissolution is especially suitable for foam applications, such as XPS or EPS, used in clam shells, (food packaging) fish boxes, but also insulation for building and construction [30].

Similar to the pros and cons of the various routes of incorporating bio-based content into polymers and other chemicals, also recycling pathways have different strengths and weaknesses, as outlined in Table 1.6 to Table 1.8.

Table 1.6 De-Polymerization Examples: Polystyrene or PMMA Cleavage Back into Their Monomers; Hydrolytic Degradation ("Solvolysis") of PET into Its Precursors

Strengths	Weaknesses/Threats	Opportunities
▪ Monomers are generated by cleavage of the polymer chain ▪ After purification and re-polymerization: basically same quality as "virgin" type polymer	▪ Some sorting and conditioning of the post-consumer waste required ▪ Purification step of monomers required ▪ Scale is lower than world-scale production of monomers from crude oil	▪ UVP for PS, PMMA due to their low T_c (ceiling temperature) ▪ Various reactor types possible

Table 1.7 Mechanical Recycling

Strengths	Weaknesses/Threats	Opportunities
▪ Cost-effective ▪ Established solution for industrial waste/off-spec. material ▪ Not much energy required	▪ No removal of contaminants like flame retardants ▪ Quality, lot-to-lot inconsistencies ▪ Material degradation possible after frequent cycles	▪ Food quality possible for polystyrene and PET in case of well sorted, washed and pure waste

Table 1.8 Dissolution Technology: A Post-Consumer Waste Is Subjected to Treatment with a Defined/Adapted Solvent. The Respective Polymer Is Selectively Dissolved, Re-Precipitated, and Recovered

Strengths	Weaknesses/Threats	Opportunities
▪ No disintegration of the polymer ▪ Moderate energy consumption due to relatively moderate temperature	▪ By-products can be significantly reduced, but mostly not completely removed ▪ Scale is lower than world-scale production of polymers from crude oil	▪ Very selective for defined plastics

In contrast to nature, technical solutions will probably never work with 100% yield and therefore not allow literally endless recycling circles, so process losses have to be replaced by new material. Even so, above mentioned recycling processes typically show a significant lower environmental impact than the production and linear use of plastics, while contributing to reducing litter at the same time.

■ 1.5 Summary

Plastics offered and still offer an unprecedented step-up in quality of life. Their beneficial effects range from reduction of food waste by extension of shelf life, affordable healthcare measures through mass production of diagnostics or infusion kits, significant CO_2 savings through lightweight applications in automotive and aircraft up to a wide range of electronic gadgets in lightweight yet stiff and tough protective housings. A replacement or even no-plastics strategy would not only affect lifestyle, but also increase the emission of greenhouse gases like CO_2 significantly.

At the same time it is clear that environmental pollution by plastic parts needs to be kept in check and hence must not continue as it has done until now. While recovery of precious metals – even from landfills – is already in place in many parts of the world, it is about time to treat plastics not as waste anymore, but as precious raw materials source. Plastics are not the problem, but rather insufficient collection schemes and recycling initiatives.

This book is more than a compilation of state-of-the-art recycling technologies: it starts with the fundamental understanding of plastics structure/property relations, deep-dives into economic, legal and political boundary conditions, and explains methods and results of life cycle analyses in order to identify the best recycling solutions for our environment. The in-depth look into the complete value chain – from collection, sorting, cleaning up to applications, design-for-recycling and new developments – provides a holistic view on this topic, beyond the pure technical perspective.

References for Chapter 1

[1] Tschimmel, U.: *Die Zehntausend Dollar Idee*, ECON, 2nd edition 1991

[2] Letcher, T. M.: *Plastics Waste and Recycling*, Elsevier 2020, p. 53 ff.

[3] INEOS Styrolution Group GmbH information

[4] PlasticsEurope, *Plastics – the Facts 2019, https://plasticseurope.org/wp-content/uploads/2021/10/2015-Plastics-the-facts.pdf*, accessed March 29, 2022

[5] Ebnesajjad, S. (Ed.): *Plastic Films in Food Packaging*, Elsevier, (2012), p. 71 ff

[6] PlasticsEurope, *Plastics – the Facts 2015, https://plasticseurope.org/wp-content/uploads/2021/10/2015-Plastics-the-facts.pdf*; accessed March 29, 2022

[7] SAS Learning Post, *https://blogs.sas.com/content/sastraining/2018/04/23/can-you-guess-which-10-rivers-produce-95-of-oceans-plastic-pollution/*

[8] "Polymerchemie - Leben mit Plastik", *Frankfurter Allgemeine Zeitung*, January 12, 2015

[9] Ivleva, N. P., Wiesheu, A. C., Nießner, R.: *Angew. Chem.*, (2017) 56 (7), pp. 1720–1739

[10] Ragaert, K.: TED talk, University of B-Ghent, *https://www.ted.com/talks/kim_ragaert_plastics_rehab/transcript*

[11] *www.european-bioplastics.org/bioplastics/feedstock/*

[12] DIN EN 14995 (2007-03): Plastics Evaluation of compostability - Test scheme and specifications

[13] Kim, H. R., Lee, H. M., Yu, H. C., Jeon, E., Lee, S., Li, J., Kim, D.-H.: Biodegradation of Polystyrene by Pseudomonas sp. Isolated from the Gut of Superworms (Larvae of Zophobas atratus). *Environmental Science & Technology*, (2020) 54(11), pp. 6987–6996; *https://doi.org/10.26434/chemrxiv.11295296.v1*

[14] CMAI Green Styrenics 2011/12

[15] Tachibana, Y., Kimura, S., Kasuya, K.-i.: *Scientific Reports*, (2015) 5 (8249), p. 2864

[16] Haveren, J., Scott, E., Sanders, J.: Bulk chemicals from biomass. *Biofuels, Bioprod. Biorefin.*, (2008) 2, pp. 41–57

[17] Harmsen, P., Hackmann, M., Bos, H.: Green building blocks for bio-based plastics. *Biofuels, Bioprod. Biorefin.*, (2014) 8, pp. 306–324

[18] Claypool, J. T., Raman, D. R., Jarboe, L. R., Nielsen, D. R.: Technoeconomic evaluation of bio-based styrene production by engineered Escherichia coli. *Journal of Industrial Microbiology and Biotechnology*, (2014) 41(8), pp. 1211–1216

[19] Guerrero-Perez, M. O., Banares, M. A.: Metrics of acrylonitrile: From biomass vs. petrochemical route. *Catal. Today*, (2015) 239, pp. 25–30

[20] ISO 14021:2016 - ISO 140213. The products under this paper are “Renewable Attributed Products”

[21] *https://standards.cen.eu/* - CEN/TC 411 – Bio-based products

[22] *https://www.bbc.com/future/article/20200616-how-iceland-is-undoing-carbon-emissions-for-good*

[23] *www.covestro.com/en/sustainability/lighthouse-projects/co2-dreams*

[24] Li, F., Thevenon, A., Rosas-Hernández, A., Wang, Z., Li, Y., Gabardo, C. M., Ozden, A., Dinh, C. T., Li, J., Wang, Y., Edwards, J. P., Xu, Y., McCallum, C., Tao, L., Liang, Z.-Q., Luo, M., Wang, X., Li, H., O'Brien, C. P., Tan, C.-S., Nam, D.-H., Quintero-Bermudez, R., Zhuang, T.-T., Li, Y. C., Han, Z., Britt, R. D., Sinton, D., Agapie, T., Peters, J. C., Sargent, E. H.: *Nature*, (2020) 577, pp. 509–513; *https://www.nature.com/articles/s41586-019-1782-2*

[25] Jiang, Y., Loos, K.: Enzymatic synthesis of biobased polyesters and polyamides. *Polymers*, (2016) 8(7), p. 243

[26] Press Release “Styrenics Circular Solutions demonstrates mechanically recycled polystyrene is suitable for food contact”, October 14, 2020

[27] *www.recyclingmagazin.de/2021/02/09/ineos-styrolution-und-polystyvert-wollen-polystyrol-recyceln/*

[28] *www.foamcycle.com*

[29] *www.creacycle.de/en/the-process.html*

[30] *thebalancesmb.com/introduction-to-eps-recycling-2877866*

2 Introduction to Major Thermoplastic Polymers in the Context of Recycling

Tristan Kolb, Andreas Schedl, Hans-Werner Schmidt

This chapter gives a basic insight into thermoplastic polymers, their architecture, polymerization mechanisms, and solid state morphology. Further, it focuses on the degeneration of the polymers during processing and use. The last part gives a brief overview of the most important thermoplastic polymer types.

Based on their origin, polymers are usually classified into biopolymers and synthetic polymers. Biopolymers are naturally occurring polymers, like proteins, polynucleotides (e.g., DNA and RNA), polysaccharides (e.g., starch, glycogen, cellulose, and chitin), and natural resins (e.g., amber, mastic, and shellac). In contrast, synthetic polymers are created by chemical reactions from petroleum-based and (increasingly nowadays) natural building blocks. In this chapter, we will focus on synthetic polymers, with their wide industrial application.

The breakthrough for industrial polymers was the discovery in 1839 of rubber vulcanization by Charles Goodyear, who obtained elastomers from the cross-linking of natural rubber with sulfur [1]. Since that time, the world of polymers has grown continuously to over 200 different polymer types today, with large differences in properties allowing a broad spectrum of applications in all fields of our daily life including packaging, construction, automotive, electronics, agriculture, and medical devices.

Polymers often replace conventional materials, for example billiard balls formerly made from ivory were later produced from celluloid. Celluloid was one of the first commercial polymers, and was prepared from a mixture of nitrocellulose and camphor. Consequently, the demand for ivory was reduced and with it the need for elephant hunting. Depending on their property profile, completely new applications have become possible, like polystyrene foam as an insulation material for buildings to reduce energy consumption for heating, or advanced food packaging to slow down food spoilage.

Key features of polymers are tailorable mechanical properties, availability in huge quantities, affordable price, and low density. The latter allows lighter parts and thus an energy conservation in their application, especially with respect to transportation. However, some further key features lead to a dilemma. The stability and durability of these materials allows on the one hand long-term usage in applications, but also causes on the other hand only very slow decomposition under natural conditions. This has provoked the current issue of plastics accumulation in the environment. This can be overcome if the polymer waste is completely collected, sorted, purified, and then treated as resource for recycling. The waste management as well as current and upcoming recycling methods are covered in the various chapters of this book.

2.1 Chemical Structure and Synthesis of Industrial Polymers

2.1.1 Polymer Architecture

Whether we consider any of the industrial polymer types, or one of the many biopolymer types, the fundamental architecture of a polymer is a long chain. These can form linear, star-like, branched, and cross-linked network architectures (Figure 2.1). The chain architecture has an enormous impact on the properties of the macroscopic material (e.g., different types of polyethylene, described in Section 2.3.2.1).

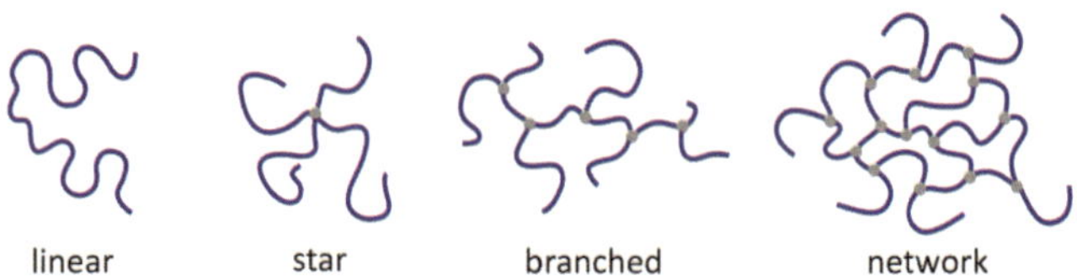

Figure 2.1 Schematic illustration of the main polymer architectures (linear, star, branched, and network)

The long polymer chains are formed from small building blocks, the so-called monomers, which are covalently bound to each other. An example is shown in Figure 2.2 for the polymerization of the monomer ethylene to the polymer polyethylene.

$$n\,H_2C{=}CH_2 \longrightarrow \left[H_2C{-}CH_2 \right]_n$$

Figure 2.2 Polymerization of the monomer ethylene to the polymer polyethylene by forming covalent bonds between the monomer units

A further important structural feature of polymer chains is their configuration (Figure 2.3). This is relevant for vinyl monomers with substituents (e.g., the phenyl group in styrene or the methyl group in propylene). The orientation of such groups along the polymer chain can be controlled during the polymerization process. Hence, selective process conditions give access to a wide variety of materials with different properties while using the same monomer (e.g., different types of polypropylene, described in Section 2.3.2.2).

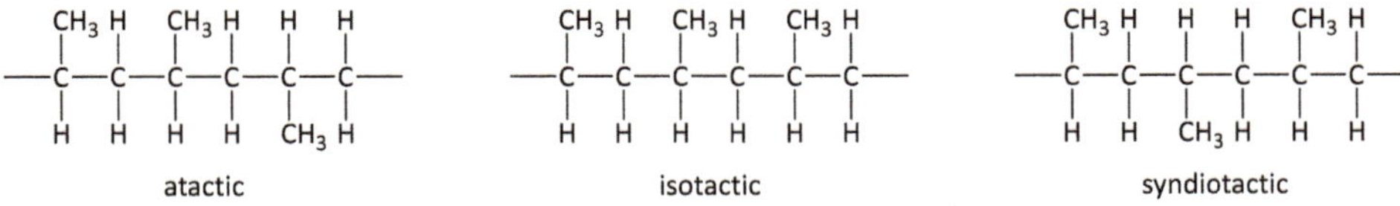

Figure 2.3 Schematic representation of potential configurations, exemplified for polypropylene: Isotactic polymers have all their substituents on the same side along the chain. In contrast, syndiotactic polymers have alternating positions of the substituents along the chain. If a chain has no regularity, it is called atactic.

So far, we have focused on polymers that are built up from only one monomer. Beyond these so-called homopolymers, copolymers can be obtained from the polymerization of two or more monomers. The arrangement of the different monomers along the polymer chain can be controlled by synthetic procedures. For two monomers, the main sequences are random, alternating, and a block fashion (Figure 2.4). In random copolymers, the different monomers are arranged without any ordering along the chain. Alternating copolymers reveal a strict alternating arrangement of both monomers along the chain. Graft copolymers are composed of a backbone of one monomer and grafts of the second monomer. In block copolymers the monomers are arranged in blocks along one chain. An AB diblock copolymer is composed of a block of monomer A directly linked via one single bond to a block of monomer B. The concept of block copolymers can also be expanded to more sophisticated architectures with an arrangement of distinct blocks, such as an ABA triblock copolymer. A special type of block copolymers are segmented $(AB)_n$ copolymers. Here, several segments of different length are connected in one chain. By introducing a third monomer block, ABC triblock terpolymers are even accessible. Notably, the arrangement of monomer units within specific segments might lead to the formation of immiscible blocks which cause microscopic phase separation on the nanometer scale.

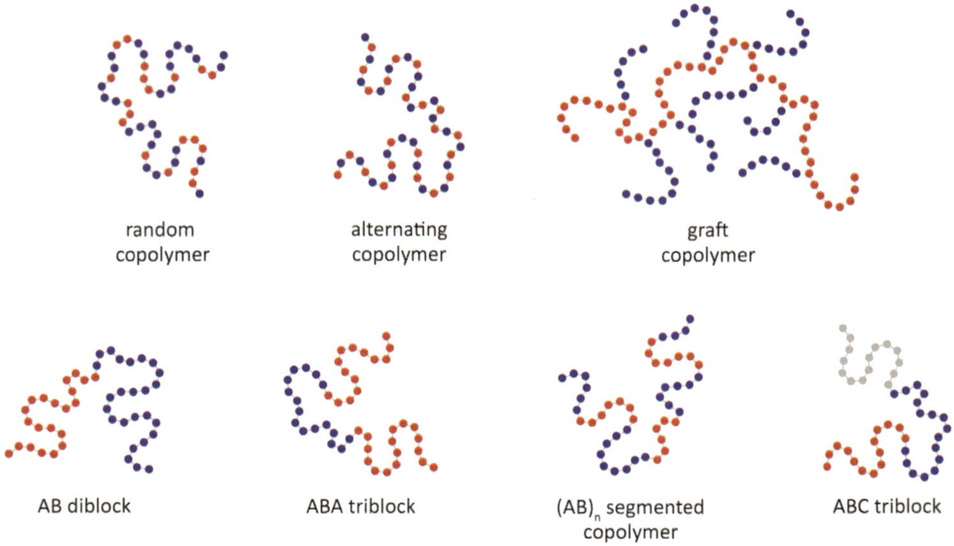

Figure 2.4 Main possibilities for the sequence of two monomers in copolymers along the chain are random, alternating, graft, AB diblock, ABA triblock, and $(AB)_n$ segmented. In case of ABC triblocks, three monomers are used.

A blend of two different homopolymers is typically not homogeneously miscible. Instead, macroscopic phase-separated areas on the micrometer scale are formed. These so-called domains consist each of only one of these polymers. Block copolymers can be employed as compatibilizer and stabilize the morphology of blended systems. This technique is widely used to combine the properties of two polymer types in one product, while maintaining a good connection between the different phases, which is, for example, crucial for good mechanical properties. The physical background for the phase separation is based on the interaction forces between the different monomer units in the polymer chains. Mostly, the interactions between chain segments based on the same monomer are stronger than interactions between chain segments consisting of different monomers. Such so-called secondary interactions originate from the weak attractive forces between the chains, which occur beside the covalent bonds of the polymer chain. The responsible intermolecular forces originate from so-called van der Waals forces and hydrogen bonds.

The different van der Waals forces originate either from interactions based on spontaneous electron fluctuations (London dispersion forces) between atoms or chain segments, or permanent polarization of functional groups based on atoms with different electronegativity in the functional groups inducing induction forces and dipole–dipole interactions. All of these have in common that the strength of the interactions is strongly dependent on the distance between the respective parts of the chain segments. Consequently, their contribution rises with closer proximity of the polymer chains.

Hydrogen bonds (H-bonds) are defined by IUPAC as "an attractive interaction between a hydrogen atom from a molecule or a molecular fragment X–H in which X is more electronegative than H, and an atom or a group of atoms in the same or a different molecule, in which there is evidence of bond formation" [2]. They are formed, for example, between N–H groups and carbonyl groups, as depicted in Figure 2.5 for polyamides and polyurethanes. Since these forces are usually stronger than van der Waals forces, they have a large contribution to the properties of the respective polymers.

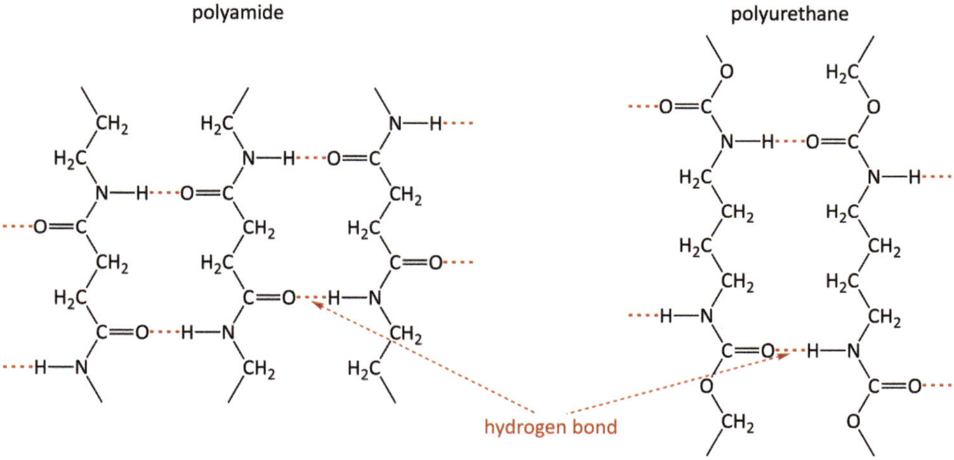

Figure 2.5 Hydrogen bonds between N-H groups and carbonyl groups in polyamides and polyurethanes

Beside the mentioned interactions, a further mechanical contribution originates from chain entanglements. Polymer chains form random coils for entropic reasons and entangle with neighboring chains. Due to that, the chains can only move to a limited degree along each other (Figure 2.6). This results, for example, in the common viscous flow of molten polymers. The number of entanglements increases with the length of the polymer chains. Consequently, grades with a high molecular weight show higher viscosity than grades with a lower molecular weight.

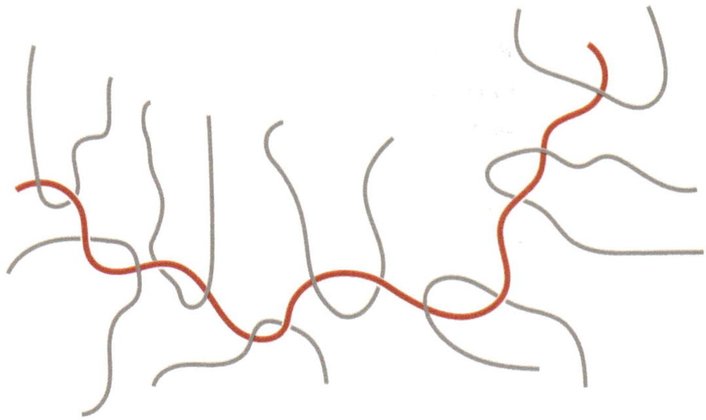

Figure 2.6 Illustration of entanglements of a polymer chain with neighboring chains

2.1.2 Polymers in the Solid State

When polymer materials are cooled down from the molten state, the mobility of the chains decreases. Polymers can solidify in a random coil-like and entangled form, which is called the amorphous state (Figure 2.7(a)). It is characterized by being disordered and having an isotropic orientation, and thus a high entropy. An important characteristic for the description of that state is the glass-transition temperature T_g. At this temperature, chain segments start to get mobile and thus the material becomes (to a certain degree) flowable and formable.

Polymers can also form ordered structures in the solid state. Hereby, the chains are partially packed in a crystalline structure with a regular orientation, and thus the entropy is reduced. Consequently, the ordering process, the so-called crystallization, must provide a sufficient energy gain to result in an overall favorable process. The energy gain for the crystallization arises from the previously discussed secondary interactions such as van der Waals forces and H-bonds. The latter require that the functional units are correctly oriented with respect to each other. Consequently, an ordered hydrogen bond pattern is formed. In contrast to small molecules, polymer chains cannot crystallize completely from the molten state. The main reason is the presence of entanglements, which restrict complete crystallization. This leads to the formation of crystalline areas, the so-called lamellae. Within these lamellae, the chain segments are well orientated and closely packed. The chains leave the lamella and statistically re-enter at a different position within the same lamella or connect to different crystalline lamellae. Tie molecules between the crystalline parts are formed, which are important for the mechanical properties. Loops, tie molecules, and disordered chains form amorphous regions. This structure is illustrated in Figure 2.7(b). Imperfections in the chain such as branches are usually present in this amorphous region, since these do not allow a high degree of ordering of the respective chains. From a more macroscopic perspective, several of these crystalline lamellae are present within an amorphous matrix, as illustrated in Figure 2.7(c), which both are crucial for the mechanical properties of the material. Due to the coexistence of amorphous and crystalline regions, the materials are called “semicrystalline”. Such materials are characterized by the degree of crystallinity, which represents the fraction of the monomers in the crystalline areas to all monomers. It depends on the polymer class but also on the microstructure, such as the amount of branching (which typically ranges from 40% to 70% for semicrystalline thermoplastic polymers). Furthermore, the amorphous regions are characterized by the glass-transition temperature. The crystalline areas are characterized by the melting temperature T_m. For temperatures above T_m, a homogeneous melt is formed.

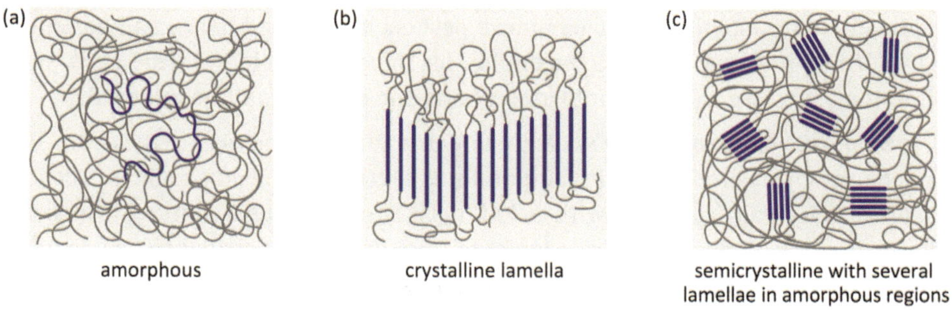

Figure 2.7 Structures of solid-state polymers: (a) In amorphous polymers, the chains form coils entangled with neighboring chains; (b) crystalline lamella of polymer chains; (c) semicrystalline polymers feature several crystalline lamellae, which are separated by an amorphous matrix

2.1.3 Categorization of Polymers

The architecture of a polymer has a direct effect on the thermomechanical properties of the material. In general, there are three main categories of polymer materials:

- Thermosets
- Elastomers
- Thermoplastics (with the subcategory of thermoplastic elastomers)

Thermosets are based on a fully cross-linked polymer network. As the different chains are directly bonded to each other, such materials cannot be melted upon heating and thus cannot be reformed without destroying the polymer architecture.

To prepare such materials, usually a reactive resin is filled into a mold and subsequently cured to result in a high degree of cross-linking. After curing, such materials can only be reshaped by subtractive manufacturing.

Elastomers are polymer networks with a relatively small number of cross-linking points and a base polymer with a glass-transition temperature below the service temperature. Like thermosets, these materials are not meltable and hence not formable upon heating. In contrast to thermosets, they show an elastic behavior. Due to the low glass-transition temperature, the polymer chains have a certain mobility and hence allow a mechanical deformation of the material by a force. Since the chains are connected to each other, the original shape is restored upon releasing the force.

Thermoplastics can be shaped at elevated temperatures. This allows repeated melting and remolding. Thermoplastics consist of polymers with a linear or branched polymer architecture but are not cross-linked. Upon heating, the chains become more and more mobile, and so can move past each other and hence can be repeatedly formed into new shapes. Depending on their morphology in the solid state, thermoplastics are classified into amorphous and semicrystalline materials.

Thermoplastic elastomers (TPEs) combine elastic and thermoplastic behavior. This is achieved by a combination of hard and soft segments. The soft segments consist of an elastic main material with a low glass-transition temperature, which are physically cross-linked by hard segments (like crystalline structures or domains with a glass-transition temperature higher than the service temperature). The physical cross-links break up at elevated temperatures and thus the entire material can be processed in a thermoplastic manner. Upon cooling, the hard segments are formed again and provide the material with physical cross-links and thus the elastic behavior. Typically, a block copolymer architecture is used for TPEs, with one domain featuring the soft segment and one the hard segment.

2.1.4 Polymerization

The term “Polymerization” covers different chemical reactions leading to covalent bonds between monomer units and simultaneously forming polymer chains or networks. In general, these reactions can be categorized into *chain-growth polymerization* and *step-growth polymerization*. These two types differ significantly in the mechanism of macromolecule formation. This is shown in Figure 2.8 for the chain length as a function of the conversion of monomers and functional groups, respectively.

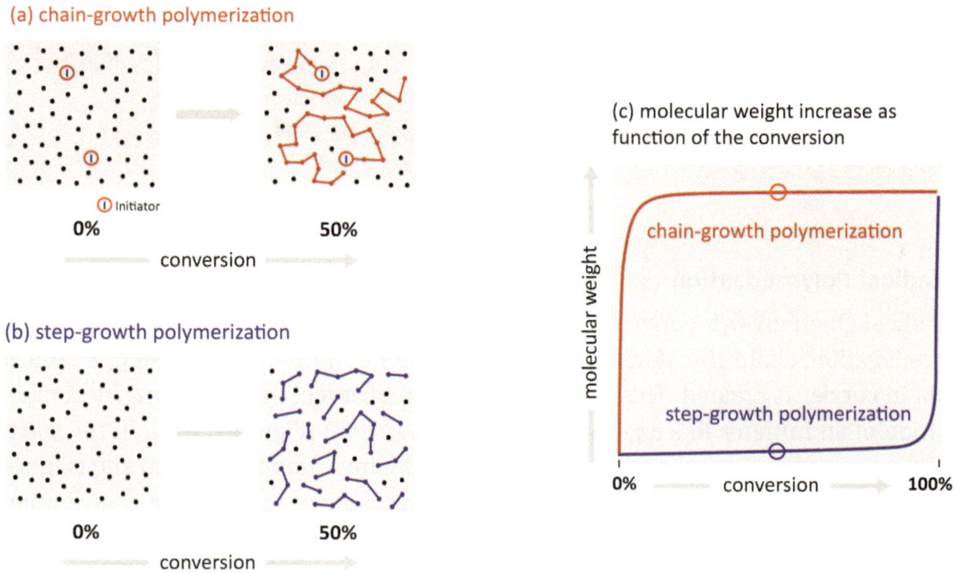

Figure 2.8 Comparison of polymerization mechanisms: (a) Chain-growth polymerization; (b) step-growth polymerization; (c) the respective increase of the molecular weight as function of the conversion

Chain-growth polymerization starts from an active chain carrier formed by an initiator. A polymer chain is formed by the subsequent addition of monomers (Figure 2.8(a)). Typically, polymer chains with high molecular weight are formed even at a low monomer conversion (Figure 2.8(c)). In contrast, systems with step-growth polymerization follow a different mechanism. Instead of reactions taking place exclusively at the active chain carriers, all monomers and reaction products formed from them have a similar probability of reacting with each other. This leads to the formation of a broad variety of dimers, trimers, tetramers, and oligomers at low conversions during the polymerization (Figure 2.8(b)). High molecular weight polymer chains are formed at a very high conversion (Figure 2.8(c)).

2.1.4.1 Chain-Growth Polymerization

Chain-growth polymerization can occur at different kinds of active sites. This has a huge effect on the polymerization process, conditions, and the resulting polymer. Table 2.1 gives an overview of possible active sites and the derived classification. These classes will be discussed below.

Table 2.1 Possible Active Sites for Chain-Growth Polymerization and the Derived Classification

Active site	Classification
Radical	Radical polymerization
Anionic or cationic species	Ionic polymerization
Transition metal-carbon bond	Coordination polymerization

Radical Polymerization

Radical chain-growth polymerization consists of the reaction steps initiation, chain propagation, chain transfer, and chain termination. In the initiation step, a radical chain carrier is created. This can take place, for example, by the thermal decomposition of an initiator like azobisisobutyronitrile (AIBN) (Figure 2.9(a)). In the starting reaction the first monomer is added while maintaining the radical site (Figure 2.9(b)). This active chain end repeatedly adds monomers in the chain propagation step, building up the polymer chain (Figure 2.9(c)). Thereby, the active center at the chain end is maintained. However, such radical-carrying chain ends are very reactive and can be subjected to different kinds of further reactions. First of all, they can react with inactive molecules (e. g., low molecular weight species or any polymer chains present) and thereby transfer the radical species to these molecules (Figure 2.9(d)). This leads to cessation of the growth of the original chain. The active radical site is now located on the new molecule and a new chain can propagate from this site. If the transfer takes place to another chain, or even another position on the same chain, long and short branches are formed. Radical-carrying species can also react with each other and thereby terminate the active centers and hence the chain growth. If two radical chain ends react by forming a new covalent bond, the molecular weight of the resultant chain substantially increases in one step. This process is called recombination (Figure 2.9(e)). In a disproportionation reaction, one hydrogen atom (marked in red) is exchanged between two active chain ends. The radical character of both chains is lost, and thereby one saturated and one unsaturated chain is formed. The latter can react with another active radical site.

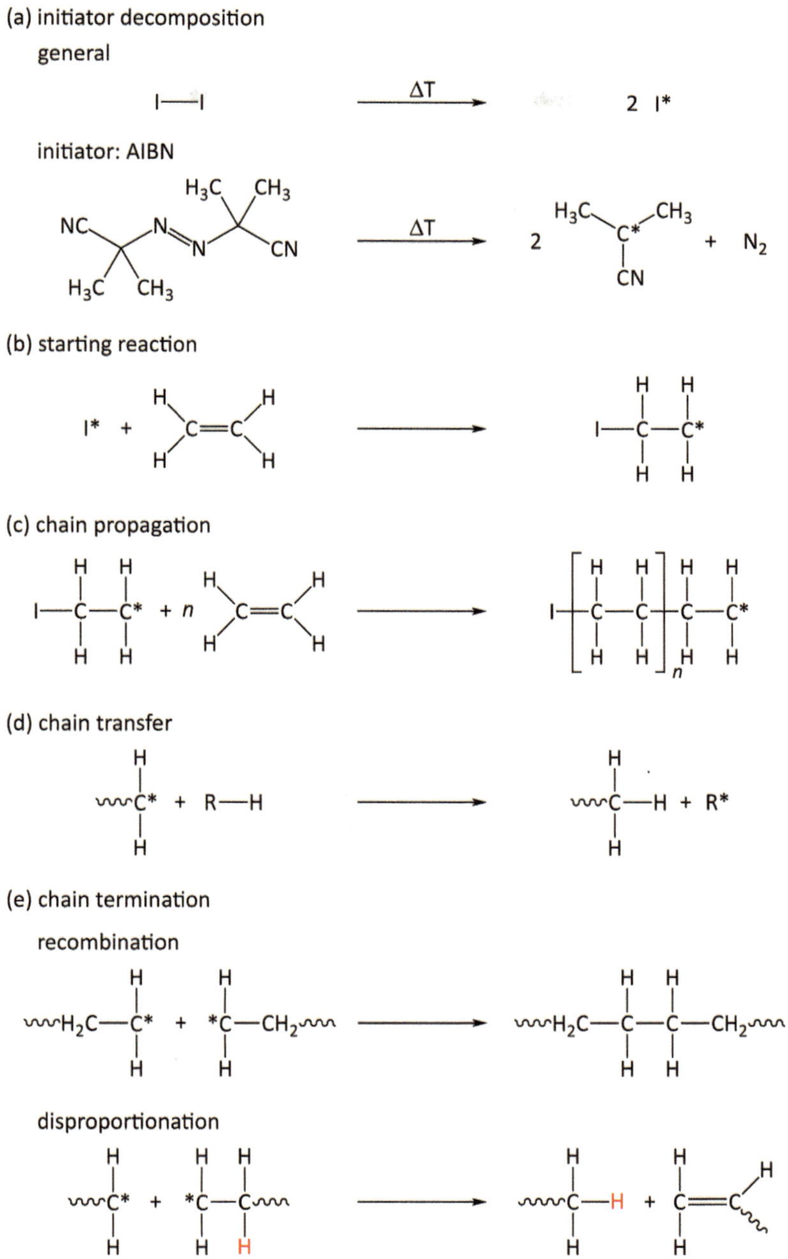

Figure 2.9 Reaction mechanism of radical chain-growth polymerization: (a) Initiation, here also shown for AIBN as an example; (b) starting reaction; (c) chain propagation; (d) chain transfer; (e) chain termination by recombination or disproportionation

All reaction types can take place, with their distribution being determined by their reaction speeds. This leads to a characteristic variety of chain lengths and thus a broad molecular weight distribution. This is usually characterized by the average molecular weight. It can be affected, for example, by the reaction temperature or the amount of initiator. Higher initiator amounts and higher temperatures result thereby in lower molecular weights.

The molecular weight is one of the key parameters influencing processing and material performance. A higher molecular weight leads to higher mechanical strength, toughness, stress crack resistance, and heat deflection temperature. However, a higher molecular weight also increases the viscosity of the material in the melt and thus decreases the flowability of the material in the molten state. Consequently, high-molecular-weight materials are more difficult to process to specific shapes than low-molecular-weight materials.

Ionic Polymerization

Ionic polymerization differs from radical polymerization by having a charged ion species or an ion pair as the active site. Depending on the charge of the active site, the reaction mechanism is further categorized into cationic and anionic polymerization.

Cationic polymerizations are initiated by a proton donor such as a strong acid (e.g., sulfuric acid) or Lewis acids (e.g., boron trifluoride) in combination with water. By reaction of the proton with the monomer, a carbenium ion is formed that can then propagate the chain by repeated reaction with further monomers. Side-reactions of the active chains can occur by a chain transfer reaction to the monomer or solvent, leading to transfer of the active center. Consequently, the growth of the original chain is stopped and a new chain can start from the active species formed. This mechanism is the main reason for the limitation of the resultant molecular weight, as termination reactions between active chain ends cannot occur due to electrostatic repulsion of the charged species. Controlled termination is usually carried out by addition of bases or water.

Anionic polymerizations are initiated by strong bases like sodium hydroxide, sodium alcoholates or organolithium reagents such as *n*-butyllithium. During initiation, they form carbanions with monomers. The carbanions can add further monomers and hence propagate the chain. Reaction of the carbanion species with polar molecules like water can take place und thus deactivate the active sites. Thus, such contaminations must be thoroughly removed prior to anionic polymerization. Analogous to the cationic polymerization, termination reactions between active chain ends cannot occur due to electrostatic repulsion of the charged specimen. Remarkably, this kind of polymerization can be performed with nearly complete exclusion of chain transfer. The active sites have a nearly unlimited lifetime and grow with a similar speed until all monomer is consumed. These “living polymers” feature the unique property of propagating further when more monomer is later added. This technique is usually applied in the production of block copolymers, where a second monomer is added to the living chain end. Polymers with a highly-defined narrow molecular weight distribution and architecture can be obtained in this manner.

Coordination Polymerization

Coordination polymerization uses catalysts based on transition metals to form the polymer chain. The monomer is added at the active center of the catalyst and then inserted into the metal–carbon bond of the growing chain. The catalyst restricts possible orientations for the coordinated monomer by the presence of its ligands around the active center. This leads to inclusion of the monomers in (primarily) a selective manner, and hence buildup of a stereoregular polymer chain is possible. The most prominent examples are Ziegler–Natta catalysts, which are based on transition metal chlorides like titanium tetrachloride, which is activated by organoaluminum compounds like triethylaluminum. Thereby, the transition metal receives an alkyl ligand, but still features a free coordination position (Figure 2.10(a)). At this position, the monomer is coordinated (Figure 2.10(b)) and inserted, via a 4-center-4-electron transition state, into the metal–carbon bond (Figure 2.10(c)). After this insertion, the chain is elongated and the active catalyst center with a free coordination position is reformed (Figure 2.10(d)). The coordination polymerization is widely used, for example, for the production of unbranched high-density polyethylene, or stereoselective polymers like isotactic polypropylene.

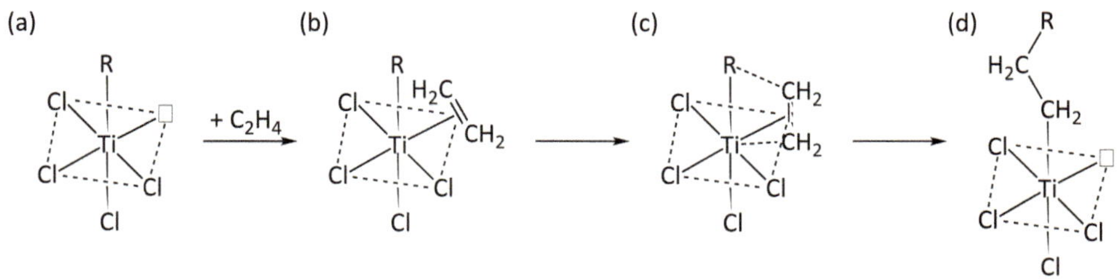

Figure 2.10 Simplified mechanism for coordination polymerization: (a) Catalyst featuring a free coordination position and a metal–carbon bond to the growing chain R; (b) coordination of the monomer in the free coordination position; (c) insertion of the monomer into the growing chain via a 4-center-4-electron transition state; (d) restoration of the free coordination position of the active catalyst

2.1.4.2 Step-Growth Polymerization

Step-growth polymerization is based on the reaction of functional groups of bifunctional monomers. In a first reaction step, a dimer is formed. This is illustrated in Figure 2.11(a) for two different bifunctional monomers with two equal functional groups F and G, respectively. The reaction is also possible with only one bifunctional monomer, which carries both functional groups F and G (Figure 2.11(b)).

(a) F—M—F + G—N—G ⟶ F—M—FG—N—G

(b) F—M—G + F—M—G ⟶ F—M—GF—M—G

Figure 2.11 (a) Formation of dimers by step-growth polymerization based on two bifunctional monomers with the functional groups F and G. (b) Formation of a dimer by a monomer featuring both functional groups

The dimers formed feature again two reactive functional groups at the ends, and thus can react with further monomers or oligomers. In general, the reactivity of the functional groups is similar regardless of their presence on monomers, dimers, trimers, or longer chains. At the outset, a large number of monomer molecules is present, and hence primarily oligomeric chains are formed at low conversion. At increasing conversion and hence higher concentrations of oligomer fragments, these increasingly react with each other, and form longer chains. Consequently, a very high conversion is needed to achieve a high-molecular-weight polymer by step-growth polymerization. Furthermore, it is essential to have equal amounts of functional groups F and G present in the polymerization, as otherwise only oligomers are formed, which will feature chain ends with the excess functional group. As these cannot react with themselves, the molecular weight is restricted. Linear polymer chains are obtained from bifunctional monomers. By the addition of trifunctional or even multifunctional monomers in different amounts, weakly or highly cross-linked networks can be synthesized.

Step-growth polymerization can be categorized into polycondensation or polyaddition reactions. In a polycondensation, two functional groups react with each other and thereby release a low-molecular by-product. In contrast, the reaction of the functional groups in a polyaddition takes place without the release of any by-product. In both cases, the main chain of the polymer typically contains heteroatoms. At such positions in some cases step-growth polymers are suitable for chemical recycling by hydrolysis of the respective unit. This differentiates them from radically prepared chain-growth polymers, which typically contain only carbon atoms in the main chain. Consequently, these polymer chains cannot be broken up by hydrolysis.

Polycondensation

In a condensation reaction, two functional groups react with each other and form a condensation product by releasing a volatile low-molecular-weight by-product (Figure 2.12). The reaction kinetics depend strongly on the equilibrium between the sum of all educts and the sum of all products. To achieve a high conversion and hence a high molecular weight of the formed polymer, the equilibrium must be shifted to the product side. This is achieved by continuously removing the low-molecular-weight by-product. Typical polycondensation reactions involve, for example, the formation of amide linkages (resulting in *polyamides*), and ester linkages (resulting in *polyesters*).

$$R{-}X + Y{-}R' \rightleftharpoons R{-}R' + XY$$

Figure 2.12 Equilibrium between educts and products in a condensation reaction by releasing a by-product

Polyamides can be formed by reaction of bifunctional monomers with carboxylic acid and amine functions while water is released as a low-molecular-weight by-product. The carboxylic acid function can also be substituted with its more active derivatives like acid halides or esters from low-boiling alcohols, resulting in formation of the respective hydrogen halide or alcohol during the reaction. The main industrial polymer classes of polyamides are aliphatic polyamides. Also, partially aromatic polyamides (named polyphthalamides) and fully aromatic polyamides are commercial products. Aliphatic polyamides are typically named according to the number of carbon atoms in the amine and the carboxylic acid monomer. Thus, polyamide 6.10 is synthesized from the 6-carbon amine hexamethylenediamine and sebacic acid, which contains 10 carbons (Figure 2.13).

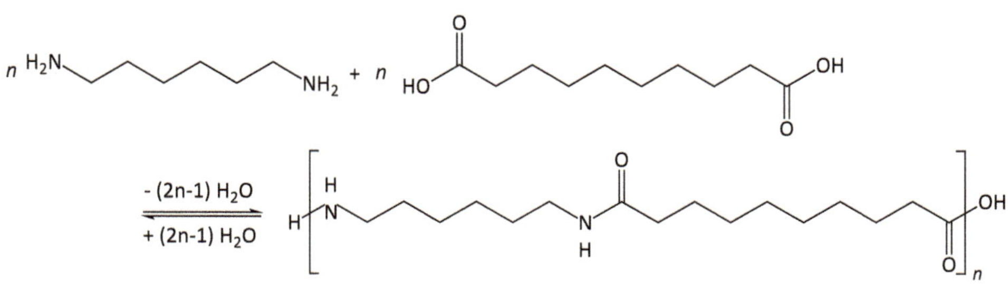

Figure 2.13 Polycondensation of hexamethylenediamine and sebacic acid to polyamide 6.10. The first number represents the number of carbon atoms in the diamine, while the latter represents the number of carbons in the diacid.

Furthermore, polyamides can be synthesized from amino carbonic acids, like PA 12 from the 12-carbon monomer ω-aminolauric acid. Alternatively, PA 12 can be prepared by anion-initiated ring-opening polymerization of laurolactam, which follows the kinetics of chain-growth polymerization.

Partialy or fully aromatic polyamides contain aromatic monomers to a certain degree. The most prominent representative for fully aromatic polyamides (so-called polyaramides) is Kevlar®, which is prepared by the reaction of *p*-phenylenediamine and terephthaloyl chloride. Kevlar is typically used as high-performance lightweight fiber.

Polyesters are synthesized from monomers with carboxylic acid and alcohol functions, with water being released as a low-molecular-weight by-product (Figure 2.14(a)). Analogous to polyamides, the carboxylic acid function can also be exchanged with its derivatives like acid halides or esters (with low-boiling alcohols), resulting in the release of hydrogen halides or alcohols, respectively. From a chemical point of view, polycarbonates can be regarded as the diesters of a carbonic acid with a difunctional alcohol. Since the reactivity of the carbonic acid is quite low, they are usually created by the reaction of phosgene with the respective difunctional alcohol (Figure 2.14(b)) or by transesterification, for example of diphenyl carbonate with the respective difunctional alcohol.

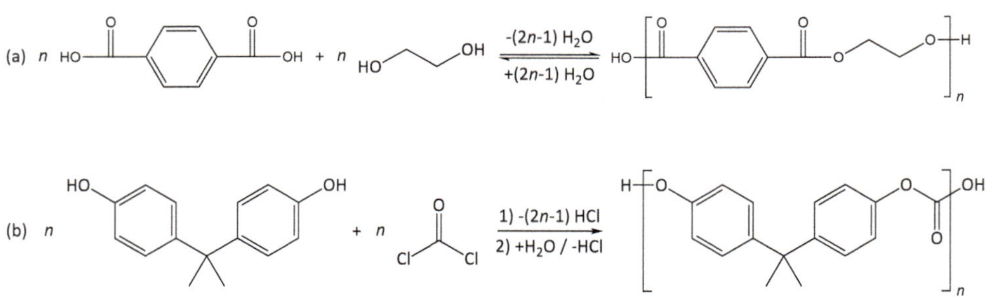

Figure 2.14 (a) Polycondensation of terephthalic acid and ethylene glycol to polyethylene terephthalate (PET). (b) Polycondensation of bisphenol A with phosgene to polycarbonate (PC)

Polyaddition

Like polycondensation, polyaddition follows a step-growth polymerization mechanism. However, the functional groups in a polyaddition react without the formation of any by-product. The most important polymer classes synthesized by this mechanism are polyureas, polyurethanes, and epoxy resins.

Polyureas are formed by the reaction of diisocyanates with diamines, resulting in the formation of urea groups (Figure 2.15(a)). In contrast, *polyurethanes* are formed by the reaction of diisocyanates with diols resulting in the formation of urethane groups (Figure 2.15(b)). Aliphatic amines, aliphatic alcohols, and aromatic or cycloaliphatic isocyanates are very reactive, and allow a rapid reaction even at room temperature. Furthermore, an anhydrous reaction procedure is required for non-foamed applications, since isocyanates are susceptible to side-reactions with water, forming carbon dioxide and amines. The carbon dioxide acts as a foaming agent and thus polyurea foams result.

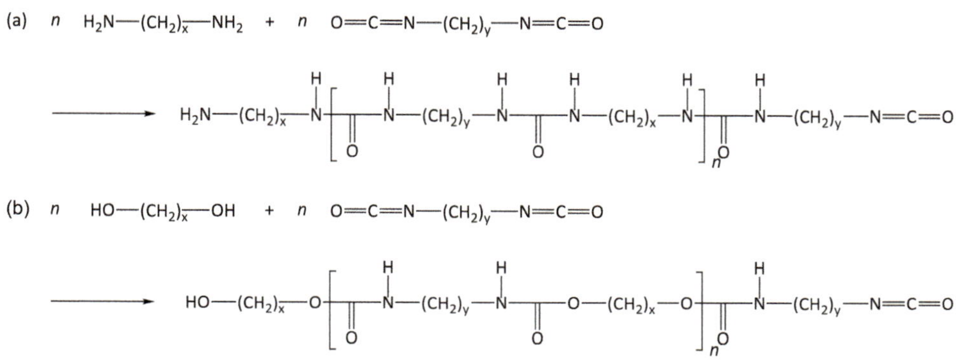

Figure 2.15 Polyaddition of diisocyanates with diamines to polyureas (a), and with diols to polyurethanes (b)

Epoxy resins belong to the class of thermosets and are highly cross-linked. They are cured by a polyaddition reaction of multifunctional monomers with epoxy functions and amine functions, respectively. Thereby, the reaction of epoxy monomers with secondary amines leads to the formation of linear connections. The reaction with primary amines (Figure 2.16(a)) leads to the formation of secondary amines, which can subsequently react with further epoxy functions and thus result in a higher cross-linking density (Figure 2.16(b)). To obtain highly cross-linked thermosets, multifunctional monomers are additionally applied.

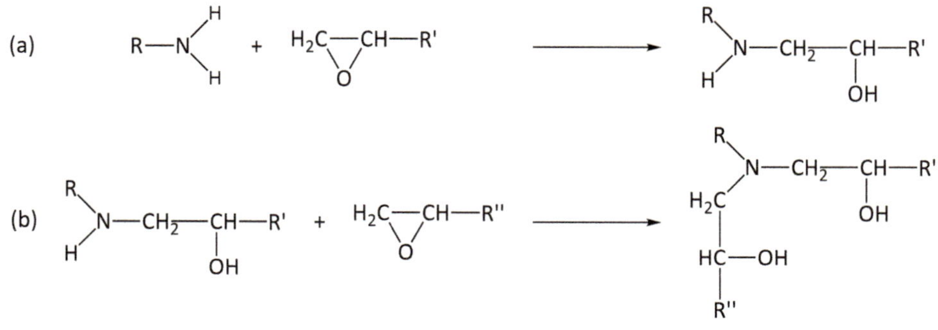

Figure 2.16 Polyaddition of epoxy groups with primary amines (a), and successive reaction with further epoxide functions to chemical cross-linkages (b)

2.2 Polymer Degradation and Recycling

Most materials are affected by processes which change the material properties during processing or during service life. These can be biological processes like the decomposition of wood in humid natural environments, chemical processes like the oxidation of metals (e. g., rusting of iron), or physical processes like the melting of ice to water. These processes influence the specific material properties in a different manner. For example, rust on the surface of an iron plate has a strong impact on the material color, but results in a negligible change of the mechanical properties of the entire material.

Polymers are also subjected to such processes, which typically lead to a change in molecular structure and deterioration of their properties. Polymer degradation becomes especially important within the context of mechanical recycling (Chapter 8), whereby the material gets a second use after remelting of the polymer waste and reprocessing into new parts. In the ideal case, the polymer is reused for a similar application. However, polymer degradation might prevent this due to alterations of the molecular structure and material properties. In the following sections, we will focus on the main physical, biological, and chemical processes leading to polymer degradation. According to DIN 50035, all processes leading to *reversible changes* of the material are termed *physical aging*, while all processes resulting in *irreversible* changes of the polymer material are classified as *chemical aging*. A special case of the latter are biological processes, which can take place in a natural environment.

2.2.1 Physical Aging

Polymers are usually processed in a quite fast manner, for example by injection molding or extrusion. This involves heating the materials up above their plasticization temperature and processing them under shear into the targeted shape. Especially for complex shapes, the polymer melt flow results in orientation of the polymer chains due to the rapid processing. This orientated state is rapidly frozen during the subsequent cooling and causes internal stress [3]. However, the energetically favored shape of a polymer chain (and thus the equilibrium condition) is a coil for amorphous materials, and a certain amount of crystallinity for semicrystalline polymers. If the energetically preferred state is not formed, reorientation processes can occur. These are hindered by a specific activation energy barrier. Consequently, they take place with low speed at low temperatures, but dramatically accelerate for temperatures at or above the glass-transition temperature.

In amorphous materials, several material properties like the specific volume, heat capacity, and mechanical properties can change during these relaxation processes [4]. This can lead to deviations in the dimensions of an object. Semicrystalline polymers do not reach usually their equilibrium crystallinity after processing. During aging, amorphous chains can rearrange themselves into a crystalline structure, and hence increase the overall crystallinity. As crystalline material has a higher density and hence a closer packing than amorphous material, local shrinkage and crack formation can occur [5]. Furthermore, a change of the crystalline structure can occur during aging. An example is PA 6, which crystallizes in its metastable nematic γ-structure during fast cooling. Annealing at 190 °C induces transformation into the more stable monoclinic α-structure [6].

Besides these rearrangement processes, migration of additives can lead to physical aging. Additives are typically applied in a polymer formulation to boost targeted material properties. For example, plasticized PVC (PVC-P) contains up to 50% of plasticizers, which give the material its soft and elastic characteristics [7]. During use (especially at elevated temperatures), additives can migrate out of the material or form domains with enriched additive and enriched polymer content, respectively. If such processes occur, the additive is no longer present in the original targeted concentration, and thus the desired property improvement is lost. In the context of plasticizers, this can lead to embrittlement and hence to failure.

From the perspective of recycling, the above-mentioned cases of physical aging just influence the chain orientation, the degree of crystallinity, or the additive content; the polymer material itself remains unchanged. Reheating of physically aged materials above their plasticization temperature can heal the effects by reorientation processes, causing the new parts to be formed in an unaged state. Similarly, the loss of additives can be compensated by respective additions during reprocessing.

2.2.2 Chemical Aging

In contrast to physical aging, processes resulting in chemical aging change the molecular architecture of the polymer. Such processes can take place at every step in the life of a polymer material. Typical steps in the polymer lifecycle are:

- Processing and modification (e. g., extrusion or compounding)
- Shaping of the part (e. g., injection molding, blow molding, extrusion)
- Service life with single or multiple use
- Potential start of a second material cycle by mechanical or chemical recycling
- End of life by waste disposal (landfill or recovery of thermal energy)

In these various steps, the material is affected by shear stress during processing or environmental influences such as temperature, level of oxygen, humidity, light radiation, mechanical stress, and the presence of living organisms. Consequently, there are a wide range of chemical reactions that can lead to chemical aging and so change the material properties. Industry has therefore found innovative ways to mitigate or even inhibit these processes with special additives.

2.2.2.1 Thermal and Thermomechanical Aging

For material processing, the polymer is heated above its plasticization temperature to make it formable. In addition to thermal stress, the material experiences mechanical stress during processing, for example by conveying with a screw in extrusion and compounding, where the material is affected by shearing between the screws and the wall of the extruder. This can result in local "hot spots" and overstretching of the chains, leading even to chain breakage. Consequently, materials with inherent instability can start to degenerate. The main degeneration mechanisms are *side-group elimination*, *random chain scission*, and *depolymerization*.

Side-Group Elimination

Upon being subject to a specific thermal stress, side-groups of certain polymers can be eliminated. This is one of the main decomposition pathways during the processing of PVC. During polymerization, tertiary or allylic chlorine atoms are formed as side-reactions. These lead to chemical aging of the material by side-group elimination at temperatures above 80 °C.

The decomposition eliminates hydrogen chloride from the polymer chain by forming a double bond (Figure 2.17(a) and (b)). The allylic chlorine next to that double bond can react accordingly (Figure 2.17(c)). In this manner, a conjugated double bond system can be formed. Upon reaching a length of six conjugated bonds, the material becomes colored and thus the damage becomes obvious [8]. Furthermore, the formed hydrogen chloride acts as a catalyst for the side-group elimination and thus accelerates the aging process. To overcome this decomposition pathway, additives are usually applied, which react with allylic chlorides by forming a stable bond. Moreover, it is important to bind the formed hydrogen chloride. Examples for such additives are organotin, phosphite, and epoxy-based stabilizers.

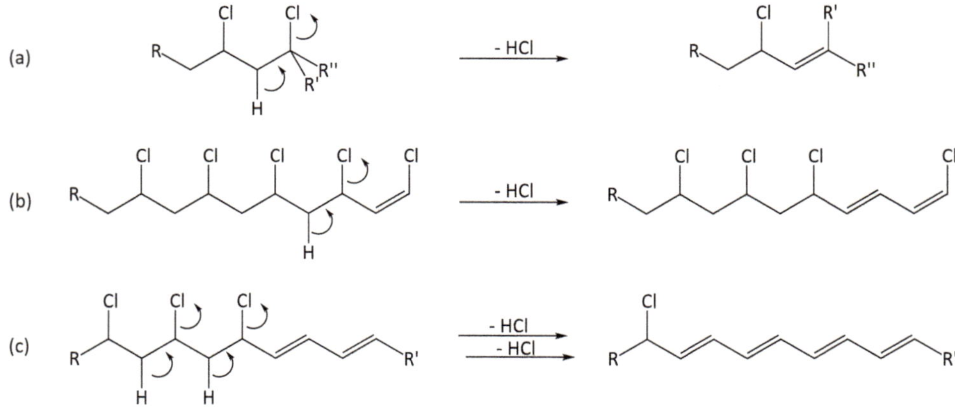

Figure 2.17 Decomposition of PVC. Hydrogen chloride is formed by side-group elimination of tertiary (a) or allylic chlorine atoms (b, c)

Random Chain Scission

Polymer chains may experience local temperature peaks and large local mechanical stresses especially in the entanglements during extrusion. If these applied forces exceed the binding force of the backbone functionality in a polymer, the chain can break homolytically into two fragments, which carry both a radical species (Figure 2.18(a)). If oxygen is present, these radicals will react with it. This so-called thermal oxidative aging will be discussed later. Without oxygen, the radicals can undergo β-cleavage (Figure 2.18(b)), recombination (Figure 2.18(c)), and disproportionation (Figure 2.18(d)). If double bonds are present, addition to these can take place (Figure 2.18(e)) [9]. While β-cleavage reduces the molecular weight, recombination and addition to double bonds increase it. Depending on the molecular architecture, either processes to reduce or to increase the molecular weight dominate. For example, PP and PS typically show a decrease of the molecular weight during processing, while PE shows an increase of the molecular weight, particularly with a large number of vinyl bonds present [9]. This aging can be overcome by reducing shear, either by changing to a lower-shear extrusion setup, or by applying lubricants.

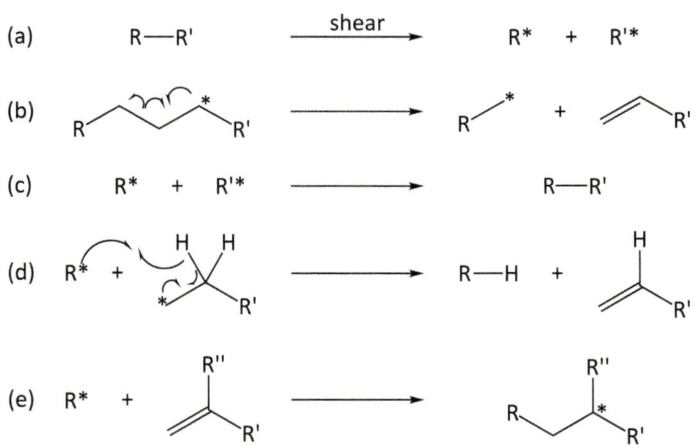

Figure 2.18 Formation of radicals by homolytic chain scission (a) and potential consecutive reactions: β-cleavage (b), recombination (c), disproportionation (d), and addition to double bond (e)

Depolymerization

The stability of the polymer chain depends on the functional groups bound to the backbone. Especially for space-demanding groups, the energy gain from the polymerization reaction is so small that the polymerization can be reversed just by heating if chain cleavage occurs. An important parameter for this is the so-called "ceiling temperature" at which the polymerization reaction has the same speed as the reverse reaction forming monomers from the polymer. Above the ceiling temperature, the depolymerization reaction becomes dominant and thus monomers are formed. Examples are poly(α-methylstyrene) with a ceiling temperature of 61 °C, and polymethylmethacrylate (PMMA) with a ceiling temperature of 164 °C [10]. Nonetheless, a low ceiling temperature for such polymers should not be viewed purely as a disadvantage, because it offers a great opportunity for chemical recycling back to the monomers simply by heating. An example of such a system is described in Section 9.4 for polystyrene with a ceiling temperature of 310 °C [11]. Notably, the literature states a range of different ceiling temperature for polymers like PS (230 °C, Section 7.3, reference 32; 390 °C, Section 9.4, reference 1). This could be attributed to different ceiling temperature measurement techniques applied.

2.2.2.2 Oxidative Aging

Degeneration in the presence of oxygen is the most important aging mechanism for polymers. Its mechanism was firstly investigated by Bolland and Gee for rubbers, which are materials containing double bonds [12]. They proposed an autoxidation by a radical chain mechanism.

The decomposition reaction is initiated by breakage of a chemical bond and the formation of a radical species. This radical formation can be induced by light (especially UV light) during the service life, shear forces or heat during processing, or other influences like impurities or residues of catalysts (Figure 2.19(a)). In contrast to the previously discussed oxygen-free radical mechanism, this radical reacts rapidly with the present oxygen by forming a peroxyl radical (Figure 2.19(b)). This is followed by hydrogen abstraction from another polymer chain to form an organic hydroperoxide (Figure 2.19(c)) and an active radical. The formed organic hydroperoxide can decompose by different mechanisms to two active radicals (Figure 2.19(d) and (e)). These reactions typically require elevated temperatures but can also take place at lower temperatures in presence of catalyzing metal ions or light. The formed hydroxyl radicals and alkoxyl radicals are more reactive than the peroxyl radicals. Thus, they easily perform hydrogen abstractions and transfer the active site to a new chain (Figure 2.19(f) and (g)). It is also possible that alkoxyl radicals perform a β-cleavage to different chain fragments and thereby lower the molecular weight of the polymer chains (Figure 2.19(h)). Active radicals can terminate either by recombination or disproportionation (Figure 2.19(i) and (j)). Recombination is possible for all mentioned radical species and is shown here for a radical-carrying chain as an example. Notably, the radicals can also cross-link different chains.

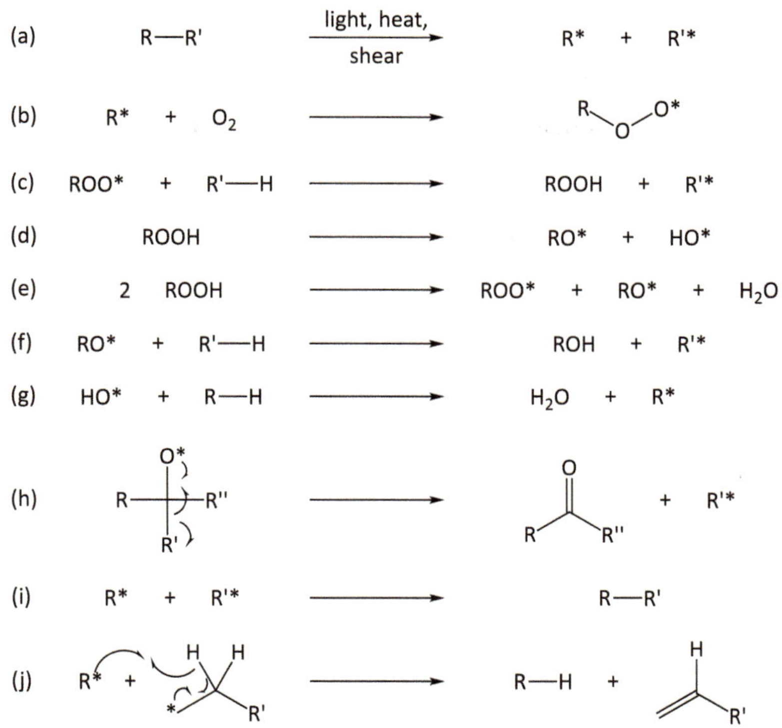

Figure 2.19 Oxidative aging of a polymer. Radicals are formed by the influence of light, heat, or shear (a). After addition of oxygen (b) and subsequent hydrogen transfer (c), an organic hydroperoxide is formed which can decay in further radicals (d, e). These radicals can transfer the active center to other chains (f, g) or eliminate a radical species by β-cleavage (h). Termination reactions are recombination (i) and disproportionation (j)

The self-acceleration of this mechanism is based on step (c), in which one radical reacts to form another radical and an organic hydroperoxide. Upon decomposition, the latter forms additional radicals and thus increases the number of active centers. The rate-determining step for this process is the abstraction of a hydrogen atom and thus the formation of an organic hydroperoxide. This abstraction is favored for easily abstractable hydrogen atoms such as allylic hydrogens or tertiary hydrogens and hence leads to a vulnerability of polymers in general to oxidative aging. However, oxidative aging is also observed for polymers where hydrogens cannot be subtracted easily. M. Coote et al. suggested that polymer defects like local double bonds play a substantial role in this oxidative aging process [13]. Additional pathways for the autoxidation of aliphatic polymers could be reactions with other atmospheric oxidants (e.g., ozone or hydroperoxyl radicals) [14]. Although there is still some research needed to study the exact mechanism for each polymer, it is undeniable that oxidative aging occurs for most polymer classes.

The effect of oxidative aging depends on the polymer architecture. It can result in a decrease of the molecular weight by chain scission or an increase of the molecular weight by branching or cross-linking. Furthermore, the oxidized material features functional groups that can affect the water uptake or the color of the material. The formed oxidized groups and decomposition products depend on the chemistry of the respective polymer. Examples are peroxides, alcohols, ketones, aldehydes, acids, peracids, peresters, and γ-lactones [15, 16]. Some of these decomposition products absorb light in the visible regime, which is observable as a yellow (or more intense) coloration of the material [17].

To overcome oxidative degeneration, different kind of stabilizers can be applied. The main groups are hydroperoxide decomposers such as phosphites, and radical-scavenging agents such as sterically hindered phenols and hindered amine light stabilizers (HALS). UV-absorbing additives can absorb the critical radiation during use, and therefore reduce the formation of the radical species and hence the degeneration processes.

The damage resulting from oxidative aging is irreversible and must be considered in recycling. Mechanical recycling of strongly oxidatively aged materials allows only applications not requiring the performance of the virgin material. Furthermore, repeated use of recycled materials accelerates oxidative aging, since decisive functional groups like hydroperoxides are already present. These functions can initiate further aging [18].

2.2.2.3 Aging by Hydrolysis and Biological Processes

In general, hydrolysis describes the splitting of a component into two or more molecules in aqueous media. It is the reverse reaction of the condensation (Figure 2.12). Like condensation, hydrolysis is also an equilibrium reaction and can be influenced by the concentration of water. While water must be removed in condensation reactions to shift the equilibrium to the product side, in contrast hydrolysis *requires* the presence of water (Figure 2.20).

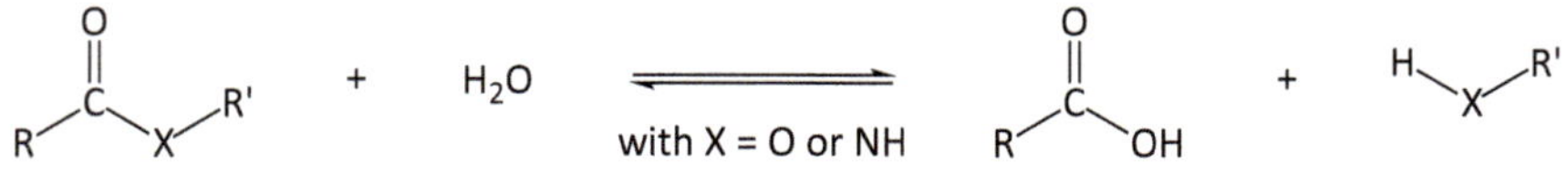

Figure 2.20 Esters and amides hydrolyze in an equilibrium reaction to carbonic acids as well as amines and alcohols, respectively.

Consequently, this kind of aging affects primarily polymers prepared by step-growth polymerization such as polyesters and polyamides. In contrast, chain-growth polymers like PE and PS are not affected since the C–C bond cannot be hydrolyzed. However, the polymerization mechanism gives a first hint of whether the material is sensitive to hydrolysis. More precisely, the probability of hydrolysis reactions occurring depends on the presence of sensitive functional groups within the material. This can be also in side-groups. For example, the chain-growth polymer PMMA features ester groups as side-groups, which can also be hydrolyzed. Polyamides synthesized by anionic ring-opening polymerization are also hydrolyzable. Oxidative aging can also lead to hydrolysis by the formation of hydrolyzable functional groups.

The reaction itself needs a certain activation energy and thus takes place very slowly during the service life of a polymer. The required activation energy can be reduced by catalysts like acids and bases, resulting in an accelerated hydrolysis. The reaction can also be accelerated by elevation of the temperature. For example, melt processing of a polyester that is not thoroughly dried results in a decrease of the molecular weight and thereby the mechanical properties. This is particularly crucial for susceptible groups present in the polymer backbone, as each hydrolysis reaction will split the polymer chain into two chain fragments. Reactions within the side-groups have a less pronounced effect on the molecular weight, but can still result in property changes such as a higher polarity or a higher absorption of oxygen and water.

In biological systems, amide and ester functionalities are the basis of all living matter. They are, for example, key functionalities in proteins and lipids. Nature has developed sophisticated enzymes for the synthesis and cleavage of such units. Due to the structural similarity of the functional units in natural and artificial hydrolyzable polymers, the same enzymes can catalyze the decomposition of such polymers. Hence, these polymers can be hydrolyzed to a certain degree within a natural environment. The speed of this biological process depends on the polymer architecture and particularly the accessibility of the functional units by the enzyme. The solubility in water is crucial for this process, since enzymes usually act in an aqueous environment. Easily affected polymers like polylactic acid [19], polybutylene succinate [20], and polyvinyl alcohol [21] are even rated as biodegradable polymers. However, decomposition of those hydrolytically susceptible polymers can take up to several years, depending on the environment.

Aging by hydrolysis is partly reversible in some cases due to the equilibrium character between hydrolysis and condensation. Consequently, if the heat treatment in mechanical recycling is conducted under anhydrous conditions, the resultant partial shift of the equilibrium leads to an increase of the molecular weight [22]. Hydrolysis is also a powerful approach for the chemical recycling of polycondensates. The monomers obtained in this way can then be purified and used again for synthesizing the same (or other) polymer materials again.

2.3 The World of Thermoplastic Industrial Polymers

2.3.1 Overview of Polymer Classes and Use

More than 200 thermoplastic polymer classes are applied in industry. So far, we have focused on the basic principles of polymer architectures, their behavior in the solid state and the general polymerization mechanisms. An application-oriented categorization according to their thermal and chemical properties is more common. Based on the market volume and application fields, polymers are categorized into commodity polymers, engineering polymers, and high-performance polymers. Illustrating them in the form of a pyramid (Figure 2.21) also indicates the increase of the market volume from the top to the bottom – for example, commodity polymers have a substantially higher volume than high-performance polymers. It should be noted that the volume of polyolefins (PE-LD, PE-LLD, PE-HD, and PP) counts for almost 50% of the total polymer market volume. The inverse direction, from the bottom to the top, represents the increasing price and better performance of the polymers. Additionally, the behavior in the solid state is considered by listing the amorphous polymers on the left and the semicrystalline polymers on the right. To complete the overview, the main polymerization mechanism of the respective polymer classes is also highlighted. Thus, polymers prepared by chain-growth polymerization are highlighted in red, while polymers prepared by step-growth polymerization are shown in blue. Typically, engineering or high-performance polymers are prepared by step-growth mechanism, whereas commodity polymers are prepared by chain-growth mechanism.

Commodity polymers are used for applications that have generally lower property requirements regarding service temperature and chemical properties (e.g., packaging and food containers). These polymers are usually produced in high volumes and are comparably inexpensive. This is possible since most of the monomers can be easily synthesized from the oil feedstock in sufficient quantities, for example by naphtha cracking. The most widely used thermoplastics, polypropylene and polyethylene, are among them. Both polymers had a combined market volume of 49.2% in Europe in 2019 (Figure 2.22).

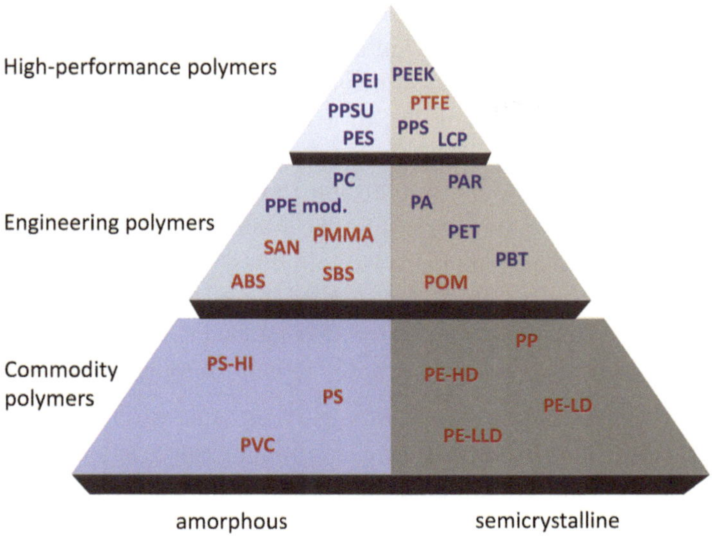

Figure 2.21 The polymer performance pyramid divided into commodity polymers, engineering polymers, and high-performance polymers, and the respective behavior in the solid state with prominent examples; red-labeled polymers are prepared by chain-growth, while blue-labeled polymers are prepared by step-growth polymerization. PVC: polyvinyl chloride (PVC has typically a low crystallinity and thus is often categorized as amorphous); PS: (general-purpose) polystyrene; PS-HI: high-impact polystyrene; PE (LLD, LD, HD): polyethylene (linear low density, low-density, high-density); PP: polypropylene; ABS: acrylonitrile-butadiene-styrene copolymer; SBS: styrene-butadiene-styrene block copolymer; SAN: styrene-acrylonitrile copolymer; PMMA: poly(methyl methacrylate); PPE mod.: modified poly(*p*-phenylene ether) (usually with PS, PS-HI or PA); PC: polycarbonate; POM: poly(oxymethylene); PBT: poly(butylene terephthalate); PET: poly(ethylene terephthalate); PA: polyamide; PAR: polyaramide; PES: poly(ether sulfone); PPSU: polyphenyl sulfone; PEI: polyether imide; LCP: liquid crystalline polymer; PPS: poly(phenylene sulfide); PTFE: polytetrafluoroethylene; PEEK: poly(ether ether ketone)

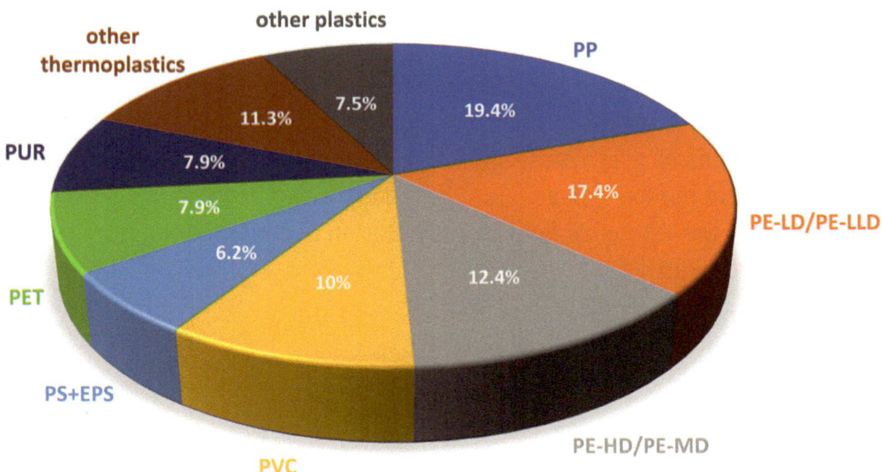

Figure 2.22 Plastic demand distribution by polymer type in 2019 in EU28+NO/CH (Adapted from [23])

Engineering polymers feature better mechanical and/or thermal properties than commodity polymers and thus are suitable for applications with higher requirements, such as in mechanical parts or construction parts. Important properties are high impact strength, good sliding and abrasion behavior, high heat deflection temperature, low deformation upon continuous load, and chemical stability as well as fire safety. Due to the very low weight and the beneficial properties, they have gradually replaced traditional materials like wood and metals in many applications. Since each polymer grade has a very unique property profile, a suitable material has to be selected carefully for each application. Their production usually requires more expensive monomers or more complex processing. Hence, these materials are only used in applications where the better property profile is needed.

High-performance polymers are used for highly demanding applications like in the aerospace sector. They usually feature continuous service temperatures above 150 °C, excellent mechanical properties, and chemical resistance. In some applications, very specific properties are also required, like the liquid-crystalline behavior of liquid-crystalline polymers (LCPs), which are widely applied in LCP-monitors and displays. Due to their typical high heat deflection temperatures, high-performance polymers require higher plasticizing temperatures than commodity and engineering polymers. Consequently, modified high-temperature equipment for temperatures above 300 °C and more complex techniques are required. Together with the high costs for the monomers, that leads to their very high price of 5–20 €/kg, in special cases also up to 100 €/kg [24].

Considering the amounts of polymers used in different industries and applications, it is obvious that commodity polymers are preferably applied within the sectors of household, leisure & sports, packaging, building & construction, and agriculture. Their percentage was over 67% of the total demand in 2019 (Figure 2.23). In contrast, engineering polymers and high-performance polymers are typically applied in the more demanding applications within the sectors automotive and electrical & electronic. However, the property profile of the pure polymer is mostly not sufficient for the targeted application. To overcome this, the property profile is adjusted by additives. Prominent examples are fillers or reinforcing fibers to improve the mechanical properties, antioxidants and UV/light stabilizers for long-time stability and outdoor usage, and processing aids such as lubricants. A more detailed view on the extensive field and application of additives can be found in the relevant literature [25].

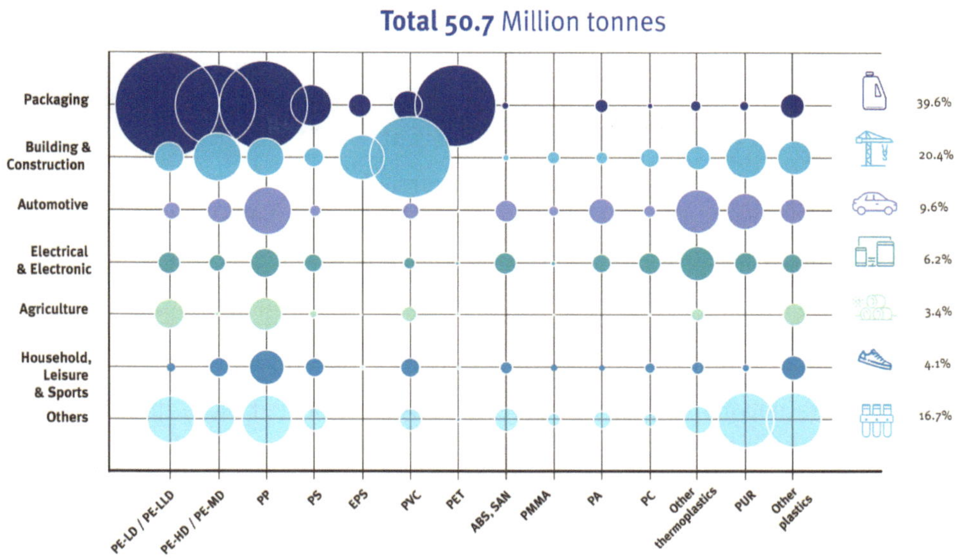

Figure 2.23 Plastic demand 2019 in Europe broken down into important industrial segments and polymer types [23]. Data for EU28+NO/CH (source: PlasticsEurope Market Research Group and Conversio Market & Strategy GmbH)

2.3.2 Profiles of Important Commercial Polymers

This section presents profiles of important commercial polymers and their characteristic features. Focus is given to the history, manufacturing, types, physical properties, processing techniques, and applications. Emphasis is also placed on recycling possibilities, which will be discussed in more detail in the corresponding chapters of this book.

2.3.2.1 Polyethylene

Structural formula:

$$-\!\!\left[H_2C-CH_2\right]_n\!\!-$$

Abbreviation according to DIN 7728: PE

Recycling code:

History [26]:

Polyethylene was first synthesized by the German chemist Hans von Pechmann by accident in 1898. His synthetic route was based on the toxic and unstable diazomethane, and did not became practically relevant.

The first industrial synthesis of PE-LD was carried out by chemists from Imperial Chemical Industries (ICI) in 1933, also by accident. They obtained a white waxy material from a mixture of ethylene and benzaldehyde at high pressures. After working on a reproductive high-pressure procedure, ICI started the production of PE-LD on a small industrial scale in 1939. A breakthrough for polyethylene was the use as insulation material for high-frequency applications during World War II.

Low-pressure polymerization was developed by Karl Ziegler et al. in 1953. Here, catalysts are used to polymerize ethylene by coordination polymerization at lower pressure.

Manufacturing:

Polyethylene is manufactured by chain-growth polymerization of ethene:

$$n\ H_2C{=}CH_2 \longrightarrow \left[H_2C - CH_2 \right]_n$$

High-pressure process: Ethene is channeled at a pressure of 1400–3500 bar and a temperature of 150–350 °C through stirring autoclaves or through tubing with a length of several hundred meters. The obtained solution of polyethylene in non-polymerized supercritical ethene is cooled and the pressure is relieved. After precipitation, *low-density polyethylene (PE-LD)* is obtained from this radical polymerization. Non-polymerized monomer is recycled.

Low-pressure process: Ethene is led through organic solvents with metal organic catalysts (e.g., Ziegler–Natta catalysts) at a pressure of up to 10 bar at normal or slightly increased temperature. After precipitation, *high-density polyethylene (PE-HD)* is obtained from this coordination polymerization. Copolymerization of ethene with α-olefins using a similar procedure results in *linear-low density polyethylene (PE-LLD)*.

Types:

- *PE-HD*: Practically linear polymer with a crystallinity of 70–80% [1] and only a few short side chains (5 per 1000 chain units). It typically has a density of 0.94–0.96 g/cm^3 and a melting temperature of 126–135 °C [7].
- *PE-LD*: Polymer with a crystallinity of 35–55% [1] and a medium number of branched side chains (8 to 40 per 1000 chain units). It typically has a density of 0.915-0.92 g/cm^3 and a melting temperature of 105–118 °C [7].

- *PE-LLD*: Polymer with a crystallinity of 55–65% [1] and manifold short side chains (up to 100 C_4 to C_8 groups per 1000 chain units). It typically has a density of around 0.935 g/cm^3 and a melting temperature of around 126 °C [7].

Processing techniques:

The most common primary shaping techniques are extrusion, injection molding, and blow molding. Post-processing techniques include adhesive bonding, welding, machining, and imprinting.

Application examples:

PE-LD is used for foils, pipes, and cable coatings. PE-LLD is typically used for foils with good optical properties. PE-HD has application for household goods, containers, and pressure pipes.

Possible recycling routes:

Purified polyethylene waste can be recycled very well by mechanical recycling, which maintains the chemical structure of the polymer. It can be directly melted to new products or processed to regranulate (Section 8.2). Typical recycled products are films, garbage bags, canisters and drums, trash cans, drinking water pipes, landfill liners, and cable insulations [27]. By chemical recycling, polyethylene waste can be transformed into a hydrocarbon feedstock for the manufacturing of new polyolefins (Section 9.2). A further route with a focus on energy recovery is thermal pyrolysis. The process is very economic and does not require sorted and purified waste fractions. However, the material leaves the circle irreversibly.

2.3.2.2 Polypropylene

Structural formula:

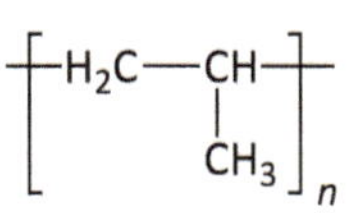

Abbreviation according to DIN 7728: PP

Recycling code:

History:

Polypropylene was presumably synthesized for the first time by J. P. Hogan and R. Banks in 1951. In 1955, G. Natta and P. Chira reported metallic catalysts to produce isotactic polypropylene (PP-I). The commercial production of PP-I started in Italy by Montecatini in 1957.

Manufacturing:

Polypropylene is manufactured by chain-growth polymerization of propene.

The synthesis is usually carried out by coordination polymerization by Ziegler–Natta catalysts or metallocene catalysts. Typical procedures are gas-phase polymerization, bulk polymerization, and slurry polymerization.

Types:

Most types of polypropylene homopolymers have an *isotactic* structure (PP-I) with a substituent on the same position along the main chain. In addition to this, *syndiotactic* PP (alternating position of the substituent along the main chain; PP-S) and *atactic* PP (random position of the substituent along the main chain; PP-A) are available. Melting temperatures for PP-S are around 138 °C and for PP-I around 186 °C, while PP-A is an amorphous material and does not have one. The density depends strongly on the crystallinity of the material, since the densities of the crystalline and the amorphous phase are different. The density for the amorphous phase of PP-I is in the range 0.850–0.854 g/cm^3, while the crystalline phase of PP-I has a density of 0.932–0.943 g/cm^3. For PP-S, it is 0.856 g/cm^3 for the amorphous and 0.91–0.989 g/cm^3 for the crystalline part. The density of PP-A is in the range of the above-mentioned amorphous phases of PP-I and PP-S [28].

Processing techniques:

The most common primary shaping techniques are extrusion (mainly fibers and foils) and injection molding (mainly thin-wall injection molding). Expanded polypropylene (EPP) can be obtained by using physical or chemical blowing agents. Post-processing techniques include adhesive bonding, welding, machining, and imprinting.

Application examples:

Polypropylene is mainly applied for automotive components, household goods, pipes, packaging, and fibers.

Possible recycling routes [29]:

Purified polypropylene waste can be recycled very well by mechanical recycling, which maintains the chemical structure of the polymer. It can be directly melted to new products or processed to regranulate (Section 8.2). By means of chemical recycling, PP waste can be transformed into a hydrocarbon feedstock for the synthesis of new polypropylene or other chemical products (Section 9.2). The thermal decomposition for energy recovery is also possible and not critical.

2.3.2.3 Polyvinyl Chloride

Structural formula:

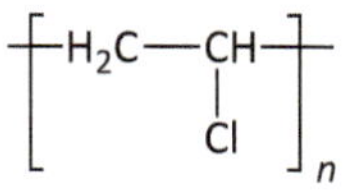

Abbreviation according to DIN 7728: PVC

Recycling code:

03
PVC

History:

F. Klatte from Griesheim-Elektron patented the polymerization and industrial production of PVC in 1912.

Manufacturing:

Polyvinyl chloride is manufactured by polymerization of vinyl chloride:

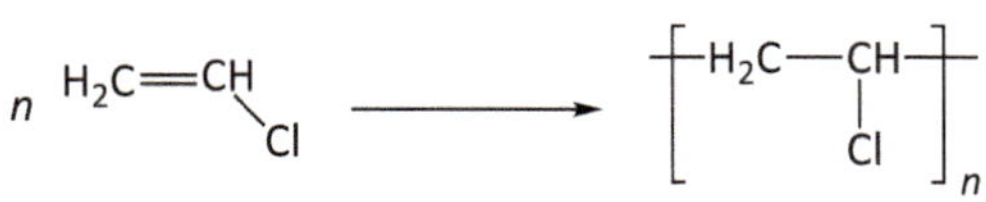

The synthesis is usually carried out by emulsion polymerization (PVC-E) and suspension polymerization (PVC-S).

Types:

Hard (unplasticized; PVC-U) and soft (plasticized; PVC-P) polyvinyl chloride are commercialized. The difference between these materials originate from the addition of plasticizers to the material. PVC-U is hard and rigid. PVC-P has an increased softness and flexibility due to an added plasticizer. Common plasticizers are based on diesters of phthalic acid with long chain alcohols. PVC-U has a glass-transition temperature of around 81 °C [30] and a density of 1.38–1.40 g/cm^3 [7]. Both values can be lowered by adding a plasticizer [31].

Processing techniques:

The most common primary shaping techniques are extrusion (mainly pipes, panels, and profiles), calendaring, and compression molding. Injection molding and blow molding is also possible. Post-processing techniques include thermal forming, adhesive bonding, and welding.

Application examples:

Typical applications of PVC are foils, window profiles, floor coverings, pipes, cable coatings, and gaskets.

Possible recycling routes [32]:

Amongst all plastics, the recycling of polyvinyl chloride has the longest history. The mechanical recycling of PVC waste is well-established (Section 8.6), and chemical recycling is also possible. Depending on the process, synthesis gas or hydrogen chloride can be obtained at high temperatures. The thermal decomposition of PVC yields energy, but less than for polyolefins. Additionally, hydrogen chloride is formed, which can be collected and used directly or as chlorine source.

2.3.2.4 Polystyrene

Structural formula:

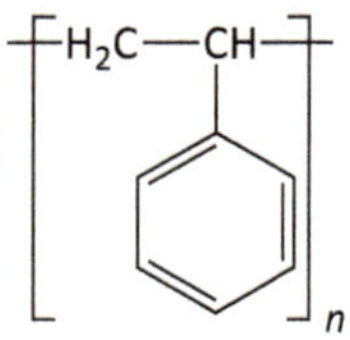

Abbreviation according to DIN 7728: PS

Recycling code:

06
PS

History:

Polystyrene was often synthesized in the past without recognizing the potential of the material (e.g., by E. Simon in 1935 and by A.W. Hofmann and J. Blyth in 1845). The technical production of polystyrene started at I.G. Farben in 1931.

Manufacturing:

Polystyrene is manufactured by polymerization of styrene:

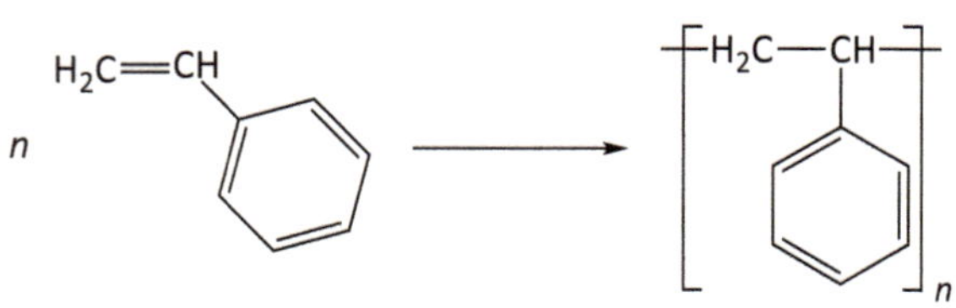

The synthesis is usually carried out by radical bulk polymerization, emulsion polymerization, and suspension polymerization. Anionic polymerization is also possible.

Types [7]:

Besides the homopolymer "general-purpose polystyrene", impact-modified versions are commercially available. "High-impact polystyrene" (PS-HI) contains additional polybutadiene rubber and has an enhanced impact strength. Copolymers, for example ABS and SAN, can be obtained from polymerization of styrene with comonomers such as butadiene and acrylonitrile. Polystyrene and its copolymers are usually amorphous polymers and thus have no melting temperature. The glass-transition temperature of general-purpose PS is around 100 °C, while it is increased by the addition of acrylonitrile as co-monomer in the respective copolymers. Other co-monomers like butadiene can also result in decreased glass-transition temperatures. The density of general-purpose PS is 1.04–1.05 g/cm^3, while it is 1.03–1.04 g/cm^3 for PS-HI and 1.03–1.07 g/cm^3 for ABS. SAN has a density of 1.07–1.08 g/cm^3.

Processing techniques:

The most common primary shaping techniques are injection molding and extrusion (mainly profiles, panels, and foils). Expanded polystyrene (EPS) can be produced by addition of carbon dioxide or a low-boiling hydrocarbon (e. g., pentane) as foaming agent. Post-processing techniques include adhesive bonding, welding, and imprinting.

Application examples:

Polystyrene is typically used for packaging, insulation, personal protection, domestic appliances, automotive parts, containers, and jewel cases.

Possible recycling routes:

Purified polystyrene waste can be recycled very well by mechanical recycling and melted to new products, or processed to regranulate. Typical recycled products are furniture, hangers, and plastic boxes (Section 8.4.1) [33]. Chemical recycling with a breakdown to the monomer is also possible, since PS depolymerizes above its ceiling temperature of 310 °C (Section 9.4) [11]. Thermal decomposition for energy recovery is also possible.

2.3.2.5 Polyethylene Terephthalate

Structural formula:

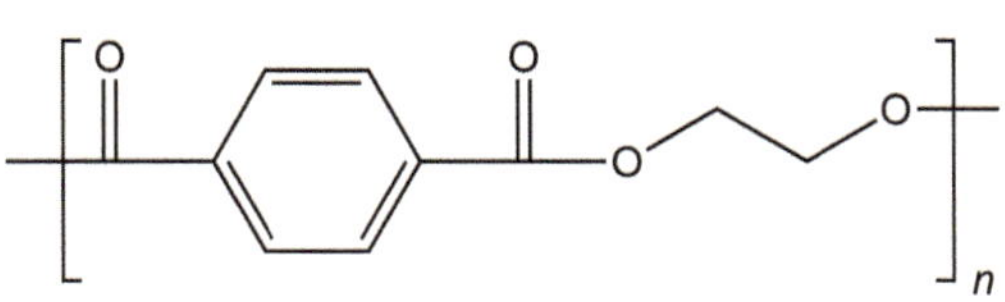

Abbreviation according to DIN 7728: PET

Recycling code:

History:

J.R. Whinfield and J.T. Dickson (Calico Printers Association) prepared textile fibers from PET in 1941. ICI started commercial production in 1953. The first PET bottle was used by Pepsi-Cola® in 1975.

Manufacturing:

Polyethylene terephthalate is manufactured by polycondensation of terephthalic acid and ethylene glycol:

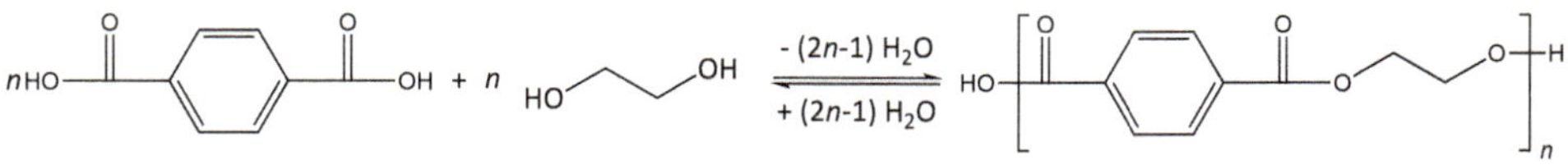

Industrial synthesis is carried out by transesterification of dimethyl terephthalate with ethylene glycol. Alternatively, melt-phase polycondensation with subsequent solid-phase polycondensation (due to low molecular weights in the first step) is possible. A further method is ring-opening polymerization, which allows high molecular weights due to the absence of condensate.

Types:

Depending on the processing method, two different materials, the semicrystalline PET-C and the amorphous PET-A, can be obtained. PET-C is formed in processes with comparably slow cooling at mold temperatures between 90 °C and 100 °C. Thereby, the material crystallizes, and the semicrystalline material is received. In processes with fast cooling and low mold temperatures between 20 °C and 40 °C, the PET solidifies in the amorphous state. If PET-A is tempered at temperatures above the glass-transition temperature, crystallization can occur, and a semicrystalline PET is formed. The glass-transition temperature is 67 °C for PET-A and 81 °C for PET-C. The density of PET-A is 1.335 g/cm^3, while PET-C has a density of 1.385–1.420 g/cm^3, depending on the crystallinity. The melting temperature of PET-C is 250–265 °C [34]

Processing techniques:

The most common primary shaping techniques are injection molding, stretch blow molding (bottles, containers), and extrusion (foils, panels, and profiles). Post-processing techniques include adhesive bonding and welding.

Application examples:

Polyethylene terephthalate is typically used for fibers, filaments, bottles, bearings, cog wheels, blister packaging, and food packaging.

Possible recycling routes:

Purified polyethylene terephthalate waste can be mechanically recycled very well, and remelted directly to new products or to regranulate. Typical recycled products are bottles, foils, and textile fibers (Section 8.3) [35]. Chemical recycling by depolymerization allows the respective monomers to be regained (Section 9.3). Thermal decomposition for energy recovery is also possible.

2.3.2.6 Polyamide

Structural formula:

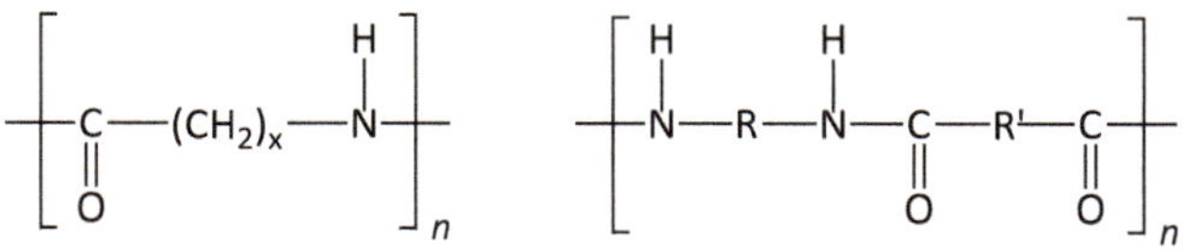

Abbreviation according to DIN 7728: PA

Recycling code:

History:

The first syntheses of polyamide 6.6 were carried out by W.H. Carothers from DuPont in 1935 and allowed the production of fully synthetic polymer fibers. DuPont started the industrial production of "Nylon 6.6" in 1939.

Manufacturing:

There are three methods to manufacture polyamide:

1. Anionic or cationic ring-opening polymerization of lactams

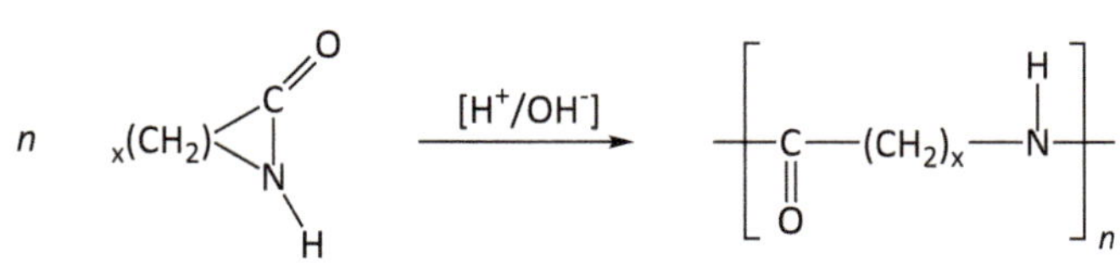

2. Polycondensation of diamines and dicarbonic acids

3. Hydrolytic chain polymerization of ε-caprolactam

ring opening

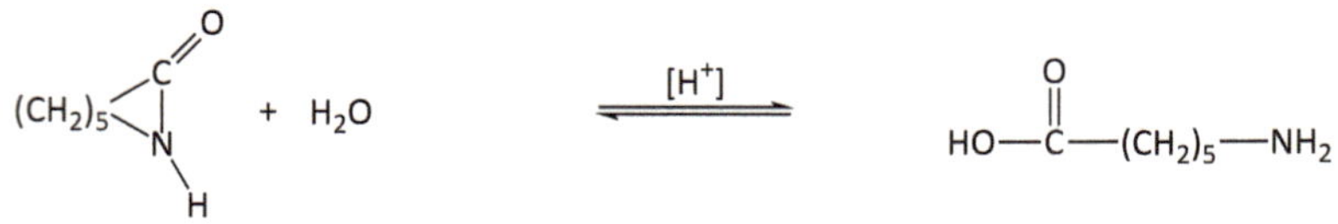

polymerization by polyaddition

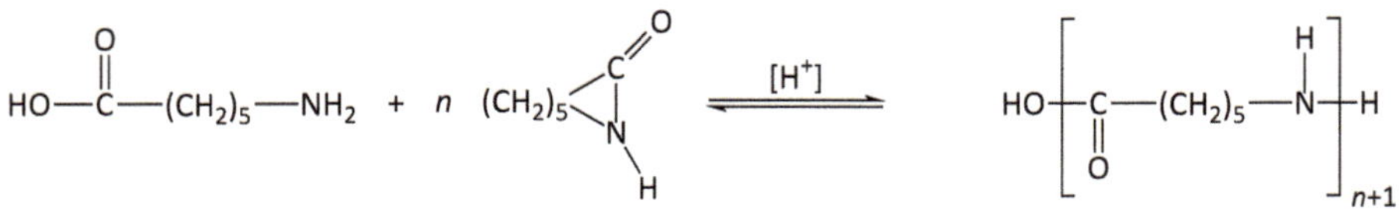

polymerization by polycondensation

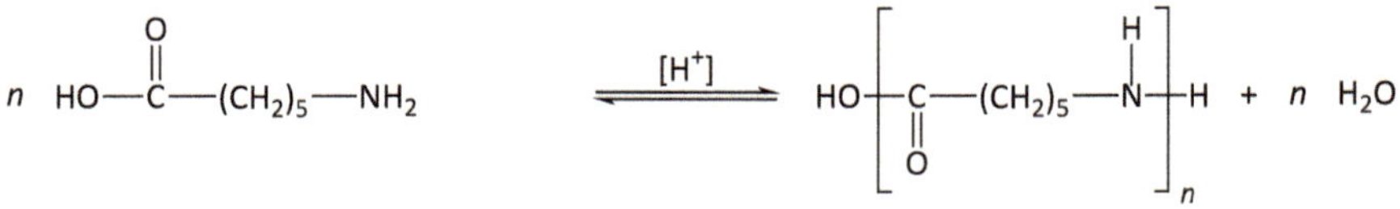

Types:

In general, three classes of polyamides are commercialized:

- Aliphatic polyamides: Building blocks can either be monomers featuring carboxylic acid and amine functions (ω-aminocarbonic acids or lactams) or a combination of diamine and diacid. In general, aliphatic polyamides are semicrystalline materials with a high polarity and thus a high water absorption. The low carbon-containing PA 4.6 shows a high melting temperature of 295 °C and a high water absorption of 3.8% under normalized conditions, while the less polar PA 12 has a melting temperature of 178 °C and a water absorption of only 0.7%. The di-monomer types have higher melting temperatures than comparable mono-monomer types (PA 6 with 222 °C and PA 6.6 with 260 °C) [7]. The density of the materials depends on the structure and on the processing. With fast cooling and thus a resultant low crystallinity, PA 6 has a density of 1.09 g/cm^3, while PA 12 has a density of 1.01 g/cm^3. For slow cooling and thus high crystallinity, the density of PA 6 is in the range 1.12–1.14 g/cm^3, while PA 12 shows a density of 1.04 g/cm^3 [36]. Glass-transition temperatures of aliphatic polyamides are around 50 °C [30].
- Polyphthalamides: These partially aromatic polyamides are based on the use of terephthalic acid or isophthalic acid as diacid component. Compared to aliphatic polyamides, the resulting polymers have a higher melting temperature (between 295 °C and 320 °C), a lower polarity corresponding to a lower water absorption, and a higher density (around 1.2 g/cm^3) [7].

- Fully aromatic polyamides: These polyamides are based on aromatic diamine and dicarbonic acid monomers. Due to the aromatic rings and directional amide bonds, linear polymer chains are achieved. As this conformation is also present in the molten state, these materials show a partly orientated behavior in melt and solution, which is called a liquid-crystalline behavior. Especially for small and linear monomers such as terephthalic acid and *p*-phenylene diamine giving the so-called Kevlar, a very high ordering with very strong hydrogen bonds is achieved. Such materials cannot be molten prior to decomposition and thus can only be processed by solvent-based processes (e. g., fiber spinning from sulfuric acid solution for Kevlar). With longer and twisted aromatic groups as well as the addition of substitutions on the aromatic rings, the ordering of such materials can be reduced and the materials can be processed by thermoplastic methods.

Processing techniques:

For primary shaping of aliphatic and polyphthalamides, all common processing techniques for thermoplastics (injection molding, extrusion, and blow molding) are possible. Foils can be prepared by the chill-roll process. Post-processing techniques include adhesive bonding, welding, and machining. Notably, pre-drying is necessary before processing.

Fully aromatic polyamides based on small monomers with linear architectures can only be processed by solvent-based processes like fiber spinning of Kevlar from a solution in sulfuric acid. With non-linear, non-planar, or partly substituted monomers, thermoplastic materials are possible, which can be processed by extrusion, injection molding, and thermoforming.

Application examples:

Aliphatic polyamide is used for technical parts (bearings, cog wheels, sealings, and cases), fibers, oil tanks, and sports equipment (ski shoes and roller-skates).

Polyphthalamides are typically used for applications at higher temperatures than aliphatic polyamides, for example technical automotive parts in engine compartments. Further applications are barrier layers in multi-layer foils.

Fully aromatic polyamides based on small monomers with linear architecture are usually used for fibers. Due to their high toughness, these materials are used in military applications such as bulletproof waistcoats. Fully aromatic polyaramides with non-linear, non-planar, or partly substituted monomers can be processed in a variety of shapes by thermoforming. Applications are usually elements exposed to dynamic load in the automotive sector like bearings or V-belt pulleys, or connectors in electronic applications.

Possible recycling routes:

Aliphatic and partially aromatic polyamides can be recycled mechanically and directly melted to new products or to regranulate (see Section 8.5.2.1). For example,

PA from the carpet industry can also be recycled chemically [37]. After splitting the polymer into its monomers and purification, new polymers can be regained (see Section 9.5.2.1). Non-recyclable material can be used for energy recovery by thermal decomposition.

Fully aromatic polyamide fibers can be recycled mechanically and used as reinforcement material for selected materials, similar to glass fibers. Other fully aromatic polyamides are usually not collected separately and recycled, since their total volume is comparably small. Consequently, they are usually thermally decomposed with the residual waste.

2.3.2.7 Polycarbonate

Structural formula:

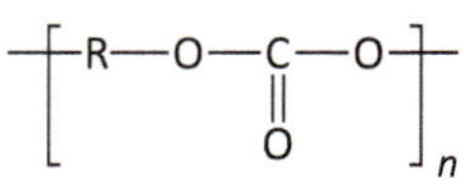

Abbreviation according to DIN 7728: PC

Recycling code:

History:

Polycarbonate was developed by H. Schnell (Bayer AG) in 1953. Commercial production started in 1958 under the brand name “Makrolon®”.

Manufacturing:

Polycarbonate is manufactured industrially by two methods. Polycarbonates with molecular weights of up to 30,000 g/mol can be prepared by melt condensation of bisphenol A and diphenyl carbonate, with basic catalysts and temperatures between 180 and 300 °C.

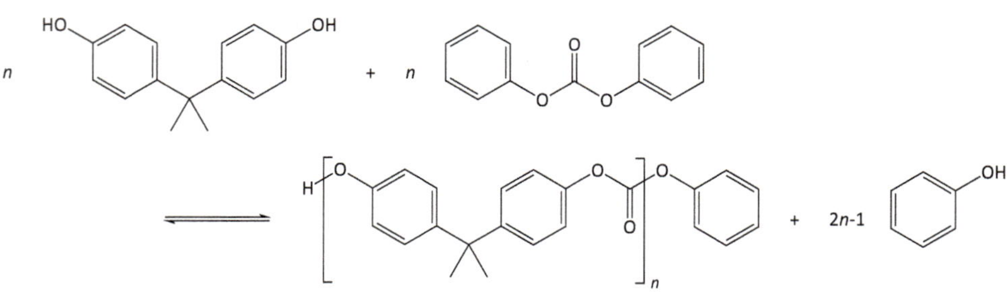

Polycarbonate with higher molecular weights of up to 200,000 g/mol is prepared by interface condensation of a caustic soda solution of bisphenol A with phosgene dissolved in an organic solvent. The reaction can by catalyzed by tertiary amines.

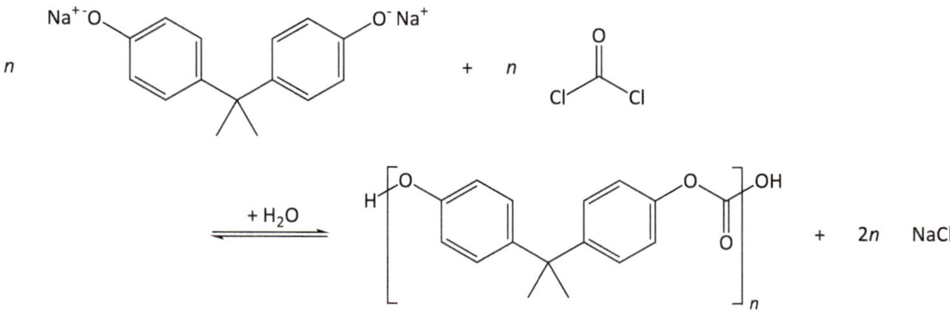

Types:

Polycarbonates are thermoplastic polyesters of carbonic acid with mainly aromatic diols. The most important type is PC based on bisphenol A (BPA). Polycarbonates are usually amorphous, polar, and highly transparent materials with a glass-transition temperature of 145 °C (BPA-PC) [30] and a density around 1.20 g/cm³ [7].

Processing techniques:

For primary shaping of polycarbonates, all common processing techniques for thermoplastics (injection molding, extrusion, and blow molding) are possible. Foils can be prepared by casting of dichloromethane solutions. Post-processing techniques include thermoforming, adhesive bonding, welding, and machining. Notably, pre-drying is necessary before processing.

Application examples:

Polycarbonates are used in optical data storage devices (e.g., CDs, DVDs, and Blu-Ray discs), glazing applications, laptop or smartphone housings, safety glasses, and helmets.

Possible recycling routes:

Polycarbonate can be recycled mechanically and directly melted to regranulate [38]. Research on chemical recycling has also been conducted, which has already identified several potential options [39]. Non-recyclable material can be used for energy recovery by thermal decomposition.

2.3.2.8 Poly(methyl methacrylate)

Structural formula:

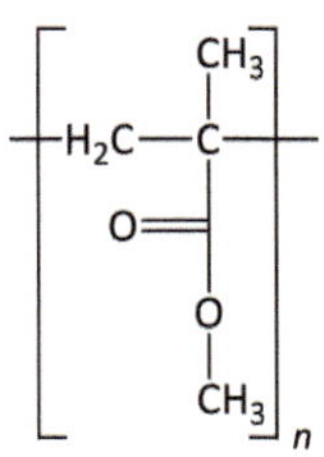

Abbreviation according to DIN 7728: PMMA

Recycling code:

History:

Otto Röhm produced the first "organic glass" by polymerization of methyl methacrylate in 1912. The industrial production of PMMA started in 1935 by Röhm & Haas under the brand name Plexiglas®.

Manufacturing:

Poly(methyl methacrylate) is manufactured by polymerization of methyl methacrylate:

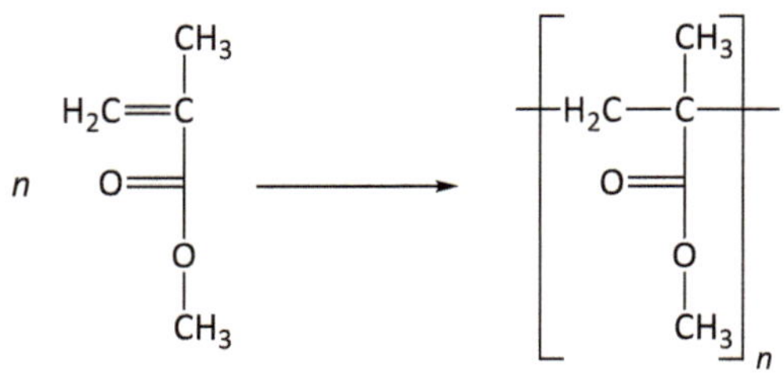

The synthesis is usually carried out by radical bulk polymerization, emulsion polymerization, and suspension polymerization. Anionic polymerization is also possible.

Types:

PMMA is an amorphous polymer with a glass-transition temperature of 105–106 °C and a density of 1.188–1.19 g/cm³ [40]. Key properties are the high transparency and a very high scratch resistance, which allows the application as acrylic glass.

Processing techniques:

The most common primary shaping techniques are injection molding, extrusion, and casting. Post-processing techniques include imprinting, adhesive bonding, and welding.

Application examples:

Poly(methyl methacrylate) can be used as a molding material and as cast resin. Examples for molding materials are sanitary equipment, glasses, lenses, wavelength filters, and optical storage devices. Examples of cast resins are glazing, encapsulations, floor coverings, road markings, and prosthodontics.

Possible recycling routes [41]:

Poly(methyl methacrylate) can be chemically recycled very well. Above the ceiling temperature of 164 °C [10], the monomer methyl methacrylate is regained, which can then be polymerized again to new polymers. Thermal decomposition for energy recovery is also possible. Mechanical recycling is not very common, as the material is not available in large amounts, and the high requirements for optical applications cannot be easily realized.

2.3.2.9 Polytetrafluoroethylene

Structural formula:

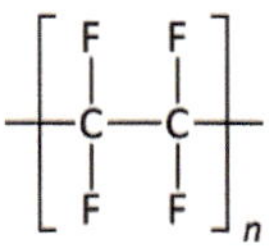

Abbreviation according to DIN 7728: PTFE

Recycling code:

History:

R.J. Plunkett and J. Rebok (DuPont) found polytetrafluoroethylene by accident during investigation of cooling agents in 1938.

Manufacturing:

Polytetrafluoroethylene is manufactured by polymerization of tetrafluoroethylene:

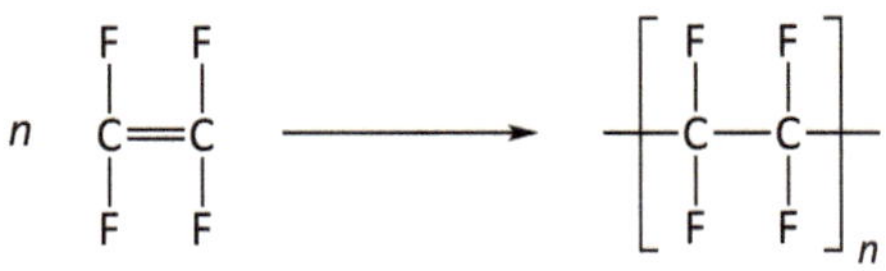

The synthesis is usually carried out by emulsion polymerization.

Types:

Polytetrafluoroethylene is a semicrystalline polymer with a density of 2.280–2.290 g/cm^3 and a melting temperature of 327 °C [42]. This high melting temperature allows continuous use at temperatures up to 250 °C. However, standard processing techniques like injection molding are not possible, since the material starts to decompose above 400 °C before a sufficient flowable melt is realized. Consequently, alternative techniques like sintering have to be applied.

Processing techniques:

The most common primary shaping techniques are sintering and powder extrusion. Post-processing techniques include adhesive bonding (after etching with solutions of alkali metals) and machining.

Application examples:

Polytetrafluoroethylene is typically used for gaskets, machine elements, crucibles, coatings, and foils.

Possible recycling routes:

Polytetrafluoroethylene can be mechanically recycled by regrinding and reprocessing. Furthermore, it is possible to obtain a lower molecular weight material by thermomechanical or light-assisted decomposition. This material can be applied as a glide agent. In both cases, the properties of the material are partly or totally different from the original material [43].

Polytetrafluoroethylene can also be chemically recycled. The monomer tetrafluoroethylene is regained by microwave-assisted depolymerization and can then be polymerized again to new polymers (Section 9.6).

References for Chapter 2

[1] Kaiser, W.: *Kunststoffchemie für Ingenieure*, Hanser, Munich (2006)

[2] Arunan, E., Desiraju, G.R., Klein, R.A., Sadlej, J., Scheiner, S., Alkorta, I., Clary, D.C., Crabtree, R.H., Dannenberg, J.J., Hobza, P., Kjaergaard, H.G., Legon, A.C., Mennucci, B., Nesbitt, D.J.: *Pure Appl. Chem.*, (2011) 83, pp. 1637–1641; DOI: 10.1351/PAC-REC-10-01-02

[3] Franck, A., Herr, B., Ruse, H., Schulz, G.: *Kunststoff-Kompendium*, Vogel, Würzburg (2011)

[4] Johlitz, M.: *Zum Alterungsverhalten von Polymeren: Experimentell gestützte, thermo-chemomechanische Modellbildung und numerische Simulation*, Dissertation, Universität der Bundeswehr München, Munich (2015)

[5] Ehrenstein, G.W., Pongratz, S.: *Resistance and Stability of Polymers*, Hanser, Munich (2013)

[6] Androsch, R., Stolp, M., Radusch, H.-J.: *Acta Polym.*, (1996) 47, pp. 99–104; DOI: 10.1002/actp.1996.010470206

[7] Oberbach, K., Baur, E., Brinkmann, S., Schmachtenberg, E.: *Saechtling Kunststoff Taschenbuch*, Hanser, Munich (2004)

[8] Schiller, M.: *PVC Stabilisatoren*, Hanser, Munich (2010)

[9] Wegmann, A., Le Gal, A., Müller, D.: In: Maier, R.D., Schiller, M. (Eds.), *Handbuch Kunststoff-Additive*, Hanser, Munich (2016), pp. 1–153

[10] Busfield, W.K.: In: Brandrup, J., Immergut, E.H. (Eds.), *Polymer Handbook*, John Wiley & Sons, New York (1989), pp. II/295-II/334

[11] Menges, G., Haberstroh, E., Michaeli, W., Schmachtenberg, E.: *Werkstoffkunde Kunststoffe*, Hanser, Munich (2002)

[12] Bolland, J.L., Gee, G.: *Trans. Faraday Soc.*, (1946) 42, pp. 236–243; DOI: 10.1039/TF9464200236

[13] Gryn'ova, G., Hodgson, J.L., Coote, M.L.: *Org. Biomol. Chem.*, (2011) 9, pp. 480–490; DOI: 10.1039/C0OB00596G

[14] Smith, L. M., Aitken, H. M., Coote, M. L.: *Acc. Chem. Res.*, (2018) 51, pp. 2006–2013; DOI: 10.1021/acs.accounts.8b00250

[15] Zweifel, H.: *Stabilization of Polymeric Material*, Springer, Berlin/Heidelberg (1998)

[16] Richaud, E., Diogo, O. O., Fayolle, B., Verdu, J., Guilment, J., Fernagut, F.: *Polym. Degrad. Stab.*, (2013) 98, pp. 1929–1939; DOI: 10.1016/j.polymdegradstab.2013.04.012

[17] Tröbs, M.: "Wrinkles" in Plastics due to Aging. *Kunststoffe Int.*, (2017) 8, pp. 45–47; *https://en.kunststoffe.de/a/specialistarticle/wrinkles-in-plastics-due-to-aging-260570*

[18] Phaendner, R.: In: Maier, R. D., Schiller, M. (Eds.), *Handbuch Kunststoff-Additive*, Hanser, Munich (2016), pp. 1203–1224

[19] Hastenreiter, L. L. G., Ramamoorthy, S. K., Srivastava, R. K., Yadav, A., Zamani, A., Åkesson, D.: *Polymers*, (2020) 12, p. 2849; DOI: 10.3390/polym12122849

[20] Yin, Y., Yang, J., Meng, L.: *J. Appl. Polym. Sci.*, (2021) 138 (51010); DOI: 10.1002/app.51010

[21] Mandal, S., Dasmahapatra, A. K.: *J. Polym. Res.*, (2021) 28, p. 269; DOI: 10.1007/s10965-021-02624-9

[22] Dimonie, D., Socoteanu, R., Pop, S., Fierascu, I., Fierascu, R., Petrea, C., Zaharia, C., Petrache, M.: In: Achilias, D. (Ed.), *Material Recycling – Trends and Perspectives*, InTech (2012), p. 85; 114; DOI: 10.5772/31841

[23] PlasticsEurope, *Plastics – the Facts 2020, https://www.plasticseurope.org/de/resources/publications/4312-plastics-facts-2020*, accessed July 28, 2021

[24] Brünglinghaus, C.: "Studie sieht Potenzial für Hochleistungspolymere", SpringerProfessional, *https://www.springerprofessional.de/werkstoffe/leichtbau/studie-sieht-potenzial-fuer-hochleistungspolymere/6561168#:~:text=Hohe%20Preise%20sind%20eine%20weitere,50%20und%203%2C50%20Euro*, accessed July 28, 2021

[25] Maier, R. D., Schiller, M.: *Handbuch Kunststoff-Additive*, Hanser, Munich (2016)

[26] Lautenschläger, K.-H., Schröter, W., Wanninger, A.: *Taschenbuch der Chemie*, Harri Deutsch, Frankfurt am Main (2007)

[27] Arbeitsgemeinschaft Verpackung + Umwelt e. V., *https://www.agvu.de/en/polyethylene-pe-467/*, accessed July 26, 2021

[28] Quirk, R. P., Alsamarraie, M. A. A.: In: Brandrup, J., Immergut, E. H. (Eds.), *Polymer Handbook*, John Wiley & Sons, New York (1989), pp. V/27-V/33

[29] Arbeitsgemeinschaft Verpackung + Umwelt e. V., *https://www.agvu.de/en/polypropylene-pp-471/*, accessed July 26, 2021

[30] Peyser, P.: In: Brandrup, J., Immergut, E. H. (Eds.), *Polymer Handbook*, John Wiley & Sons, New York (1989), pp. VI/209-VI/277

[31] Hitachi High-Tech Science Corporation, *https://www.hitachi-hightech.com/file/global/pdf/products/science/appli/ana/thermal/application_TA_065e.pdf*, accessed July 26, 2021

[32] VinylPlus, *https://vinylplus.eu/uploads/2015-12-08_Recycling-Technologies-German.pdf*, accessed July 26, 2021

[33] Arbeitsgemeinschaft Verpackung + Umwelt e. V., *https://www.agvu.de/en/polystyrene-474/*, accessed July 26, 2021

[34] Lawton, E. L., Ringwald, E. L.: In: Brandrup, J., Immergut, E. H. (Eds.), *Polymer Handbook*, John Wiley & Sons, New York (1989), pp. V/101-V/106

[35] Arbeitsgemeinschaft Verpackung + Umwelt e. V., *https://www.agvu.de/en/polyethylene-terephthalate-pet-469/*, accessed July 27, 2021

[36] Pflüger, R.: In: Brandrup, J., Immergut, E. H. (Eds.), *Polymer Handbook*, John Wiley & Sons, New York (1989), pp. V/109-V/116

[37] Mihut, C., Captain, D. K., Gadala-Maria, F., Amiridis, M. D.: *Polym. Eng. Sci.*, (2001) 9, pp. 1457–1470; DOI: 10.1002/pen.10845

[38] Plastic Expert, *https://www.plasticexpert.co.uk/plastic-recycling/polycarbonate-recycling/*, accessed July 27, 2021

[39] Kim, J. G.: *Polym. Chem.*, (2020) 11, pp. 4830–4849; DOI: 10.1039/c9py01927h

[40] Wunderlich, W., In: Brandrup, J., Immergut, E. H. (Eds.), *Polymer Handbook*, John Wiley & Sons, New York (1989), pp. V/77-V/80

[41] Kikuchi, Y., Hirao, M., Ookubo, T., Sasaki, A.: *Int. J. Life Cycle Assess.*, (2014) 19, pp. 120–129; DOI: 10.1007/s11367-013-0624-y

[42] Sperati, C. A., In: Brandrup, J., Immergut, E. H. (Eds.), *Polymer Handbook*, John Wiley & Sons, New York (1989), pp. V/35-V/56

[43] Industrieverband Halbzeuge und Konsumprodukte aus Kunststoff e. V., *https://www.pro-kunststoff.de/assets/Merkbl%C3%A4tter%20und%20Co/FP%20TM-10-Recycling-von-Fluorkunststoffen.pdf*, accessed July 27, 2021

3 Legislation

Michael Ulbrich, Sören Hörnicke

Governments and policy makers are taking action to address environmental challenges of wasteful and polluting plastic consumption, and to increase resource efficiency. Public awareness about marine litter and ocean plastic in particular has caused many legislators in recent years to become more active. In addition, a recognition of the environmental benefits of recycling such as reduced greenhouse gas emissions or reduced fossil feedstock usage have stimulated new legislation, often following an impetus from agreements such as the Paris Agreement or the European Green Deal.

Historically, as a means to reduce the impact of waste, in the waste treatment hierarchy recycling has been considered superior to energy recovery (incineration) and to disposal (landfilling), but less preferable than a reduction and prevention of waste. Today, recycling is increasingly integrated into a holistic approach to transitioning towards a circular economy and to keeping resources in the loop for as long as possible (Figure 3.1). While legislation is aimed at all levels of the waste hierarchy – for example, it supports waste prevention and reduction through a facilitation of shared economy models, of reuse, repair and product durability – in this chapter we will focus on regulatory action supporting recycling.

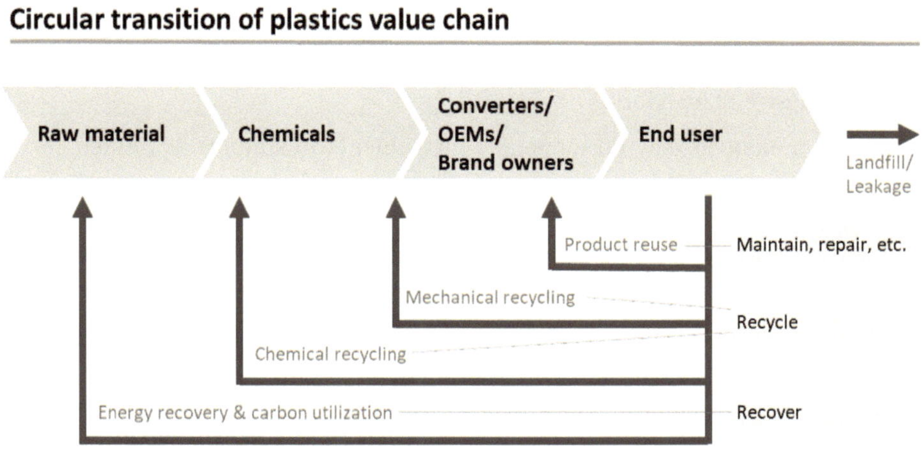

Figure 3.1 Recycling as part of a holistic transition to a circular economy [1]

3.1 A Categorization of Regulatory Approaches to Promote Recycling

Policy makers in political unions, governments, and municipalities are deploying a variety of approaches to promote plastics recycling, which are often intertwined with motivations such as advancing a circular economy, addressing environmental challenges, increasing resource efficiency, etc. This variety has resulted in a heterogeneous landscape of policies across world regions, countries, and municipalities. Generally, the policies can be distinguished into legal and financial measures (Figure 3.2).

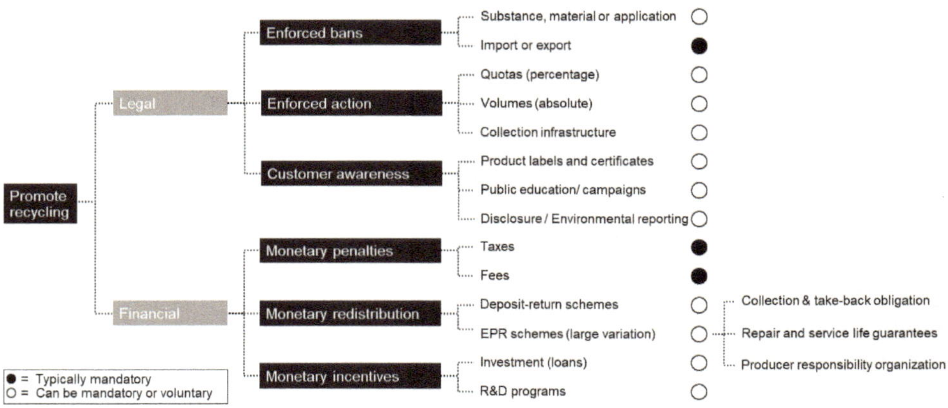

Figure 3.2 Legislation: categorization of approaches [1]

Legal measures include the prohibition or enforcement of certain behaviors, but also regulations to raise the awareness to encourage certain behaviors indirectly, including guidelines of how to set the desired behaviors into action. While such policies also have a financial impact on affected stakeholders, it is typically not possible for individual market participants to buy one's way into the right to continue non-desired behaviors.

Financial measures rely on monetary instruments to support or punish certain behaviors without making them illegal. They depend on incentives and market mechanisms rather than legal enforcement to lead the industry in the desired direction. The promoted actions and corresponding business models gain economic attractiveness and may allow for overcoming competitive disadvantages versus established business models that benefit from legacy value chains, mature technologies or inherent advantages or disadvantages.

In addition to regulation by policy makers, in recent years there has been a trend of **voluntary commitments** ("pledges") by market players to achieve certain targets or discontinue certain practices, often in an attempt to preempt regulatory action or to cater to recognized new customer preferences and thereby realize price premiums. These self-commitments follow similar mechanisms as some of the regulatory approaches, as indicated in Figure 3.2. They are typically introduced with the intention to create a competitive advantage, either for the company or the whole industry, which may unfold in shorter or longer time frames. The Ellen MacArthur Foundation's Global Commitment [2] for circular plastics, the voluntary phase-out of additives in PVC, or the voluntary commitment [3] to use 10 million tons of recycled plastics per year by 2025 by the Circular Plastic Alliance [4] are prominent examples.

While this chapter provides a possible categorization of regulatory approaches to promote recycling and discusses the characteristics of the individual mechanisms, it is worth noting that in practice, regulation can come in the form of broader programs that utilize a combination of the described mechanisms.

3.1.1 Legal Measures

Legal policies can be classified into enforced bans, enforced action, and raising customer awareness.

Enforced Bans

Enforced bans restrict the usage of certain **substances**, **materials** or **applications** as well as limit their permitted content in end products. Examples include the restricted substances under the REACH framework of the European Union [5], where listed substances are either limited in their use or banned altogether in the European Union due to environmental or health concerns, or the various bans on plastic bags or single-use plastics applications globally. The impact of such bans on recycling can be beneficial, but it some cases it can also be detrimental. For example, recycling is promoted if a substance, material or application that cannot be handled by available technology is banned, such as a mixture of multiple additives, or certain multi-layer products. Conversely, bans impede recycling if a substance gets banned that is present in legacy plastics, e.g., lead or cadmium in PVC [6]. In such cases, a trade-off arises whether to phase out the non-desired additives and forego opportunities to circulate the plastic material, or whether to keep the plastic material in a loop and accept unwanted additives. Bans on materials or applications may cut into the supply for recycling operations and may also support substitutes, e.g., glass, metal or paper instead of plastic, which can create unwanted consequences such as when the substitute material has a more adverse life-cycle impact.

Further, **imports and exports can be banned**. The import restrictions on low-quality and contaminated plastic waste, which were first introduced by China and consecutively by other South-East Asian countries to address local pollution and incorrect recycling methods, are a prominent example [7]. However, under this approach, there is a risk that waste streams get redirected, simply cascading the problems of waste imports to other regions that may have a less suitable and appropriate waste treatment infrastructure. Conversely, under the Basel Convention among the OECD member states, there is an agreed restriction on exporting certain hazardous wastes including contaminated plastic waste. A localization of waste treatment with appropriate environmental and health standards and the corresponding expansion of local recycling instead of exporting can be a desired effect.

Enforced Actions

Enforced actions describe a goal and actions to implement that goal. Targets are typically defined as quotas or volumes, as well as regarding the establishment of the necessary prerequisites such as a suitable collection infrastructure.

Quotas are expressed in percentage of a total volume and usually refer to a minimum or maximum fixed share of the total volume. On the other hand, targets can also be defined in terms of absolute **volumes**. Quotas and volumes can be set at various stages of the value chain, for example, for the collection of particular waste streams, for recycled volumes, or for the usage of recyclate in consumer products. For example, there are quotas on the collection of particular waste streams such as end-of-life vehicles or waste of electric and electronic equipment (WEEE) in the EU. The mandatory use of recyclate in beverage containers in California [8] is another example of a quota.

An advantage of quotas and volumes is that they clearly define a goal at a certain point in time. This offers policy makers and market players a timeline and the perspective to introduce measures in order to reach this goal. Targets for collection, recycling, and recyclate usage, by definition, are supporting more recycling activity.

However, quotas and volumes can also result in downcycling if there are no sufficient quality standards, which may create an incentive to deploy low-cost and inefficient low-value options to achieve the targets. What's more, quotas and volumes may suffer from weak definitions such as the point of calculation or import and export balances. For example, in the EU, the recycling rate used to be measured as the volume collected for recycling and not which amount of that was actually recycled. Process losses in separation, sorting and washing result in significant yield losses. With that definition under the Directive (EU) 94/62/EC, the recycling rate for plastic packaging in 2018 was 42%, whereas with a new definition planned under Directive (EU) 2018/852, materials entering pelletization and extrusion molding to obtain the same volume of recycled material will be reported as having a recycling rate of only 29% [9]. Further, exports with the purpose of recycling were counted towards the recycling rate without any assurance that the exported waste really was recycled appropriately. Moreover, the recycling rate used to be defined in relation to the collected waste stream, while only a certain share of the consumed volume in fact gets turned into collected volume of any given category. For example, 50.7 million tons of plastic have been consumed by converters in the EU28 plus Norway and Switzerland in 2019, but only 29.1 million tons were collected as post-consumer waste [10]. The difference is accounted for by the accumulation of plastic material in the anthroposphere, e.g., in buildings, but also due to leakage into the environment or import and export balances.

Enforced action can also refer to the implementation of a **collection infrastructure** to collect waste in a defined manner. Prominent examples include the obligation to introduce separated collection of different waste streams in the EU under the Directive 2008/98/EC for at least paper, metal, plastic and glass, and in the amending Directive (EU) 2018/851 for bio-waste, hazardous waste produced by households, and textile waste in all member states. A dedicated collection infrastructure facilitates recycling by providing pre-sorted waste streams with fewer contaminants. This reduces the required subsequent separation effort and increases the yield of valuable recycled plastic material.

Customer Awareness

Another legal regulation method revolves around raising **customer awareness**. The reasoning behind providing information is to build knowledge for the general public and specific groups such as companies or officials to act in favor of promoting recycling.

Product certification and labeling provide information about products and might be defined by law or certified by a third-party organization. Certificates or labels can be included directly on the packaging or advertisement of a product or can be issued as a document in paper or online where there is no packaging, e.g., for online services or for buildings. They may indicate the recyclability, the recyclate content, the correct end-of-life treatment of a product, or similar details. Typical examples include the labeling of certain recyclate content in packaging or the well-known recycling codes, which indicate what the material is made of. The certification of buildings which includes, inter alia, criteria on their sustainability and environmental impact, is less known but equally important (Figure 3.3). In that context, construction materials made of recyclate or being easily recyclable support a high score.

BREEAM United Kingdom 1990	**LEED** United States 1993	**LBC** United States 2006
DGNB Germany 2007	**Nordic Swan** The Nordics 2009	**WELL** United States 2014

Figure 3.3 Selected building certificates

Product certificates and labels drive recycling rates by helping buyers and consumers to adapt their purchase behaviors in favor of recyclable products or products containing recyclate. In addition, they facilitate a better sorting of end-of-life products and, consequently, the creation of more homogenous waste streams, which then allows for more and better recycling. Certificates and labels therefore create an incentive for market players to incorporate recycling in their product design, either by improving recyclability or by using more recyclate. However, some potential downsides need to be considered, too. Production certificates and labeling can be overwhelming and their criteria and meaning can remain unclear to end users and consumers. For example, the recycling codes for particular plastic types might not be understood by end consumers. Moreover, the labels only name the plastic types, but don't give an indication about the recyclability or the correct sorting for disposal. Finally, there is a challenge around trust and potential greenwashing. A certifier of a label might not be independent or might not be perceived as such, so that as a result, consumers may not pay attention or conclude adverse behaviors in the desired ways.

Public education and campaigns are targeting the general public to inform and create awareness about correct waste disposal, sorting and recycling options, etc. These can be courses in public education facilities, field trips to waste treatment sites, or broad campaigns using various channels such as information brochures, advertisements on TV, radio and online, or public signs explaining how to correctly dispose of waste. These methods are usually used by the government as large-scale campaigns and, especially by local municipalities and waste treatment authorities, to educate their public and industrial customers about local systems for waste disposal and treatment. Plastic recycling can benefit from public education and campaigns if consumers follow the advised sorting methods so that collected volumes grow and waste streams are better pre-sorted. Also, public awareness might drive the demand for products that contain recyclate or are easier to recycle, as it helps to address greenwashing as well as misconceptions and prejudices about the life-cycle impact of plastics and recycling.

Disclosure and environmental reporting support transparency about the activities and products of industry players. The partially voluntary reporting on environmental, social and governance (ESG) criteria and reports that are mandatory under the Non-Financial Reporting Directive [11] and the EU taxonomy [12] are examples. Consumers as well as investors are increasingly rating companies based on ESG criteria to mitigate liabilities resulting from poor ESG performance. Just like with certifications and labeling, however, efforts are required to avoid weak definitions, unclear information and lacking trust to avoid greenwashing and confusion.

3.1.2 Financial Measures

Financial measures can be classified into penalties, redistribution, and incentives.

Penalties

Penalties can be imposed in the form of **taxes** or **fees**. Both are common regulatory instruments to influence private and commercial behaviors, as they impact the economic attractiveness of certain products or actions. They can be designed in a way to compensate for competitive disadvantages of recycling business models compared to those building on virgin feedstock. For example, the UK will introduce a tax on plastic packaging with less than 30% recyclate content of about £200 per ton – both for domestically produced and imported packaging material [13]. But taxes and fees can also have a negative influence or none at all if the cost impact is too low compared to non-recycling options.

Redistribution Schemes

In **redistribution schemes**, the monetary value is reallocated in order to promote recycling.

Deposit-refund schemes are a straightforward mechanism: When a product is bought, the buyer pays a deposit which will be refunded once the product is returned correctly. If the product is not returned correctly, the deposit remains with the producer or is passed on to a private or public organization, which might use the money to fund dedicated environmental efforts. This is commonly known for bottles made from glass or polyethylene terephthalate (PET). Due to deposit-return schemes on PET bottles in several European countries, PET recycling rates are significantly higher than for other plastic types. A deposit-return scheme minimizes random disposal methods such as dumping items in the environment. What's more, this enables well pre-sorted, homogenous, low contaminant, large-volume waste streams that are easy to recycle. And further, the return of deposit items may become so lucrative that non-professional actors start to collect incorrectly disposed of items from the environment or trash to return them and harvest the deposit. That way, even leaked waste is retrieved. On the flipside, deposit-return schemes may incentivize consumers to avoid the products covered by the scheme and choose less recycling-friendly options to avoid being troubled with the system; and market players may choose alternative, non-regulated packaging options to counter low demand for regulated packaging as well as to avoid the costs associated with additional infrastructure and operating deposit-return schemes.

Extended producer responsibility (EPR) schemes have a wide variety of implementations but are all based on the same idea: The responsibility of a producer for their products is extended beyond production, shipping and selling to the end-of-life phase. While there are various ways of accomplishing this, all schemes have in common that an additional fee has to paid by one value chain participant to fund the end-of-life management and waste treatment. These additional financial resources positively impact plastic recycling business cases, and also incentivize manufacturers to consider end-of-life criteria such as recyclability of products already during design, as well as encourage them to implement high-value end-of-life treatment options such as reuse, repair and recycling.

- **Collection and take-back obligations** are common EPR schemes. In these approaches, producers or retailers must take back their own products or the product category. In the EU, for example, this applies to waste electrics and electronic equipment [14]. Retailers of a certain size must take back end-of-life equipment at their stores. Battery collection at retailers is another prominent example. Collection and take-back obligations enable reliable sourcing of material for recycling and pressure the obligated manufacturers and retailers to find the most valuable waste management options. This way recycling can become more profitable than landfilling or incineration.
- **Producer responsibility organizations** represent another form of EPR schemes that take over the responsibility of end-of-life management of a product from the producer or seller. Producer responsibility organizations can be private or public and can be voluntary to comply with the requirements of a law or mandatory by law. The producers of an industry may fund an own, private organization together to manage the correct disposal of their products. Another option is that a public organization collects fees for sold products and then finances the waste treatment. One example for such an EPR scheme organization is Eco-Mobilier in France for furniture. When purchasing a product, a certain fee has to be paid on top of the product price which is later used to collect and recycle the furniture at its end of life. Another example is the Duales System in Germany, a public organization which collects fees for sold packaging to fund their waste treatment.

Incentives

Incentives support an organization or action financially. Governments and financial institutions may support **investments** to facilitate the introduction and implementation of recycling. Examples include the European Investment Bank [15] or the German state-owned KfW Bank [16]. In addition, **R&D** is another pillar for financial incentives to promote recycling. Public grants can finance research projects at public institutions or in cooperation with the industry. PUReSmart is an example of a European plastic recycling project that is developing recycling for polyurethane, funded by the EU with about €6 million. Loans from public institutions with low interest rates can propel the build-up of capital-intensive waste treatment and recycling facilities.

3.2 Approaches in Various World Regions

Policies, regulations and legislations for recycling vary globally and for each region, with very limited global coordination. However, the global community has understood that recycling and corresponding regulations are a crucial aspect, and awareness about the global character of the challenges around plastics is on the rise. Global organizations such as the WEF, WBCSD, or UN, cooperate and support initiatives to promote recycling as well as the collection of statistical data on waste management and knowledge transfer. Examples include the UN Global Partnership on Marine Litter [17] and the action plans put forward by the G7 and G20 [18].

There are significant differences across regions in legislation regarding recycling with a few frontrunner regions such as Europe or certain states in the US like California or Illinois. On their heels, there are the historically strong regions like Japan and South Korea. Further, there are rallying regions like Canada and China. And then there are still a large number of regions with little or no significant legislation regarding recycling.

3.2.1 Global Regulation

The only global regulation with significant direct impact on recycling is the Basel Convention. The Basel Convention regulates the export and import of hazardous waste between participating nations. It roots from the intention to limit the export of hazardous waste from developed to developing countries to stop environmental challenges in the latter. Since the adaption of the Plastic Waste Amendments in 2019, plastic waste is presumed to be hazardous, and only plastics that are free of certain additives (e. g., halogens, fluorinates) and contaminants may be traded [19] with the objective to be recycled. To date, the Plastic Waste Amendment of the Basel Convention has been signed by 188 states. The United States of America is a notable exception, at least as of early 2021. The Plastic Waste Amendment is proof that the challenges of plastic and thereby also plastic recycling have been understood to be a global challenge. Moreover, it localizes and introduces quality standards to waste treatment if the waste is destined to be exported.

3.2.2 Europe

Europe has historically pioneered plastic recycling as a preferred concept of waste treatment, and today is a frontrunner for regulations supporting recycling. While early regulation focused on recycling as a preferred option to treat waste instead of landfilling and incineration in the waste hierarchy framework, today's regulation on plastic recycling is increasingly embedded in the general concept of the circular economy. Although the EU is driving a coordinated and harmonized approach to policy making, historically there have been, and still are, differences between member states in the pace, scope, concept and implementation of regulations regarding plastics recycling.

The earliest endeavors to support recycling via policies were pursued in the 1980s and 1990s. The first European Directive 75/442/EEC on Waste from 1975 already supported recycling to treat waste and the concept of the waste hierarchy was first introduced in the Dutch parliament in 1979 [20, 21]. Since then, the Directive 75/442/EC on waste has been consecutively further amended and complemented by additional directives to incorporate aspects like separated collection schemes, reporting of recycling rates, the waste hierarchy, target rates for certain waste categories, and extended to further products and waste segments (Figure 3.4).

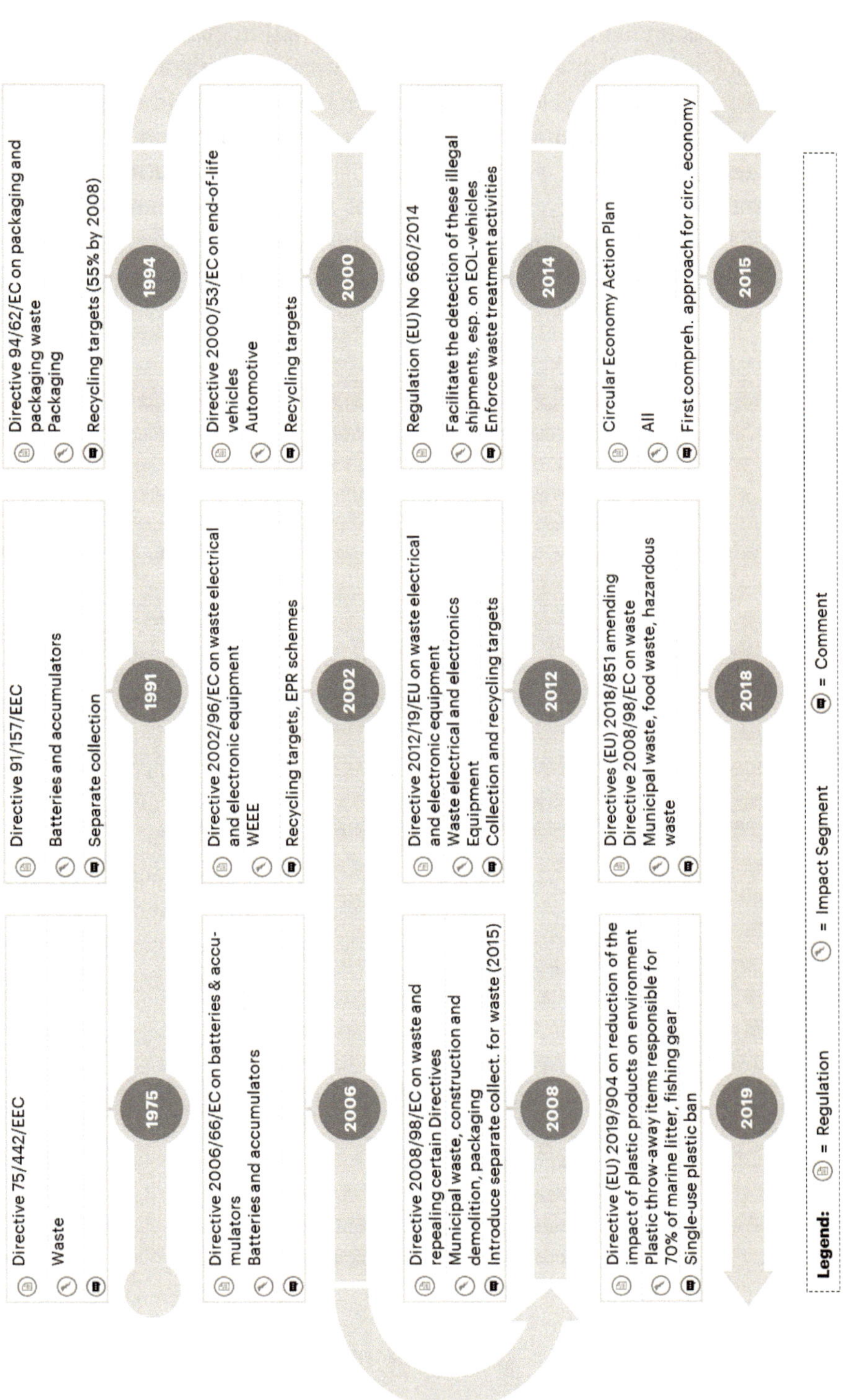

Figure 3.4 Overview of regulations in the European Union with impact on recycling (not exhaustive)

The first EU Action Plan for the Circular Economy [22] in 2015 marks the first holistic approach to establish a harmonized framework of regulations and initiatives on the whole product life cycle to promote a circular economy in Europe. It covers firstly production and consumption, secondly waste management and the market for secondary raw materials, and thirdly a revision of the legislation on waste (Table 3.1). Understanding recycling as a part of the circular economy has several advantages: Design for recycling and collection schemes can address typical issues around plastic recycling and allow for more valuable and larger volume recycling.

Table 3.1 EU Action Plan 2015 [22]

Value Chain Steps	Measures
Production	▪ Circular product design (durability, repairability, reuse and recyclability); esp. for WEEE ▪ Production processes (sustainable resource sourcing, resource efficiency, industry symbioses for by-products)
Consumption	▪ Counteract greenwashing with own label initiatives, particularly labels for energy efficiency ▪ Incentives and taxation based on environmental footprint ▪ Product legal guarantee periods ▪ Reuse and repair programs ▪ Reduce waste production via information campaigns and regional-level funding and cooperation ▪ Collaborative economy (shared infrastructure and services) ▪ Green public procurement
Waste treatment	▪ Promote the waste hierarchy for all waste types ▪ New legislative proposals on waste to provide a long-term vision for increasing recycling and reducing the landfilling of municipal waste ▪ Recycling targets for packaging materials ▪ EPR schemes for waste collection and sorting with minimal conditions, extend to textiles and furniture ▪ Update of the method to calculate recycling rate
Secondary raw materials	▪ Single market for secondary raw materials, especially chemical tracking in products ▪ Raw material information system
Focus areas	▪ Plastics ▪ Food waste ▪ Critical raw materials ▪ Construction and demolition ▪ Biomass and bio-based products

Plastics had already been identified as a key topic of the first circular economy action plan, and based on that, a specific plastics strategy [23] was developed. It recognizes the important role of plastic in our society and its versatile and beneficial technical applications, but demands sustainable handling of plastic and makes a strong case for a circular economy for plastics. The goal is to create a new plastics economy, "where the design and production of plastics and plastic products fully respect reuse, repair and recycling needs and more sustainable materials are developed and promoted." In the context [24] of recycling, the most important goals are that all plastic packaging shall be recyclable by 2030 and single-use items will be banned. New rules for ten throw-away plastic items, which have been identified to account for 70% of marine litter, became effective in 2021, including a single-use plastic ban for cotton buds, cutlery, plates, straws, drink stirrers and sticks for balloons [24]. Also collection rates for single-use plastic bottles shall be introduced. Moreover, a minimum recycled content in beverage bottles of 30% and a corresponding methodology for the calculation and verification of such is pressed for in the context of the latest directive on single-use plastics.

The latest major legislation efforts in the EU regarding plastic recycling are incorporated in the new circular economy package of 2020. The most prominent initiative promoting plastic recycling is the Circular Plastics Alliance [25]. About 260 signatories support the voluntary goal to use 10 million tons of plastic recyclate annually by 2025. These signatories include companies along the plastics value chain from converters, brand-owners to waste managers and recyclers, industry associations, NGOs and public institutions, which are involved in plastic recycling. In addition, the European Commission wants to develop a framework to regulate bio-based and degradable plastics. Both can be a competitor or a complement to plastics recycling. The regulation framework shall clarify the environmental benefits of bio-based plastics and the labeling of bio-based and degradable plastic in order to reduce misleading information, which may lead consumers to plastic littering.

Legislative pressure on plastic is increasing in Europe in- and outside the EU. For example [13], from April 2022 on, a tax on plastic packaging with less than 30% recyclate content of £200 will become effective in the UK.

In conclusion, in Europe plastic recycling is promoted through a wide variety of regulations and legislation as well as additional initiatives, exploring the whole range of possible methods. Plastic recycling is seen as a part of the circular economy and is advanced to gain a competitive advantage by increasing resource efficiency, decarbonization and independence from resource imports or waste exports. Hence, a further increase in regulations and legislation to support plastics recycling in Europe can be expected.

3.2.3 APAC

There are historical leaders in the APAC region like Japan and South Korea, with China becoming a fast adopter.

Japan has historically taken the lead in the APAC region regarding regulations and focus on EPR schemes, which not only include the producers, importers or retailers of products, but also put sorting and collection duties on consumers and municipalities. The first and most prominent law for plastics is the Containers and Packaging Recycling Act from 1995 [26]. The act establishes a producer responsibility organization, the Japan Containers and Packaging Recycling Association, which collects fees from all businesses using defined packaging and therefore operates recycling operations and orchestrates sorted waste collection. In this way, the act combines the responsibilities for all involved stakeholders. Businesses pay fees to operate recycling facilities, while consumers are responsible to sort their waste, and municipalities are in charge of collecting it. Over the years, the act and corresponding organization have evolved to cover more waste streams. For example, in 2006 the amendment included promotion of thin and lightweight products and reduction of less recyclable materials leading to the phase-out of green-colored PET bottles [27]. In 2009 it started to fund municipalities as well. Other laws like the Home Appliance Recycling Act (1998) [28] and the End-of-Life Vehicle Recycling Law (2002) [29] also distribute the responsibilities between the different stakeholders: Consumers need to transport the end-of-life products to a retailer, who has an obligation to take them back. The manufacturer or importer then has to recycle them. Delegating the responsibility for the recycling activity back to the producers encourages them to include recycling criteria in their product design. As a result, for example, automotive producers in Japan have included a “3R evaluation” system and life-cycle assessment to react to the End-of-Life Vehicle Recycling Law [27]. In order to address the relatively low recycling rates in the corporate sector, policy makers are planning to enforce a separate collection of plastic waste and to make recycling mandatory [30]. However, while these efforts generally promote a reduction of landfilling end-of-life products, the existing definitions are not fully incentivizing to choose the most valuable way of closing the loops: Incineration is considered recycling, too, including the usage of end-of-life plastic packaging as fuel in blast furnaces, as syngas or as a coal alternative in coke ovens. That definition is limiting the attractiveness of potentially higher value mechanical recycling or closed-loop chemical recycling methods.

South Korea is another historical leader in recycling regulations and puts a strong focus on EPR schemes [31], which results in one of the highest recycling rates globally at 44.8% for plastic waste in 2018. As in Japan, energy recovery is considered recycling, too, and recycling rates for mechanical recycling of plastic were around 22.7% in 2017 [32]. Similar to Japan, the South Korean waste management system relies on very strict pre-sorting of waste by consumers and on payment of waste treatment by producers and manufacturers [33].

Chinese policy makers started to include circular economy principles in the 2000s as pilot projects and embedded recycling prominently in the 12th five-year plan in 2011. Higher targets to reuse solid waste as resources were set in the Circular Economy Development Strategic Action Plan in 2013 and consecutively in the 13th five-year plan in 2016. Beside general targets for reuse rates, Chinese policy makers focus on promoting certain pilot and model regions, cities, or industrial parks to introduce circular economy strategies. In addition, some local governments are spearheading with recycling regulations. For example, in Shanghai companies and individuals have to pre-sort their household waste for better waste collection and treatment [31]. In addition, a national ban on single-use plastic items like bags is planned [34]. The Chinese "National Sword" operation, which became effective in 2018, banned the import of contaminated and mixed plastic waste beside other solid waste to counteract the local mismanagement of imported plastic waste. Before this ban, the majority of plastic waste from Europe, the USA and other developed countries had been shipped to China. The Chinese ban shifted these imports to other Asian countries, which also reacted with import bans. The ban of plastic waste imports has shifted the associated problems back to where the plastic waste is generated and re-localized the waste treatment responsibility. In the shorter term, in some regions that creates problems as local infrastructure is often overwhelmed with the large quantities of waste piling up, so that in order to handle those large volumes, fast and scalable, but non-sustainable options such as incineration or landfills are being used instead of recycling. In the longer term, however, the necessity to process waste locally supports the building up of an effective local waste treatment and recycling infrastructure. The recent keen actions of the Chinese central government might indicate a trend of more policies supporting plastic recycling and this might get a more prominent role in the next five-year plan.

3.2.4 NAFTA

In the NAFTA region, regulation and legislation regarding plastic recycling is very fragmented and far less distinctive than in Europe or in the developed nations in APAC. Despite the high consumption of plastics in the USA, there are only individual policies in certain states or cities regarding plastics recycling. Some states are more advanced, while the majority lack any measures. For example, San Francisco was the first city in California to ban plastic bags at checkouts followed by many other municipal governments until plastic bags were banned everywhere in the state in 2016. In total, 31 states include cities or municipalities which either ban or charge for plastic bags. In addition, there are fragmented landfill bans or fines for disposing certain items for different states. About 19 states ban WEEE from landfills and 13 states ban or fine the disposal of certain plastic items like plastic containers in landfills [35]. Single states like Illinois or California are even introducing regulation to the same extent as other frontrunner regions. Illinois is supporting curbside collection and recycling with targets as well as EPR schemes with targets for WEEE for manufacturers [36]. California introduced an enforced target that beverage containers need to incorporate recyclate [37] and even set an aspiring target of 75% recycling of solid waste till 2020 [38].

The same, fragmented picture of policies regarding recycling historically applies to Canada. Certain cities like Montreal (2016) and six municipalities across Alberta, Manitoba, and Quebec (2007–2010) have imposed plastic bag bans [39]. However, recently policy makers in Canada raised their ambition on plastic recycling dramatically and in 2018 announced that Canada is set to achieve zero plastic waste by 2030. First actions include the planned banning of single-use plastic items in 2021 – a whole action plan is underway. This regulation and legislation effort in Canada would make it the leading actor in plastic recycling in North America.

The future trend for NAFTA might be a further differentiation into certain frontrunner or fast adopters on the one hand and regions that are not putting any effort into recycling legislation on the other.

3.2.5 Rest of the World

Most other countries – if at all – have only limited regulations in place to support plastics recycling. Many countries, especially in Africa, are banning or charging fees for lightweight plastic bags in order to stop pollution (Figure 3.5). Similar actions have been imposed on single-use plastic items. For example, Costa Rica planned to be the first country to be free of single-use plastics by 2021 [40]. Bans or charges on plastic bags or single-use items may help to eliminate the environmental challenges of plastic pollution, but do not have any directly supporting effects for plastic recycling. Banning certain items might even decrease the volume of recyclable content, while driving the demand for items from recycled content at the same time. For example, lightweight plastic bags might be substituted with thick, multi-use plastic bags from recycled content. Finally, some regions have legislation in place that is influencing plastics recycling. For example, Chile, Argentina [41], and South Africa [42] use EPR schemes to recycle certain waste streams.

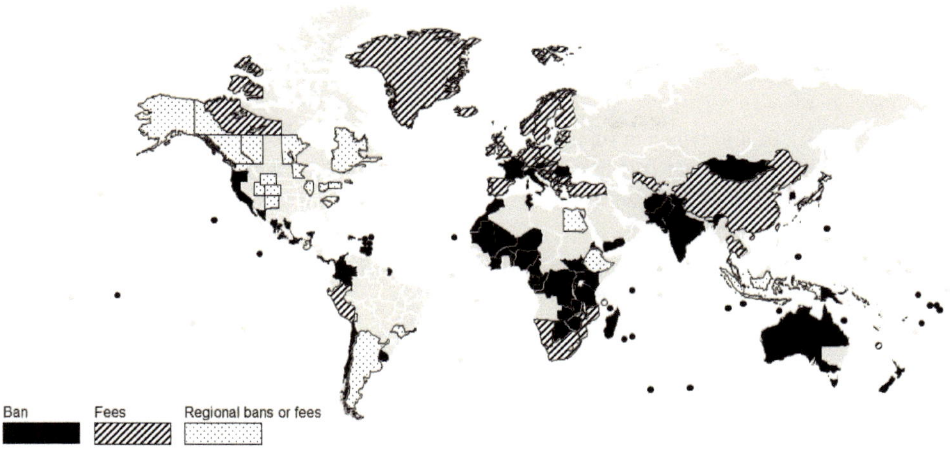

Nations and sub-national regions with bans or fees on lightweight plastic bags

Figure 3.5 Nations and sub-national regions with bans or fees on lightweight plastic bags [1]

References for Chapter 3

[1] Accenture

[2] Ellen MacArthur Foundation, Global Commitment (2021), *https://www.newplasticseconomy.org/projects/global-commitment*, United States

[3] VinylPlus, "The European PVC industry's experience in replacing lead and cadmium-based stabilisers" (2014), *https://www.stabilisers.eu/wp-content/uploads/2015/11/VinylPlus_Contribution-Cefic_Eu-Industry.pdf*, Brussels

[4] European Commission, Circular Plastics Alliance (2020), *https://ec.europa.eu/growth/industry/policy/circular-plastics-alliance_en*, Brussels

[5] European Commission, REACH (2020), *https://ec.europa.eu/environment/chemicals/reach/reach_en.htm*, Brussels

[6] European Parliament, "Parliament objects to lead in PVC to protect public health and the environment" (2020), *https://www.europarl.europa.eu/news/de/press-room/20200206IPR72018/parliament-objects-to-lead-in-pvc-to-protect-public-health-and-the-environment*, Strasbourg

[7] Brooks, A.L., Wang, S., Jambeck, J.R.: The Chinese import ban and its impact on global plastic waste trade, United States, (2018)

[8] California Legislative, "AB-793 Recycling: plastic beverage containers: minimum recycled content" (2020), *https://leginfo.legislature.ca.gov/faces/billNavClient.xhtml?bill_id=201920200AB793*

[9] PlasticsEurope, *The Circular Economy for Plastics – A European Overview*, 2019) Brussels (2019)

[10] PlasticsEurope, *The facts 2019 – An analysis of European plastics production, demand and waste data*, Brussels (2019)

[11] European Commission, Non-financial reporting – Directive 2014/95/EU (2020), Brussels

[12] European Commission, EU taxonomy for sustainable activities (2020), Brussels

[13] UK Governance, "Introduction of Plastic Packaging Tax" (2020), *https://www.gov.uk/government/publications/introduction-of-plastic-packaging-tax*

[14] European Commission, "WEEE responsibilities" (2020), *https://europa.eu/youreurope/business/product-requirements/recycling-waste-management/weee-responsibilities/index_en.htm*

[15] European Investment Bank, "EUR 10 billion to support the Circular Economy in the EU" (2019), *https://www.eib.org/en/press/all/2019-191-eur-10-billion-to-support-the-circular-economy-in-the-eu*

[16] KFW Bank, "EUR 10 billion to support the Circular Economy in the EU" (2019), *https://www.kfw.de/KfW-Group/Newsroom/Latest-News/Pressemitteilungen-Details_531840.html*

[17] UN, "Addressing Land-Based Pollution" (2020), *https://www.unep.org/gpa/what-we-do/global-partnership-marine-litter*

[18] G7, "Leaders' Declaration: G7 Summit" (2015), June 7 to 8, 2015, *https://www.g7germany.de/Content/EN/_Anlagen/G7/2015-06-08-g7-abschluss-eng_en.html*

[19] Basel Convention, "Plastic Waste Amendments" (2021), *http://www.basel.int/Implementation/Plasticwaste/PlasticWasteAmendments/Overview/tabid/8426/Default.aspx*

[20] Parto, S., Loorbach, D., Lansink, A., Kemp, R.: Transitions and institutional change: The case of the Dutch waste subsystem, United States (2007)

[21] Parto, S., Herbert-Copley, B.: In: *Industrial innovation and environmental regulation: Developing workable solutions*, New York: United Nations University Press, United States (2007)

[22] European Commission, "First Circular Economy Action Plan" (2015), *https://ec.europa.eu/environment/circular-economy/first_circular_economy_action_plan.html*

[23] European Commission, "A European Strategy for Plastics in a Circular Economy" (2018), *https://ec.europa.eu/environment/circular-economy/pdf/plastics-strategy.pdf*

[24] European Commission, "Single-use plastics: New EU rules to reduce marine litter" (2018), *https://ec.europa.eu/commission/presscorner/detail/en/IP_18_3927*

[25] European Commission, "Circular Plastics Alliance" (2020), *https://ec.europa.eu/growth/industry/policy/circular-plastics-alliance_en*

[26] JCPRA, "What is the Containers and Packaging Recycling Act?" (2020), *https://www.jcpra.or.jp/english/tabid/614/index.php#Tab614*

[27] OECD, "The State of Play on Extended Producer Responsibility (EPR): Opportunities and Challenges" (2014), *https://www.oecd.org/environment/waste/Global%20Forum%20Tokyo%20Issues%20Paper%2030-5-2014.pdf*

[28] Japanese Governance, "Law for the Recycling of Specified Kinds of Home Appliances (Home Appliance Recycling Law)" (2020), *https://www.env.go.jp/en/laws/recycle/08.pdf*

[29] Japanese Governance, "Law for the Recycling of End-of-Life Vehicles (End-of-Life Vehicle Recycling Law)" (2020), *https://www.env.go.jp/en/laws/recycle/11.pdf*

[30] Kyodo News, "Japan to make plastic recycling mandatory for large corporate users" (2020), *https://english.kyodonews.net/news/2020/10/dd18d86cd307-japan-to-make-plastic-recycling-mandatory-for-large-corporate-users.html*

[31] BBC – Kerry Allen, "Shanghai rubbish rules: New law sends Chinese city into frenzy" (2019), *https://www.bbc.com/news/world-asia-china-48847062*

[32] Statista, "Distribution of plastic waste disposed in South Korea in 2017, by method type" (2020), *https://www.statista.com/statistics/1100696/south-korea-plastic-waste-disposal-by-method/*

[33] Global Recycling, "South Korea: The Aim Is a Resource-Circulating Society" (2021), *https://global-recycling.info/archives/3205*, South Korea

[34] Reuters, "China moves to phase out single-use plastics" (2020), *https://www.reuters.com/article/us-china-environment-plastic/china-moves-to-phase-out-single-use-plastics-idUSKBN1ZI0CV*

[35] Earth911, "What's Banned in Landfills: A State-by-State Guide" (2020), *https://earth911.com/business-policy/landfill-bans/*

[36] IEC, "Zero waste is an achievable and necessary goal" (2020), *https://ilenviro.org/waste-reduction/*

[37] California Legislative, "AB-793 Recycling: plastic beverage containers: minimum recycled content" (2020), *https://leginfo.legislature.ca.gov/faces/billNavClient.xhtml?bill_id=201920200AB793*

[38] CA Governance, "California's 75 Percent Initiative Defining the Future" (2020), *https://www.calrecycle.ca.gov/75percent*

[39] Xanthos, D., Walker, T. R.: International policies to reduce plastic marine pollution from single-use plastics (plastic bags and microbeads): A review, *Marine Pollution Bulletin*, (2017) 118, Issues 1-2, pp. 17-26, May 15, 2017

[40] The Costa Rica News, "Businesses in Costa Rica Must Adapt to the New Laws that Regulate Single-use Plastic and Polystyrene" (2020), *https://thecostaricanews.com/businesses-in-costa-rica-must-adapt-to-the-new-laws-that-regulate-single-use-plastic-and-polystyrene/*

[41] Valpak, "How Argentina is following Chile's lead on EPR Regulations in Latin America" (2021), *https://www.valpak.co.uk/news-blog/blog/how-argentina-is-following-chile-s-lead-on-epr-regulations-in-latin-america*

[42] Business Insider South Africa, "Here's how SA's radical new recycling laws will affect you" (2020), *https://www.businessinsider.co.za/how-the-recycling-laws-will-affect-south-africans-2020-11*

4 Economics

Michael Ulbrich, Marvin Stiefermann

With plastic waste more and more becoming a threat to marine life and the environment, it is becoming increasingly evident that the level of plastics recycling needs to be massively enhanced.

An important factor to attract market players to venture into the recycling of plastics is the likelihood that they will be able to extract an economic profit from those activities. For most plastics types, recycling business cases are still a challenge. In this chapter, we explore the fundamental cost drivers of recycling plastics and the scale disadvantage of recycled raw materials compared to the usage of fossil raw materials. For the purpose of this discussion, we neglect broader aspects such as a societal perspective on preferable recycling methods when considering, for example, the greenhouse gas footprints of different recycling routes, and instead focus on the perspective of an individual potential market entrant purely interested in generating an economic profit. We conclude the discussion with the implications for new types of economic collaboration along the emerging circular value chain, as well as for potential new business models for the value chain actors involved.

4.1 Economic Challenges of Recycling Plastics

While some materials such as glass and steel, at recycling rates exceeding 86% globally [1], are already used in a very circular way, the recycling rate for polymers collected in Europe has struggled for years to exceed about 30% [2] and has only recently started to grow. It is difficult to come up with global numbers, but the most optimistic current assumptions are that about 12% are being recycled [3]. There are three major reasons for recycling rates being significantly lower for plastics than for glass or steel:

1. **Energy:** Energy need is a major cost driver in glass and steel production. By recycling those materials, energy savings of about 67% or more are realized in comparison to the production from virgin raw materials. In contrast, the energy consumption for plastics recycling in most cases is higher than for the production of new plastics from virgin feedstock.
2. **Technology:** Recycling of glass and steel is rather simple from a technological perspective. By melting these materials, they can be processed into the desired application without any degradation or decrease in quality. For most plastics, on the other hand, this process is far more challenging. Some polymers, e.g., polyethylene (PE), polypropylene (PP), or polyethylene terephthalate (PET) can be melted multiple times, however this often leads to a degradation and decrease in quality. Other plastics require more sophisticated technologies in order to be recycled.
3. **Waste stream purity & density:** If sorted by color, glass can be collected with a very high degree of purity and effortlessly recycled after cleaning. Similarly, steel is sorted according to its composition and can be easily processed afterwards into the desired end product by melting the required quantities of scrap steel. Plastic, on the other hand, is an inhomogeneous waste material. Plastic waste is composed of a large variety of different polymers, requiring separation in order to be able to recycle these materials. Mixtures or contaminants in recycling mostly result in poor material properties, rendering it unsuitable for many applications. In order to achieve a high degree of purity, plastic waste has to be sorted by material type, which is often challenging due to similar properties of different materials or nearly impossible if waste materials are a composite of different polymers.

As a result, plastic recycling has not yet evolved to a similar extent as the recycling of steel, paper, or glass. To make matters worse, the production of virgin feedstock from crude oil benefits from much more favorable economies of scale in comparison to plastic waste recycling. Plastic waste is typically scattered over large geographic areas in fragmented, non-controlled specifications and quantities, and therefore requires the deployment of complex collection and sorting systems in order to gain the required feed both in terms of volume and quality. In contrast, crude oil is extracted in high quantities from concentrated spots and subsequently cracked in world-scale steam crackers with capacities of up to 1 million tons per year. Even in metropolitan regions with a high waste density, recycling plants cannot be scaled up to similar dimensions as the size of the required catchment areas would lead to prohibitively high transportation costs.

However, the growing problems associated with plastic waste, such as ocean waste affecting all marine life forms, or microplastic contamination through the food chain resulting in deposits in the human body, have resulted in both a rising demand for plastic recycling by consumers, brand owners and investors, and the initiation of a process to create new economic incentives that may propel recycling business cases in the near future.

4.2 Framework for Recycling Business Cases

The typical cost drivers in the production of recycled feedstock are depicted in Figure 4.1. They depend on a variety of input parameters such as availability and purity of the material to be recycled, the deployed recycling process, or geographical location of the plant.

- **Collection and pre-treatment:** The cost of waste material is strongly affected by the geographical location of the recycling plant. Collection costs are lower when local waste density is high and the time to fill one truck load as well as the actual transportation distance are relatively short. This is typically the case in metropolitan areas with a high population density. Second, the cost depends on the type of waste. Some plastics make up for a higher percentage of the waste collected than others and can therefore be collected in large amounts in a short time. And third, there is the cost of sorting which is linked to the plastic type, too. Some materials can be easily sorted, while others require more expensive technology to be filtered out in sufficient purity.
- **Recycling process:** The type of recycling process has a strong influence on the total cost of recycling. Further, cost is strongly affected by the plant capacity – generally, with better fixed cost dilution as plant capacity increases.

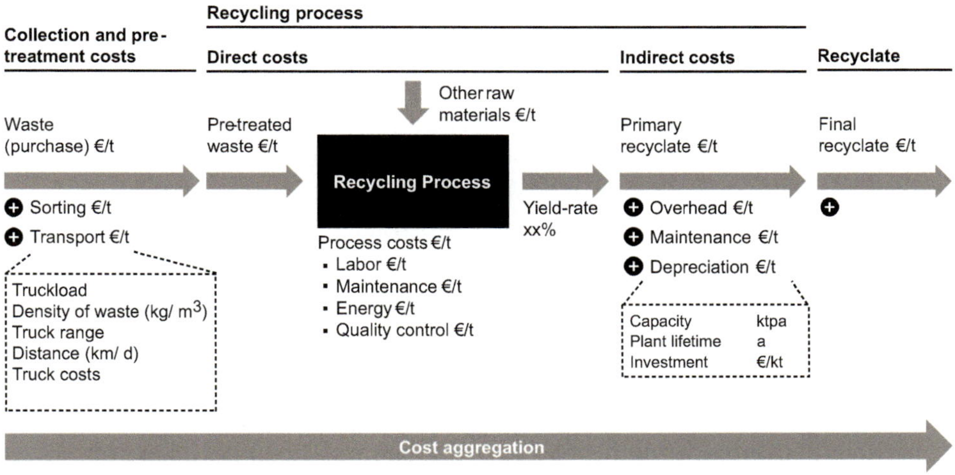

Figure 4.1 Cost framework for plastics recycling from collection to recyclate

With its potential to make further waste streams accessible for processing, chemical recycling – next to mechanical recycling – may play an important complementary role in reaching the ambitious material recycling targets. As the associated technologies such as pyrolysis or solvolysis have largely not been scaled up yet, there are only limited data points available from practice so far. In the following, we explore the economic viability of these technologies by applying the outlined cost framework to selected examples. We are assuming fictitious recycling plants and are making numerous assumptions about waste density, baled waste plastic purchase price (in itself highly volatile), cleaning costs, converting costs, personnel costs and energy costs, and also attainable yields and other input factors in orders of magnitude that are typical for Western Europe. By their nature, such calculations can provide directional guidance but may look quite different for actual, specific projects.

4.2.1 Illustration 1: Chemical Recycling of Polyolefins

More than 40% of all plastic waste in Germany is composed of PE and PP [4]. Some of that material is not processable through mechanical recycling. Pyrolysis to produce synthetic oil can be a solution for such materials. Pyrolysis oil can be used as a substitute for crude oil or as fuel in marine diesel engines. However, while converting plastics to fuel may be preferable over just depositing or combusting plastic waste, it is not typically considered recycling by authorities as it does not displace any fossil-based plastic usage. An advantage of this process is that it does not require the same level of waste stream purity as mechanical recycling. A different ratio of raw materials results only in a shift in the composition of pyrolysis oil. Therefore, the cleaning and sorting effort – often a large cost driver – can be kept at a minimum for chemical recycling.

Pyrolysis is an energy-intensive process, as the PE and PP waste material must be heated up to over 400 °C at atmospheric pressure. In addition, the complete baking cycle has to be carried out in an oxygen-free atmosphere to avoid the oxidation of the hydrocarbons. Hence, energy consumption is a major cost driver for chemical recycling plants.

In our business case, we assess the economic feasibility from the perspective of a pure recycling plant operator, who has to procure the cleaned and sorted material stream from a predecessor in the value chain that absorbs any potential gate fees paid by the waste collector and takes care of sorting and baling. We use the average market price of PE and PE bales from 2015–2020 as an assumption for the raw material purchase cost, and a plant capacity of 100 kilotons per year [5] – still a quite ambitious capacity at this point in time; however, first pilot plant projects are already underway with even larger capacities. For the other parameters, due to a lack of actual values from scaled-up chemical recycling plants, we apply typical values for chemical plants located in Western Europe as a proxy. We assume that only specific types of PE/PP waste will become accessible for pyrolysis plant operators that cannot be recycled mechanically. Summing up all cost items and allowing for a profit mark-up of 15%, we arrive at a total price of €532 per ton of pyrolysis oil (Figure 4.2).

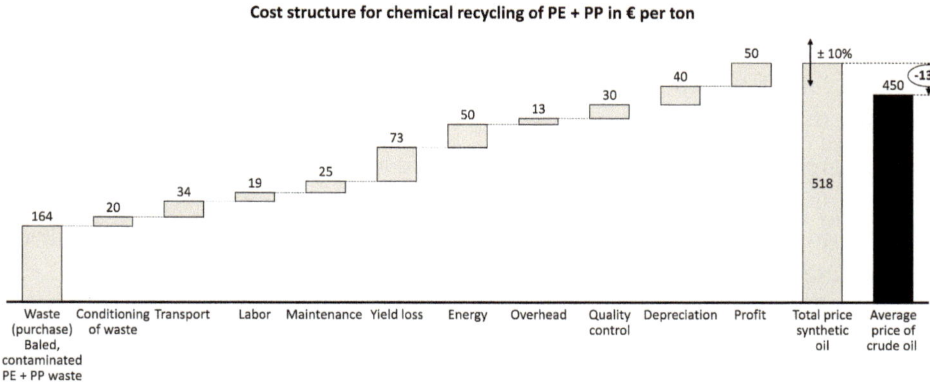

Figure 4.2 Exemplary cost structure of a PE/PP pyrolysis plant in Europe

The comparison of the hypothetical pyrolysis oil market price with the average market price for crude oil over the last 10 years indicates that economic viability is within reach, and even more so as oil prices are on the rise again. However, the extent to which pyrolysis and crude oil can be directly compared is limited due to their different composition. The extra in price could be compensated through the usage for higher-value applications for synthetic oil. The calculation also shows that the business case is strongly sensitive to the total yield of the chemical recycling process. If the overall yield can be increased further, the economics improve. Additionally, the purchase price of PE + PE bales is strongly linked to gate fees paid by the waste stream collector for each ton. To ultimately achieve a positive business case, regulatory support through mechanisms such as bans on landfilling, bans or heavy taxation of incineration, and bans on end-of-life (EoL) material exports will be inevitable.

In conclusion, the business case for pyrolysis of PE and PE is strongly linked to external factors such as gate fees, the price for crude oil, and regulatory incentives. Other factors such as the yield can be improved under the control of the operator with additional research and development to reduce the overall loss of material during the process.

4.2.2 Illustration 2: Chemical Recycling of PET

End-of-life PET is one of the largest waste streams - about 7.9% [6] of plastic waste in Europe. PET is a good water and moisture barrier with a high tensile strength and a nice gloss, which is why it is the standard polymer for bottles for water and carbonated drinks. PET can easily be retrieved from EoL bottles, which in some countries are collected via dedicated deposit schemes, resulting in a relatively high purity of PET waste streams.

For PET, different recycling technologies exist with several already being operated at large scale. Mechanical recycling of PET has been used for a long time and is currently the most common process.

Another process for PET which avoids the downsides of mechanical recycling such as rapid degradation is the chemical recycling into unsaturated polyester (UP). Through this process, the conventional precursor for UP, orthophthalic anhydride, is substituted by PET at only a fraction of the cost.

Since PET is available as a comparatively pure waste stream and the total costs for collection and sorting are comparatively low, the raw materials required for UP production represent the major cost driver [7].

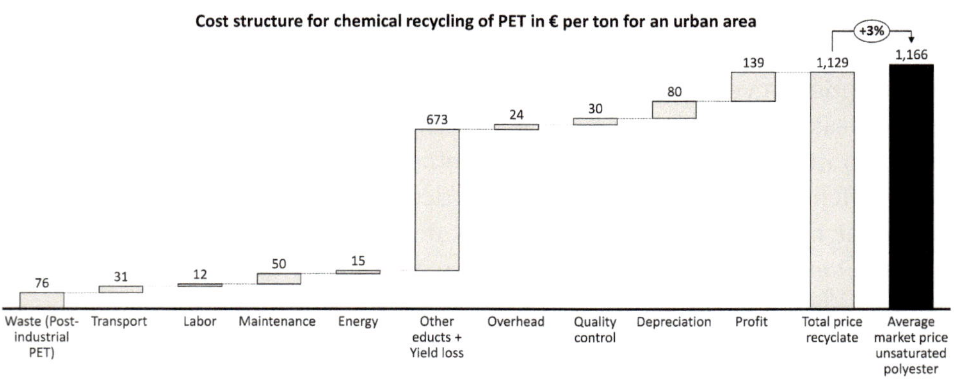

Figure 4.3 Exemplary cost structure of chemical PET recycling in an urban area

Figure 4.3 shows that with an average virgin UP market price of €1166 per ton [7] over the last four years, chemical recycling of PET in an urban area can be economically viable.

If the economic parameters are modified slightly to reflect a more rural area in which the average EoL PET transportation distance doubles compared to an urban location, the business case turns negative (Figure 4.4).

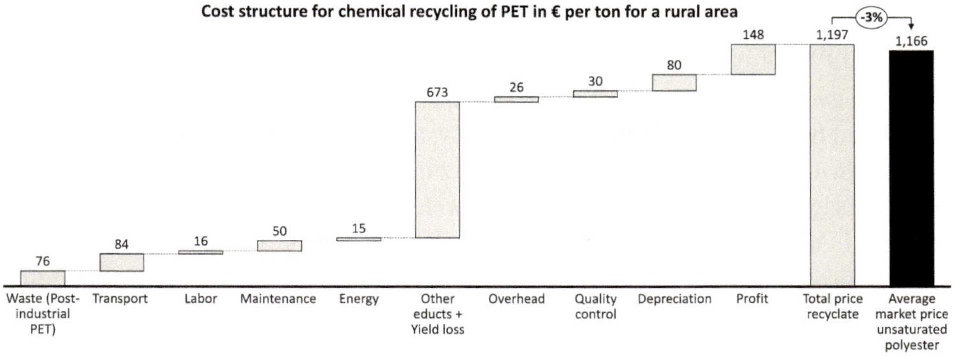

Figure 4.4 Exemplary cost structure of chemical recycling of PET in a rural area

However, the transformation into UP does not loop PET back into the cycle. This is why other technologies are currently being tested, which could enable a closed PET recycling loop and therefore could significantly affect the amount of virgin PET feedstock required.

Unlike in the PE/PP example, where pyrolysis is a rather complementary technology for recycling materials that cannot be processed mechanically, in this example of chemically recycling EoL PET into UP, the technology might directly compete for the waste stream with conventional mechanical PET recycling, and standard economic parameters such as cost of goods sold or attainable market price may turn the needle for a purely economics-driven potential market entrant.

4.2.3 Illustration 3: Microwave-Assisted Depolymerization of Styrene

Polystyrene (PS) is the polymerized version of styrene, which makes up about 6.9% [6] of the total European plastic production per year. PS is widely used for food packaging, e.g., dairy products, or in the form of foamed or extended polystyrene (EPS) as thermal insulation for takeaway food or for buildings; it is also used as transport protection for sensitive goods such as electronics. Due to a low material price in combination with the extremely low density of EPS and its straightforward way to be brought into shape through vacuum forming, it has become the standard material for protective packaging. With a continuously growing demand for protective packaging, the need for polystyrene recycling is growing, too.

Generally, where sufficient to meet customer specifications and other requirements, mechanical recycling is the preferred option due to lowest process cost and lower energy consumption. In case of higher requirements, dissolution is a more advanced available technology; and in case of high purity needs and broad application flexibility, depolymerization with the use of microwaves is one of the most promising technologies currently applied. This process enables recycling into monomers without any degradation or loss of material properties and is therefore an ideal process when it comes to closing the loop for styrene applications.

Technology startups are successfully exploring this space. Amsty and Agilyx have pioneered depolymerization of post-use polystyrene through a 50/50 joint venture located in Oregon (US). The company is converting end-of-life polystyrene back into its original components, on a commercial scale [8]. Or another example: As styrene is an important ingredient in synthetic rubbers, which is in large scale used in tires, Michelin®, together with Pyrowave® [9], has announced the building of a demonstrator for styrene recycling by 2023. An indicative business case [7, 10] for a small-scale recycling facility utilizing microwaves for depolymerization is shown in Figure 4.5.

It can be seen that even in small-scale recycling facilities, the microwave-assisted depolymerization of styrene can be an economically attractive alternative to virgin feedstock application. This opens up the possibility of smaller styrene recycling facilities being next to municipal collection points, which would additionally positively impact the transportation cost of styrene waste to the recycling facility [7].

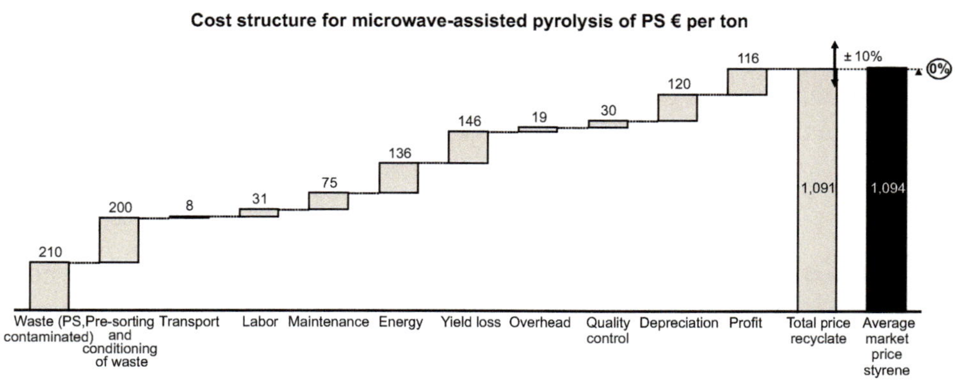

Figure 4.5 Exemplary cost structure of microwave-assisted depolymerization of PS

Another study, by contrast, has determined that for small pilot plants, the product obtained through depolymerization of styrene may realistically cost up to €500 per ton more than fossil-based styrene, based on a detailed evaluation of incremental costs from raw material purchase and yield losses, e. g., at the processing of raw styrene through distillation [11]. A positive business case may become achievable upon full industrial scale-up and completion of the learning curve as soon as the technology is more broadly applied.

The two presented cost positions mark the bandwidth of current economics around PS recycling from constrained lab/pilot scale to fully scaled up in the future and indicate an interesting upside potential.

■ 4.3 Implications and Outlook

As discussed, several novel recycling processes are emerging that may soon be ready to be operated in an economically viable way under favorable conditions: waste stream densities must be sufficiently high and sorting efforts low. In turn, this means that recycling processes for waste streams from less common plastic types and for waste from rural areas will not break even with virgin alternatives on their own for the foreseeable future [12].

As it is difficult to forecast technological progress for recycling processes, the best bet to raise plastic recycling volumes and close the loops will be through tailored regulation that facilitates the allocation of funds to improve the economics of recycling business cases for operators, and through the education and gradual shifting of consumer behavior and buying preferences.

As none of today's market players have all the capabilities and experience needed to bring fully closed loops to life on their own, it will be increasingly vital to collaborate across the value chain and forge new alliances in order to take advantage of the opportunities offered by increasingly attractive business cases. A number of emerging collaboration models can be observed in the market.

- **Chemical company/waste manager cooperation:** Some chemical and waste management companies have started to establish joint frameworks and infrastructure for economical recycling processes. The waste managers contribute a reliable and sufficiently large waste supply with the needed composition and quality for economical recycling processes. The chemical companies provide the necessary knowledge for the manufacturing technology to develop and set up a chemical recycling process or to use the recyclate. Established collaborations on the market so far focus on recycling PE, PP, and high-density and low-density PE for various application industries.
- **Chemical company/waste manager partnership, joined by a chemical recycling technology provider:** One chemical company is actively cooperating with technology startups to find solutions for the chemical recycling of PS through pyrolysis-based and microwave-assisted depolymerization. Their cooperation could provide an opportunity to test and gather experience with different chemical recycling technologies, and on how to handle the yielded monomers in manufacturing new PS.
- **Cooperation of chemical company, converter, brand owner and waste manager:** This emerging type of cooperation bundles expertise in material properties, product manufacturing and product requirements. In one joint development project, the partners are testing a solvent-based process to recover complex multi-layer packaging to gain virgin feedstock-like low-density PE for packaging.

In another example, synthetic pyrolysis oil is produced as a substitute for crude oil, using a thermo-anaerobic process on mixed plastic waste.

- **Direct chemical company/brand owner cooperation:** This model is well suited for selected niches. In one example, the partners are developing a textile fiber-reinforced composite for infinite recycling loops that is reprocessed into new shoes. Other examples include the production of recycled carbon fibers for use in composite materials in computers, and the production of cosmetic packaging from a cellulose-based thermoplastic engineering bioplastic. It is worthwhile noting that in this collaboration model, the role of the waste manager is bypassed by the retailer and/or the manufacturer as they have direct access to pure waste streams.
- **Direct cooperation of a technology startup and waste manager with a brand owner:** In this model, the traditional chemicals producer is bypassed. In one example, a producer of carbonated soft drinks has financed a startup to build a chemical recycling plant for colored and contaminated PET.

These partnerships are increasingly coalescing into circular ecosystems, which will be further propelled by an increasing transition from the firm domain boundaries of the established industry associations towards outcome-driven cross-value chain associations such as the Alliance to End Plastic Waste which was established in early 2019.

Finally, from an economic perspective, the growing attractiveness of business cases and emerging circular ecosystems will facilitate novel business models. These could include more obvious extensions of current models such as the provision of development services for recyclate applications or the certification of recycled materials and products; but also more progressive business models such as operating EoL material platforms to match scattered supply with concentrated demand; or even somewhat futuristic models such as "circular polymer as-a-service", where the producer of a plastic molecule would remain its owner throughout its lifecycle, making a profit by managing its conversion and performance and ultimately assuming the responsibility to circle it back into the value chain after its use life.

References for Chapter 4

[1] *https://www.steelsustainability.org/recycling*

[2] *https://www.plasticseurope.org/application/files/5716/0752/4286/AF_Plastics_the_facts-WEB-2020-ING_FINAL.pdf*

[3] OECD, Improving Plastics Management: Trends, policy responses, and the role of international co-operation and trade, *https://www.oecd.org/environment/waste/policy-highlights-improving-plastics-management.pdf*

[4] Umwelt Bundesamt, Kunststoffabfälle, *https://www.umweltbundesamt.de/daten/ressourcen-abfall/verwertung-entsorgung-ausgewaehlter-abfallarten/kunststoffabfaelle#kunststoffvielfalt*, accessed January 15, 2021

[5] Plasticker, Raw Materials & Prices, *https://plasticker.de/preise/preise_monat_multi_en.php*, accessed January 15, 2021

[6] *https://www.plasticseurope.org/application/files/5716/0752/4286/AF_Plastics_the_facts-WEB-2020-ING_FINAL.pdf*, accessed January 15, 2021

[7] Source: Accenture research

[8] Agilyx, "A cooperative venture that is changing the way we recycle polystyrene plastic", *https://www.agilyx.com/regenyx/*, accessed October 5, 2021

[9] Michelin, "Michelin and Pyrowave join forces to industrialize an innovative plastic waste recycling technology", *https://www.michelin.com/en/press-releases/michelin-and-pyrowave-join-forces-to-industrialize-an-innovative-plastic-waste-recycling-technology/*, accessed January 15, 2021

[10] Discussions with a polystyrene recycling startup, based on small industrial-scale plant

[11] Project ResolVE funded by the German Federal Ministry for Education and Research; findings represent the consolidated views of a broad expert group. Nießner, N., Nosić, F., Wilhelmus, B., et al.: TIBKAT (2020): ResolVe - Recycling von Polystyrol mittels rohstofflicher Verwertung: Projektlaufzeit 20172–020. Available from the German National Library of Science and Technology: *https://www.tib.eu/de/suchen/id/TIBKAT:1748492039/ResolVe-Recycling-von-Polystyrol-mittels-rohstofflicher?cHash=6eae3ce01d861a107b20ab29bdada904*, accessed August 4, 2021

[12] Based on our publication: "*Winning in a circular economy*", *https://www.accenture.com/_acnmedia/PDF-117/Accenture-Winning-In-A-Circular-Economy-Executive-Summary.pdf*

5 Political Boundary Conditions – Plastics Waste Management

Klaus Wittstock, Victoria Wessolowski[1]

This chapter describes boundary conditions for plastics recycling topics in legislation and policies. Recycling is not a standalone concept but should be seen in the context of the broader and more complex area of waste management. The chapter thus starts by explaining why plastics recycling has been one of the key focus areas of environmental legislation for decades. It follows with an overview of the existing legislation in the selected geographies: worldwide, Europe, Germany, the USA, and China. In conclusion, the chapter looks into the current developments in legislation on plastics recycling.

5.1 Plastics Recycling and Its Political Relevance

Plastics are one of the newest and most widespread materials that were invented. Since the end of the 19th century, different types of plastics with novel features have been discovered with an immense uptake of plastics production as of the 1950s. To the public, plastics had an unambiguously positive image and were generally regarded as versatile, benign, sanitary, safe, and superior to many other materials, offering distinct characteristics and greater convenience. As plastics were increasingly produced in mass and could be broadly used in automotive parts, building insulation, food packaging and electronic and medical applications, the material became cheaper and could enter new markets [1]. For the last 70 years, global plastics production has shown continuous growth, increasing from 1.5 million tons in 1950 to ~367 million tons in 2020, with China being the leading producer of plastics globally (Figure 5.1) [2].

[1] Special thanks to Sigrid Meyer, Uwe Blumenstein, and Nan-Hua Nadja Yang.

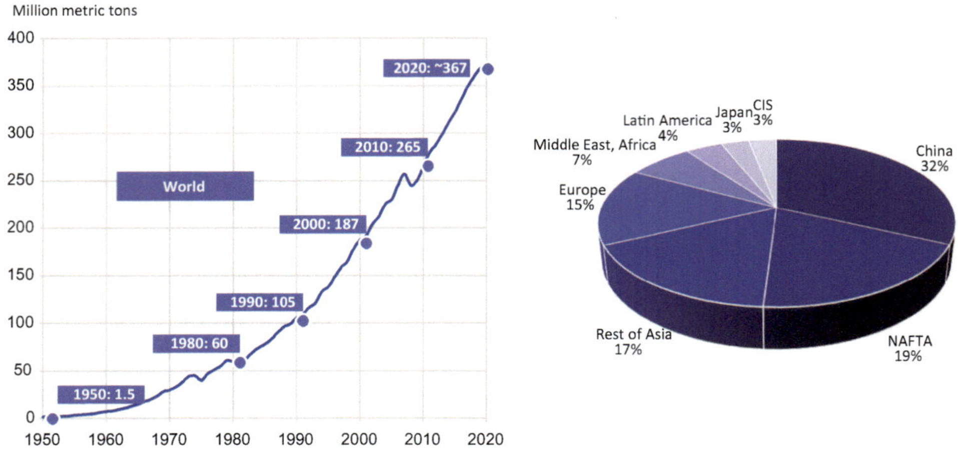

Figure 5.1 Global plastics production 1950–2020 and its regional shares in 2020 (source: PlasticsEurope Market Research Group (PEMRG)/Conversio Market & Strategy GmbH)

In the second half of the 20th century, the reputation of plastics gradually started to suffer. Most concerns revolved around the industrial production of the raw materials, health issues and pollution [3]. The image of plastics worsened as environmental concerns increased. By that time, plastics production was booming, with ever-increasing consumption, while the waste management infrastructure was lagging. Existing waste management facilities are inadequate to deal with our current levels of plastic production and consumption. Most plastic wastes generated today are not collected or adequately managed, and recovery once they have entered the waterways is virtually impossible, as is the case for microplastics. This asymmetric development has resulted in uncontrolled landfilling and leakage into the environment: Of 250 million tons of generated waste in 2018, 20% were sorted for recycling, 20% were energetically recovered, 29% landfilled and 31% were disposed of under improper conditions (e.g., unsanitary landfills, improper burning or burial) or leaked into the soil or aquatic environment. Asia and the Oceania region account for the most plastic waste produced worldwide (more than 40%), along with China with 60 million tons, but this is below United States in terms of per capita waste generation (Figure 5.2, Figure 5.3, Figure 5.4) [4]. Under the business-as-usual scenario, global plastic production "will account for 20% of total oil consumption and 15% of the global annual 2 °C carbon budget by 2050" [5].

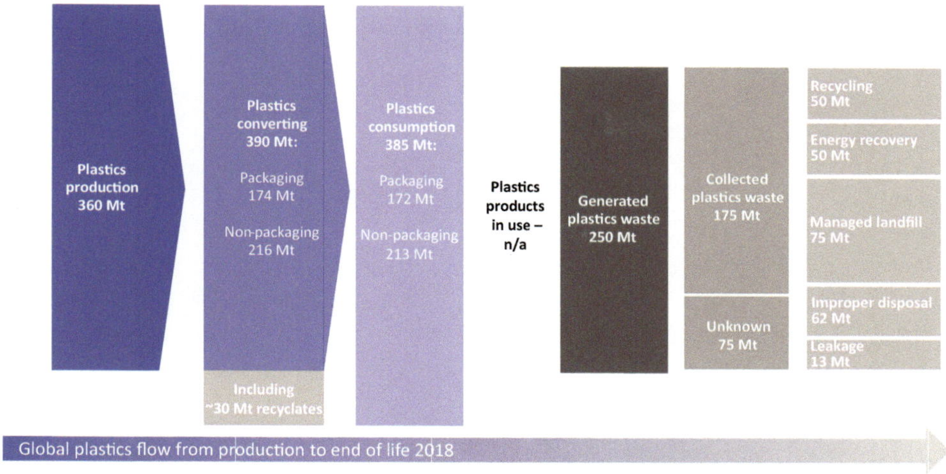

Figure 5.2 Global plastics flow from production to end of life 2018 (source: Conversio Market & Strategy GmbH, Global Plastics Flow 2018, 2/2020)

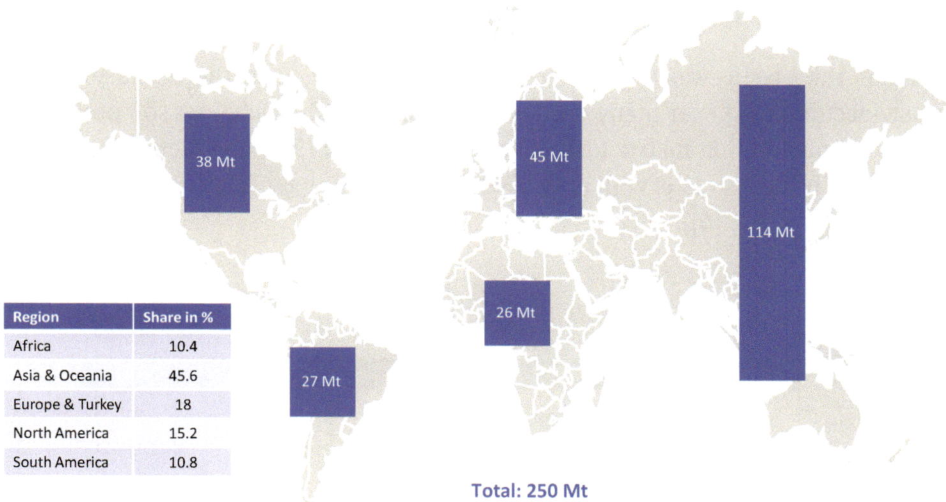

Region	Share in %
Africa	10.4
Asia & Oceania	45.6
Europe & Turkey	18
North America	15.2
South America	10.8

Figure 5.3 Post-consumer plastics waste generation by region 2018 (source: Conversio Market & Strategy GmbH, Global Plastics Flow, 2/2020)

Country	Generated plastics waste in Mt	Collected for treatment in Mt
Europe	30	29.1
Turkey	3.6	1.84
United States	34.5	33.8
Canada	3.27	3.24
Brazil	7.2	4.77
Colombia	1.40	0.95
China	59.1	40.1
India	9.5	6.89
Japan	9.03	9.03
Indonesia	5.70	2.63
Philippines	2.0	0.83
Thailand	2.50	1.70
Malaysia	1.80	1.11
South Africa	1.70	1.12
Ghana	0.43	0.08

Figure 5.4 Post-consumer plastics waste generation by selected countries 2018 (source: Conversio Market & Strategy GmbH, Global Plastics Flow, 2/2020)

Besides the plastic waste generated during a specific year, there is also plastic waste accumulated over many decades. Since the 1950s, global plastic production exceeded nine billion metric tons. More than half of it (5 billion metric tons) is estimated to become waste, with only seven percent being recycled during the entire period (Figure 5.5, [6]).

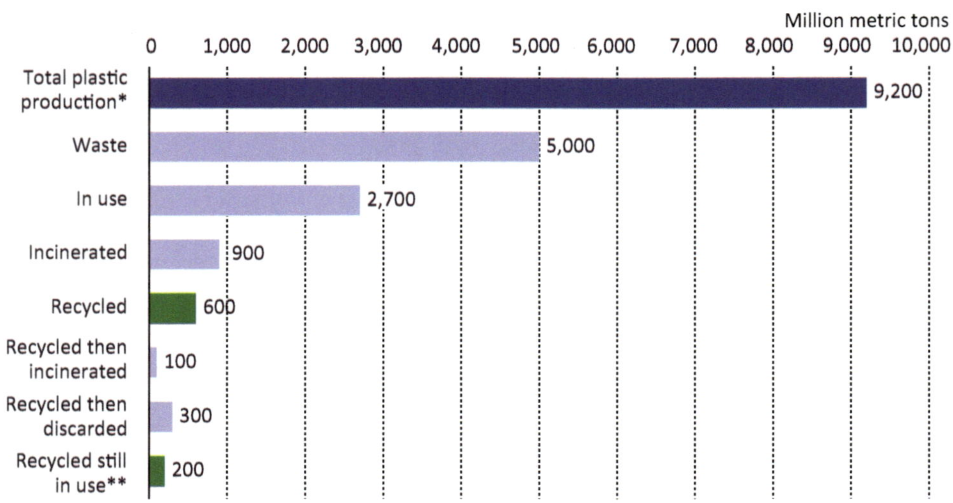

Figure 5.5 Global plastic materials flow 1950–2017 (source: Statista 2021; Heinrich-Böll-Stiftung, 10/2019)

In the 1980s, the plastics industry took the lead in encouraging municipalities to include recyclable (i. e., able to be collected, sorted and recycled with the existing infrastructure and technologies) plastics into their collection and recycling systems [7]. A slow upward trend in recycling rates can be observed in most countries over the course of the last 30 years [8]. While some countries already have relatively high rates, others still recycle almost none of their plastic waste: For example, the United States' plastics recycling rate was about 2% in 1990 and rose to over 9% in 2019. While South Korea had a plastics recycling rate of more than 40% during the decade up to 2021, this development has also not been steady [9].

Meanwhile, the awareness of plastic waste in soil and aquatic environments has massively increased. The first marine debris was found in the 1960s [7]. In the following decades, more environmental impacts of discarded plastic waste were identified, bringing the topic of mismanaged plastic waste to the wider public debate and policy developments. A new wave of research activities began in the early 2000s when ubiquitous microplastics were discovered in the ocean [10]. Since then, public attention to the negative environmental impacts of plastic leakage has increased exponentially, with the intensified activities of, among others, the media, and non-governmental organizations (NGOs).

To prevent more plastic waste leaking into the natural ecosystems, numerous countries have introduced landfill restrictions. For example, Germany has had landfill restrictions for municipal waste since 1993. Austria and the Netherlands took landfills out of operation in 1996. Switzerland banned combustible waste from landfilling in 2000. And Sweden has had bans on landfilling of sorted combustible waste since 2002 [11]. These European countries have in common that they have excessively increased their recycling and energy recovery capacity to balance out the waste management infrastructure via landfills. This clear correlation between the restriction of landfilling and the increase of recycling and energy recovery of plastics is obvious. Hence, the more landfill restrictions are imposed nationally and internationally, the more recycling efforts are induced as depicted in Figure 5.6 and Figure 5.7 for Europe [12]. The concept of "diverting waste from landfill" is thus key to increasing plastics recycling.

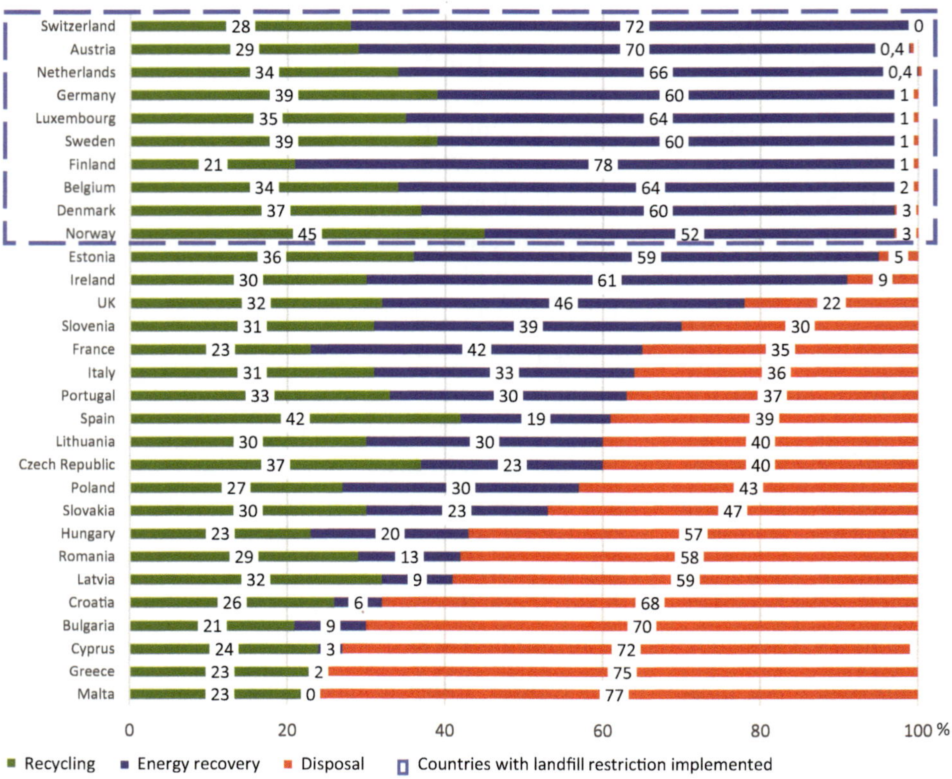

Figure 5.6 Post-consumer plastic waste treatment in percentage in EU28+2 countries in 2018: Correlation between landfill restrictions and recycling and energy recovery of plastics (source: PlasticsEurope Market Research Group (PEMRG)/Conversio Market & Strategy GmbH)

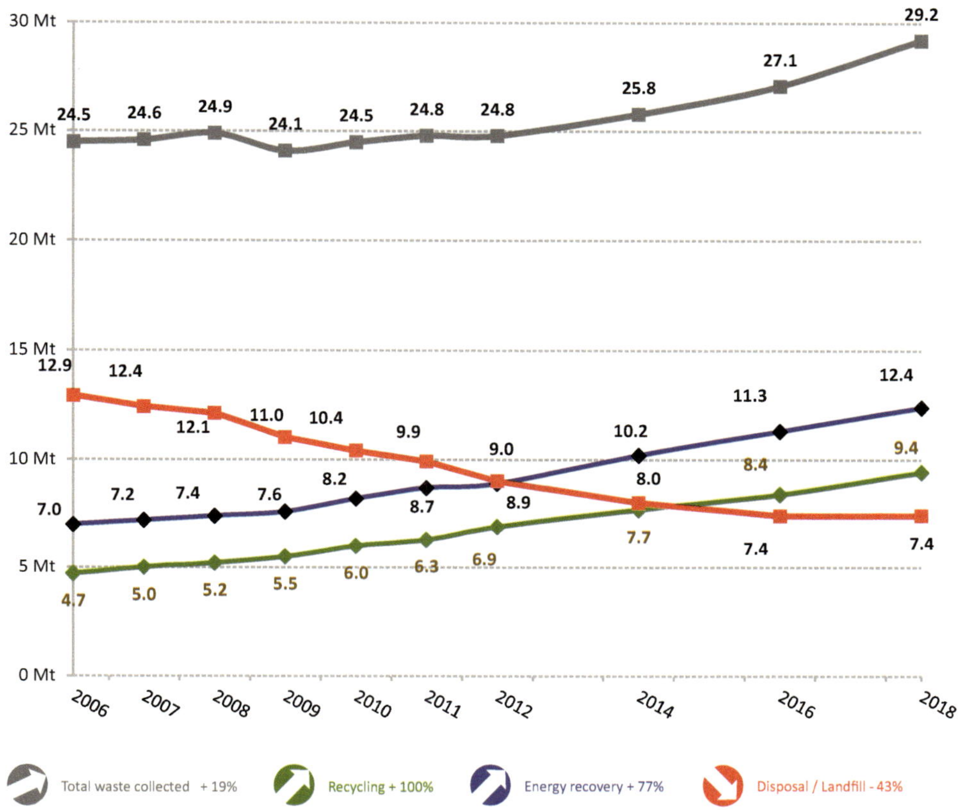

Figure 5.7 Post-consumer plastic waste treatment in EU-28+2 countries 2006–2018 (source: PlasticsEurope Market Research Group (PEMRG) / Conversio Market & Strategy GmbH)

Landfilling and improper disposal are still cheaper both in CAPEX and OPEX than energy recovery and recycling, and most countries are nowadays aware of the negative long-term consequences (i. e., external costs of improper waste management). However, as establishing a proper recycling and energy recovery infrastructure is time-consuming, cumbersome, and costly, many countries cannot keep up with the rapid pace of plastics production. In many countries, the notion of preventing waste as such thus evolved as the key principle of waste management. For instance, Germany set out a waste prevention program in 2013 and Japan adheres to the 3R principles (Reduce, Reuse, Recycle). The anticipated and predicted growth of virgin plastics production is perceived as a big concern which will only aggravate the problem of plastic pollution even further. For example, over one-third of plastic production goes into single-use applications, with 95% of its material value lost to the economy after a short first use. Packaging is considered a major contributor to plastic pollution. Only 14% of plastic packaging is collected for recycling globally and, when additional sorting and reprocessing losses are factored in, only 5% of material value is available for subsequent use, with most being recycled into lower-value applications that are subsequently not recyclable. Recycling alone cannot solve the plastic pollution problem. Production capacity is expected to increase by almost 40% (approximately 100 million tons) for ethylene and propylene - the key building blocks for cheap and disposable plastic applications - undermining efforts to reduce consumption, keep up with the waste management infrastructure, and end the plastic waste problem [13]. Faced with such a massive increase in plastic production, improved recycling and waste management alone would at the very best only stabilize the amount of plastic ending up in the ocean and the environment. There are no waste management mechanisms in place that can outcompete overproduction of plastic, which is supported by massive private investments, nor even curb the amount of plastic that ends up in the environment. While improved waste management approaches such as zero waste are still necessary, any intervention that only addresses waste management will fail to prevent plastic pollution unless plastic production is also reduced. Furthermore, existing solutions to manage plastic at the end of its initial-use phase, if and where it is adequately collected, are still largely insufficient to address impacts of plastic throughout its life cycle. Especially, incineration, "waste-to-energy", and "waste-to-fuel" methods undermine impact reduction efforts as a one-way use of fossil fuels in plastic [14].

While restrictive legal requirements on single-use items only affect a smaller fraction of plastic products, the demand for most products can still not be prevented, or many products just cannot be reused completely as of today. Moreover, much of the plastic waste generated in the past still needs to be treated. The recycling rate of plastics in general is even lower; plastics intended for more durable applications and/or enhanced properties may be manufactured with additives to improve the material properties. These include plasticizers to soften the product, coloring agents, or agents and chemicals to confer UV-resistance and flame-retardation. There is some criticism that existing and proposed technology solutions to deal with plastic waste, such as incineration, waste-to-energy, plastic-to-fuel, plastic-to-road, plastic-to-brick, chemical and mechanical recycling are incapable of adequately and safely dealing with the toxic legacy of most of the plastic produced, used and discarded [14].

Considering the well-known waste hierarchy with prevention as the preferred option, increased reuse, recycling, and recovery measures seem to be inevitable to avoid the last option of disposal (Figure 5.8 and Table 5.1) [15]. This means that, first, every effort should be made to design products that need the fewest resources or materials possible while providing the functions required during these products' use. Only when the products cannot be reused, and it is unavoidable that they become waste, should the decision be made which recycling (mechanical, chemical, or organic) or recovery route best fits the specific plastic waste stream. Landfilling should be the last resort, if previous options are not available, while littering and any leakage of plastic waste into the environment should be halted.

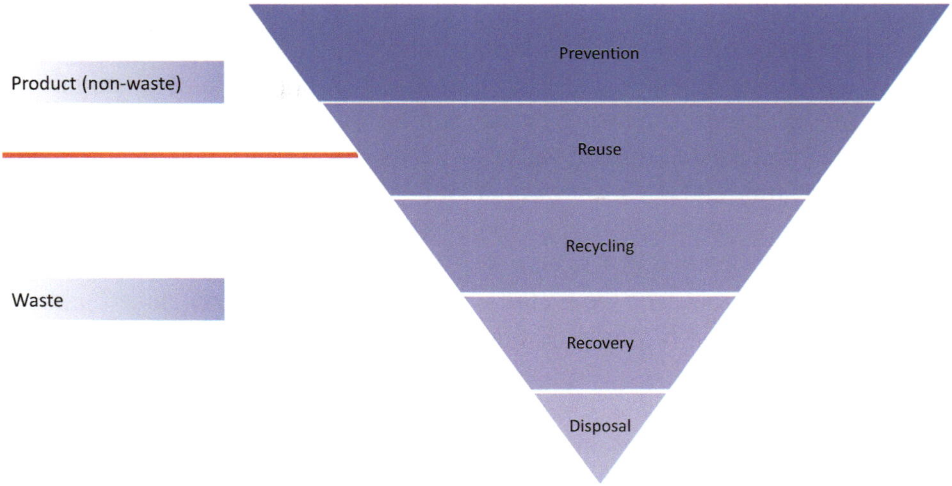

Figure 5.8 Waste hierarchy

Table 5.1 Waste Hierarchy: Definitions of the Terms Used in the EU Waste Framework Directive (2008/98/EC)

- **Waste:** Waste is any substance or object which the holder discards or intends or is required to discard.
- **Waste management:** Waste management means the collection, transport, recovery and disposal of waste, including the supervision of such operations and the aftercare of disposal sites, and including actions taken as a dealer or broker.
- **Prevention:** Measures taken before a substance, material or product has become waste, that reduce:
  - The quantity of waste [...]
  - The adverse impacts of the generated waste on the environment and human health
  - The content of harmful substances in materials and products
- **Reuse:** Any operation by which products or components that are not waste are used again for the same purpose for which they were conceived
- **Preparing for reuse:** Checking, cleaning, or repairing recovery operations, by which products or components of products [...] can be reused without any other pre-processing
- **Recycling:** Any recovery operation by which waste materials are reprocessed into products, materials or substances whether for the original or other purposes; it includes the reprocessing of organic material but does not include energy recovery and the reprocessing into materials that are to be used as fuels or for backfilling operations.
- **Recovery:** Any operation, the principal result of which is waste serving a useful purpose by replacing other materials which would otherwise have been used to fulfil a particular function, or waste being prepared to fulfil that function, in the plant or in the wider economy
- **Disposal:** Any operation which is not recovery, even where the operation has as a secondary consequence the reclamation of substances or energy

As plastic waste is sold and traded across the globe, its management has important implications. Predominately, high-income countries retain high-quality waste they can manage cost-efficiently and export waste of lower-cost value to middle- or low-income countries. Importing countries often lack the proper waste management infrastructure and discard waste if it doesn't meet sufficient requirements. Collectively, China and Hong Kong imported more than 70% of globally traded plastic waste [16]. A major turning point in global plastic waste management was China's amended legislation at the end of 2017 prohibiting imports of 24 kinds of solid waste, including eight types of plastic, and lowering the contamination tolerance threshold to 0.5% (Chinese "National Sword" initiative). Many recycling facilities worldwide cannot reach this standard. While the ban was aimed at reducing foreign waste to help protect China's environment and improve public health, it was also part of a strategy to boost the domestic recycling capacity to 350 million tons across various waste resources by 2020, and help the country move towards a more circular economy. The short-term impact of this ban will likely lead to a demand for disposal services, such as incineration and landfill, in other markets [17].

Until 2018, China was the largest importer of plastic waste, which amounted to 56% of the global market (Figure 5.9) [18]. Most imports came from the United States, Germany, Japan, Australia, the United Kingdom and Canada [19]. For exporting countries, it was commonly more cost-efficient to export their plastic wastes rather than to recycle, recover or dispose of the materials on site. Yet China used to consider the plastic wastes as valuable resources to manufacture new recycled products. However, much of the imported waste was non-recyclable and contaminated, which mostly led to illegal landfilling and littering. The remaining recyclable goods could mostly only be downcycled and the manufacturing often came along with health or environmental issues.

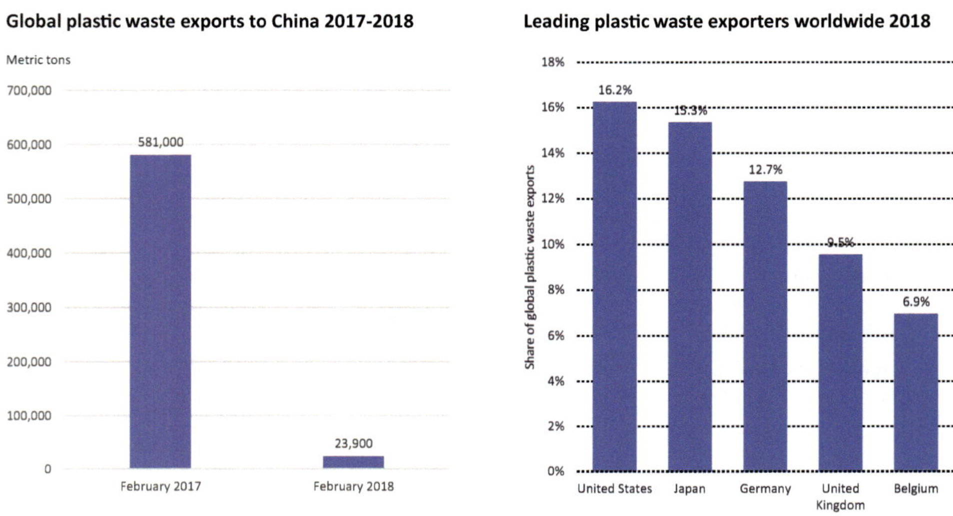

Figure 5.9 Global plastic waste exports (source: Statista, 2021)

By 2030 an estimated 100 million tons will therefore be displaced due to the ban, thus requiring domestic handling or exporting to another country [16]. Exporting countries might improve their domestic recycling infrastructure and generate internal markets or even reduce primary plastic production to handle less waste, if this infrastructure is missing. The new situation can also change the nature of the global plastic trade as such by harmonizing standards or promoting best practices on waste management. Those countries considering importing may, for example, introduce taxation to fund the development of the needed infrastructure to handle this waste [20]. Due to the de facto plastic import ban by the biggest plastic waste importer, many countries and players tried to open new export destinations for their plastic waste as this would still be the most cost-efficient option. The top alternative destinations, for example for Europe, became Malaysia, Turkey, Hong Kong, Vietnam, and Indonesia (Figure 5.10) [21]. As more illegal import activities have been detected, many of these alternative export countries also started to impose restrictions and bans on low-quality plastic waste imports by setting higher-quality standards [22]. This overall shift in global waste dynamics has caused export countries to start reconsidering their plastic waste strategies, inevitably improving the domestic waste management infrastructure by enhancing recycling and recovery processes.

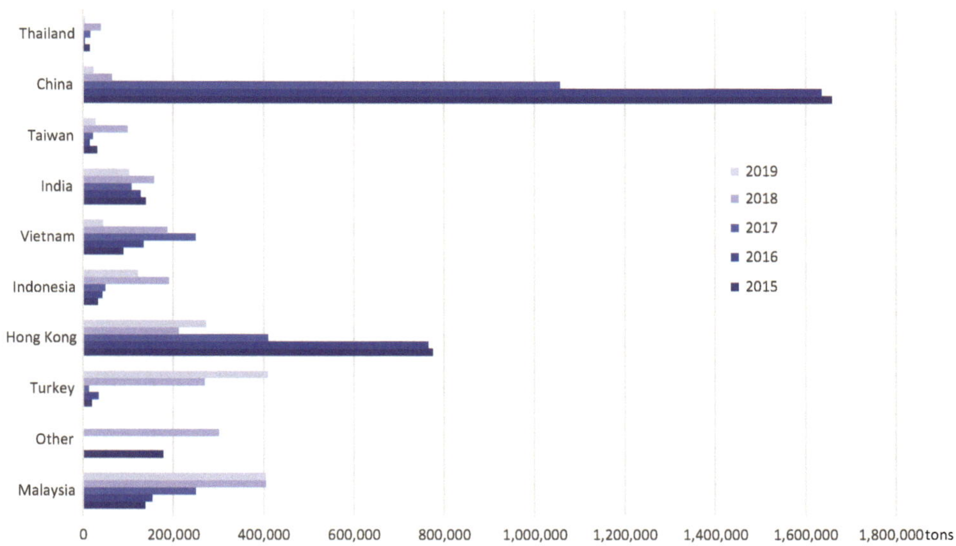

Figure 5.10 Receiving countries of plastic waste exported from the EU 2015–2019 (source: Eurostat COMEXT, Statista, 2021)

Another significant trigger for enhanced plastics recycling and related legislation is the ambition to transition towards a circular economy.

- The concept of a circular economy cannot be traced back to one specific event or year. The Ellen MacArthur Foundation, a non-profit organization and the most prominent advocate of the circular economy, states that the concept gained momentum in the 1970s [23] when the architect Walter Stahel introduced the concept of "Cradle to Cradle". According to Rizos et al. [24], the first publications about the circular economy are attributed to the environmental economists Pearce and Turner, who stated that the current linear economy could not last forever and must be replaced with a circular system [25].
- The concept of a circular economy can be defined in multiple ways. For example, "a circular economy is an alternative to a traditional linear economy (make, use, dispose) in which we keep resources in use for as long as possible, extracting the maximum value from them whilst in use, then recovering and reusing products and materials" [26].
- One of the primary goals of the circular economy by the Ellen MacArthur Foundation is "to decouple economic growth from natural resource depletion and environmental degradation." Consequently, no or minimal waste shall be generated in a circular economy [23].
- The idea of a circular economy is based on the insight that resources on the earth are finite. Today's linear economic waste flow as shown in Figure 5.2 ends in a significant amount of incineration and landfilling. Following, it was recognized as sensible to maximize the reuse or recycling of resources.
- Especially in this and the last decades, this concept has gained significant importance. The reasoning behind this is the general insight that it is crucial to handle and process finite resources in a sustainable manner.

While in Germany the circular economy and waste law adopted in 1996 only focuses on the end-of-life processing of waste, the circular economy strategy of China also encompasses the optimization of consumption behavior and improvements during the design and production phase [27]. Leading non-profit organizations like the Ellen MacArthur Foundation have developed comprehensive descriptions of a circular economy where no waste leaks into the natural ecosystem. This concept of zero waste implies that all materials are viewed from the resource perspective which leads to enhanced resource circulation. As such, the circular economy should not be reduced to mere waste management efforts, but should be commonly designed as a holistic approach, ranging from business model changes to product re-designs to recycling and recovery [23].

Among other recent developments, the European Commission introduced Circular Economy Action Plans in 2016 and 2020 with a first-of-its-kind European-wide strategy on plastics in 2018 as part of the transition towards a circular economy. The strategy includes measures such as:

- improving the economics and quality of plastics recycling by designing for recyclability, boosting demand for recycled plastics, and improving separate collection and sorting
- curbing plastic waste and littering by introducing Extended Producer Responsibility (EPR) schemes and plastic bans
- driving innovation and investment towards circular solutions
- harnessing global action [28]

Lastly, EPR schemes have been set out in multiple countries to encourage or make producers take more responsibility for the end-of-life phase of their products. Germany, for example, imposed the Packaging Law in 1991 to direct the responsibility of proper endoflife treatment away from the public/communal level and towards the private sector. This should encourage producers to consider the holistic lifecycle responsibility for their products.

The 21st century has been marked by numerous practices and legislation on plastics, as a material class in production and use, as well as plastic waste to prevent further negative environmental and health impacts. Worldwide, the introduction of bans on single-use products made of plastics has risen exponentially. Resource efficiency policies, such as the German ProgRess program, have also been put forward. Circular economy strategies have been widely accepted and adopted; these include not only the end-of-life treatment, but target the whole value chain of plastics. Based on the European Green Deal, the European Union recently committed itself to bolster the circular economy and reduce plastics leakage to a minimum while increasing plastics recycling to a maximum. EPR legislation further stimulates the recycling of plastics waste and boost the development of innovative technologies. Not only Europe but also other regions around the world are working on more efficient ways to deal with their plastic waste in a circular manner. As plastic waste is nowadays viewed from the resource perspective, the developments towards a more controlled and sustainable treatment of plastic waste are promising.

5.2 Overview of Political Measures Related to Plastic Waste Management

Most countries have experienced similar developments in plastic waste management in the past decades, but started them at different points in time. As plastics production was increasing rapidly and individual consumption of plastic products was rising, plastic waste first landed in formal or informal landfills. However, when those landfills could not bear more waste and health and environmental issues arose, legislation had to step in to set up a framework and support formalized waste management infrastructure in the form of recycling, composting and incineration plants as well as better-organized landfills. Today's political landscape has changed: Next to policy makers, these are also consumers, NGOs, designers, industry, retailers, investors, and shareholders demanding and shaping the policies and strategies to minimize risks and impacts associated with plastic. Most policies and regulations have focused on plastic packaging, but the focus is now shifting to other plastics applications in transportation, construction, and consumer sectors.

The legislation and number of regulations on plastics and plastic waste has been increasing across regions, countries, cities, and municipalities (Figure 5.11). These policies define measures on the use of plastics to reduce consumption and on disposal of plastic waste to improve its management, thus impacting entire value chains up- and downstream. Germany and Japan were one of the pioneers in initiating the first laws on plastic waste. In 1972, Germany was one of the first countries that introduced a Waste Disposal Act to mitigate uncontrolled landfilling [29]. Europe followed with the EU Waste Framework Directive in 1975 (75/442/EEC). Over the years, more and more legislation on plastics has been steadily set out in different countries around the world. Each piece of legislation provides valuable insights into the policy dynamics around this topic; the mere breadth and complexity would be enough to cover in a separate book. This chapter provides an excerpt from international, regional, and national legislation on plastic waste management shaping the legal framework for recycling, without claiming to be comprehensive.

Figure 5.11 Plastics-related legislation in various countries 2021 (source: Principles for Responsible Investment: 2021 Mapbox – OpenStreetMap)

5.2.1 Global Plastic Waste Management

One of the global instruments on managing plastic waste is the **International Convention for the Prevention of Pollution from Ships ("MARPOL")** that addresses pollution of the marine environment from ship-based discharge of harmful substances or effluents containing such substances. Since its adoption in 1973, MARPOL has included Annex V on the prevention of pollution of the seas by garbage from all kind of ships. Since the revision of Annex V in 2011, it generally prohibits the discharge of all kinds of domestic and operational garbage, all plastics, cargo residues and fishing gear into the sea. If plastic is mixed with other garbage, it must be treated as if it is all plastic and is thus subject to the most stringent procedures for handling and discharge. Every ship must carry a garbage management plan and record books with written procedures for minimizing, collecting, storing, processing, and disposing of garbage. These plans should be cost-effective and environmentally sound and use a combination of complementary techniques, including reduction at source, reusing or recycling, onboard processing, and discharge to a port reception facility. In addition, governments should provide adequate port reception facilities for garbage. Since 2016, the adopted amendments have included e-waste. Seafarers are encouraged to recover persistent garbage from the sea during routine operations and retain the material for discharge to port reception facilities [31, 32].

Another and the most prominent legislation on the international level is **Basel Convention on the Control of Transboundary Movements of Hazardous Wastes and their Disposal**, which was signed by more than 50 countries and has been joined by more than 180 parties since 1989. It was the public pressure after the discovery of deposits of toxic wastes in Africa and other developing countries imported from abroad that led to the development of this regulation. The objective of the Convention was "to protect human health and the environment against adverse effects of hazardous wastes." The scope includes a wide range of wastes based on their origin and/or composition and characteristics. The Convention is intended to reduce the generation of hazardous waste, to restrict their movement (in case of unsound practices), promote environmentally sound waste management, and establish regulation for transboundary shipments. Exports to Antarctica, to a state that is not party to the Convention, or to a party that has banned the import of hazardous waste are forbidden, but bilateral and multilateral agreements are allowed under certain conditions. In principle, the movement of waste takes place only under prior informed consent between the exporting, importing, and transit countries [30].

The Basel Convention introduced new rules in 2019 on the export and import of plastic waste. The key objective was to protect the low-developed countries with no recycling capabilities and no infrastructure for residuals of plastic waste streams which might otherwise end up in uncontrolled landfills and finally in waterways. The newly introduced waste categories and requirements are:

- for hazardous plastics waste A3210: prior informed consent procedure (PIC)
- for mixed, contaminated plastic waste Y48 (like household waste): PIC
- only for sorted non-mixed, clean, ready for recycle plastic waste B3011: no PIC

This global regulation has, for example, huge consequences for waste owners and recyclers inside the European Union (waste transport between EU Member States) and between EU and non-EU countries. The EU Waste Shipment Regulation 1013/2006/EC regulates in detail the transboundary transport of waste within, from and to the European Union and implements updates of the Basel convention into EU legislation. Most important is the difference between green-, amber-, and red-listed waste. Green-listed means that only a notification to the authorities of the sending, the receiving and the countries passed are necessary. Amber-listed means that a complex prior informed consent process is necessary where every involved state of export and import must approve the transport. Red-listed means a ban of transport. Table 5.2 shows the adoption of the Basel categories and the translation to the intra-EU situation. In the end, the export of plastic waste to low-developed countries is almost banned, to industrial OECD countries and for the intra-EU trade of plastic waste somewhat allowed.

Table 5.2 Adoption of the BASEL CATEGORIES (Green-, Amber- and Red-Listed Waste)

Plastic waste	Intra-EU	OECD outside EU	Non-OECD
EU3011	Green: Import/Export (Art. 18)	N/A	N/A
B3011	N/A	Green: Import/Export (Art. 18)	Amber: Prior notification and consent procedures by default, unless indicated differently in Commission Regulation (EC) 1418/2007 as amended
Mixtures of plastic waste in § 4 of Annex IIIA (PET, with PE/PP)	Green: Import/Export (Art. 18)	Amber: N/A (Y48 applies)	Amber: N/A (Y48 applies)
EU48	Amber: Prior notification and consent procedure	N/A	N/A
Y48	N/A	Amber: Import/Export: prior notification and consent procedure	Red: Export: prohibition Amber: Import: prior notification and consent procedure

Plastic waste	Intra-EU	OECD outside EU	Non-OECD
AC300 (the shipment of hazardous waste to and from OECD countries and intra-EU; identical with A3210 for Non-EU AND Non-OECD countries)	Amber: Prior notification and consent procedure	Amber: Import/ Export: prior notification and consent procedure	N/A
A3210	N/A	N/A	Amber: Export: prohibition Import: prior notification and consent procedure

There is a clear need to recycle waste in the best way and the best place possible. On the other hand, regional solutions are preferable to minimize waste shipment. A circular economy can only be efficiently established when product and waste processes are complementarily aligned without compromising on human health and environment, neither on the product nor on the waste level.

Some countries started to call for an international, legally binding agreement on plastics and plastic waste at **the third session of the United Nations Environmental Assembly (UNEA-3)** in 2017. Actions approved by 193 UN member states included such measures as:

- prevention of waste generation by binding reduction targets
- caps on production and consumption of the unnecessary use of plastic
- increase in reuse and recycling
- loss prevention
- emphasizing the need for EPR [33]

While several studies estimate some countries in Asia to contribute to ocean leakage, 85% of the top 20 fast-moving consumer goods companies and 95% of the top 20 plastics corporations are based in the U. S. and Europe [34]. Therefore, international policy should address the geopolitical barriers caused by product design and plastic production made in countries other than those currently in the spotlight for ocean pollution. Policy emphasis must be placed on the sources of the plastic pollution problem. According to the UNEA's Ad Hoc Open-Ended Expert Group on Marine Litter and Microplastics, addressing the challenge of marine plastic pollution should start with prevention by reducing rapidly increasing plastic production, and the use of chemicals of concern such as Endocrine Disruptive Chemicals and Persistent Organic Pollutants (POPs) involved in plastic production, as well as exploring other waste management options. This requires broadening the scope from marine litter and microplastics to a broader one on plastic waste management [35]. In Section 5.3.1, we will address the current developments around UNEA.

In the last couple of years, plastic products designed for single use (such as bags, bottles, plates, cutlery, and straws) became one of the most obvious and preventable polluters, constituting over a third of total plastic production in 2017 [36] and frequently found discarded in the environment. Multiple countries reacted quickly by introducing, for example, bans, consumption reduction targets or taxation on the use of these products. Denmark was one of the first countries that introduced a tax, and India was one of the first that introduced bans on plastic bags (local ban) in 1998 [37]. Nowadays, a vast amount of countries have introduced restrictions or bans on various single-use plastic products, and the number is rapidly increasing. Figure 5.12 shows by way of example the current status quo of plastic-bag bans in the world in 2020 [38]. The increasing popularity of bans is not only visible in Europe: Kenya banned single-use plastic bags in 2017 nationwide and all single-use plastic products in 2020. Zimbabwe outlawed polystyrene food containers in 2017. The legal framework in the U.S. is more complicated due to its federalized system: While some states and cities banned different types of single-use plastic products, there is no federal ban. Similarly, India did not pass a nationwide regulation but rather encouraged the states to address the issue [39]. China prohibited plastic bags and other disposable plastic products in all major cities in 2020 with all municipalities and markets to follow in the next five years [40]. Although it has been reported as an outright single-use plastics ban, Canada's 2021 environmental regulation amendment in fact enables a more targeted mitigation approach which could lead to prohibition at a later point [41].

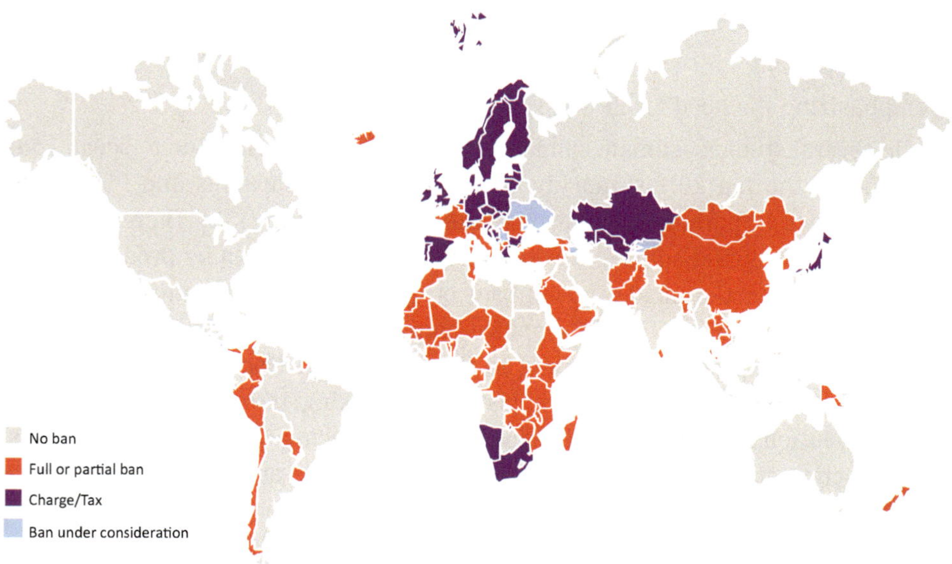

Figure 5.12 Countries banning or restricting plastic bags in 2020 on the national level (source: Statista, 2021; United Nations)

As plastic waste varies in its composition and timely generation across regions and countries, the solutions and legislation to its management are complex. A concerted action by different stakeholders at the national, regional, and global levels is needed to define a holistic and dynamic multilateral approach to plastic waste management. Although the topic of plastic is already being discussed in various international for a, and the need for long-term, encompassing, and global action is recognized, there is no international agreement so far which is fit to deal with plastics waste management in its complexity.

5.2.2 Plastic Waste Management in Europe

The European Union wields an extraordinary external influence on legislation across the globe. Therefore, it might be interesting to take a closer look at the legislation on plastic waste management the European Union has set out in the last couple of years to direct the various EU Member States.

With its 2019 **EU Green Deal** (COM/219/640), the European Commission addressed the UN 2030 Agenda. The Green Deal enshrines the goal of a climate-neutral Europe by 2050. One of the measures presented to achieve the objectives is the establishment of a CO_2-neutral circular economy with major implications for the industry. A series of actions planned include the EU Industrial Strategy (published in 2020) and the continuation of the Circular Economy Action Plan, focusing on the construction, textiles, electronics, and plastics sectors, with the development of measures to deal with microplastics. Envisioned actions also include following up on the measures and objectives of the EU Plastics Strategy communicated in 2018, demanding all plastic packaging to become reusable or recyclable by 2030, and calling for industry's pledge to use 10 million tons of plastic recycled material across the EU by 2025. In addition, more attention will be paid to a "circular-oriented design." It remains to be seen how the announced actions will affect the development of legislation at both EU and its Member State levels [42].

The first **EU Circular Economy Action Plan** (CEAP 1.0, COM/2015/614) was launched in 2015 announcing the development of a strategy to address the aspects of plastics along their entire life cycle. Several measures and policy revisions on waste were announced:

- Increasing high-quality recycling, also by a more ambitious EU-wide target on the recycling of plastic packaging
- developing quality standards for secondary raw materials
- clarifying existing rules on the "end-of-waste" criteria to determine when secondary raw materials should no longer be considered "waste"

- revising the Directive on the Port Reception Facilities 2000/59/CE and cross-border trade in recyclable plastics [43]

The new CEAP 2.0 (COM/2020/98) published in 2020 emphasized the role of eco-design. It is perceived as a good way for a circular economy to avoid waste generation and improve the durability, repairability, reusability, and recyclability of products. The EU envisages increasing the possibilities of high-quality and efficient recycling. It also aims at setting rules on mandatory recycled content for certain materials or components in packaging, construction, vehicles and electrical and electronics equipment in 2021/2022, also establishing the rules for the safe recycling into food-contact applications other than PET. In addition, EU Directives on packaging, single-use plastic products, eco-design and end-of-life vehicles will potentially be accordingly amended. Although the EU Ecodesign Directive (2009/125/EC) has historically focused on energy-related products, it is most likely that the EU will expand the scope and include non-energy-related aspects (e. g., recycled content, recyclability) and all kinds of products [44].

In 2019, the **EU Taxonomy for Sustainable Finance**, i. e., a classification system for environmentally sustainable economic activities, was released. To reduce CO_2 emissions from the plastics sector, it would be relevant, for example, to promote materials' recirculation. The manufacturing of plastics should be based on secondary raw materials which can significantly reduce life-cycle emissions – in the manufacturing process and the embedded ones. Recycled material can thus be produced by mechanical and chemical recycling processes. This objective is currently only considered under "do no significant harm" by the mitigation and adaptation objective. The technical screening criteria for the objectives is expected to be developed further by scrutinizing business activities conceived "green" in their contribution to the circular economy [45].

The **EU Waste Framework Directive (2008/98/EC) (WFD)** regulates the management of waste and is to be considered as the framework for all relevant laws and regulations on waste in the European Union, to be subsequently transposed on the EU Member States level. It was revised as part of the 2018 Waste Package (Directive 2018/851/EU). The definitions of waste were harmonized (Art. 3), EPR tightened (Art. 8a), higher reuse and recycling quota were defined and a new method of calculation for the recycling rate was established (Art. 11 and 11a). Moreover, the EU set the target of reducing the amount of non-recyclable municipal waste by at least 50% by 2030. During the next revision as of 2022, it is expected that further mass reduction targets will be set for certain waste streams, but it is not yet known for which specifically [46, 47].

The **EU Packaging & Packaging Waste Directive (94/62/EC) (PPWD)** regulates the handling of packaging and packaging waste and was also amended as part of the 2018 Waste Package (Directive 2018/852/EU). Amendments include the harmonization of definitions and terms. For example, the term "packaging waste" should be defined in a general way. In addition, in the future Member States will be obliged to promote measures (e.g., deposit schemes) that increase the reuse of packaging. The new changes set recycling targets for plastic packaging waste at 50% by 2025 and 55% by 2030 (Art. 6) as well as defined rules on the calculation of the attainments of targets. It was also announced that the essential requirements listed under Annex II should be reviewed, and, if necessary, amended to enforce them. Particular attention should be given to the prevention of packaging waste generation, reducing (over)packaging and material complexity, but also to the design for reuse and a high-quality recycling. The EU Commission will thus develop requirements to ensure that all packaging in the EU market is reusable or recyclable in an economically viable manner by 2030. Ensuring the functioning of the internal market, avoiding obstacles to trade and distortion and restriction of competition are central to the new provisions in the same way [48, 49].

The **Directive on reducing the impact of certain plastic products on the environment (2019/904/EU)**, also known as the "Single-Use Plastics" Directive, aims to reduce pollution of the environment by certain products frequently found on EU beaches. Those products that can be easily replaced by alternatives should be banned; those which cannot fall under consumption reduction targets or may be impacted by other measures. For beverage bottles with a capacity for up to three liters, each EU Member State shall establish separate collection (77% by 2025, 90% by 2029) and ensure that

- from 2025, beverage bottles manufactured from polyethylene terephthalate (PET) as the major component contain at least 25% recycled plastic, calculated as an average for all PET bottles placed on the market on the territory of that EU Member State, and
- from 2030, beverage bottles contain at least 30% recycled plastic, calculated as an average for all such beverage bottles placed on the market on the territory of that EU Member State.

By 2022, the EU Commission shall adopt implementing acts laying down the rules for the calculation and verification of these targets which might potentially influence other legislative processes on mandatory use of recycled content. In addition, EPR should be implemented for fishing gear and all listed single-use plastic products, adding the clean-up costs on top of the partly established collection, sorting and treatment costs [50].

The **EU End-of-Life Vehicles Directive (2000/53/EC)** regulates the handling, recovery, and disposal of end-of-life vehicles (ELV). In all EU Member States, reuse and recovery is expected to be at least 95% of the average vehicle weight per year since 2015 and at least 85% are to be reused and recycled. This Directive was also amended as part of the 2018 Waste Package (Directive 2018/849/EU). Since 2019, the EU Commission has been reviewing the ELV Directive in terms of its feasibility aiming to increase the effectiveness, efficiency, relevance, and coherence of the Directive, as well as its compliance with EU values. In the coming years, aspects to be considered in a revision of the Directive are the shipment of old vehicles, the proportion of hazardous substances and plastics recovery and recycled content [51, 52].

The **EU Waste Electrical and Electronic Equipment Directive (2012/19/EU) (WEEE Directive)** was amended as part of the 2018 Waste Package (2018/849/EU). EU Member States were encouraged to develop new recovery, recycling, and treatment technologies for WEEE as well as incentivize the application of the waste hierarchy. From 2019, the minimum collection rate for WEEE to be fulfilled has been 65 wt% per annum of EEE placed on the market of the concerned EU Member State in the three preceding years. The new CEAP foresees the creation of a Circular Electronic Initiative in 2020/2021 to improve the collection and treatment of WEEEs and announced the introduction of plastic recycled content [53].

In summary, apart from the legislation on waste prevention, reuse and landfilling, European legislation is in a process of setting and implementing recycling-related targets on various plastic applications to develop circular economy. Starting with plastic packaging, the focus is gradually shifting to other applications from the automotive, construction and various consumer sectors, with partial adoption of similar provisions. The targets are defined either for a class of products (e.g., packaging) or specific applications (e.g., bottles or bags) and aim to bolster recycling by increasing recycling quota for waste, encouraging and, in some cases, mandating recyclability or design for recycling and recycled content as well as EPR for products (Figure 5.13).

Application \ Targets	Recycling	Recyclability	Recycled content	Extended Producer Responsibility
Plastics	n/a – falling under the general municipal waste targets	n/a	Call on industry to use 10 million tons by 2025 in different applications	n/a
Packaging	▪ 50% by 2025 ▪ 55% by 2030 ▪ EU member states' contribution for non-recycled waste	All types of packaging by 2030	▪ 25% rPET in bottles by 2025 ▪ 30% rPlastics in bottles by 2030 ▪ Mandatory requirements announced	▪ In place for household packaging in almost ever EU member state ▪ Mandatory for all types by 2025
Automotive	▪ Mandatory dismantling of large plastic parts ▪ 85% car recycling ▪ 95% car recovery	Stricter requirements expected as of 2022	Mandatory requirements under consideration	In place
Building & Construction	▪ 70% of construction waste to be recovered by 2020 ▪ Material-specific target under consideration	Introduction under consideration	Mandatory requirements under consideration	n/a
Electrical & Electronic	Depending on the appliance 55-80% from 2018	Circular initiative	n/a	In place
Textiles	n/a – falling under the general municipal waste targets	Circular design	Uptake	To be introduced
Fishing gear	n/a	EU standard on circular design in development	n/a	To be introduced

Figure 5.13 Recycling-related targets in European legislation as of 2020

5.2.3 Plastic Waste Management in Germany

In 2019, Europe represented 16% of global plastics production, and Germany accounted for 25% of the European plastics market. In 2018, 5.3 million tons of plastic post-consumer waste were collected in Germany through official EPR schemes. From 2006 to 2018, the volumes for recycling increased by 80%, energy recovery increased by 73% and landfill decreased by 80%. Germany has a strong track record of implementing innovative waste legislation, including the 1991 Packaging Waste Ordinance, the first application-related plastics waste legislation in Europe setting recycling targets for plastics packaging. As Germany is the strongest economy in Europe, its waste legislation is influencing EU and global plastics waste laws. Therefore, it might be interesting to take a closer look at the German legislation for waste of specific plastic application sectors (the following data are from [54]).

In countries like Germany, most of the waste ended up in informal trash dumps until the late 1960s. By then, 50,000 informal trash dumps existed next to only 130 formal landfills, 16 composting sites and 30 incineration plants that could only handle approximately 37% of household waste in line with environmental standards [29]. The German magazine *DER SPIEGEL* stated that each year a 3,000 meter high mountain of waste was growing out of the Olympiastadion in Munich [56]. According to the Environmental Program of 1971 of the German Parliament (Bundestag), more than 90% of all waste ended up in informal trash dumps. However, the public realized that this waste was causing the steady poisoning of the ground water. Due to such growing health and environmental concerns, a central department for waste disposal was introduced in 1965.

A turning point in Germany's waste history was the **Waste Disposal Act** of 1972. Worldwide it was one of the first legal frameworks for waste disposal. Many informal trash dumps were closed, and more formal waste disposal facilities were introduced. By 1984, the 50,000 informal trash dumps were reduced to 385 official landfill sites. Meanwhile, 16 additional incineration plants were built by 1987. Despite gas filters, such plants caused increased and toxic gas emissions. Simultaneously, 28 composting plants were built by 1987 in Germany. However, as a large amount of waste was not compostable, this process led to increased environmental harm [56].

Starting in 1975, the **"Bundesmodell Recycling"** was implemented as a first large-scale attempt to formalize communal recycling in the southwest of Germany. Germany funded a demonstration plant in Steinlachtal, which was intended to show that mechanical separation and recycling of waste was possible on a larger scale. The realization of the project was rather difficult. At the beginning, the stakeholders in the project disagreed on the technical components and the establishment of the plant was postponed for several years. Starting in 1981, the pilot plant ran for

10 years backed by government subsidies. The project was eventually a failure because the recycled products were not demanded on the market. Furthermore, the project was very costly and energy-intensive despite continuous technical optimizations. Last but not least, the unpleasant odor caused much local resistance so that the plant was shut down in 1991 [56].

Germany was also one of the first countries to introduce a **Packaging Ordinance** in 1991, which aimed to prevent and reduce waste, reuse packaging, and recycle packaging waste [3]. With this ordinance, producers and retailers had also to take on responsibility for the waste since they can steer many optimizations along the development and production of packaging products. Hence, the responsibility for collection and disposal of packaging waste, which amounted to most of the municipal waste, shifted to the producers (see also Section 5.3.6). As a matter of fact, the Dual System Germany AG was formed which conducted its licensees' take-back and recycling obligation [56].

Despite all product-specific regulations like the Packaging Ordinance, the landfill ban on carbon-rich waste (effective 2005) was key to raising the recovery rate for plastics or plastic-rich waste. As waste flows to the cheapest price, recycling and energy recovery options that are more expensive than landfill can only be attractive if the cost benchmark is shifted to a higher level (e.g., by putting a tax or ban on landfill). The effect of boosting plastics recovery rates within a few years thanks to a "divert-from-landfill" concept in Germany is shown in Figure 5.14.

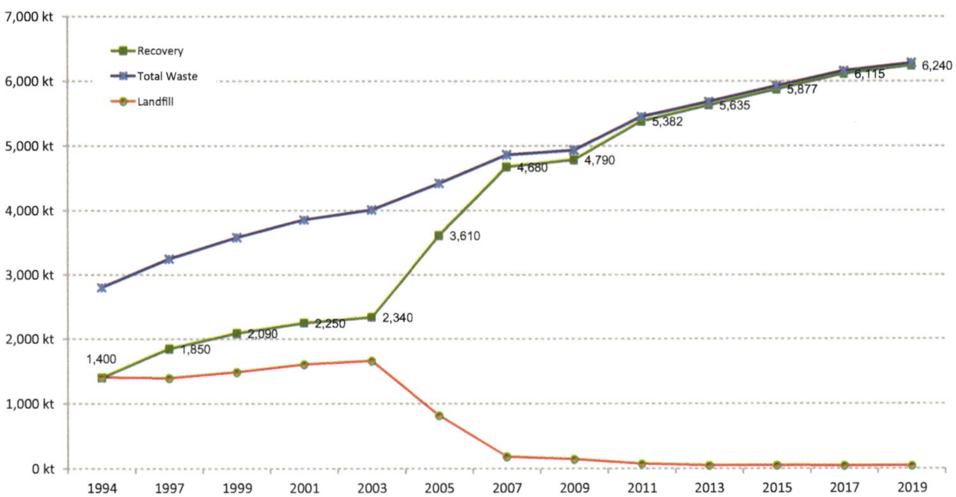

Figure 5.14 Development of post-consumer plastic waste treatment in Germany 1994–2019 (source: BKV; Conversio Market & Strategy GmbH)

The Circular Economy Law (Kreislaufwirtschaftsgesetz/**KrWG**) of 2012 serves as a general basis for the handling of waste and its management. It is the German implementation of the EU Waste Framework Directive (EU WFD) and forms the legislative framework for the management of waste. The revised KrWG is from the October 2020. Amendments include the harmonization of definitions. Important newly inserted and/or revised definitions refer to the terms “recycling”, “municipal waste”, “waste management” and “end-of-waste status”. It is, in particular, the definition of “end-of-waste status” that plays a decisive role in recycling, since this is the interface with the chemical and substance law and therefore leads to a change in the handling of substances and mixtures. In § 5, the waste status of a substance or article ends if it has undergone a recovery process and there is a market or demand for it; it is used for certain purposes; it complies with the technical requirements and legislation for which it is used; and has no harmful effects on humans or the environment. Newly produced substances and articles from waste are to comply with the requirements of laws applicable for substances and articles in accordance with § 7a. Interestingly, the EU WFD’s provision for all producers and importers of articles to submit information on those containing substances of very high concern (SVHC) (according to REACH) to the SCIP database (managed by the European Chemicals Agency, thus making this information available for recyclers and consumers), was implemented in Germany in the chemical law (ChemG, § 16f) and not in the KrWG.

As in the WFD, product responsibility or EPR will be further expanded in the KrWG (§ 23–25). In addition to the directly adopted requirements from the WFD, the German Federal Government goes even further in its draft law. It wants to oblige manufacturers, processors, and first-time polluters to contribute to the costs of cleaning up the environment in addition to the usual cost sharing. They should also be obliged to act in an informative and advisory capacity on waste management.

Other obligations under the draft law in the context of product responsibility are:

- Products should be designed to be more durable, repairable, and reusable.
- Recycled materials should be used primarily to produce products.
- Products with critical raw materials should be labeled.
- Reuse and reuse systems should be strengthened and supported.
- The obligation to take back and treat waste, as well as the duty of care for manufacturers, is also crucial. The duty of care is intended to maintain the suitability for use of products in the sense of waste prevention.

§ 14 of the KrWG reflects the reuse and recycling targets set out in § 11 of the EU WFD, which were directly adopted, with the only difference that, since 2020, 50% of municipal waste is to be used in preparation for reuse and recycling. Another planned change in § 24 of the KrWG concerns the obligation of manufacturers to provide information. This includes the labeling of substances used, the prevention of waste, the possibility of return, possible recovery, and disposal methods.

In 2019, most plastics in Germany were processed in packaging (about 30.7% of the total amount). Of 3.224 million tons consumed, 3.160 million tons incurred as waste: Households produced 2.091 million tons and industry – 1.069 million tons of waste. 1.559 million tons of plastic packaging were recycled. Approximately 474 kt of plastic recyclates were used in the production of plastic packaging, whereby about 255 kt came from waste in the post-consumer sector. In addition to the KrWG, the **Packaging Act (VerpackG)** has regulated the consumption and waste management of packaging and packaging waste since 2019, which replaced the packaging ordinance in force until then. A key change that has been in effect is the registration requirement for system-participating packaging at the Central Agency Packaging Registry (ZSVR). This includes the registration of the company itself, the quantities of packaging in circulation as well as the EPR scheme used. Since a large part of packaging is made of plastics or contains plastics, the plastics industry is addressed directly by law as a manufacturer or distributor within the EPR framework.

The requirements for preparation for reuse or recycling of plastic packaging in accordance with § 16 (VerpackG) are:

- At least 90 wt% to be recycled
- At least 58.5 wt% material recycling
- At least 63 wt% material recycling from 2022

The EU Plastics Strategy and the legislative package show that the EU is currently focusing strongly on plastics and plastic packaging aiming to curb marine litter. The related amendments to the EU WFD and PPWD led to the revisions of the KrWG and the VerpackG. The revisions of the KrWG regarding the new calculation methods for recovery rates and the extension of the EPR also affect the packaging industry. In the future, the calculation should relate not to the input but to the output quantities of sorting systems. In addition, manufacturers and those who bring the product to the market for first time are obliged to cover cleaning costs in public spaces. The 2020 revision of the VerpackG includes a ban on plastic carrier bags. Further revisions are planned to transpose the requirements from the EU "Single-Use Plastics" Directive into national law in a 1:1 approach, as well as further requirements on the modulation of EPR fees based on packaging's recyclability and recycled or renewable content.

Due to the ever-increasing quantities of (plastic) packaging and other (plastic) disposable products in the environment, the Federal Government published a **5-Point Plan** in 2018, which set out objectives and measures for a "way out of the throw-away society" to reduce the littering of the environment. The 5-Point Plan includes the EU objectives of the EU Plastics Strategy. These include avoiding unnecessary products and packaging, eco-friendly product and packaging design and promoting fabric cycles through high-quality recycling.

The handling and treatment of construction and demolition waste is regulated in German legislation by the **Industrial Waste Ordinance** (GewAbfV). The new GewAbfV entered into force in 2017. There is a pre-treatment obligation for "predominantly non-mineral" construction and demolition waste (§ 9, GewAbfV), which includes, among other things, the sorting out of plastics. Plastic packaging, which is produced on construction sites, is subject to the VerpackG. A sorting rate of mixed commercial waste of 85 wt% has been achieved in pre-treatment plants since 2019. The recycling rate to be fulfilled in accordance with § 6(5) of the GewAbfV is at least 30 wt%. In 2019, 3.58 million tons of plastics were processed for the construction industry in Germany. The share of the total amount of plastic demand for construction is 25.2%, making construction the second largest user of plastics. In 2019, the amount of recycled plastics used in the construction sector was 834 kt. This corresponds to approximately 43% of the total amount of recyclates produced in Germany (1.9445 million tons). The volume of plastic waste from the construction industry in 2019 was approximately 0.522 million tons, or 0.23% of the total amount of construction and demolition waste.

In 2019, approximately 1.5 million tons of plastics were installed in vehicles. In total, about 0.33 million tons of plastic waste from vehicles were generated. With 10.6% of the total amount of plastic processed in Germany, the automotive industry is the third largest industry in which plastics are processed and consumed. Following the European legislation, the recycling target is 85% and the recovery target 95% under the **ELV Ordinance (AltfahrzeugV)**. Interestingly, shredder and dismantling companies together must fulfill the targets, meaning that a cooperation between these actors is required.

The legal framework for the treatment of WEEE is the **Electrical and Electronic Equipment Act (ElektroG)**, which was amended in 2015. Distributors have been obliged to document the collection and waste management of electrical and electronic equipment via the Foundation EAR (Stiftung Elektroaltgeräte). Since 2019 electrical and electronic products placed on the German market must also be registered at the Foundation EAR. § 10 of the ElektroG sets the target that the quantity of waste collected must correspond to at least 65% of the quantity placed on the market in the last three years. The recycling and recovery targets are defined individually for each group of collected products. The quantities are recorded at the Foundation EAR. Among other things, it also coordinates the collection of WEEE at the respective collection points in the German municipalities. The total amount of collected WEEE across all device categories was 853,124 tons in 2018 [55]. The total amount of plastic waste generated from the post-consumer WEEE sector in 2019 was approximately 316 kt. According to the Federal Environment Agency, Germany achieved a recovery rate of 43.11% in 2018, which was the minimum collection rate at the time. In 2019, a total of about 881,000 tons of plastics were processed from WEEE.

5.2.4 Plastic Waste Management in China

Since the 1950s, China has initiated several pieces of legislation on plastics waste management on the national and province levels.

In 2009, the **Circular Economy Promotion Law** came into effect aiming to realize sustainable development, promote a circular economy, protect and improve the environment, and raise resource utilization efficiency. In this case, the term "circular" is synonymous with "sustainable" and covers activities along the products' life cycle. It means that the use of resources and waste generation should be reduced in production, consumption, and circulation. Waste should be reused directly as product, after repair, renovation, or reproduction, or, wholly or partly, as parts of other products. Recycling refers to the direct use of wastes as raw material or after waste regeneration. Production safety shall be guaranteed to ensure quality standards are complied with and secondary pollution is avoided. By providing funding, the state should encourage and support the research, development and promotion of circular economy-related science, technology, and infrastructure, in urban and rural areas. It should also develop a national circular economy plan and set indicators for resource output capacity, waste reutilizing rate and recycling rate. Certain businesses must be responsible for the recycling products they place on the market, either themselves or by entrusting other organizations. In the design of products, the preference should be given to using fewer materials to avoid wasting resources as well as materials which are recyclable, dismountable, degradable, less harmful or poisonous [57].

As the quality of recyclable materials imported to China over the past two decades gradually declined and was mainly mixed with food, garbage, and other pollutants, China implemented a **"Green Fence"** policy in 2013. It temporarily restricted plastic imports with significantly less contamination by setting a limit of 1.5% of "allowable" contaminant in each bale of imported recyclables. Chinese regulations had long banned excessive levels of contamination in imports of recyclables, but these had rarely been enforced. By forcing foreign companies to increase the quality of the waste they sell to China, the Chinese government aims to reduce the environmental damage caused by small Chinese firms using improper recycling techniques that pollute the country's environment [58].

The above requirements were tightened in 2017 by introducing the **"National Sword"** campaign, as was described in Section 5.1. These new restrictions tightened importing recyclable materials and other scrap for processing and increased inspections. This program was launched as a key component of the overarching plan to reduce plastic waste in China. It banned the import of plastic scrap from other regions, resulting in a significant reduction thereof. Licenses to operate domestic recycling plants were either cancelled or suspended as they failed to comply with stricter rules. The intention was to enforce stricter environmental regulation and invest in developing domestic recycling infrastructure to offset the decline in imports [59].

The **Solid Waste Law** originally adopted in 1995 was revised in 2020. It covers the prevention and control of pollution from industrial, household, construction, agricultural, and hazardous waste. Among other things, the government should establish EPR systems for electrical and electronic products. Producers are required to establish a recycling system for the volume of products placed on the market and to raise awareness of their recycling systems. Manufacturers, sellers and importers of products and packaging that are included in the compulsory recycling catalogue should finance the collection and recycling of end-of-life products, either individually or via a third party. Retailers, e-commerce, express delivery, and fast-food providers have to report the use and recycling of single-use plastic products to the authorities [60].

In 2020, the Chinese government additionally announced a new set of stringent regulations under **China Plastic Pollution Control (CPPC)** to reduce plastic consumption in different industry sectors and set specific goals for 2020, 2022 and 2025. By 2020, the production, sale and use of certain plastic products in some pilot areas and sectors in China were prohibited and a total ban on plastic waste imports was imposed. By 2022, the consumption of disposable plastic products should be significantly reduced, and alternative products promoted. By 2025, a management system for the production, consumption, and circulation of plastic products should be established. Affected products are disposable foamed tableware, beauty and personal care products containing microbeads, items provided at hotels, plastic packaging, and non-biodegradable plastic bags. It also prohibits the production and sale of plastic shopping bags less than 0.025 mm thick and polyethylene agricultural film less than 0.01 mm thick. Recycling and converting plastic medical waste into new products is prohibited as well. Another strategy is to charge citizens for waste production by establishing a pricing system for municipal solid waste. Along with this, a nationwide campaign to promote waste separation was initiated [61, 62].

In addition to reducing the volume of imported waste, China also enacted policies to reduce waste generation within the country. To decrease waste volumes significantly, China is increasingly shifting from landfilling to incineration. In 2016's "Thirteenth Five-Year Plan", it targeted >50% of the waste being incinerated rather than landfilled or disposed of by other means. As China generates an increasing amount of waste, one policy measure by the Chinese government is a plan to develop zero waste cities, as announced in 2018. This strategy reflects, for instance, how one can maximize recycling, including the utilization of different types of waste as secondary carbon feedstock for production as well as help build up a better functioning domestic infrastructure [63].

5.2.5 Plastic Waste Management in the United States of America

Since the 1970s, several hundred bills and resolutions on plastic waste-related topics were introduced at the federal, state or local level in the United States, with two under discussion for the national legislation and several hundred in more than forty states as of the beginning of 2021. Especially the latest developments in international trade, changes in the material flow and economics, and concerns about the domestic waste infrastructure have spurred the U. S. Congress to initiate several waste and recycling-related bills. It seems that a bipartisan consensus prevails on the importance of tackling the plastic waste problems. In 2020, only the Save Our Seas Act 2.0 made its way into the law; however, other bills are also back on the table in 2021. These bills might change the way how the U. S. deals with recycling in the coming years, be it investments in infrastructure, EPR, bans on certain products or awareness raising.

The bills are frequently initiated or supported by the industry associations and recently also NGOs. A strong focus lies on developing roadmaps and building alliances, such as Advanced Recycling Alliance for Plastics. Plastic packaging applications are at the center of current legislative provisions. There is no national law that mandates recycling, and state and local governments often introduce their own recycling requirements. Similarly, there has been no federal law governing EPR in the U. S. yet. The often-preferred term used in the U. S. for practices related to EPR is "product stewardship". For example, the focus of most product stewardship programs on vehicles at the national level in the U. S. has been on voluntary measures to address contaminants of particular concern or to develop recycling [64].

The **Save Our Seas Act 2.0** (S.1982) passed in the Senate at the beginning of 2020 and requires the improvement of domestic infrastructure to prevent marine debris through new grants for and studies of waste management, mitigation, and new markets for recycled plastics. It also strengthens the U. S. response capability and enhances global engagement to combat marine debris. For example, the collection, removal, and recycling of the derelict fishing gear should be supported. An international cooperation to combat marine debris should help, among others, strengthening systems for reducing the generation of plastic waste and recovering, managing, reusing, and recycling. Advanced technologies to capture value from post-consumer materials and municipal solid waste streams through mechanical and other recycling technologies should be deployed and made accessible.

The **Break Free From Plastic Pollution Act** (H.R.5845), introduced in 2020 and reintroduced in 2021, sets requirements on waste and recycling collection systems for a variety of products and materials, including plastics. It is considered an amendment to the Solid Waste Disposal Act in order to reduce the production and use of certain single-use plastic products and packaging; to improve the responsibility of producers in the design, collection, reuse, recycling, and disposal of their consumer products and packaging; and to prevent pollution from consumer products and packaging from entering into animal and human food chains and waterways. It makes packaging product producers take responsibility for collecting, managing, and recycling or composting waste after consumer use and removes the burden from the cities, states, and taxpayers. In addition, the bill enacts minimum percentages of products that must be reused, recycled, or composted. Mandatory post-consumer recycled content for plastic beverage containers must gradually increase from 25% by 2025 to 80% in 2040. Requirements for other products are to be set by the Environmental Protection Agency (EPA). Beginning in 2022, a variety of single-use products should be phased out or their use discouraged. Consumers' behavior should be encouraged by establishing programs to refund consumers for returning beverage containers (e.g., a nationwide 10-cent deposit program) and by establishing a tax on carryout bags. A temporary three-year moratorium on new or expanded permits for facilities that manufacture plastics and chemical recycling plants is set until regulations are updated to address pollution from the facilities. The EPA must publish guidelines for a national standardized labeling system for recycling and composting receptacles. Producers must include labels on their products that are easy to read and indicate whether the products are recyclable, compostable, or reusable. Finally, the bill establishes limitations on the export of plastic waste to other countries [65].

The **Plastic Waste Reduction and Recycling Act** (H.R.7228) was introduced in June 2020 and is expected to be reintroduced in 2021. It announced the establishment of a "Plastic Waste and Reduction and Recycling Program". The federal government is called to develop a strategic plan for waste reduction, recycling, and remediation. Key pillars should be improvement of the global competitiveness of the U.S. plastics recycling industry, development of standards for plastics recycling and related technologies, as well as investments in recycling technologies. For this, funding should be authorized for five years and $85 million invested in 2021 as well as public-private partnerships facilitated. It should bolster plastic sorting infrastructure, including that for biobased plastics, research, and development of chemical and "bio-inspired" recycling and their environmental implications. In addition, research and other activities on advanced recycling technologies "optimized for recyclability", remediation, and health impacts of microplastics should be driven forward. The Break Free From Plastic Pollution Act and the Plastics Waste Reduction and Recycling Act were the first attempts to establish a federal plastics waste legislation [66].

The **CLEAN Future Act** introduced in 2020 and reintroduced in 2021 aims to achieve net zero greenhouse gas emissions by 2050, by bringing them down 50% from 2005 levels. It calls for various environmental, economic, and job-related efforts in recycling. This bill aims to establish standards on labeling and collection of recyclables and post-consumer recycled content for certain products, as well as implement a national bottle deposit program. A task force to establish EPR system for certain products should also be set up. The EPA should develop grants for zero-waste initiatives, recycling and waste reduction education, and composting or anaerobic digestion projects. A "temporary pause" on permitting new and expanded plastic production facilities has been put in place to give time for the EPA to update clean air regulations for facilities by addressing environmental justice impacts on affected communities [67].

Introduced in 2019 and reintroduced in 2021, the **RECOVER Act** (H.R.2357, formerly H.R.5115) aims to establish a Recycling Infrastructure Program with the EPA to award financial assistance to states, local, and tribal governments. Infrastructure projects should be boosted with up to $500 million in federal grants to improve curbside collection, processing infrastructure, and education [68, 69].

The **RECYCLE Act**, introduced in 2021, targets improving the effectiveness of residential and community recycling programs through increased public education and outreach by authorizing up to $15 million a year in grants for states and local governments, tribes, and nonprofit organizations. The EPA has been asked to update its guidelines for products containing recycled content on a more regular basis [70, 71].

Solid waste programs are mainly managed at the state and local level. Various states are gradually taking up plastic recycling in **state-level legislation**, whereby California is the most active one in proposing and implementing laws. For example, California supports products becoming reusable, recyclable, or compostable by 2030. It also introduced mandatory targets on the recycled content for single-use applications: 20% as of 2020, 40% by 2028, and 72% by 2030. One of the bills aims to achieve a 75% source reduction, recycling, and reuse goal by making businesses financially responsible for waste management. Nearly a dozen states introduced EPR for packaging, shifting the recycling costs away from taxpayers to packaging producers. Some of the states mandate take back for selected products, such as electronics, and are considering including other products. Although many states' own bills were postponed in 2020, more substantive changes are expected to happen on this level.

The EPA has drafted a **National Recycling Strategy**, which prioritizes the reduction of contamination in the recycling stream, improvement of the domestic market for recyclables and an increase in processing efficiency. The national recycling rate should be increased to 50% by 2030. It is, however, too early to say how this strategy will influence the change. The federal government, as a big purchaser of products, could update its purchasing requirements, including products containing recycled content.

The political disposition on recycling in the U.S. is quite divergent, varying from total opposition to recycling in general to strongly supporting it. More than a dozen states have initiated bills supporting the relatively new advanced recycling technology; others, in turn, have even called for a total ban on the building of such facilities. Whereas some states have initiated bills supporting the relatively new advanced recycling technology, others, in turn, have even called for a total ban on the building of such facilities. It is expected that other recycling-related legislation will evolve, and that Congress might become more engaged on this topic, such as funding expansion of the existing recycling infrastructure. Some may call for increasing incentives to expand capabilities in domestic recycling, either taken up in its own bill or as part of a broader infrastructure package. Even if the bills might not become law in the short term, lawmakers remain motivated to keep recycling on their agenda, because it has become essentially relevant for their constituents by providing employment and other benefits. In addition, environmental justice to reduce harm from climate change and environmental contamination affecting communities of color and low-income areas might shape the legislation on recycling [72].

5.3 Developments and Political Trends in Plastic Waste Management

The legislation on plastics recycling has been evolving for many decades. This chapter will reflect on some of the prominent developments shaping the current political and legal landscape around plastic waste management.

5.3.1 Harmonization of Legal Aspects

Achieving a world free from (unrecovered) plastic waste, where plastic products are valued for their contribution to the UN Sustainable Development Goals (SDGs) (Figure 5.15), are sustainably produced and used, and recovered after use into increasingly circular systems is the vision of both the plastics industry as well as regulators. One measure to achieve this could be an international agreement on plastics at the UN level analogous to the Minamata Convention on mercury, the Paris Agreement on climate action or the Montreal Protocol on CFCs, HCFCs. Although the details of such an agreement would be subject to negotiations by the UN Member States, an effective UN treaty could be decisive in addressing the key barriers to successful waste management of plastics: the fragmentation and lack of scale in voluntary initiatives, misalignment of regulations, lack of data and structural capability gaps [73].

Figure 5.15 UN Sustainable Development Goals (United Nations, 2021)

The United Nations Environment Assembly (UNEA) resolution UNEP/EA.3/L.20 already emphasized environmentally sound waste management as the highest priority in reducing the leakage of plastic debris into the ocean. Nonetheless, present efforts are still insufficient to solve the issue of plastic litter. Addressing this global challenge requires increased coordination among all actors and prioritization of investments by, e.g., multinational development banks, sovereign wealth funds, or institutional investors. While strengthening existing multilateral agreements is needed, a global architecture will enable rapid scaling of innovations and investments. Such massive efforts will help achieve the UN 2030 Agenda and the SDGs. In the case of plastics recycling, particular focus should be on SDG 12–ensuring sustainable consumption and production, including environmentally sound waste management. Within this goal, certain targets also promote a circular economy by calling to "substantially reduce waste generation through prevention, reduction, recycling and reuse by 2030" [74].

The **fifth UNEA session** in February 2021 focused on *Strengthening Actions for Nature to Achieve the Sustainable Development Goals*. Before this meeting, various stakeholders from NGOs, industry and more than 40 countries renewed calls for an international treaty to address plastic waste pollution in a harmonized way. There is a need for governments to commit to coordinated actions and policies to tackle plastic waste problems and advance a circular economy at scale. Such a framework would help businesses and governments move in the right direction. For example, the suggested elements of the treaty should integrate:

- agreed harmonized standards and definitions for defining products and processes to be applied across markets
- a common international policy framework with national targets, action plans and minimum requirements to deliver on the treaty's overarching objective
- common reporting metrics and methodologies across the plastic value chain supervised by the intergovernmental scientific review panel
- coordinated investment approaches to support innovation and infrastructure development in key markets [73]

Various institutions and organizations are now working to advocate for a binding treaty to address plastic pollution on a global scale, looking at plastics along their entire value chain. In addition, the call urges openly involving all relevant stakeholders in the development of this policy. In 2022, a mandate to create an Intergovernmental Negotiating Committee to form such an agreement is likely to be established [75]. Even though the regional and national instruments are clearly an important part of an integrated approach against pollution, they have serious limitations which can be addressed by a new global convention. "Such an agreement would harmonize efforts and provide an overarching strategy to achieve common objectives – including developing minimum standards, shared indicators, sustainable product design, global liability, and compensation for pollution by plastic, shared indicators, and global reviews and assessments" [75].

Due to the COVID-19 pandemic and a decision by UN Member States and stakeholders, UNEA-5 is taking place in a two-step approach: The first session was held virtually in February 2021. Substantive matters that require in-depth negotiations will be postponed to a second, in-person session in February 2022. A global framework will most likely have a clear target like the Osaka Blue Ocean Vision goal of zero discharge of plastic waste into the marine environment by 2050. The framework should encourage governments to develop and then submit national solid waste management plans (e.g., the EU CEAP and its national implementations) to address leakage of plastic waste into the environment.

In the same way, plastic waste is on the agenda of other high-level political, economic, and climate fora such as the World Economic Forum (WEF), G7, G20, OECD, and the UNFCCC, – all having in common the uptake of similar tenets on plastic waste management.

5.3.2 Restrictions on Microplastics

The debates about littering and microplastics are everywhere in the news and one of the main reasons for legislators to work on bans on plastics ("avoiding the problem at the source") and recycling legislation ("recycled is not littered"). Therefore, it is important for the plastics recycling discussion to understand what microplastics are, where they come from and how are they regulated. Microplastics and their influence on human health and the environment have been heavily debated since 2010. Since then the scientific literature has been developing exponentially. The main reasons are monitoring data which show that microplastics can be found almost everywhere. They are ubiquitous in the air, water, and soil, and in urban and remote areas like the Arctic.

The definitions of microplastics vary a lot. It has been agreed that the upper limit is 5 mm, whereas the lower limit is 1 µm or even lower. Some stakeholders put the lower limit down to 1 nm so that the microplastics discussion has an overlap with the one on nanoplastics. At that low nanometer level the size of a single polymer molecule is reached at moderate molecular weights. However, in the latter case, the definition covers a range of more than six orders of magnitude. At the beginning of the microplastics discussion, typical packaging thermoplastics like PE, PP or PET were the focus. In the meantime, the understanding of what a microplastic is has become much broader. Common to all definitions is that a solid microplastic consists of, or incorporates, a high molecular synthetic polymer which is not biodegradable (also known as very persistent according to regulatory terminology). The upcoming European REACH restriction on intentionally added microplastics has this definition of microplastic:

- a particle of any composition that is solid, i.e., it is neither liquid nor gas, as there is no definition or standard of solid applicable, and
- has 100 nm – 5 mm in size of any shape, including fibers, and
- either has a continuous polymer coating of any thickness, or
- contains more than 1 wt% of any polymer (REACH definition) which is not natural or biodegradable.

The restriction applies at >100 ppm polymer content in a mixture (as MP).

That means that, apart from the well-known microbeads in cosmetics and secondary microplastics from degraded macroplastics (e.g., littered packaging), also resin-coated seeds or fertilizers are microplastics as are encapsulated fragrances in washing powder. Resins, rubber (e.g., tire abrasion), and paints also belong to the universe of microplastics.

Essentially, there are three categories of microplastics:

- Intentionally produced microplastics as described in the REACH definition above, including lost plastic beads (see Operation Clean Sweep®, *www.opcleansweep.org*)

- Secondary microplastics which are produced by abrasion during use from tires, shoes, or textiles (wearing and washing) or paints
- Secondary microplastics which get fragmented after use like plastic marine litter or coated fertilizer (Figure 5.16, [76])

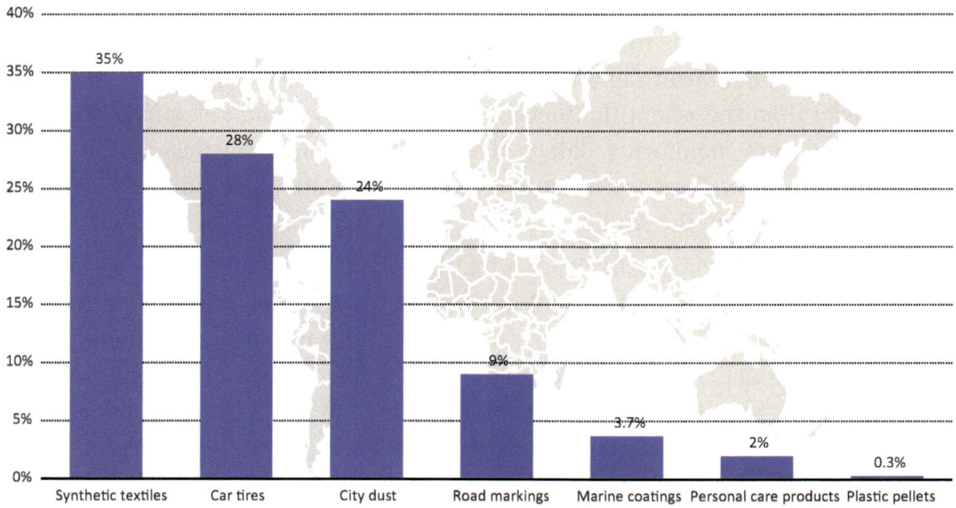

Figure 5.16 Where the oceans' microplastics come from: Distribution of microplastics' sources in the world's oceans (source: Statista GmbH, 2021)

The greatest concern is the increasing accumulation in the environment together with potentially unknown effects caused by:

- The particles themselves with the questions: Do they influence the feeding of small organisms? Do they penetrate the cells of organisms and cause harm to the organs or metabolism?
- Additives: Are the particles carrying additives or soaking substances which might harm the environment or human health?
- Surface: Are they carriers for harmful algae or bacteria?
- In general: Where are the sources and sinks?

As research by academia, authorities, and the industry on microplastics and their effects is steadily evolving, legislators are forced to act. More and more authorities in OECD countries want to phase out intentionally-added microplastics. They should be substituted with biodegradable or natural alternatives, or the application should even cease. The European Union is striving to mitigate the release of secondary microplastics during use, such as in tires, textiles, and plastic beads. The emissions by fragmentation is mainly sought by reducing the roots of marine litter, whether by better waste management infrastructure (collection and recycling) or reducing waste and the use of plastics, especially for single-use packaging. California might introduce the first microplastic threshold in clean (drinking) water, despite of a lot of shortcomings in enforcement, relevant adverse effects, and missing standards.

5.3.3 Transparency on Chemicals

With the European Green Deal, the EU Commission intends to make the EU's economy more sustainable. The Chemicals Strategy for Sustainability (CSS) published in October 2020 aspires to a "toxic-free environment" and is set as a focus area with the same level of commitment as others, e.g., CO_2-neutraltity, a resource-efficient circular economy, and financial transformation/taxonomy. In principle, virgin and recycled materials should respect the same limit value for hazardous substances. Therefore, the question of how recycled material can fulfill the requirements of chemical regulations is of utmost importance for recyclers. A deep understanding of chemical regulation is necessary for bringing recycled materials to the market and compliance with all kinds of legislation - not only with waste legislation setting recycling targets. The CSS includes an action plan consisting of a list of legislative measures and will likely result in a re-opening of REACH. A resolution from the European Parliament on the CSS supports a more protective, more preventive and more hazard-driven approach to managing chemicals.

In Europe, the public is increasingly concerned (outcome of the EU Commission's survey) about the negative effects of chemicals, including from marine littering, microplastic generation, use of endocrine disruptors, cocktail/mixture effects, and a loss of biodiversity. NGOs and likeminded regulators in Europe have taken up this development and are aiming for stricter regulations and measures, including bans of chemical substances (e.g., contained in plastics) that supposedly harm the environment or human health. The CSS depicts a paradigm shift for the assessment and management of chemicals in the EU and challenges established practices in the scientific and regulatory community. In particular, the following measures have been proposed:

- Broadening the definition of "hazardous chemicals" which leads to restrictions and bans of a significant number of substances, likely including a significant number of substances related to plastic materials such as additives, monomer residuals, and other impurities
- Applying a preventive approach, referred to as a "generic approach to risk management" used for CMR (Carcinogenic, Mutagenic, and toxic for Reproduction) substances as the default option for the regulation banning hazardous chemicals in consumer products

- Using grouping approaches to ban entire groups of substances (e.g., PFAS [per- and polyfluoroalkyl substances]) if individual molecules in a group have hazardous properties; the goal is to simplify and speed up regulatory measures for a broad class of chemicals and to avoid regrettable substitution

Only substances for which an essential use for society can be demonstrated are exempted from such bans. While a definition of "essential uses" is not yet available, one suggestion is to take into consideration the definition of the Montreal Protocol (*www.unep.org/ozonaction/who-we-are/about-montreal-protocol*) on substances that deplete the ozone layer, which would make it very difficult for exemptions to get approved.

The assessment of chemical substances has shifted away from classical risk assessment and risk management towards generic approaches to risk management. In effect, this results in a purely hazard-based chemical management approach, neglecting levels of exposure to people and the environment. Any unwanted effect of a substance (hazard), independent of its probability (exposure) through the life cycle of the product will lead to bans of the substance and related chemicals in consumer products. This shows clearly how difficult it will be to fulfill those requirements in the case of mechanical recycling of post-consumer waste.

Important exemplary trends which can be observed apart from the paradigm shift away from risk towards more hazard approach are:

I. Endocrine Disruptors (EDs): The Commission proposal will lead to the introduction of new hazard classes (human health and environment) for substances based on their endocrine disrupting properties, on the basis of the definition of the World Health Organization and on the criteria already developed for pesticides and biocides. The new hazard classes will then apply to all legislations. Both classes will have categories like CMR substances. Of great concern is the "transitional" weakest category of suspected EDs, where, when a low bar for proof has been set: essentially only the endocrine activity is measured by in-vitro (cellular assays) methods. There is the risk that many substances might get an ED label which leads to a precautionary deselection by OEMs and brand owners due to its highly emotional significance. Plastics are very vulnerable to that process, as they depend on a lot of substances for performance reasons (additives) and for production needs. The latter are always a matter for impurity discussions. Both have an influence on recycling when "non-toxic material circles" are demanded.

II. Persistency: Persistency is another key issue which will be addressed by the EU Commission, also as part of the zero-pollution ambition. REACH already address persistent, bioaccumulative and toxic substances (PBT and vPvB) by nominating them as substances of very high concern (SVHC) with the aim of banning all uses of those substances (authorization route). With such restrictions, even manufacturing of those substances on EU level can be banned.

III. Mixture/Cocktail Effect: There is an assumption that in a mixture of substances the toxicity of each substance is not independent but additive or, even worse, multiplicative. As it is not possible to check all substances in all combinations, there will be an attempt to address that uncertainty by introducing a mixture assessment factor of 10. With that factor, all safe threshold values derived from animal studies will be lowered by an additional safety margin of 10, where those values already have safety factors between 10–1,000. This might have dramatic consequences on allowed concentrations and emissions of substances. The CSS has announced that the EU Commission will introduce (via the Sustainable Products Initiative [SPI] as a framework and the EU Ecodesign Directive as an implementing act) legal requirements for the presence of substances of concern in products, giving priority to textiles; packaging, including food packaging; furniture; electronics and ICT; and construction and building. It will also ensure information requirements on chemical content and safe use in 2021–2022 via a digital product passport, which might come into conflict with confidential business interests. The SPI will also reflect on categories for recyclability, recycling content, reparability, upgradability, durability and, in the case of chemicals, will be strongly influenced in the highest category by ecolabel criteria. In addition, the EU Commission will ensure that the authorizations and derogations from restriction for recycled materials under REACH are exceptional and justified.

Among other things, the digital product passport will contain important information about:

- materials and substances present in the product and their quantities
- recycled content of each material present in the product
- whether hazardous chemicals are present, and if so, their location

5.3.4 Eco-Design, Design for Recycling and Design for Sustainability

Final products such as packaging, buildings or cars often contain different types of materials, including plastics, to deliver the required functionality. It is thus about the functional unit's (e.g., car's) life cycle and not the material's (e.g., plastic's) life cycle where the design of a new product should ensue, i.e., the design must always refer to the specifically intended function of the final product to decide which material or material combination should be used. Different materials or material combinations can frequently enable various functions and be resource-efficient along their life cycle.

In the last decades, the concept of eco-design in its various forms, like Design for Recycling (DfR), has enjoyed great popularity. Eco-design is defined as a systematic and comprehensive design approach for products to reduce environmental impacts throughout their life cycle. It is thus considered as complementary to the classical product development requirements for functionality, safety, ergonomics, and cost-benefit ratio with environmental compatibility [77]. Eco-design has already been practiced on several applications. It was even the subject of an ISO Technical Report in 2002 (*ISO TR 14062:2002: Environmental management – Integrating environmental aspects into product design and development*) and of an ISO standard published in 2020 (*ISO 14006:2020: Environmental management systems – Guidelines for incorporating ecodesign*). At the same time, the environmental assessment of products over the entire life cycle is well covered in the ISO standards 14040:2006 and 14044:2006 for Life Cycle Assessments. Such standards provide useful processes in guiding decisions on product design. They help in anticipating and identifying the environmental aspects of a product throughout its life cycle. Such standards furthermore explain how a product's environmental impact over its life cycle must also be balanced against other factors, such as the product's intended function, performance, safe use, quality, legislative compliance, cost, and marketability. Nowadays, by adding social aspects to this list [78, 79], the whole sustainability of products can be assessed by stakeholders along the entire value chain [80]. An efficient eco-design process should be applied during the early product design planning stage to be most beneficial. It should also be "iterative and flexible, promoting creativity and maximizing innovation and opportunities for environmental improvement" and "competitiveness" [81].

Taking meat packaging as example, the savings from avoiding food waste by improving the protective properties of the packaging are 10 times higher than the carbon footprint of the total packaging. The plastic material and the material composition chosen for the differently packed products demanded by society will differ depending on the requirements to contain, protect and preserve these wide ranges of products. It is therefore important to recognize that it is the type of packed product and its production and consumption patterns that govern life-cycle resource efficiency rather than only the materials used to make the packaging. Similar results for the Design for Sustainability (DfS) life-cycle analyses have been reported for the protection effect of packaging for TVs, the lightweight effect of plastics in cars, and the insulation effect of organic foams in building and construction. Coming back to the example of meat packaging, it is self-evident that protecting and preventing degradation of the meat has a far greater impact on the environment than the actual packaging. The priority is therefore to protect the food in the most resource-efficient way, even if this would be at the expense of today's limited ability to recycle such packaging. This is where innovation plays a significant role, i.e., to produce the most resource-efficient packaging to protect and save the food and as such reduce the overall environmental impact. Other sustainability aspects which have a higher priority than only recyclability are health (e.g., food-contact material approval) and safety. For instance, a safety belt or an air bag in a car are plastic products, difficult to recycle but making the car lighter (leading to less fuel consumption), ensuring the driver's safety and thus overruling design for recyclability considerations. It is, however, important to continue to search for innovations to eco-efficiently recycle today's not-recycled plastics. Chemical recycling offers the most promising way to close the loop for such plastic applications.

Experience has shown that the best solutions are specific to the product. Products are complex and diverse. Knowledge and technology today evolve at an ever-increasing rate, thereby improving the ability to create and apply novel solutions that previously would not have been considered possible. Examples of some popular eco-design shortcuts are DfR, for biodegradability, for durability, and for recycled content. They may all be good or bad options, potentially leading to sub-optimum designs, and must be assessed holistically. This means assessing the environmental, social, and cost aspects during the entire life cycle of the product, from cradle to grave, and could be at best called a design for sustainability (DfS). Plastic materials offer diverse possibilities in contributing to more sustainable products thanks to their potential for innovation and competitive costs. DfS of products must be a multi-criterial approach based on life-cycle assessments to lead to the most environmentally friendly, cost-effective, and societally beneficial solution. It should be ensured that resource efficiency is optimized across the entire life cycle of a system and not only for one part. Increasing recyclability by DfR, durability or biodegradability should always be balanced against the actual environmental impact of the required final product in its specific application. As eco-design, DfR or DfS are increasingly more integrated into legislation (e.g., EU Ecodesign Directive for energy-related products, EU ELV Directive, EU PPWD), it is important that eco-design policies focus not only on a single aspect of the life cycle, such as end of life, but support solutions which are resource-efficient, have a reduced carbon footprint and improved sustainability along the entire life cycle.

5.3.5 Economic Instruments

In addition to the mandating statutes, policymakers often make use of economic instruments like EPR, deposit refund schemes, taxation mechanisms, Sustainable Public Procurement or ecolabeling as a lever for policy. By sending market signals, these instruments cause change or influence the behavior of market players. They can be designed in different ways and are used as means for the consideration of external costs, i.e., "costs to the public incurred during production, exchange or transport of various goods and services", which can include natural resource depletion, environmental degradation, health impacts, and social impacts. In the following, EPR and taxes are highlighted as the most prominent instruments used to steer recycling efforts [82].

Extended Producer Responsibility (EPR)

With the waste export market slowly shrinking, governments are increasingly looking into EPR and deposit return schemes to reduce the amount of domestic waste going to landfill or being incinerated. Although EPR is not a new concept for waste legislation, there is an increasing perception of the need to move from voluntary to mandatory implementation of EPR. One of the guiding principles here is that producers take responsibility for the full cost and impact of their products over the life cycle and redesign or innovate better materials and systems. EPR legislation holds the producers of products responsible for any negative environmental externalities and associated costs. It should offer manufacturers incentives for designing resource-efficient and low-impact products, and for facilitating effective end-of-life collection and treatment of collected products for reuse and recycling. While plastic waste is primarily governed by regulations focused on packaging, EPR legislation also exists for vehicles and electrical and electronic equipment, bringing some plastic items under these rules as well. EPR policies exist in the U.S., Canada and the EU and are on the rise in other countries around the world.

To create a basic understanding of EPR and its intentions and goals, a study from Bio by Deloitte found:

"The goal of EPR systems could be stated as follows: Extend the producer's physical and financial responsibility for a product to the post-consumer state of a product's life cycle, in order to internalize the end-of-life management costs according to high environmental standards and provide an incentive for producers to take environmental considerations into account along the products' life from the design phase to their end-of-life." [83]

Based on this definition, EPR has two main goals: to improve the recycling and reuse of waste and to-become-waste products as well as to provide an incentive to producers to create more sustainable and environmentally-friendly products. Compared to the traditional solid waste management approach, EPR involves a shift in responsibility (administratively, financially and/or physically) from governments or municipalities, and thus taxpayers, to the entities that make and market the products that are destined to become waste (Figure 5.17). To this extent, EPR still constitutes the implementation of the polluter-pays principle (PPP), but induces a change in the definition of the "polluter." Whereas in the classical version of the PPP the polluter was the individual directly causing pollution (in general, the consumer), within the EPR framework the polluter is the economic agent who can play a decisive role in avoiding pollution, for example, through eco-design efforts [83].

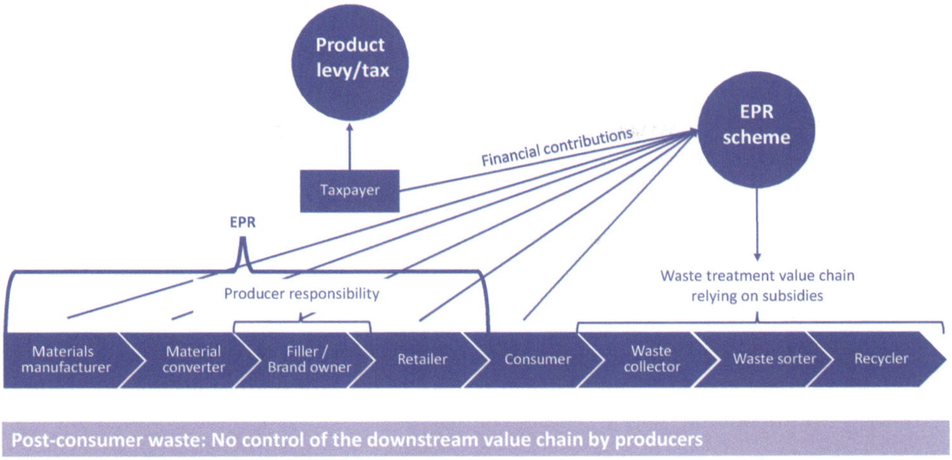

Figure 5.17 Extended producer responsibility

The emerging responsibility for obliged producers to implement a national EPR scheme may be financial or organizational. EPR schemes are often described as being "financial" when the responsibility for waste management is left to municipalities and the financial responsibility is left to producers. "Organizational" EPR schemes are those schemes where the physical responsibility of waste management is transferred to the producers. There is nowadays a big variety of schemes, and the boundary between these two models is indistinct. The producers' responsibility within an EPR scheme may be defined as follows [83]:

- **Sole financial responsibility:** Producers have no obligation except to finance the existing waste management channels. Mere financial responsibilities may offer less incentive to producers in terms of improving waste management and the eco-design of their products as their burden is only monetary.
- **Financial responsibility through contracts with municipalities:** Producers establish contracts with municipalities to collect and manage waste. The producers' motivation to improve waste management depends on the type of contract and on the dialogue with municipalities. The financial contribution of producers can be modulated, e. g., based on collection or recycling rate, or a quality check of the waste stream.
- **Financial responsibility and partial organizational responsibility:** Some activities fall under the responsibility of municipalities and backed financially by producers, while other activities are the responsibility of the producer. For example, public waste collection operators may collect the waste, but the sorting or reselling of recovered materials may be done by producers.
- **Financial responsibility and full organizational responsibility:** The producers subcontract professional waste collection and treatment activities, or even own part of the collection and treatment infrastructure.

It must be considered that EPR is primarily an individual obligation on the part of each producer who places products on the market subject to EPR legislation. In most industries and branches, the exercise of the individual obligations is likely to be executed by a collective scheme where all individually obligated bodies participate. Individual national implementations may differ a lot from each other. The following definitions list the players on the market rather generally:

- **Manufacturer/Producer/Retailer:** Manufacturers, producers and retailers are the firms that are held responsible for bringing waste to the market for the first time. To fulfill their responsibility, producers and retailers must ensure that all waste they might directly or indirectly produce and that is subject to EPR legislation is recycled or recovered in a manner that legislation and national implementations are fulfilled.
- **Producer Responsibility Organization (PRO):** A PRO is a collective entity set up by producers, or through legislation, which is responsible for meeting the recovery and recycling obligations of the individual producers. PROs are established if producers assume their obligations would be better fulfilled by collective implementation of their individual responsibilities. Depending on national implementations, individually responsible firms and persons might be obligated to participate in a collective scheme and thus to contract with a PRO, or they might have the choice of fulfilling their responsibilities on their own. PROs potentially have three main functions: 1) financing the collection and treatment of the product at the end of its life in the targeted waste stream by collecting fees and redistributing the corresponding financial amounts; 2) managing the corresponding data; and 3) organizing and supervising these activities.
- **Clearinghouses:** The clearinghouse is a central organization that works independently of PROs and fulfills mainly surveillance and monitoring tasks. The tasks of the clearinghouse may include:
  - Verifying compliance (identifying free riders) together with public authorities in charge of enforcement, ensuring that all competing PROs work on a level-playing field by verifying that all requirements are met, calculating market shares and ensuring a fair determination of the PRO’s individual objectives
  - Centralizing and aggregating data reported and inspection of data quality and completeness as a “register’s role”
  - Cost sharing related to certain operations (e. g., reimbursement of local authorities, national communication campaigns) through common agreements with public local authorities or through common calls for tenders
  - Managing common communication and R&D activities
  - Managing a common communication fund for raising public awareness to support the understanding of the system, fed by each PRO’s communication budgets

- **Waste management industries:** Any firm or branch that is responsible for the collection, sorting, recycling and/or recovery of the collected waste. These operators may be internalized in the firm structure of the PRO or they may be independent establishments with whom the PROs are contracting.
- **Municipalities:** The role of municipalities highly depends on the national implementation and historic development of the countries' waste management. Theoretically, municipalities can be completely excluded from an EPR scheme or they can be the PRO of the system.
- **End-Customer:** The end-customer can be identified as the consuming unit that utilizes and creates the waste and whose "task" is to be aware of where to dispose the waste one is producing and how one participates in the system so that one's waste is collected. Many systems and schemes differentiate between private and non-private end-customers.
- **National/State-level authorities:** All public-law authorities that are obliged, in any manner, to control, monitor or cooperate with the participants of an EPR scheme. Roles and responsibilities of the authorities strongly depend on the way the EPR scheme is implemented. Tasks may include monitoring of the system and PROs, as well as prosecution if any player commits an illegal action.
- **Free riders:** Any producer who is subject to EPR but does not participate in the EPR scheme, whether by the non-fulfilment of the individual tasks or by not cooperating with a PRO [83].

EPR schemes extend the producers' responsibility to meeting recycling and recovery targets specified in national legislations. However, no one, including producers, should be made responsible for acts, misbehaviors, or happenings they cannot control. It is therefore important to define the areas where the producer cannot be held liable for emerging circumstances. For example, packaging associations demand that the cost for litter cleanup be excluded from EPR schemes, arguing that littering is an improper behavior of individuals for which the industry cannot take responsibility. However, proper disposal behavior campaigns could be supported by the industry and municipalities [84]. This example also illustrates that it is not always easy to conclude which costs should be covered by EPR schemes and which ones should be exempted. In this context, the overarching associations of each industry that was subject to EPR have recently called for an "an explicit evaluation and re-allocation of responsibilities consistent with the PPP" [85, 34].

Taxation

Several countries around the globe have already introduced or are considering the introduction of taxation mechanisms on certain (or all) plastic packaging materials and applications. In parallel, the European Union has decided to introduce an Own Resource (or EU-wide levy on non-recycled plastic packaging waste) on the regional level, as part of the COVID-Recovery Package under the EU Multiannual Financial Framework for the period 2021–2027. The EU Member States can decide from which domestic sources they contribute their portion. For 2021, some EU Member States will cover the costs from the domestic budget without imposing the burden on the industry. This may be later changed or even translated into new taxation schemes at the national level, thus potentially doubling the costs for the industry.

It is questionable whether levies, taxes and further fiscal measures are the most efficient tool to drive the innovation and investments that are needed to meet the intended circular economy and recycling policy objectives (e.g., of the EU Green Deal), whereas EPR fees are mainly invested in the waste management infrastructure (CAPEX and OPEX). For example, packaging users or other stakeholders are already paying fees under EPR schemes, which are set to boost recycling (e.g., via fee modulation based on product's recyclability). In particular, the uptake of sustainable (e.g., recycled) material should be encouraged and rewarded by EPR systems to close the loop of circularity. In contrast, the announced EU Own Resource from non-recycled plastic packaging waste is considered to cover the costs of COVID and Brexit recovery and is not earmarked for funding the recycling infrastructure or other environmental measures.

Fiscal measures on materials and applications, whether at the national or regional level, risk hampering efforts to move towards a circular economy and improve waste management systems across countries and regions. In the absence of earmarking revenues, there is no guarantee that the resources collected through such fiscal instruments will be reinvested into better collection, sorting and recycling infrastructures needed to increase the amount of waste that is effectively recycled.

5.3.6 New Recycling Technologies

There are also developments in the way plastic waste can be recycled, and legislators are looking at how to avoid landfill and incineration and close the carbon loop. Chemical (or feedstock, or advanced as in the U.S.) recycling is a relatively new technology which offers an opportunity to recover a wider range of plastic waste than mechanical recycling, while maintaining the same quality the original feedstock initially had (Table 5.3). It is currently the most prominent development, and the question is how to integrate it into the existing legislation which has historically been written with a focus on mechanical recycling as a single long-standing recycling technology.

Table 5.3 Terms and Definitions of Chemical/Feedstock/Advanced Recycling

Chemical or feedstock recycling:
▪ ISO 15270: 2008 (Note: Definition uses the terms "feedstock recycling" and "chemical recycling" interchangeably). A conversion to monomer or production of new raw materials by changing the chemical structure of plastics waste through cracking, gasification or depolymerization, excluding energy recovery and incineration ▪ ISO 18601: 2013 Process of recovering valuable chemical substances by chemical treatment of used packaging by hydrolysis, glycolysis, methanolysis, catalytic reaction, thermal reaction, and other chemical processes – process of substituting used packaging for natural resources ▪ European view: Chemical recycling is indirectly defined in the EU Waste Framework Directive as "recycling", meaning any recovery operation by which waste materials are reprocessed into products, materials, or substances whether for the original or other purposes. It includes the reprocessing of organic material, but does not include energy recovery and the reprocessing into materials that are to be used as fuels [47].
Advanced recycling:
▪ Advanced recycling technologies are innovative manufacturing processes that convert post-use plastics into their original building blocks, specialty polymers, feedstocks for new plastics, fuels, waxes, and other valuable products [86] ▪ According to the U.S. view, the terms advanced recycling (and recovery, transformational technologies) and chemical recycling are interchangeable. All these terms help differentiate advanced recycling from the more widely-known recycling processes that use mechanical technologies to recycle used plastics. However, under ISO and European legislation the production of fuels is explicitly excluded from the recycling definitions.

For example, several U.S. states have passed legislation which defines advanced recycling as manufacturing, enabling more investments in this technology and, in turn, a more circular plastic value chain [86]. Contrary to those individual U.S. jurisdictions, the European Union is deciding how chemical recycling should be further accepted under the EU legislative framework. Although in Europe chemical recycling is already accepted by different legislations, a discussion started during the revision process for PPWD and WFD about whether mechanical and chemical recycling are on the same level in the waste hierarchy. Linked to this is the question of whether end-of-life products entering a chemical recycling process are "recyclable" and whether the resulting products from chemical recycling plants are recognized in the same way as recycled materials from mechanical recycling plants. The current discussion about the status of chemical and mechanical recycling on the EU level will most likely lead to an inconsistent definition on the EU Member States level. In Germany, chemical recycling is fully accepted in all kinds of waste legislation except for a plastics recycling target in the packaging legislation [87]. Although Germany has a strong history of using chemical recycling as complimentary to mechanical recycling for plastic (packaging) waste [88], in this specific case the existing legislation for plastics packaging recycling (revision is planned in 2022) explicitly demands "material recycling".

The absence of appropriate and consistent national and global standards across thousands of recycling systems hinders the growth of recycling initiatives and related efforts to reduce plastic waste in landfills and oceans. Waste legislation should always be interpreted with an openness to different technologies, so that chemical recycling – complementary to mechanical recycling – can contribute to a circular economy of plastics.

5.3.7 Replacement of Virgin by Recycled Content

Several obstacles are holding back a decisive breakthrough in the use of recycled materials: the quality of mechanically recycled plastics is lower than that of virgin plastics due to inevitable damage of the polymer structure during the mechanical recycling process. Consequently, recycled plastics are often used in applications with less stringent requirements for performance, safety, etc. (e.g., park bench, flowerpot, garbage sack). The price of virgin plastics is linked to the oil price, whereas the costs of recycled materials are determined by the process costs. Low oil price makes a switch to recycled material unattractive. Food contact is a prerequisite for a lot of packaging and sensitive applications, but also for some non-packaging applications (e.g., drawers in fridges). To achieve food-contact, material approval has been difficult, if not impossible, for mechanical recycling processes up until now (exception: PET drinking bottle-to-bottle recycling).

Although the effect of voluntary commitments for recycled content sometimes leads to a shortage (e.g., in 2020 for rPET), the supply is most often higher than the demand for recycled plastics materials due to the above-mentioned reasons. This is a limitation of theoretical recycling potentials leading to some suggestions by legislators for how to overcome this limitation:

- Incentivizing the use of recycled materials
- Offering tax reductions (e.g., Italian plastic tax of 80 €ct/kg)
- Mandating the use of recycled content on a
  - country level (e.g., x% of all plastics used in a specific country)
  - product level (e.g., x% of all packaging)
  - application level (e.g., x% of all bottles)
  - individual product level (e.g., x% of each PET bottle)
  - the level of plastics manufacturers (e.g., x% of virgin material placed on the market)

An example for mandated product-specific recycled content targets is defined in the EU "Single-Use Plastics" Directive as 25%/30% for PET/beverage bottles by 2025/2030. Recycled content targets for specific products could lead to a consumption shift from non-regulated recycling markets to regulated without increasing the total amount of recycled material on the market. Material and polymer-specific recycling targets on the level of plastics manufacturers could lead to an increase of cheap thick-wall, low-end recycled materials replacing concrete, wood and steel, which often does not deliver the full environmental potential of recycling gained by replacing virgin plastics. In addition, it might be impossible for some virgin polymers to be replaced by recycled if only mechanical recycling technologies are allowed. In contrast to mandating the use of recycled content, incentivization via modulated EPR fees, tax exemptions, Sustainable or Green Public Procurement, or ecolabeling could be a flexible economic approach using market mechanisms.

For mechanical recycling, the quality of recycled materials is key to higher uptake. Chemical recycling is another option for overcoming the quality issue of mechanical recycling. Complementing mechanical recycling, chemical recycling could lead to higher recycling rates of plastic waste and higher recycled content in plastics.

5.3.8 Keeping Carbon in the Loop

Whereas plastics recycling is a quite well-established and regulated field, carbon (CO_2) emissions from energy production, housing or cars are regulated by individual standards and legislation. However, there is carbon in biomass, paper, and many other materials. There are many options to make cycles for carbon in the technosphere and biosphere a reality. Manifold processes and strategies at different levels offer the opportunity to close cycles in various ways. This is made possible primarily by new forms of circular management through mechanical and chemical recycling and the use of biomass, CO_2 (carbon cycles) and CO_2-free hydrogen production. Already now, there are examples of cycles in all these fields. The consistent promotion and further development of technologies and processes can bring about complete carbon cycle management in the end.

The goal of keeping carbon in the loop and the broader view of integrating all carbon (from fuel to materials such as plastics, from the atmosphere to coal and oil) into this picture leads to other ways of optimization than political recycling targets for plastics. Plastics can, for instance, be seen either as a carbon sink or an incineration plant as preparation for reuse of carbon, positioning the incineration process much higher in the waste hierarchy. Most importantly, the integration of more than one industry sector in the optimization demands competitiveness in every phase of development and favorably-priced renewable energy in large volumes, combined with legal framework conditions that are open to a wide range of different technologies and innovation-friendly – so that carbon can be fully run in cycles in the future and complete greenhouse gas neutrality is achieved.

A holistic approach which considers all partial aspects of climate protection and a circular economy could be the guiding principle for achieving both resource efficiency and reducing greenhouse gas emissions [89, 90]. For example, Europe has set itself the goal of a net-zero carbon emission (climate-neutral) circular economy by 2050. Waste management and a circular economy are thus one lever – but an extremely important one – for achieving net-zero carbon emissions.

5.3.9 Conclusion

Better waste management systems which facilitate higher waste collection and recycling rates would allow plastics wastes to be captured before they begin creating problems in the natural environment. Effective management of remaining plastic waste requires the relocation of some plastic waste processing domestically and the strengthening of domestic secondary waste markets. Suppliers and buyers of recycled plastics would both benefit from larger and more liquid markets for recycled plastics, but neither party has strong incentives to act alone. In turn, improved market outcomes could, to some extent, become self-fulfilling as scale efficiencies are captured and a more widespread consumer acceptance develops. These factors provide a clear rationale for policy intervention, as well as potential insights into how to do it effectively. Policies are likely to be more effective if they jointly address the challenges – market failures, policy misalignments, and status quo biases – on both the supply and demand sides of recycled plastics markets. Put differently, an effective policy framework would address challenges across the entire plastics life cycle, from plastics and product design through to end-of-life management and recycled plastic production [91].

In conclusion, a certain pattern can be recognized across countries, regions and globally when it comes to legislating plastics recycling (Figure 5.18):

- Recycling quota or targets for specific plastic waste streams have been set and gradually increased over the years.
- To achieve these high recycling targets, Design for Recycling rules or recyclability requirements are considered prerequisite.
- To really close the loop with recycling, mandatory recycled content for new products (including regulated ingredients) is used as a lever.
- Economic instruments (e.g., extended producer responsibility, taxes or levies, public procurement programs or ecolabeling) help to define responsibilities, steer investments, finance waste management and operations, and set the general process in motion.

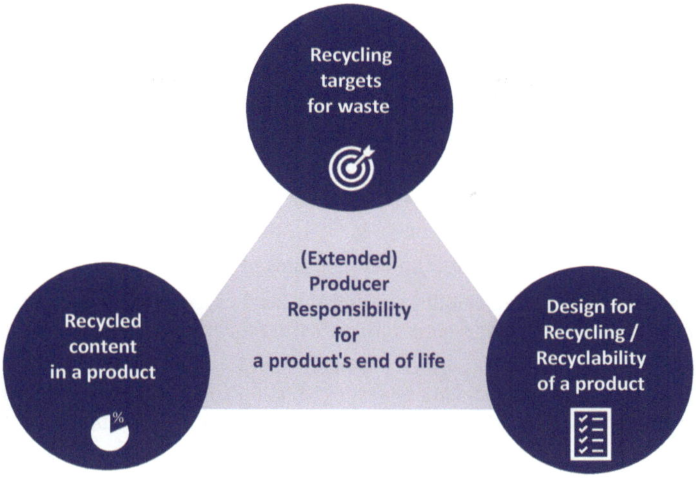

Figure 5.18 Key aspects of recycling in legislation

The plastic waste legislation impacting the recycling industry is a complex environment with a myriad of different regulations which are updated every few years, so it is not easy to keep up with it. In addition, these regulations vary in each country, region, and sometimes even locality [89, 90, 92]. In general, the legislation on plastic waste uses preventive and restrictive measures, such as bans on certain plastic applications like single-use items and microplastics or obligations and regulations on waste imports, exports, and handling. Starting with plastics packaging, other plastic applications in automotive, building and construction, or textiles are likewise increasingly becoming the focus of legislation and will see similar adaptations of legal provisions as in the case of packaging. A visionary view of the life cycle of plastics is that products would be 100% recyclable, would be in fact 100% recycled and would be made with 100% recycled feedstock, which ultimately would only be possible if the loop is extended from plastics to carbon.

References for Chapter 5

[1] Fuhr, L., Buschmann, R., Freund, J.: *Plastikatlas - Daten und Fakten über eine Welt voller Kunststoff*, Heinrich-Böll-Stiftung, BUND (2019)

[2] PlasticsEurope Deutschland (2021), *Business Data and Charts 2020 - World Plastics Demand*

[3] Mayne, N.: Current Trends in Legislation and Plastics Recycling Technology in Europe. (2006). In M Kozlowski, M.: *Plastics Recycling in Europe*, Wroclaw: GS Media

[4] Conversio Market & Strategy GmbH (2/2020), *Global Plastics Flow 2018*

[5] World Economic Forum, *The New Plastics Economy: Rethinking the Future of Plastics*, (1/2016), p. 7

[6] Statista GmbH (2021), "Flow of plastic materials worldwide between 1950 and 2017", *https://www.statista.com/statistics/959995/flow-plastic-materials-worldwide/*

[7] Science History Institute (2020, 08/06), "History and Future of Plastics", *https://www.sciencehistory.org/the-history-and-future-of-plastics*

[8] Our World in Data (2021), "How do we dispose of our plastic?", *https://ourworldindata.org/plastic-pollution*

[9] EPA (United States Environmental Protection Agency) (2021), "Plastics: Material-Specific Data", *https://www.epa.gov/facts-and-figures-about-materials-waste-and-recycling/plastics-material-specific-data*

[10] Ryan, P.: A Brief History of Marine Litter Research. In: Bergmann, M., Gutow, L., Klages, M. (Eds.): *Marine Anthropogenic Litter*, Springer (2015) pp. 1–28

[11] Herczeg, M.: Municipal waste management in Switzerland, European Environment Agency (2013)

[12] Conversio Market & Strategy GmbH, Circular Economy of Plastics 2018 EU28+2 (9/2019)

[13] Carbon Tracker, "The Future's Not in Plastics: Why plastics demand won't rescue the oil sector" (2020, *https://carbontracker.org/reports/the-futures-not-in-plastics/*

[14] Soroptimist International, "Joint position statement on the 1st meeting of the Ad Hoc Open-Ended Expert Group on Marine Litter and Microplastics" (05/2018), *https://www.soroptimistinternational.org/wp-content/uploads/2018/04/2018.04.16-Joint_Position-Paper-with-signatories.pdf*

[15] Wittstock, K., Meyer, S., Krähling, H., Schlotter, U., Cieplik, S.: *Verwertung von Kunststoffen. Makromolekulare Chemie*, Birkhäuser Basel (2010) pp. 503–518; DOI: *https://doi.org/10.1007/978-3-7643-8891-1_8*

[16] Our world in data, "Plastic Pollution" (2021, 04/15), *https://ourworldindata.org/plastic-pollution*

[17] XINHUANET "China to recycle 350 mln tonnes of waste resources by 2020" (2017, 01/28), *http://www.xinhuanet.com/english/2017-01/28/c_136018178.htm*

[18] Statista GmbH, "Global plastic waste exports to China in February 2017 and February 2018" (2021), *https://www.statista.com/statistics/1131780/global-plastic-waste-exports-to-china/*

[19] Statista GmbH, "Distribution of plastic waste exports worldwide by country between January and November 2018" (2021), *https://www.statista.com/statistics/1127258/plastics-waste-export-share-by-country-worldwide/*

[20] Brooks, A. L., Wang, S., Jambeck, J. R.: "The Chinese import ban and its impact on global plastic waste trade". *Science Advances*, (2018) 4 (6), eaat0131, *https://advances.sciencemag.org/content/4/6/eaat0131*

[21] Statista GmbH, Eurostat COMEXT-Datenbank. "Volume of plastic waste exported from the European Union (EU) in 2015 and 2018, by receiving country (in tons)" (2021), *https://www.statista.com/statistics/1074454/extra-eu-plastic-waste-exports-by-receiving-country/*

[22] Time, "Southeast Asia Doesn't Want to Be the World's Dumping Ground. Here's How Some Countries Are Pushing Back" (2019, 06/03), *https://time.com/5598032/southeast-asia-plastic-waste-malaysia-philippines/*

[23] Ellen MacArthur Foundation (2013), *Towards the Circular Economy – Economic and business rationale for an accelerated transition*

[24] Rizos, V., Behrens, A., Van der Gaast, W., Hofman, E., Ioannou, A., Kafyeke, T., Flamos, A., Rinaldi, R., Papadelis, S., Hirschnitz-Garbers, M., Topi, C.: Implementation of circular economy business models by small and medium-sized enterprises (SMEs): Barriers and enablers. *Sustainability*, (2016) 8, 1212

[25] Bennett, J., Pearce, D., Turner, R.: Economics of Natural Resources and the Environment. American *Journal of Agricultural Economics*, (1991, 02/01)

[26] Mitchell, P., James, K.: Economic growth potential of more circular economies. WRAP. (2015/09)

[27] Feng, Z., Yan, N.: *Putting a circular economy into practice in China. Integrated Research System for Sustainability*, Science and Springer (2007, 01/27), pp. 95–101

[28] Commission, E. (2018), A European Strategy for Plastics in a Circular Economy. European Commission

[29] Bilitewski, B., Härdtle, G., Marek, K.: *Abfallwirtschaft: Eine Einführung*, Springer (1990)

[30] Basel Convention, "Parties to the Basel Convention on the Control of Transboundary Movements of Hazardous Wastes and their Disposal" (2021), *http://www.basel.int/Countries/StatusofRatifications/PartiesSignatories/tabid/4499/Default.aspx*

[31] International Maritime Organization, "Final Act of the International Conference on Marine Pollution" (1973), *https://wwwcdn.imo.org/localresources/en/KnowledgeCentre/ConferencesMeetings/Documents/MARPOL 1973 - Final Act and Convention.pdf*

[32] Bundesamt für Seeschifffahrt und Hydrographie (BSH), "Simplified overview of the discharge provisions of the revised MARPOL Annex V which entered into force on 1 January 2013" (2021, 04/15), *https://www.bsh.de/EN/TOPICS/Shipping/Environment_and_shipping/MARPOL/_Module/Akkordeon/Anlage_5/Einleitbedingungen_AnlageV.pdf?__blob=publicationFile&v=3*

[33] Center for International Environmental Law, "Plastic global law & policy" (2021, 04/15), *https://www.ciel.org/issue/plastic-global-law-policy/*

[34] Ellen MacArthur Foundation; McKinsey & Company, "The New Plastics Economy – Rethinking the future of plastics" (2016), *https://www.ellenmacarthurfoundation.org/assets/downloads/EllenMacArthurFoundation_TheNewPlasticsEconomy_Pages.pdf,* accessed April 15, 2021

[35] Ad Hoc Open-Ended Expert Group on Marine Litter and Microplastics (2018), Joint Position Statement. First AHOEGG Meeting, May 2018, *https://papersmart.unon.org/resolution/uploads/2018.05.25_yi_oeeg-1_final_joint_position_paper_with_signatories_updated.pdf*

[36] BP (2021), BP Energy Outlook 2019 edition, *https://www.bp.com/content/dam/bp/business-sites/en/global/corporate/pdfs/energy-economics/energy-outlook/bp-energy-outlook-2019.pdf*

[37] Giacovelli, C., Zamparo, A., Wehrli, A., Alverson, K.: *Single-Use Plastics – A Roadmap for Sustainability*, UN Environment (2018)

[38] Statista GmbH, "The Countries Banning Plastic Bags" (2021), *https://www.statista.com/chart/14120/the-countries-banning-plastic-bags/*

[39] World Economic Forum, "As Canada bans bags and more, this is what's happening with single-use plastics around the world" (2021), *https://www.weforum.org/agenda/2020/10/canada-bans-single-use-plastics/*

[40] Lexology, "Single use plastics in China: an evolving landscape for downstream participants" (2021), *https://www.lexology.com/library/detail.aspx?g=59a44161-9d5b-4464-ab3f-95523feb9609&utm_source=lexology+daily+newsfeed&utm_medium=html+email+-+body+-+general+section&utm_campaign=lexology+subscriber+daily+feed&utm_content=lexology+daily+newsfeed+2021-03-0*

[41] Lexology, "Grasping at Straws: Regulatory and Trade Considerations Regarding Canada's Proposed Single Use Plastics Policy" (2021), *https://www.lexology.com/library/detail.aspx?g=0e901d76-721a-42cb-a7e7-1cdc521068f4&utm_source=Lexology+Daily+Newsfeed&utm_medium=HTML+email+-+Body+-+General+section&utm_campaign=Lexology+subscriber+daily+feed&utm_content=Lexology+Daily+Newsfeed+2021-02-0*

[42] The European Commission, "The European Green Deal" (2020), *https://eur-lex.europa.eu/resource.html?uri=cellar:b828d165-1c22-11ea-8c1f-01aa75ed71a1.0002.02/DOC_1&format=PDF,* accessed April 15, 2021

[43] EU Commission, "Closing the loop – an EU action plan for the Circular Economy" (2015), *https://eur-lex.europa.eu/resource.html?uri=cellar:8a8ef5e8-99a0-11e5-b3b7-01aa75ed71a1.0012.02/DOC_1&format=PDF*

[44] European Commission, "A new Circular Economy Action Plan for a cleaner and more competitive Europe" (2020), *https://eur-lex.europa.eu/resource.html?uri=cellar:9903b325-6388-11ea-b735-01aa75ed71a1.0017.02/DOC_1&format=PDF*

[45] EU Technical Expert Group on Sustainable Finance, "Taxonomy: Technical Report" (2019). *https://ec.europa.eu/info/sites/info/files/business_economy_euro/banking_and_finance/documents/190618-sustainable-finance-teg-report-taxonomy_en.pdf*

[46] European Parliament (2018, 11/19), Directive 2008/98/EC of the European Parliament and of the Council of November 19, 2008 on waste and repealing certain Directives, *https://eur-lex.europa.eu/legal-content/EN/TXT/PDF/?uri=CELEX:02008L0098-20180705&from=EN*

[47] Official Journal of the European Union (2018, 05/30), Directive (EU) 2018/851 of the European Parliament and of the Council of May 30, 2018 amending Directive 2008/98/EC on waste, *https://eur-lex.europa.eu/legal-content/EN/TXT/PDF/?uri=CELEX:32018L0852&from=EN*

[48] Official Journal of the European Communities (1994, 12/20), European Parliament and Council Directive 94/62/EC of December 20, 1994 on packaging and packaging waste, *https://eur-lex.europa.eu/legal-content/EN/TXT/PDF/?uri=CELEX:31994L0062&from=ENhttps://eur-lex.europa.eu/legal-content/EN/TXT/PDF/?uri=CELEX:32018L0852&from=EN*

[49] Official Journal of the European Union (2018, 05/30), Directive (EU) 2018/852 of the European Parliament and of the Council of May 30, 2018 amending Directive 94/62/EC on packaging and packaging waste, *https://eur-lex.europa.eu/legal-content/EN/TXT/PDF/?uri=CELEX:32018L0852&from=EN*

[50] Official Journal of the European Union (2019, 06/05), Directive (EU) 2019/904 of the European Parliament and of the Council of June 5, 2019 on the reduction of the impact of certain plastic products on the environment, *https://eur-lex.europa.eu/legal-content/EN/TXT/PDF/?uri=CELEX:32019L0904&from=EN*

[51] Official Journal of the European Communities (2000, 09/18), Directive 2000/53/EC of the European Parliament and of the Council of September 18, 2000 on end-of-life vehicles, *https://eur-lex.europa.eu/resource.html?uri=cellar:02fa83cf-bf28-4afc-8f9f-eb201bd61813.0005.02/DOC_1&format=PDF*

[52] Official Journal of the European Union (2018, 05/30), Directive (EU) 2018/849 of the European Parliament and of the Council of May 30, 2018 amending Directives 2000/53/EC on end-of-life vehicles, 2006/66/EC on batteries and accumulators and waste batteries and accumulators, and 2012/19/EU on waste electrical and electronic equipment, *https://eur-lex.europa.eu/legal-content/EN/TXT/PDF/?uri=CELEX:32018L0849&from=EN*

[53] Official Journal of the European Union (2012, 07/04), Directive 2012/19/EU of the European Parliament and of the Council of July 4, 2012 on waste electrical and electronic equipment. (2012), *https://eur-lex.europa.eu/legal-content/EN/TXT/PDF/?uri=CELEX:32012L0019&from=DE*

[54] Conversio Market & Strategy GmbH, "Stoffstrombild Kunststoffe in Deutschland 2019" (2020). Studie

[55] Umweltbundesamt (2020, 07/20), Elektro- und Elektronikaltgeräte. Sammlung und Verwertung von Elektro- und Elektronikaltgeräten: Drei Kennzahlen zählen: *https://www.umweltbundesamt.de/daten/ressourcen-abfall/verwertung-entsorgung-ausgewaehlter-abfallarten/elektro-elektronikaltgeraete#sammlung-und-verwertung-von-elektro-und-elektronikaltgeraten-drei-kennzahlen-zahlen*, accessed April 26, 2021

[56] Forstner, T., Georgi, M., Knebelspieß, S., Schindler, R.: *40 Jahre Umweltbundesamt. Umweltbundesamt* (2015)

[57] Chinalawinfo Co., Ltd. "Circular Economy Promotion Law of the People's Republic of China" (2008, 08/29), *http://www.lawinfochina.com/display.aspx?id=7025&lib=law*

[58] MBA Polymers Inc. "How is China's 'Green Fence' affecting the global recycling trade?" (2013, 07/31), *https://mbapolymers.com/news/news-green-fence/*

[59] Cowie, Lesley. Everblue. From: China's Green Fence: Impacts on U.S. Recycling (2019, 12/18). *https://everbluetraining.com/us-recycling-impacts-green-fence-initiative/: https://everbluetraining.com/us-recycling-impacts-green-fence-initiative/*

[60] Sphera EC4P. "Revision to China's Solid Waste Law" (2020, 06/08), *https://ec4p.com/resources/news/revision-to-china-solid-waste-law*

[61] CGTN. "China tightens rules to control plastic pollution" (2020, 01/19), *https://news.cgtn.com/news/2020-01-19/China-tightens-rules-to-control-plastic-pollution–NnoPKShEHu/index.html*

[62] IHS Markit, Morales, J.: "Impact of New Plastics Regulation in China" (2020, 02/03), *https://ihsmarkit.com/research-analysis/impact-of-new-plastics-regulation-in-china.html*

[63] Oxford University Press (OUP), "Sustainable waste management for zero waste cities in China: potential, challenges and opportunities" (2020, 10/06). (V. 4.-2. Clean Energy, Ed.), *https://academic.oup.com/ce/article/4/3/169/5918339*

[64] EPA (United States Environmental Protection Agency), "Recycling and Reuse: End-of-Life Vehicles and Producer Responsibility" (2016, 02/21), *https://archive.epa.gov/oswer/international/web/html/200811_elv_directive.html*

[65] (n. d.). "A Bill Break Free From Plastic Pollution Act", *https://d12v9rtnomnebu.cloudfront.net/diveimages/BFFPPA_2021_bill_language.pdf*

[66] Resource Recycling, Inc., "Plastics recycling bill introduced to Congress" (2020, 06/16), *https://resource-recycling.com/recycling/2020/06/16/plastics-recycling-bill-introduced-to-congress/*

[67] (n. d.). Bill CLEAN Future Act. Climate Leadership and Environmental Action for our Nation's Future Act (2021), *https://energycommerce.house.gov/sites/democrats.energycommerce.house.gov/files/documents/CFA%20Bill%20Text%202021.pdf*

[68] Congress.gov. (n. d.). "Realizing the Economic Opportunities and Values of Expanding Recycling Act". From: To establish the Recycling Infrastructure Program within the Environmental Protection Agency, and for other purposes: *https://www.congress.gov/bill/117th-congress/house-bill/2357/text?r=1&s=1*

[69] (n. d.) "A BILL - To establish the Recycling Infrastructure Program within the Environmental Protection Agency, and for other purposes" (2019, 11/09), *https://d12v9rtnomnebu.cloudfront.net/diveimages/RECOVERText.pdf*

[70] Congress.gov. (n. d.). Congress.gov., *https://www.congress.gov/search?searchResultViewType=expanded&pageSort=dateOfIntroduction:desc&q={%22congress%22:[%22117%22],%22source%22:[%22legislation%22],%22search%22:%22plastic%22}*

[71] WasteDive, "Tracking the future of US recycling policy in Congress" (2021, 04/21), *https://www.wastedive.com/news/tracking-the-future-of-us-recycling-policy-in-congress/570778/*

[72] Quinn, M.: "Environmental efforts key for Biden, but industry expects limited recycling policy movement in 2021" (2021, 01/20), *https://www.wastedive.com/news/biden-harris-congress-epa-recycling-plastic-climate-pfas/593525/*

[73] WWF, Ellen MacArthur Foundation, Boston Consulting Group, The business case for a UN treaty on plastic pollution" (2020), *https://f.hubspotusercontent20.net/hubfs/4783129/Plastics/UN%20treaty%20plastic%20poll%20report%20a4_single_pages_v15-web-prerelease-3mb.pdf*

[74] United Nations, "Sustainable Development Goals" (2021), *https://www.un.org/sustainabledevelopment/news/communications-material/*

[75] Center for International Environmental Law (CIEL), "Convention on plastics pollution - Toward a new global agreement to address plastic pollution" (2020, 06), *https://www.ciel.org/wp-content/uploads/2020/06/Convention-on-Plastic-Pollution-June-2020-Single-Pages.pdf*

[76] Statista GmH, "Where do the oceans' microplastics come from?" (2021), *https://www.statista.com/chart/17957/where-the-oceans-microplastics-come-from/*

[77] Umweltbundesamt, "Reducing pollution through product design" (2016, 08/18), *https://www.umweltbundesamt.de/en/reducing-pollution-through-product-design*, accessed April 20, 2021

[78] UNEP, SETAC & Life Cycle Initiative, "Guidelines for Social Life Cycle Assessment of Products" (2009), *https://www.lifecycleinitiative.org/wp-content/uploads/2012/12/2009%20-%20Guidelines%20for%20sLCA%20-%20EN.pdf*

[79] Roundtable for Product Social Metrics, "Handbook for Product Social Impact Assessment 5.0" (2020), *https://product-social-impact-assessment.com/wp-content/uploads/2021/04/20-01-Handbook2020.pdf*

[80] UNEP, SETAC & Life Cycle Initiative, "Towards a Life Cycle Sustainability Assessment – Making informed choices on products" (2011), *https://www.lifecycleinitiative.org/wp-content/uploads/2012/12/2011%20-%20Towards%20LCSA.pdf*

[81] (n. d.). ISO 14006:2020: "Environmental management systems – Guidelines for incorporating product eco-design"

[82] World Health Organization, "Economic instruments as a lever for policy" (2021), *https://www.who.int/heli/economics/econinstruments/en/,* accessed April 23, 2021

[83] Deloitte, Bio by, Development of Guidance on Extended Producer Responsibility" (2014). Brussels: Deloitte

[84] Europen, *Factsheet – Extended Producer Responsibility for Used Packaging,* Europen (2016)

[85] ETRma, EBPA, LighteningEurope, AIM, Common waste stream views on EPR in Europe. Europen (2015)

[86] ACC (American Chemistry Council, Inc.), "Plastic Makers Applaud Governor Wolf for Signing Advanced Recycling Legislation into Law" (2021), *https://www.americanchemistry.com/Media/PressReleasesTranscripts/ACC-news-releases/Plastic-Makers-Applaud-Governor-Wolf-for-Signing-Advanced-Recycling-Legislation-into-Law.html*

[87] Umweltbundesamt (Vogel, J., Krüger,F., Fabian, M.), "Chemical recycling" (12/2020), *https://www.umweltbundesamt.de/publikationen/chemical-recycling*

[88] Buttker, B., Giering, R., Schlotter, U., Himmelreich, B., Wittstock, K.: *Full-Scale Industrial Recovery Trials of Shredder Residue in a High Temperature Slagging Gasifier in Germany,* PlasticsEurope report 8043/GB/06/2005

[89] Hoppe, W., Bringezu, S., Thonemann, N.: "Comparison of global warming potential between conventionally produced and CO_2-based natural gas used in transport versus chemical production" *Journal of Cleaner Production,* (2016) 121, pp. 231–237, *https://doi.org/10.1016/j.jclepro.2016.02.042*

[90] Hoppe, W., Bringezu, S.: "Vergleichende Ökobilanz der CO_2-basierten und konventionellen Methan- und Methanolproduktion" uwf UmweltWirtschaftsForum, (2016) 24, pp. 43–47, *https://doi.org/10.1007/s00550-016-0389-4*

[91] OECD, "Improving Plastics Management: Trends, policy responses, and the role of international co-operation and trade (Environment Policy Paper NO. 12)" (2018), *https://www.oecd.org/environment/waste/policy-highlights-improving-plastics-management.pdf*

[92] Universität Kassel – Center for Environmental Systems Research (CESR) (03/2020), Zukünftige Nutzung von CO_2 als Rohstoffbasis in der deutschen Chemie- und Kunststoffindustrie, *https://kobra.uni-kassel.de/bitstream/handle/123456789/11483/ZukuenftigeNutzungVonCO2AlsRohstoffbasis.pdf?sequence=4&isAllowed=y*

6 Life Cycle Assessment (LCA)

6.1 Introduction: What Is Life Cycle Assessment?

Peter Saling

6.1.1 History and General Aspects

Sustainable development (SD) has been defined as the balance of economic success, environmental protection, and social responsibility. To effectively manage sustainability, companies must be able to measure or otherwise quantify sustainability aspects. For the environmental impacts, life cycle assessment (LCA) has been applied for decades, and continues to be today. To get a holistic view on the environmental impacts of products throughout the life cycle, LCA is a powerful method creating much important information for decision-making processes, both for the company and for consumers deciding to buy a product or to whether to consider an alternative.

The assessment of the environmental impacts of products compared to alternatives became important when public debate started about environmental preferences. The first studies were published as partial assessments in the late 1960s and early 1970s, focusing on energy analysis. In the early 1990s, studies focused on packaging alternatives was an important topic.

"Jute instead of plastic?" was a question asked 40 years ago, and an attempt was made to use LCA to answer it. It was stated that renewable materials are inherently more sustainable than materials made from fossil-fuel-based sources, because four times as much energy was used to manufacture a single plastic bag as to produce a jute sack. Today we know better that it is necessary to consider the whole life cycle, including use phases and recycling steps, if results of an LCA study can be interpreted and used effectively. Today, many more indicators are addressed, beyond the energy consumption for the production phase. Formerly, widely diverging approaches and results were linked to a lack of international scientific discussion and exchange platforms for LCA, as well as standards that can be used for the preparation of LCA studies. Because of the lack of this expertise, several life cycle impact assessment methods, were developed [1].

Today, LCA is widely used, as a "cradle-to-grave" approach for evaluating products and services. This also includes the assessment of processes, including recycling options for products, which is needed specifically for materials made from plastics. A cradle-to-grave process begins with the extraction of materials from the earth (or from recycling processes) to manufacture a product used in a defined application over a defined time. It ends at the point when all materials are ultimately returned for disposal. LCA links the various environmental impacts at all stages of the product's life.

Further assessment tools may include the economic or the social dimension of sustainability. Eco-efficiency analyses (EEAs) incorporate both economic and environmental aspects and are used by research organizations and companies in different sectors to include sustainability aspects in decision-making processes. Quantification of sustainability may be utilized in R&D, marketing, sales, strategic planning, and communications. The SEEbalance® covers all three dimensions of sustainability in an integrative approach following holistic assessment principles [2].

Eco-efficiency is generally measured by the ratio of a useful output divided by useful input. The monetary value created by the business can be (but is not limited to) a useful input. The monetary value created, justified for environmental external costs, can be the useful output [3]. It is a common concept designed to drive the decoupling of economic growth from environmental deterioration [4]. But it must be pointed out that eco-efficiency has no direct link to sustainability per se. There is still a need for a consistent definition of the term. A "strong" eco-efficiency definition implies improvement in both dimensions - environment and economy [5].

The empirical facts and developments that have been achieved in the field of eco-efficiency are able to support decision-making on a micro- and macro-level, but the relationships between economy and environment are not self-evident. By clarifying the "why and what", eco-efficiency is a first and important step towards decision-making support including relevant aspects of SD [6].

6.1.2 LCA Principles Defined by ISO Standards

6.1.2.1 Definitions

The international standards ISO 14040:2006 [7] and ISO 14044:2006 [8] are the relevant standards used today by every practitioner in the world. ISO 14040 describes the principles and framework for life cycle assessment (LCA) including: definition of the goal and scope of the LCA, the life cycle inventory (LCI) analysis phase, the life cycle impact assessment (LCIA) phase, the life cycle interpretation phase, reporting and critical review of the LCA, limitations of the LCA, the relationship between the LCA phases, and conditions for use of value choices and optional elements. ISO 14040 covers LCA studies and LCI studies. It does not describe the LCA technique in detail, nor does it specify methodologies for the individual phases of the LCA. The intended application of LCA or LCI results is considered during definition of the goal and scope, but the application itself is outside the scope of this international standard.

Linked with ISO 14040, ISO 14044 specifies requirements and provides guidelines for LCA including: definition of the goal and scope of the LCA, the LCI analysis phase, the LCIA phase, the life cycle interpretation phase, reporting and critical review of the LCA, limitations of the LCA, relationship between the LCA phases, and conditions for use of value choices and optional elements.

Eco-efficiency analysis (EEA) was established in recent decades, by combining economic and environmental concepts. Eco-efficiency is defined as aspect of sustainability relating to the environmental performance of a product system to its product system value. Within this context, the environmental performance relates to measurable results concerning environmental aspects. Environmental aspects are defined as elements of an organization's activities, products, or services that can interact with the environment. A product system value has worth (or desirability) ascribed to it, and may be expressed in monetary terms or other value aspects. A product system reflects a collection of unit processes with elementary and product flows, performing one or more defined functions, and which models the life cycle of a product.

Eco-efficiency is covered by ISO 14045 [9] and has also subsequently been linked to ISO 14040 and 14044. EEA is defined within these standards as a quantitative management tool which enables the consideration of the life cycle environmental impacts of a product system, alongside its product system value to a stakeholder. EEA shares with LCA many important principles such as life cycle perspective, comprehensiveness, functional unit approach, iterative nature, transparency, and priority of scientific approach [10].

6.1.2.2 Goal and Scope of an LCA Study

According to the ISO standards, life cycle assessment needs to be conducted in four phases (Figure 6.1): the description and definition of the goal and scope (Phase 1), followed by the inventory analysis (Phase 2), the impact assessment (Phase 3), and the interpretation (Phase 4). LCA is an iterative approach, in which these four phases always interact with each other, and during which they shall be used as necessary until the study is finished. The goal is the generation of a meaningful study with reliable results reflecting as well as possible the environmental impacts of alternative products or processes.

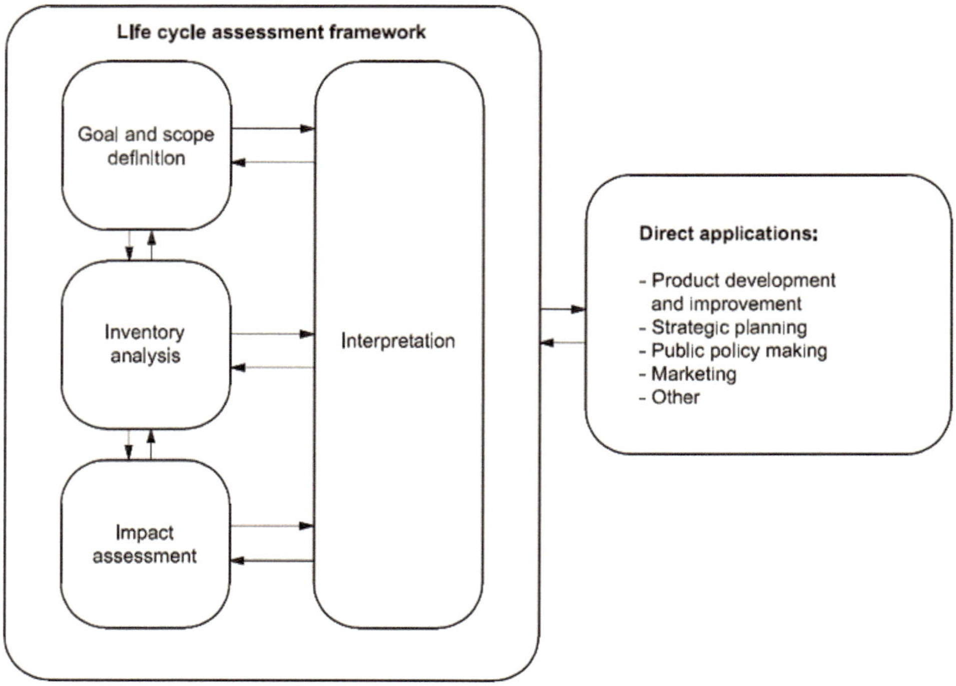

Figure 6.1 Phases of an LCA study

The first phase, "**goal and scope**", is important for the definition of the purpose of an LCA study. The functional unit is a quantified performance of a product system for use as a reference unit. It needs to be defined carefully and should consider all relevant aspects of the study.

According to the aims, data assessment needs to be defined including levels of details and qualities addressed. It has also to be decided whether and how an external report (critical review) needs to be added. Specific requirements are needed if a comparative assertion disclosed to the public is intended. Comparative assertion is defined as environmental claim regarding the superiority or equivalence of one product versus a competing product that performs the same function. All alternative products addressed in an LCA study need to be able to fulfill the functional unit in the same way, otherwise they cannot be compared. In this first phase of a study, a clear functional unit together with a meaningful definition of alternatives is needed as a basis. The following gives an example of this::

Functional unit: Packaging and distribution of 1000 L mineral water for the household market with a distribution circle of 200 km.

Alternatives: PET bottles, single-use or with re-use opportunities with sizes of 1 L or 1.5 L. In addition: glass bottles with 0.7 L content.

This example shows that the comparison of bottle sizes, materials, uses and re-uses (including different recycling options for the household market) generates different alternatives than (for example) the convenience market, which might have other needs.

In the next step the "**definition of the system boundaries**" needs to be defined. In this step all relevant process steps need to be addressed throughout the whole life cycle. The system boundaries need to be determined according to the defined goals of the LCA. The temporal framework, and the spatial and technical spheres, need to be defined and included accurately. In this step all contributions need to be challenged. On one hand, the system boundaries need to be complete, while on the other hand, full completeness will realistically never be achieved. To keep the efforts on an acceptable level and to be comprehensive enough, cut-off criteria need to be applied. For example, very often discussed in this context are the investment goods. In many LCA studies, they are excluded because they have a very low impact on the results of the study. In this case, the production of the filling machine or the production of trucks can be excluded, due to the fact that compared to the other materials this impact is low. However, this needs to be decided and cross-checked in every study. If a contribution of a life cycle step is unclear, an overview calculation is done, and the effect will be interpreted. However, in some cases, the contribution of a life cycle step to one impact category is low, but to another impact category is high. That needs to be considered in the design of the study, and probably adjusted during the study, so as to not underestimate important impacts. In Figure 6.2 an example for a packaging LCA for mineral water is given. It was shown by calculations that the production of the trucks needed for distribution does not need to be considered here, because the contributions to the relevant impact categories were very low.

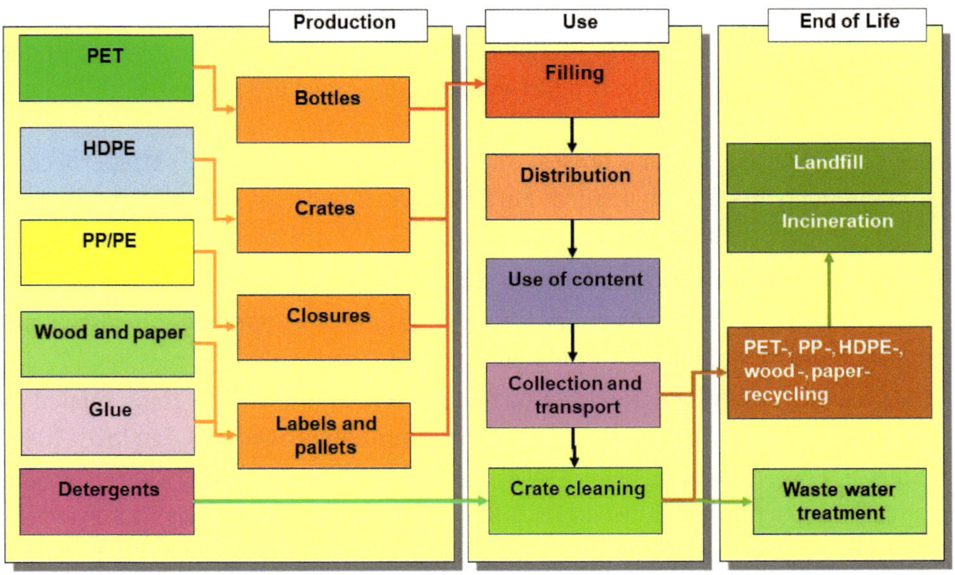

Figure 6.2 System boundaries of mineral water packaging systems for a single-use PET bottle

6.1.2.3 Life Cycle Inventories

In the next step the "**life cycle inventory**" collects all data needed for the step of the impact assessment. Material and energy flows are gathered and listed during their entire life. The material and energy flows are determined as input/output figures for every process step, building a so-called unit process. The partial processes and the links between the modules are used to generate a meaningful basis for the assessment. The mass and energy balances need to be closed on the inventory level of the whole system. All material and energy streams which pass the system borders are listed as quantities in physical units. The data refer to the reference flow of the functional unit. Many life cycle inventories can be derived from commercially available databases, information from open sources, association data platforms, or by direct contact to suppliers. Very often, new life cycle inventories need to be generated by the practitioners, based on primary or secondary data sources. To exchange LCI between different data owners requires the use of data formats that can be used by as many users as possible, which is often a challenge even when data formats (such as ILCD) exist. It was recognized that the naming of the so-called flow quantities is difficult and not harmonized, leading to the incomparability of data sets, or mistakes when flows are not covered accurately. Therefore, during the life cycle inventory phase it is necessary to select the data sets carefully, because they are the basis for the subsequent work in the LCA.

6.1.2.4 Processes with Co-Products

If considering a process where co-products are generated (defined as any of two or more products coming from the same unit process or product system), a specific treatment of their environmental flows is needed. In ISO 14044 clear guidance is given that practitioners need to follow. The exercise is partitioning the input or output flows of a process or a product system between the product system under study and one or more other product systems. For recycling processes, this situation is observed very often. Recycling processes can generate different materials, energies, and wastes in one process step, and therefore the different approaches for co-products need to be applied. Furthermore, inputs can be generated by processes with co-products as well. The procedure is defined as follows:

- **Step 1:** Wherever possible, allocation should be avoided by:
  - Dividing the unit process to be allocated into two or more sub-processes and collecting the input and output data related to these sub-processes, or
  - Expanding the product system to include the additional functions related to the co-products, taking into account the requirements described in ISO 14044.
- **Step 2:** Where allocation cannot be avoided, the inputs and outputs of the system should be partitioned between its different products or functions in a way that reflects the underlying physical relationships between them; i.e. they should

reflect the way in which the inputs and outputs are changed by quantitative changes in the products or functions delivered by the system.

- **Step 3:** Where a physical relationship alone cannot be established or used as the basis for allocation, the inputs should be allocated between the products and functions in a way that reflects other relationships between them. For example, input and output data might be allocated between co-products in proportion to the economic value of the products.

Some outputs may be partly co-products and partly waste. In such cases, it is necessary to identify the ratio between co-products and waste, since the inputs and outputs will be allocated to the co-products part only. Allocation procedures shall be uniformly applied to similar inputs and outputs of the system under consideration. The inventory is based on material balances between input and output. Allocation procedures should therefore approximate as much as possible such fundamental input/output relationships and characteristics.

6.1.2.5 Allocation in Recycling Processes

The allocation principles and procedures mentioned in the previous chapter also apply to reuse and recycling situations. Changes in the inherent properties of materials are considered. In addition, particularly for the recovery processes between the original and subsequent product system, the system boundary shall be identified and explained, ensuring that the allocation principles are observed as described in ISO 14044.

Additional elaboration is needed in this context for the following reasons:

- Reuse and recycling (as well as composting, energy recovery and other processes that can be assimilated to reuse/recycling) may imply that the inputs and outputs associated with unit processes for extraction and processing of raw materials and final disposal of products are to be shared by more than one product system
- Reuse and recycling may change the inherent properties of materials in subsequent use
- Specific care should be taken when defining system boundary regarding recovery processes.

Note that:

- A closed-loop allocation procedure applies to closed-loop product systems. It also applies to open-loop product systems where no changes occur in the inherent properties of the recycled material. In such cases, the need for allocation is avoided, since the use of secondary material displaces the use of virgin (primary) materials. However, the first use of virgin materials in applicable open-loop product systems may follow an open-loop allocation procedure outlined in b).

- An open-loop allocation procedure applies to open-loop product systems where the material is recycled into other product systems and the material undergoes a change to its inherent properties.

In the recently published Amendment 2 to ISO 14044, an informative annex is included that describes allocation guidance for recycling. In the case of recycling, one way to avoid allocation can be by calculating a recycling credit based on the technical substitutability of the secondary material(s), i. e. taking into account any changes to the inherent properties and quality of the secondary material versus the substituted primary material. If the secondary material X from the product system under study is substituted for a primary material Y, then the recycling credit corresponds to subtracting the inventory associated with the acquisition of the primary material Y from the inventory calculated for the product system under study. If an input to a product system is a recycled material that has previously implied a credit to the product system that the recycled material comes from, such recycled material carries the credit as a potential environmental impact to the product system that it enters. In general, if several allocation procedures appear to be permissible, a sensitivity analysis must be carried out.

6.1.3 Impact Assessment

A substance causes several environmental impacts. These impacts need to be transferred from the inventory (stage) with a characterization model to different environmental impacts. This process is called impact assessment. Impacts considered in a LCIA can include (in a typical set-up) climate change, ozone depletion, eutrophication, acidification, human toxicity, photochemical ozone formation, land use, and resource depletion. In addition, respiratory inorganics, ionizing radiation, ecotoxicity, and other impacts might be addressed. The emissions and resources are assigned to each of these impact categories. They are then converted into indicators using impact assessment models.

In LCA in general two different approaches are used in the impact assessment. They are often defined as "midpoint" and "endpoint" characterization models. Although the terms have yet to be rigorously defined, midpoints are considered to be a point in the cause-effect chain (environmental mechanism) of a particular impact category, prior to the endpoint, at which characterization factors can be calculated to reflect the relative importance of an emission or extraction in an LCI (e. g., global warming potentials defined in terms of radiative forcing and atmospheric half-life differences) [11]. In the LCIA, the data of a LCI of the considered products defined in the system boundaries of a product system are transformed into understandable impact categories that represent the impact on the environment.

Various impact assessment models have been developed, such as TRACI, CML, ReCiPe and LIME [12–15]. New developments have either been finished or are still ongoing to develop additional assessment methods for specific impact categories, such as the WULCA group for water use [16, 17], the UNEP-SETAC land use group [18], the developments for marine litter treatment in LCA with the MariLCA project [19], and the developments dealing with marine litter impacts [20] and the ProScale project as a new approach for human toxicity and ecotoxicity assessments [21]. In addition to the extensive work on assessment models, recommendations for aggregation of environmental impacts have been developed [22].

6.1.3.1 Environmental Footprint of the EU

The most wide-ranging recent initiative is the European Product Environmental Footprint (EU PEF), which has been developed as part of the "single market for green products", and represents a harmonized approach for the life cycle assessment of products [23]. Industry initiatives have developed sector-specific guidelines to ensure consistent accounting and reporting in an industrial sector and to ensure a practicable approach [24].

In the environmental footprint of the EU, both the product-related view and the company-related view are covered. The final "Product Environmental Footprint Category Rules" (PEFCRs) and "Organization Environmental Footprint Sector Rules" (OEFSRs) can be used for calculating the environmental footprint profile for products and organizations in scope. Final PEFCRs and OEFSRs are uploaded when finalized. The PEFCRs and OEFSRs were developed according to version 6.3 of the PEFCR guidance and the OEFSR guidance. The EU PEF method covers 15 environmental impacts. The European Commission provided different types of information, highlighted the possible applications, and answered general questions [25].

PEF and OEF methods could be used for a variety of applications, including:

- In-house improvement of product environmental performance (design for environment)
- Communication (B2B and B2C) of the environmental performance of a product or organization (without comparisons against similar products or organizations)
- Improvement of production processes along the life cycle

However, whenever the intention is to make a comparison against other products or organizations, some further methodological features need to be developed.

Circular Footprint Formula

In the environmental footprint, among other methodological developments, a **circular footprint formula (CFF)** was developed, which needs to be applied in this program to recycling processes. This is very relevant for plastics recycling processes and also their LCA evaluation. It gives guidance on how different impacts need to be addressed, including market shares of products, and was developed within the PEF programme of the European Commission [26]. It is stated as:

The PEF is an LCA-based method to quantify the environmental impacts of products (goods or services). It builds on existing approaches and international standards. The overarching purpose of PEF information is to enable to reduce the environmental impacts of goods and services considering supply chain activities (from extraction of raw materials, through production and use to final waste management). This purpose is achieved through the provision of detailed requirements for modelling the environmental impacts of the flows of material/energy and the emissions and waste streams associated with a product throughout its life cycle.

It answers the question of how recycling and energy recovery processes, as well as the use of secondary materials and energy, can take into account the benefits and burdens of these processes. For reproducibility and fairness, guidance is needed on:

- Where is the boundary between the first and the second product system?
- How should the benefits and burdens of generating and of using recycled material be shared between the first and the second product system?
- As generated secondary materials and energy carriers avoid primary materials and energy carriers being produced, how should one select the specific primary material and/or energy that is avoided?
- How to handle downcycling, i. e. differences in quality between secondary material or energy and the primary material or energy?
- How to avoid double counting or gaps of benefits and burdens?

In this context, different examples were developed to transfer technical information from recycling processes to LCA-based definitions, ending in detailed calculations of the environmental impacts of different types of recycling or end-of-life options or combinations of them.

From secondary materials input and energy flows, the system boundaries can be defined, and the relevant impacts can be transferred to the CFF illustrated by Wolf [27, 28] in the following figures (Figure 6.3, Figure 6.4).

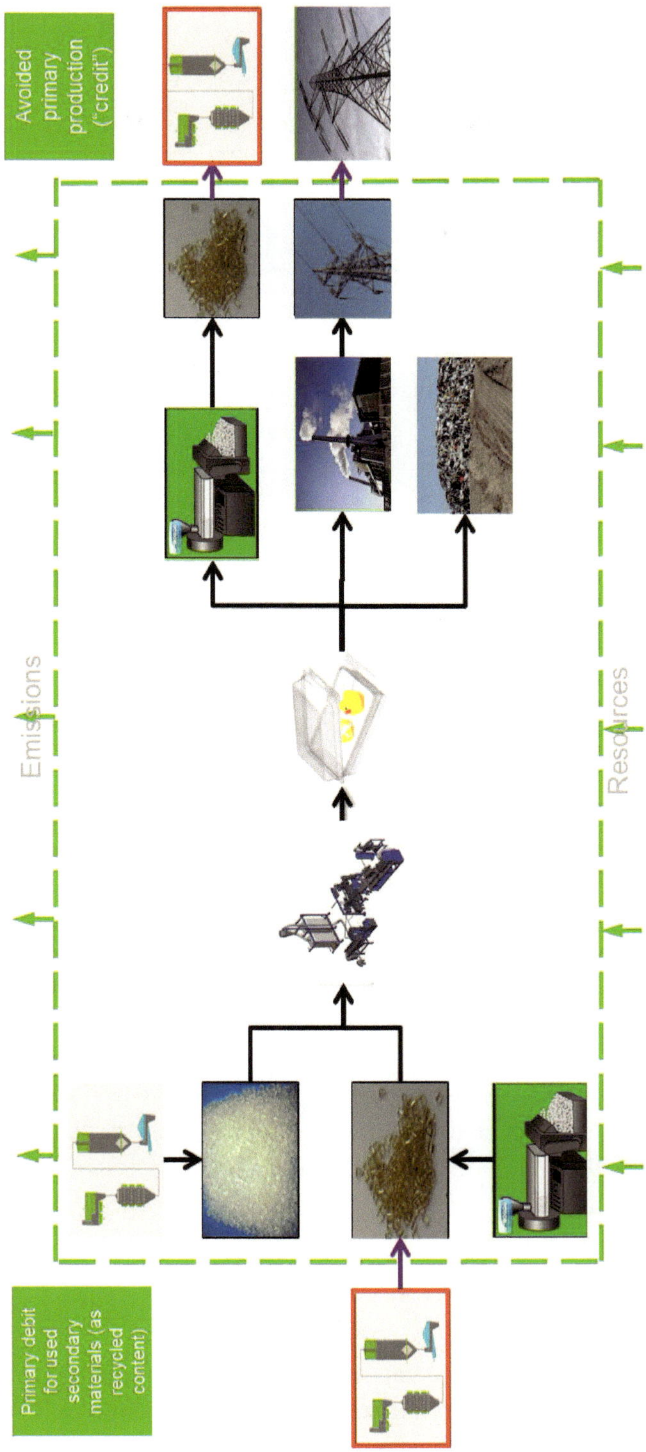

Figure 6.3 System boundaries of the recycling of a PET blister package

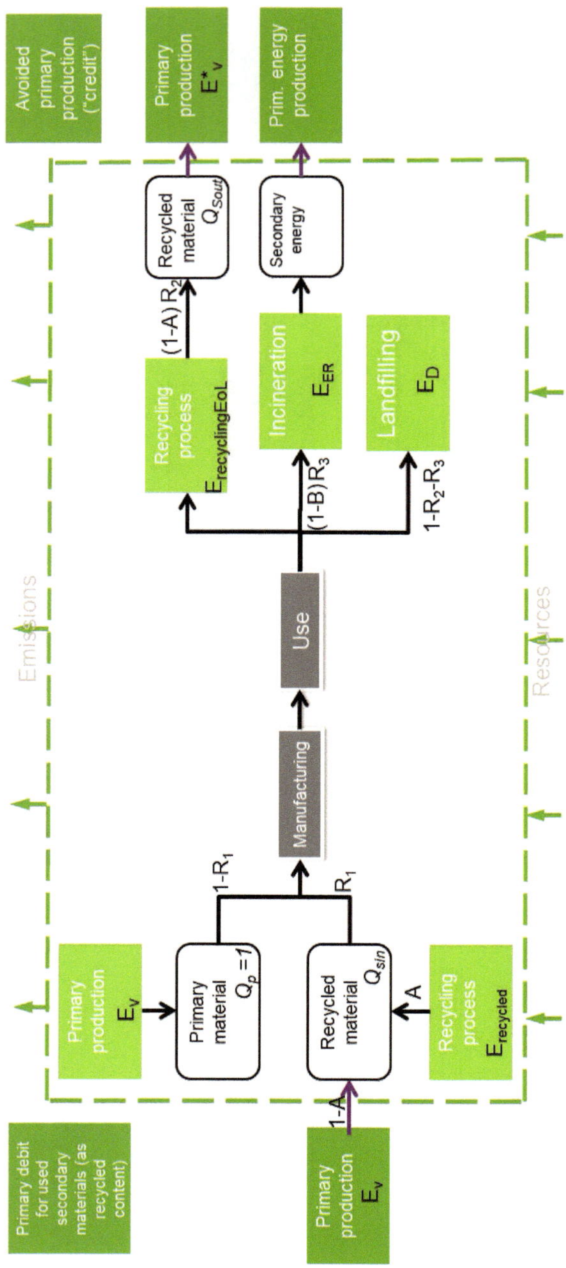

Figure 6.4 Flow description of the LCA model within the CFF

To cover all relevant flows and impacts, a CFF was developed that can be used based on the flow definitions of Figure 6.4 and that covers all relevant flows of a recycling system (Figure 6.5). Factor *A* is used for sharing recycling burdens and benefits among connected life cycles, and here is either 0.2, 0.5, or 0.8. Factor *B* is used for sharing energy recovery burdens and benefits among connected life cycles, and is always 0. For different types of materials, *A* and *B* allocation factors are fixed in the first instance, and can be changed if market data support that.

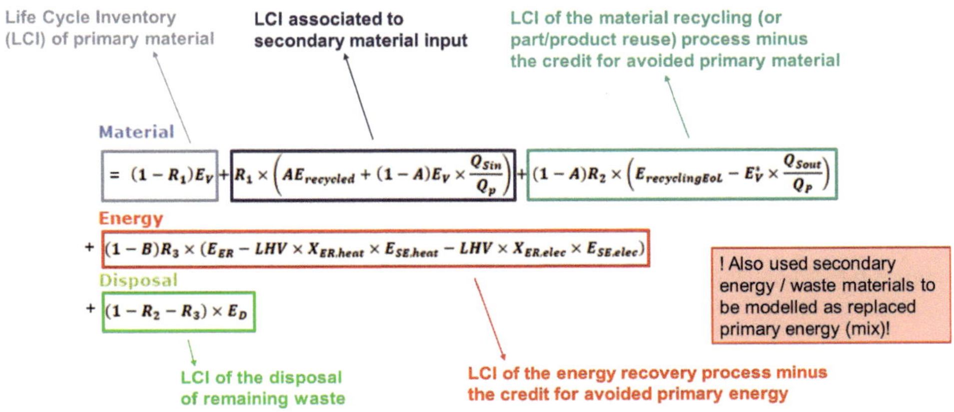

A	Allocation factor of burdens and benefits (jointly: "credits") between supplier and user of recycled materials.
B	Allocation factor of energy recovery processes. It applies both to burdens and benefits.
Q_{Sin}	Quality of the ingoing secondary material, i.e. the quality of the recycled material at the point of substitution.
Q_{Sout}	Quality of the outgoing secondary material, i.e. the quality of the recycled material at the point of substitution.
Q_P	Quality of the primary material, i.e. quality of the virgin material.
R_1	Proportion of material in the input to the production that has been recycled from a previous system.
R_2	Proportion of the material in the product that will be recycled (or reused) in a subsequent system. R2 shall therefore take into account the inefficiencies in the collection and recycling (or reuse) processes. R2 shall be measured at the output of the recycling plant.
R_3	Proportion of the material in the product that is used for energy recovery at EoL.
$E_{recycled}$	Specific emissions and resources consumed (per functional unit) arising from the recycling process of the recycled (reused) material, including collection, sorting and transportation process.
$E_{recyclingEoL}$	Specific emissions and resources consumed (per functional unit) arising from the recycling process at EoL, including collection, sorting and transportation process.
E_v	Specific emissions and resources consumed (per functional unit) arising from the acquisition and pre-processing of virgin material.
E^*_v	Specific emissions and resources consumed (per functional unit) arising from the acquisition and pre-processing of virgin material assumed to be substituted by recyclable materials.
E_{ER}	Specific emissions and resources consumed (per functional unit) arising from the energy recovery process (e.g. incineration with energy recovery, landfill with energy recovery, etc.).
$E_{SE,heat}$ and $E_{SE,elec}$	Specific emissions and resources consumed (per functional unit) that would have arisen from the specific substituted energy source, heat and electricity respectively.
E_D	Specific emissions and resources consumed (per functional unit) arising from disposal of waste material at the EoL of the analyzed product, without energy recovery or other usable product output.
$X_{ER,heat}$ and $X_{ER,elec}$	The efficiency of the energy recovery process for both heat and electricity.
LHV	Lower heating value of the material in the product that is used for energy recovery.

Figure 6.5 The CFF includes all relevant flows of a recycling system

The PET blister recycling calculation shows the separate steps in the CFF. Positive and negative contributions for CO_2-eq along the whole life cycle are displayed, along with the total (Figure 6.6). All other indicators can be treated and calculated with different impact methods in the same way.

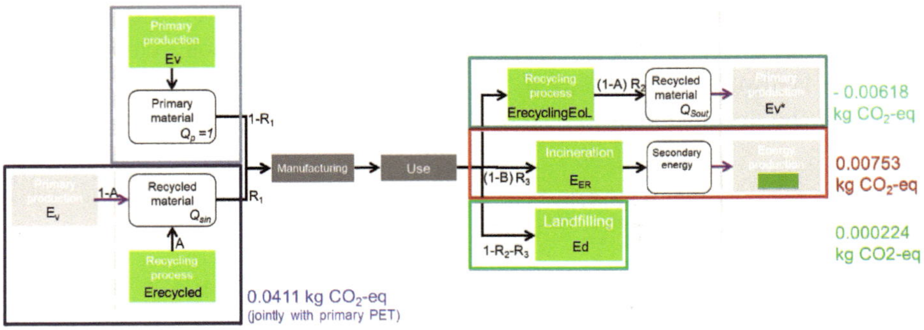

= 0.0411 – 0.00618 + 0.00753 + 0.000224 = **0.0427** kg CO_2-eq per 1 PET blister packaging from CFF calculation

! The PEF study shall report all parameters and variable values used, the source and – if not default – provide evidence.

Figure 6.6 Carbon footprint results for a PET blister system following the CFF

6.1.3.2 Impact Categories Used by BASF LCA Methods

It is often necessary to reduce the number of indicators to a reasonable number. In this context product category rules (PCRs) are very helpful for defining a consensus level of indicator sets for specific products or processes.

In our assessments, we use mainly two sets of indicators, depending on the relevance check we do in conjunction with a specific LCA study. The reduced set of indicators often is good enough to cover many LCA studies. If the relevance check suggests a higher number of indicators, we will include them (Table 6.1). Detailed information on the methods and the relevant citations is available [26].

Table 6.1 Overview of Impact Categories for LCA Applied by BASF

Reduced set of impact categories	Extended set of categories	Impact assessment method
Resource use *Minerals and metals* *Fossil fuels*	ADP (abiotic depletion potential) *Minerals and metals* *Fossil fuels*	ADP of EU PEF [13, 29]
Climate change	Climate change	GWP of EU PEF [30]
Photochemical ozone formation, human health	Photochemical ozone formation, human health	POCP of EU PEF [31]
Acidification	Acidification	AP of EU PEF [32, 33]
Eutrophication (freshwater, marine)	Eutrophication (freshwater, marine)	EP of EU PEF [32–34]
Human toxicity potential	Human toxicity potential	BASF [35]
	ODP/ozone depletion potential	ODP of EU PEF [36]
	Water consumption	Available WAter REmaining (AWARE) [37]

Reduced set of impact categories	Extended set of categories	Impact assessment method
	Ecotoxicity, freshwater	EU PEF [38]
	Land use	LANCA of EU PEF [39, 40]

6.1.4 Interpretation

To finalize an LCA study, the fourth element in this order is the interpretation. A meaningful interpretation of the results is key for understanding the messages developed. It supports decision-making processes, allows policy makers to take the outcomes from a study and introduce them to legislation, and allows producers to initiate meaningful claims for business (as well as helping them to improve products throughout the life cycle). The identification of significant issues based on the results of the environmental and product system value assessment phases, aspects and description of completeness, sensitivity, uncertainty and consistency of data and results is needed. This will support the formulation of conclusions, limitations, and recommendations to the recipient or client of such a study.

Normalization is an (optional) impact assessment step for better understanding the results and to compare the different impact categories of an environmental footprint study. The normalized impact categories provide insight into the magnitude of the impact categories relative to a reference unit (e.g. CO_2 emissions of a product relative to the CO_2 emitted by an EU citizen). The normalization step is not mandatory when using the PEF method, and could simply be used to identify the most relevant environmental impacts. Normalization factors were published by the Joint Research Centre and will be used in the environmental footprint of the EU.

When comparisons of LCA results between product systems or within the same product system are made, they are based in the first analysis on the same indicator. That allows clear comparison without further issues. If the comparison covers all indicators assessed, it is then necessary to put the results on the same level. In an LCA this normalization step generates dimensionless factors linked to specific normalization factors. These need to be generated using the same approach and with the same level of references. For example, normalization factors for Europe should not be mixed up with those from China. If the normalization is done accurately, different indicators can be identified contributing on different levels of environmental impacts of a defined region. That process also helps to identify and discuss trade-offs that might be visible and that might influence the overall interpretation.

Decision-makers often want to get an overall result of an LCA study, because several indicators might deliver different results that are very difficult to compare and to summarize. The aggregation of a set of indicators to one single result needs normalization, aggregation, and weighting steps.

Weighting is subjective and based on value choices. It results in the potential for possible manipulation of results or influence by pre-conditions set by decision-makers. However, if there are clear preferences stated for some of the indicators, it is helpful to address them because that increases the acceptance of the outcome of a study. The creation of weighting schemes can be done in different ways. One option is a distance-to-target approach, where targets set by authorities for reaching specific goals of environmental impacts can be used as a basis. Furthermore, so-called panel weighting factor approaches can be applied, where different stakeholders are asked for their preferences. In an ideal way, statistical methods can be used to determine the stakeholder asked for their opinion to assure a high level of representativeness.

The weighting should also follow specific guidance which is given in ISO 14040 and ISO 14045. However, weighting is not allowed according to ISO 14040 and ISO 14044 if comparative assertions are made about the environmental superiority of one of the systems compared. In such a case, a critical review must be done according to the panel method [41].

6.1.5 Examples of LCA Studies in the Recycling Context

6.1.5.1 Recycling of PET

An example for recycling of PET materials compared domestic and transboundary recycling scenarios between Japan and China, and disposal scenarios from the viewpoints of greenhouse gas (GHG) emission and fossil resource consumption [42]. The results showed that all the domestic and transboundary recycling scenarios had smaller GHG emissions and fossil resource consumptions than the incineration scenario, and that the chemical recycling scenarios had larger GHG emissions and fossil resource consumptions than the other recycling scenarios. The landfilling scenario had the largest fossil resource consumption, while it was better than the incineration scenario and slightly better than the chemical recycling scenarios from the viewpoint of GHG emission. This example shows that different indicators can deliver different results and need to be carefully addressed in the interpretation phase.

The application of different assessment approaches for recycling of polyester materials was published last year [43]. The authors applied cut-off, waste evaluation, and system expansion approaches to show the different impacts of the basic decisions on the assessment system. The system boundary of the cradle-to-factory gate was followed by analysis of the cut-off, system evaluation, and waste evaluation methods. The use of rPET enables the reduction of the GWP from 25 to 75%, depending on the type of technology. It was found that most of the environmental categories are reduced by the impact of bottle-to-fiber recycling based on all three techniques. Chemical recycling offers the highest product quality, but has higher impact on the environment than the mechanical and semi-mechanical routes. The overall conclusion was that rPET offers net reduction of emission of CO_2, carbon monoxide, acids and their gases, particulate matters, heavy metals, and dioxins. This is due to the reduction of the landfill as well as emissions associated with the manufacturing an equivalent amount of virgin PET.

The primary purpose of another study was to understand the environmental impacts of recycled PET fiber compared to virgin PET [44], and several studies have reported the environmental impacts of PET recycling cited in this article. In these studies, PET recycling treated as a post-consumer waste management option was compared with other options such as landfilling and incineration. The goal of the study was to understand the environmental impact of making recycled PET fibers and to address different allocation methods for this open-loop recycling case.

A common practice is to follow the "cut-off" principle, which distinguishes the first life (virgin product) and the second life (recycled product) as separate systems. This method, in contrast to the environmental footprint (EF) approach, has been used very often in different applications. The advantage of this approach is that no further steps nor allocation of burdens and credits in the downstream are needed. The disadvantage is a kind of "unfair" treatment of users of recycled materials that are not within the system boundaries of the cut-off system. The cut-off method is considered simple and easy to apply for using recycling materials, because no data of the first life are needed. In the overview of this topic [44] alternative methods were introduced to further develop the methodology for open-loop recycling. The first alternative method is the "waste valuation" method, which follows the principle of economic allocation. The second alternative method is the "system expansion" method, in which the entire system (cradle-to-grave) is analyzed. Four PET recycling cases are investigated in the study, including mechanical recycling, semi-mechanical recycling, back-to-oligomer recycling, and back-to-monomer recycling. In most of the impact categories, the virgin material had disadvantages compared to the recycling alternatives.

6.1.5.2 Recycling of WEEE

Waste from electrical and electronic equipment (WEEE) includes a large range of devices such as computers, fridges, and mobile phones at the end of their life. This type of waste contains a complex mixture of materials, some of which are hazardous. These can cause major environmental and health problems if the discarded devices are not managed properly. In addition, modern electronics contain rare and expensive resources, which can be recycled and re-used if the waste is effectively managed. The EU [45] has introduced the "WEEE Directive" to deal with the growing amount of WEEE. This Directive focusses on measures to protect the environment by reducing overall impacts of resource use and improving the efficiency of such use in accordance with the concept of sustainable development. In this context, recycling plays an important role and LCA can support the decision-making processes used.

In this context, an LCA study was worked out to generate an in-depth life cycle-based information of the recycling of mixed, plastics-rich residues from WEEE treatment in an operational, state-of-the-art plastics recycling plant. It was investigated how different recovery and disposal routes for plastics-rich WEEE treatment residues originating from the mechanical treatment of WEEE perform from an environmental point of view. Furthermore, it was investigated how post-consumer recycled (PCR) plastics from this plant perform relative to the production of the respective virgin plastics from an environmental point of view.

The LCAs performed in this study focused on the comparison of an existing recycling process with alternatives producing a similar reference flow of materials. The environmental impacts of the recycling process were compared with the impacts of a "final disposal" of the same residual material through incineration in a state-of-the-art European municipal solid waste incineration (MSWI) plant. In a second step, the environmental impacts of the plastics recycling process were compared with those resulting from primary production of an identical amount of (plastics) material. Attributional LCA models using a common "basket of products" through system expansion were used in both cases. In view of having an identical product output or basket of products, respectively, the systems were each expanded with additional processes where necessary [46]. This example shows that this approach can be applied as a good solution for multifunctionality situations by subdividing the unit process into mono-functionality processes or by expanding the product system to include the additional functions of the co-products in the functional unit, which is illustrated by Heijungs [47]. This type of system expansion is a good approach to ensure that all environmental burdens and credits are taken into account [48] and is shown in Figure 6.7.

It was found in the WEEE example that with two exceptions, the incineration in an MSWI plant results in impacts that exceed those of recycling by a factor of 4 and more. The exceptions are the factors "FEP" (freshwater eutrophication potential), with a difference of only factor 2 between the recycling and the MSWI option, and "ODP" (ozone depletion potential), with the impact for recycling exceeding the corresponding score for the MSWI option by about 25%. 70% of this impact is due to the additional "heat & power production" (caused to a large extent by Halon 1211 and Halon 1301 releases in the fossil fuel supply) required to establish the common basket of products. Further details and scenarios can be found in the publication [49].

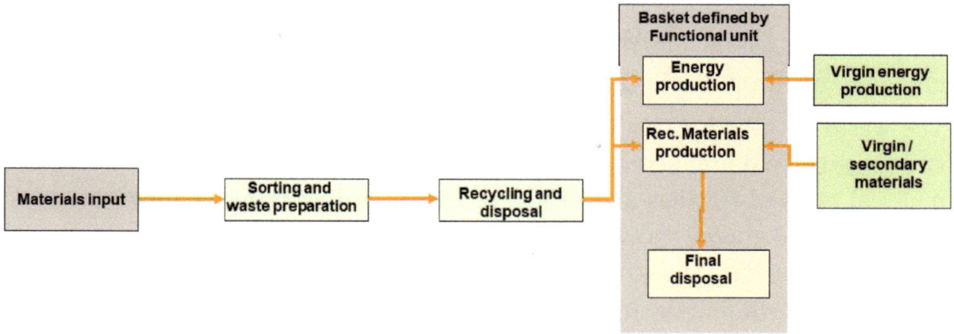

Figure 6.7 General approach of system boundaries to cover all relevant functions of a large product system

6.1.5.3 ChemCycling™ of Plastic Waste from Packaging

For mixed and contaminated plastics, it may be impossible (or very inefficient) to recycle them mechanically. This means that they are most likely incinerated, and the resulting energy is used to generate steam or electricity. In this case, chemical recycling is the better option. Chemical recycling complements mechanical recycling and can be a more sustainable solution than incineration or landfill of plastic waste. It thus contributes to a circular economy and the responsible use of resources. In this context, BASF developed a process together with partners – the ChemCycling™ process – and analyzed it with an LCA study.

Three separate LCA studies were considered, to compare the different options and to generate information for decision-makers as well as for the public discussion. The alternative studies cover firstly the "waste perspective", which compared pyrolysis with the incineration of mixed plastic waste. In addition, a "product perspective" addressed the comparison of plastics based on pyrolysis oil and conventional plastics from primary fossil resources (naphtha). Thirdly, the "plastics quality perspective" compared the life cycle of 1 t of virgin plastics with three end-of-life options.

The LCA studies prepared by Sphera on behalf of BASF [50] for all these three cases compared different indictors of plastics produced with pyrolysis oil under a mass balance approach with conventional plastics made from fossil naphtha. The general approach analyzes the pyrolysis of plastic wastes as the thermal cracking of polymeric chains into a mixture of short-chain hydrocarbons, versus competing end-of-life options such as mechanical recycling and incineration. Moreover, it compares the production of chemically recycled plastics to the production of conventional plastics made from primary fossil resources (naphtha). For the LCA, real process data were used for the pyrolysis process, and common practices of waste treatment were taken as the basis for incineration and mechanical recycling. Several scenarios and situations were calculated and interpreted. System expansions were mainly used by integrating relevant and meaningful substitutes, linked with credits for their production or the respective avoided production.

The waste perspective compared pyrolysis with an incineration process. In this case, the efficient production of oil as a feedstock for the chemical industry with a material yield of 70% has almost no need for an external thermal energy supply due to internal energy recovery. The incineration generates electricity and steam, replacing electricity and steam of productions based on light fuel oil and natural gas. It was shown that the pyrolysis has a significantly lower carbon footprint than the incineration. The difference was 1 kg/kg of plastic waste (Figure 6.8).

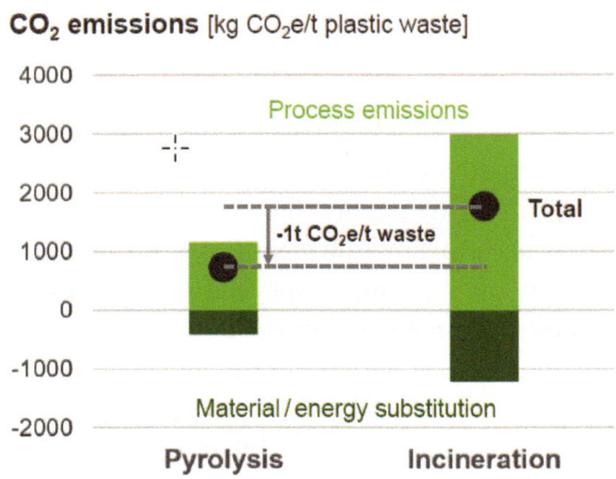

Figure 6.8 Pyrolysis of 1 t mixed plastic waste against incineration in a future scenario

In the second study it was investigated whether replacing virgin materials by plastic material based on waste pyrolysis oil results in lower CO_2 emissions. This perspective gives more insights into which direction the materials generated should be used. The LCA study showed that the production of a reference plastic (LDPE) from pyrolysis oil under a mass balance approach emits 2.3 t less CO_2 than 1 t LDPE produced from fossil naphtha, based on an upstream system expansion approach including the energy substitution of alternative waste incineration. The pyrolysis can offer (via the mass balance approach) a high-quality material with significantly lower CO_2 emissions. That material can be used for the same applications as virgin materials, and offers producers the potential to reduce their carbon footprint (Figure 6.9).

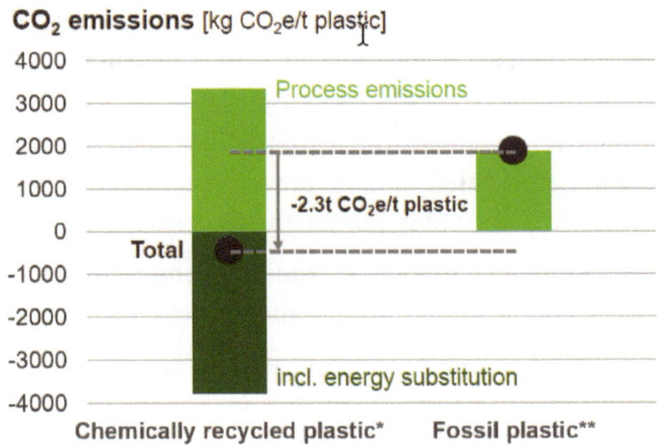

Figure 6.9 Production of 1 t of PE based on pyrolysis compared with the production of a virgin material based on fossil fuels

Overall, it was concluded that chemically recycled plastics result in significantly lower CO_2 emissions than those produced from primary fossil resources. The lower emissions result from avoiding the incineration of mixed plastic waste. Moreover, the LCA study found that manufacturing of plastics via either chemical recycling (pyrolysis) or mechanical recycling of mixed plastic waste results in similar CO_2 emissions. It was considered that the quality of chemically recycled products is the same as that for the virgin material, and that usually less input material needs to be removed during the sorting process than for mechanical recycling.

6.1.5.4 Eco-Efficiency Analysis

The comparison of EEA results is a method using LCA as the assessment basis for the environmental impacts and a life cycle costing approach covering the life cycle costs of a whole system. To derive messages and initiate improvements of product systems, the interpretation is a very important element.

The multi-criteria LCA approach delivers different types of information in an EEA. This makes it comprehensive but often not easy to interpret, even for specialists.

Therefore, to achieve better understanding of the outcome of an EEA, a normalization and weighting step is applied as an additional step in the LCA [51]. In this sense it is very helpful to aggregate single pieces of information to give an overall result. The requirements and recommendations in ISO 14044 should also apply for the interpretation of the EEA. In addition, the interpretation should consider the relationship between environmental results and product system value results. ISO 14045 describes how to accurately define and describe a product system by different stakeholders, who may encounter different values for the same product system. For instance, the product system value to the consumer may be different from the product system value to the producer, and in turn different to the investor.

Therefore, the interpretation should describe the stakeholder's value(s), the type of value(s), and the methods to determine the product system value(s) to be used in the assessment. The value(s) should be quantifiable with reference to the functional unit according to the goal and scope of the EEA. The types of product system values may be a functional value, a monetary value, or other values. The specific EEA method of BASF [52] was validated for the first time by TÜV Rheinland [53].

For the choice of eco-efficiency indicators, the following requirements of ISO 14045 should be considered and modeled to show higher eco-efficiencies:

- Increasing efficiency at the same product system represents improved environment
- Increasing efficiency at the same environmental impact represents improved product system value

Life cycle impact category indicator results, as determined according to ISO 14044, may be used for EEAs. Such data will typically result in an eco-efficiency profile, where several environmental aspects are considered in parallel. All single results should be reported separately with meaningful information. They should be displayed so that life cycle impacts are easily understandable, the main impacts can be detected, and improvement potentials can be identified as well.

The BASF methodology was used in several hundred cases [54, 55] and after each significant update of the method, it underwent an external validation process by NSF [56]. All major changes in the future will be followed by an updated validation process, which ensures a fully validated methodology and toolbox. Most of the impact assessment methods were adopted and harmonized with the environmental footprint methods of the EU [41].

The EEA has used an eco-efficiency portfolio for the display of aggregated results for 20 years [57]. The figure gives clear information on which alternative is (within the assessment) the most sustainable alternative, positioned in the upper right corner of the portfolio with the highest distance from the diagonal. Figure 6.10 shows the final and aggregated result of the previously described packaging system for mineral water. The environmental impact is the lowest for the 1.0 L PET refillable bottle, closely followed by the 1.5 L PET single-use bottle with deposit. In combination with the costs of the alternatives, these two alternatives be equal, because the single-use bottle is cheaper. Both systems have the same eco-efficiency. Using the person time concept, we also apply here a final aggregation to a single score, showing the result on the right-hand side of Figure 6.10.

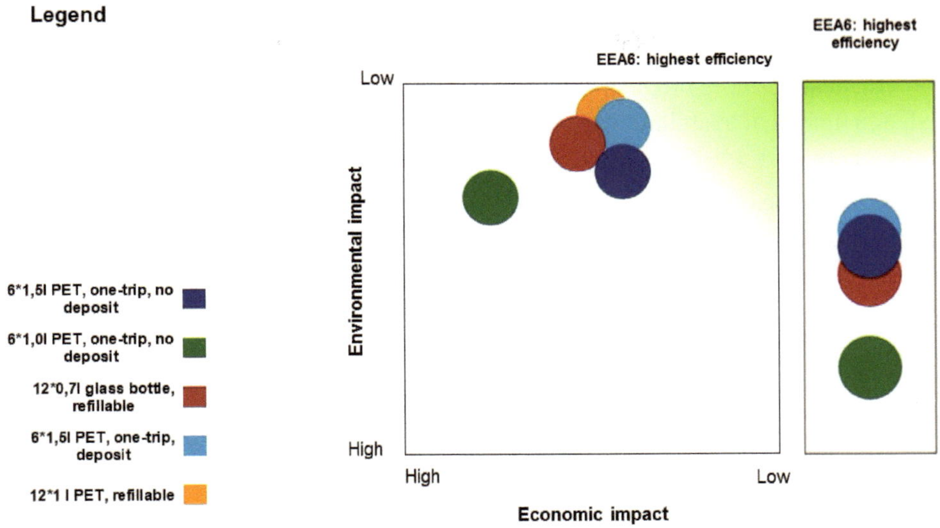

Figure 6.10 Eco-efficiency results of packaging alternatives for mineral water in the household market

Sensitivity analysis is a procedure to determine how changes in data and methodological choices affect the results of the EEA. A sensitivity analysis may provide additional information on data choice(s). In an EEA, several different methods for the determination of environmental and product system value indicators may be used. Therefore, a sensitivity analysis should be conducted to assess the consequences on the EEA results of different choices of methodology and data. An uncertainty analysis should be conducted to determine how uncertainties in data and assumptions affect the reliability of the results of the EEA.

6.1.6 Summary

Recycling of different types of materials will become much more important in the future. The circular economy is one key aspect of the green deal of the European Commission, but is also well implemented in a lot of strategies in countries and companies [58]. The new Action Plan announces initiatives throughout the entire life cycle of products, targeting (for example) their design, promoting circular economy processes, fostering sustainable consumption, and aiming to ensure that the resources used are kept in the EU economy for as long as possible. To make products fit for a climate-neutral, resource-efficient, and circular economy, to reduce waste, and to ensure that the performance of front-runners in sustainability progressively becomes the norm, the Commission will propose a sustainable product policy legislative initiative.

The core of this legislative initiative will be to widen the Ecodesign Directive beyond energy-related products to make the Ecodesign framework applicable to the broadest possible range of products and make it deliver on circularity. To establish a principle of sustainability, it is proposed to develop procedures on the following aspects:

- Improving product durability, reusability, upgradability, and reparability, addressing the presence of hazardous chemicals in products, and increasing their energy and resource efficiency
- Increasing recycled content in products, while ensuring their performance and safety
- Enabling remanufacturing and high-quality recycling
- Reducing carbon and environmental footprints
- Restricting single-use and countering premature obsolescence
- Introducing a ban on the destruction of unsold durable goods
- Incentivizing product-as-a-service or other models where producers keep the ownership of the product or the responsibility for its performance throughout its life cycle
- Mobilizing the potential of digitalization of product information, including solutions such as digital passports, tagging and watermarks
- Rewarding products based on their different sustainability performance, including by linking high performance levels to incentives

Priority will be given to addressing product groups identified in the context of the value chains featuring in this action plan, such as electronics, ICT, and textiles but also furniture and high-impact intermediary products such as steel, cement, and chemicals. Further product groups will be identified based on their environmental impact and circularity potential.

Certainly, plastics also have high potential for circular opportunities. Economic considerations have already established many different approaches to support a circular economy, with plastic materials having been recycled for decades now. However, to achieve more sustainable solutions, methods and processes need to be implemented that can measure the sustainability of products and processes supporting decision-making processes. LCA and sustainability assessment systems as EEA and SEEbalance have been used and many case studies have been prepared to generate meaningful information on life cycle approaches. These assessments will become even more important in the future, because many more different technologies, recycling business models, designs of materials for a circular economy, etc. will be developed and need clear information on sustainability aspects. Standards help to define conditions, and how these types of studies can be performed.

Recycling options can be analyzed and developed by focusing on their sustainability potentials. An additional opportunity to use this type of assessment is to generate specific tools that can generate a high number of scenarios in a short time frame, based on automatic systems that support practitioners very efficiently. Different types of such tools have been developed and successfully applied as well.

The engagement of all relevant stakeholders in supply chains linked to actors along the whole life cycle of products is based on the provision of meaningful information to the decision-making processes. Sustainability assessments will support this engagement in a very powerful manner, generating meaningful information to reach the goal of improved solutions with high positive contributions to sustainability.

References for Section 6.1

[1] Guinee, J.: Life cycle assessment: Past, present and future. International Symposium on Life Cycle Assessment & Construction, Nantes, 2012

[2] Saling, P., Alba-Perez, A., Kölsch, P., Grünenwald, T.: Generation, calculation and interpretation of social impacts with the social analysis of SEEbalance®, in: Traverso M., Petti L., Zamagni A. (eds): Perspectives on Social LCA. Springer Briefs in Environmental Science. Springer, Cham, 2020, *https://doi.org/10.1007/978-3-030-01508-4_8; ISBN 978-3-030-06564-5.*

[3] Steen, B.: Eco-efficiency of a wind farm: A case study. Environmental Systems Analysis CPM – Center for Environmental Assessment of Product and Material Systems, Chalmers, 2009, *http://www.cpm.chalmers.se/document/reports/09/2009_1%20EE%20of%20Borkum.pdf*, accessed July 9, 2012

[4] Carlson, R.: Eco-efficiency – The conceptual model, the concept model and an operational data structure, Chalmers, 2009, *http://lifecyclecenter.se/wordpress/wp-content/uploads/2011/01/2009_2.pdf*, accessed July 9, 2012

[5] Finkbeiner, M.: On the sustainability of eco-efficiency. LCA VIII – Seattle 2008. TU Berlin, Sustainable Engineering, 2008, *http://lcacenter.org/LCA8/presentations/ss-ecoefficiency-Finkbeiner.pdf*, accessed July 9, 2012

[6] Huppes, G., Mansanobu, I.: Eco-efficiency in industry and science, Quantified eco-efficiency: An introduction with applications, Vol. 22, Springer, London, 2007

[7] ISO 14040 (2006), Environmental management – Life cycle assessment – Principles and framework, Geneva

[8] ISO 14044 (2006), Environmental management – Life cycle assessment – Requirements and guidelines, Geneva

[9] ISO 14045 (2012), Environmental management – Eco-efficiency assessment of product systems – Principles, requirements and guidelines, Geneva

[10] Saling, P.: Eco-efficiency assessment, 2016, in: Finkbeiner, M. (eds): Special Types of Life Cycle Assessment. LCA Compendium – The Complete World of Life Cycle Assessment, 115–178; *http://doi.org/10.1007/978-94-017-7610-3_4*, Klöpffer, M., Curran, M. A. (series Eds), Springer, Dordrecht, 2016, ISBN 978-94-017-7608-0

[11] Bare, J. C., Hofstetter, P., Pennington, D. W., Udo de Haes, H. A.: Midpoints versus endpoints: The sacrifices and benefits. *International Journal of Life Cycle Assessment*, (2000) 319, *https://doi.org/10.1007/BF02978665.*

[12] Goedkoop, M., Heijungs, R., Huijbregts, M., De Schryver, A., Struijs, J., van Zelm, R.: ReCiPe 2008: A life cycle impact assessment method which comprises harmonized category indicators at the midpoint and at the endpoint level. Report I: Characterisation (1st edition), 2009

[13] Guinee, J. B., Gorree, M., Heijungs, R., Huppes, G., Kleijn, R., de Koning, A., van Oers, L., Wegener Sleeswijk, A., Suh, S., Udo de Haes, H. A., de Bruijn, H., van Duin, R., Huijbregts, M.: IIb: operational annex, in: Guinee, J. B. (Ed.): Handbook on life cycle assessment. Operational guide to the ISO standards, Kluwer Academic Publishers, Dordrecht, 2002, ISBN 1-4020-0228-9

[14] Itsubo, N., Inaba A.: A new LCIA method: LIME has been completed, *International Journal of Life Cycle Assessment*, (2003) 8, p. 305, *https://doi.org/10.1007/BF02978923*

[15] Bare, J.: Tool for the Reduction and Assessment of Chemical and Other Environmental Impacts (TRACI) v2.1. EPA/600/R-12/554, 2012

[16] Boulay, A., Bare, J., Benini, L., Berger, M., Klemmayer, I., Lathuilliere, M., Loubet, P., Manzardo, A., Margni, M., Núnez, M., Ridoutt, B., Worbe, S., Pfister, S.: Building consensus on a generic water scarcity indicator for LCA-based water footprint: preliminary results from WULCA, 2014, in: 9th International Conference LCA of Food, San Francisco USA, (2014) pp. 8–10

[17] Boulay, A., Bare, J., De Camillis, C., Doll, P., Gassert, F., Gerten, D., Humbert, S., Inaba, A., Itsubo, N., Lemoine, Y., Margni, M., Motoshita, M., Núnez, M., Pastor, A. V., Ridoutt, B., Schencker, U., Shirakawa, N., Vionnet, S., Worbe, S., Yoshikawa, S., Pfister, S.: Consensus building on the development of a stress-based indicator for LCA-based impact assessment of water consumption: outcome of the expert workshops, *International Journal of Life Cycle Assessment*, (2015) 20, pp. 577–583, *https://doi.org/10.1007/s11367-015-0869-8*

[18] Teixeira, R., de Souza, D., Curran, M., Anton, A., Michelsen, O., Mila i Canals, L.: Towards consensus on land use impacts on biodiversity in LCA: UNEP/SETAC Life Cycle Initiative preliminary recommendations based on expert contributions, *Journal of Cleaner Production*, (2015) 112, pp. 4283–4287, *https://doi.org/10.1016/j.jclepro.2015.07.118*

[19] Boulay, A.-M., Woods, J., Verones, F., Vazquez, I.: Marine impacts in life cycle assessment, Poster at SETAC conference, Helsinki, 2019, *https://doi.org/10.1016/j.jclepro.2020.120792content/uploads/2019/05/PosterSetac2019MARILCA.pdf*, accessed March 9, 2021

[20] Saling, P., Gyuzeleva, L., Wittstock, K., Wessolowski, V., Griesshammer, R.: Life cycle impact assessment of microplastics as one component of marine plastic debris, *International Journal of Life Cycle Assessment*, (2020) 25, pp. 2008–2026, *https://doi.org/10.1007/s11367-020-01802-z*

[21] Kalberlah, F., Schmincke, E., Saling, P., de Hults, Q.: Performance indicator for potential health impact analysis within LCA framework and for environmental product declaration (EPD), *International Journal of Life Cycle Assessment*, (2019) 24, pp. 181–190, *https://doi.org/10.1007/s11367-018-1513-1*

[22] Huppes, G., van Oers, L.: Evaluation of weighting methods for measuring the EU-27 overall environmental impact. *JRC Scientific and Technical Reports*, (2011), JRC - IES 2011. Report EUR 24985 EN

[23] EU Commission, Single market for green products initiative, 2020, *https://ec.europa.eu/environment/eussd/smgp/index.htm*, accessed March 18, 2021

[24] WBCSD (2014), Global chemical companies release guidance for environmental footprint of products, *https://www.wbcsd.org/Projects/Chemicals/News/Global-Chemical-Companies-Release-Guidance-for-Environmental-Footprint-of-Products*, accessed March 18, 2021

[25] EU commission (2018), Q+A for the PEF and OEF, *https://ec.europa.eu/environment/eussd/smgp/pdf/q_a.pdf*, accessed March 12, 2021

[26] Zampori, L., Pant, R.: Suggestions for updating the Product Environmental Footprint (PEF) method, *JRC Technical Reports*, (2019), European Union, ISBN 978-92-76-00654-1, ISSN 1831-9424, *https://doi.org/10.2760/424613*, *https://eplca.jrc.ec.europa.eu/permalink/PEF_method.pdf*, accessed March 9, 2021

[27] Wolf, M. A., Hofstra, U., Vroege, G. J., De Schryver, A., Zampori, L., Wolf, K.: Circular footprint formula webinar; Environmental footprint (EF) transition phase, 2019, *https://ec.europa.eu/environment/eussd/pdf/Webinar%20CFF%20Circular%20Footprint%20Formula_final-shown_8Oct2019.pdf*, accessed March 9, 2021

[28] Wolf, M. A., Partl, H., Wolf, K., Boyano-Larriba, A., Thellier, L.: The circular footprint formula (CFF) and its practical application, 2020, *https://ec.europa.eu/environment/eussd/smgp/pdf/Training CFF%20Circular%20Footprint%20Formula10Nov2020_final_corr.pdf*, accessed March 29, 2021

[29] Van Oers, L., de Koning, A., Guinee, J. B., Huppes, G.: Abiotic resource depletion in LCA. Road and Hydraulic Engineering Institute, Ministry of Transport and Water, Amsterdam, 2002

[30] Intergovernmental Panel on Climate Change – IPCC: IPCC Good practice guidance for land use, land-use change and forestry, Intergovernmental Panel on Climate Change, Hayama, 2003

[31] Van Zelm, R., Huijbregts, M. A. J., Den Hollander, H. A., Van Jaarsveld, H. A., Sauter, F. J., Struijs, J., Van Wijnen, H. J., Van de Meent, D.: European characterisation factors for human health damage of PM10 and ozone in life cycle impact assessment. *Atmospheric Environment*, (2008) 42, pp. 441–453

[32] Seppälä, J., Posch, M., Johansson, M., Hettelingh, J. P.: Country-dependent characterisation factors for acidification and terrestrial eutrophication based on accumulated exceedance as an impact category indicator, *International Journal of Life Cycle Assessment*, (2006) 11, pp. 403–416

[33] Posch, M., Seppälä, J., Hettelingh, J. P., Johansson, M., Margni, M., Jolliet, O.: The role of atmospheric dispersion models and ecosystem sensitivity in the determination of characterisation factors for acidifying and eutrophying emissions in LCIA, *International Journal of Life Cycle Assessment*, (2008) 13, pp. 477–486

[34] Struijs, J., Beusen, A., van Jaarsveld, H., Huijbregts, M. A. J.: Aquatic eutrophication, chapter 6, in: Goedkoop, M., Heijungs, R., Huijbregts, M. A. J., De Schryver, A., Struijs, J., Van Zelm, R.: ReCiPe 2008 – A life cycle impact assessment method which comprises harmonised category indicators at the midpoint and the endpoint level. Report I: Characterisation factors (1st edition), 2009

[35] Landsiedel, R., Saling, P.: Assessment of toxicological risks for life cycle assessment and eco-efficiency analysis, *International Journal of Life Cycle Assessment*, (2002) 7, pp. 261–268

[36] World Meteorological Organisation (WMO): Scientific assessment of ozone depletion: 2014, Global Ozone Research and Monitoring Project Report No. 55, Geneva, Switzerland, 2014

[37] UNEP: Global guidance for life cycle impact assessment indicators, vol 1, ISBN: 978-92-807-4143 3630-4, 2016, *http://www.lifecycleinitiative.org/life-cycle-impact-assessment-indicators-and-characterization-factors/*

[38] Fantke, P., Bijster, M., Guignard, C., Hauschild, M., Huijbregts, M., Jolliet, O., Kounina, A., Magaud, V., Margni, M., McKone, T. E., Posthuma, L., Rosenbaum, R. K., van de Meent, D., van Zelm, R.: USEtox® 2.0 Documentation (Version 1), 2017, *http://usetox.org*, *https://doi.org/10.11581/DTU:00000011*

[39] Beck, T., Bos, U., Wittstock, B., Baitz, M., Fischer, M., Sedlbauer, K.: LANCA land use indicator value calculation in life cycle assessment – Method report, Fraunhofer Institute for Building Physics, 2010

[40] Bos, U., Horn, R., Beck, T., Lindner, J. P., Fischer, M.: LANCA® – Characterisation factors for life cycle impact assessment, Version 2.0, 978-3-8396-0953-8, Fraunhofer Verlag, Stuttgart, 2016

[41] Grosse-Sommer, A. P., Grünenwald, T. H., Paczkowski, N. S., van Gelder, R. N. M. R, Saling, P. R.: Applied sustainability in industry: The BASF eco-efficiency toolbox, *Journal of Cleaner Production*, (2020) 258, 120792, *https://doi.org/10.1016/j.jclepro.2020.120792*

[42] Nakatani, J., Fujii, M., Moriguchi, Y., Hirao, M.: Life-cycle assessment of domestic and transboundary recycling of post-consumer PET bottles. *International Journal of Life Cycle Assessment*, (2010) 15, pp. 590–597, *https://doi.org/10.1007/s11367-010-0189-y*

[43] Periyasamy, A. P., Militky, J.: LCA (life cycle assessment) on recycled polyester, in S. Muthu (ed.), Environmental footprints of recycled polyester, textile science and clothing technology, 2020, *https://doi.org/10.1007/978-981-13-9578-9_1*

[44] Shen, L., Worrell, E., Patel, M.K.: Open-loop recycling: A LCA case study of PET bottle-to-fibre recycling, *Resources, Conservation and Recycling*, (2010) 55, pp. 34–52, *https://doi.org/10.1016/j.resconrec.2010.06.014*

[45] EU commission (2012), Waste from electrical and electronic equipment (WEEE), *https://ec.europa.eu/environment/topics/waste-and-recycling/waste-electrical-and-electronic-equipment-weee_en*, accessed February 25, 2021

[46] Fleischer, G., 1994, Methodik des Vergleichs von Verwertungs-/Entsorgungswegen im Rahmen der Ökobilanz, *Abfallwirtsch.*, 6 (1994) 10, pp. 697–701

[47] Heijungs, R.: Ten easy lessons for good communication of LCA, *International Journal of Life Cycle Assessment*, (2014) 19, pp. 473–476, DOI 10.1007/s11367-013-0662-5

[48] Schrijvers, D.L., Loubet, P., Sonnemann, G.: Developing a systematic framework for consistent allocation in LCA, *International Journal of Life Cycle Assessment*, (2016) 21, pp. 976–993, DOI 10.1007/s11367-016-1063-3

[49] Wäger, P.A., Hischier, R.: Life cycle assessment of post-consumer plastics production from waste electrical and electronic equipment (WEEE) treatment residues in a Central European plastics recycling plant, *Science of the Total Environment*, (2015) 529, pp. 158–167

[50] BASF: Project ChemCyclingTM Environmental Evaluation based on Life Cycle Assessment (LCA), 2020, *https://www.basf.com/global/de/who-we-are/sustainability/we-drive-sustainable-solutions/circular-economy/mass-balance-approach/chemcycling/lca-for-chemcycling/download-study.html*, accessed January 21, 2021

[51] Uhlman, B.W., Saling, P.: Measuring and communicating sustainability through eco-efficiency analysis. *Chemical Engineering Progress*, 106 (2010) 12, pp. 17–26

[52] Saling, P., Kicherer, A., Dittrich-Krämer, B., Wittlinger, R., Zombik, W., Schmidt, I., Schrott, W., Schmidt, S.: Eco-efficiency analysis by BASF – The method, *International Journal of Life Cycle Assessment*, (2002) 7, pp. 203–218, *https://doi.org/10.1007/BF02978875*

[53] TÜV Germany, ID-Nr. 5711150561: BASF Aktiengesellschaft, Ökoeffizienz-Analyse, Geprüfte Ökoeffizienz-Analysenmethode, 2002, *https://www.certipedia.com/quality_marks/5711150561/public_documents?locale=de&q[s]=document_type_name+asc*, accessed March 16, 2021

[54] Saling, P., Hofer, R.: Metrics for sustainability, in: Clark, J., Kraus, G. (Eds.): Sustainable solutions for modern economies, *RSC Green Chemistry Series*, (2009) 4, pp. 25–37, *https://doi.org/10.1039/9781847552686-00025*

[55] Uhlman, B.W., Saling, P.R.: The BASF eco-efficiency toolbox: Holistic evaluation of sustainable solutions, in: Abraham, M.A. (ed): Encyclopedia of sustainable technologies (1st edition), Elsevier, Amsterdam, 2017, pp. 131–144, *https://doi.org/10.1016/B978-0-12-409548-9.10042-9*

[56] Uhlman, B., Laginess, T., Green, D., Saling, P., Grosse-Sommer, A., Paczkowski, N., van Gelder, R.: NSF protocol 352, Validation and verification of eco-efficiency analysis, Part A, BASF's eco-efficiency analysis methodology, USA, 2019, *https://info.nsf.org/Certified/Sustain/ProdCert/C0023412-002.pdf*, accessed March 16, 2021

[57] Saling, P.: The BASF eco-efficiency analysis – A 20-year success story, BASF Verlag, Ludwigshafen, 2016, ISBN 978-3-00-054273-2

[58] EU commission: Circular economy action plan for a cleaner and more competitive Europe; COM/2020/98 final, 2020, *https://eur-lex.europa.eu/legal-content/EN/TXT/?uri=CELEX:52020DC0098*, accessed March 10, 2021

6.2 Life Cycle Assessment for Different Recycling Routes of Polystyrene

Regino Weber, Thomas Neumeyer

6.2.1 Background of the Study

A life without plastics would be unimaginable today, as they are used in all manner of applications. Thus, in terms of responsible use of resources and the environment, the circular economy will become much more important for polymer materials in the future. Plastics such as polystyrene (PS) currently mainly undergo carbon emitting end-of-life options like incineration or are sent to landfill. However, different technologies are under development to enable a circular economy for this type of polymer.

The overall plastics converters demand in Europe almost reached 51 Mt in 2019. Out of these, the packaging sector is by far the largest market, taking a share of nearly 40%. Resins from styrenic polymers like PS, but also expanded PS (EPS), are often used in applications like food packaging (dairy, fishery), building insulation, and in electrical and electronic equipment. The main segment for PS is packaging, as can be seen in Figure 6.11. The share of PS and EPS of all resin types is about 6.2% [1].

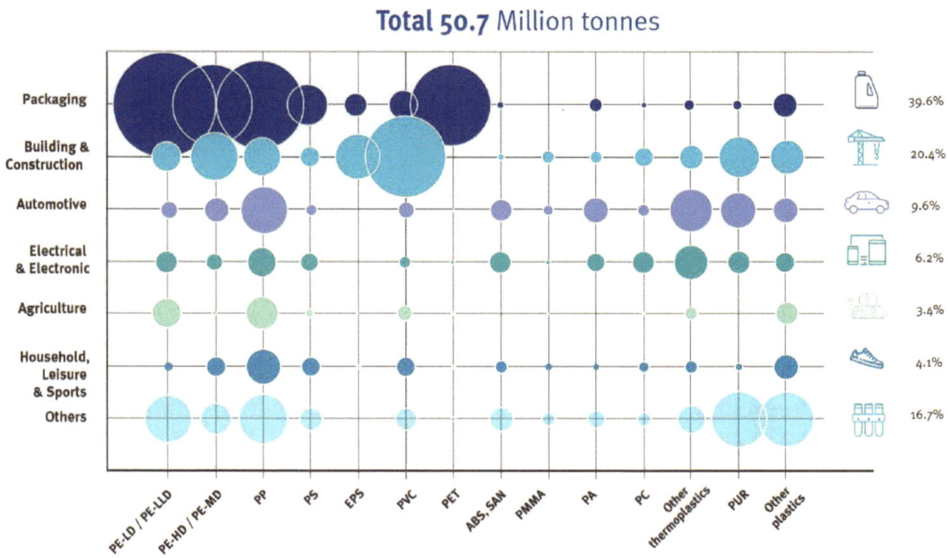

Figure 6.11 Plastics demand by segment and polymer type in 2019 [1]

Overall, the amount of plastic waste sent to recycling has nearly doubled since 2006, but 25% of plastic post-consumer waste was still sent to landfill in 2018. In order to reach net-zero emissions, the recycling rates in Europe need to increase further in the future. To meet this need, more recycling technologies need to be implemented, which also need to be environmentally viable, and efficient options for the treatment of plastic waste to close the loop. The very good sortability and recyclability make styrenic polymers an exceptional material for a circular economy [1].

This study focuses on the environmental impacts of three different recycling technologies for PS: high-purity mechanical recycling, depolymerization, and dissolution, which are all capable of converting PS waste into new granules for food-contact applications. The main objective of this study is to evaluate the environmental impacts of these end-of-life options and to compare them with widely used incineration. The effects of the replacement of fossil-based PS by recycled PS on the environmental impacts will also be explored.

6.2.2 Different Recycling Technologies for Polystyrene Waste

PS is very suitable for various recycling technologies. Styrenics show an excellent sortability from mixed waste, due to their specific signal easily recognizable by near-infrared (NIR), and have 'low diffusion' characteristics, enabling high suitability (for example) for food contact status. Furthermore, there are solvents available for specifically dissolving PS. PS can also be processed back to its monomer without additional steam cracking (as necessary for polyolefins). Thus, the following three technologies are currently used (albeit on different scales) to convert PS waste:

- Mechanical recycling
- Dissolution
- Depolymerization

Using **mechanical recycling**, impurities in the waste stream are sorted out and any organic residues are removed in a washing step. Subsequently, the feedstock is melt-filtered and regranulated to create PS of very high purity. Using high-purity mechanical recycling, the molecular structure of the polymer is not changed (see Chapter 8 and especially Section 8.4).

The **dissolution** process (see also Chapter 10) is another recycling possibility for PS, which does not degrade the polymer chains. Instead, the waste feedstock is treated by a solvent, which specifically dissolves PS. The resulting solution is filtered and PS is precipitated and granulated.

Another option is the **depolymerization** of the polymer material (for PS see Section 9.4), which converts post-consumer PS waste back into monomers. The monomer styrene can subsequently be re-polymerized into new products. During the depolymerization process other valuable substances like ethylbenzene, toluene, and α-methylstyrene are created, which can be used as a replacement for fossil-based chemicals.

All these technologies are contributing to close the material loop for products made out of styrenic polymers, which is important to meet set environmental guidelines. It has to be stated that these processes can still be considered as emerging technologies, thus there is still an immense potential for optimization and increase in efficiency to be expected in the future.

6.2.3 Goal and Scope of the Study

The study presented here was carried out in 2020 and 2021 in cooperation with Styrenics Circular Solutions (SCS), which is a joint value chain initiative with members acting all along the value chain for PS.

Application of the Study

This life cycle assessment was done to evaluate the environmental impacts of different recycling routes for PS as an end-of-life option, and should demonstrate whether there is an environmental benefit of these recycling technologies compared to incineration.

The environmental assessment of the above-mentioned recycling technologies is also intended to provide an overview of available recycling technologies for polystyrene based on comprehensive LCA data and results, which also provide a clear understanding of the environmental impacts of the different recycling possibilities and their opportunities to improve the sustainability of PS and enable a circular economy.

This study was performed according to ISO 14040:2006 and ISO 14044:2006, and focuses on three technologies:

- Technology 1: High-purity mechanical recycling
- Technology 2: Dissolution
- Technology 3: Depolymerization

Within the study it is considered that these technologies are not only end-of-life (EoL) options for PS, but are also capable of producing new feedstock material, which can replace fossil-based PS.

Functional Unit

The function of the product system is defined as the treatment of PS waste from municipal plastic waste. This includes the collection of the waste, initial sorting of the waste (waste composition is described later in this chapter), as well as different EoL-treatment technologies (mechanical recycling, dissolution, and depolymerization). As a reference scenario the incineration of the PS waste fraction is assessed.

Two separate perspectives were considered. One perspective comprises the treatment of waste containing PS objects, the other perspective evaluates the production of HIPS granules from recycling compared to HIPS based on a fossil feedstock (Figure 6.12). The PS objects within the municipal plastic packaging waste are defined as the baled output of an initial sorting step. PS objects contain 70% of actual PS polymer.

To account for that, there are also two separate functional units defined:

- Treatment of municipal plastic waste containing 1 t of PS objects
- Production of 1 t of HIPS granules

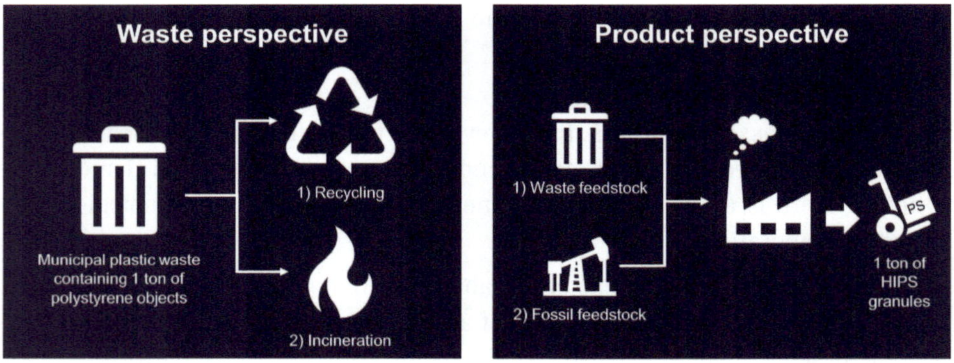

Figure 6.12 The two perspectives considered within the study. The waste perspective represents the treatment of municipal plastic waste containing 1 t of PS objects and the product perspective the production of 1 t of HIPS granules

Within the study, the avoided incineration and also the virgin material which can be replaced by recycling are considered. The emissions savings associated with diversion of waste from incineration has been considered within the waste perspective. In the product perspective the recycling technologies are compared to the production of fossil-based counterpart. Besides that, a combined functional unit is assessed: The treatment of municipal plastic waste containing 1 t of PS objects and substitution of the corresponding amount of fossil-based HIPS granules are considered. As every recycling technology has a different yield, the amount of substituted HIPS granules varies accordingly between the different technologies.

System Boundaries

This study follows a cradle-to-gate approach: it does not include the complete life cycle, but focuses on the processes involved in the recycling of the polystyrene, as can be seen in Figure 6.13.

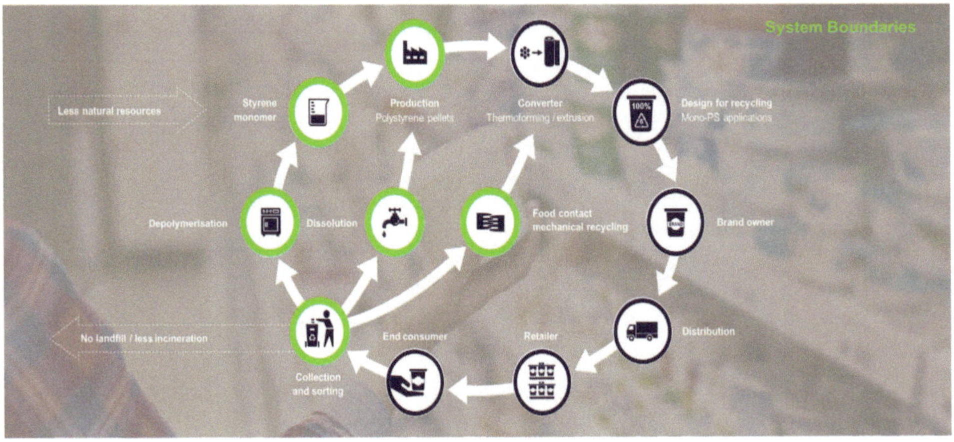

Figure 6.13 System boundaries of the study representing a cradle-to-gate approach starting from municipal plastic waste

The system boundaries (Figure 6.14 and Figure 6.15) include the collection and initial sorting of mixed plastic packaging waste, as well as the three different conversion methods to create recycled granules. The figures show all relevant processes and material flows, as well as the system expansion due to material and energy substitution. The impacts relating to construction of the factory, employee, and facility, as well as the infrastructure are not included, as they should not have a significant impact on the overall results, due to allocation to the high quantities of waste being converted.

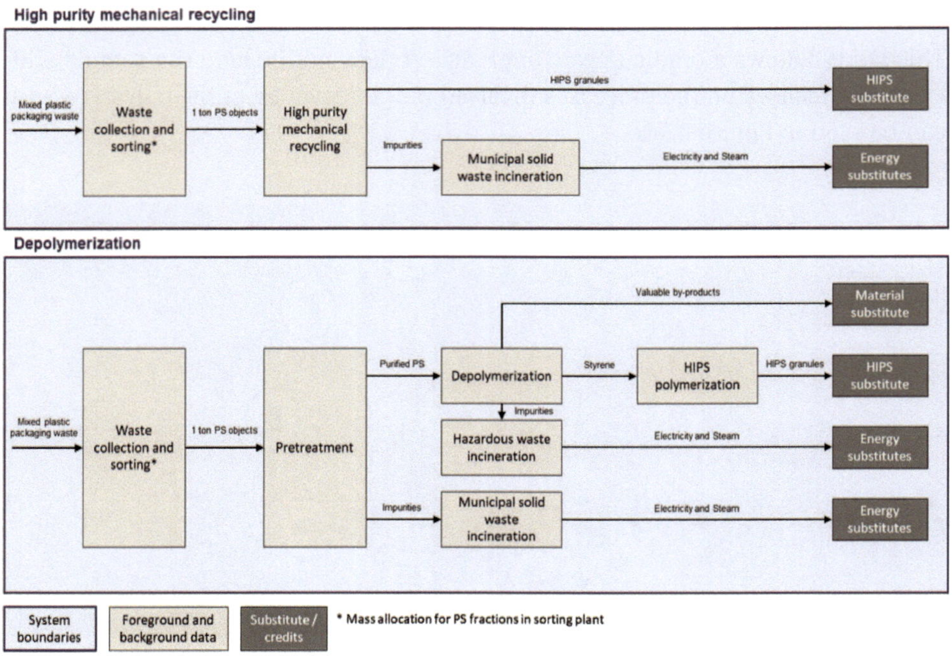

Figure 6.14 System boundaries of high-purity mechanical recycling and depolymerization

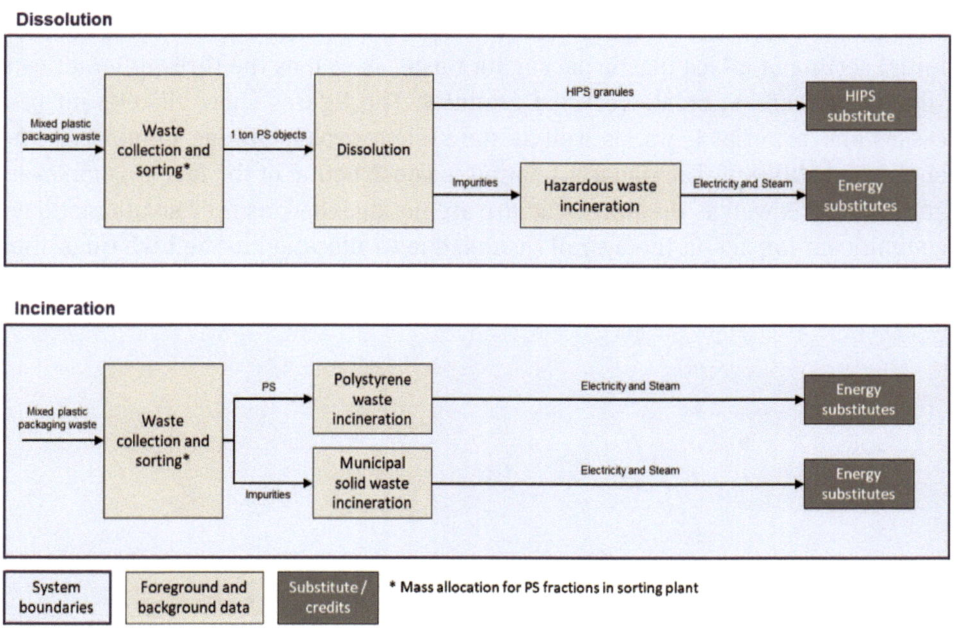

Figure 6.15 System boundaries of the dissolution and municipal waste incineration

In order to create comparable results for all the examined end-of-life options, the following criteria apply:

- Amount and type of waste: Starting point for all options is mixed plastic waste containing 1 t of PS objects
- System boundaries: Start with the collection of the waste at households and end with products or energy. Material or energy substitution is credited by system expansion.
- Applied data: Foreground and background data are used equivalently for all technologies.
- Material output: For all recycling technologies the recycled HIPS granules are defined to have a polybutadiene content of 4%.

All three technologies as well as the assessed reference scenario are described in Figure 6.16, as well as in Table 6.3.

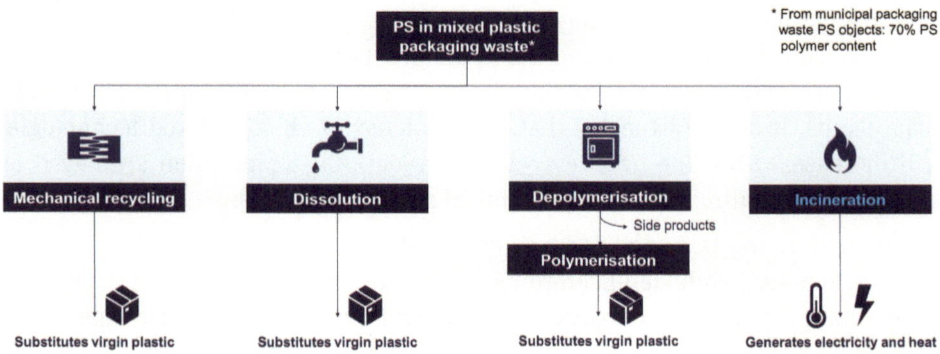

Figure 6.16 Overview of the assessed technologies (mechanical recycling, dissolution, depolymerization, and incineration)

Table 6.3 Technologies and Reference Scenario Assessed in the Study

Product system	Description
High-purity mechanical recycling	The defined feedstock is treated in a mechanical recycling plant, where it is sorted, washed, and regranulated. The resulting HIPS granules have food-contact quality and can be used as feedstock for the production of new PS products.
Depolymerization	The defined feedstock is treated in a depolymerization plant. An extra sorting step is necessary to meet the purity requirements for the depolymerization technology. The resulting PS oil is distilled to extract PS monomer, which is then re-polymerized. To create a comparable output material, a defined amount of polybutadiene is added to create HIPS. The resulting HIPS granules have food-contact quality and can be used as feedstock to produce new PS products.

Table 6.3 Technologies and Reference Scenario Assessed in the Study *(continued)*

Product system	Description
Dissolution	The defined feedstock is treated in a dissolution plant, where it is converted to HIPS granules. These granules have food-contact quality and can be used as a substitute for fossil based HIPS material.
Incineration	The defined feedstock is treated in a municipal solid waste incineration plant. The recovered energy (electricity and thermal energy) can be used as a substitute for the EU-28 electricity and thermal energy mix.

All input and outputs of the assessed recycling technologies are considered in this investigation. Thus, all required energy inputs, resources, electricity, and chemicals are included, as well as the disposal of all process waste. For all recycling technologies, primary data provided by commercial manufacturers has been used. For high-purity mechanical recycling, the data were provided by TOMRA (Asker, Norway). For depolymerization, data were provided by Agilyx (Tigard, USA), and for dissolution from Polystyvert (Montréal, Canada). Some of the provided process data are not based on European locations. In order to meet the defined goal and scope, all consumptions were modeled on datasets based on EU-28 data. Data on dissolution and depolymerization in particular represent early-stage, intermediate-scale plants. It can be expected that the efficiency of all considered technologies will increase within the next few years. The data for incineration are based on datasets representing the current technical circumstances.

All the underlying LCA models were done within the LCA-software GaBi 9, developed by Sphera (Leinfelden-Echterdingen, Germany). The background data to calculate the environmental impacts were sourced from the GaBi LCI database.

Definition of the Covered Products

The starting point for this study is mixed plastic waste collected from households, containing mainly PO, PS, and PET. In the EU a total amount of approx. 5000 kt of plastics were collected in separate collection in 2017. This amount contained about 200 kt of PS and thus a PS share of 4% was considered within the study. After the separate collection, the waste is transported to a sorting center, where it is sorted into bales, containing 700 kg PS polymer content per tonne (Figure 6.17).

Figure 6.17 Feedstock and initial processes converting it into presorted bales as a starting point for the recycling processes

Output HIPS Granules – Quality Factor

Considering the defined goal and scope of this study for all of the assessed technologies, the output material is HIPS granules with a polybutadiene content of 4%. The quality of the material also plays an important role in the results of this study. By definition, HIPS granules that have approval to be used for food-contact applications have equal quality compared to virgin material, and thus receive a quality factor of 1.

Data Collection and Quality

In order to create representative, complete, consistent, and precise results, high-quality data are used in the study. In this regard the primary data used for the study are sourced from measured data, calculated data, and literature values. The secondary data are based on industry averages. With regards to completeness, all relevant data for each process are included.

In terms of time coverage, all primary process data represent the current status of processes in 2020. Also for the background data, the newest available datasets were used, which represent the current state. As geographical reference the European countries were chosen and therefore, for all datasets, the EU-28 mix was used. The technology reference covers current end-of-life options for PS waste in mixed plastic packaging waste, by waste incineration, mechanical recycling, dissolution, and depolymerization.

Further technologies covered within this study are the sorting of mixed plastic waste containing PS, to generate bales containing 70% of PS objects, as an input for the different recycling technologies.

Allocation

As stated in ISO 14044 (Section 4.3.4.2), the most suitable allocation rule for the corresponding processes needs to be applied and documented.

For the first sorting of mixed plastic waste a mass allocation for the PS fraction has been applied. As approximately 4% of the considered waste is PS, 4% of the environmental impacts of this sorting step are allocated to the PS fraction.

Regarding the end-of-life, the following allocation criteria used energy recovery and substitution (the "avoided burden approach"). Accordingly, for all waste streams sent to a waste incineration plant, datasets were used that account for the electricity and thermal energy created during the incineration process. These are credited as a replacement for:

- Recovered electricity from incineration, as a substitute for electricity from the average EU-28 grid mix
- Recovered thermal energy from incineration, as a substitute for thermal energy generated in the EU-28

Impact Assessment and Impact Categories

In this study, the life cycle impact assessment (LCIA) was performed based on the ReCiPe 2016, v.1.1 [2] - Hierarchical (H) midpoint indicators.

6.2.4 Data for Life Cycle Inventory (LCI)

The life cycle inventory is mainly based on primary data provided by companies that have implemented the different technologies, and consists of mass balances and consumption values of the different processes. Wherever no primary data were available, secondary data were collected from various literature sources.

The inventory data are further detailed in the following, split into different stages, starting with the collection of the mixed plastic waste and the initial sorting, followed by the different recycling technologies and energy recovery.

Mixed Plastic Waste Collection

The starting point of this study is mixed plastic waste, which is collected from households as part of the "yellow bag" system in Belgium. These yellow bags are collected and transported to a sorting plant. The collection vehicle is powered by diesel fuel. The amount of diesel for the collection of 1 t of waste is given in Table 6.4. For the background system, a GaBi dataset for diesel fuel is used.

The transport distances to the sorting plant, as well as later on to the recycling plant, were estimated by experts to be 200 km each. Secondary data from GaBi were used to model the transport processes. Datasets for transport vehicles and fuels were implemented, which are representative for the transport processes. A Euro 6 diesel-powered truck-trailer was used for all transport processes within the model.

Table 6.4 Inputs and Outputs of Mixed Plastic Waste Collection

Type	Flow	Value	Source
Inputs	Diesel	3.24 kg/t	Based on literature data from [3]
	Mixed plastic waste	1 t	Allocation
Outputs	Mixed plastic waste	1 t	Allocation

Mixed Plastic Waste Sorting

The mixed plastic waste is then sorted into different fractions. The electricity needed for this sorting process is based on literature values, which are shown in Table 6.5. The fraction of interest for this study is the polystyrene fraction. Due to the used mass allocation, the consumptions are allocated accordingly. The output of this initial sorting step are baled PS objects, which contain 70% of actual PS polymer. These bales are transported to the corresponding recycling plants.

Table 6.5 Energy Use and Mass Balance of Mixed Plastic Waste Sorting

Type	Flow	Value	Source
Inputs	Electricity	173 MJ/t	Based on an average of literature data [4, 5]
Outputs	Mixed plastic waste	1 t	Allocation
	Baled PS objects	1 t	Allocation

High-Purity Mechanical Recycling

High-purity mechanical recycling consists of a sequence of processes to sort out as much of undesirables as possible before degranulating the material. In total, five subprocesses are needed to treat the baled PS objects: deep sorting, hot washing, flake sorting, super-cleaning, and extrusion. The resulting material has a purity of minimum 99.9% PS and can also be used for food-contact applications. Due to the food-contact approval, a quality factor of 1 is assumed for the study.

The total electricity, gas, and water consumption, as well as the material inputs and outputs, are based on calculated data. Secondary data from the GaBi database were used for the fuel and energy inputs, which are suitable for the goal and scope of the study.

The impurities removed as part of the above-mentioned processes are transported to a municipal solid waste incineration plant, where they are incinerated, with recovery of the generated heat and electricity. Within the model, secondary data from GaBi were used for the incineration process.

Depolymerization

As the purity of the baled PS objects is not high enough to directly enter the depolymerization process, an additional sorting and cleaning step is needed, consisting of removal of non-plastic materials, as well as a washing process. The corresponding electricity and water consumption are based on calculated data.

Depolymerization consists of many different processes to split up the polymer chains to produce recycled styrene monomer (see Section 9.4). Some of the byproducts of this process are valuable substances, which can replace feedstock chemicals. A valorization of 50% of the byproducts was estimated by industry experts. The byproducts ethylbenzene, toluene, and benzene, as well as chars are considered using a system expansion approach, as they can replace virgin products. They have been credited accordingly. For these credits, secondary data from the GaBi database were used. The steam for the purification of the styrene monomer and other valuable substances was considered to be burden-free based on the assumption that the recycling plant will be part of an integrated site and thus steam will be available.

Primary data (electricity, gas, and water consumptions, as well as the material inputs and outputs) are calculated on basis of a commercial plant. Secondary data from the GaBi database were used for the fuel and energy inputs, which are suitable for the goal and scope of the study.

After the depolymerization, a polymerization step is necessary to produce HIPS granules again. The feedstock as well as further required materials and energy is based on literature values [6]. Due to the fact that the recycled HIPS is polymerized from styrene monomer, the produced granules are virgin-like and thus also suitable for food contact. Therefore, a quality factor of 1 was considered for the depolymerization process.

The impurities of this process are transported to a municipal solid waste incineration plant, where they are incinerated, with recovery of the generated heat and electricity. Within the model, secondary data from GaBi were used for the incineration process.

Dissolution

The dissolution consists of distinct processes to dissolve, filter, and precipitate the polymer prior to regranulation. Solvent and antisolvent are recycled by distillation and are reused for the process over and over again. Depending on the purity of the feedstock entering the process, the amount of solvent and antisolvent wasted differs, but overall, very small quantities are lost. These do not have a significant impact on the overall results.

Primary data have been calculated on the basis of a technical-scale plant. Secondary data from the GaBi database were used for the fuel and energy inputs, which are suitable within the goal and scope of the study.

After the dissolution process the PS granules are also suitable for food contact. Based on this fact, a quality factor of 1 was considered for the dissolution process [7].

The impurities resulting from this process are transported to a municipal solid waste incineration plant, where they are incinerated, with recovery of the generated heat and electricity. Within the model, secondary data from GaBi were used for the incineration process.

Background Processes

For all described GaBi datasets, the corresponding documentations can be found online on the GaBi/Sphera website.

For the purpose of the study and with regards to the set goal and scope, GaBi datasets for the European average (EU-28) electricity grid mix, for thermal energy from natural gas as well as for natural gas, were used. The corresponding information about the used datasets can be found in Table 6.6.

Table 6.6 Energy, Key Material and Process Datasets Used Within the Study

Location	Dataset	Provider	Reference Year
EU-28	Electricity grid mix	Sphera	2016
EU-28	Thermal energy from natural gas	Sphera	2016
EU-28	Natural gas	Sphera	2016
EU-28	Diesel	Sphera	2016
EU-28	Waste incineration (commercial waste)	Sphera	2019
EU-28	Waste incineration (hazardous waste)	Sphera	2019
RER	Polybutadiene	Sphera	2002
EU-28	Ethylbenzene	Sphera	2005
EU-28	High-impact polystyrene	Sphera	2010
GLO	Truck-trailer, Euro 6	Sphera	2019

6.2.5 Life Cycle Impact Assessment (LCIA) Results

The results of the life cycle impact assessment are presented below for the three recycling technologies. For each of them, the global warming potential is discussed from three perspectives – the waste perspective, the product perspective, and a combined perspective. The waste perspective represents a functional unit "treatment of municipal plastic waste containing 1 t of PS objects" whereas the product perspective represents the functional unit "production of 1 t of HIPS granules". The combined functional unit includes the conversion of 1 t of PS objects as well as the avoided production of the same amount of fossil-based PS (as recycled PS is produced by the respective technology at an input of 1 t of PS objects). For each technology four impact categories have been considered (fossil depletion, terrestrial acidification, freshwater eutrophication, and freshwater consumption).

High-Purity Mechanical Recycling

In the case of high-purity mechanical recycling, the results of the climate change impact for the three above-mentioned perspectives are shown in Figure 6.18. The figure clearly indicates that the climate change impact of the recycling process is significantly lower for each perspective. The result for the treatment of waste with 608 kg CO_2-eq is about 65% lower than the incineration option. The production of 1 t of recycled HIPS granules emits 918 kg CO_2-eq, and thus more than 60% below the production of virgin HIPS from fossil resources. Taking into account the combined functional unit, the benefit of recycling and thus avoiding production of fossil-based material becomes even more evident, as the recycling route emits over 80% less.

The overall value for mechanical recycling can be further divided into collection, initial sorting, and the recycling process itself. The main contributor is the high-purity mechanical recycling process, contributing 568 kg CO_2-eq to the overall number of 608 kg CO_2-eq. Within the recycling process the driving force to climate change is the incineration of residuals, which are separated during the process, taking a share of more than 50% of the CO_2-eq emissions.

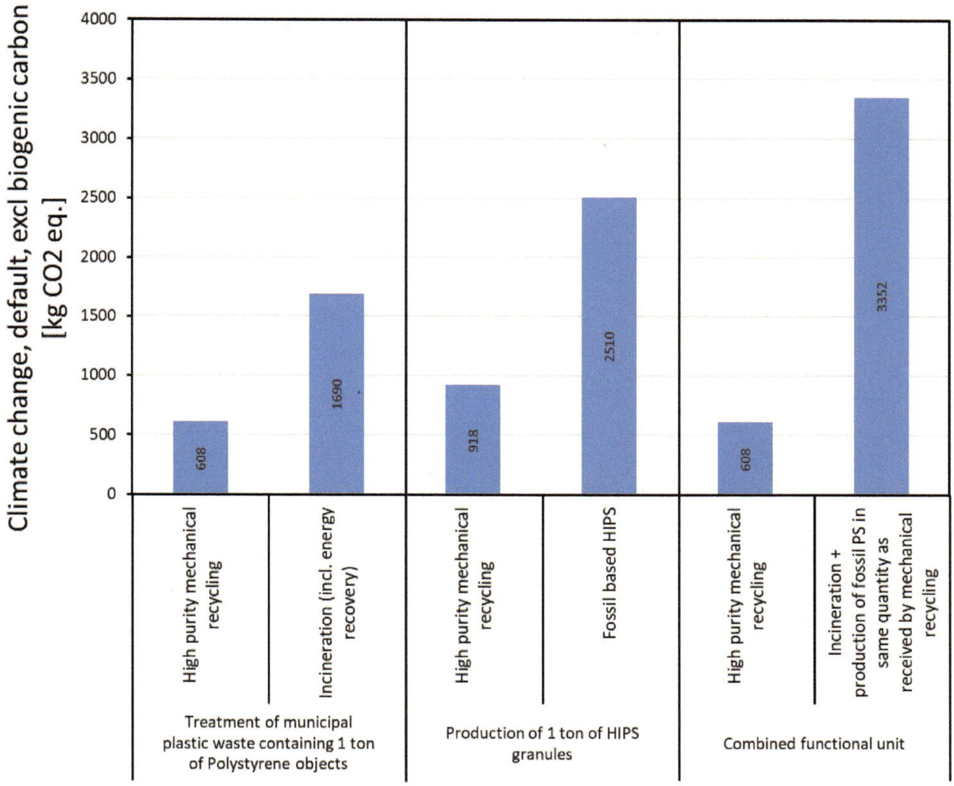

Figure 6.18 Climate change impact of high-purity mechanical recycling of PS in comparison with incineration for three functional units

As shown in Figure 6.19, high-purity mechanical recycling also has lower impact in other impact categories. With regard to fossil depletion, it is a significantly better option, as, with 86 kg oil-eq, it depletes about 90% less than the incineration and virgin material production. Also with regard to terrestrial acidification and freshwater consumption, the recycling process is preferred. However, with respect to freshwater eutrophication, the mechanical recycling performs slightly worse. This is mainly due to relatively high credits for incineration (for recovering energy).

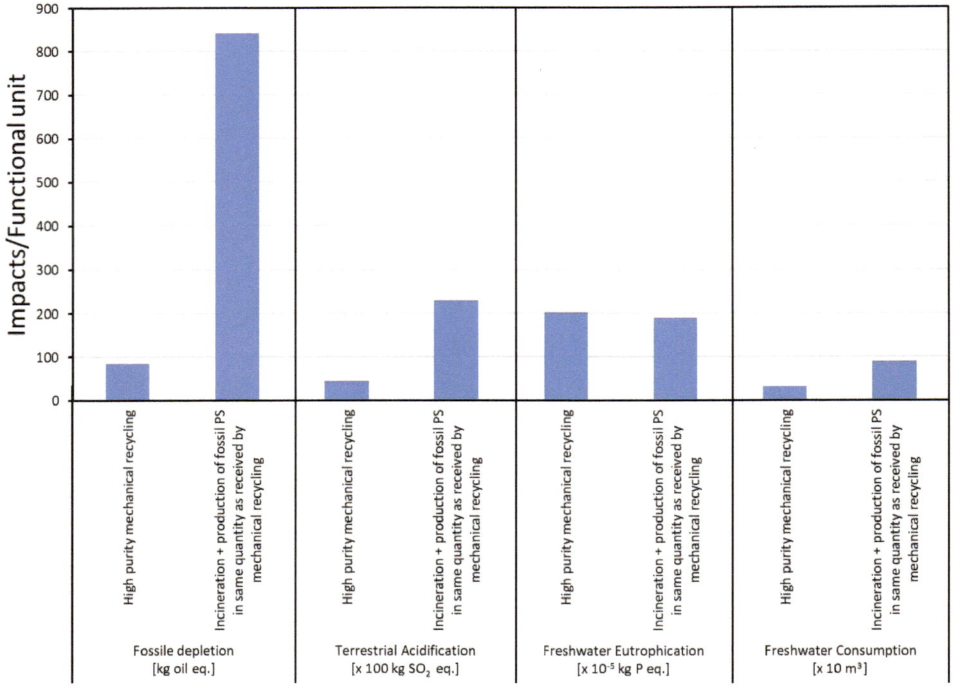

Figure 6.19 Environmental impacts of high-purity mechanical recycling in comparison with material substitution and avoided incineration. Divide by the factor shown to obtain the original values

Depolymerization

In this section, the impacts of the depolymerization process are compared to incineration and virgin material production. Results shown in Figure 6.20 reveal a lower global warming potential for all assessed perspectives. The treatment of municipal waste containing 1 t of PS objects has a significantly lower climate change impact compared to the incineration of the same waste fraction: 658 vs 1690 kg CO_2-eq per tonne of PS objects. Furthermore, 1 t of granules generated via depolymerization and repolymerization emit 1620 kgCO_2-eq and thus about 35% less compared to fossil-based HIPS. Combining the depolymerization process with the avoided incineration and the replaceable virgin material results in 70% reduction in CO_2-eq. As in the combined scenario, the yield of the recycling processes is considered. Thus, only the emissions from the production of fossil PS in the specific depolymerization yield is added to the impact of incineration. The emissions of the reference combined scenario is therefore different for the three recycling technologies.

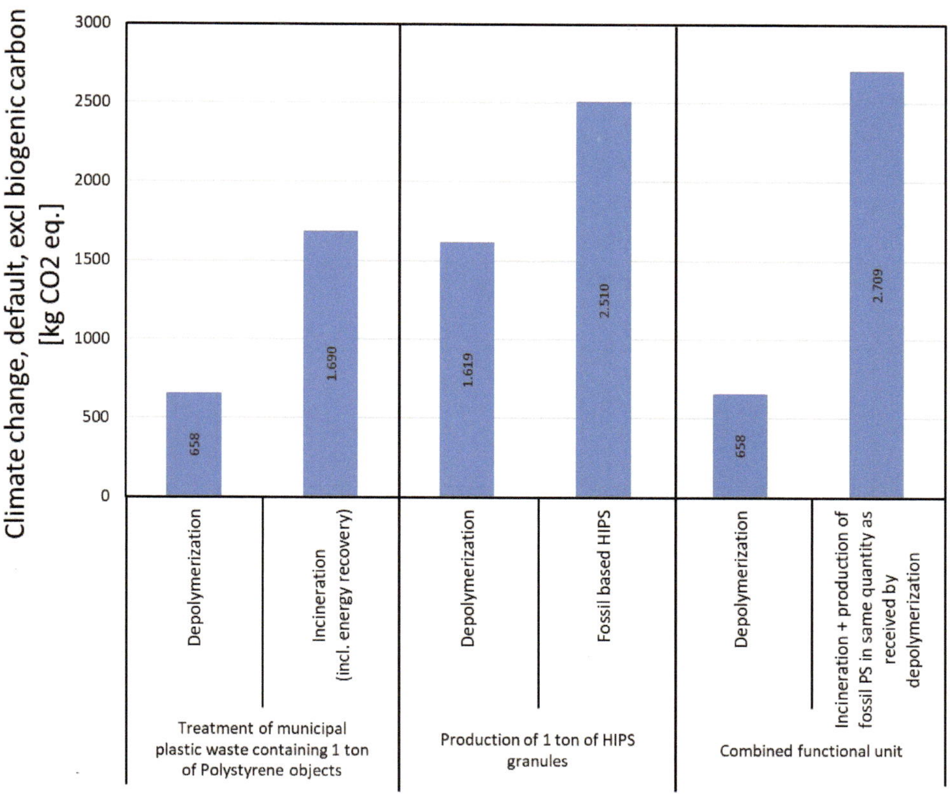

Figure 6.20 Climate change impact of depolymerization of polystyrene in comparison with incineration for three functional units

The overall value for depolymerization can be further divided into collection and initial sorting, extra sorting, depolymerization, and re-polymerization. The main contributors are the pretreatment process and the depolymerization process, adding 343 kg CO_2-eq and 167 kg CO_2-eq, respectively, to the overall number of 658 kg CO_2-eq. Within the pretreatment step, the driving force to climate change is the incineration of the residues separated during the process, taking a share of more than 50%. By refinement and advancement of the technology, the yields can be further enhanced and thus residues as well as the corresponding emissions can be lowered. In the depolymerization process, the valorization of some of the byproducts like ethylbenzene and toluene is credited, contributing positively to the overall result. The valorization results in a reduction of the climate change impact of about 130 kg CO_2-eq. The valorization of by-products is based on an average assumption. Depending on how the technology is integrated into a chemical site, this may vary.

Also with respect to other impact categories, the depolymerization technology performs significantly better compared to incineration and virgin material production, as can be seen in Figure 6.21. For all four categories, the recycling process shows less environmental impact. Looking for example at fossil depletion, the overall value of 85 kg oil-eq is approximately 75% below the reference scenario. Also much less water is consumed during the process. With regards to freshwater eutrophication, the assessment results in a negative value for depolymerization, due to the credits for valuable substances.

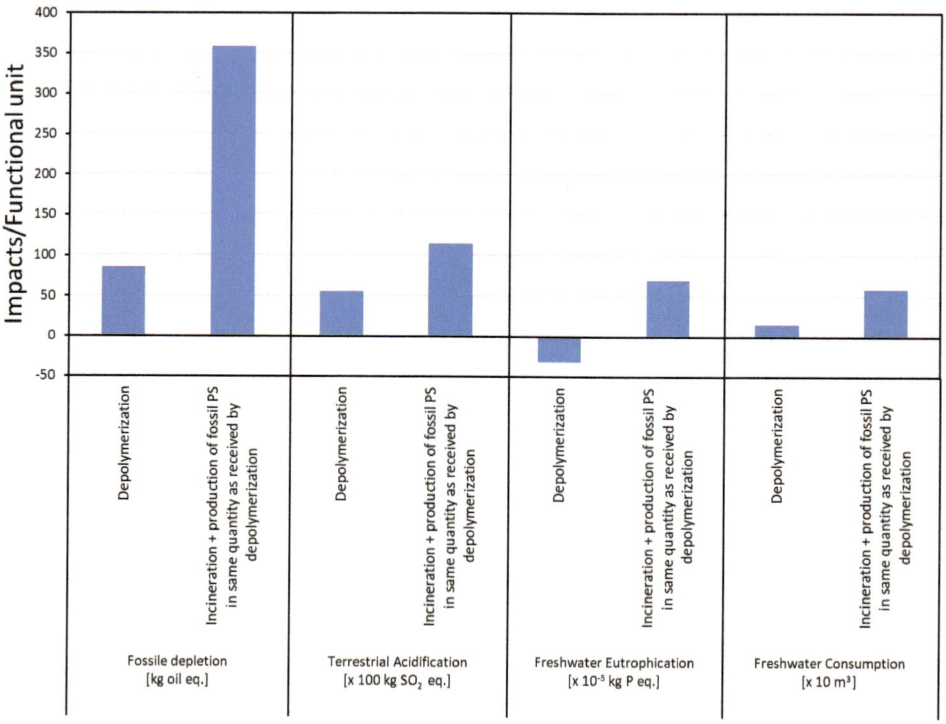

Figure 6.21 Environmental impacts of depolymerization in comparison with material substitution and avoided incineration. Divide by the factor shown to obtain the original values

Dissolution

In this case, the dissolution technology to convert municipal plastic waste into new HIPS granules is compared to the waste treatment using incineration and the production of polymer granules from a fossil feedstock. As shown in Figure 6.22, dissolution has beneficial effects on the global warming potential from all considered perspectives. The reduction for both waste perspective and product perspective for climate change impact is about 50%. With regards to the combined functional unit (treatment of waste containing 1 t of PS objects/production of HIPS granules) dissolution has a much lower overall impact: 888 vs 3394 kg CO_2-eq.

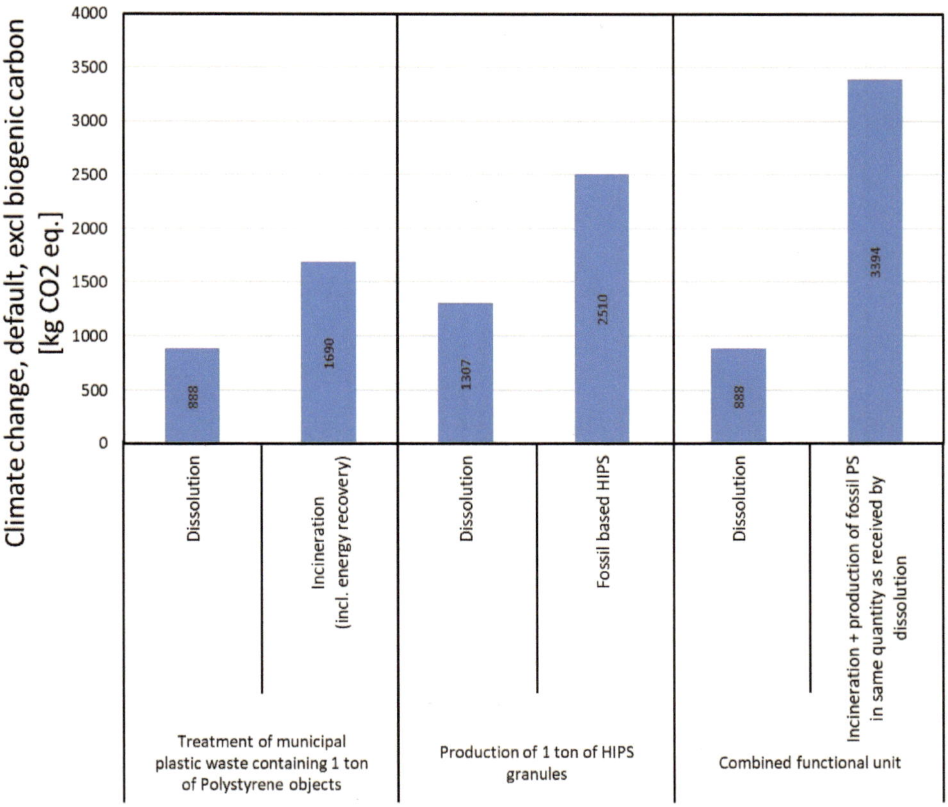

Figure 6.22 Climate change impact of dissolution of PS in comparison with incineration for three functional units

The overall value for dissolution can be further divided into collection, initial sorting, and the dissolution process itself. The main contributor is the dissolution process, contributing 848 kg CO_2-eq to the overall number of 888 kg CO_2-eq. The main contributors to the climate change potential within the dissolution process are the required electricity and thermal energy for the process, with 444 and 195 kg CO_2-eq, respectively.

Figure 6.23 shows a better performance of dissolution compared to incineration and material substitution. With regards to fossil depletion, the result is about 70%, and for terrestrial acidification approximately 60% lower. For freshwater eutrophication and freshwater consumption, the reduction in the impacts are about 35% and 45%, respectively.

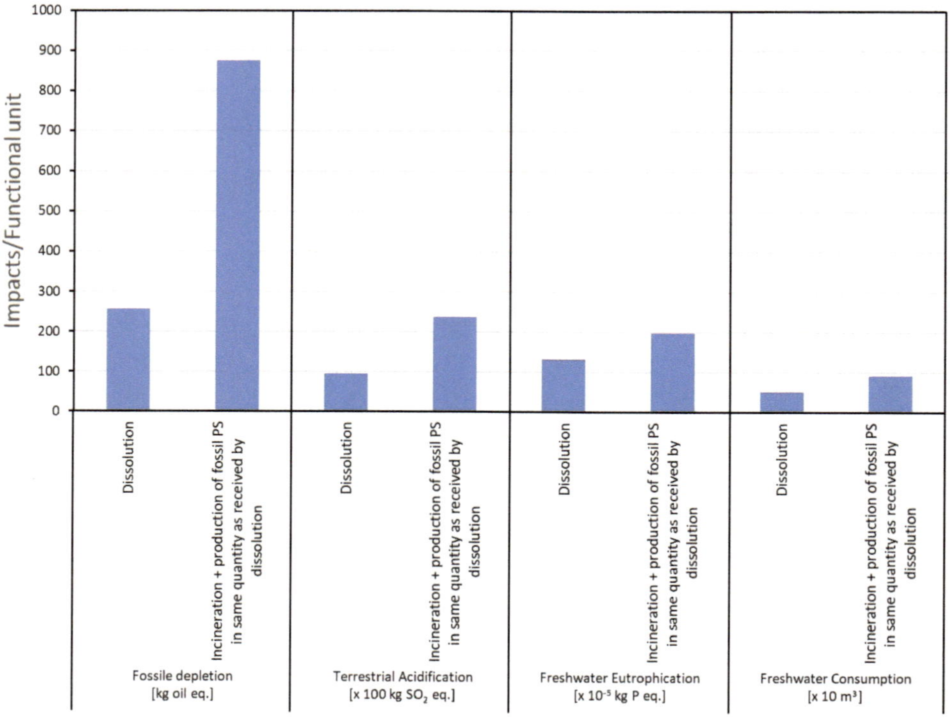

Figure 6.23 Environmental impacts of dissolution in comparison with material substitution and avoided incineration. Divide by the factor shown to obtain the original values

6.2.6 Analysis and Interpretation

Relevant Findings

The goal of the study was the evaluation of the environmental impact of different recycling technologies for PS in comparison to incineration as currently used alternative. Waste treatment was considered starting from the waste collection and ending with the generated product (material or energy). Thus, this study does not cover the whole life cycle, but relevant parts (cradle-to-gate approach).

All three assessed recycling technologies clearly show beneficial performance regarding the climate change impact, when compared with incineration. From the perspective of the functional unit "treatment of 1 t of municipal plastic packaging waste", high-purity mechanical recycling, depolymerization, and dissolution emit significantly less CO_2 compared to incineration. For other impact categories the incineration by itself partly results in lower values, as credits are allocated for the avoided burden of electricity and heat produced. However, recovery of product loss through waste incineration must be taken into account.

The functional unit "Production of 1 t of HIPS granules" shows quite interesting results, as all three recycling technologies have significantly better performance with regards to climate change. In this context, high-purity mechanical recycling achieved the best performance with respect to climate change impact, followed by dissolution and depolymerization. For many other impact categories too, the recycling technologies provide clearly superior performance compared to virgin material production.

When evaluating the defined dualistic functional unit, the significance of the comparison becomes even clearer. Comparing the recycling technologies with the emissions caused by the incineration of the same amount of waste, plus the emissions avoided by not generating HIPS granules from a fossil feedstock, all recycling technologies show a significantly better performance in climate change potential, fossil depletion, freshwater consumption, and terrestrial acidification. For freshwater eutrophication, high-purity mechanical recycling shows slightly worse results.

Since the feedstock has a PS polymer content of 70% after the initial sorting process, the remaining 30% cannot be converted into recycled HIPS granules and thus have to be extracted and treated. The treatment of these residues is done by incineration, which contributes a significant share of the overall emissions.

Assumptions and limitations

Within this study three different recycling technologies are assessed as well as incineration and fossil-based HIPS production as references. For the recycling options a conservative approach was used for data such as polymer content, yields, and treatment of residues.

A relevant limitation of the study, especially for the comparison of the recycling technologies with incineration and HIPS production, is the development state and scale. Processes such as incineration and HIPS production have existed for decades and thus are very well optimized. In comparison, all three recycling technologies (high-purity mechanical recycling, depolymerization, and dissolution) are quite young. Hence, the data available for this study are based on intermediate-scale technical plants and thus the processes are not yet fully optimized. Future developments of these processes will reduce the energy demand, and so will have a positive impact on the emission results.

The composition of the waste feedstock will have a significant impact on the overall results, as the treatment of the residues does have a major impact on the overall results for all the assessed technologies. The data used are based on a study representing an EU average, and should be quite representative of the current state.

Another limitation of this LCA is the fact that the data for dissolution and depolymerization are based on plants not located in Europe. All included energy and material consumptions were then added by using EU-28 datasets to evaluate a scenario corresponding to the defined goal and scope.

For some of the required material inputs, no datasets or other usable data were available. Therefore, proxies or datasets of chemicals with similar molecular structures were used in the models. As these materials are only used in very small quantities, the overall impact on the final results is quite small, and thus the results are still representative.

Reliability of the Results

In this study, both industry data as well as background LCA data from the GaBi database were used to generate reliable results. The datasets from GaBi are widely used for LCA studies and provide a data source meeting requirements of industrial and scientific applications worldwide. The quality of foreground data used within this study is considered to be good. All the relevant data were provided by owners of the corresponding technology in form of measured or calculated process information. All background data have been sourced from the GaBi database. Completeness of the provided data was cross-checked before using it for this study; no data were knowingly left out. Due to these facts, the data completeness can be considered as very good. All data were gathered with the same amount of detail, in order to ensure consistency in the collected data. Because much of the used data provided by the technology owners are confidential, the reproducibility is limited. This study represents the current state of the art, and is based on datasets representing the latest energy and fuel mixes. All primary and secondary data have been collected and/or adapted to the defined region (the EU-28). Wherever possible, region-specific datasets were used, but otherwise, proxies were applied. In this regard, the geographical representativeness is considered to be good. Technology representativeness is considered to be high, as the model is based on technology-specific primary and secondary data. All relevant process steps for each product system were considered and modelled to represent each specific technology. The process chain data are sufficiently complete and detailed with regards to the goal and scope of this study. All the assumptions, methods, and data are consistent and in-line with the study's goal and scope. System boundaries, allocation rules, and impact assessment methods have been applied consistently throughout the study.

6.2.7 Conclusion

The results of this study clearly show that all three assessed recycling technologies are very viable options for the treatment of PS within a municipal plastic packaging waste stream. With respect to climate change potential and fossil depletion, these processes have a significant advantage over incineration. It can be concluded that high-purity mechanical recycling, depolymerization, and dissolution are advantageous compared to currently implemented end-of-life treatments: not only are emissions from incineration reduced, but fossil-based materials can be replaced.

One of the biggest influences on the overall result for the incineration processes is the avoided burden of electricity and heat. The credits for electricity take a significant share, as they reduce the overall impact values significantly. This is mainly due to the implemented current grid mix. The increasing use of green electricity will therefore have a significant impact in the future, resulting in even bigger advantages for the recycling technologies, because their overall impact will decrease while the impact of incineration will increase (due to reduced credits).

Overall, this study provides an overview on environmental impacts caused by different end-of-life options for PS waste. As this study is a cradle-to-gate assessment, it does not provide impacts of specific products and application cases. Depending on the application and waste fractions, all three technologies are viable, but there are also complementary solutions. All three recycling technologies are able to deliver substantial environmental improvements compared to incineration.

References for Section 6.2

[1] PlasticsEurope, *Plastics – The Facts*, Brussels, 2020

[2] Huijbregts, M. A. J., et al.: ReCiPe 2016: A harmonized life cycle impact assessment method at midpoint and endpoint level, RIVM Report, 2016

[3] Boss, A.: Life cycle assessment of a gas-electric hybrid waste collection vehicle – Comparison with conventional waste collection vehicles, 2005, *http://cpmdatabase.cpm.chalmers.se/DataReferences/ESA_2005-7.pdf*

[4] Dehne, I., Oetjen-Dehne, R., Siegmund, N.: Stoffstromorientierte Lösungsansätze für eine hochwertige Verwertung von gemischten gewerblichen Siedlungsabfällen, 2015, *https://www.umweltbundesamt.de/sites/default/files/medien/378/publikationen/texte_18_2015_stoffstromorientierte_loesungsansaetze_fuer_eine_hochwertige_verwertung_0.pdf*

[5] Dehoust, G., Stahl, H., Gebhardt, P.: Klima- und Ressourcenbilanz für die Abfallentsorgung des Landkreises Kassel: Vergleich des Status Quo mit vier Entwicklungsszenarien, 2014, *https://docplayer.org/13415518-Klima-und-ressourcenbilanz-fuer-die-abfallentsorgung-des-landkreises.html*

[6] Nexant: Process Evaluation/Research Planning (PERP) report, San Francisco, 2016

[7] Mady, N.: Intertek 3rd Party Certification – Polystyvert's Recycling Process, Intertek, Boca Raton, 2020

6.3 Life Cycle Assessment of a New Monomer Recycling Process for PET

Mandy Paschetag, Hannes Schneider, Stephan Scholl

Background of the Study

In the EU, a total of more than 4.3 Mt of PET plastic is produced annually, of which approximately 96% is used for food packaging. From the latter only 45% of the packaging materials used are collected and sorted for recycling from a range of schemes across the EU-28. Various deposit return schemes account for up to 35% of collected, recycled and then reused PET, for plastics with contact to foodstuffs [1]. In Germany, for example, the recycled PET is primarily obtained from the deposit system of beverage bottles. Except for these bottles, the majority of the PET-containing packaging for foodstuffs, such as PET trays, is downcycled (for lower-grade reuse) or thermal utilization. Currently, several chemical recycling technologies are in the development pipeline, and may be a suitable alternative to recycle the complex composite materials for high-quality reuse.

These emerging end-of-life technologies are listed at the third level of measures according to the European waste prevention and management legislation and policy [2] and the German law of Circular Economy [3] – before other recycling (e. g. thermal recycling) on the fourth level and lastly waste disposal. The first level comprises the avoidance of waste, followed by the preparation of waste for its reuse. Currently, no established recycling technology is available to process contaminated or mixed PET plastic waste with inhomogeneous quality as well as multilayer plastic trays, e. g., made of PET and PE layers. The latter are often employed as packaging for e. g., fruit, vegetables, cheese, meat, or salad. For these complex waste streams, chemical recycling technologies seem to be a promising way to enable recycling for a high-grade reuse, e. g., back to the food sector. This would create a sustainable solution with increased resource efficiency. For this reason, economic feasibility as well as the environmental benefits of these innovative recycling technologies have to be investigated and proven [4].

In comparison to conventional state-of-the-art recycling technologies, such as the mechanical recycling of beverage bottles, chemical recycling – such as monomer recycling – is complementary to mechanical recycling, and could help to achieve an energy- and resource-efficient circular economy for PET plastics. For instance, currently the waste flows for mechanical recycling processes suffer from a gradual decrease in polymer chain length; they therefore cannot be re-processed indefinitely, and have to be thermally utilized after several cycles. As an extension of this recycling path, additional recycling options via monomer recycling or other chemical recycling processes can therefore contribute to a real circular economy for PET plastics.

One of these pipeline technologies is the revolPET® process, which depolymerizes PET into its monomers terephthalic acid (TA) and ethylene glycol (EG) by hydrolysis. The efficient separation of contaminants ensures high purity of the resulting monomers. The resulting secondary raw materials are of primary quality and can be used entirely for the production of new high-grade PET polymers. Thus, on the one hand a circular economy is facilitated and on the other the PET production via the fossil route can be reduced significantly. Consequently, a comparison of both the recycling and the fossil route, is implemented in this study to quantitatively assess the environmental footprint of the monomers obtained via the revolPET process vs the fossil route. To design a new technology to be as environmentally friendly as possible, the integration of environmental aspects in the process development is indispensable. The impact of such investigations has been shown (see, for example [5–10]). Compared to retrospective LCA, which assesses an existing process, the assessment during process development is more challenging. On the one hand there is a lack of available data during process development and on the other hand a thorough comprehension of the background data is necessary to predict the market developments up to the point where the technology reaches the final technology readiness level (TRL). Equivalent to the economic situation, the specific environmental expenditures per product unit strongly depend on the TRL and typically decrease with increasing technical maturity of the process. This is especially true for processes still on the lab scale or pilot plant development phase, where the investigation focus gradually shifts over time from feasibility to process and equipment optimization. Despite that, the integration of prospective LCAs in the development stage offers the possibility to integrate environmental considerations into the design phase, and supports decisions in regards to greater sustainability. The so-called ex-ante LCAs [11–13] are future-oriented assessments, trying to determine the environmental impact of emerging technologies on an industrial scale, with predicted background data.

For the life cycle analysis of the revolPET process, holistic investigations were performed to identify parameters with a strong influence on the environmental impact. Being well aware that the current TRL of the technology under development holds a lot of potential for improvement, different studies are necessary to detect hotspots and opportunities that must be considered in the ongoing process development and design. The working progress can be supported by detailed material flow net modeling to assess and analyze the operation of a single process step, apparatus, or unit operation, and the whole process as well as the site infrastructure.

6.3.1 The revolPET® Process

The revolPET technology keeps the complexity within the process as low as possible. It is based on selective depolymerization of the PET from composite materials or mixed fractions by means of alkaline hydrolysis of the PET macromolecule's ester bonds.

Figure 6.24 illustrates the revolPET process as a combination of its process tasks. In the first step, depolymerization, approximately 97% of the PET reacts with a metal hydroxide to form a salt of the terephthalic acid (rTA) and ethylene glycol (r-EG); the other, mostly polyolefin components (e. g., polyethylene, polypropylene) of the input material remain chemically unchanged.

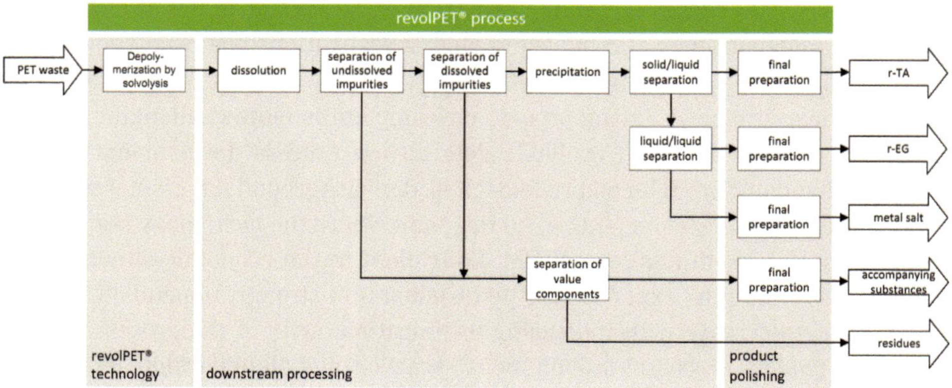

Figure 6.24 Block flow diagram of the revolPET process

The resulting mixture is then dissolved in pure water. All inert impurities, including possible byproducts that have not reacted with the metal hydroxide, are separated in a filtration cascade. The purified solution contains the metal salt of r-TA and r-EG. Through the addition of an acid, the TA is precipitated. Afterwards, the remaining r-EG monomer, as well as the metal salt, are extracted by a solid-liquid separation step. Both the r-EG and the metal salt are purified separately by product polishing steps after an additional liquid–liquid separation. Accompanying substances like polypropylene or polyethylene could be purified by additional downstream processing [14].

6.3.2 Goal and Scope of the Study

The presented study is performed based on the framework of DIN EN ISO 14040/44 [15, 16] and was carried out within the research project "Development of a recycling technology for waste plastic from multilayer and other waste composites –

revolPET" funded by the German Federal Ministry of Education and Research (BMBF) within the funding program "Plastic in Environment – Source, sinks, solutions".

Within the project framework, the revolPET process has been assessed with regard to quantifying environmental impacts as well as to support the design of a sustainable recycling process.

Application of the Study

The environmental assessment of the revolPET process was performed continuously, starting from the earliest stage and accompanying the process development and design. The results intend to:

- Provide valuable information for decision-makers in research and development to guide the process design in regards to an environmentally friendly direction
- Identify suitable operational boundary conditions for an environmentally friendly production setting

Furthermore, the results of the study are of interest for other stakeholders of the recycling and plastics value chains, e. g., PET producers, as well as for NGOs, policy makers, and society. The study and its results may support the development of policy tools to reinforce R&D incentives, as well as the integration of innovative technologies for the goal of a circular economy. The current status of the study offers initial insights into the environmental impacts of the investigated recycling process, and builds a first step towards a comparative study about the integration of the technology in the whole value chain in the context of a circular economy.

Functional Unit

For the assessment of recycling processes, a dual functional unit exists, with the perspective of a recycling company on the one hand and of a material supplier on the other hand. The former perspective considers the disposal of PET-containing waste as the main goal, while the latter considers the consumption of and expenditures for the monomers TA and EG to produce a valuable product. For the environmental assessment of a monomer recycling technology shown here, the supplier perspective is chosen. The reason is the low data availability for a depiction of the whole PET value chain due to many unanswered questions, such as the necessary purity of the waste flows for the different available technologies, or the ability to handle the different qualities of recyclates in the framework of a life cycle assessment. An approach to take the quality into account for the recycling of the plastic polylactide is shown in [17]. However, for PET plastics no similar consideration exists.

For this study, the production of 1 kg recycled terephthalic acid (r-TA) is used as the functional unit. The reference feed flow to the revolPET process is represented by a post-consumer fraction of PET-waste with a share of roughly 3% polyethylene (PE).

System Boundaries

This study is a process-LCA with a focus on the production phase of an emerging recycling process. The environmental assessment follows a cradle-to-gate approach. The system boundaries include expenditures related to the recycling process, except for the impact of service and maintenance and the plant and connected infrastructure itself (see Figure 6.25). Further processing of the produced monomers as well as additional life cycle phases in the circular value chain are not part of this study. In addition, the transports of the reactants, auxiliaries etc. are not considered due to the low data availability. In other investigations of chemical production processes, these impacts have shown a minor role within the ranges 2–4% (organic), 4–5% (inorganic) or 5–8% (pharmaceutic) [18–20] of the overall environmental impact.

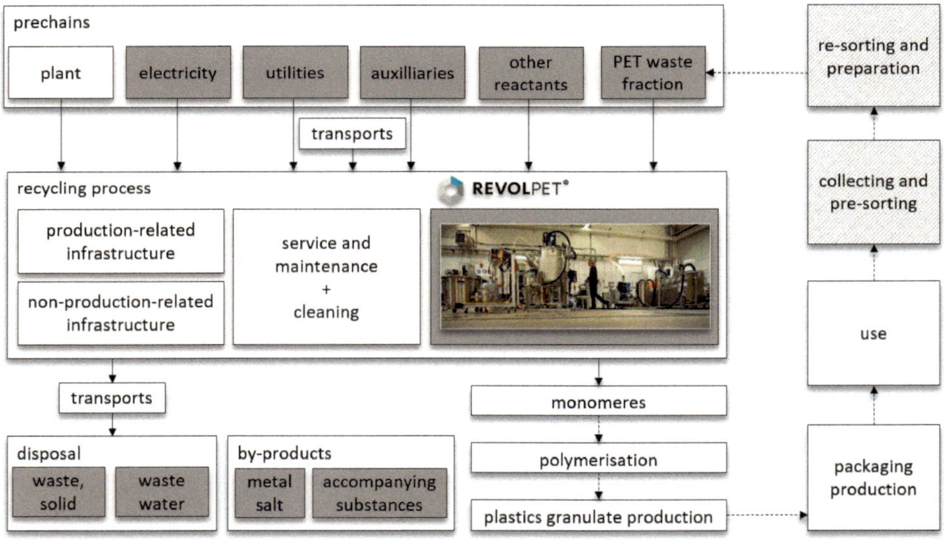

Figure 6.25 System boundaries of the revolPET LCA study: fields with gray background show the considered phases of the life cycle; life cycle phases in dashed fields are part of the generic dataset which represent the PET waste fraction [21]

Following [22], all inputs and outputs of the recycling process are considered in the investigation. This covers the expenditures caused by the supply of reactants (PET waste fractions and others), utilities, and auxiliaries, as well as for the electricity, process emissions, and the disposal of waste. For the life cycle analysis of the recycling technology, the same recording approach is applied as for production processes in the process industries.

The background data to calculate up to the elementary flows between ecosphere and technosphere are represented by generic datasets of the ecoinvent database v3.7.1 [23]. For this study no flows are cut off within the foreground data regardless of their share of the overall material and energy flows. Furthermore, no substitutions are issued and no credits are calculated for the recycled monomers or byproducts, to ensure a comprehensive process-related assessment of the new recycling process.

Data Collection and Quality

The inventory of the revolPET process is mainly based on primary data collected from experiments and measurements, complemented with calculations. For the calculation of the elementary flows for the life cycle inventory, generic datasets from the ecoinvent database with the system model APOS (allocation at the point of substitution) are implemented.

- *Time-related reference*: The revolPET primary process data were captured during 2020–2021. The incorporated generic datasets for the LCI calculation refer to a recording period of 2000–2017. In this context, it has to be further noted that the datasets of the ecoinvent database are often extrapolated from earlier recording periods to a period of 2015–2020 by adjustment of the uncertainties to validate them to the specified period.
- *Geographic-related reference*: The primary data were directly collected at the technical lab plant in Braunschweig, Germany. For the input electricity, a generic dataset representing the German national mix is used. The same applies for other inputs and outputs: if the German geographical reference region is available in the ecoinvent database, it was chosen as the most suitable alternative. Otherwise, datasets representing the European market are used.
- *Technology-related reference*: For the assessment of the revolPET-technology, the collected data are based on the consumption measurements at the plant with a "technology readiness level" (TRL) of 4 [24]. This level implies a working technology in a relevant production environment with a near-time perspective of a functional demonstrator. The technology of the fossil production route of TA (virgin TA, v-TA) is categorized at TRL 9, the highest TRL.

Multifunctionalities and Allocation

As many other recycling processes, the monomer recycling is a multi-output process (see Figure 6.26), with the three main value components (r-TA, r-EG, and a metal salt), as well as various accompanying substances depending on the processed waste fraction. In this first LCA study, an attributional approach was chosen to assess exclusively the environmental impact of the recycling process detached from a consideration of the complete PET value chain. This approach usually proceeds from the status quo and considers the material and energy flows related to the product system, here the revolPET process. The product system is considered isolated from the market and the environmental impacts are usually divided into the products and byproducts using allocations.

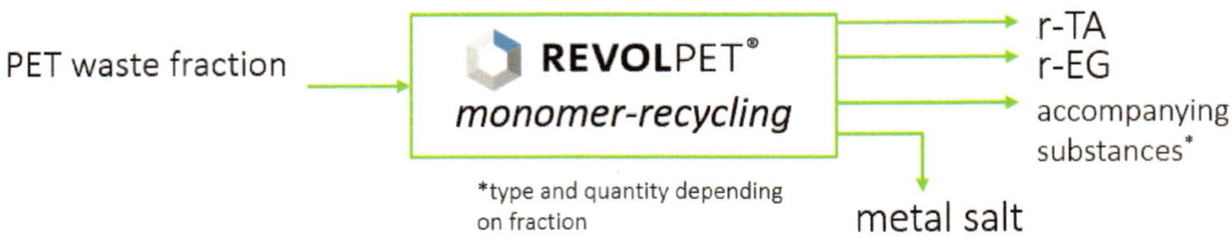

Figure 6.26 revolPET process as a multi-output process with the value components r-TA, r-EG, and a metal salt, as well as (depending on the waste fraction) accompanying substances as possible additional value components

Another approach is the consequential LCA that focuses on the consequences for the market caused by the implementation of an innovative and new technology or a decision taken. For such investigations, allocations are replaced by an extension of the system boundaries. For byproducts, the substitution of their production by other production routes is assumed, and therefore environmental credits are issued [23].

In this study, an allocation by mass is carried out at the points of separation in the recycling process (see Figure 6.24). Therefore, the inputs and outputs of the respective process steps "separation of undissolved impurities" as well as "solid/liquid separation" and "liquid/liquid separation" are allocated according to the mass balance of the single value components of these process steps.

Alternative allocation approaches (e.g., monetary based), are currently under investigation and may be implemented as a consolidated data basis becomes available.

Impact Assessment and Impact Categories

In this study, the life cycle impact assessment (LCIA) was performed using mainly the characterization models described in the ILCD handbook [25]. Those 16 categories were complemented by the ReCiPe 2016 [26] midpoint characterization model with a hierarchical perspective for the impact category "fossil depletion". The reason for this deviation is the more descriptive impact indicator "unit of oil equivalents" in comparison to the ILCD-recommended unspecific "Megajoule" for this category. The oil equivalents describe a reference of crude oil as feedstock in the ground with a caloric value of 42 MJ/kg [27].

Sensitivity Studies

As described above, the revolPET process is currently classified as being at technology readiness level (TRL) 4. The life cycle analyses carried out in this early stage accompany the process development and design, and are used for the prediction of potential environmental impacts. Through the implementation of sensitivity studies, the potentials of a technology can be analyzed as well as quantified – an approach typically used for ex-ante LCAs. During the early stage of development, the research focus typically rests on the proof of feasibility of the investigated process or technology. Being located in a technical lab, the revolPET process is not yet integrated into a functional production infrastructure. This means that typical synergies of an industrial-scale production or even an industrial park, such as heat integration or energy supply by a large steam network with an efficient large-scale generator, are not yet utilized. As such, potentials of a more efficient energy supply will be the first aspect to be investigated in the following studies. Secondly, the environmental impact will be estimated in regards to an increased technology readiness level based on the experimental data from the pilot plant combined with initial scale-up investigations. The third study combines the previous investigations and is then compared against the fossil-based production of virgin TA (v-TA).

Sensitivity study 1: "Operational infrastructure"

This study investigates the influence of the production site on the environmental impact of the process. Therefore, a change of the site infrastructure is assumed. Instead of the current technical lab, the revolPET process is operated within the infrastructure at a typical chemical industrial park. Environmental impact indicators for the heat and power generation from the industrial park Hoechst (Frankfurt/Main, Germany) were used to represent this change in location for the utility supply. The data were taken from a 2019 report of the industrial park [28].

In the technical lab, the required heat energy is supplied by electrical power from a local energy provider. In comparison, over 70% of the heat at the industrial park is provided by cogeneration and the remainder by waste heat from the production and the substitute fuel power plant. The electricity is supplied by a mix of cogeneration (52%), the substitute fuel power plant (16%) and bought-in (31%). In this sensitivity study, only the potential environmental expenditures for the impact category climate change could be determined because only CO_2-eq indicators were available. For heat the indicator is 0.283 kg CO_2eq/kWh and for electricity 0.293 kg CO_2-eq/kWh [28].

Sensitivity study 2: "Technology readiness level"

The second study addresses the technology readiness level of the technology. To investigate the changes in the environmental performance resulting from process improvement, the changes between TRL 4 and 7 are determined. The increase to TRL 7 represents the next scheduled development step and will be finalized with mini-plant scale equipment as a demonstration plant to produce customer samples. A mini-plant scale typically represents the operation of all process steps including closed recycle streams in a reproducible continuous operation, and is fully scalable.

In detail, the increase of the TRL along the development path is expected to lead to operational improvements like the recirculation of process water, heat integration measures, improved equipment design bases, an optimization of operational points, as well as the implementation of a full process control concept. The effects of these prospective considerations have a direct impact on the utility consumption and energy demand, which enable an increase of efficiency of 20–70% in regards to utility consumption, based on short-cut calculations and expert estimations.

Improvements on the process side (e.g., the most appropriate and minimum mass of acid for precipitation, or the most suitable adsorbents for the separation of dissolved impurities), are not yet accounted for in this study. In order to determine these associated potentials for improvement, various experimental investigations are underway.

In this second study, the influence of the site infrastructure is also taken into account. As already described for study 1, two operation locations are investigated by using the CO_2-eq indicators related to the Hoechst industrial park. Similar to study 1, only the impact category climate change is considered here.

Sensitivity study 3: "r-TA vs v-TA"

Since the revolPET process provides the PET monomers r-TA and r-EG in a primary quality, a comparison with the fossil production route as a benchmark is carried out in this study. Through integration of the new recycling technology into the value chain of PET plastic, a circular economy is enabled. Thus, a downcycling of post-consumer PET waste as well as its inevitable end-of-life expenditures can be almost completely avoided.

In this sensitivity study, the revolPET process is compared to the fossil production route of vTA through the implementation of generic process data [29]. For this purpose, a dataset is chosen that represents the mix of commercial TA production technologies in Europe. The dataset is based on data collected between 2011 and 2013 from plant data of all PlasticsEurope members. The data are representative for 79% of the overall production capacity in Europe at the current technology level. The result of studies 1 and 2 will be included to also consider the boundary conditions (operational area and technology readiness level) assessed so far. The assumptions made for both previous studies apply unchanged.

6.3.3 Data for Life Cycle Inventory (LCI)

The life cycle inventory is mainly based on the mass and energy balance acquired from consumption measurements at the experimental pilot plant (TRL 4). It considers all expenditures in regards to the continuous processing of 14 kg/h of PET/PE waste fraction including all reactants, auxiliaries, utilities, electrical energy, residues, and emissions. These consumption data were further complemented by calculations. Not included in the investigations are the expenditures resulting from the provision and disposal of the production plant itself as well as correlated maintenance expenditures for the equipment. Table 6.7 shows a summary of the necessary inputs and resulting outputs of the recycling process.

Table 6.7 Process Inputs and Outputs of the revolPET® Process (TRL 4)

Inputs		Outputs	
Reactants	PET waste fraction	Value components	r-TA
	Metal hydroxide		r-EG
Auxiliaries	Acid		Metal salt
	Adsorbent	Residuals	Waste water
	Deionized water		Solid waste
Utilities	Nitrogen		
	Cooling water		
	Heat		
Energy	Electricity		

For this study, the production data for the removal of a post-production waste fraction of PET/PE with a total throughput of 14 kg/h and a resulting recovery of approximately 10 kg r-TA/h as well as 3.8 kg r-EG/h and 9.3 kg metal salt/h were taken into account.

Reactants: The only reactants used in the revolPET process are shredded PET waste fractions and a solid metal hydroxide. To gather the elementary flow data of the reactants for the life cycle inventory, corresponding generic datasets from the ecoinvent database v3.7.1 [23] for the depiction of the background data were chosen. The data for the solid metal hydroxide can only be covered by a dataset for a dissolved form of the substance. Due to a lack of primary data, the additional necessary heat energy theoretically needed to transform the dissolved form into its solid state was calculated by considering substance data [30].

Auxiliaries: In the study all auxiliaries have been taken into account. For the acid and the deionized water, generic data sets were used that record their provision, whereas for the adsorbents the treatment for reactivation is taken into account. How often it has to be replaced is currently under investigation. The current laboratory data show high-quality reactivation over the long-term, indicating a high adsorbent lifespan and therefore long exchange intervals. If a solid database is available the replacement of the adsorbents will be implemented as well, whereby no significant influence is expected. For the gathering of the inventory data of the auxiliaries, datasets from the ecoinvent database v3.7.1 [23] could be used without any adaptations or extensions.

Utilities: During operation, the supply of cooling water and nitrogen (for inertization) was determined based on measurements. The heat demand during operation was determined by the electricity demand of the thermostats. Datasets from the ecoinvent database v3.7.1 [23] could be used for the inventory data of the utilities without any adaptations or extensions. For the electricity a dataset considering medium voltage was chosen.

Energy: The data relative to the demand during operation were measured during production tests and represent an average case. For the electricity a generic dataset from the ecoinvent database v3.7.1 [23] considering medium voltage was chosen.

Residuals: Both accruing waste water and solid waste were recorded in detail and included in the assessment. For the inventory data of the residuals generic dataset from the ecoinvent database v3.7.1 [23] representing the treatment of waste water and municipal solid waste are chosen.

Future process improvements are expected to significantly reduce the resource consumption and energy demand. The influence of the current development stage and potential for further development will be investigated in Sensitivity study 2, as described above.

6.3.4 Results of the Life Cycle Impact Assessment (LCIA)

As described above, the potential environmental impacts of the revolPET process at the current development stage were assessed in all 16 recommended ILCD impact categories [31].

The results were subdivided, on the one hand, into a part which is significantly influenced by the technology and, on the other hand, into a part which is mainly influenced by the type of energy supply employed, correlating with the available site infrastructure (cf. Figure 6.27). In the following, the term "technology" is always applied to the entire revolPET process, as shown in Figure 6.24.

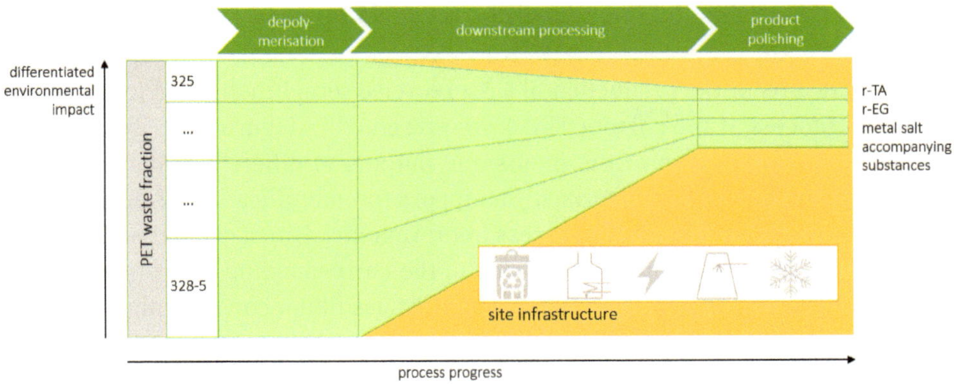

Figure 6.27 Approach for representing the results of the life cycle impact assessment

Additionally, the total PET-content, as well as accompanying substances and contaminations, differ considerably between the individual PET waste fractions. Therefore, it should be considered that the influence of the waste fraction composition will be reflected in the impacts induced by the process parts *depolymerization* and *downstream processing*. The differences originate from the varying necessary auxiliary quantities, as well as the expenditures for the separation of the contaminations and accompanying substances to isolate and purify the products. Afterwards, the environmental impacts for the *product polishing* are independent from the initial waste fraction, since at this point the starting conditions of the isolated product mix are identical. In this initial study, the potential environmental impact during the processing of fraction 325 (PET bottles, transparent) [32] were determined. To estimate and assess the resulting influence, data for the other fractions must be collected and the corresponding datasets have to be edited.

This structure of the presentation of results is chosen to particularly present the influence of the TRL and the site infrastructure on the environmental assessment of new technologies with focus on a recycling process.

Not included in the assessment are credits for the substitution of fossil-based monomers or the consideration of the thermal recycling of the process waste, since their integration into the value chain of PET was not part of the life cycle analysis. The consideration of those environmental impacts would be of interest for a producer of PET, who wants to replace the fossil terephthalic acid in their products.

In the second part of this section, the numerical results as well as hotspot analyses and sensitivity studies are presented and discussed comprehensively.

Potential Environmental Impacts of the revolPET Process at the Actual Stage of Development

The LCA results of the production of 1 kg recycled terephthalic acid (r-TA) by the revolPET process at the current TRL 4 are presented in Figure 6.28. By deploying the chosen classification of the inputs and outputs to different material classes, hotspot analyses can be performed to support the identification of hotspots and thus key aspects for the start of process improvements. Direct comparison between the impact categories is not feasible due to the corresponding differing reference substances as well as varying environmental issues under consideration.

	Impact categories	Unit	Utilities	Reactants	Electricity	Auxiliaries	Residuals
	Climate change	kg CO_2-eq	52%	17%	20%	9%	2%
Ecosystem quality	Freshwater and terrestrial acidification	mol H^+-eq	14%	10%	6%	68%	2%
	Freshwater ecotoxicity	CTU	16%	19%	6%	44%	15%
	Freshwater eutrophication	kg P-eq	54%	6%	21%	3%	16%
	Marine eutrophication	kg N-eq	21%	16%	8%	9%	46%
	Terrestrial eutrophication	mol N-eq	38%	25%	15%	16%	6%
Human health	Carcinogenic effects	CTUh	27%	31%	9%	20%	13%
	Ionizing radiation	kg ^{235}U-eq	47%	18%	18%	16%	1%
	Non-carcinogenic effects	CTUh	14%	7%	5%	52%	22%
	Ozone layer depletion	kg CFC-11-eq	22%	61%	9%	7%	1%
	Photochemical ozone creation	kg NMVOC-eq	32%	22%	12%	31%	3%
	Respiratory effects, inorganics	Disease incidence	8%	15%	3%	70%	4%
Resources	Dissipated water	m^3 water-eq	12%	17%	4%	54%	13%
	Land use	points	57%	9%	23%	10%	1%
	Minerals and metals	kg Sb-eq	19%	27%	7%	45%	2%
	Fossil depletion (ReCiPe midpoint (H) V1.13)	kg oil-eq	43%	13%	17%	27%	1%

Legend: ≤ 10%, ≤ 20%, ≤ 30%, ≤ 40%, ≤ 50%, ≤ 60%, ≤ 70%

Figure 6.28 Results of the hotspot analysis of the environmental assessment of the revolPET process at the current development stage (TRL 4)

In half of the impact categories, the *utilities* account for the main impact. This is due to the thermal energy needed for the downstream processing as well as final preparation.

In six categories, the *auxiliaries* have the highest percent share of the impact (e.g., *freshwater* and *terrestrial acidification*, *respiratory effects*, *inorganic* or *dissipated water*). This is mostly due to the deployment of the acid for the precipitation, which has a high impact in those categories. In general, the residues have a low percent share of the overall impacts, due to their low accrued quantities. In view of these results, further process development is needed to recover as many flows as possible and return them to the value chain, with the goal of reducing the need to dispose of flows.

The impact share of the *reactants* ranges between 6–61% of the overall impact. In this class a low optimization potential is expected, since the required reactant quantities are determined by the stoichiometric ratio and henceforth any reduction is fairly limited.

The *electricity* demand amounts from a medium up to a high share of the overall impacts in all categories. Because thermal energy is supplied with electrical heating, which has the main impact of the utilities, the environmental impact (EI) ratio of $EI_{energy}/EI_{utilities}$ is close to 0.38 in all categories. Hotspot analyses of the overall impact in a single category attributed to the input and output classes, as well as to the technology and infrastructure aspect, were performed to identify the main influence(s).

Figure 6.29 shows results of a hotspot analysis for the impact category *climate change* and *dissipated water*. The results in the two categories *climate change* and *dissipated water* show a variety of trends with respect to the assignment of the potential environmental impacts to the technology itself and the site infrastructure. For the category *climate change*, more than 50% of the 0.82 kg CO_2-eq expenditures of the technology part stem from the reactants, in this case the PET waste fraction and the sodium hydroxide. The lowest share, with 2%, belongs to the utilities, here nitrogen for inertization. For the site infrastructure, 7% of the 1.6 kg CO_2-eq results from the utilities, in this case the expenditures for the provision of heat energy on different temperature levels, whereas 29% is caused by the electricity demand of the equipment.

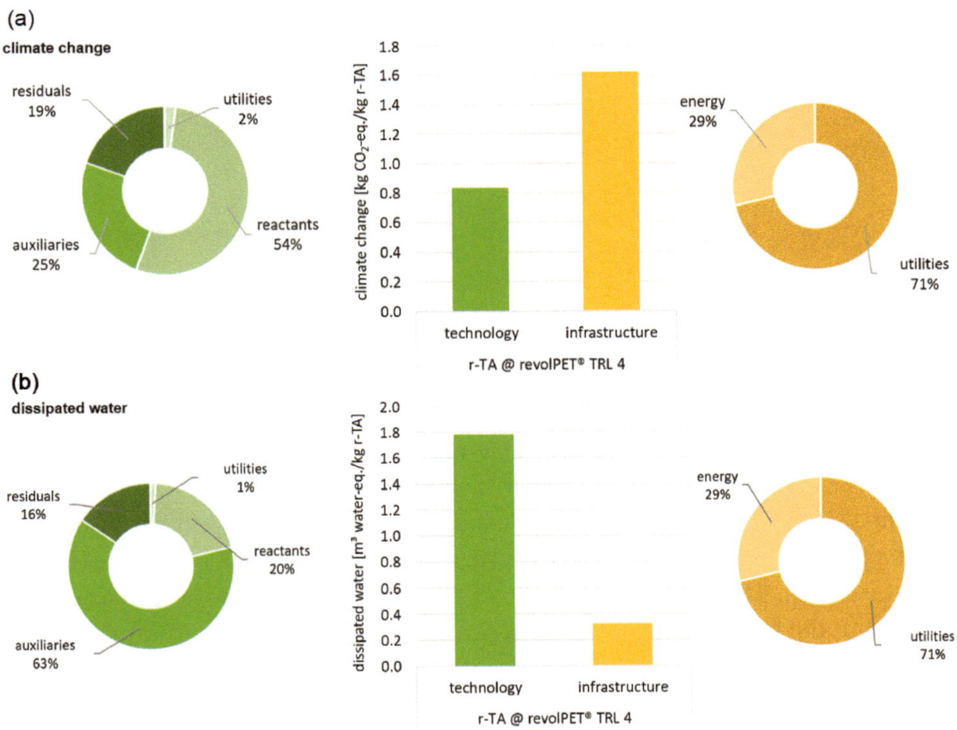

Figure 6.29 Excerpt of the results of the life cycle assessment, here a hotspot analysis of the impact categories *climate change* (a) and *dissipated water* (b) in regards to input and output classes

In the category *dissipated water*, the share of the site infrastructure is much lower, with only 15% of the overall potential impacts of 2.1 m^3 water-eq. The hotspot analysis in regards to the electricity and utilities for heating and cooling show for the assessment of the current development stage an identical percent assignment in all impact categories. This results from two facts: firstly, at the lab scale both the electricity and the heat energy are provided by electrical power and, secondly, the volume of the tap water for cooling processes is very low. Therefore, the water consumption has no significant influence on the results. The main impact of the technology part is caused by the auxiliaries, with a share of 63%. Compared to the reactants' share in the category climate change, their contribution is much lower, with 20% of the 1.78 m^3 watereq.

Additional hotspot analysis with an assignment of the environmental impacts to the process parts *reaction*, *downstream processing*, and *product polishing*, as well as to the technology and the site infrastructure, also present diverging trends across the investigated impact categories. Figure 6.30 shows the findings for the categories *climate change* and *dissipated water*.

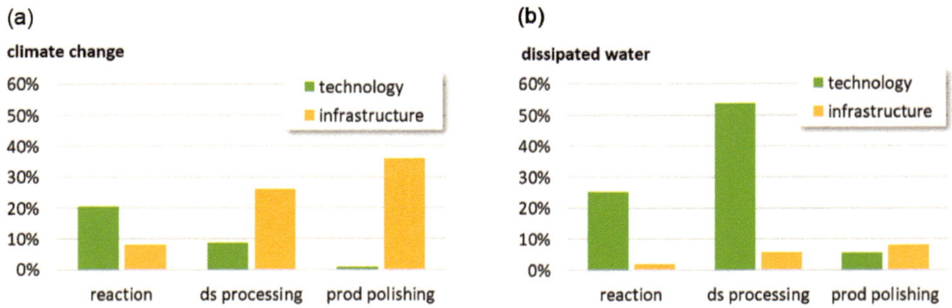

Figure 6.30 Hotspot analyses of the potential environmental impacts assigned to the process stages *reaction*, *downstream (ds) processing*, and *product polishing*, as well as to the technology and infrastructure parts, here exemplified by the impact categories *climate change* (a) and *dissipated water* (b)

For the category *climate change*, the potential impacts are allocated almost evenly among the process stages within a narrow range of 28–37% each. It is noteworthy that in a similar proportion as the influence of the technology on the environmental impact decreases over the process progress, the influence of the infrastructure part increases. The *reaction* stage shows the lowest impact of the three process stages but the highest percent share of the technology part, with 20%. In comparison to the third stage *product polishing*, the impact related to the infrastructure caused by expenditures for the heating steps is the main driver, with an overall share of approx. 36%. The inputs and outputs in this stage relevant for the technology on the other hand (e. g., waste water and auxiliaries) make up for less than 1% of the overall impacts.

For the category *dissipated water*, the earlier investigations have already demonstrated a high environmental impact stemming from the auxiliaries. In the results shown in Figure 6.31, this fact is again clearly reflected for the second process stage, *downstream processing*. In this category the percent share of the overall impact caused by the auxiliaries is much higher than the proportional distribution in the category *climate change*. These environmental expenditures are primarily caused by the acid for the precipitation of the r-TA followed by the adsorbent for the removal of dissolved impurities. The technology share of the impacts for the stage *reaction* stems for both categories *climate change* and *dissipated water* mainly from sodium hydroxide, one of the reactants. Like in the category *climate change*, the impact share of the infrastructure is the highest for the *product polishing* and the share of the technology the lowest compared with other process stages.

Sensitivity Studies

With Sensitivity studies 1 and 2, the influence of boundary conditions is investigated to identify parameters that are highly dependent on the actual development stage. In Sensitivity study 3, the results of both previous studies is compared with the established fossil-based production route of v-TA for an initial evaluation of the revolPET process prior to an assessment over the whole value chain.

Sensitivity study 1: "Operational site"

In this study, the influence of the operational site is investigated. Therefore, the revolPET process was assessed as if located at two different production sites: the technical lab in Braunschweig and at the industrial park Höchst, Frankfurt/Main. In this context only the site infrastructure part of the potential impacts of the revolPET process was adopted, specifically the mode of energy generation to cover the calculated demand (see Figure 6.31). The analysis shows a reduction of approx. 0.8 kg CO_2-eq/kg r-TA for the infrastructure expenditures, resulting in an overall reduction of about 33% of the environmental impact of the technology at TRL 4.

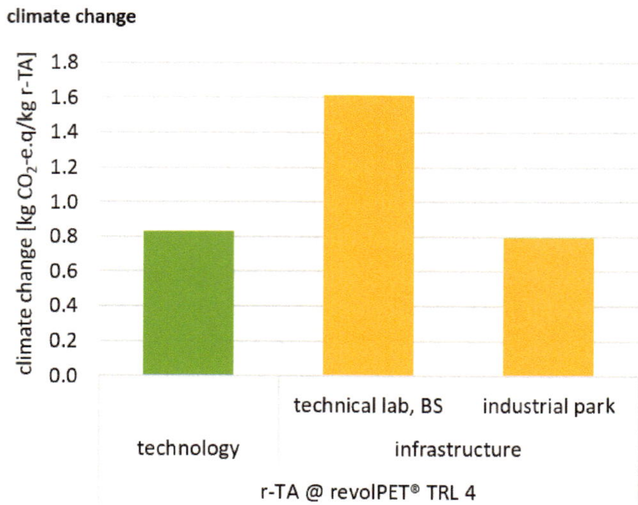

Figure 6.31 Results of Sensitivity study 1 in the impact category *climate change* for the revolPET technology at TRL 4, conducted at two different production sites

The results of this study show that without any direct adaption or improvement of the process itself at the current stage of development, the environmental impact can be reduced significantly by choice of another mode of energy generation, as available at a typical chemical production site. While the annual electrical energy production in Germany has an impact indicator value for the category *climate change* of 0.582 kg CO_2/kWh, the direct supply of heat energy with steam has only an impact indicator value of 0.372 kg CO_2/kWh [23]. This offers a significant reduction potential by switching from electrical to direct heating. In the future, this might change due to a greater reliance on renewable energies in the electricity production mix and thus a decreasing environmental impact per kWh. While the generic datasets are based on the market mix in a specific year, the real impact is often lower due to the implementation of better and more efficient energy generation technologies compared to the common average. For example the Höchst industrial park published a specific impact for heat of 0.283 kg CO_2/kWh [28], which is 25% lower than the generic dataset. The potential environmental impact differs even more significantly for electricity, where the generation at the Höchst industrial park is related to an impact of only 0.293 kg CO_2/kWh, which is only about 50% of the generic dataset value of ecoinvent (v3.7.1) [23]. In general, by choosing a suitable energy generation method available at a production site, and combining it with efficient and state-of-the-art infrastructure, a significant reduction of the overall impact can be achieved.

Sensitivity study 2: "Technology readiness level"

In study 2 the current technology readiness level of the revolPET process was investigated to identify potentials from an environmental point of view, in regards to the future development of the technology. Instead of TRL 4, a TRL 7 was assumed for both cases: the production at the technical lab site and at the industrial park. For the investigation, only the improvements of equipment and operational aspects are taken into account.

For the mass and energy balance, the following consequences could be quantified: With regard to the higher development stage of the revolPET process (TRL 7), the overall thermal energy demand is reduced by 60–78% per kg PET. The reduction is mainly due to the process improvements through heat recovery and optimized process operation. The recovery of the process water also reduces the amount of used water and subsequently of the generation of waste water by 20–70%. Figure 6.32 presents the results of the study.

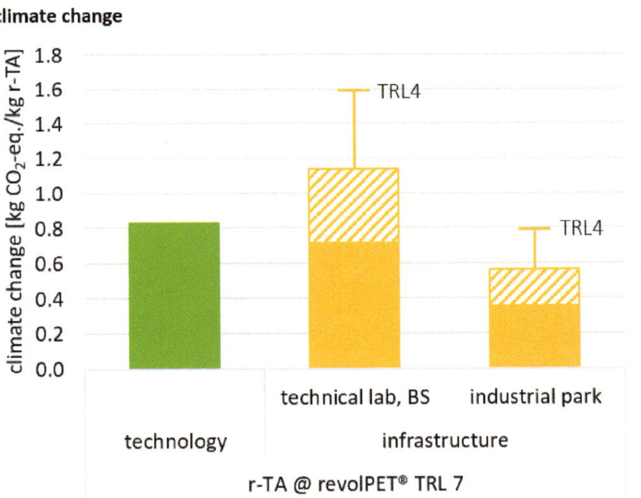

Figure 6.32 Results of Sensitivity study 2, revolPET technology at TRL 7 compared to TRL 4 without improved process parameters, here for the impact category *climate change*

Since no improvement possibilities of the process side were considered, the potential environmental impacts of the technology part remain unchanged at 0.82 kg CO_2-eq. However, the impacts in the site infrastructure part are considerably reduced. The direct comparison between the technology at TRL 4 and TRL 7 results in a reduction of the environmental impacts by 30–55% (see the dashed region (yellow) infrastructure bars. In relation to the technical lab, a decrease of 0.5–0.9 kg CO_2-eq/kg r-TA is indicated. For a production site at an industrial park a reduction amounting to 0.2–0.4 kg CO_2-eq/kg r-TA is recognizable.

The results of this study indicate an overall reduction potential for the operation at the technical lab site between 19–37% and for the industrial park between 14–27% influenced by the infrastructure part of the revolPET process. Furthermore, potential improvements that may exist on the process side have to be detected and assessed as well. With the latter, the overall reduction potentials by increasing the technology readiness level from 4 to 7 will be fully assessable.

Sensitivity study 3: "r-TA vs v-TA"

In the third study, the fossil route of TA production (v-TA) is compared with the provision of TA by the revolPET process (r-TA) from PET waste fractions. In this investigation the results of both previous studies are integrated. Therefore, the boundary conditions of the infrastructure site as well as the technology readiness level are varied and all data sets are compared to a data set of v-TA. The latter is represented by a generic data set "purified terephthalic acid production [RER]" from the ecoinvent database [23], which depicts the expenditures of the common European average of the v-TA production.

Figure 6.33 presents the environmental impacts as quantified for the category *climate change* for both production routes, using fossil resources (gray bar) and the revolPET route (yellow/green bars). For the revolPET route, the technology contribution and the site infrastructure contribution are depicted as stacked bars. The dash area of the bars represents the expected reduction potential of the environmental impact by increasing the TRL. A reduction of the environmental impact due to technology improvements in the revolPET process has not been considered here. The corresponding green bar in Figure 6.33 therefore shows a constant height.

The environmental impact of the production of 1 kg terephthalic acid (v-TA) by the fossil route is 1.78 kg CO_2-eq. At present, the impact induced from the revolPET process at the current development stage at the technical lab in Braunschweig is 37% higher. However, these results are mainly due to the energy supply on the operation side. By operation of the revolPET process with TRL 4 at an industrial park, a reduction of approx. 30% to 1.63 kg CO_2-eq is achieved. This is already approx. 9% lower than for v-TA produced by the fossil route. Additionally, the increase of TRL 4 to TRL 7 at the technical lab will provide a benefit compared to the fossil route, recording the assumption of an improvement of equipment and operational parameters from up to 70%. The highest environmental improvement potential of the scenarios investigated here will be generated by operating the revolPET technology at TRL 7 in an industrial park. In a best-case scenario, the environmental impacts for the production of 1 kg r-TA with the revolPET process operated in an industrial park is approx. 33% lower compared to 1 kg v-TA via the fossil route applying today's state-of-the-art technology. Future improvements of the revolPET process will therefore enlarge these benefits even further.

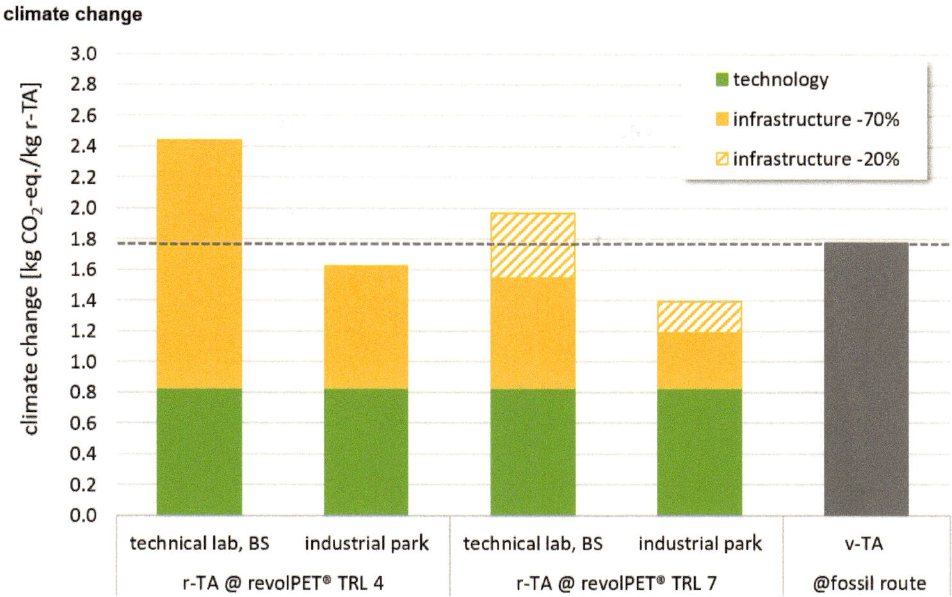

Figure 6.33 Environmental impact in the impact category *climate change total* for the production of v-TA and r-TA, including variation of the infrastructure as well as the TRL contribution

Just as for the assessment of the current technology development stage for this study, the ILCD-recommended 16 impact categories are considered for a multi-criteria assessment. Figure 6.34 shows an excerpt of the multi-criteria analysis. Here, the data set "TRL 7, site infrastructure of the technical lab in Braunschweig, increase of efficiency by 70%" is presented.

The results for this data set indicate a reduction of 20% in the impact category *climate change* compared to the fossil route of the TA production (v-TA). Further potential benefits for the recycling route via the revolPET technology are seen in the impact categories *fossil resources*, with a reduction of 51%, and *non-renewable energy resources* (minus 62%), as well as *photochemical ozone creation* with 26%. In the categories *dissipated water*, *carcinogenic effects*, and *freshwater ecotoxicity* the revolPET technology causes more potential environmental effect than the fossil route of TA production. Particularly noticeable are the significantly higher effects in the impact category *ecotoxicity*. Detailed hotspot analyses identified the sulfuric acid with 33% as the main contributor to this impact category. Another considerable flow is the municipal solid waste stream with 20% share in the overall impact.

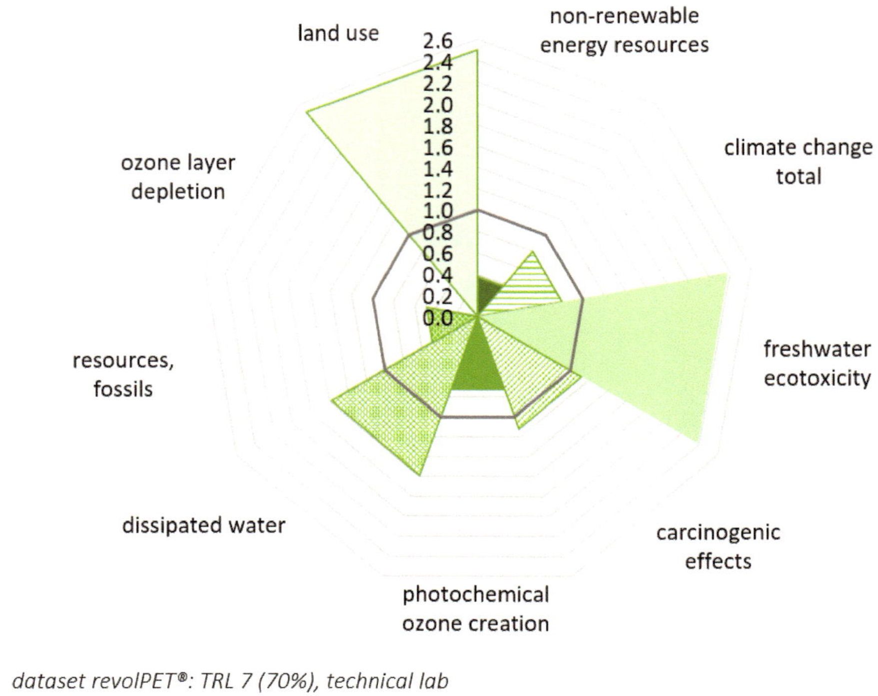

Figure 6.34 Multi-criteria assessment of the environmental impact of r-TA produced by revolPET at the technical lab in Braunschweig at TRL 7 (cf. Sensitivity study 2) with the v-TA produced by the fossil route as represented by the gray polygon (excerpt of results arbitrary units)

Overall, the hotspot analysis identifies the auxiliary *sulfuric acid*, used as precipitant, as an input with a significant influence on the LCA results. In 5 of the 16 investigated categories the acid caused more than 50% of the total environmental impact, in half of them more than 20%. Another significant influence parameter is the *electricity demand*. Here, the share of the overall potential environmental impacts are higher than 50% in three categories and still more than 20% in seven categories.

For the process development these identified hotspots and boundary conditions will be addressed in future technology development steps aiming to continuously reduce the environmental impacts of the revolPET technology.

6.3.5 Analysis and Interpretation

The presented LCA of the revolPET technology at the current development stage targets the assessment of the potential environmental impacts as well as the identification of influence parameters that deserve primary attention in the ongoing development and process optimization. Therefore, a holistic assessment including most of the ILCD-method recommended impact categories supplemented by hotspot analyses as well as sensitivity studies was carried out. Furthermore, this analysis led to a database available for further investigations in the framework of process development and design.

Reliability of the Results

The defined system boundaries enable a complete compilation of a process-LCA regarding all related inputs and outputs, except for the expenditures induced by the plant itself. The chosen level of detail enables the identification of key parameters for improvement, in regards to the inputs and outputs as well as to unit operations or process parts.

It can be expected that the integration of the plant expenditures will slightly increase the overall potential environmental impacts of the technology. Here the approach by the European Commission [33] with a production plant lifetime of approximately 20 years will result in a corresponding annual depreciation which has to be added to the potential environmental impacts.

Most of the used primary data are directly collected through measurements at the technical lab in Braunschweig. Only the product polishing is calculated and based on data sheets from the employed machines. Overall, the data quality of the foreground systems can be evaluated as "very good". For the consideration of upstream chains, generic datasets from the ecoinvent data base (v3.1.7) were used to record the background data. Their quality score in time-related reference ranges in accordance with the ecoinvent pedigree matrix between 1 and 5. In terms of geographic correlation, the score ranges between 1 and 3, depending on the generic dataset. The technology correlation score is between 1 and 3 as well [34]. Consequently, the overall data quality of the background systems can be evaluated as "good".

For the life cycle analysis of the current development state of the revolPET technology, all recommended impact categories by the ILCD were assessed complemented by fossil depletion by ReCiPe2008. The same applies for the investigations in Sensitivity study 3. However, for Sensitivity studies 1 and 2 only the category climate change was considered. For the provision of heat and electricity at the industrial park only information about the CO_2-emissions regarding the supply of the energy sources are available. For a fully comprehensive life cycle assessment, a dataset comprising mass and energy balances for the supply of electricity and steam at an industrial park are necessary. The collection of such data will be part of future work, and will be integrated into the investigations as well.

Evaluation of the Results and Recommendations

The obtained results show that an ex-ante LCA study can very well support the technology development concerning the identification of hot spots and improvement potentials. On the one hand the present study provides an initial idea in terms of the environmental relief potentials associated with the new recycling technology. On the other hand, the study indicates influence factors that have to be addressed in the progressive development of the revolPET process.

Special attention will be given to the selection of an optimum production site and infrastructure for the operation of the revolPET process. The available site infrastructure, respectively the way of thermal and electrical energy supply, exerts a substantial influence on the environmental impact resulting from process operation. For the technology at the current TRL of 4 the direct comparison of the technical lab and an industrial park shows a difference of 0.82 kg CO_2-eq/kg r-TA, representing 34% of the overall impacts in the category *climate change*. The results of Sensitivity study 2, investigating a progressive TRL, has detected a reduction by 1.26 kg CO_2-eq/kg r-TA, with an overall impact reduction of 51%. Combining the results of Sensitivity studies 1 and 2, the site infrastructure appears as a major enabler for an environmentally friendly technology design.

However, this improvement potential is only partially influenced by the revolPET technology, and may only be leveraged as developed production sites become available. To reduce this influence, it should be evaluated whether a self-sufficient energy supply can be integrated in the technology concept. This could be realized, for example, by joint ventures with companies that offer technologies for the provision and storage of renewable energies. It would also support the applicability of the revolPET technology at locations with little or almost no technical infrastructure available but severe waste plastic issues, such as developing countries.

Furthermore, it can be expected that the whole environmental impacts caused by the revolPET technology will be reduced significantly by progress of the development stage. Regarding the current TRL of the revolPET technology and the investigated TRL 7 (see Sensitivity study 2), the potential impacts in the category *climate change total* are reduced by 0.47–0.90 kg CO_2eq/kg r-TA for the operational area of the technical lab. In most of the other investigated categories the improvements range between 10–40% compared with TRL 4.

For an environmentally friendly process performance, the next steps of the technology development should, among others, focus on the identification of heat integration possibilities, as well as the reuse and recirculation of process water and cooling media. However, it must be taken into account that Sensitivity study 2 considers only an improvement of equipment (here, the scale-up effect of the extruder regarding to the energy demand), and operational aspects like heat integration. Ongoing investigations at the technical lab will provide indications of further potential on the process side as well. The integration of the improvement potentials regarding the parameters on the process side will additionally influence the current results. After their integration, more detailed hotspot analyses are necessary to identify further environmentally relevant parameters of the technology.

The comparison of the revolPET process with the fossil production route of terephthalic acid (see Sensitivity study 3) already shows environmental benefits in the current development stage of the revolPET technology. As stated above, the energy supply is one of the main influencing factors. Between the operation sites technical lab and industrial park the differences for TRL 4 and 7 are approximately 34% and 23% respectively. The differences between TRL 4 and 7 at an industrial park are approximately 14% and 27%.

Based on these results, it is expected that the influence resulting from the energy supply depending on the production infrastructure will tend towards a minimum value corresponding to an increasing TRL. In further studies this minimum will be addressed in detail. Initial calculations applying an estimated energy mix of 2040 [35] show that the integration of a higher part of renewable energy carrier (approximately 70%) results in an overall reduction potential (compared with the fossil route) of 42% for TRL 7 and 39% for TRL 4, both located at an industrial park.

Building on solid data to estimate the environmental impact of the revolPET technology at TRL 7, further studies have to be carried out to investigate the influence of the revolPET technology on the circular economy of PET plastics. In particular, this includes the effects on the entire value creation network and influences of expected changes of the boundary conditions emerging as a result of efforts by the EU and existing legislation. In this context, the increase in the recycling quota as well as the climate protection and (with the latter) the accompanying changes in the energy supply both in the industrial and the private sectors will play a stronger role.

However, in this current early stage of development, the proof of concept is focused. The integration of the technology in a real chemical production infrastructure as well as the realization of circular process flows or the optimization of equipment operating points are still under way. Furthermore, scale up considerations of the relevant plant equipment are addressed in the future. The continuous integration of the environmental assessment during the process development and design phase will ensure that the development of the revolPET recycling process is as sustainable as possible.

Future Role of Monomer Recycling

In view of a comprehensive circular economy for plastics and PET in special, a multitude of different recycling technologies is already available, and even more are under development. Based on the available feed waste material and the intended use of the products of the recycling processes, different technologies will find their place based on technical applicability, economic competitiveness, and ecological profitability. In this overall context, PET monomer recycling, such as the revolPET technology, is seen as complementary technology to alternative recycling technologies, such as mechanical recycling or pyrolysis. These technologies offer complementary process paths to recycle PET or other plastics fractions depending on feed material quality and recycling product application. PET monomer recycling via the revolPET technology offers the opportunity to recycle complex PET-containing waste material, like multilayer trays or colored and opaque packaging. As polyolefin constituents of the processed mixed plastic waste, such as PE or PP, remain inert and may be separated, this by-product of PET monomer recycling may serve as feed material for other recycling technologies.

There still is a long way to go to develop the different approaches for monomer recycling. The decisive step will be the scale-up to industrial scale and the resulting economic and environmental impacts. Particularly the economic figures are crucial for the ultimate competitiveness of a monomer recycling technology. As monomer recycling ensures the preservation of non-renewable resources, their positive environmental impact is undoubted. A transparent and fact-based LCA with virgin monomers produced via the fossil route as the reference subject will quantify and clarify the sustainability profile of recycled monomers in the future.

References for Section 6.3

[1] Lugal, L., Grant, A., Cordle, M., Fletcher, D.: PET Market in Europe – State of Play: Production, Collection and Recycling Data, 2020

[2] Directive 2008/98/EC of the European Parliament and of the Council of November 19, 2008 on waste and repealing certain Directives, European Parliament, Council of the European Union, 2008

[3] Gesetz zur Förderung der Kreislaufwirtschaft und Sicherungder umweltverträglichen Bewirtschaftung von Abfällen(Kreislaufwirtschaftsgesetz – KrWG), Deutsche Bundestag, 2012

[4] Vogel, J., Krüger, F., Fabian, M.: Chemisches Recycling: Hintergrundpapier, 2020

[5] Patel, A. D., Meesters, K., den Uil, H., de Jong, E., Blok, K., Patel, M. K.: Sustainability assessment of novel chemical processes at early stage: Application to biobased processes, *Energy & Environmental Science*, 5 (2012) 9, p. 8430, 10.1039/c2ee21581k

[6] Curzons, A. D., Jiménez-González, C., Duncan, A. L., Constable, D. J. C., Cunningham, V. L.: Fast life cycle assessment of synthetic chemistry (FLASC™) tool, *International Journal of Life Cycle Assessment*, 12 (2007) 4, pp. 272–280, 10.1065/lca2007.03.315

[7] Biwer, A., Heinzle, E.: Environmental assessment in early process development, *Journal of Chemical Technology & Biotechnology*, 79 (2004) 6, pp. 597–609, 10.1002/jctb.1027

[8] Grundemann, L., Kuschnerow, J. C., Brinkmann, T., Scholl, S.: Using ecological assessment during the conceptual design phase of chemical processes – A case study, in: Globalized Solutions for Sustainability in Manufacturing, pp. 617–622, Springer, Berlin/Heidelberg, 2011

[9] Kuschnerow, J. C., Wesche, M., Scholl, S.: Ökobilanzielle Bewertung des Einsatzes rezyklierter ionischer Flüssigkeiten als Umesterungskatalysatoren, *Chemie Ingenieur Technik*, 83 (2011) 10, pp. 1582–1589, 10.1002/cite.201100097

[10] Paschetag, A., Wesche, M., Nieder, H., Scholl, S.: Integration of ecological aspects during process development and design – A case study of batch to continuous production, *Procedia CIRP*, (2016) 48, pp. 532–537, 10.1016/j.procir.2016.03.154

[11] Buyle, M., Audenaert, A., Bilen, P., Boonen, K., Van Passel, S.: The future of ex-ante LCA? Lessons learned and practical recommendations, *Sustainability*, 11(19), p. 5456, 10.3390/su11195456

[12] Cucurachi, S., van der Giesen, C., Guinée, J.: Ex-ante LCA of emerging technologies, *Procedia CIRP*, (2018) 69, pp. 463–468, 10.1016/j.procir.2017.11.005

[13] van der Giesen, C., Cucurachi, S., Guinée, J., Kramer, G.J., Tukker, A.: A critical view on the current application of LCA for new technologies and recommendations for improved practice, *Journal of Cleaner Production*, (2020) 259, 120904, 10.1016/j.jclepro.2020.120904

[14] Biermann, L., Brepohl, E., Eichert, C., Paschetag, M., Watts, M., Scholl, S.: Development of a continuous PET depolymerization process as a basis for a back-to-monomer recycling method, *Green Processing and Synthesis*, 10 (2021) 1, pp. 361–373, 10.1515/gps-2021-0036

[15] Deutsches Institut für Normung e. V., DIN EN ISO 14040:2021-02, Umweltmanagement- Ökobilanz-Grundsätze und Rahmenbedingungen (ISO 14040:2006 + Amd 1:2020); Deutsche Fassung EN ISO 14040:2006 + A1:2020. Beuth Verlag GmbH, Berlin

[16] Deutsches Institut für Normung e. V., DIN EN ISO 14044:2021-02, Umweltmanagement- Ökobilanz-Anforderungen und Anleitungen (ISO 14044:2006 + Amd 1:2017 + Amd 2:2020); Deutsche Fassung EN ISO 14044:2006 + A1:2018 + A2:2020. Beuth Verlag GmbH, Berlin

[17] Maga, D., Hiebel, M., Thonemann, N.: Life cycle assessment of recycling options for polylactic acid. *Resources, Conservation and Recycling*, (2019) 149, pp. 86–96, 10.1016/j.resconrec.2019.05.018

[18] Wernet, G., Mutel, C., Hellweg, S., Hungerbühler, K.: The environmental importance of energy use in chemical production, *Journal of Industrial Ecology*, 15 (2011) 1, pp. 96–107

[19] Jiménez-González, C., Curzons, A.D., Constable, D.J.C., Cunningham, V.L.: Cradle-to-gate life cycle inventory and assessment of pharmaceutical compounds, *International Journal of Life Cycle Assessment*, 9 (2004) 2, pp. 114–121, 10.1007/BF02978570

[20] Gasafi, E., Meyer, L., Schebek, L.: Using life-cycle assessment in process design, *Journal of Industrial Ecology*, 7 (2003) 3–4, pp. 75–91, 10.1162/108819803323059415

[21] Kägi, T.: Waste polyethylene terephthalate, for recycling, sorted: Europe without Switzerland, ecoinvent 3.7.1 dataset documentation, 2020

[22] Wesche, M., Häberl, M., Kohnke, M., Scholl, S.: Ökologische Bewertung von Produktionsprozessen in Mehrproduktbatchanlagen, *Chemie Ingenieur Technik*, 87 (2015) 3, pp. 203–215, 10.1002/cite.201400086

[23] Wernet, G., Bauer, C., Steubing, B., Reinhard, J., Moreno-Ruiz, E., Weidema, B.: The ecoinvent database version 3 (part I): Overview and methodology, *International Journal of Life Cycle Assessment*, 21 (2016) 9, pp. 1218–1230, 10.1007/s11367-016-1087-8

[24] HORIZON 2020: Work programme 2018–2020, European Commission

[25] International reference life cycle data system (ILCD) handbook: Recommendations for life cycle impact assessment in the European context, Publications Office, Luxembourg, 2011

[26] Huijbregts, M.A.J., Steinmann, Z.J.N., Elshout, P.M.F., Stam, G., Verones, F., Vieira, M., Zijp, M., Hollander, A., van Zelm, R.: ReCiPe2016: A harmonised life cycle impact assessment method at midpoint and endpoint level, *International Journal of Life Cycle Assessment*, 22 (2017) 2, pp. 138–147, 10.1007/s11367-016-1246-y

[27] Goedkoop, M., Heijungs, R., Huijbregts, M., De Schryver, A., Struijs, J., van Zelm, R.: ReCiPe 2008: A life cycle impact assessment method which comprises harmonised category indicators at the midpoint and the endpoint level (1st edition, version 1.08), 2013

[28] Erfolgsfaktor Energieeffizienz Carbon Footprint-Bericht Infraserv Höchst 2010, Infraserv GmbH & Co. Höchst KG, Frankfurt/Main, 2011 (2017)

[29] Fröhlich, T.: Purified terephthalic acid production: RER (Europe), ecoinvent 3.7.1 dataset documentation, 2020

[30] Kurt, C., Bittner, J.: Sodium hydroxide, in: Ullmann's Encyclopedia of Industrial Chemistry, Wiley-VCH, Weinheim, 2000

[31] International reference life cycle data system (ILCD) handbook: General guide for life cycle assessment: Provisions and action steps. Publications Office, Luxembourg, 2011

[32] Product Specification 03/2018, Fraction-No. 325: Sorting fraction: PET bottles, transparent, DSD – Duales System Holding GmbH & Co. KG

[33] European Commission, Commission Recommendation of 9 April 2013 on the use of common methods to measure and communicate the life cycle environmental performance of products and organisations, 2013

[34] Ciroth, A., Muller, S., Weidema, B., Lesage, P.: Empirically based uncertainty factors for the pedigree matrix in ecoinvent, *International Journal of Life Cycle Assessment*, 21 (2016) 9, pp. 1338–1348, 10.1007/s11367-013-0670-5

[35] Pfluger, B., Testeegen, B., Franke, B.: Langfristszenarien für die Transformation des Energiesystems in Deutschland. Modul 3: Referenzszenario und Basisszenario. Fraunhofer Institut für System- und Innovationsforschung (ISI)/Consentec GmbH/Institut für Energie- und Umweltforschung Heidelberg gGmbH, 2017

PART B

Major Recycling Technologies

7 Overview of Major Recycling Technologies

Christian Hopmann, Rainer Dahlmann, Martin Facklam

It is fundamental for the establishment of a circular economy to consider the various material, product, and waste flows as well as various recycling technologies and the resulting recycling products [1]. This chapter provides an overview of the major recycling technologies.

By way of introduction, the current market situation of plastic products is first described and the changes concerning the handling or utilization of plastic waste are discussed. Based on this, the current situation of plastic recycling in Europe is derived. This is followed by an overview of the recycling technologies that are increasingly taking on a leading role in plastic waste management or closing the material loops. A distinction is made between mechanical recycling and feedstock recycling. In context of feedstock recycling, alternative utilization routes are also outlined and distinguished from recycling in terms of their relevance and material use of the resulting products.

In general, the focus in this chapter is on thermoplastics. Thermosets, elastomers and fiber-reinforced plastics are not discussed in detail.

7.1 Introduction

Plastic products are often characterized by low weight, whereby the choice of materials and the adaptability of the material performance result in high-quality usage properties. At the same time, high volumes can be realized at low cost and with a high degree of design freedom [2]. Plastics are therefore often regarded as an energy- and resource-saving alternative to other materials such as glass, paper, or metal. Consequently, everyday life is shaped by an extremely versatile range of applications for plastics covering numerous industries and products.

Regardless of the material, the topic of waste recycling is nowadays receiving significantly more attention in view of environmental and socio-political developments. The evident environmental impacts caused by improper handling of plastic waste are offset by the great ecological and economic usefulness and indispensability of plastic products [1]. However, the global reputation of plastics has deteriorated due to the generally inadequate recycling of plastic products after use [3]. In addition, the handling of plastic waste is currently being scrutinized much more intensively. It is increasingly becoming apparent that there is a great need for action in the field of waste management and that a fundamental change from a linear to a circular economy is essential. Up to now, valuable material flows have not often been appropriately recycled after use.

Plastics have undergone tremendous development since the 1950s, both in terms of their variety of grades, processing technologies, markets, and applications, as well as margins. The current global plastic demand is approximately 368 Mt, of which Europe accounts for about 16% [4]. Packaging dominates the market in terms of the plastic demand, followed by construction, automotive, electrical and electronics, household, leisure and sports, agriculture, and other industries [4]. Table 7.1 provides an overview of the various types of plastics on the market, their relationship to recycling codes, and the respective application examples and annual demand.

Table 7.1 Plastics Overview, Application Examples and Annual Plastic Demand for EU28 + NO + CH (2019) [4]

Symbol	Designation	Abbreviation	Application examples	Plastic demand [Mt/a]
01 PET	Polyethylene terephthalate	PET	Plastic bottles, textile fibers, films, food packaging etc.	4.00
02 PE-HD	Polyethylene (high-density)	PE-HD (or HDPE)	Gas and water pipes, cleaning agent containers, fuel containers, household items, toys etc.	6.23
03 PVC	Polyvinyl chloride	PVC	Window profiles, shutters, cable sheathings, foils, flooring, hoses etc.	5.07
04 PE-LD	Polyethylene (low-density)	PE-LD (or LDPE)	Garbage bags, shrink films and agricultural films etc.	8.82
05 PP	Polypropylene	PP	Food packaging, bottle caps, suitcase trays, bumpers, infusion containers etc.	9.84
06 PS	Polystyrene	PS	Food packaging, sound and heat insulation, packaging etc.	3.14
07 O	Other plastics (ABS, PA, PC, PMMA, PUR, SAN etc.)	O	N/A	13.6

Especially in the packaging sector, product life cycles are often very short (Figure 7.1). The lifetime of flexible packaging (e.g., films) or rigid packaging (e.g., cups or bottles) is often only a few days or weeks. The volume of waste is therefore particularly high due to the fact that the ratio of waste to production volume is constantly increasing as a result of economic growth and new applications. However, in the majority of the industries, products remain in use for many years. For example, the life cycles in the construction sector are the longest, averaging 35 years. For particularly durable products such as plastic pipes made of polyethylene, a lifetime of 100 years is expected [5].

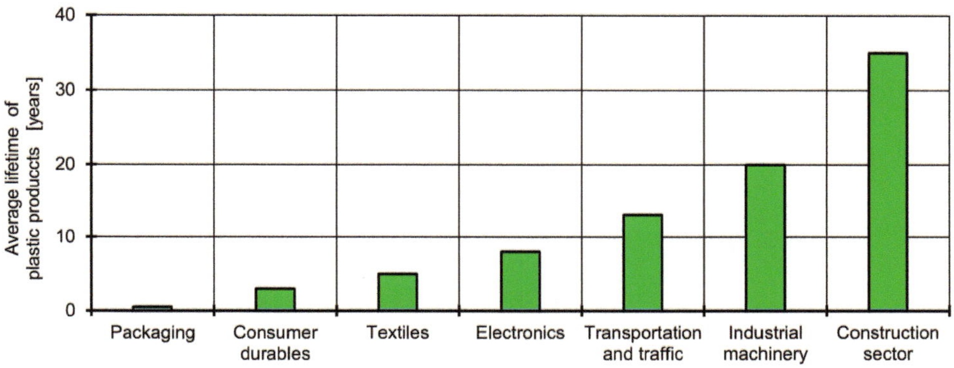

Figure 7.1 Average lifetime of plastics by industry in 2019 [Creative Commons (CC-BY-4.0), see: L. Fuhr, R. Buschmann and J. Freund, *Plastikatlas 2019 – Daten und Fakten über eine Welt voller Kunststoffe*, Möller Druck und Verlag GmbH, 2019]

The wide variety of products and different product life cycles ultimately lead to widely differing waste streams, which vary greatly in terms of their quantity and material composition, as well as over the course of the year. Responsible use of the available material resources is thus challenging, but may also develop into an innovation driver for the plastics industry, in the wake of changing climate and environmental policy objectives due to the constantly tightening legal requirements [1].

Plastic waste is often divided into three categories with regard to its origin. Accordingly, industrial waste from semi-finished or component production, such as sprues, scrap parts (injection molding), or edge trimmings (film extrusion), which are not recycled and reused at the respective production site, are considered *post-production waste*. This waste can be collected by type without great effort, and recycled as materials in an economically lucrative manner. Waste streams generated during product distribution (e.g., transport packaging) are referred to as *post-industrial waste*. Here, the degree of impurities often increases noticeably through use, and the plastic waste is increasingly less sorted. Finally, *post-consumer waste* includes all remaining products whose life cycle has come to an end. This waste flow is characterized by a high level of impurities and a large variety of materials with varying plastic content. In summary, the material composition is often undefined, and needs to be sorted and cleaned, before the fraction of recoverable plastics can be reused to produce new products.

The approaches to enable the efficient recycling of plastic products and to reprocess waste streams to a much greater extent than before are anything but new (for example, see ref. [6]). However, it was not until 2018 that the EU adopted the policy framework for dealing with plastic waste, with the European Strategy for Plastics in a Circular Economy as part of its action plan [7].

The various routes for dealing with plastic products and waste are briefly illustrated in Figure 7.2. Best practice is to avoid and reduce plastic waste. However, this approach offers only limited room for improvement, due to economic growth, new products, and high product requirements, among other things. On the other hand, the economic incentives for waste prevention are also limited.

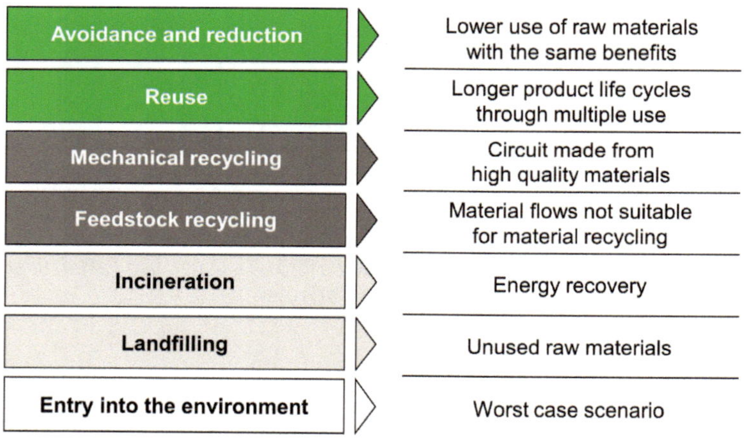

Figure 7.2 Routes for dealing with plastic waste [6, 8, 9]

The reuse of products aims at individual solutions and requires a sufficiently high market share (e. g., plastic bottles). Once the possibilities for avoiding and reducing waste or for reusing the products have been exhausted, various approaches exist to enable efficient circular material flows. For example, products can be optimized for recycling in the development phase (design for recycling), and targeted configurations of material composites (material design for recycling) means that the recyclability of waste is considered from the outset [1].

Material cycles for plastic products can be realized in different ways, as shown in Figure 7.3, based on the principle of the EU waste hierarchy. Depending on the variant, different path lengths for the cycles result. Also the product requirements and materials must always be considered in detail [3].

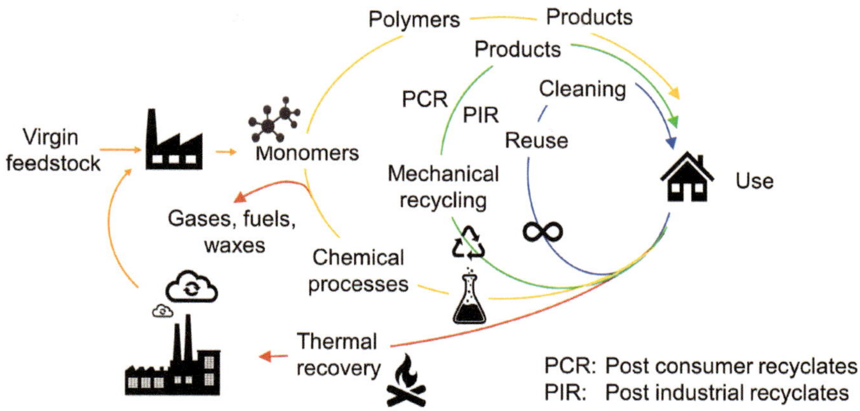

Figure 7.3 Illustration of the possibilities for recycling and recirculation of plastics [3]

A distinction is made at the material level between mechanical and feedstock recycling to enable closed material cycles (see Figure 7.3). In general, the waste is collected and, depending on the quality, sorted before it is recycled. Mechanical recycling is a multi-stage process that essentially comprises the steps of shredding, sorting, cleaning, drying, and regranulation. In this process, the polymer structure of the plastic should not be affected at all, or only slightly, so that high-quality products can be manufactured again from the recyclate [1].

In feedstock recycling, the polymer structure is (for example) broken down into monomeric units and finally rebuilt into a polymer by a synthetic process. This recycling path is considered if the energetically more advantageous mechanical recycling is not possible [1].

The following consideration of the current market situation for recycling of plastic waste can be assumed as an approximation for the EU. It should be noted that data collection differs from country to country, and is often not very accurate. It can be assumed that the true amount of plastic waste recycled is significantly lower than that indicated by the data. Another potential source of inaccuracy is that waste streams exported from the EU are included in the recycling statistics, but their proper recycling is not ensured.

The total European plastics converter demand was approximately 50.7 Mt in 2019. In the previous year, the amount of collected post-consumer waste with a high polyolefin content related to the plastic packaging was about 29.1 Mt, which includes EU exports of about 11.3 Mt. The material recycling accounts for 32.5% (9.6 Mt) of the collected post-consumer waste. The remaining share of 42% is incinerated and 24.9% is landfilled [4]. However, feedstock recycling represents less than 1% in terms of volume.

How plastic waste is handled also varies significantly from country to country. Some European countries only landfill low volumes of plastic waste. Safe storage of plastic waste does result in a reduction in mass and volume, making landfilling a simple form of waste disposal. However, the recycling potential is completely lost. In particular, the high proportion of organic hydrocarbons is considered critical [9]. Due to legal restrictions or high taxes, landfilling has now been completely eliminated in some other countries (e. g., Austria, Belgium, Germany, Denmark, Finland, Luxembourg, Netherlands, and Sweden). Thermal recycling or use as a substitute fuel currently accounts for the majority of waste utilization. The high energy content of petrochemical-based plastics allows them to be used instead of primary fuels (such as coal, oil, and gas) for energy generation [1, 9]. Other European countries, on the other hand, do not have the technological infrastructure for material recycling, with the result that a large proportion of plastic waste is sent to landfill.

At the same time, political pressure on European converters is increasing, as recycling targets are increasingly being defined in the various plastics sectors. For example, 50% of plastic products in the packaging sector must be used to create new products by 2025, and 55% by 2030 [10]. However, in order to enable industry-wide implementation, the technologies of mechanical recycling must be significantly further developed and recycling capacities drastically increased [1].

In addition, various obstacles exist to establish material cycles. First, existing thermal or energy recovery is widespread and efficient. In Germany, for example, the business of waste management is worth approximately €50 billion per year [11]. Plant operators are also not keen to give up their monetary share of the disposal business. Furthermore, the tracking of material flows along the value chain has not yet been established. These are just a few examples that show the need for action.

It is apparent that there is a backlog of investment, as plastics recycling has received insufficient attention for many years since the 1990s. Investments are urgently needed to create recycling capacities or to increase recycling rates. This also applies to providing high-quality, valuable recyclate streams, which in some cases only inadequately meet the high product requirements and can also exhibit fluctuating material properties

In summary, it can be stated that from a recycling infrastructure perspective, there is an increasing need for development, in order to serve the growing secondary raw materials market as well as to be able to cover the increasing demand for high-quality recycling material flows in the future. At the same time, the value chain of plastic products (e. g., material producer, converter, distributors) as well as waste management involves many different stakeholders, some of whom pursue very different goals. As long as there is no merging of interests, the systematic return of plastic waste to the value chain will be hindered. Further developments in the various recycling technologies and the development of new recycling routes are also required, in order to return far greater quantities of plastics to the secondary raw material cycle in the future.

Figure 7.4 provides an overview of the most important technological approaches used for recycling and thermal recovery of plastic waste. Essentially, a distinction can be made between physical and chemical process paths, which in turn cover mechanical recycling, solvolytic processes, chemical recycling, thermolysis, and biodegradation. In the following sections, the fundamental differences, the types of plastics used in each case, and the special process engineering features will be discussed, among other things.

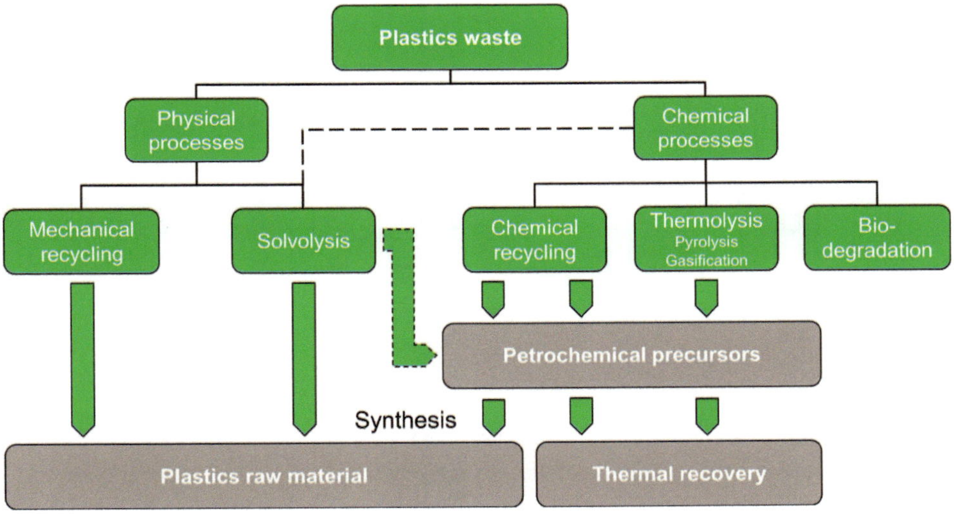

Figure 7.4 Overview of physical and chemical processes for the recycling or utilization of plastic waste

7.2 Mechanical Recycling

Mechanical recycling describes a type of recycling where the macromolecular base structure of plastics remains mainly unchanged. Material recycling can be subdivided into primary and secondary mechanical recycling. *Primary recycling* is mostly applied on post-industrial and therefore high-quality plastic wastes. Plastics entering primary recycling are brought back into the production process after shredding, for example by mixing it into the virgin material or into an intermediate layer in the case of multilayer semi-finished products. In *secondary recycling*, the plastics are shredded, washed, dried, and granulated or compounded again. Secondary recycling usually processes post-consumer waste that is collected mixed with different types of plastics. Due to the limited compatibility of different plastics and other interfering substances (e.g., organic contaminants, labels, fillers), mechanical recycling often leads to a significant deterioration of the material and optical properties.

Figure 7.5 shows the life cycle of a packaging product using the example of a yoghurt cup. Cup forming can be achieved by injection molding or by film extrusion followed by thermoforming. The overview illustrates the multitude of factors influencing the process steps and the use phase of the recyclate on the product quality in this application.

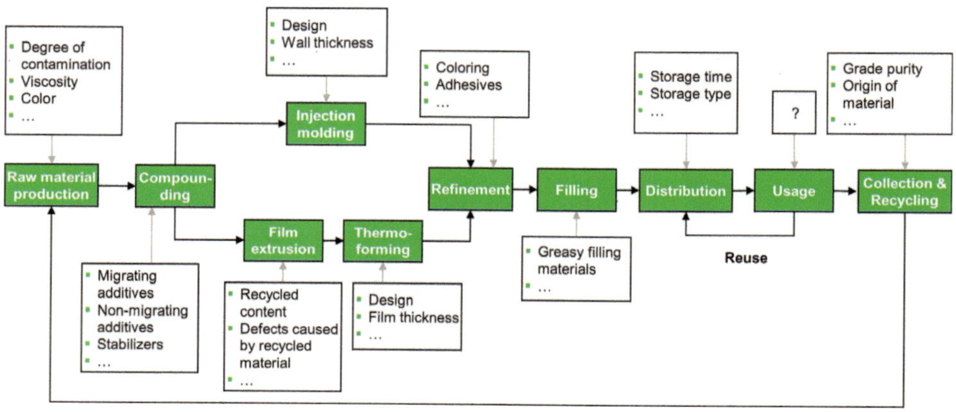

Figure 7.5 Sorting and recycling processes for lightweight packaging, including relevant characteristics for separation [12, 13]

In addition to the interfering substances in the recyclate, the degradation of the plastics during their processing, use, and reprocessing also negatively affects the subsequent product properties. Degradation results from thermal and mechanical stresses as well as exposure to oxygen and UV light [14, 15]. Polymer chains can be cut off or re-cross-linked by radical chain reactions [16]. This changes the macromolecular constitution [17]. If the molecular weight distribution (significant change in the length of the polymer chains) is changed, the viscosity of the thermoplastic also changes [15]. Accordingly, chemical degradation has a significant impact on mechanical recycling and secondary product quality. Consequently, the potential applications for such recycled plastics in new products are currently limited compared to primary plastics. Conversion of a product to a lower-quality end product due to the influences mentioned is often described as *downcycling*. A high degree of purity of the material streams to be recycled is therefore one of the decisive prerequisites for the mechanical recycling [15, 18].

Collection, sorting and pre-processing (recycling) are the major steps of the mechanical recycling path, shown in Figure 7.6. The recycling process depends on the type of waste stream. Waste is sometimes mixed and collected separately according to the recyclable materials present. How collection systems are designed varies by country or region. In most countries, mixed forms of recycling exist. One plastic waste stream that is highly relevant for recycling in terms of volume results from the collection of lightweight (sales) packaging. For example, in Germany, more than 30% of the plastic related to packaging, which has a comparatively short service life compared to products from the construction or automotive sectors [19]. In Germany, for example, lightweight packaging materials like aluminum cans, tin cans, multilayer packaging, liquid cartons, and packaging plastics (PP, PE, PET, and PS), are collected from the curb-side in the so-called "yellow bag or bin", which is in the responsibility of the dual public/private systems [20].

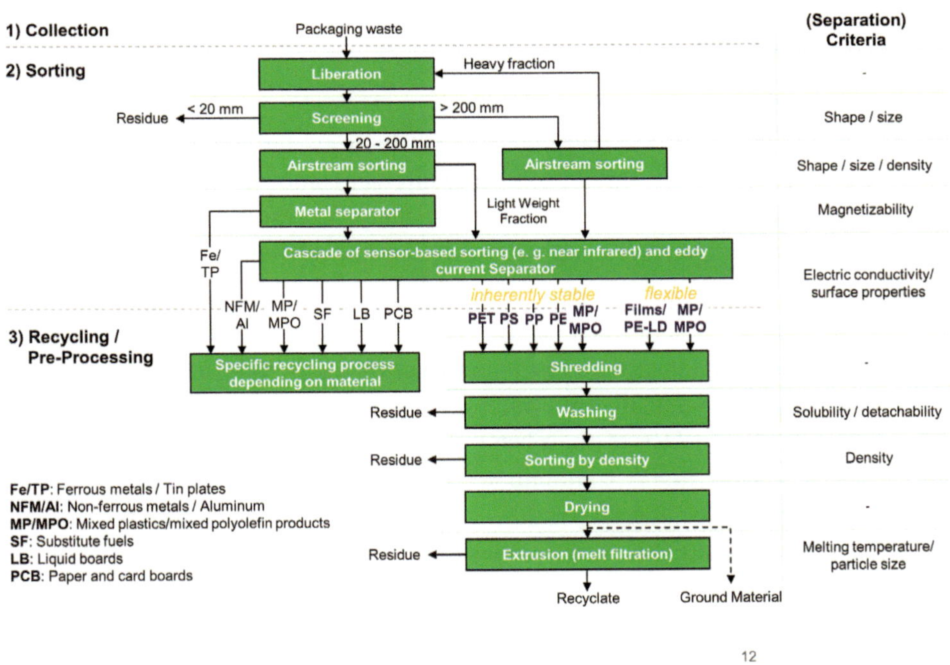

Figure 7.6 Sorting and recycling process of lightweight packaging including relevant characteristics for separation [12, 13]

A plant for sorting of lightweight packaging consists of several units for screening, airstream sorting, magnetic metal separation, non-ferrous metal separation, and sensor-based sorting to remove other recyclables (paper, ferrous and non-ferrous metals, etc.) and foreign materials from the thermoplastics waste flow [20, 21]. To achieve this, particle size and various physical properties like density, wettability, electrical conductivity, susceptibility, and spectroscopic properties are used for classifying and sorting [18, 22]. Due to the overall low content of recyclable plastic in lightweight packaging material and the limitations of applied sensor-based sorting technologies, the plastic output to be recycled further is only around 30% by weight of the plant's input [21, 23]. For example, (carbon) black plastics or pieces of plastic film smaller than DIN A4 are not detected by common near-infrared sensors, and therefore ejected as mixed plastics or substitute fuel [21]. Some of these mixed plastics are used in plant pots, benches and poles, but about 85% of them are incinerated [23].

Upon entering the pre-processing stage – the main process of mechanical recycling – the purities of the pre-sorted standard packaging plastics (PE, PP, PS, and PET) are around 94% [12, 20]. The pathway in Figure 7.6 shows the basic processing steps (shredding, washing, sorting by density, drying and extrusion) for standard plastics. Depending on the type of plastic, further steps can be added upstream or downstream [13]. For example, the pre-processing of PET usually includes more distinct sorting by color, hot alkaline washing, as well as drying and post-processing before extrusion [12, 13].

Starting with shredding, the thermoplastic products are cut most efficiently using a cutting mill, single-shaft shredder, rotary cutter, or comb-shaft shredder. Stresses on thermoplastics due to cutting lead to most adequate results. A uniform particle size distribution and particle shape is required for subsequent sorting [18, 22].

Afterwards the approximately equally sized plastic cuttings go into the washing process. In subprocesses of soaking/dissolving, detaching, and separation, all types of contaminants and adhesives can theoretically be removed from the thermoplastic pieces [18]. While usually only water is used due to the presence of sufficient surfactants on the surface of the plastic pieces [13], friction through stirring is applied to clean more highly contaminated plastics [18, 20]. However, most barrier layers, bonding agents, and large-scale organic contamination cannot be removed by common techniques [13]. The final sorting by plastic type is performed by applying a combination of density sorting, float–sink processing, centrifuging, electrostatic separation, selective dissolution, sensor-based sorting, and perhaps additional separation by color (in the case of high-quality requirements) [18, 22]. One scenario would be to transfer the cleaned plastic pieces into a float–sink separation unit to allow sorting by density, followed by (a cascade of) near-infrared sensor-based sorting units to further purify the output stream [20]. Even plastics with similar physical properties (like density in the case of PE and PP (0.895–0.97 g/cm^3) [13, 24]) can be adequately separated by sensor-based sorting.

The origin of the material, as well as the separation techniques used for waste sorting, have an influence on the recyclate quality. One factor in terms of material origin is the presence of contamination of the plastics. Paper labels, non-polyolefin plastic parts, barrier layers, and bonding agents are examples of foreign material, and cannot easily be removed from the thermoplastic stream [13]. Furthermore, plastic packaging contains, embedded in the polymer base matter, additives like color pigments, antistatics, antioxidants, lubricants or other processing agents, and filler materials like minerals [25]. Greater imprints on plastics are also rather detrimental to mechanical recycling [13]. Labels or radio-frequency identification (RFID) tags lead to negative effects on color, mechanical and thermal properties when not excluded from secondary plastics [12]. Remnants of adhesives lead to (for example) discoloring of recyclates, and can compromise melt filtration and extrusion [12]. Remnant odors in recycling products is another challenge yet to be overcome. Volatile organic compounds like carboxylic acids form a major share of odorous substances attached to, for example, secondary plastic films, which are formed by microbial activity within the waste, or are products of the volatile degradation of polymers and contaminants [26].

Melt filtration is a necessary processing step in mechanical recycling to remove solid contaminants of foreign material or fragments of degraded material before extrusion. The right filtration method to be applied depends on the type and quantity of impurities and the targeted recyclate quality. Types of contaminants vary in size, morphology, density, and elasticity. For example, to remove sand grains a different filtration technique is required than to remove paper label residues. Filtration systems can be discontinuous or continuous in terms of their operation, self-cleaning, and can use different media and mesh sizes/pore space [27]. Degassing offers the possibility to remove low-molecular components, gases, or air from the melt during the extrusion process. However, the substances to be degassed vary considerably in some cases due to the special composition of the recycling materials, as documented in the literature more than 25 years ago [18].

In accordance with the above-mentioned possibilities of recyclate quality deterioration and as shown in Figure 7.6, the proportion of recyclate applied is an important factor affecting extrusion. Moreover, additives for better processability or mixability with other plastic types are added depending on the recyclate quality. Common attempts to reduce negative effects of lower-quality recyclates in new products are to mix them with virgin polymers.

The efficiency and quality affect every step of the process, starting from preparation of materials, the manufacture of plastic products, the life cycle in the field, through to valuable material separation and recycling. More specifically, obstacles to mechanical recycling of post-consumer plastic waste are, on the one hand, the complexity of sorting due to inhomogeneous material flows and the technological limitations to full separation, and on the other hand, impurities or contaminants like ink particles, dirt, other plastics, or polymer degradation products requiring extensive pre-processing and increased additive application.

It should be noted that the ability to maintain the remaining level of material input plus energy input and external impacts on the environment determines the ultimate sustainability of the overall system. In general, mechanical recycling of sorted materials is more sustainable than incineration and should therefore be preferred [28, 29]. However, there are significant variations in energy consumption, water, and emissions during the different recycling processes depending on the material and the purity of the fractions. This is also due to the fact that the industry is in a state of change, which still requires a lot of research and investment.

The sustainability of the recycling route depends on the amount of treatment required and is therefore also determined by the type of material stream. The more polluted the waste is, the higher is the energy input (and costs) needed to process the material to a defined quality level. Material reprocessing therefore only makes sense if the energy expenditure of the recycling does not exceed the process energy requirement for the production of the polymer [18, 28].

■ 7.3 Feedstock Recycling

Multi-layered material composites that cannot be mechanically separated from one another, and plastic waste that has been severely damaged or contaminated by use, are two reasons why mechanical recycling fails to reach its process engineering limits. Also, it is not yet possible to use recyclates for the production of food packaging without further effort, since such packaging must verifiably meet high quality requirements.

However, feedstock recycling can be an alternative measure for closing the material cycle, and has therefore been receiving noticeably more attention in recent years. The aim is now no longer to convert the plastic waste into products by remelting it. Instead, the aim is to use technologies to achieve targeted interventions in the polymer structure. For many types of plastics, the degradation of the macromolecular polymer structure to smaller chain segments, or to the basic chemical units (depolymerization), has already been investigated, and was patented shortly after polymerization on an industrial scale [18]. One example is the in situ polymerization of poly(methyl methacrylate), which succeeded in the 1930s. Shortly after, the first patent for the depolymerization was submitted [30].

The major technological processes available for feedstock recycling are described below, starting with chemical recycling followed by solvolysis, thermolysis, and biodegradation.

7.3.1 Chemical Recycling

Chemical recycling describes the polymer degradation back to the raw material, which can be used to re-polymerize plastics that are qualitatively comparable to virgin material. However, this is not suitable for all types of plastics. The thermally induced depolymerization of poly(methyl methacrylate) (PMMA) or polystyrene (PS), for example, allows the recovery of the respective monomers. The plastic waste is converted into a valuable feedstock for the renewed polymer synthesis. In contrast, the resulting raw feedstock for most types of plastics may also consist of longer-chain hydrocarbon compounds (gases, oils, or waxes). This applies, among other things, to polyolefins (PE, PP), which are mostly degraded at high temperatures by means of a pyrolytic process (see Section 7.3.3). Most of the degradation residues are used as fuel or in the chemical industry for manufacturing purposes. A small residue is then available as pyrolysis oil for renewed polymer synthesis.

Degradation is characterized by the cleavage of the repeating unit in the main chain of the polymer. In this process, a polymer molecule can be broken down into its components by different mechanisms. On the one hand, degradation can take place through successive chain degradation at the end of the chain. A high energy input at a random point in the chain can trigger another mechanism. This breaks the bond between the repeating units at a random point. Another possibility of chain degradation is offered by weak points, which are randomly distributed along the polymer chain. These have a lower binding energy and have a higher probability of being broken compared to normal bonds. The first mechanism is a chain reaction (which is also called depropagation, unzipping, or chain depolymerization). The second and third mechanism is a stepwise reaction. All mechanisms can occur separately or in combination with each other, and influence the product yield of the depolymerization. Degradation can be triggered by ultrasonic irradiation, or by thermal, photochemical, or mechanical energy input [31, 32].

A chemical reaction, such as polymer degradation, can only take place spontaneously if the free molar energy of reaction ΔG - as defined by the Gibbs equation of state (Equation 7.1) - is negative [33, 34].

$$\Delta G = \Delta H - T\Delta S \qquad \text{Equation 7.1}$$

The molar enthalpy of reaction describes ΔH, which represents the difference of the energy contents of products and starting materials. The molar entropy of reaction ΔS stands for the difference between the entropies of the reaction products and the entropies of the starting materials. During the reaction of monomers to polymers, the state of order of the system increases, as the number of unbound and thus free particles present decreases. The entropy consequently decreases during the polymer formation reaction and ΔS is negative. Also ΔH is negative, so that ΔG is also negative, and the reaction proceeds spontaneously by itself. The activation energy for this reaction varies depending on the plastic involved. For polystyrene, an activation energy of 230 kJ/mol is needed [31]. If the temperature rises to the ceiling temperature, the polymerization rate (k_p) is in equilibrium with the depolymerization rate (k_{dp}) (Equation 7.2) [35, 36]. Figure 7.7 shows this equilibrium as a function of temperature for both reaction rates.

$$M_n^* \underset{k_p}{\overset{k_{dp}}{\rightleftharpoons}} M_{n-1}^* + M \qquad \text{Equation 7.2}$$

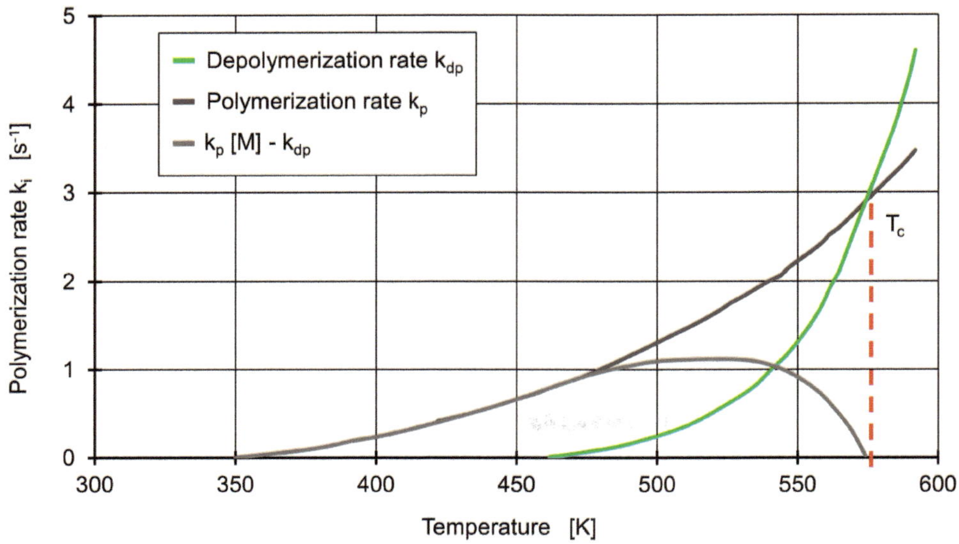

Figure 7.7 Polymerization and depolymerization rate as a function of temperature for polystyrene at a given concentrate [36, 37]

Due to the equilibrium between the depolymerization and polymerization rates, the free molar energy of the reaction ΔG is 0. As a result, the ceiling temperature T_c can be described by Equation 7.3 [34, 36].

$$T_c = \frac{\Delta H}{\Delta S} \qquad \text{Equation 7.3}$$

If the temperature rises above T_c, depolymerization is the dominant reaction, which means that the polymer chains are broken. Therefore, polymers that naturally have a low molar enthalpy of reaction ΔH and low ceiling temperature T_c produce higher monomer yields during thermal depolymerization. This is illustrated by Table 7.2, where w_a is the monomer yield by thermal depolymerization at a temperature of 300 °C to 500 °C in vacuum, v_v is the volatilization per minute at a temperature of 350 °C, T_d is the decomposition temperature, T_h is the temperature at which a polymer loss of 50 wt% is achieved in 40–50 min, and T_c is the ceiling temperature.

Table 7.2 Depolymerization Characteristics of Vinyl Polymers [32]

Polymer	w_a [wt%]	v_v [%]	T_d [°C]	T_h [°C]	T_c [°C]
Poly(methyl methacrylate)	92–100	5.2	330	283	220
Polystyrene	42	0.24	360	364	230
Poly(α-methylstyrene)	95–100	230	290	287	7
Polyethylene, linear	0.03	0.004	400	415	400
Polypropylene	0.2–2	0.069	380	387	300
Polyisobutylene	18–32	2.4	340	348	50
Polytetrafluoroethylene	97–100	2×10^{-6}	510	509	580

At this stage we want to refer to literature references stating higher ceiling temperatures for polymers like PS (390–395 °C, see e.g. Section 9.4, Ref [1]. The authors assume that different T_c measurement techniques are the root cause; possibly those techniques measure the thermodynamic equilibrium between monomer and polymer, plus the kinetic hindrance to physically depolymerize the polymer chain.

The depolymerization of plastics like polystyrene can be carried out in a continuously operating screw reactor. A co-rotating intermeshing twin screw extruder is suitable for this purpose, as large amounts of energy can be transferred into the melt via the screws. The mechanical energy input ensures that the plastic is melted and then depolymerized. The extruder barrels are also heated. The residence time can be adjusted specifically to the plastic and the product equilibrium, which is defined by the time that the material needs to pass through the extruder. A vacuum system removes the gaseous reaction products from the extruder, and these can then be liquefied in a condenser. Impurities and partially depolymerized plastic are discharged from the extruder. Only plastics with a relatively high zip length and low ceiling temperature, like polystyrene or poly(methyl methacrylate), are suitable for continuous depolymerization in a screw reactor [31, 38–41]. The operating temperature of this reactor is between 200 °C and 500 °C.

The screw reactor is already effectively used in the field of research to continuously depolymerize polystyrene and poly(methyl methacrylate) into its monomers. Michaeli et al. widely analyzed and researched the depolymerization of poly(methyl methacrylate). In these investigations, monomer yields of up to 99% were achieved. Due to economic reasons and lack of available feedstocks, industrial scale-up was not carried out [39–41]. In the research project with the short designation "ResolVe", which was funded by the German Federal Ministry of Education and Research

(No. 033R194C), a screw reactor was used to depolymerize polystyrene waste economically and continuously [38]. It was possible to depolymerize post-consumer waste with condensate yields of more than 60%. Furthermore, the reaction equilibrium could be specifically adjusted by the process parameters temperature, vacuum pressure, and residence time in the reactor [38, 42, 43]. INEOS Styrolution Group GmbH (Frankfurt am Main, Germany), and Americas Styrenics LLC (AmSty) (Texas, USA), two of the largest global producers of polystyrene, have also announced plans to construct a joint 100 t/d facility in Channahon, Illinois, that will utilize the Agilyx advanced recycling technology to recycle post-consumer polystyrene products back into virgin-equivalent styrene monomer [44].

Another example is the thermal depolymerization of polyamide 6 (PA 6), which is carried out at temperatures between 260 °C and 340 °C. The optimal process temperature is influenced by the residence time and the resulting reaction products at this temperature. Decarboxylation and deamination lead to a variety of secondary products, which influence the quality of the produced caprolactam during depolymerization. Due to the complexity of the reaction kinetics, the rate of monomer formation for polyamide 6 is very low. Similar to polyethylene terephthalate, hydrolysis can be used for an increased yield in the depolymerization of PA 6 (see Section 7.3.2). This is only possible with unmixed PA 6 waste, as the additives required for the application are technologically difficult to control. Contamination results in process instabilities [45, 46].

7.3.2 Solvolytic-Based Recycling

In this section, solvolytic approaches for the recycling of plastic waste are described, first the chemical solvent processes, followed by the physical dissolution of individual plastic waste fractions.

Chemical Solvent Processes

Chemical dissolution is aimed at plastics synthesized by means of polycondensation reactions (chemical equilibrium reactions), such as polyesters (polyethylene terephthalate (PET) or polycarbonate (PC), as well as polyamides (PA)) and polyurethanes (PUR) synthesized by polyaddition. Due to the complex reaction kinetics in polycondensation and the low monomer yield by thermal depolymerization, these plastics cannot be economically degraded solely by the introduction of thermal energy. Therefore, degradation agents must be used for chain scission [18]. The chemical dissolution involves the use of reagents (e.g., alcohols, alkalis, acids, or amines) which, with the simultaneous addition of heat and, if necessary, catalytic support, allow selective decomposition of the polymer structure into its monomer components (depolymerization) [9, 47]. The polymer degradation thus takes place through the reverse reaction of the polymer synthesis. Since polyamides have a higher chemical resistance than polyesters, the reverse degradation does not proceed as easily [18]. In general, however, a high purity of the waste fractions to be recycled is required to avoid undesirable side reactions and thus a reduction in the quality of the monomers [47]. The resulting monomer components can either be used to re-produce the same plastic or serve as a petrochemical feedstock for other chemical syntheses [9]. The following section therefore focuses on the plastic PET as an example.

Chemical depolymerization of polyesters by solvolysis is mainly used for PET, and a variety of methods have been known for many years [6, 18]. According to Vollmer et al. [48], the number of publications on chemical depolymerization of PET is greater by far than for any other polymer. One explanation is that a collection infrastructure for PET has been in place for some time, and this plastic has therefore attracted the attention of both the public as well as the research community [48].

PET is mostly used in the form of fibers, films, sheets, foams, or bottles. For many applications, a high molecular weight or sufficiently fast crystallization of the PET is important for achieving a good mechanical property profile in the application. Once the polymer is synthesized, it is very difficult to remove impurities [49]. If material recycling is not an option, chemical recycling by means of solvolysis offers a promising alternative - the advantage being that, depending on the process, pure monomers can be recovered from the waste for renewed plastics synthesis. At the same time, the recycling route allows the production of high-grade recyclates that have a quality comparable to virgin material and can be used, for example, in the food sector for packaging applications.

Figure 7.8 shows some solvolytic process variants which differ with respect to the reagents as well as the process control and the resulting degradation products. Under the action of water (and sulfuric acid or sodium hydroxide), glycol, or methanol in excess, the chemical equilibrium often shifts to the monomer (TS) or the oligomers (BHET, DMT) at temperatures above 200 °C [6]. Catalysts allow the reaction to proceed and be accelerated at lower temperatures.

However, hydrolysis has not yet been able to establish itself on an industrial scale for the recycling of PET, since the terephthalic acid obtained can only be purified from particles (e.g., pigments, fine dust adhering to the plastic waste) or catalysts at considerable expense, and the degradation reaction is quite slow. In addition, a large amount of chemicals and salts are required to neutralize the alkalis or acids, and their disposal also hinders technical implementation [18].

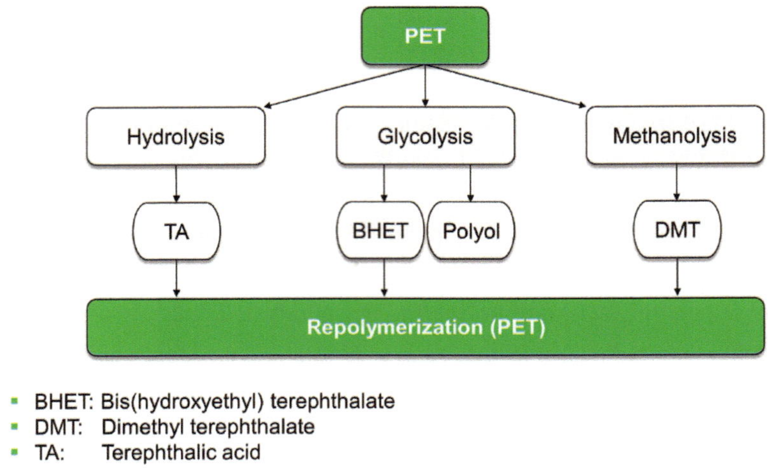

Figure 7.8 Solvolytic process variants for depolymerization using the example of PET [6] (figure: ref. [8])

The Swiss start-up gr3n has been pursuing a novel approach to PET recycling since 2011, using an alkaline hydrolysis process in which the reaction is supported by microwave technology [50]. The patented technology, called DEpolymerization by MicrowavE TechnolOgy (DEMETO) allows PET flakes, powder, or fibers to be depolymerized in a continuous process within 10 min. Contaminants (colorants, titanium oxide, metals) can be removed using a purification step after the depolymerization process [50]. A pilot plant will demonstrate the concept of the microwave-assisted recycling technology on an industrial scale. Currently, the plant is in the construction phase, and was designed for a PET input flow of 60 kg/h to produce monomers, terephthalic acid, or monoethylene glycol for the production of virgin PET [51].

The glycolysis of PET with ethylene glycol results in transesterification, which starts the depolymerization [52]. For a long time, purification of the oligomer melt after depolymerization was also considered hardly possible in glycolysis. Only coarse impurities could be removed by filtration, and finer particles (as well as catalyst or stabilizer residues) could only be separated with considerable effort. Separation of co-monomers, thermal-oxidative degradation products, and low-volatility by-products, among others, was practically impossible [18]. In the meantime, DuPont has developed a more advanced process variant for industrial use [53]. Mechanically recovered PET flakes are successfully degraded to the oligomer BHET (see Figure 7.8), which is chemically indistinguishable from primary monomer. Impurities are removed during the glycolysis process by a combination of monomer and polymer filtration units and with the aid of vacuum extraction, which runs for several hours at temperatures between 270 °C and 300 °C. The base monomer is repolymerized and then processed into PET film [53].

One advantage of methanolysis is that it is relatively insensitive to impurities of all kinds, additives, modificators, and color pigments. However, in the past, yields of only 80–85% were possible. In addition to the high process temperatures, pressurization (20–40 bar) is also necessary, which makes it difficult to introduce the feedstock into the reactor, especially in the continuous process [6, 9]. Another challenge is the investment cost, particularly relating to the strict safety requirements resulting from the use of superheated methanol. [9].

Industrially, methanolysis is promoted by the company Eastman Chemical Company, Kingsport, Tennessee, USA, which since 2019 has been conducting a technical feasibility study with the aim of implementing the construction of a commercial-scale methanolysis plant [54]. The company now plans to build a $250 million PET depolymerization facility at its company headquarters by the end of 2022 [55]. The methanolysis technology will be used to recycle 100 kt/a of PET waste from sources such as carpet fibers and PET packaging, which can then be processed as a secondary raw material to manufacture plastic products such as refillable water bottles. According to the company, this approach is 20–30% less greenhouse gas intensive than using primary raw materials. Also, other companies, such as Loop Industries Inc., are engaged in the industrialization of PET solvolysis [55].

Physical Solvent-Based Processes

The separation of different polymer types by their different solubility in (mostly) organic solvents is the basic principle of solvent processes. This process variant dates back to patents in the 1990s and has so far been little represented on the market [47]. The dissolution of the plastic is a physical process and aims to ensure that the basic chemical structure of the polymer chains remains unaffected. Chain degradation, as occurs in mechanical shredding or in chemical depolymerization, can be largely avoided if a suitable polymer-specific solvent is selected and the process is adequately controlled. Physical solvent-based recycling is thus distinct from both mechanical and chemical recycling (see Figure 7.4), although chemicals are necessarily part of these processes as solvents. In addition, the process offers the advantage that additives or fillers that are insoluble in the solvent can be separated to leave the purified polymer. In order to obtain the required processing properties or functional properties of the material, it is therefore necessary to use additives again before reprocessing into plastic products.

The majority of plastics on the market are polyolefins, which are frequently used in packaging applications (among other things). Due to the high chemical resistance of these materials, chemical solvolysis is not a technologically and economically useful approach to tapping secondary raw materials in the context of a circular economy. At the same time, numerous plastic products, such as PE films or some injection-molded parts, are made of multi-material composites, which are also not recyclable by the chemical recycling route. Alternatively, for some plastic waste streams, physical dissolution of plastics offers a promising alternative to increase recycling rates. Typical plastics considered for physical solvent recycling are therefore PE, PS, PA, PBT, and PVC, whereby metallic composite components such as aluminum barrier layers or electroplated layers can also be separated from the polymer [47, 56, 57].

Figure 7.9 shows the so-called CreaSolv® process as an example of solvent-based recycling. The development is the result of cooperation between the Fraunhofer Institute for Process Engineering and Packaging (IVV) (Freising, Germany) and CreaCycle GmbH (Grevenbroich, Germany) [56, 57]. The target plastic is physically dissolved by a selective solvent and foreign materials or impurities are separated. The selection of economical and label-free solvent formulations with a high flash point plays an important role for safe plant operation. After the polymer and solvent have been separated, the solvent can be recovered for reuse by means of a processing step [57]. Especially for large-scale applications, reprocessing is extremely important from an economic as well as an environmental point of view. After precipitation and drying, a high-purity polymer remains, which is processed into plastic granules by extrusion.

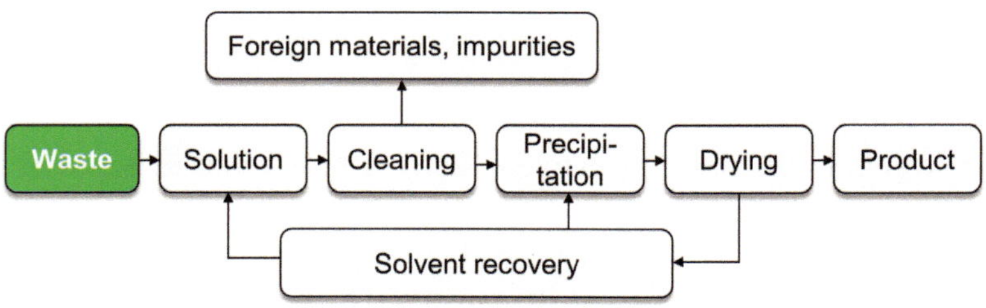

Figure 7.9 Plastics recycling using the solvent-based CreaSolv® process [56, 57]

There are several collaborations with industry partners worldwide that have resulted in pilot plant projects for recycling various plastic waste streams through an adaptation of the CreaSolv process [58], four of which are presented below:

- Unilever PLC (London, UK), operates a plant in Indonesia that recovers pure PE from multi-layer sachets comprising 60% PE, enabling consumers in developing countries to use small quantities of quality products in these sachets without the associated risk of the plastic getting into the environment. The plant currently processes 3 t of waste per day, and is able to produce recycled polyethylene resin for one-sixth of the energy cost of producing virgin resin [59].
- The closed-loop recycling of PS foam waste, which is used as an insulation material in the construction sector and contains a brominated flame retardant, is being driven by European Initiative Polystyrene Loop B.V. (Terneuzen, Netherlands). The pilot plant is to be operated with a capacity of 3 kt/a and went into operation in June 2021 [60].

- A similar approach to solvent-based recycling is being taken by the APK AG (Merseburg, Germany). With its Newcycling technology, 12 kt/a of mixed plastic waste and multi-layer packaging are recycled. This includes post-consumer composite packaging waste containing PE or PA in the material composite, which acts as the target plastic [61].
- The PureCycle process developed by Procter & Gamble Corp. (Cincinnati, Ohio, USA) enables recovery of high-quality polypropylene from composite waste such as old carpets or packaging. The PureCycle process allows removal of colors, odors, and impurities [47, 62], and the plant in Ohio has a capacity of 50 kt/a [47].

In summary, high-quality plastic recyclates can be recovered from plastic waste by both chemical and physical processes, which can even be used in the food sector and have a quality comparable to primary plastics. The processes used are limited to selective material streams, which must be available in sufficient quantities and are often not qualitatively suitable for material recycling. Currently, recycling technologies still play a minor role on the market, but development activities are increasing. Innovation barriers and market entry barriers result from high investment costs and long development times for the processes, which must offer "added value" both environmentally and economical to compete with thermal recovery and material recycling. The process engineering costs are generally high, with material losses during the process and the use of chemicals (e.g., solvents) also having a disadvantageous effect on market penetration. In particular, the process requires chemicals that are economically advantageous (as well as being largely harmless toxicologically and environmentally), and a safe plant operation. At the same time, the disposal of chemicals is also cost-intensive, so that a purification and multiple use is aimed for.

7.3.3 Thermochemical Processes

In case of heavily contaminated plastic waste or composite materials, material recycling is often not economically feasible and results in downcycling. A solution is provided by thermochemical processes, with which feedstock recycling can be achieved [18]. In thermochemical processes the polymer bonds are broken by thermal energy, and the process runs in an inert atmosphere or under the influence of reactive gases [47]. Gases, petrols, oils, and waxes can be recovered from plastic waste, from which important raw chemicals can be obtained. This recovery process has its largest application for polyolefins (PE and PP), but there are also emerging developments for recycling PS, PET, and PMMA. Thermochemical feedstock recycling currently plays a minor role in the handling of plastic waste, accounting for less than 1% of the waste stream, although this value is predicted to rise to 13% by 2030 [48].

Pyrolysis and gasification are the main thermochemical methods and are described in more detail below. Pyrolysis is a thermal process and runs in inert gas such as nitrogen. Gaseous hydrocarbons, oils and waxes can be obtained as products. The gasification is based on a thermal oxidative recycling, so that the process takes place in oxygen or water steam. This results in a synthesis gas (e.g., carbon monoxide (CO) and hydrogen (H_2)) as the degradation product.

Pyrolysis

In pyrolysis, polymer chains from waste plastics can be broken down to shorter hydrocarbon components by residence at temperatures above 300 °C in an inert gas. This process is characterized by a broad product spectrum, which is influenced by the starting material fed in, the process conditions, and the type of reactor. Consequently, the resulting pyrolysis oils and gases can be used in a variety of ways, but must be subjected to petrochemical separation processes [9].

Pyrolysis is characterized by the fact that sorting residues from material recycling of different quality can be fed into the pyrolysis reactor. The addition of PVC-rich plastic waste is avoided, which leads to problems in downstream processes and requires special plant technology and the addition of lime or ammonia [18].

Thermal degradation proceeds at high temperature, and on the basis of the process temperature a differentiation is made between slow pyrolysis (at less than 400 °C), moderate pyrolysis (from 400 °C to 600 °C) and fast pyrolysis (above 600 °C). The temperature influences the residence time of the plastic in the pyrolysis reactor and varies from a few seconds to several hours [9]. Furthermore, the thermal process control also influences the resulting products. At high reactor temperatures, lighter products such as gas and coke are produced. At low temperatures, a higher proportion of waxes and oils are produced, as shown in Figure 7.10 [63]. Atmospheric pressure is usually used, although the influence of pressure is still being researched [64].

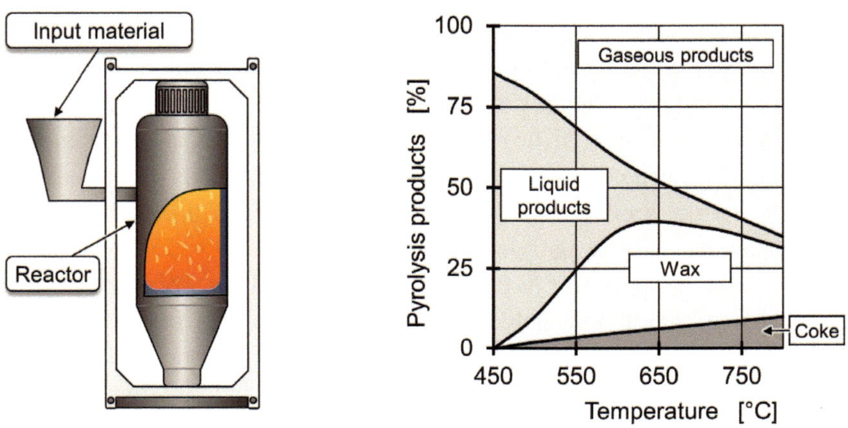

Figure 7.10 Isobaric fractional change in the degradation of polyolefins as a function of the applied temperature [63] (figure: ref. [8])

If the isobaric fraction change during the degradation of polyolefins from Figure 7.10 is compared with an existing pyrolysis plant, it can be seen that at temperatures of 800–900 °C only about 25% of the total amount are oils and waxes. The RT7000 plant from Recycling Technologies Ltd (Swindon, UK), is used as an example for the commercial pyrolysis of plastic waste. Their aim is the conversion of sorting residues of plastic waste (PE and PP), whereby proportions of PET, PC, and PS are also tolerable. In contrast, plastic containing PVC, PU, and PA does not give a high-quality pyrolysis oil. The plant produces 5200 t/a of oils and waxes, which can (in part) be used for the indirect production of new plastics [65]. By means of further process steps, monomers for the production of plastics can be recovered in a similar way to conventional crude oil, for example in an ethylene cracking plant. In addition, the partially unsaturated hydrocarbons can be converted by hydrogenation into more stable substances for use as fuels or chemical precursors [9].

Figure 7.11 shows the distribution of the distillates depending on the carbon distribution. Only 70% of the oils and waxes could theoretically be used for the production of new plastics. However, many of these products are used for petrochemicals and fuel production. In general, pyrolysis is characterized by a low proportion of recyclable products that can be used for polymer production, and therefore offers only a small potential for the circular economy of plastics.

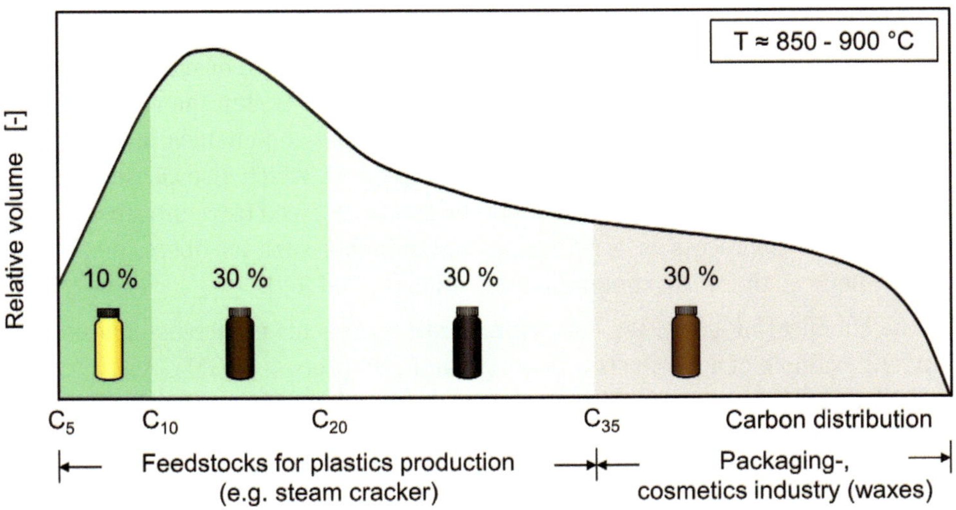

Figure 7.11 Total volume output of pyrolysis oils and waxes correlated to carbon distribution of the RT7000 plant of Recycling Technologies Ltd (UK) [65] (figure: ref. [8])

Considering the thermal and rheological properties of plastics as well as the composition of plastic waste streams or sorting residues, the selected reactor type also has an important effect on the composition of the degradation products. From a process engineering point of view, the aim is also to achieve the most favorable and reliable mode of operation in terms of energy, in order to open up a market opportunity for the pyrolysis of plastic waste in both environmental and economic terms.

Different plant designs and reactor types have been developed, including melting pots, autoclaves, tubular reactors, rotary kilns, and fluidized-bed reactors. Rotary kilns are the simplest and most stable plant technology and widespread in the market. Another possibility to pyrolyze plastics on an industrial scale is a closed batch reactor. In this process, the plastic is placed in a reactor filled with inert gas, which is hermetically sealed from the environment. Due to the high process temperatures (over 700 °C) and long residence times, products with a low molecular weight are preferentially produced. In addition, the amount of gaseous products increases with rising temperature [66]. The disadvantages of batch reactors are the slow heating rates and long residence times at high temperatures. For this reason, closed batch reactors are primarily used for the pyrolysis of polyolefins.

A further development of this technology is a flow reactor. The main difference between these reactors is that an inert gas flows continuously through the reactor. In this way, the vaporized products are removed and condensed in a subsequent fractionation process outside the reactor. As a result, the reaction products are exposed to high temperatures for a comparatively short time, which reduces subsequent reactions [67].

The fluidized-bed reactor is characterized by good heat and mass transfer, by increasing the surface area of the plastic. Compared to the batch reactor, the fluidized-bed reactor offers a significantly shorter reaction time. For increased energy input, a heated inert gas is bubbled through quartz sand within the reactor. Under these conditions, the sand behaves similarly to a fluid and enables rapid heat release to the material. Subsequently, the inert gas flow, which also carries the gas- and vapor-phase reaction products out of the reactor, is freed from entrained quartz sand [68]. This process is sensitive to contaminants such as fibers, metal, and other fillers, as these can contaminate the quartz sand [69].

In the chemical industry, the conventional stirred reactor is increasingly used. A current example of a pyrolysis process is the ReOil process of OMV AG (Burghausen, Germany). This works with a solvent to create better heat transfer. The pilot plant can process 100 kg of mixed plastic such as polyolefins or polystyrene from the packaging sector plastic in 1 h and produce 100 l of synthetic crude oil [70].

In general, the pyrolysis is characterized by a poor life cycle assessment, but it is still better than incineration with energy recovery. With the introduction of regulations for recycling and a CO_2 tax, there is an increased potential for the use of pyrolysis in the field of plastic recycling [48].

It has long been known that pyrolysis results in 20-fold reduction in the amount of flue gases, which leads to significant savings in gas cleaning [18]. The life cycle assessment can be improved by reactor design and by catalyst input. Pyrolysis offers the possibility to recycle mixed waste streams, but due to the thermally insulating behavior of the polymers, they have to remain in the plants for a long time. The investments for a recycling plant are very high and result in high product prices. The conventionally produced plastics from crude oil are cheaper than recycled plastics from pyrolysis products [48]. Also, only a small proportion of the degradation products are suitable for polymer synthesis and can be recovered. From some of the oils, gases, and waxes, new plastics with a quality comparable to virgin material can be produced in cooperation with polymer producers. As an example, SABIC AG (Riab, Saudi Arabia), is working with Plastic Energy Ltd (London, UK) and Petronas AG (Kuala Lumpur, Malaysia) to promote a circular economy [71]. ChemCycling by BASF SE (Ludwigshafen, Germany) has been producing commercial polyolefin granules from pyrolysis products since 2020 [72]. Products that can be made from the re-synthesized plastics include pharmaceutical boxes, protective transport packaging, and food packaging. Close cooperation is maintained with pyrolysis partners in the course of feedstock recycling [72].

Gasification

Gasification technology is a further feedstock-based system for the production of chemical precursors and fuels. It can break down virtually all carbon-based feedstocks such as heavy oil, hazardous waste, and sewage sludge into their constituents, and was originally developed for coal and refinery residues [9, 18]. In addition to pyrolysis, the resulting products are reused as basic raw materials in the chemical industry. The process is a partial oxidation of the carbon-rich materials with an insufficient amount of oxygen for complete incineration. Gasification proceeds using air, oxygen, and steam, or mixtures of these, and usually at temperatures between 700 °C and 1600 °C and pressures between 10 bar and 90 bar [6].

In contrast to pyrolysis, there is no splitting back into short-chain hydrocarbons. The quality of the feedstock supplied to the reactor is decisive for the decomposition products. Starting from high-quality synthesis gases for the petrochemical industry with a high carbon monoxide and hydrogen content, via reduction gases for blast furnace processes, fuel gases with widely differing calorific values are obtained for heat and power generation [18].

The gasification of plastics to synthesis gas does not, strictly speaking, fall within the concept of a circular economy, since no monomers and oligomers are obtained that can be reprocessed to new plastics [48]. However, a major advantage of gasification of plastic waste is the purity requirements of the input material. As a result, the sorting effort is significantly lower compared to pyrolysis or material recycling [73]. In addition, it is also possible to process mixed materials, such as sorting residues of household waste. A major part of research and development activities is focused on this process variant [9].

The variety of process variants for gasification is large, with more than 18 differentiations described as early as 1995 [18]. As one of the few existing examples of plastic waste processing by means of gasification, the plant of the company Showa Denko K.K. (Kawasaki, Japan), with an annual processing capacity of 70 kt, is described in more detail below. The plant has been operating since 2003, and Figure 7.12 shows the process flow [74].

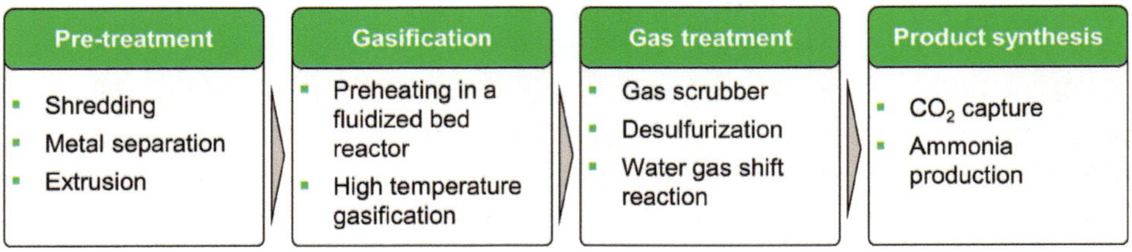

Figure 7.12 Process flow of the plastic waste gasification of Showa Denko K.K. [74]

The plastic waste is first shredded and metals are extracted. This is followed by degassing of highly volatile components, drying, and compacting using an extruder with 140 hole-type dies. Gasification takes place in two steps. A mixture of steam and oxygen is used as gasification medium. In the first step, preheating is done in a fluidized-bed reactor at 600–800 °C, and in the second step, high-temperature gasification is done in a shaft furnace at 1300–1500 °C. The resulting syngas is used for the production of ammonia and has to be upgraded to pure hydrogen in several steps. It is fed to an alkaline scrubber for cleaning. Then the hydrogen content is increased by a water-gas shift reaction and other components are separated. For ammonia production, the CO_2 present must be removed from the gas, by cooling and pressurizing to liquefy the CO_2, enabling it to be separated and sold on as dry ice. The remaining hydrogen is used as feedstock for ammonia synthesis. Slagging problems in both reactors and the disposal of residues (metals, sulfur, slag) due to poor quality are a challenge for plant operations [74].

Although gasification does not allow direct feedstock recycling, it can achieve long-term success in commercial operations. However, this technology is only expected to be economical for large capacities above 100 kt/a [47].

7.3.4 Biodegradation

In addition to hydrolysis, depolymerization, and other degradation mechanisms, which are described earlier in this chapter, there is another degradation mechanism based on the biodegradability of plastics. Since this degradation mechanism only occurs in small proportions compared to all the other mechanisms mentioned, it will only be briefly discussed below.

Microorganisms (such as bacteria, fungi, and algae [75]) play an essential role in biodegradation [76]. The degradation of conventional plastics by these organisms (biocorrosion) can lead to undesirable changes in material properties (for example, discoloration or brittleness). Often, only part of the plastic material is attacked by the microorganisms. However, if microorganisms should be used to dispose of plastic waste, the "biodegradation" must not simply consist of physical degradation of the material to small fragments that are no longer visible [76]. Instead, according to the definition, the entire polymer chain of a biodegradable plastic must be able to be cleaved by these microorganisms via enzymes and completely degraded to carbon dioxide, water, methane, mineral salts, and biomass. Only in this case can a plastic can be called "biodegradable" [16, 77, 78]. But the degradation products must be captured and recycled so that the CO_2 is not released into the atmosphere. Without this measure, biodegradation in the environment is not a solution. In addition, from a technological point of view, it is disadvantageous that biodegradation interrupts the material cycle and the energy stored in the material may remain unused.

Process

Since plastics are generally not water-soluble and their polymeric components would be too large to pass through the cell membranes of microorganisms anyway, the organisms must excrete enzymes from the cells and first reduce the chain length of the polymers until they are water-soluble. The dissolved intermediates can then be taken up into the cells for further conversion. Since the enzymes, in turn, cannot enter the plastics due to their size, biodegradation is an interfacial process, and the material is gradually removed from the surface. In the case of some plastics, the initial breakdown – usually of non-water-soluble and non-toxic plastics into small, soluble, and thus mobile fragments – can under certain circumstances lead to environmentally problematic effects. However, parts of these fragments may have such a structure that they are not further converted by microorganisms, and so accumulate in the environment [76].

Whether plastics are attacked by enzymes is basically determined only by the structure of the polymers and not, for example, by the origin of the raw materials. Thus, chemically unmodified natural polymers such as cellulose, starch, or proteins are basically biodegradable. However, chemical modifications (e.g., esterification to cellulose triacetate) or polymerization of natural monomers (e.g., lactic acid) can produce polymers that can no longer be broken down by enzymes. On the other hand, a number of demonstrably biodegradable plastics exist that are produced entirely from petrochemical substances (for example, various polyesters, polyester amides, and polyester urethanes) [76]. However, most oil-based plastics (polyethylene (PE), polypropylene (PP), and polyethylene terephthalate (PET)), are not biodegradable. This is in contrast to poly(ε-caprolactone) (PCL), polybutylene succinate/adipate) (PBS/A), and polybutylene adipate-co-terephthalate) (PBA/T) [75, 78]. These polymers have ester bonds that are degradable by certain enzymes secreted by microorganisms, so they are classified as petroleum-based biodegradable plastics [75].

Influence of Polymer Structure on Biodegradation

In practice, it has been shown that an essential prerequisite for the enzymatic cleavage of polymer chains is the presence of heteroatoms such as oxygen or nitrogen. Chains consisting exclusively of carbon (as in polyethylene or polypropylene) are not attacked, or are attacked only extremely slowly, by the enzymes secreted by microorganisms (polyvinyl alcohol is an exception here). The corresponding (and technically interesting) biodegradable plastics are mainly polyesters, polyamides, polyester urethanes, and modified polysaccharides [76]. However, though enzymes may in principle be able to split the ester, ether, amide, and urethane bonds, other factors play an important role. For example, biological polyamides (proteins) are known to be biodegradable, while technical polyamides (for example, nylon polymers) show no measurable enzymatic attack. In the case of technical polyamides, strong attractive forces (hydrogen bonds) exist between the polymer chains, so that the polymer chains are not mobile enough to fit into the catalytically active center of the enzymes. Similar conditions exist in aromatic polyesters (such as polyethylene terephthalate (PET)). The stiffening of the polymer chains by the aromatic rings prevents biodegradation, while polyesters without aromatic rings (aliphatic polyesters) are often quickly cleaved enzymatically [76]. However, other properties of the polymers, such as the length of the polymer chains, any branching or crosslinking, the distribution and arrangement of disordered (amorphous) and ordered (crystalline) regions in the plastics, and surface properties can also have a significant influence on the biodegradation behavior [76].

Measurement of Biodegradation

For the detailed investigation of biodegradation processes on plastics, it is necessary to install special optimized laboratory test systems. This often involves the isolation of individual microorganism strains from natural mixed populations that are particularly effective at degrading certain plastics [76, 79]. With the aid of these special strains, or their extracellular enzymes, degradation tests are then carried out in the laboratory under very defined conditions and within a reasonable time [76]. The production of carbon dioxide as the end product of mineralization is a generally accepted parameter in standardized laboratory methods for determining the complete biodegradability of a material in an aerobic environment. Under aerobic conditions, CO_2 production correlates closely with oxygen consumption, which is recorded as a metric in some test procedures [80]. Since degradability and degradation rate depend on the environmental conditions, correspondingly adapted test methods have been developed for the different habitats such as compost, soil, and aqueous environments, and are available as standardized methods [80].

Technological Processes

Since biodegradable plastics are supposed to decompose themselves, there are no direct technological processes. Thus, a distinction is only made between "industrial composting", "garden composting", and "fermentation".

Industrial composting basically consists of three steps: preparation, intensive rotting, and post-rotting. In the preparation stage, the input material is shredded and homogenized. This is followed by intensive rotting. Under a controlled oxygen and moisture supply, the organic and green waste is decomposed, resulting in fresh compost. In the subsequent post-rotting stage, the fresh compost is stabilized and converted into finished compost [80]. Importantly, organic and green waste includes biodegradable plastics used for waste bags for the collection of organic waste, and waste from agriculture, horticulture, pond management, forestry, hunting, and fishing, such as mulch films. According to the BioAbfV, packaging made of bioplastics, in contrast, does not belong to the waste that is permitted for recycling in an industrial composting or fermentation plant [80].

Composting biodegradable plastics on one's own compost in the garden is possible in principle, provided that the products are certified as "home compostable" or "garden compostable", and biodegradable plastic waste is permitted for home composting according to the relevant waste statutes [80]. In terms of the CO_2 balance, however, this is not advisable, especially if the polymer is of fossil origin.

In contrast to industrial composting, the fermentation of organic and green waste produces biogas as well as compost. Here, too, the waste is processed before biological treatment. The actual degradation process takes place under oxygen exclusion in a closed container, at either mesophilic (35–40 °C) or thermophilic (50–55 °C) temperatures. Biogas is formed, which can be used for energy purposes or synthesis of new polymers [80].

References for Chapter 7

[1] Dahlmann, R., Facklam, M.: Kunststoffverwertung erneut im Wandel – Forschungsprojekte zur Kreislaufwirtschaft. *RWTH Themen,* (2020) 1, pp. 36–41

[2] Dahlmann, R., Haberstroh, E., Menges, G.: *Menges Werkstoffkunde Kunststoffe*, Hanser Verlag (2021)

[3] Dahlmann, R.: Closing loops: Opportunities for a plastics circular economy. In: *30th International Colloquium on Plastics Technology 2020*, Shaker Verlag (2020)

[4] *Plastics – The Facts 2019 – An Analysis of European Plastics Production, Demand and Waste Data*, Plastics Europe, Association of Plastics Manufacturers (2020)

[5] Hessel, J.: 100-years service-live for polyethylene pipes – Review and prospects. *3R International* 46 (2007) 4, pp. 242–246

[6] Aguado, J., Serrano, D.: *Feedstock Recycling of Plastic Wastes*, The Royal Society of Chemistry (1999)

[7] *Circular Economy Action Plan – For a cleaner and more competitive Europe*, European Union (2020)

[8] Facklam, M.: "Rohstoffliches Recycling von Kunststoffabfällen". Promotionsvortrag. RWTH Aachen University (2019)

[9] Lechleitner, A., Schwabl, D., Schubert, T., Bauer, M., Lehner, M.: Chemisches Recycling von gemischten Kunststoffabfällen als ergänzender Recyclingpfad zur Erhöhung der Recyclingquote. *Österreichische Wasser- und Abfallwirtschaft*, 73 (2020), pp. 47–60

[10] Directive (EU) 2018/852 of the European Parliament and of the Council. *Official Journal of the European Union*, (2018), pp. 141–154

[11] *Prognostizierte Umsatzentwicklung in der Branche Abfallwirtschaft in Deutschland in den Jahren von 2012 bis 2024, https://de.statista.com/statistik/daten/studie/247964/umfrage/prognose-zum-umsatz-in-der-abfallwirtschaft-in-deutschland/*, accessed June 30, 2021

[12] Kreibe, S., Pitschke, T., Berkmüller, R., Bokelmann, M., Förster, A., Stramm, C.: *Umweltbezogene Bilanzierung von "intelligenten" und "aktiven" Verpackungen hinsichtlich der Recyclingfähigkeit und Durchführung eines Dialogs mit Akteuren der Entsorgungs- und Herstellungsbranchen*, Umweltbundesamt (Ed.), Dessau-Roßlau (2017)

[13] *Prüfung und Testierung der Recyclingfähigkeit: Anforderungs- und Bewertungskatalog des Institutes cyclos-HTP zur EU-weiten Zertifizierung (CHI-Standard)*, Institut cyclos-HTP GmbH (2019)

[14] Potthoff, P.: Alterung und Recycling von HDPE für Flaschentransportkästen. *Kunststoffe* 75 (1985) 8, pp. 481–486

[15] Rudolph, N., Kiesel, R., Aumnate, C.: *Understanding Plastics Recycling – Economic, Ecological, and Technical Additional Profit Aspects of Plastic Waste Handling*, Hanser Verlag (2020)

[16] Ehrenstein, G., Pongratz, S.: *Beständigkeit von Kunststoffen*, Hanser Verlag (2007)

[17] Pongratz, S.: *Alterung von Kunststoffen während der Verarbeitung und im Gebrauch*, [Dissertation], Erlangen-Nürnberg University (2000)

[18] Brandrup, J., Bittner, M., Michaeli, W., Menges, G.: *Die Wiederverwertung von Kunststoffen*, Hanser Verlag (1995)

[19] *Stoffstrombild Kunststoffe in Deutschland 2019*. Study, Conversio Market & Strategy GmbH (2020)

[20] Tschachtli, S., Pitschke, T., Kreibe, S., Martin, A.: *Recyclingfähigkeit von Verpackungen – Konkretisierung, Untersuchungsrahmen und Kriterienkatalog*, Bifa Umweltinstitut GmbH (2019)

[21] Feil, A., Coskun, E., Bosling, M., Kaufeld, S., Pretz, T.: Improvement of the recycling of plastics in lightweight packaging treatment plants by a process control concept. *Waste Management & Research*, 37 (2019) 2, pp. 120–126

[22] Kranert, M.: *Einführung in die Kreislaufwirtschaft: Planung - Recht - Verfahren*, Springer Vieweg Verlag (2017)

[23] Wagner, J, Günther, M., Rhein, H., Meyer, P.: *Analyse der Effizienz und Vorschläge zur Optimierung von Sammelsystemen der haushaltsnahen Erfassung von Leichtverpackungen und stoffgleichen Nichtverpackungen auf der Grundlage vorhandener Daten*, Umweltbundesamt (Ed.), Dessau-Roßlau (2018)

[24] *Polyolefins. https://plasticseurope.org/plastics-explained/a-large-family/polyolefins-2/*, accessed January 17, 2022

[25] Nentwig, J.: *Kunststoff-Folien*, Hanser Verlag (2006)

[26] Strangl, M., Ortner, E., Fell, T., Ginzinger, T., Buettner, A.: Odor characterization along the recycling process of post-consumer plastic film fractions. *Journal of Cleaner Production*, 260 (2020) 121104, pp. 1–11

[27] Dreu, W.: Schmelzefilter für die Polymeraufbereitung. *Kunststoffe*, 94 (2004) 2, pp. 67–72

[28] Hopewell, J., Dvorak, J., Kosior, E.: Plastics recycling: Challenges and opportunities. *Philosophical Transactions of the Royal Society B: Biological Sciences*, 364 (2009) 1526, pp. 2115–2126

[29] *Resources SAVED by recycling*, Fraunhofer-Institut für Umwelt-, Sicherheits- und Energietechnik (UMSICHT) (2020)

[30] Strain, D.E.: U.S. 2 030 901: Process for depolymerizing alpha substituted acrylic acid esters. U.S. Patent and Trademark Office (18.02.1936)

[31] Lechner, M.D., Gehrke, K., Nordmeier, E.H.: *Makromolekulare Chemie - Ein Lehrbuch für Chemiker, Physiker, Materialwissenschaftler und Verfahrenstechniker*, Springer-Verlag (2014)

[32] Sawada, H.: Depolymerization. In: Mark, H.F. (Ed.): *Encyclopedia of Polymer Science and Technology*, John Wiley & Sons (2007)

[33] Callen, H.B.: *Thermodynamics and an Introduction to Thermostatistics*, John Wiley & Sons (1998)

[34] Kaitz, J.A., Lee, O.P., Moore, J.S.: Depolymerizable polymers: preparation, applications, and future outlook. *MRS Communications*, 5 (2015) 2, pp. 191–204

[35] Koltzenburg, S., Maskos, M., Nuyken, O.: *Polymere: Synthese, Eigenschaften und Anwendungen*, Springer-Verlag (2014)

[36] Odian, G.: *Principles of Polymerization*, John Wiley & Sons (2004)

[37] Cook, R.E., Dainton, F.S., Ivin, K.J.: Effect of structure on polymerizability: Olefin polysulfone formation. *Journal of Polymer Science*, 29 (1958) 120, pp. 549–556

[38] Niessner, N., Nosić, F., Wilhelmus, B., Nottbohm, J., Bittner, T., Schäfer, P., Kolb, T., Schedel, A., Feil, A., Rolf, K.: *ResolVe - Recycling von Polystyrol mittels rohstofflicher Verwertung*, Final report on the German Bundesministerium für Bildung und Forschung (BMBF)-project (No. 033R194C) (2021)

[39] Michaeli, W., Breyer, K.: Chemisches Recycling von PMMA-Depolymerisation durch Extrusion. *Achema Mag*, 97 (1997) 53, pp. 52–56

[40] Michaeli, W., Breyer, K.: Chemisches Recycling PMMA – Die degradative Extrusion mit Doppelschnecken. *Plastics-Special*, (1997), pp. 52–58

[41] Menges, G., Michaeli, W., Lackner, V.: Die degradative Extrusion – Stand und Aussichten. *Monatshefte für Chemie*, 127 (1996) 8–9, pp. 821–839

[42] Schäfer, P., Hopmann, C., Facklam, M., Hollerbach, L., Kolb, T., Schedl, A., Schmidt, H.-W., Nosić, F., Wilhelmus, B.: Continuous chemical recycling of polystyrene with a twin-screw extruder. In: Hopmann, C., Dahlmann, R. (Ed.): *Advances in Polymer Processing 2020 – Proceedings of the International Symposium on Plastics Technology*, Springer-Verlag (2020), pp. 37–49

[43] Schäfer, P., Hopmann, C., Facklam, M., Dauber, N.: Increasing the efficiency of the continuous depolymerisation of polystyrene. *Proceedings of the 79th Annual Technical Conference (ANTEC) of the Society of Plastics Engineers (SPE)*, (2021)

[44] "INEOS Styrolution and AmSty announce joint facility for advanced recycling of polystyrene in Channahon, Illinois, using Agilyx technology", *https://www.ineos.com/businesses/ineos-styrolution/news/ineos-styrolution-and-amsty-announce-joint-facility-for-advanced-recycling-of-polystyrene-in-channahon-illinois-using-agilyx-technology/*, accessed July 02, 2021

[45] Braun, D., Disselhoff, R., Guckel, C., Illing, G.: Rohstoffliches Recycling von glasfaserverstärktem Polyamid 6. *Chemie Ingenieur Technik*, 73 (2001) 3, pp. 183–190

[46] Meys, R., Frick, F., Westhues, S., Sternberg, A., Klankermayer, J., Bardow, A.: Towards a circular economy for plastic packaging wastes–the environmental potential of chemical recycling. *Resources, Conservation and Recycling*, 162 (2020) 105010, pp. 1–10

[47] Schlummer, M., Fell, T., Mäurer, A., Altnau, G.: Die Rolle der Chemie beim Recycling – Physikalisches und chemisches Kunststoffrecycling im Vergleich. *Kunststoffe*, 110 (2020) 6, pp. 51–54

[48] Vollmer, I., Jenks, M.J.F., Roelands, M.C.P., White, R.J., van Harmelen, T., de Wild, P., van der Laan, G.P., Meierer, F., Keurentjes, J.T.F., Weckhusysen, B.M.: Beyond Mechanical Recycling: Giving NewLife to Plastic Waste. *Angewandte Chemie, International Edition*, 59 (2020), pp. 15402–1542

[49] Sinha, V., Patel, M.R., Patel, J.V.: PET Waste management by chemical recycling: A review. *Journal of Polymers and the Environment*, 18 (2010), pp. 8–25

[50] Crippa, M., Morico, B.: PET Depolymerization: A novel process for plastic waste chemical recycling. In: Basile, A., Centi, G., De Falco, M., Iaquaniello, G. (Ed.): *Studies in Surface Science and Catalysis*, 179 (2020), pp. 215–229

[51] *DEMETO – Newsletter April 2021*, DEMETO (2021)

[52] Carta, D., Cao, G., D'Angeli, C.: Chemical recycling of poly(ethylene terephthalate) (PET) by hydrolysis and glycolysis. *Environmental Science and Pollution Research*, 10 (2003) 6, pp. 390–394

[53] "DuPont Teijin Films launches chemical recycling LuxCR™ depolymerization process", *https://www.petcore-europe.org/news-events/227-dupont-teijin-films-luxcr.html*, accessed June 09, 2021

[54] "Eastman offers innovative recycling technology for polyesters", *https://www.eastman.com/Company/News_Center/2019/Pages/Eastman-offers-innovative-recycling-technology-for-polyesters.aspx*, accessed June 09, 2021

[55] Tullo, A.H.: "Eastman will build a $250 million plastics recycling plant", *https://cen.acs.org/environment/recycling/Eastman-build-250-million-plastics/99/web/2021/02*, accessed June 09, 2021

[56] Menz, V., Levèvre, J., Schlemmer, M.: Werte wiedergewinnen – Recycling. *Kunststoffe*, 102 (2012) 7, pp. 72–75

[57] Schlummer, M., Kappich, F.: *Recycling von Metall-Kunststoffverbunden und Hybridwerkstoffen*, Abschlussbericht zum Forschungsvorhaben im Projektverbund ForCycle – Ressourcenstrategie für Bayern unter besonderer Berücksichtigung von Sekundärrohstoffen. Fraunhofer Institut Verfahrenstechnik und Verpackung IVV, 2016

[58] "CreaCycle GmbH – Innovative Plastic Recycling Technologies", *https://www.creacycle.de/en/creasolv-plants.html*, accessed June 12, 2021

[59] "Our solution for recycling plastic sachets takes another step forward", *https://www.unilever.com/news/news-and-features/Feature-article/2018/our-solution-for-recycling-plastic-sachets-takes-another-step-forward.html*, accessed June 12, 2021

[60] "Technology – One of the key options for dealing with the growing volumes of end of life construction PS foam waste streams in the decades to come is the closed-loop recycling of PS foam waste." *https://polystyreneloop.eu/technology/*, accessed June 12, 2021

[61] "Newcycling® – aus komplexen Abfällen werden hochwertige Kunststoffe", *https://www.apk-ag.de/newcycling/*, accessed June 12, 2021

[62] "PureCycle", *https://purecycletech.com/*, accessed June 12, 2021

[63] Grzybowski, P.: Transformation of the polyethylene wastes into fuel products. *Chemik*, 66 (2012) 7, pp. 725–734

[64] Schubert, T., Lehner, M., Karner, T., Hofer, W., Lechleitner, A.: Influence of reaction pressure on co-pyrolysis of LDPE and a heavy petroleum fraction. *Fuel Processing Technology*, 193 (2019), pp. 204–211

[65] Recycling Technologies Ltd. *https://recyclingtechnologies.co.uk/technology/the-rt7000/*, accessed November 15, 2019

[66] Onwudili, J.A., Insura, N., Williams, P.T.: Composition of products from the pyrolysis of polyethylene and polystyrene in a closed batch reactor: Effects of temperature and residence time. *Journal of Analytical and Applied Pyrolysis*, 86 (2009) 2, pp. 293–303

[67] Kim, Y.S., Hwang, G.C., Bae, S.Y., Yi, S.C., Moon, S.K., Kumazawa, H.: Pyrolysis of polystyrene in a batch-type stirred vessel. *Korean Journal of Chemical Engineering*, 16 (1999) 2, pp. 161–165

[68] Liu, Y., Qian, J., Wang, J.: Pyrolysis of polystyrene waste in a fluidized-bed reactor to obtain styrene monomer and gasoline fraction. *Fuel Processing Technology*, 63 (2000) 1, pp. 45–55

[69] Kaminsky, W., Predel, M., Sadiki, A.: Feedstock recycling of polymers by pyrolysis in a fluidized bed. *Polymer Degradation and Stability*, 85 (2004) 3, pp. 1045–1050

[71] "Plastik zu Plastik. Mit ReOil schließt sich der Kreis im Kunststoffrecycling", *https://www.omv.com/de/blog/plastik-zu-plastik-mit-reoil-schliesst-sich-der-kreis-im-kunststoffrecycling/*, accessed June 22, 2021

[72] "Sabic and customers launch certified circular polymers from mixed plastic waste", *https://plasticenergy.com/sabic-and-customers-launch-certified-circular-polymers-from-mixed-plastic-waste/*, accessed June 22, 2021

[73] "ChemCycling – From plastic waste to virgin-grade products", *https://www.basf.com/global/en/who-we-are/sustainability/we-drive-sustainable-solutions/circular-economy/mass-balance-approach/chemcycling.html*, accessed July 2, 2021

[73] Lopez, G., Artetxe, M., Amutio, M., Alvarez J., Bilbao, J., Olazar, M.: Recent advances in the gasification of waste plastics. A critical overview. *Renewable and Sustainable Energy*, 82 (2018) 1, pp. 576–596

[74] Quicker, P.: Evaluation of recent developments regarding alternative thermal waste treatment with a focus on depolymerisation processes. *Waste Management*, 9 (2019), pp. 361–370

[75] Iwata, T.: Biologisch abbaubare und biobasierte Polymere: Die Perspektiven umweltfreundlicher Kunststoffe. *Angewandte Chemie*, 127 (2015), pp. 3254–3260

[76] Müller, R. J.: Bioverfahrenstechnik – Biologisch abbaubare Kunststoffe. *Biologie in Unserer Zeit*, 30 (2000) 4, pp. 218–225

[77] Grigat, E., Selewski, K.: Biologisch abbaubar in 60 Tagen. *Kunststoffe*, 87 (1997) 1, pp. 63–65

[78] Tokiwa, Y., Pranamuda, H.: Microbial degradation of aliphatic polyesters, In: *Biopolymers (Vol. 3b)*, Steinbchel, A., Doi, Y. (Eds.), Wiley-VCH (2002) pp. 85–104

[79] Kleeberg, I., Hetz, C., Kroppenstedt, R. M., Müller, R.-J., Deckwer, W.-D.: Biodegradation of aliphatic-aromatic copolyesters by *Thermomonospora fusca* and other thermophilic compost eluates. *Applied and Environmental Microbiology*, 64 (1998) 5, pp. 1731–1735

[80] Burgstaller, M., Potrykus, A., Weißenbacher, J., Kabasci, S., Merrettig-Bruns, U., Sayder, B.: Gutachten zur Behandlung biologisch abbaubarer Kunststoffe. *Umwelt Bundesamt, Texte*, 57 (2018), Projektnummer 93711

8 Mechanical Recycling

8.1 Basic Principles

Achim Schmiemann

The main objective of plastic recycling is to produce a high-quality recyclate. A distinction is made between purely mechanical recycling and feedstock or chemical recycling. A possible third route is to use the plastics as long-chain hydrocarbons and thus as an energy source. In addition, there are the possibilities of reusing plastic components on the one hand and the (outdated) possibility of landfilling them on the other. Figure 8.1 gives an overview of the different possibilities.

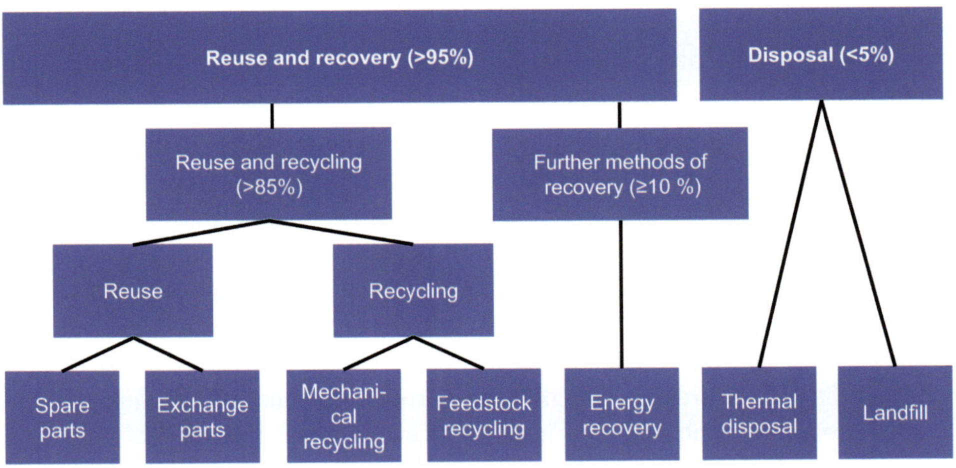

Figure 8.1 Different recycling possibilities (terms and definitions)

In this chapter, only mechanical recycling will be discussed in more detail. The starting point for the discussion is the wide diversity of consumer and economic goods, which consist of plastics and/or which contain plastics. A closer look at individual products reveals that individual polymer are rarely used in the absence of other plastics or materials. For consumer products in particular, it is the case that they are typically distributed over large areas across the country, across the continent, and across the world. Collection and related logistics are of critical importance. Thus, before mechanical recycling can begin, sufficiently large volume streams must be collected and transported to the point of recycling.

Upon closer examination of the collected waste, two other technical issues immediately come to the forefront, in addition to cleaning the materials. These are shredding and sorting of the collected materials.

In fact, these two basic principles of technical processing are essential for mechanical plastics recycling. In the process, the following three tasks can be performed simultaneously within the scope of a (frequently multi-stage) shredding process:

- Reduction of volumes
- Achieving a particle size suitable for the processes of mechanical recycling
- Material separation

"Material disintegration" is perhaps a better name for material separation. This disintegration can be achieved selectively by careful selection of grinding techniques, or by ensuring that the grinding processes result in a product that is sufficiently fine.

In addition to the separation of foreign materials such as metals, sorting of the plastics will also have to take place if high-quality recyclates are to be the goal of mechanical recycling. Plastic sorting plays a major role for thermoplastics, as many plastic melts are not compatible with each other. Tables such as the one in Figure 8.2 provide information on which plastics are compatible, conditionally compatible, or incompatible with each other [1].

The incompatibilities can have various causes such as an immiscibility of the polymer melts or even a mutual chemical degradation of the melts. In this respect, the grade purity of plastics is of great importance for mechanical recycling. For plastics sorting, the classical methods of ore processing have been used since the 1970s or, to an increasing extent, methods of optical sorting. However, since in all cases of mechanical recycling of plastics, 100% purity can never be achieved when more than one plastic is present in a material stream, Figure 8.2 will indicate the effect of the (admixture) component in the surplus component. The results are of a qualitative nature, since a scientifically exact evaluation is lacking.

In addition to pre- and post-processing measures such as material feeding or material drying, shredding and sorting are the central processing steps in the mechanical recycling of plastics. Figure 8.3 shows an example of the structure of a mechanical recycling plant for plastics.

	ABS	ASA	PA	PBT	PBT/PC	PC	PC/ABS	PC/PBT	PE	PET	PMMA	POM	PP	PPO	PPO/PS	PS	PVC	SAN	TPU
ABS	+	+	0	+	+	+	+	+	0	0	+	0	0	0	0	0	+	+	+
ASA	+	+	0	+	+	+	+	+	0	0	+	0	0	0	0	0	+	+	+
PA	0	0	+	0	0	-	-	-	0	0	0	0	0	-	0	0	-	0	+
PBT	+	+	0	+	+	+	+	+	0	0	0	0	0	0	0	0	-	+	0
PBT/ PC	+	+	0	+	+	+	+	+	0	0	0	-	0	0	0	0	-	+	+
PC	+	+	-	+	+	+	+	+	0	+	+	-	0	0	0	0	-	+	0
PC/ ABS	+	+	0	+	+	+	+	+	0	+	+	0	0	0	0	0	-	+	+
PC/ PBT	+	+	-	+	+	+	+	+	0	+	+	0	0	0	0	0	-	+	+
PE	-	-	0	-	-	0	-	-	+	-	-	-	+	-	0	-	0	-	0
PET	+	+	0	+	+	+	+	+	0	+	0	0	0	0	0	0	-	+	0
PMMA	+	+	0	0	0	+	+	+	0	0	+	-	0	0	0	0	0	+	0
POM	0	0	0	0	0	-	-	-	0	0	-	+	0	0	0	0	0	0	0
PP	-	-	0	-	-	-	-	-	0	-	-	-	+	-	0	-	0	-	0
PPO	0	0	0	0	0	0	0	0	0	0	0	0	0	+	+	+	-	0	0
PPO/ PS	0	0	+	0	0	0	0	0	0	0	0	0	0	+	+	+	-	0	0
PS	0	0	0	0	0	0	0	0	0	0	0	0	0	+	+	+	0	0	0
PVC	+	+	-	-	-	-	-	-	0	-	+	+	0	-	0	0	+	+	+
SAN	+	+	0	+	+	+	+	+	0	0	+	0	0	0	0	0	+	+	0
TPU	+	+	+	0	+	+	+	+	0	+	+	+	0	0	0	0	+	+	+

+ Good compatibility over wide mixing ratios
0 Conditional compatibility in small quantities
- Incompatibility

Figure 8.2 Compatibility of plastics

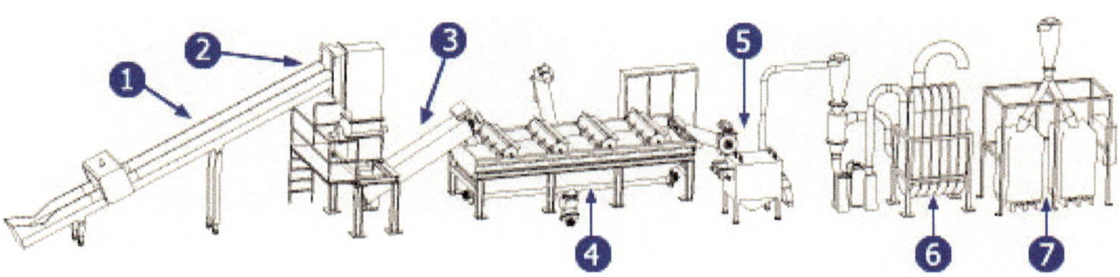

Figure 8.3 Example for a plastics recycling plant (Courtesy of ZERMA). 1 Conveyor belt with metal detection bridge, 2 Wet granulator, 3 Friction washer, 4 Float-sink container, 5 Mechanical dryer, 6 Thermal dryer, 7 Big-bag filling station

There are many conceivable ways to crush components and materials. The basic physical principles are pressure, impact, shock, or shear. In 1965, Rumpf defined four types of stress during comminution (see Figure 8.4) [2].

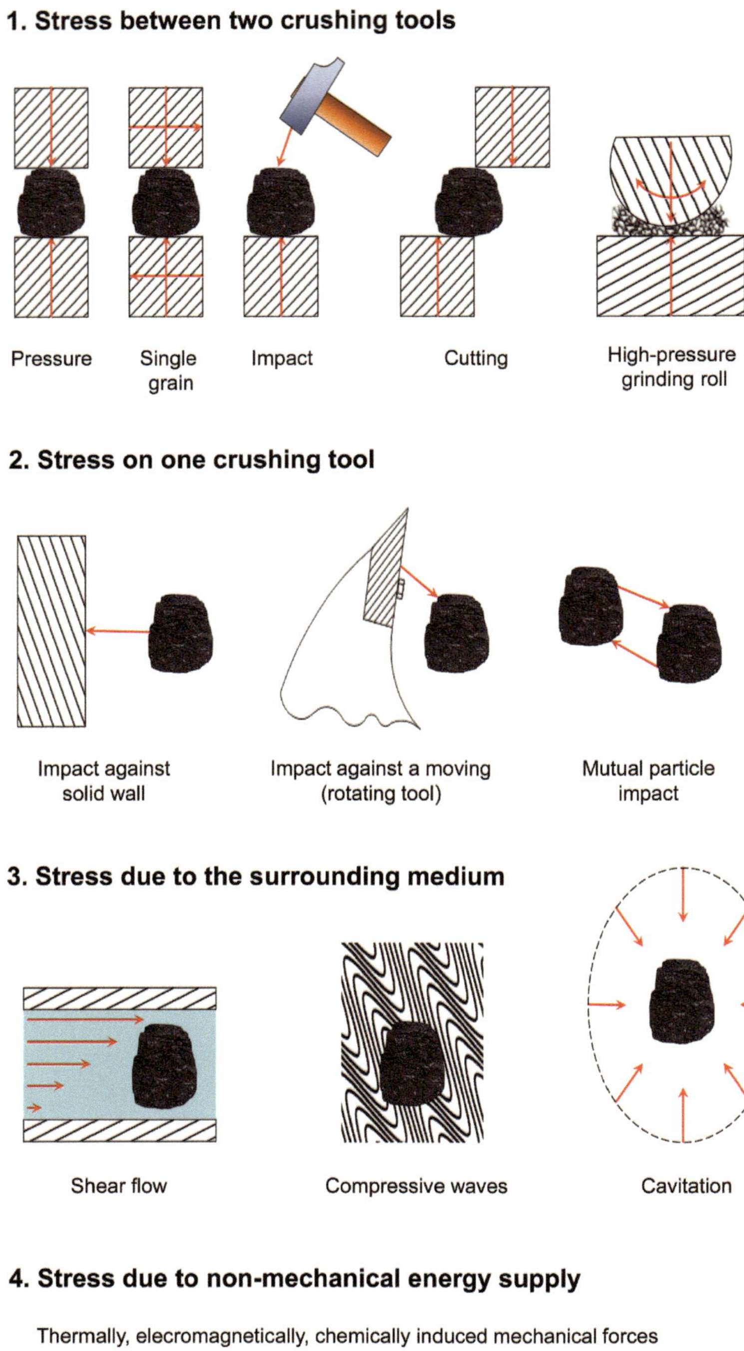

Figure 8.4 Specific types of stress according to Rumpf [2]

In purely qualitative terms, elastomers and most thermoplastics can be regarded as tough-elastic materials. Tough-elastic behavior can certainly no longer be said to exist in the case of highly filled thermoplastics or thermoplastics reinforced with high proportions of glass fibers. These modifications can be described as brittle, as indicated by the low elongation at break. Thermoset moldings or moldings made from compounds with a thermoset matrix behave in exactly the same way. Table 8.1 gives a classification of material properties and suitable stress types during comminution.

Table 8.1 Relationships Between Material Property and Stress Type

Material property	Stress type				
	Beating	Impact	Pressure	Cutting	Shearing
Rigid	++	++	++	–	–
Brittle	++	++	++	–	–
Medium hard	++	++	++	–	+
Soft	+	+	+	++	++
Elastic	–	–	–	++	–
Tough	–	–	–	++	–
Fibrous	+	+	–	++	+
Heat-sensitive	–	++	–	+	–

++ Good resistance, + Moderate resistance, – Not resistant

In general, cutting comminution is the best way of shredding plastics. In general, when shredding, a distinction is made between:

- Coarse (particle size >20 mm)
- Medium (particle size 1–20 mm)
- Fine or ultra-fine (particle size <1 mm)

8.1.1 Coarse or Preliminary Shredding

To achieve coarse shredding, shearing or cutting using slow-running rotor shears is used. The name of the rotor shears is derived from the significant shear stress in a narrow gap between the overlapping rotors (in the case of multi-rotor designs) or between the rotor blade and stator (in the case of single-rotor designs). The cutting blades, which are mounted on two parallel cutting shafts, mesh with each other at different speeds so that there is a gusset area similar to that of household shears (Figure 8.7). Versions with four or six cutting shafts are also available. In this case, the shafts mounted above the actual cutting rotors are used only for feeding (Figure 8.5, right).

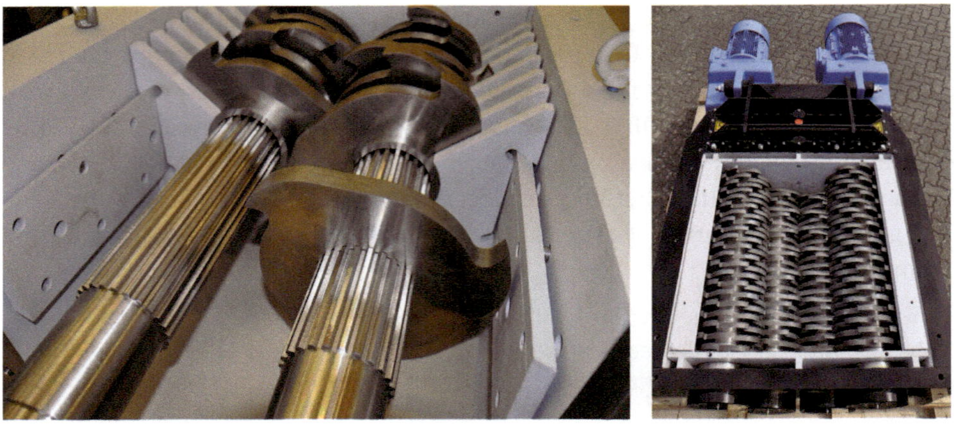

Figure 8.5 Rotary shearing setup. Left: Shaft assembly; Right: Four-rotor design (Courtesy of Bomatic)

The cutting shafts are equipped with high-strength cutting blades, which are individually slid onto the shafts and can therefore be replaced separately in the event of breakage or wear. Additional ripping teeth are mounted on the cutting knives. These grip the material and pull it in. They also shorten the strips of plastic produced by the cutting blades. The result of the shredding process is strip-shaped parts, with the strip width corresponding to the blade width, and the strip length corresponding to the distance between the ripping blades on the circumference of the discs (Figure 8.6).

Figure 8.6 Shredded polypropylene bumper (Courtesy of WIPAG)

Automatic overload protection and a reversing circuit are standard equipment on these machines. If the cutting rollers cannot rotate without interference after a set number of reversing operations, the machine is automatically stopped. Forced feeding of particularly bulky, large-volume parts can be achieved by mechanical tamping devices, such as feed slides or screws, so that bridging is eliminated and continuous size reduction is ensured.

The throughput rate and the shape of the material to be ground depend to a large extent on the material and on technical aspects of the machine. The size of the individual pieces can be changed by retooling with cutting discs of different width and tooth number.

In smaller series, the drive is usually provided by an electric motor via a planetary gear to the two cutting shafts. In these designs with only one electric motor, it is a disadvantage that the fast-rotating shaft applies less torque than the slow-rotating one. This can be remedied by using two electric motors, which drive each shaft with the same torque.

When driven by a hydraulic motor, the first cutting shaft is driven directly and without intermediate gearing, and the second cutting shaft is driven by the first shaft via a gear reduction unit. This drive concept offers advantages in terms of more uniform power development and torque characteristics. With regard to the design of the twin-shaft shredders, a number of points are important, e.g. the overlapping of the cutting contour, as shown in Figure 8.7.

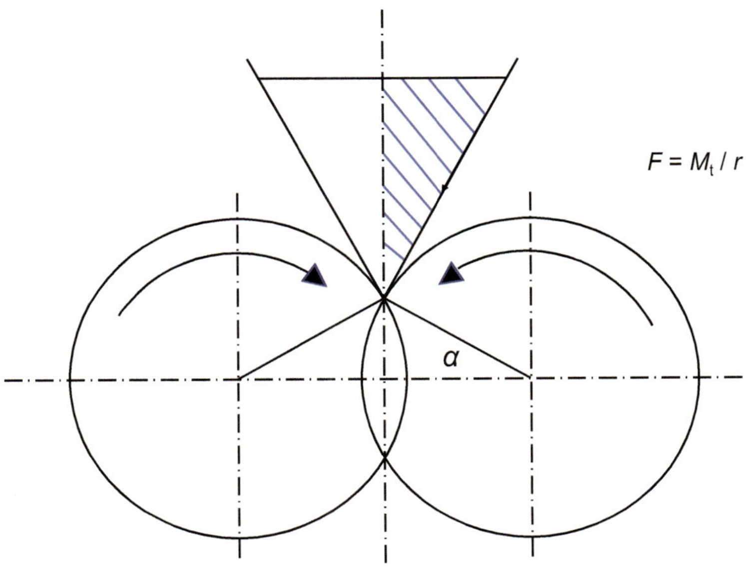

Figure 8.7 Cutting contour overlap

With regard to the size of the overlap, there are two points to consider:

1. With increasing overlap, the pull-in behavior becomes worse. The indentation conditions $\mu = \tan \alpha$ can – depending on the material pairing – no longer be fulfilled.
2. With decreasing overlap – i.e. with decreasing angle α, the radial force on the shafts of the shredding machine becomes greater at a constant circumferential force $F = Mt/r$. The shafts are then subjected to high bending stresses.

High drive torques increase the circumferential forces on the outer edge of the knife, while small knife diameters give less favorable feed behavior. In particular, material dimensions that are larger than the shaft distance require a holding pressure device (Figure 8.8).

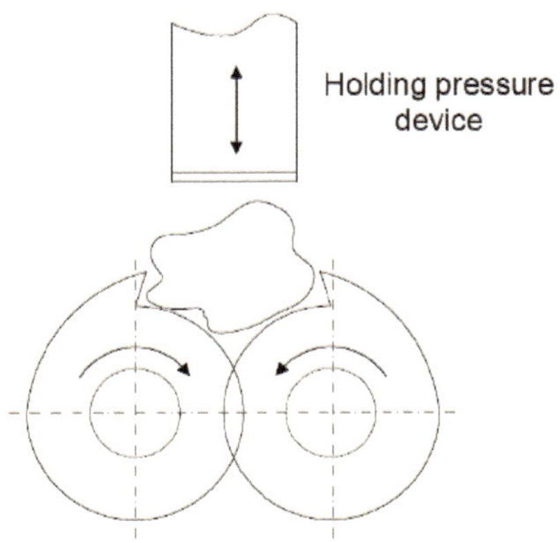

Figure 8.8
Holding pressure device

The influences of tangential and radial force demand that great care is taken when selecting a connection between shaft and cutting tool. Various connections are possible. Because of the ease of interchangeability, positive locking should be preferred to frictional locking. Figure 8.9 shows variants.

For good mountability and interchangeability, the fit selection is limited to a sliding fit, which at the same time, means that micromovements are permitted. With every half revolution, the knife shifts by the fit clearance. Reversing frequency, temperature influence, etc. can increase this clearance, so that, for example, with the hexagonal shaft there is a risk of "running in". The splined shaft profile also results in a high surface pressure and is not considered to be rotationally stable during reverse operation. Further information on design parameters and specific energy requirements can be found, for example, in refs. [4–7].

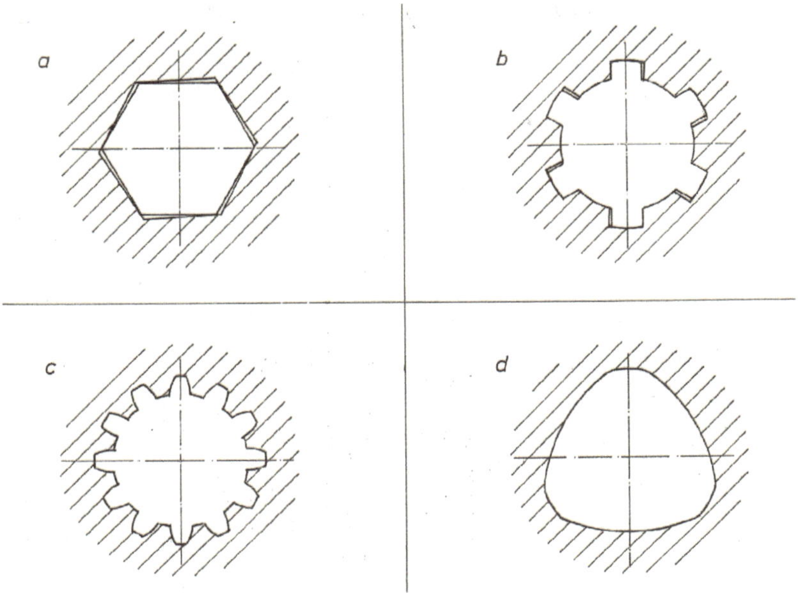

Figure 8.9 Torque transmission options. (a) Hexagon shaft profile; (b) Splined shaft profile; (c) Tooth shaft profile; (d) Polygon shaft profile

In addition to the frequently used two-shaft shredders, which are used in different variants (e.g. also as a four-shaft unit, where two of the four shafts just perform a feeding function), a number of other units are in use. For example, guillotines are used in the plastics sector for the pre-shredding of film bales and film remnants. Screw shredders have also been proposed. However, these machines, which are familiar from the wood sector, are only useful for brittle material such as glass-fiber-reinforced thermosets. The comminution process in such machines is not a pure cutting process. A combination of cutting, tearing, and breaking between the paired shafts limits the range of applications. Other shredding machines for pre-shredding such as saws, shears, and disc chippers are used to shred production waste (e.g. from profile extrusion) and return it to the production process.

In addition to the radial rotary shears described above, axial rotary shears as shown in Figure 8.10 have become increasingly important. The advantage of this machine is that the blades screwed onto the profile rotor can be turned up to eight times. The so-called indexable inserts can therefore be used several times at low cost.

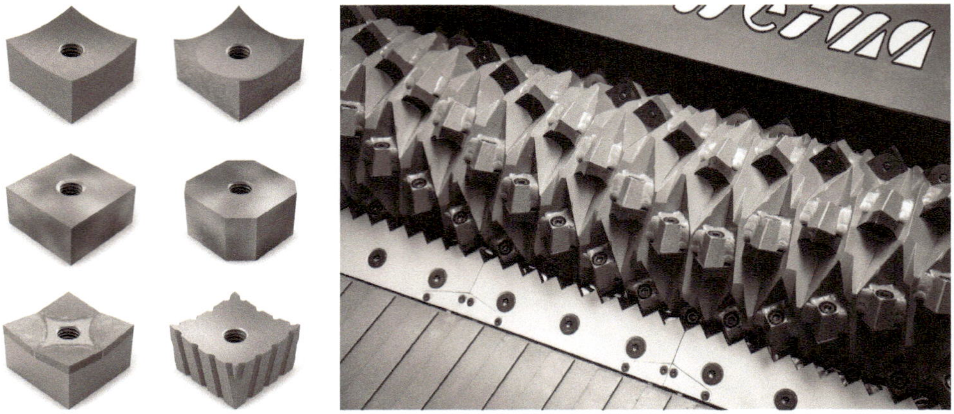

Figure 8.10 Axial rotary shears with reversible cutting blades (Courtesy of WEIMA)

These axial cutters are usually designed as single-shaft machines. A hydraulic pressure device provides the material feed. Screen inserts limit the particle size. The areas of application in plastics recycling are quite diverse. Cutting drum shredders are used, for example, for films, various production wastes, cable waste, and electronic scrap containing plastics. Compared to high-speed granulators, these machines, which also operate on the cutting principle, are less sensitive to adhering or entrained foreign materials such as metals.

In plastics processing, granules in cylinder or cube form with a particle size smaller than 10 mm are standard. Cylindrical pellets are produced by feeding extruded strands uniformly to a cutter via a roller feeder (strand pelletizer). If a strip is extruded instead of the strands, it is first cut into longitudinal strips with staggered rolls, and subsequently cut directly crosswise into cubes (strip pelletizer).

Residual thermoplastic materials are produced during the manufacture of plastic products, and can be reprocessed directly, because they are already sorted and uncontaminated. Post-consumer materials pass through several process steps until the required grade purity is achieved, which is necessary for the manufacture of quality products. For economic reasons, pure material from production has long been reground to a particle size of less than 10 mm and then added to the virgin material in various concentrations, or it has been reprocessed. An advantage at this point is that the material here does not suffer from excessive thermal damage. Furthermore, changes in properties as a result of repeated recycling cycles must be taken into account here, depending on the plastic.

8.1.2 Medium Size Shredding

Cutting mills are generally used almost exclusively to reduce thermoplastic materials to a medium size. The granulators developed in the USA after World War II are so-called pile cutters, which achieve a ground material with non-uniform particle shape and particle size distribution (polydisperse ground material).

The basic design of granulators has hardly changed in recent years. Instead, progress has been made through the development of modified cutting geometries, numerous sizes, adapted special or custom designs, and specific accessories. This also applies to sound insulation and the ease of maintenance or safe handling of granulators. The drive powers of each series are available in close increments, so that the processor of plastic residues can select the machine that best suits the application.

The schematic design of a granulator is shown in Figure 8.11 (see also refs. [10, 11]). The feed material falls through the feed shaft into the cutting chamber and is repeatedly comminuted by cutting between the rotor and stator knives until it passes the built-in sieve as ground material. The sieve enclosing the lower half of the grinding chamber is replaceable. Thus, the sieve perforation can be adapted to the required particle size of the material to be comminuted. The throughput capacity and the grinding material quality achievable by a granulator are largely dependent on the design elements and parameters described below, which must be adapted to the requirement profile of the feed material.

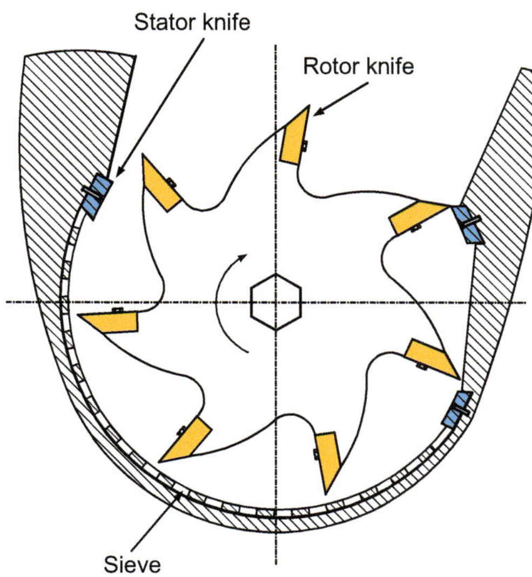

Figure 8.11
Structure of a classic cutting mill

The number of stator and rotor blade rows has a decisive part in the throughput capacity of a granulator. Usually, the number of rotor knife rows is between one and ten, and that of stator knife rows between one and five.

In contrast to the rotor knives, which are arranged evenly around the circumference of the rotor, the stator knives can only be arranged unevenly in the grinding chamber. The feed opening and the built-in sieve hinder regular placement of the stator knives. The number of stator knives is therefore less than that of the rotor knives. However, the ratio of stator knives to rotor knives should still be about 2:3 or 4:5, since this allows a high cutting performance per revolution (number of stator knives times number of rotor knives) to be achieved with fewer rotor knives. This results in a better rotor knife pitch and an optimum material intake. It also achieves uniform loading and wear.

Compared to continuous arrangements, segmented stator and rotor knives can simplify handling when changing knives. In the event of damage, it is usually only necessary to replace individual segments. Likewise, knife breakage due to distortion is prevented if the contact surface is not clean. Knives with a rectangular cross-section allow multiple uses by turning until regrinding.

Not every rotor is suitable for every shredding task. Thus, different rotor variants have been developed over the course of time. A basic distinction is made between closed and open designs (Figure 8.12).

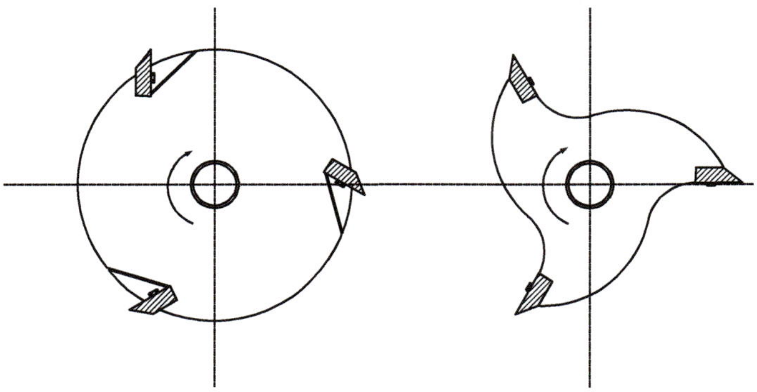

Figure 8.12 Different rotor types. Left: Enclosed rotor; Right: Open rotor

Closed rotors are the preferred choice for shredding compact, heavy plastic lumps. The rotor knives are attached to the circumference of a solid roller and have a diameter of 4 mm to 15 mm. In this case, the compact material to be ground cannot penetrate very deeply between the rotor and stator knives. It is "nibbled off", thus avoiding impact loads. Blocking of the rotor by compacted material is avoided.

For bulky, thin-walled plastic parts such as foils, foams, etc., open rotors are used, which have better feed characteristics than the closed version, and cut with reduced noise. This allows the open rotor to penetrate deep into the material to be shredded and so cut it up. Large impact loads do not occur here, as the parts do not have a large mass.

To counteract thermal damage to the material and increase throughput, the residence time of the material to be ground in the grinding chamber must be kept short. The use of closed rotors is more advantageous in this respect than open ones, since the material to be ground is held better on the circumference of the rotor and is comminuted more frequently between the rotor and stator knives.

8.1.3 Comminution of Bonded Materials

In the case of bonded materials (Figure 8.13), it is often necessary to use comminution methods that enable composite disintegration. Frequently, conversions or modifications – or even developments of new comminution units – are then required. A basic idea here is often to comminute one of the composite partners and to leave one or more others as intact as possible. Another way is to comminute the composite partners, releasing the interface between the composite partners. Overall, the following possibilities are distinguished with regard to the release of components from bonded materials:

- Summative comminution
- Comminution along the interface
- Selective comminution

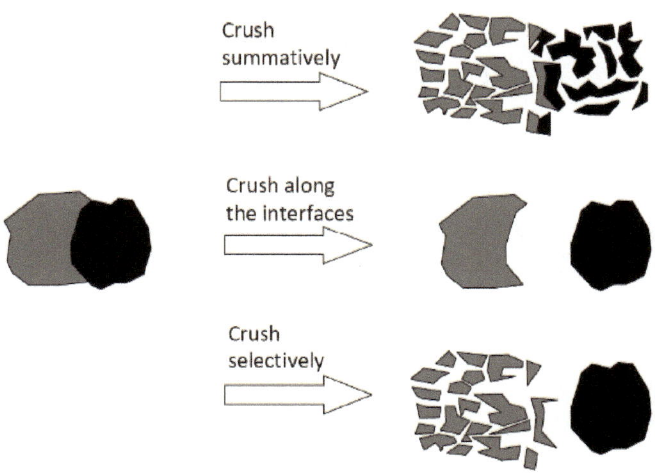

Figure 8.13 Possibilities for a composite disintegration

In contrast to selective comminution and interfacial separation, summative comminution divides the material independently of its structure. In this process, a certain degree of compound disintegration is also achieved. Two types of particles are produced: bonded and exposed particles.

In contrast to mineral composites, the interface of composite waste from industrial production is relatively small. There are no interpenetrations or intergrowths. In this respect, the simplifying assumption that (random) fragmentation along the interface can be neglected in the case of summative comminution is permissible. Then the complete interface is preserved in the connected particles.

For a measurable interface A and volume V of a composite part, a characteristic length x can be derived according to [8, 9]:

$$x = \frac{V}{A}$$

For different, simple geometries, the characteristic lengths can then be easily determined (Figure 8.14).

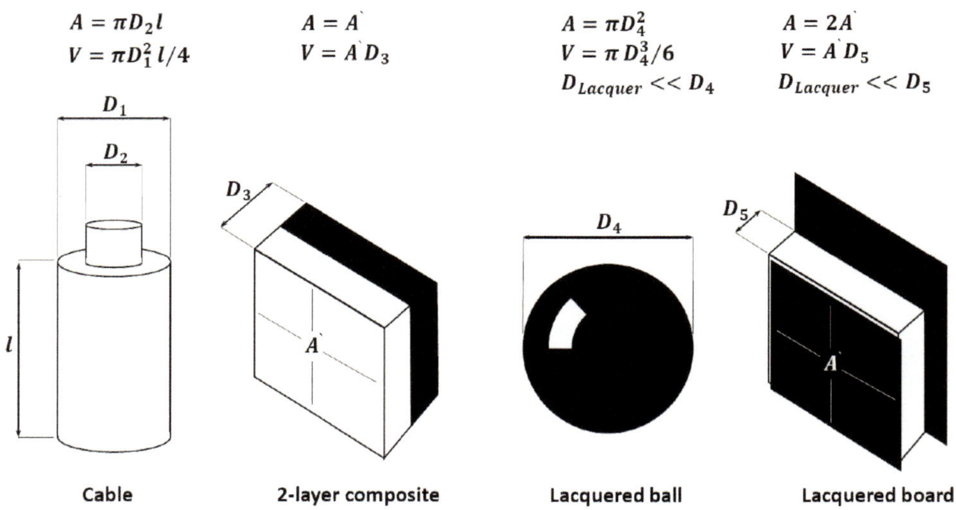

Figure 8.14 Composites of simple geometry

For example, for a cable of diameter D_1 and inner wire diameter D_2, the characteristic length is x_C:

$$x_C = \frac{D_1^2}{4D_2}$$

After crushing to the average particle size d, the total volume of composite particles is:

$$V' = Ad$$

The exposed material thus has a volume of L, resulting in an exposure or disintegration rate of:

$$L = \frac{V - V'}{V} = \frac{V - Ad}{V} = 1 - \frac{d}{x}$$

Converted, for a known characteristic length x, the required particle size d is given by:

$$d = x(1 - L)$$

Figure 8.15 shows a diagram for calculating the disintegration during summative comminution for various bonded materials.

If the core diameter (inner diameter) D_2 is the same as the sheath thickness of the cable under consideration, the result is, for example, x_C:

$$x_C = \frac{9D_1^2}{4}$$

For a comminution ratio D/d of 3, a material disintegration of 80% is already achieved according to Figure 8.15. If the non-exposed particles are subsequently treated again in a second stage, exposure rates of >90% can be achieved.

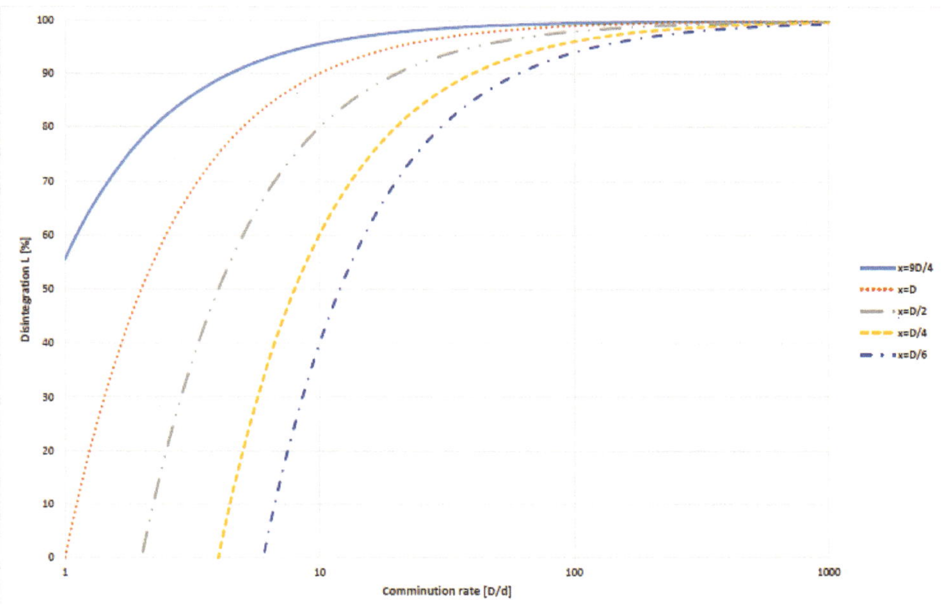

Figure 8.15 Disintegration L with different composites and different comminution ratios D/d

8.1.4 Classifying and Sorting

Classification is the separation of a mixture according to external particle properties such as size or shape. Sorting, on the other hand, is the separation of a particle mixture according to material properties such as density or color. Despite the low densities of plastics compared to other materials (Table 8.2), e.g. metallic materials, density sorting methods are also predominantly used for sorting plastics in material recycling [12-14]. It is certainly helpful that polyolefins - unless heavily modified - have a density of less than 1.0 g/cm³ and thus float on water. A simple separation of the polyolefin fraction from a batch stream is easily achieved in a water bath.

Table 8.2 Polymers and Their Raw Densities

Polymer	Abbreviation	Raw density [g/cm³]
Polypropylene	PP	0.91
Polyethylene	LDPE	0.92
Polyethylene	HDPE	0.96
Polyamide	PA	1.14
Polystyrene	PS	1.05
Poly(methyl methacrylate)	PMMA	1.18
Poly(vinyl chloride)	PVC	1.39
Polytetrafluoroethylene	PTFE	2.20

The physical principles underlying both separation tasks are identical. In Figure 8.16, the fundamentals of separation on a single grain are derived. First, the forces acting on a single particle in the earth's gravitational or centrifugal field are discussed. To simplify matters, a stationary process is assumed and the Coriolis force is not considered. The force equilibria resulting from the sinking velocity in the earth's gravitational field (or in the centrifugal field) are shown.

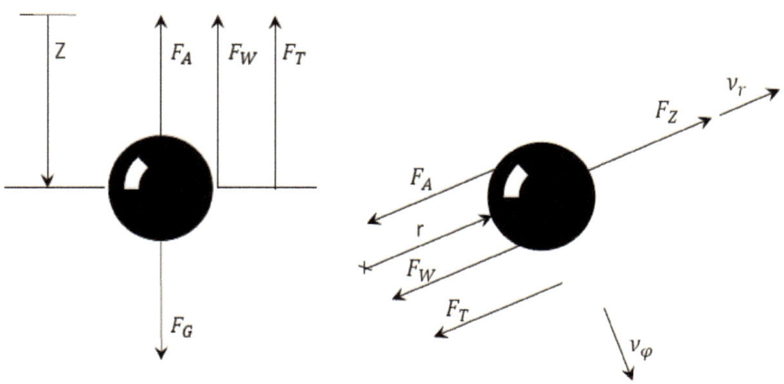

Figure 8.16 Forces on particles in (left) a gravity field and (right) a centrifugal field

with:

$$F_G = \rho_P V_P g \qquad\qquad F_z = \rho_P V_P r\omega^2$$
$$F_A = \rho_L V_P g \qquad\qquad F_A = \rho_L V_P r\omega^2$$
$$F_W = \frac{\rho_L}{2} v_z^2 A_s c_w(\text{Re}) \quad \text{and} \quad F_W = \frac{\rho_L}{2} v_r^2 A_s c_w(\text{Re})$$
$$F_T = f\left(\frac{dv}{dt}\right) = 0, \text{stationary} \qquad\qquad F_T = f\left(\frac{dv}{dt}\right) = 0, \text{stationary}$$

The force equilibria in the z- and r-directions are:

$$v_z^2 = \frac{4}{3}\frac{\rho_P - \rho_L}{\rho_L}\frac{gd}{c_w(\text{Re})} \quad \text{and} \quad v_r^2 = \frac{4}{3}\frac{\rho_P - \rho_L}{\rho_L}\frac{dr\omega^2}{c_w(\text{Re})}$$

Consideration of the Stokes approach $c_w = \frac{24}{\text{Re}} = \frac{24\eta}{vd\rho_L}$ for viscous flow around the particles provides:

$$v_z^2 = \frac{\rho_P - \rho_L}{18\eta} gd^2 \quad \text{and} \quad v_r^2 = \frac{\rho_P - \rho_L}{18\eta} r\omega^2 d^2$$

Simplifying strongly, a viscous (Stokes) flow is assumed. It is evident that the sinking velocity depends on the density difference between particle and surrounding fluid, as well as on the square of the particle diameter. Likewise, a correspondingly high rotational speed can increase the sinking speed substantially. Of course, in real-life applications (plants) the mutual particle obstructions as well as friction phenomena have to be taken into account. In the vast majority of cases, turbulent flow will also be observed.

Nevertheless, two conclusions can be drawn from these simple considerations:

1. In density sorting, classifying effects always take place.
2. Density sorting with the highest yield requires a uniform particle geometry.

The density of the separation medium must be chosen so that it lies midway between the densities of the substances to be separated. This is done either by dissolving substances of lower or higher density than water, such as alcohols or salts, or by suspending fine-grained solids of higher density than water, such as kaolin or ferromagnetic minerals. The latter separation fluid is also referred to as a heavy slurry. Density adjustment by means of solution has the advantage of a constant density but entails disposal problems. Density adjustment by means of suspension involves the problem of segregation, but has the advantage of easy disposal and concentration by sedimentation tanks.

The aggregates used in this form of density sorting are also called (static) float-sink separators. Continuous separators have proven successful here, consisting of one or more open basins filled with separating liquid and equipped with charging and discharging devices (Figure 8.17).

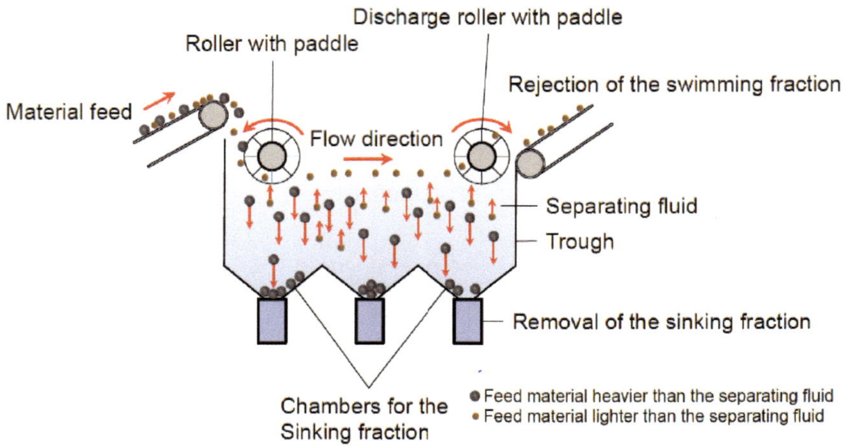

Figure 8.17 Continuous float-sink separator

The regrind from the granulator is fed into the front part of the basin and brought below the level of the separating liquid level by a roller with paddles. Heavy particles sink and light ones float up. False discharges can occur, as air bubbles can become attached to individual particles if they are not sufficiently wetted, and these can then float up, even though they are heavier than the floating material (the geometry of the feed plays an important role here). Wetting agents can also be added to the separating liquid to improve wetting of the plastic particles. To ensure a sufficient flow rate in the plant, the mass flow drifts through the basin at a horizontal speed of about 0.1–0.4 m/s. During their residence time in the basin, the particles can float or sink in the release liquid according to their density. In order for this process to take place unhindered, the flow in the basin should be kept as vortex-free as possible. This is also the reason for the low throughput capacity of the float-sink separators. However, this disadvantage can be eliminated by connecting several float-sink stages in parallel.

The discharge of the floating material at the end of the basin is carried out by means of discharge paddles or finger rollers, with which the material can be skimmed off with as little turbulence as possible. The material sinking to the bottom of the basin is discharged hydraulically, usually discontinuously. Dividing the basin bottom into several chambers has the advantage of creating three sink fractions, the first fraction containing materials with high density and sinking velocity (e.g. mineral and metallic contaminants), while the last fraction contains materials with lower density and sinking velocity, such as paper and plastic regrind.

Various calculation models have been developed to calculate the separation efficiency, separation performance, and retention time in float-sink density sorting. Depending on the type of suspension removal, a distinction is made between suspension-tapping and suspension-dividing models. The design features in static float-sink separators result in tapping of the suspension at the water level and at the bottom of the basin, so that the suspension-tapping model can provide a basis for calculation.

The float-sink separators in use are designed almost exclusively for polyolefin separation. In addition, an almost loss-free separation of already dissolved contaminants and heavy foreign bodies (stones and metal parts) is achieved. When processing certain types of waste, for example used PE hollow bodies from households, the addition of wetting agents can be dispensed with due to the surfactants adhering to them.

In order to significantly increase the attacking field force above the level of the earth's gravity, various processes exist which make use of the centrifugal field of a rotary design. In addition to various centrifuge designs in which the rotation is introduced mechanically from the outside via a drive, there are also stationary apparatuses in which the rotation of the separating fluid is generated by the flow energy of the fluid itself. An example of such a flow-energy-driven centrifugal separator is the hydrocyclone (Figure 8.18).

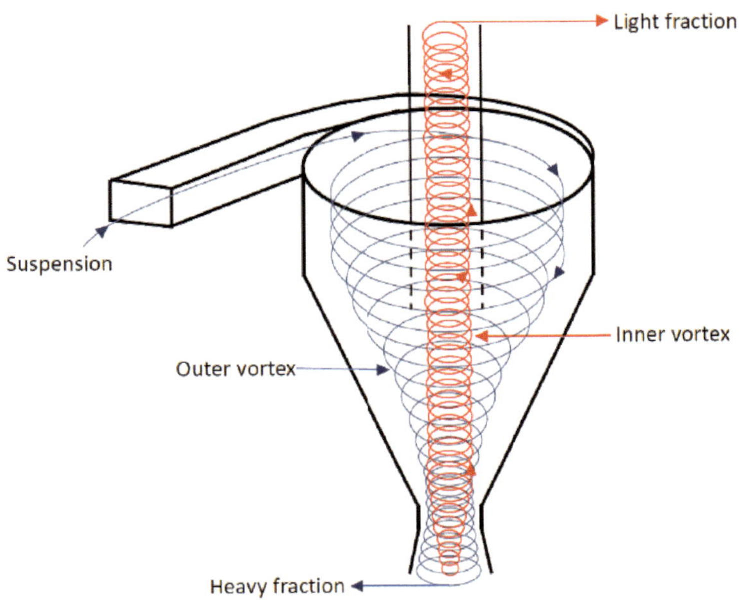

Figure 8.18 Operating principle of a hydrocyclone

The separating liquid (loaded with plastic particles) flowing under pressure into the tangential feed of the apparatus is deflected by the geometry into a circular motion, resulting in a vortex. The circular motion generates centrifugal accelerations that are greater than the acceleration due to gravity by a factor of $K = r^2/g$. Common hydrocyclones reach values up to 250 times the acceleration due to gravity.

This acceleration now drives the particles that are heavier than the fluid further outward in the vortex, while particles that are lighter than the fluid move toward the center of the vortex. An air surface is located in the center of the vortex. The separation process is similar to the static float-sink separator, but takes place with a multiple of the acceleration due to gravity and not in a vertical but in a radial direction.

A pipe, the so-called vortex finder, protrudes into the vortex of the hydrocyclone from above. The inner part of the vortex, i.e. the buoyant material, is extracted through this pipe. The outer part of the vortex, which contains the heavier sinking fraction, runs off through the downward-tapering cone. Here, the geometry of the hydrocyclone determines its separation characteristics.

In addition to the hydrocyclone design with a low cone angle of about 10°, known as the cone cyclone, designs have also been investigated where the cyclone transitions from full diameter (without a cone) to the smaller diameter of the bottom run. These designs, with a cone angle of 90°, have a flat bottom and are therefore also referred to as flat-bottom cyclones. This special design makes it more difficult for the heavy material to be discharged in terms of flow, and some of the heavy material ends up in the overflow. This makes it possible to achieve separation using the cyclone at a density that is higher than that of the separation medium. In tests on the separation of a PS/PVC mixture in a flat-bottom cyclone using water as the separating liquid, the PS almost completely passed into the overflow, so that a PVC fraction of up to 99.95% by weight purity was obtained in the underflow. The flat-bottom cyclone therefore saves operating costs and avoids the need for equipment required to prepare and adjust a separating liquid with density between PS and PVC.

Hydrocyclones achieve a much higher throughput than static float-sink separators. In technical use, there is a proportionality between the applied pressure and the square of the achievable throughput. They also have impressively stable separation behavior. Due to their shape, the flow that forms remains stable even in the event of pressure fluctuations in the feed.

The solids content of the feed suspension is limited on the one hand by economic efficiency and on the other hand by the risk of clogging of the plant. The solids content is uniformly adjusted in a mixing tank. From this, a pump draws off the suspension and forces it into the hydrocyclone at pressures between 1 and 2 bar. Experience has shown that safe operation is guaranteed up to 16 or 20 g of plastic per liter of separating liquid. In retrospect, this means that, depending on the purity of the feed material, about 10–15 m^3 of separating liquid must be recirculated per ton of plastic recovered. For this reason, the released liquid is recirculated. After leaving the hydrocyclone, it is separated, processed, and recycled in a more or less complex manner, depending on the plant.

A sorting centrifuge, as shown in Figure 8.19, is a double-cone full-jacket screw centrifuge which separates two solid phases and one liquid phase from each other [14, 15]. The centrifuge is partially filled with the separating liquid, and due to the fast rotation (up to 1500 times the gravitational acceleration) a liquid ring is formed. The mixed plastics are fed axially into the centrifuge as a suspension through a stationary tube, where they meet the liquid ring rotating at high speed. Due to the strong turbulence, the plastic particles are separated and freed from any dirt still adhering to them. Particles whose density is greater than that of the liquid are flung radially outwards towards the centrifuge shell, while the lighter particles float inside. Both fractions are transported to one of the conical ends of the centrifuge by screws. There they are discharged separately.

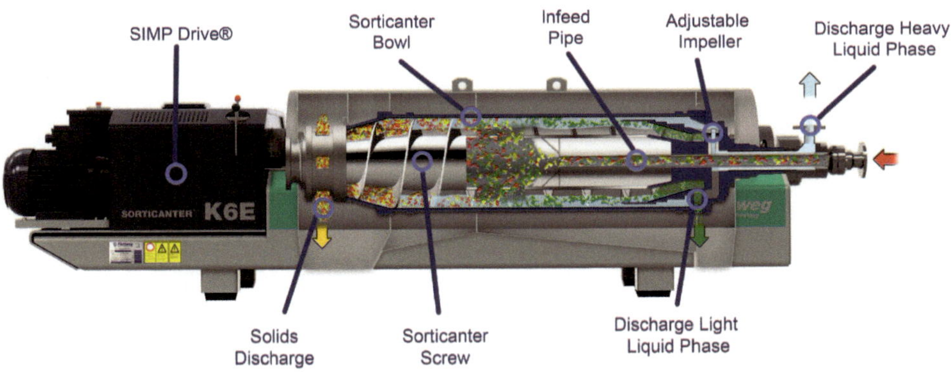

Figure 8.19 Sorting or decanter centrifuge (Courtesy of Flottweg)

8.1.5 Extrusion

Following reprocessing of used plastics, it is usually necessary to regrind the material into uniform-sized granules, in order to further process it into a finished part, or more generally make it into a saleable product. This process, which is carried out in extruders, is known as compounding, and can be combined with additional processing steps on the melt, to improve the properties, for example by adding suitable additives or intimately mixing them with virgin material. In all cases it has to be considered whether the added value created by the additional extrusion is greater than the associated costs.

Basically, an extruder performs a number of tasks, each of which is critical to the proper functioning of the extruder. These tasks or functions are:

- Pumping
- Plasticizing (melting)
- Mixing

Depending on the application, an extruder can also perform additional tasks (with additional equipment if necessary), including:

- Dosing
- Degassing
- Filtration
- Shaping

Single-screw extruders are often used in plastics processing - an example is shown in Figure 8.20. The molding compound is fed via a funnel to the screw, where the plastic is melted by friction and heat conduction, and optionally degassed, homogenized, and compressed.

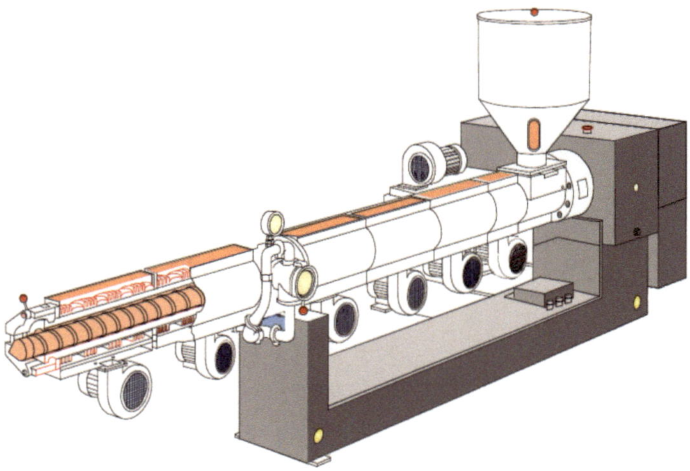

Figure 8.20 Single-screw extruder

The basic structure of an extrusion screw with feed, transition, or compression zone and discharge or pumping zone is shown in Figure 8.21. The screw geometry is adapted to the plastic to be processed. Further information can be found, for example, ref. [16].

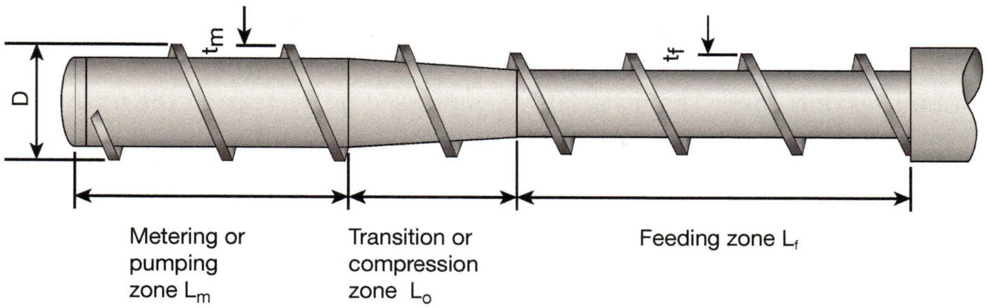

Figure 8.21 Three-zone screw

Single-screw extruders - often also referred to as plasticizing extruders - are primarily suitable for processing granules, and are not usually used for compounding tasks. The process part is built in one piece, because the pressure build-up for downstream equipment is high. It is available in a wide variety of sizes and with a very wide throughput range (from around 5 kg/h to 30 t/h).

Twin-screw extruders are more often used for compounding tasks. A distinction is made between *co-rotating* twin-screw extruders and *counter-rotating* twin-screw extruders. Depending on the center distance of the two worm shafts and the direction of rotation, the following subdivision applies:

- Tangential co-rotation
- Closely intermeshing co-rotation
- Tangential counter-rotation
- Closely intermeshing counter-rotation

The extruder design used for a given application depends on the required degree of task fulfillment (see above). Counter-rotating twin-screw extruders introduce little shear into the material and protect the melt because they induce low thermo-mechanical damage. They are therefore used, for example, in the processing of temperature-sensitive thermoplastics such as PVC. In addition to their good intake behavior, co-rotating twin-screw extruders have their strengths in their good dividing and distribution effect. They are also self-cleaning, and their process parts are modular and therefore easy to handle and adapt to various tasks. As for single-screw extruders, they are available with a large throughput range.

The core assembly is the cylinder with the screws inside, in which all process engineering operations take place. This section is usually heated electrically or using hot air saturated with water vapor ("saturated steam") (for very large units) and can be cooled with water (by evaporation) or air. The main areas are:

- **Feed zone:** This is where the material (regrind in the case of recycling) is drawn in and conveyed into the system.

- **Melting zone:** Here the ground material is melted and any additives are incorporated.
- **Compression zone:** The material is compressed by decreasing the pitch of the screws (i. e. increasing their core diameter).
- **Mixing zones:** The mixing performance is increased by special mixing elements attached to the screw.
- **Degassing zone:** In order to allow volatile constituents to escape from the melt, the filling level in the screw cylinder must be below 100%, otherwise the melt would be forced into the degassing nozzle.
- **Discharge zone:** The discharge zone ensures the necessary pressure build-up around the flow resistance of the downstream equipment (e. g. melt filter).

The co-rotating twin screws have two axially parallel screws in a cylinder with an eight-shaped bore. The screws rotate with the same angular speed and the same direction of rotation. The geometry of the elements, based on the "Erdmenger" system, results in mutual "stripping" of the screws with a sufficiently close clearance between the screws [17]. The so-called sealing profile can be seen in Figure 8.22. The multi-thread conveyor elements are shown in combination with the so-called kneading blocks, which have the same profile in the frontal section as the conveyor elements.

Figure 8.22 Sealing profile of a co-rotating twin-screw (Courtesy of Coperion)

For decades, the entire process section (cylinder and screws) has had a modular structure and can therefore be adapted very flexibly to different process tasks. The "worms" often have modified toothing, multi-spline or polygon geometries for torque transmission. Because of the necessary safety when transmitting very high torques, feather key connections are no longer used in practice. For the same reason, aerospace materials that have very high strengths are sometimes used for the worm shafts. The lengths of screws and cylinders are given in relation to the external screw diameter (*L*/*D*). Usual housings are 3*D* or 4*D* long, and in some cases up to 8*D* are used for economic reasons [18].

Due to a wide range of application tasks, extruders are specifically equipped for each application. In addition to scales, dosing aids may be necessary for continuous, precise, and safe dosing. Especially when substances that are difficult to dose, such as light fibers or thin-walled flakes, have to be fed into an extruder, specially designed tamping units (Figure 8.23) or side feeders are used.

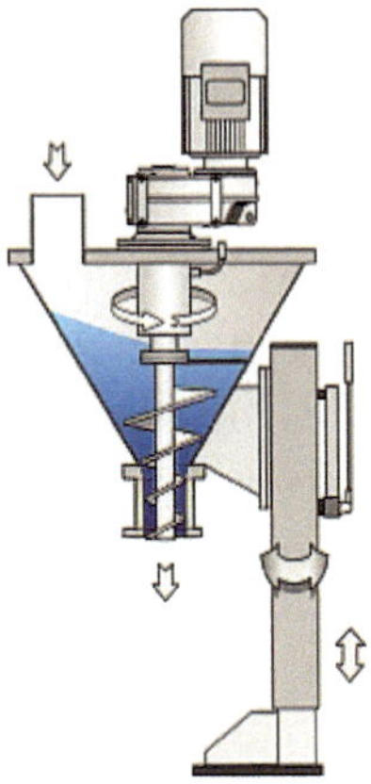

Figure 8.23
Special feed hopper (Courtesy of Kreyenborg)

Last but not least, the extrusion of regrind from used plastics offers an important opportunity to be able to remove the last remaining impurities from the melt flow. A melt filter is attached to the extruder for this purpose. Melt filtration is a mechanical separation process for separating both solid and flexible (gel-like) particles from a polymer melt with the aid of a metallic filter. Detailed information can be found in ref. [19]. Backwashable screen changers are recommended for large amounts of soiling (compared to new goods). These are system components that have to be operated continuously, such as the rotating screen changer (Figure 8.24).

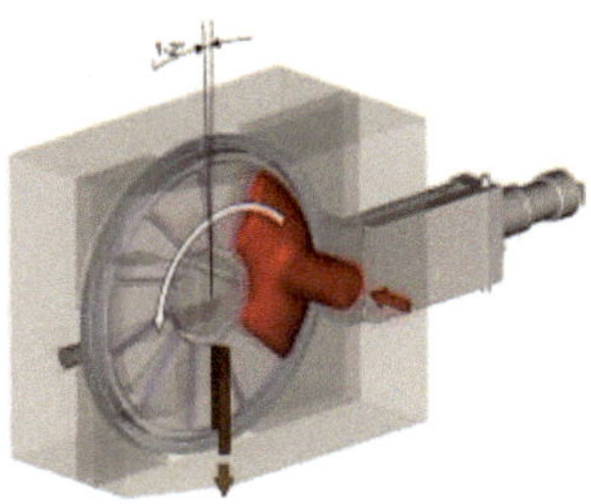

Figure 8.24
Rotary filtration system (Courtesy of Gneuß)

A rotary screen changer essentially consists of three parts: an inlet block, an outlet block, and a screen disc rotating in between. Ring-shaped sieve elements are arranged on the sieve disk and move through the melt channel. When the plastic melt flows through the sieve surface, the dirt particles are deposited on it, increasing the differential pressure slightly. The control reacts to this increase in pressure and, if necessary, allows the sieve disc to cycle by around 1 to 2°. As a result, contaminated screen surface is continuously moved out of the melt channel, and fresh screen surface is moved into the channel without changing the active screen surface.

With this mode of operation, the filtration system works as a constant-pressure, continuous process. From the very beginning, the development of the rotary screen changer was aimed at achieving the longest possible service life. For this purpose, the circular arrangement of a certain number of sieves was chosen. The dirty screens are cleaned shortly before they re-enter the melt channel, and the dirt cake is removed by high-pressure segment backwashing. In order to do this, already filtered melt is slowly brought from the outlet channel of the filtration system into a hydraulically driven "shooting pot" and then shot under high pressure (approx. 30–80 bar) in a reverse direction through the sieve disc and into the inlet block, from where it is discharged in a defined manner. Only a small segment (approx. 1% of the screen area) is shot free. In this way, the sieves are completely cleaned and, depending on the filter fineness, can be reused several hundred times. It must always be taken into account that, as a result of the backwashing process, in addition to the dirt that has been screened off in this way, there is always a certain proportion of "good product" in the filter cake.

References for Section 8.1

[1] Rink, M., Meyer, H., Orth, P., Schauf, D.: *Kunststoffe*, 85 (1995) 2, pp. 245–249

[2] Rumpf, H.: *Chem.-Ing.-Tech.*, 37 (1965) 3, pp. 187–202

[3] Michaeli, W., Bittner, M., Wolters, L.: *Stoffliches Kunststoff Recycling*, (1993), Hanser

[4] Jäckel, H.-G., Schubert, G.: *Chem.-Ing.-Tech.*, 69 (1997) 5, pp. 640–648

[5] Schubert, G.: *Aufb.-Tech.*, 43 (2002) 9, pp. 6–23

[6] Woldt, D., Schubert, G., Jäckel, H.-G.: *Chem.-Ing.-Tech.*, 75 (2003) 12, pp. 1860–1863

[7] Woldt, D.: Zerkleinerung nicht-spröder Stoffe in Rotorscheren und -reißern [Dissertation] (2004), TU Freiberg

[8] Löhr, K., Melchiorre, M., Kettemann, B.-U.: *Aufb.-Tech.*, (1995) Hanser

[9] Melchiorre, M., Güldenpfennig, M., Löhr, K., Zürn, J.: *Chem.-Ing.-Tech.*, 66 (1994) 5, pp. 661–670

[10] Konieczka, R.: *Kunststoffe*, 79 (1989), pp. 589–601

[11] Leschonski, K. in: Dietz, P. (Ed.): *Konstruktion Verfahrenstechnischer Maschinen*, (2000), Springer

[12] Bahr, A., Vogt, V., Djawadi, H.: *Sortierung von Kunststoffabfällen*, Dissertation (1980), TU Clausthal-Zellerfeld

[13] Bilitewski, B. in: Tiltmann, K. O.: *Handbuch Abfall Wirtschaft und Recycling*, (1993), Springer

[14] Martens, H., Goldmann, D.: *Recyclingtechnik*, (2016), Springer

[15] Anlauf, H.: *Chem.-Ing.-Tech.*, 90 (2018) 12, pp. 1929–1938

[16] Baur, E., Brinkmann, S., Osswald, T., Rudolph, N., Schmachtenberg, E.: *Saechtling Kunststoff Taschenbuch*, (2013), Hanser

[17] Kohlgrüber, K.: *Der Gleichläufige Doppelschneckenextruder*, (2016), Hanser

[18] Limper, A.: *Verfahrenstechnik der Thermoplastextrusion*, (2013), Hanser

[19] Grimm-Bosbach, T. in: Kohlgrüber, K. (Ed.): *Polymer-Aufbereitung und Kunststoff-Compoundierung*, (2019), Hanser

8.2 Mechanical Recycling of Polyolefins

Marco Amici

8.2.1 Introduction – Market Size and Recycling Challenge

Polyolefins (POs), such as polypropylene (PP), polyethylene (PE), low-density PE (LDPE), linear low-density PE (LLDPE), and high-density PE (HDPE), are the most ubiquitous of plastics worldwide [1–3]. In Europe alone, POs represent more than 50% of all the plastics used.

While figures vary from region to region [3], it is safe to assume that recycling rates for PO have ample margins for improvement [2–4]. Recent legislation, known as Directive 2018/852 – Packaging and Plastic Packaging Waste Directive (PPWD) [5], has set ambitious targets for plastics recycling in the EU, with 50% of collected plastics packaging waste recycled by 2025, rising to 55% by 2030. PO packaging waste [4] in 2018 was 12 Mt, of which 40% was recycled (4.8 Mt, slightly lower than overall figure for plastics packaging waste of 42%). Considering the target of 50% of collected plastics packaging waste recycled by 2025, this would imply an overall additional 1.4 Mt of plastics to be recycled, of which PO would need to be 1.2 Mt (to reach at least 50%, i. e. 6 Mt from 4.8 Mt currently).

These staggering numbers alone portray a very clear strategy for the global PO industry: how to evolve a vast, linear offering of resins into an even larger set of circular solutions. While advanced and chemical recycling have many merits, mechanical recycling is still a suitable solution in many cases, and offers the possibility to process and upgrade large volumes of PO waste from packaging and other sources. This section will therefore focus therefore on mechanical recycling of PO, in particular from packaging (which constitutes the vast majority of PO applications).

8.2.2 Main PO Applications

POs can be found in essentially all applications for plastics [1], and hence there are many possible ways to segment them. A usual method is to distinguish between *durable* applications (like automotive bumpers or pipes) and *non-durable* (like single-use packaging or some agricultural films). While figures vary for regions and specific PO polymer, non-durable applications represent on average ca. 60% of the total PO market.

Due to the nature of the product cycle, non-durable applications are those where recycling developments have historically focused, both for the sheer volumes involved (around 15 Mt just in Europe [1]) and for the challenges associated. For the sake of clarity, in this section we will focus on packaging PO (with some examples from durable applications such as pipes), and segment PO applications into three groups:

1. **Rigid:** These include detergent bottles, caps, construction pipes, and cosmetics containers. These are mostly PP and HDPE.
2. **Flexible (films) for industrial applications (secondary and tertiary packaging):** These include palletization films (stretch), protective films, and shrink films (also collation). Some agricultural films, such as those used for silage, could be considered in this segment too. These are mostly LDPE and LLDPE.
3. **Flexible (films) for food and medical packaging (primary packaging):** These are usually multilayered structures, combining all POs plus tie layers, barrier resins (such as for instance aluminum, polyamide PA, or ethylene vinyl alcohol (EVOH)), printing inks, etc. Some examples are coffee pouches, lidded trays for meat and cheese, blister packaging for tablets, and stand-up pouches for dishwasher tabs.

8.2.3 Common Processes for Mechanical Recycling of PO

There is a vast literature on mechanical recycling of POs [7–35]. Recycling usually starts with recyclate collection, where the usual distinction is between post-industrial recyclates (PIR) and post-consumer recyclates (PCR) [23]: mixed plastics waste from households are considered in the latter, together with secondary/tertiary packaging collected by retailers.

Collection processes and rates for packaging vary significantly by country. In general, sources are either households (municipal waste), logistics (retailers for instance) and special waste (such as for instance bins and crates used for transporting chemicals).

After collection, plastics waste is usually sorted, with near-infrared and density differential being quite common techniques. It is important to note that, while combining more technologies can effectively differentiate the various plastics materials, sorting methods cannot yet easily distinguish food-contact-approved plastics from other recyclates. Technologies for appropriately marking packaging for easier selection are being developed: an example is "Holy Grail" tagging [24] which combines chemical tracers and digital watermarks to facilitate identification of specific plastics for each packaging type, hence improving sorting.

Different PO films used for logistics packaging, for instance stretch wrap or collation shrink films, can be either separately or jointly collected, usually at the retailer level (the so-called "back store"): in the former case the process is lengthier and requires manual intervention, although it does deliver PE recyclates of higher quality, as polymer selection is very accurate.

A commonly used multilayer packaging is the liquid packaging board, which comprises paper board, aluminum and various layers of PE. From a recycling point of view, liquid packaging boards can often be handled using a separate, efficient process, based on separation of the paper fraction from the “PolyAl” (PE/aluminum laminate) [25].

After sorting, plastic waste is generally shredded and washed, followed by an initial extrusion step to deliver more homogeneous, higher-quality recyclate granules by reducing volatiles (odors), gels, contamination, etc. A secondary extrusion step is sometimes required to either blend in other components (for instance additives or virgin resins), or to further reduce impurities and contamination from recycled materials.

Similar to other polymers, PO usually undergoes severe physico-chemical changes during the mechanical recycling process, and yet, the most critical step, in terms of overall impact on PO structures and subsequent application of recyclates, is undoubtedly extrusion. As we will see in Section 8.2.4 and Section 8.2.5, different POs experience various physico-chemical changes during this step, and we will discuss these individually.

As POs are used in a vast array of applications, PO recyclates can present very different types of contamination:

- **Gels** can be addressed using recently developed extrusion machinery [26] – for instance, by screen filtering and on-line measurements.
- **Volatiles and odors** formed during extrusion are important considerations for POs, as they are used for a very broad range of content, from meat to milk and detergents. The array of by-products is quite broad, and includes alcohols, aldehydes, short-chain hydrocarbons, and ketones [19]. Hence, proper venting is needed to extract these volatiles and reduce their negative impact on odor for PO recyclates.
- **Oxygen**, as well as the presence of certain contaminants like paper fibers or pigments, can effectively increase degradation during extrusion of PO.

It has been noted [13] that higher molecular weight and/or branching in polymers before extrusion leads to a higher impact of the recycling process on the final recyclate's performance, likely due to higher melt viscosity (and hence shear stress). In this respect, all other things being equal, POs used for injection molding will be less susceptible to shear-induced degradation than their counterpart used for blown film production.

The most challenging films to sort and recycle are probably the “multimaterial laminates” – for example, those used to package meat, cheese, coffee and even dried food. This is because they combine materials of an entirely different nature, which are immiscible when re-processed together, and locked in structures in semi-permanent ways by either co-extrusion or adhesive lamination. Recyclates from multimaterial laminates present phase separation and large inhomogeneities, lowering drastically their value. For this reason, these PO-rich types of packaging are the object of several developments to improve the quality of final recyclates [12, 16].

Recycling PO has to ensure that final recyclates are a viable solution for real-life applications, hence mechanical properties (or mechanics in short) will be discussed for the various cases.

8.2.4 Mechanical Recycling of Rigid PO: Processes and Physico-Chemical Changes

As mentioned earlier, we will focus our attention on the extrusion step of the mechanical recycling process, because of its impact on the physico-chemical features of recycled PO. This section will discuss first the generic phenomena associated with PO, then examine in greater detail the changes occurring in (1) HDPE, (2) PP and (3) HDPE/PP blends, the most used POs for rigid packaging. We will also discuss some examples of recycled PO applications.

During PO extrusion, the simultaneous action of thermal (heat) and mechanical (shear) stress generates shear-induced chain scission, branching, or crosslinking. PO extrusion generates radicals along the aliphatic polymer chain by (A) oxygen-induced peroxy radical and (B) H-abstraction due to thermal stress [7]. The radicals formed can provoke β-scissions of PO chains, forming (macro) radicals, which the heat and shear will intensify: by recombining and separating, these (macro)radicals can transfer within the same macromolecule (intramolecular radical transfer) and subsequently produce shorter, branched, or crosslinked polymer chains [18].

While the exact extrusion setup (e.g. screw profile, screw length, filters, degassing) and conditions (e.g. temperature profile, screw rotation, output) have an influence on the changes occurring to the PO, two contrasting physico-chemical phenomena of chain scission and chain branching/crosslinking will occur, often simultaneously and in competition with each other. It is worth noting that each PO class (HDPE, LDPE/LLDPE, PP) has different prevailing mechanisms, as represented in Figure 8.25 [7].

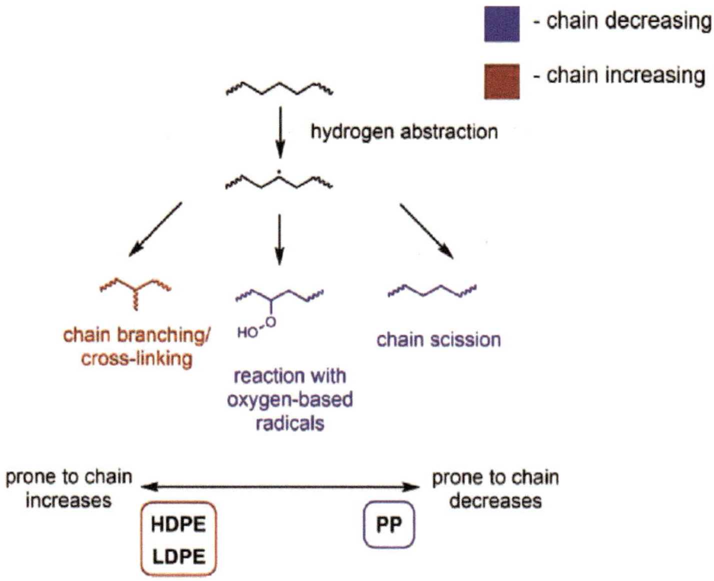

Figure 8.25 Chain scission vs. chain branching/crosslinking mechanisms for the three main classes of PO during extrusion (heat and shear stress) (Reproduced with permission from Schyns and Shaver [7])

8.2.4.1 High-Density PE (HDPE)

HDPE molecules are highly linear, forming packed crystals in the solid state. Once molten, shear-induced degradation will first cause chain scission, whereby shorter, linear, aliphatic macromolecules are formed by radicals.

These shorter polymer chains are on one hand less susceptible to the shear forces, yet, on the other hand, they will eventually be attacked by radicals and macroradicals. Long chain branches will then be formed by these (longer) macroradicals attacking the double-bonded carbons on the unsaturated (shorter) linear chains. It appears that extruder screw speed has large influence on the chain scission and branching for HDPE [11, 21]: a lower extrusion (screw) speed might create lower shear stress, hence lower branching and crosslinking as a whole, yet molecular weight distribution would play a major role in this respect.

Oxygen concentration and distribution in the melt seems to be an important factor to determine whether HDPE will encounter primarily chain scission or chain branching during recycling [18]. In a high-oxygen environment, for instance due to trapped air or oxygen adsorbed on the surface of contaminants (like paper fibers), HDPE chains will undergo scission, to ultimately form more stable carbonyl species. Where available oxygen is lower, or in areas inside the extruders with lower oxygen concentration (due for instance to lower contamination), these stable carbonyl species will not be formed after chain scissions, and chain scission will generate two reactive chain ends (macroradicals). These radicals in turn might react with the (terminal) double bonds of other chains, likely to be shorter chains less subject to mechanical-stress degradation, with the potential of branching.

Therefore, what started as a very linear HDPE polymer with a defined molecular weight (MW) distribution will be degraded, in the presence of oxygen, into carbonyl-terminated shorter chains and branched, higher MW chains. The presence of a significant level of vinyl unsaturation in some HDPE grades (like those produced with chromium catalysts) can also affect the sensitivity to crosslinking. A way to nudge HDPE degradation towards cross-linking is the combined use of a peroxide initiator with a branched chemical species with multiple unsaturated carbon bonds as terminal group [21]. These co-agents [21] prevent chain scission and favor the crosslinking reaction initiated by the peroxide itself.

Predicting mechanical properties of recycled HDPE is not an easy exercise, as it depends on many factors which are very specific to the virgin HDPE and its applications, and the exact recycling conditions (contamination, oxygen concentration, and distribution, shear/temperature profile in extruder, additives such as peroxide or stabilizers).

In general terms, recycled HDPE (as well as blends with virgin HDPE) could exhibit a decrease of tensile strength [7], flexural or tensile modulus [11, 27], and long-term properties such as resistance to cracking due to environmental stresses, as determined by, for example, the Pennsylvania edge-notch tensile test (PENT) [11]. Yield strength is less affected by the recycling process [11, 27]: this might be possibly due to the presence of carbonyl-terminated short chains, which act as in-situ plasticizers during the slow deformation in the tensile or flexural test.

8.2.4.2 Polypropylene (PP)

With reference to Figure 8.25, it is fair to say that the mechanisms for degradation of polypropylene (PP) are in contrast to those for HDPE. While recycled HDPE, properly processed, will show more chain branching (and hence an increase in MW), recycled PP will generally undergo chain scission (and hence a decrease in MW).

The onset of PP degradation is similar to HDPE, i. e. abstraction of hydrogen atoms by peroxy radicals. Subsequently, radicals encounter tertiary carbons and methine groups [17]. Radicals on methine functional groups are more reactive, and hence form more peroxy-radicals, while the tertiary carbons stabilize the radicals on the aliphatic backbone [7, 17]. A combination of both mechanisms creates favorable conditions for increased chain scission, and hence a reduction of MW for recycled PP. Notably, it has been observed that oxidation starts in the amorphous phase of PP, which is usually higher at the spherulite boundaries [17]: these boundaries therefore become points of weakness for the overall PP structure.

The mechanical properties of a polymer are in general a good indicator of the macromolecular structure: recycling PP (with the associated reduction of MW) will decrease tensile strength and impact strength. These shorter PP macromolecules could potentially increase crystallinity by facilitated crystalline arrangement [28], and accordingly, this has been reported to increase modulus and lower elongation at break of recycled PP [7].

Differently from PE, the crystalline structure of PP (which is spherulitic) apparently plays a big role in how degradation during recycling unfolds for PP. Stabilization of PP for recycling is therefore even more relevant than PE stabilization. Due to the amorphous nature of the spherulite boundaries [17], it is common that stabilizers (such as phenols) are present in higher concentrations in these areas. Therefore, it can be deduced that, to avoid over-loading PP with stabilizers to ensure a minimum critical concentration across the whole PP matrix, it is essential to minimize the spherulite size (for instance using nucleating agents).

8.2.4.3 PE/PP Blends

In municipal waste it is not uncommon to find PP and PE (more precisely HDPE) mixed together after recycling. In fact, HDPE and PP have similar density and source (bottles, canisters, crates, etc.), and cannot therefore easily be separated at source. Besides, PP and PE, despite resembling each other chemically, are generally immiscible: it is therefore key to understand how PP and PE can be possibly blended in the same final compound, so we will start from the basis of PE and PP blends.

Given the differences in their chemical structures, blends of PE and PP show high interfacial tension (IFT) [13], with poor stress transfer between the two polymers, caused by the extra energy stored at the interphase boundaries, which effectively pushes PE and PP apart during melt blending and subsequent crystallization.

There are some differences between PE and PP products synthesized using Ziegler-Natta (ZN) catalysts (which are heterogeneous) and those synthesized using metallocene (m) catalysts (which are site-specific) – in particular on how the PE and PP chains align and fold at the interface. The four mechanisms are shown graphically in Figure 8.26, and can be summarized as:

- **Entangled crystals** (mPE (mLLDPE) and mPP): Given the high crystallinity and very narrow molecular weight distribution of both polymers, the prevalent structure at the PE/PP interface is highly crystalline, hence creating a network of entangled crystals.
- **Easy chain pullout** (mHDPE and mPP): A low content of amorphous (oligomeric) material creates pockets of HDPE lamellae, showing cross-interface crystallization with PP.
- **Kinetically trapped entanglements** (mHDPE and mPP): Very slow crystallization results in cross-interface entanglements.
- **Excess oligomers** (ZN PE and ZN PP): An amorphous portion of either polymer (oligomers) separates itself at the interface between PE and PP, creating a "layer" hindering physical connection between the crystalline phases of either polymers.

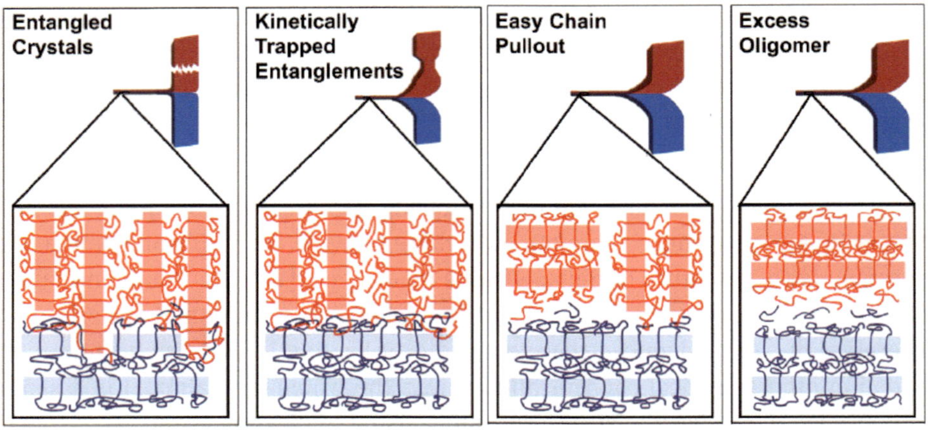

Figure 8.26 Schematic representation of the four classes of PO interfaces (PE: red, isotactic PP: blue) (Reprinted with permission from *Macromolecules*, copyright (2021) American Chemical Society) [20]

It is to be remembered that a weak interface (i.e. poor PE/PP adhesion) has a strong, negative impact on mechanical properties of PE/PP blends [13, 20]. In more practical terms, when recycling blends of PE and PP, the specific HDPE and PP being recycled might have a very large impact on the final performance of the recyclates. We will see later in this section how to improve the mechanical performance, by improved interfacial adhesion, of PE/PP blends.

PE/PP blend morphology is also highly influenced by the specific extrusion conditions. A co-continuous morphology is more favorable to increase synergies between PE and PP phases, hence delivering superior performance [13]. While under shear, PE and PP blends can form droplets that elongate into filaments (or threads), followed by fragmentation to droplets, or coalescence of droplets into larger ones. When PE and PP are immiscible in the melt, for instance if no compatibilizer is added, the high IFT promotes the coalescence of the minor phase (i.e. the polymer with lower concentration) into a matrix of the other polymer. Blends of HDPE and PP show [13] finely dispersed PE droplets at low PE concentration, with their size increasing with PE concentration until they reach a range of PE vs. PP concentrations where a co-continuous morphology is obtained. Beyond this range, PP will form large droplets into the (now) HDPE matrix, and the size of these droplets will decrease with decreasing PP percentage.

So far, we have analyzed the quasi-ideal case of linear PE (HDPE) and PP. In the presence of polymer branching, for instance when blending LDPE and PP, the interface is very poorly entangled, as the long-chain branching of LDPE and its low crystallinity are very different from the crystalline arrangement of PP.

For this reason, LDPE/PP blends have been reported with properties inferior to the "rule of mixture" of the two separate polymers. LLDPE, with its absence of long-chain branching in favor of (usually) short-chain branching, has a crystalline structure more similar to PP, and hence results in better compatibility and properties of the blend. It therefore follows that highly linear HDPE is therefore the PE that is easier to blend with PP.

From the point of view of mechanical properties, PE/PP blends exhibit different behavior depending on the exact blend composition, extrusion conditions, presence and concentration of stabilizers and compatibilizers, as well as crystallization conditions. In general terms, it has been shown [9] that blends of the same PE and PP prepared before (virgin) or after recycling (recycled) have very similar relation between their mechanical properties (tensile modulus, strength and maximum strain) vs. the content of either component. In other words, knowing how virgin PE and PP blends behave mechanically can be used as a predictive tool for the properties of recycled PE and PP blends.

It is well-established [7, 9, 13, 22] that HDPE will give better mechanics than LLDPE (and even more so than LDPE) in blends with PP. This is in line with the mechanisms of interfacial entanglement-crystallization reported in Figure 8.26. Small contaminations of LDPE (for instance from cosmetics tubes) into the HDPE/PP stream can therefore decrease the mechanical properties of the recycled PE/PP blend in a way that does not reflect their proportions.

Slower crystallization rates seem to provide better mechanics for the PE/PP blends [9, 20]. In practical terms, this might suggest that, when molding a recycled PE/PP article, using a slower cooling rate (for instance by increasing the mold temperature) will results in better mechanics overall for the new molded article. Tensile modulus and strength increase with the content of PP, either to reach a peak value with a subsequent drop [13] or following a more linear relationship [9]: HDPE has seemingly lower (negative) impact on PE/PP blends than LDPE.

More complex is the situation of elongation, or maximum tensile strain, which is usually very dependent on crystalline morphology and the presence of oligomers. Small amounts of LDPE have been reported to almost double the maximum elongation of recycled PP/PE blends [9].

As interfaces, both in terms of morphology [7, 20] and energy [9, 13], are key in determining the properties of PE/PP blends, compatibilizers are regularly used to improve the quality of recycled PE/PP blends by acting specifically on the polymer interfaces. One can distinguish between reactive and non-reactive compatibilizers. The latter usually have one side miscible with PE and the other one miscible with PP. Ethylene–propylene rubbers (EPRs), as ethylene–propylene–diene terpolymers (EPDM), are one example of non-reactive compatibilizers. EPRs are widely used in the industry, as they can increase elongation at break while not affecting tensile strength and modulus [7, 13].

EPRs effectively act to stabilize the PE/PP interface, promoting more entanglements and hence facilitating stress transfer, and consequently their influence on

PE/PP blend mechanics. The positive influence of EPRs has been reported to peak at a certain concentration [13, 29, 30], and therefore it is crucial to set the PE/PP blend composition first, then proceed to add the EPR compatibilizer (as well as for reasons of extra costs and added manufacturing complexity). Block copolymers, for instance octene–ethylene, because they have a more organized and homogeneous structure, are more effective than random copolymers in compatibilizing, as they are able to interact strongly with the interface [29, 30].

As rigid PO waste (HDPE and PP) is widely available, it has found many routes into everyday products: from vases to crates, bottles, and automotive components. The quality of the recycled HDPE and PP can vary greatly. For instance, in addition to the mechanical properties described above, the organoleptic profile can be very different for the same base (virgin) PP or HDPE, depending on the original application, washing step, and de-gassing during extrusion. It is also very tricky to separate colored artifacts from natural ones: it is therefore common to have "Natural recycled PP" and "Jazz recycled PP", for instance, with the term "Jazz" used to describe a mixture of underlying colors.

There are different routes to minimize the disadvantages of using recycled PO in everyday applications: for instance, in the HDPE bottles used to contain home -detergents, the recycled HDPE is co-molded (blown) in the middle layer of a three-layer structure, effectively "sandwiching" it with virgin resins both inside and outside the bottle.

In this way, the exact color specification for the recycled PO is not relevant (as it is hidden from the consumer), there is also a much lower (or zero) impact of the odor profile on the detergent (as the inner virgin layer acts as a barrier to the odors, at least partly), and the packaging design requires only minimal adjustment.

An alternative, rather more advanced solution is to use the so-called "hybrids". Hybrids are blends (either dry-blends or compounded) of at least one recycled PO and one virgin PO. Products like "Recycl-IN" from INEOS belong in this category. Hybrids use highly engineered virgin PO to raise the performance of the overall blend to the level of virgin-like PO. It is clear that this virgin PO has a limited influence on the color, organoleptic and contamination properties of the recycled PO and the overall blend. Yet it can be very effective to improve other properties, for instance mechanical or stress-cracking resistance (ESCR), to the level of virgin PO, hence resulting in a final PO product with a high proportion of recycling material (50% for instance) and with a performance on a par with virgin material. This hybrid approach has been adopted for industrial products, such as INEOS Recycl-IN rHD5402, suitable for detergent bottles with medium to high specifications for ESCR.

Table 8.3 and Table 8.4 summarize the main relationships between the various types of PO (either pure or in blends), their degradation mechanisms, and the resulting properties. The information in this table is not exhaustive: contamination, process conditions and exact recycling material composition can (and will) have a large influence on the final performance of the recycled PO.

Table 8.3 Main Relationships between Pure POs

	Primary degradation mechanism	Oxygen (O_2) influence	MW	Gels	Crystallinity	Stiffness (Young's or flexural modulus)	Strength (tensile or flexural)	Elongation at break	Other properties	Remarks	Ref.
LDPE & LLDPE	Chain branching/cross-linking	High O_2: Chain scission Low O_2: Chain branching	Increase	Increase	Slight decrease	Slight decrease	–	Slight increase	Haze slightly increases	–	[7, 31, 34]
HDPE	Chain branching/cross-linking	High O_2: Chain scission Low O_2: Chain branching	Increase	Increase	Slight decrease due to cross-linking (hindering folding of lamellae)	Decrease	Decrease	Stable or slight increase	ESCR decrease	Peroxide can favor branching	[7, 11, 18, 21, 27]
PP	Chain scission	–	Decrease	–	Slight increase due to lower-MW chains (easier folding)	Increase (resulting from higher crystallinity)	Decrease	Decrease	Impact strength decreases	Oxidation starts at boundary between amorphous phase and spherulites	[7, 17, 28]

Table 8.4 Main Relationships between PO Blends

	PO Interface arrangement	Effect of crystallization rate	Stiffness (Young's or flexural modulus)	Strength (tensile or flexural)	Elongation at break	Remarks	Ref.
LDPE/PP	Very poor entanglement, as for highly branched LDPE vs. linear PP	–	Large decrease with increasing LDPE content	Decrease with increasing LDPE content	Increase with increasing LDPE content	Massive increase of elongation when LDPE content is >25%	[9, 13]
ZN LLDPE/ZN PP	Excess oligomers, high interface energy (poor interface cohesion)	–	Large decrease with increasing LLDPE content	Decrease with increasing LLDPE content	Slight increase with increasing LLDPE content	–	[9, 13, 20]
mLLDPE/mPP	Entangled crystals, highly interpenetrated	Slow rate favors entanglements	Decrease with increasing LLDPE content	Stable vs. LLDPE content	Slight increase with increasing LLDPE content	–	[9, 13, 20, 29]
HDPE/PP	ZN HDPE and ZN PP: Excess oligomers, high interface energy (poor interface cohesion) mHDPE and mPP (regular crystallization): Easy chain pullout, alternating entangled lamellae with pockets of loose interfaces mHDPE and mPP (slow crystallization): Kinetically trapped entanglements	Slow rate favors entanglements	Decrease with increasing HDPE content	Stable vs. HDPE content	Slight increase with increasing HDPE content	–	[7, 9, 13, 22]

8.2.5 Mechanical Recycling of Flexible PO: Processes and Physico-Chemical Changes

This section will first focus on the challenges of mechanically recycling industrial films, and will then move on to discuss recycling of multi-material (multilayer) flexible packaging.

8.2.5.1 Industrial films

Industrial (logistics) flexible packaging is usually PO-based, and is usually classified as:

- **Secondary:** This is mostly based on LDPE, for its shrinkability and excellent optics, and include shrink-wrapped films for water bottles.
- **Tertiary:** This is mostly based on LLDPE, for its mechanics (resistance to puncture) and stretchability, and include stretch wrap films for pallet wrapping.

Protective films, either PP- or PE-based, also belong to the industrial films category, though their overall volume is definitely smaller than secondary or tertiary packaging. Ultra-low gel content is one of their main features.

Post-consumer LDPE- and LLDPE-based films are usually collected in different environments: secondary packaging (mostly LDPE films) are disposed of by consumers (households) as well as by operators of the "HoReCa" (hotel, restaurant, catering) industry, less frequently by retailers. Tertiary packaging is usually collected at retailer sites (for instance when unbundling a pallet of other goods such as confectionery) and in more industrial environments (for instance at construction sites where logistics films are used to wrap and secure bags of cement).

As industrial films are used in a variety of environments, it is difficult to predict possible contamination: it is generally safe to assume that films collected at retailers have lower contamination (like dust) than those collected in industrial operations. Yet, accidents involving breaking of primary packaging (for instance beer bottles or oil canisters) can also contaminate logistics films, with a strongly negative impact on their recyclability.

While LDPE and LLDPE films share many similarities, it would be advisable to collect and process them separately as far as possible, to extract maximum value from each resulting recyclate. When LDPE and LLDPE films are processed together (as is very commonly the case), it is more difficult to re-use the recycled material for its original applications. There are essentially two reasons for this:

- **The different degradation modes of LDPE and LLDPE:** The long-chain branched structure of LDPE [7, 31] favors crosslinking (or further chain branching) during degradation. The presence of oxygen presence will increase chain scission, while oxygen scarcity will lead to further branching. The more branched a LDPE is, the less chain scission and more chain branching will happen. Virgin

LDPE is also hardly stabilized, which adds an additional risk of gel creation to the recycling process.

The short-chain branched structure of LLDPE, on the other hand, is more prone than LDPE to chain scission, yet its exact behavior is harder to predict than LDPE. This is because LLDPE has a larger array of macromolecular structures than LDPE, depending on the α-olefin copolymer, density, short chain branching and molecular weight distribution.

- **The different intrinsic properties of LDPE and LLDPE:** The presence of large amounts of LDPE in LLDPE-rich blends will result in lower mechanical properties, in particular toughness and elongation at break, which are key for stretch-wrap or stretch-hood applications. This is due both to the poorer mechanics of LDPE compared to LLDPE (long chain branches create broader stress fields within the polymer matrix), as well as the larger amounts of gels (due to chain branching happening for LDPE during recycling).

  On the other hand, the presence of large amounts of LLDPE in LDPE-rich blends will reduce the orientability of the polymer chains in the melt, with a direct negative impact on the shrinkability in final application. The optical properties, such as haze or transparency, of the LDPE-rich blend will also decrease due to the presence of LLDPE.

8.2.5.2 Multilayer Flexible Packaging

When considering food and medical packaging, sometimes called consumer packaging, mechanical recycling cannot effectively deal with the complex, multilayered and multi-material film structures. In these multilayer films, POs represent a key component, but are far from being the only polymers.

For instance, in many pouches, the internal (sealing) layer can be either LLDPE or PP, with more PO in the core layers and on the lamination (or skin) layer too (which is generally LDPE). The outer layer, which is often reverse-printed, can be either biaxially-oriented PET (BoPET) or PP (BoPP), depending on the specificity of the packaging.

Yet, as much as PO contributes to the sealing/abuse/structural/lamination performance of the packaging, other polymers (or aluminum) will help with the gas-aroma barrier (PA, EVOH, PVDC), inter-layer adhesion (tie layers, usually PO-g-MAH) and printability/gloss (the already mentioned BoPET and BoPP). Lamination usually occurs by polyurethane adhesive or extrusion lamination (much less common). These structures are already very complex, even excluding additional printing inks, coatings, surface treatments (like AlO_x or SiO_x), as well as labels and their adhesives.

While POs are often the main component of multilayer packaging films, mechanically recycling these films is difficult at least, with final recyclates having very poor performance and consistency due to the various layer compositions and variability.

Studies have been conducted on how to improve the recyclability of PO/PA or PO/EVOH structures [12]. Generally, once recycled, these PO-rich multilayer films produce immiscible polymer blends, with very coarse phase morphologies and poor mechanics (owing to poor interfacial adhesions). Usage of compatibilizers has also been studied: PE grafted with maleic anhydride (PE-g-MAH) has been found to be an effective compatibilizer for PE/PA and PE/PET structures [12, 32, 33], at least in a certain concentration range. Elongation at break and impact strength are generally improved across the whole concentration range and processing conditions [32].

More complex is the evolution of stiffness (indicated by tensile modulus), that was shown to improve only when PA microfibers were formed in the PE matrix due to shear elongation [32]. Ionomers have also been proved effective in compatibilizing LDPE/LLDPE/PA blends [33], with a final stiffness exceeding the one of the "rule of mixture" of the virgin polymers.

These approaches have found applications in the real-life recycling of multilayer flexible films, with PE-g-MAH being added to the virgin structures (blended within other PO layers) or directly in the recycling extruder. The resulting recyclates show better morphology and mechanical properties, on both elongation and modulus, than uncompatibilized blends.

What was discussed in Section 8.2.4 and Section 8.2.5 is applicable also to repeated recycling cycles, where chain scission and branching can sometimes alternate throughout a certain number of cycles [8, 34]. As more recyclates will be included in packaging and other fast-turnover applications, more research needs to be put into how to keep the POs recyclable for multiple cycles.

8.2.6 Food Contact Regulations for Mechanically Recycled PO

As it stands today, with one exception that we will discuss, mechanically recycled PO cannot be used for any application requiring food contact approval. This also limits pharmaceutical, medical and cosmetic application, not because of regulatory limitations but because of precautionary, habitual practices in the market.

In a PO waste stream containing at least 99% food-contact-compliant PO packaging [14] (for instance HDPE for milk or juice bottles), contaminants in the recycled PO can be reduced to a level deemed suitable by the USA Food and Drug Administration (FDA) to prevent these contaminants entering the food in concentrations higher than 0.5 ppb. This is achieved by selection of the PO waste stream and performing a rigorous cleaning/extrusion process [14]. If this concentration limit is met, a non-objection letter (NOL) provided by the FDA can be issued, enabling this recycled PO to be re-used in applications in direct contact with food.

This approach is common practice for instance in the UK, where generally 30% of HDPE for milk bottles is recycled HDPE, in compliance with the FDA-NOL upper contamination limit. This recycled quota can be increased, potentially up to 100% [14], provided the upper limit for contaminant is respected.

The European Food Safety Agency (EFSA) does not have yet an equivalent to the FDA-NOL, though there is a drive to promote recycled HDPE in applications demanding very high purity of recycled materials, for instance cosmetics, where historically food-contact levels of contaminants have been considered a “golden rule”.

An open question is whether LDPE or LLDPE, having lower T_g and much faster diffusion of contaminants within the polymer matrix [35], could ever meet food contact requirements (even under FDA-NOL) once recycled: a challenge test, similar to those ran for HDPE, would need to be identified and performed, targeting not only low molecular weight contaminants but also higher MW ones (which would anyhow diffuse at speed within the more elastomeric matrix of LDPE or LLDPE compared to the more crystalline one of HDPE).

Regulations might soon evolve, and the reader is recommended to keep an eye on new legislations and regulations being approved in next years.

References for Section 8.2

[1] Plastics Europe, *Plastics – The Facts, plasticseurope.org*, accessed January 2021

[2] Duffy, J., Argus Media: *Plastics Recycling – Challenges and Opportunities*, video presentation to the CEFIC LOSG/APA sector meeting, June 2020

[3] Hundertmark, T., Mayer, M., McNally, C., Simons, T.J., Witte, C.: “How plastics waste recycling could transform the chemical industry”, *https://www.mckinsey.com/industries/chemicals/our-insights/how-plastics-waste-recycling-could-transform-the-chemical-industry*, accessed January 2021

[4] Polyolefins Circular Economy Platform PCEP: PCEP commitment, *https://pcep.eu/our-commitments*, accessed January 2021

[5] EU Law: “Packaging and Packaging Waste”, *https://eur-lex.europa.eu/legal-content/EN/TXT/?qid=1595838206165&uri=LEGISSUM:l21207*, accessed December 2020

[6] European Plastics Pact: *https://europeanplasticspact.org/wp-content/uploads/2020/03/European-Plastics-Pact-FINAL.pdf*, accessed December 2020

[7] Schyns, Z. O. G., Shaver, M. P.: *Macromolecul. Rapid Commun.*, (2020) 2000415

[8] Al-Salem, S. M., Behbehani, M. H., Al-Hazza'a, A., Arnold, J. C., Alston, S. M., Al-Rowaih, A. A., Asiri, F., Al-Rowaih, S. F., Karam, H.: *J. Mater. Cycles Waste Manage*, (2019) 21, pp. 1106–1122

[9] Aumnate, C., Rudolph, N, Sarmadi, M.: *Polymers*, (2019) 11, p. 1456

[10] Delva, L., Cardon, L., Ragaert, K.: *Env. Eng. Manage. J.*, (2018) 2, pp. 427–434

[11] Juan, R., Dominguez, C., Robledo, N., Paredes, B., Garcia-Muñoz, R. A.: *J. Cleaner Prod.*, (2020) 276, 124081

[12] Jönkkäri, I., Poliakova, V., Mylläri, V., Anderson, R., Andersson, M., Vuorinen, J.: *J. Appl. Polym. Sci.*, (2020), p. 137

[13] Graziano, A., Jaffer, S., Sain, M.: *J. Elast. Plast.*, (2019) 51, pp. 291–336

[14] Wöss, C., Kitzberger, C.: *Kunstoffe Int.*, (2020) 5, pp. 38–40

[15] D'Ambrières, W.: *Field Act. Sci. Rep.*, (2019), pp. 12–21

[16] Maris, J., Bourdon, S., Brossard, J.-M., Cauret, L., Fontaine, L., Montembault, V.: *Polym. Degrad. Stab.*, (2018), pp. 245–266

[17] Knight, J. B., Calvert, P. D., Billingham, N. C.: *Polymers*, (1985) 26, pp. 1713–1718

[18] Pinheiro, L. A., Chinelatto, M. A., Canevarolo, S. V.: *Polym. Degrad. Stab.*, (2004) pp. 445–453

[19] Hoff, A., Jacobsson, S.: *J. Appl. Polym. Sci.*, (1981), pp. 3409–3423

[20] Jordan, A. M., Kim, K., Soetrisno, D., Hannah, J., Bates, F. S., Jaffer, S. A., Lhost, O., Macosko, C. W.: *Macromolecules*, (2018), pp. 2506–2516

[21] Kim, K. J., Kim, B. K.: *J. Appl. Polym. Sci.*, (1993), pp. 981–986

[22] Klimovica, K., Pan, S., Lin, T.-W., Peng, X., Ellison, C. J., LaPointe, A. M., Bates, F. S., Coates, G. W.: *ACS Macro Lett.*, (2020), pp. 1161–1166

[23] Environmental labels and declarations – Self-declared environmental claims, ISO 14021:1999

[24] Holy Grail: Pioneer Project, *newplasticseconomy.org*, accessed January 2021

[25] Post-Consumer Recycling, Tetra Pak: *https://www.tetrapak.com/sustainability/planet/recycling*, accessed January 2021

[26] Feichtinger, K., Hackl, M., Wendelin, G.: US Patent 9 808 963 B2 (2017)

[27] Alzerreca, M., Paris, M., Boyron, O., Orditz, D., Louarn, G., Correc, O.: *Polym. Testing*, (2015), pp. 1–8

[28] Aurrekoetxea, J., Sarrionandia, M. A., Urrutibeascoa, I., Maspoch, M. L.: *J. Mater. Sci.*, (2001), pp. 2607–2613

[29] Choudary, V., Varma, H. S., Varma, I. K.: *Polymers*, (1991), pp. 2541–2545

[30] Kock, C., Aust, N., Grein, C., Gahleitner, M.: *J. Appl. Polym. Sci.*, (2013), pp. 287–296

[31] Dostá, J. I., Kašpárková, V., Zatloukal, M, Muras, J., Šimek, L.: *Eur. Polym. J.*, (2008), pp. 2652–2660

[32] Li, Z. M., Yang, M. B., Huang, R., Yang, W, Feng, J. M.: *Polym. Plast. Technol. Eng.*, (2002), pp. 19–27

[33] Kaiser, K., Schmid, M., Schlummer, M.: *Recycling*, 2018, pp. 1–9

[34] Yin, S., Tuladhar, R., Shi, F., Shanks, R. A., Combe, M., Collister, T.: *Polym. Eng. Sci.*, (2015), pp. 2899–2909

[35] Palkopoulou, S., Joly, C., Feigenbaum, A., Papaspyrides, D., Dole, P.: *Trends Food Sci. Technol.*, (2016), pp. 110–120

8.3 Mechanical Recycling of PET

Thomas Schröder

This section is not intended to describe new data but rather to collect and compile existing material. The fields of application for PET and the relevance of this material to the global and European plastics market are considered. The focus lies on PET bottles and their recycling possibilities, especially bottle-to-bottle recycling. Relevant processes of different market participants are presented that enable the production of food-grade R-PET by applying a deep cleaning process. The licensing procedure for bottle-to-bottle recycling processes is also discussed, as well as the limits of PET recycling.

8.3.1 Introduction

As a consumer, polyethylene terephthalate (PET) is ubiquitous. This thermoplastic is currently the raw material for a wide range of applications, especially in the packaging sector. PET can be processed into plastic bottles for the beverage industry, stable outer packaging trays for fruit and vegetables, films for food and non-food applications, and textile fibers. Today, PET is usually produced by direct esterification of terephthalic acid with monoethylene glycol (MEG). It is therefore the result of a reaction between acid and alcohol and belongs to the polyester family. The structural formula of PET is shown in Figure 8.27 [1, 2].

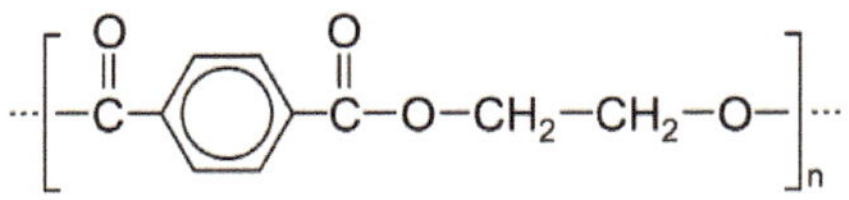

Figure 8.27
Structure of polyethylene terephthalate (PET) [2]

While terephthalic acid is currently separated exclusively from the aromatic fraction of crude oil, MEG can be produced both from naphtha and from renewable raw materials. For the latter, ethanol is obtained from sugar- or starch-containing plants, which serves as a starting point for the production of bio-MEG. If MEG from renewable raw materials is used in the synthesis of PET, it is referred to as "bio-PET 30" because 30% of the carbon atoms it contains are of biogenic origin. Since part of the petrochemical raw material components are completely substituted by biogenic raw materials here, without deviating from the established synthetic pathway of the plastic, it is referred to as a "drop-in" solution. The chemical structure of bio-PET 30 therefore corresponds to that of a petro-based PET. Thus, both polymers can achieve a very similar property profile, be used in established processing methods, and be recycled together in a recycling process. Nevertheless, the market penetration of bio-PET is currently still very low [1, 3, 4].

For the esterification of terephthalic acid, antimony or alternatively germanium catalysts are used. The resulting product is pretreated in a vacuum reactor and then processed via melt polycondensation to amorphous PET granules (A-PET). This material is characterized in particular by high transparency, but has a low molecular weight. For technically more demanding applications with food contact, solid-state polycondensation (SSP) is carried out downstream, which is shown schematically in Figure 8.28.

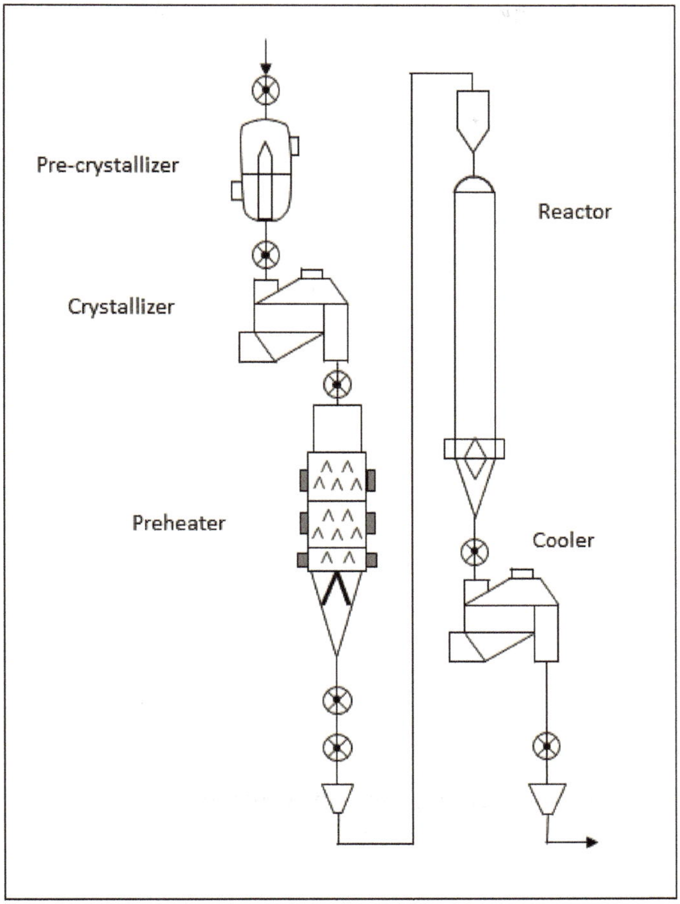

Figure 8.28 The process of solid-state polycondensation (SSP) used in PET production [5]

In the first step of the process, crystallization of the supplied A-PET is initiated by heating the polymer. The product is then passed through a preheater in which it is further heated during a defined residence time and thus prepared for the subsequent reactor. In this reactor, the actual polycondensation reaction takes place below the melting point of PET. Inert gas, such as nitrogen or helium, flows through the reactor, in which by-products such as water, ethylene glycol, or acetaldehyde accumulate and are removed by gas purification. SSP allows the molecular weight of the polymer to be increased and the acetaldehyde concentration in the polyester granules, which is critical for food packaging, to be reduced. The chain length of the polymer can be adapted very precisely to the application-specific requirements. The intrinsic viscosity (IV value) in deciliters per gram (dl/g) is used as a measure of the molecular weight. The IV value can be determined by solution viscometry or calculated from melt flow/volume rate (MFR/MVR) values using a correlation relationship [1, 2, 5–8].

Table 8.5 shows typical grades of PET, the IV values required for them, and their applications.

Table 8.5 Intrinsic Viscosity of PET for Different Applications [1]

Grade	IV [dl/g]	Example application
Fiber grade	0.40–0.70	Textile fiber
	0.72–0.98	Fibers for technical applications
Film grade	0.60–0.70	Extruded PET films
	0.70–1.00	PET films for thermoforming
Bottle grade	0.70–0.78	Bottles for non-carbonated drinks
	0.78–0.85	Bottles for carbonated drinks
Monofilament	1.00–2.00	Fishing lines, medical devices

In addition to the IV value, the (physical) properties of a semi-crystalline PET are largely determined by the degree of crystallinity. Depending on the processing conditions, a degree of crystallinity of 30–40% can be achieved with semi-crystalline PET. The crystallization rate of PET spherulites, 10 µm/min, is significantly lower than that of other polymers (e.g. 400 µm/min for POM, and 5000 µm/min for HDPE). In addition, the maximum growth rate can only be achieved in a temperature range of 200–230 °C. However, due to the high enthalpy of the melt, mold temperatures of about 140 °C are sufficient, for example in injection molding, to achieve a high degree of crystallization. Nucleating agents are also used to shorten the induction time [2].

8.3.2 Market Overview and Fields of Application

In 2019, global plastic production was 368 Mt. For the same year, the consulting company SBAcci indicated a PET demand of approx. 27 Mt in its analysis of the development of the PET market. PET therefore accounts for slightly more than 7% of global plastics production. A similar picture emerges when looking at European (EU28+NO/CH) plastic consumption in isolation. This is stated at 50.7 Mt for 2019. Of this, just over 4 Mt is PET, which corresponds to a share of just under 8% of European plastic demand. As can be seen in Figure 8.29, PET is primarily used in Europe for the production of packaging [9, 10].

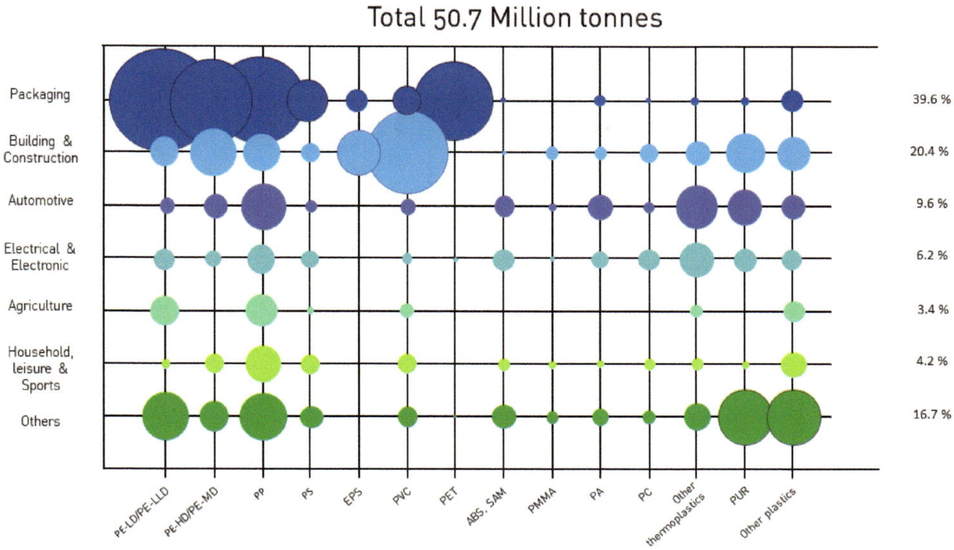

Figure 8.29 European plastics demand in 2019, divided into application areas and polymers [9]

A driving factor for the application of PET as packaging is that the crystallinity of PET can be reduced by incorporating co-monomeric building blocks to such an extent that molded parts made from it remain transparent even with high wall thicknesses. Due to its molecular structure, PET has good material and processing properties, which make it suitable for multi-stage production processes that require repeated thermal stressing of the material. Combined with the good barrier properties for oxygen, carbon dioxide etc., this means that PET can be used as packaging for several liquids. Another point that should not be underestimated is the possibility, described above, of adjusting the IV value of PET at any time using SSP. Due to this fact, packaging made of PET has a higher recyclability than packaging made from other standard plastics. Figure 8.30 shows the global PET market for 2016 broken down by product segment [1, 2].

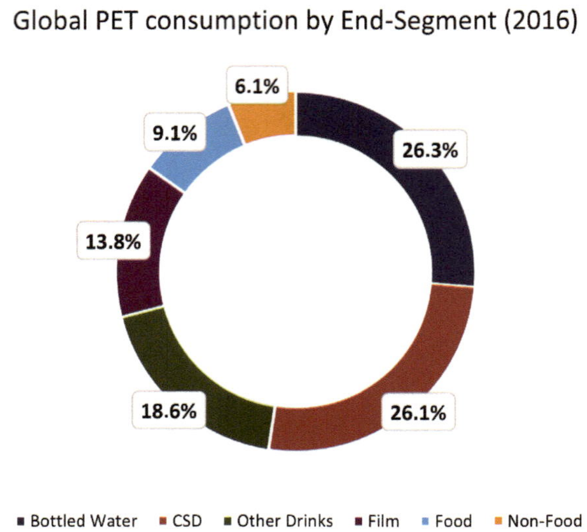

Figure 8.30 Global PET consumption in 2016 [11]

As shown above for the European market, PET is also primarily used in packaging applications on a global level. The PET market (by product segment) is dominated by water bottles and carbonated soft drinks (CSDs). Overall, more than 70% of the PET produced in 2016 was used in bottles. These thin-walled hollow packaging bodies are manufactured in the so-called "stretch blow molding" process. Within this multi-stage production process, PET granulate is injection-molded to produce a blank (preform), which is then biaxially formed into the finished PET bottle by stretch blow molding, with the addition of heat using compressed air. This process allows the preforms to be produced centrally, and not formed into bottles until the filling operation. This makes it possible to achieve tighter packing during transport, which increases the efficiency of the transport process from an environmental and economic point of view. In addition, PET bottles are lighter than glass bottles. For example, a disposable bottle weighing 25–30 g is about 25 times lighter than a glass bottle with comparable contents [1, 12].

8.3.3 Circular Economy of PET Bottles

Due to the increasing demand for PET in the packaging industry for many years (especially as a raw material for bottles), and the growing awareness of environmental protection and resource efficiency, coupled with technical innovations in recycling technologies, closed-loop recycling has established itself as a fixed pillar of the PET value chain. The closed-loop or circular economy is a model of production and consumption in which materials taken from natural sources are used as intensively as possible in order to save resources and conserve disposal capacities by extending the life cycle of products (or materials). This principle is identified, for example, in the EU's plastics strategy as a key approach to reducing the amount of plastic waste in the long term. Measures mentioned include the reuse of plastic products and an increase in the recycling rate. Across Europe, around 57% of all PET bottles on the market in 2014 were collected, sorted, and recycled after being used. This high recycling rate is directly related to the good recyclability of PET bottles. Recyclability is defined here as an individual and stepwise suitability of a product (or the materials it contains) to replace chemically identical primary material in typical applications, after passing through industrially available recovery processes [12–16].

The increased recyclability of PET bottles results, among other things, from the following points [17]:

- Established grade-specific collection systems, which greatly relieve subsequent digestion and sorting processes, and increase the quality of the secondary material
- Due to the market penetration of PET bottles, they are globally available, resulting in a high attractiveness for recycling companies and machine manufacturers.
- The use of removable caps and labels made of non-PET materials such as polyolefins enables uncomplicated separation of these interfering materials from the recyclable material stream via density separation.
- The bottles, which are usually transparent or only slightly colored, can be fractionated via color sorting.
- A subsequent increase in the IV value of a PET recyclate to that of virgin material can be achieved by using SSP (or comparable processes).
- PET is a comparatively inert polymer.

Used PET bottles thus form an ideal starting material for bottle-to-bottle recycling. The recyclate obtained in the mechanical recycling process can therefore be reused in the same or similar applications and thus be kept in a closed loop. In addition to conserving fossil resources, the use of recyclates also leads to an improvement in the environmental impact of a plastic product. The influence that the use of a recycled content has on the carbon footprint (CF) of a plastic product is shown in Figure 8.31 using the example of a PET tray [18].

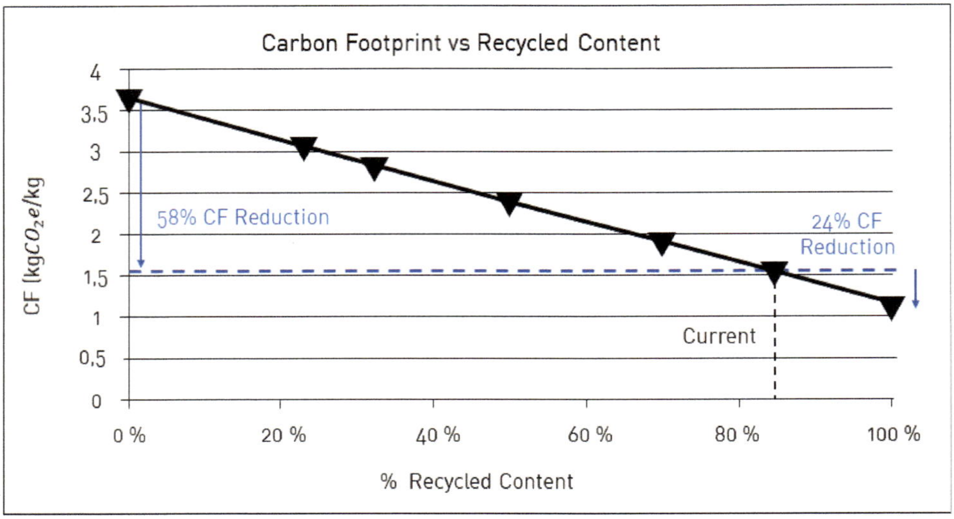

Figure 8.31 Influence of the recycled content on the carbon footprint of a plastic product made of PET [19]

The figure shows that a recycled content of 80% reduces the CF of the product by 58%. If the virgin material is completely supplemented by a secondary material, the CF could even be reduced by 82%. The difference in greenhouse gas emissions is mainly due to the saving of energy-intensive processes during plastic production when using a recyclate. These processes include extraction of the raw material base, refining into intermediates, recovery of the monomers, and their polymerization. The high energy and heat requirements of these processes are almost exclusively covered by the combustion of fossil fuels [20, 21].

8.3.4 Bottle-to-Bottle Recycling of PET

8.3.4.1 Bottle-to-Flake Processes

Regardless of the material used, plastic bottles accumulate either in separate collection systems (e.g. a deposit return system) or within a mixed plastic or waste fraction. If the latter is the case, manual or (automated) mechanical pre-sorting of the plastic bottles must take place. This process step is usually carried out on unshreddable material. Afterwards, the plastic bottles are usually available as pressed bales and can be transported to a recycling plant. Here, in the first step, the bales are broken up again and fed to a separation mechanism without further shredding. The individual objects are then passed by various sensors (X-ray sensor, NIR sensor, color sensor, shape recognition sensor). At the discharge station, the identified plastic objects are separated by controlled air blasts and transferred to the sort-specific collection containers. This process step allows the fed plastic bottles to be separated according to the grades PVC, PET and HDPE of different coloration. If, on the other hand, PET bottles are collected via a separate system, they are already largely unmixed. Therefore, no more complex separation into different plastic fractions is carried out, only a separation of metals and foreign plastics takes place. The bottles are then sorted by color as before [12, 17, 22].

Regardless of the collection logistics used (separately or as part of a waste fraction), the second step of the recycling process involves fine sorting of the PET bottles. This consists of shredding using a granulator (10–30 mm), magnetic sorting, float-sink sorting to separate the polyolefin (PO) fraction, and washing (usually hot washing) with subsequent drying. In addition, the degree of purity of the PET recyclate and thus its quality can be further increased by the addition of near-infrared (NIR) sorting for the further separation of foreign plastics. The output material of the bottle to flake process stage is thus a surface-cleaned PET flake that can be processed into regranulate in an extruder. After completion of this process step, a "non-food grade" PET recyclate (R-PET) is available, which can be used for textile fibers, films, or as core material for multilayer bottles. To obtain a food-grade R-PET that can be used in a bottle-to-bottle recycling process, additional deep cleaning of the recovered R-PET is required [1, 22].

8.3.4.2 Deep Cleaning Processes for PET Flakes

The decisive quality requirements for the production of food-grade R-PET are derived from the international food legislation and regulations of the European Food Safety Authority (EFSA) and the US Food and Drug Administration (FDA). The requirements for the recycling process are defined on the basis of a maximum permissible residual concentration of contaminants (e. g. genotoxic substances), which are derived from the results of an exposure scenario. The cleaning efficiency of the recycling process (deep cleaning) is determined in a defined test scenario. The result is then applied to the normalized "worst-case" contamination of returned PET bottles. It was also defined that the input stream of used PET bottles may consist of a maximum of 5% non-food packaging.

Exposure is assessed on the basis of three scenarios for different age groups (young children, adolescents, and adults), from which the maximum migration values for a mineral water bottle or its bottle wall are then derived. If the maximum loads determined by means of exposure calculation are below the maximum concentrations of the R-PET, the corresponding recycling process receives approval. The R-PET is then suitable for food contact and may therefore be used in bottle-to-bottle recycling. There are various deep cleaning processes on the market, which have been accepted by EFSA [1].

The individual process steps within the procedures vary depending on the plant manufacturer and the plant configuration, as well as the operator. One of these processes is vacuum decontamination in combination with melt filtration. One of the market leaders for this recycling technology is EREMA. With their VACUREMA® and VACUNITE®, they offer systems for decontamination, drying and IV treatment of PET flakes. In the case of the VACUNITE systems shown in Figure 8.32, the core is a vacuum reactor with integrated nitrogen flushing, which is directly coupled to a single-screw extruder with an underwater pelletizer. The system is supplemented by vacuum-assisted SSP under nitrogen.

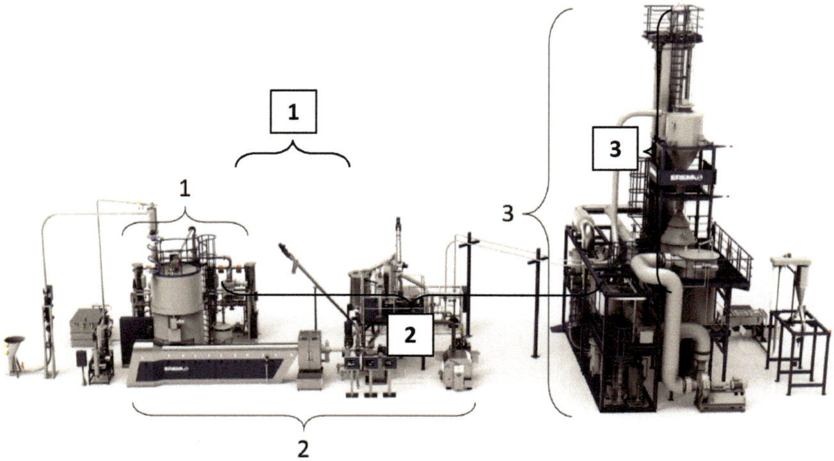

Figure 8.32 VACUNITE system from EREMA: (1) Vacuum reactor, (2) single-screw extruder with underwater pelletizer, and (3) SSP [23]

The single-screw extruder used is equipped with a fine filter system that removes even the smallest impurities such as aluminum or steel particles from the PET melt. The filter system is characterized by a large active filtration area. This lowers the pressure in the extruder and thus prevents solid impurities from being pressed through. Homogenization in the extruder ensures a stable output IV value in the material, even with different input IV values. The extruder is fed by the vacuum reactor, which prepares the material for the subsequent process. A special agitator creates three functional zones in the reactor. Within the first zone, the external moisture of the washed flakes (up to 1%) is eliminated at around 20 °C by reducing the evaporation point under vacuum. In the second zone, temperatures up to 190 °C prevail, causing the polymer structure to open. This marks the beginning of the process to remove migration substances as well as the internal moisture of the flakes, which is about 0.3%. This process gains intensity due to the steady temperature increase in the third functional zone. The total residence time in the reactor is between 1½ and 2 hours, depending on the throughput.

After passing through the two process stages, the PET melt is cut under hot water and conveyed into a centrifuge, where the granules are crystallized by latent heat crystallization from the inside out, without the input of external energy. Due to the process arrangement and temperature control of the VACUNITE process, crystallization rates of more than 40% can be achieved without additional steps. Subsequently, the produced food-grade R-PET granules from the inline crystallization can be fed to an SSP. Here, the IV value of the material is raised again to the values suitable for preform production according to the principle described in Section 8.3.1 [1, 23].

An alternative to this deep cleaning process is chemical surface treatment, which was developed by the United Resource Recovery Corporation (URRC). Veolia holds the process patent for the European market. The first process steps of this process correspond to the bottle-to-flake process described in Section 8.3.4.1. The washed flakes are then wetted with diluted caustic soda in a special device. After the material is dried, it is fed into a 26 m long and 3 m thick mixing screw called a rotary kiln, which is shown in Figure 8.33.

Figure 8.33 Rotary kiln for deep cleaning of R-PET from Veolia [24]

Under vacuum, the adhering caustic soda splits the top PET layer into the monomeric building blocks monoethylene glycol and terephthalic acid, which dissolves any impurities from the surface layer of the flakes. In addition, mechanical abrasion and thermal treatment of the material take place in the rotary kiln. As a result of this process, the flakes form a rough and thus enlarged surface, which enables the migration of impurities such as aromatics, residual monomers, and degradation products from the material. As a by-product of the reaction, a salt is formed, which is sieved off and washed out. In addition, neutralization of the material takes place. To ensure the color purity of the food-grade R-PET, a further sorting step is then carried out with the aid of optical sensors. In this process step, any PVC remaining in the material stream is also separated, as this has taken on a brownish to black color as a result of treatment with caustic soda [24, 25].

Another deep cleaning process that does not require the PET flakes to be melted is the approach developed by Krones AG. Here, a multi-stage flake washing process for surface cleaning is combined with a vacuum decontamination process. The cleaning step of the material surface goes beyond that of the conventional bottle-to-flake process. Within the MetaPure W-PET washing module, a pre-treatment, a caustic wash, and a hot post-wash result in high-purity PET flakes with optimum usability. To achieve food-grade quality, the material is then treated in a decontamination module, the MetaPure S. This is shown in Figure 8.34.

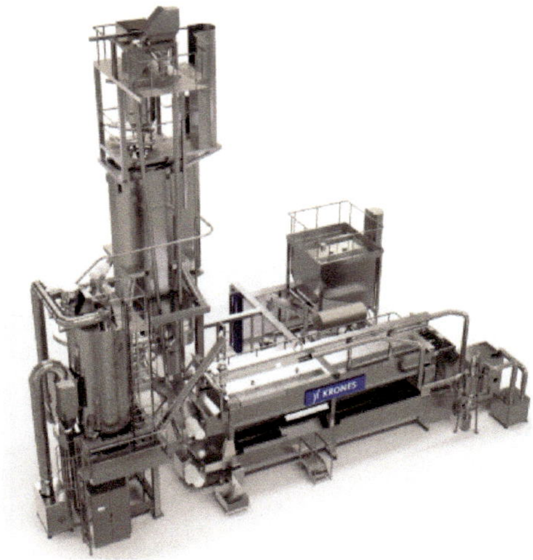

Figure 8.34
MetaPure S decontamination module from Krones AG [26]

In this module, the flakes are first dried and heated under atmospheric pressure in an inert environment. Subsequently, the removal of ingredients that have migrated into the material during use takes place under vacuum. In addition, the decontamination process can be followed by SSP and further color sorting of the flakes [26, 27].

8.3.5 Barriers to the Use of Recycled PET

As shown in the previous chapters, regardless of the prevailing collection logistics, PET bottles can be processed into a high-quality secondary material suitable for renewed contact with food by applying surface and deep cleaning processes. Nevertheless, there are a number of obstacles, mostly process-related, which can make the recovery and use of R-PET difficult or even impossible.

A first problem area is the sortability of the packaging after collection. According to Krones AG, for the production of high-quality R-PET, it is necessary to sort not only the flakes after surface cleaning, but also the bottles themselves, regardless of the prevailing collection logistics. The sorting methods here work with optical or physical principles of action. Due to the broad spectrum of identifiable plastic types, NIR is particularly widespread as an optical sorting method for packaging waste. Here, the plastic is identified by measuring the reflection radiation of excited molecules. For successful sorting of PET bottles, this means that the material used must be identifiable by its surface. For this reason, labels made of foreign plastic that cover more than 30% of the product surface can prevent sorting success. This includes, in particular, shrunk-on sleeves (whole bottle labels) that cover the majority of a bottle. For successful sorting of the flakes, density sorting in the form of a float-sink process is used in addition to NIR. This method separates plastics from each other and impurities with densities greater than 2 g/cm^3 on the basis of their density differences. It must therefore be possible to identify the substances to be separated on the basis of their material-specific density. In addition to sortability, the presence of impurities can also prevent high-quality mechanical recycling of PET bottles. Critical interfering materials that can qualitatively degrade R-PET to the point of unusability include polyvinyl chloride (PVC), non-soluble adhesives, and partially non-magnetic metals [15, 22].

Another obstacle to bottle-to-bottle recycling of PET bottles are additives added during processing to chemically bind oxygen or reduce the acetaldehyde concentration. These can discolor the R-PET yellowish, for example under the temperature conditions of a melt filtration process. Since such discolorations may not be accepted by the consumer, especially for food contact packaging, chemical lightening can be achieved by adding contrasting colors. Even if this does not affect the recycling process of a PET bottle in the further course, the additional addition of contrasting colors results in additional environmental and economic costs in the recovery of the material. The accumulation of the additives mentioned is also regarded as problematic for long-term use of the material. While the recycled content of a PET bottle is currently around 26%, the increasing expansion of recycling capacities could lead to higher average recycled content in the bottles. This would favor the accumulation of additives in the material, which is why regulatory measures are increasingly being discussed [1].

In this section, it has been shown that the quality of the recovered R-PET as well as the costs and the success of the recycling process directly depend on the properties of the recycled product. This correlation describes the product development strategy "Design for Recycling", which aims to improve the ease with which products can be recycled at their end of life (EoL). Once the requirements for sorting and the elimination of contamination are fully taken into account during the design phase of a PET bottle, it will become possible to achieve successful closed-loop recycling of PET using bottle-to-bottle recycling [28].

References for Section 8.3

[1] Kauertz, B., Detzel, A.: "Verwendung und Recycling von PET in Deutschland. Verwendung von PET und PET Rezyklaten aus Verpackungen in Deutschland. Eine Kurzstudie im Auftrag des NABU - Naturschutzbund Deutschland e.V.", 2017, *https://www.nabu.de/imperia/md/content/nabude/veranstaltungen/171025-nabu-01b_studie_verwendung-und-recycling-pet-deutschland.pdf*, accessed January 12, 2021

[2] Henning, F., Diemer, J., Sandoz, B.: Polyethylenterephthalat (PET). In: Domininghaus, H., Elsner, P., Eyerer, P., Hirth, Th. (Eds.): *Kunststoffe. Eigenschaften und Anwendungen*, Springer, Heidelberg, 2012, pp. 785–820

[3] Hochschule Hannover; IfBB: "Biopolymers facts and statistics. Production capacities, processing routes, feedstock, land and water use", 2020, *https://www.ifbb-hannover.de/files/IfBB/downloads/faltblaetter_broschueren/f+s/Biopolymers-Facts-Statistics-2020.pdf*, accessed January 12, 2021

[4] Endres, H.-J., Siebert-Raths, A.: *Technische Biopolymere. Rahmenbedingungen, Marktsituation, Herstellung, Aufbau und Eigenschaften*, Hanser, Munich, 2009

[5] Ruf, A., Borer, C.: Festphasenpolykondensation (SSP) von Polyester; Simulation zum Scale up und zur Optimierung [online]. *Chemie Ingenieur Technik*, 72 (2000) 9, p. 1032, *https://www.readcube.com/articles/10.1002%2F1522-2640(200009)72%3A9%3C1032%3A%3AAID-CITE10321%3E3.0.CO%3B2-S*

[6] Göttfert: Die Ermittlung der Grenzviskosität (Intrinsische Viskosität), *https://www.goettfert.de/fileadmin/assets/Downloads/DE/PDF/Anwendungen/Rheo_Info/GFT_RHEO-INFO_Intrinsische-Viskositaet_d.pdf*, accessed January 13, 2021

[7] Schumann, H.-D., Thiele, U.: DE19503053: Verfahren zur direkten Herstellung von Polyester-Verpackungen

[8] Schröder, T.: *Rheologie der Kunststoffe. Theorie und Praxis*, 2. Aktualisierte und erweiterte Auflage, Hanser, Munich, 2020

[9] PlasticsEurope: "Plastics – the Facts 2020. An analysis of European plastics production, demand and waste data", *https://www.plasticseurope.org/de/resources/market-data*, accessed January 14, 2021

[10] PETnology: Month in charts. PET insights 07, *https://www.petnology.com/competence-epaper.html*, accessed January 18, 2021

[11] Plastics Insight: "Polyethylene terephthalate (PET): Production, price, market and its properties", *https://www.plasticsinsight.com/resin-intelligence/resin-prices/polyethylene-terephthalate/*, accessed January 18, 2021

[12] Welle, F.: "Verpackungsmaterial aus Polyethylenterephthalat (PET)", *https://www.dlg.org/fileadmin/downloads/lebensmittel/themen/publikationen/expertenwissen/lebensmitteltechnologie/2016_4_Expertenwissen_PET.pdf*, accessed January 18, 2021

[13] European Parliament: "Kreislaufwirtschaft: Definition und Vorteile", *https://www.europarl.europa.eu/news/de/headlines/economy/20151201STO05603/kreislaufwirtschaft-definition-und-vorteile*, accessed January 19, 2021

[14] European Commission: "A European strategy for plastic in a circular economy", *https://ec.europa.eu/environment/circular-economy/pdf/plastics-strategy-brochure.pdf*, accessed January 18, 2021

[15] Institut Cyclos-Htp GmbH: "Prüfung und Testierung der Recyclingfähigkeit", 2019, *https://www.cyclos-htp.de/leistungen/pr%C3%BCfung-testierung/*, accessed January 19, 2021

[16] Stiftung Zentrale Stelle Verpackungsregister.: "Mindeststandard für die Bemessung der Recyclingfähigkeit von systembeteiligungspflichtigen Verpackungen gemäß Section 21 Abs. 3 VerpackG", *https://www.verpackungsregister.org/fileadmin/files/Mindeststandard/Mindeststandard_VerpackG_2020.pdf*, accessed January 19, 2021

[17] Welle, F.: "Twenty years of PET bottle to bottle recycling – An overview". *Resources, Conservation and Recycling*, 55 (2011) 11, pp. 865–875, *https://doi.org/10.1016/j.resconrec.2011.04.009*

[18] World Economic Forum, Ellen Macarthur Foundation and McKinsey & Company: "The New Plastics Economy - Rethinking the future of plastics", *http://www.ellenmacarthurfoundation.org/publications* , accessed January 19, 2021

[19] Dormer, A., Finn, D.P., Ward, P., Cullen, J.: "Carbon footprint analysis in plastics manufacturing". *Journal of Cleaner Production*, (2013) 51, pp. 133–141, *https://doi.org/10.1016/j.jclepro.2013.01.014*

[20] Joosten, L.A.J.: "The industrial metabolism of plastics. Analysis of material flows, energy consumption and CO_2 emissions in the lifecycle of plastics". Universiteit Utrecht (Faculteit Scheikunde), 2001

[21] Franklin Associates: "Cradle-to-gate life cycle inventory of nine plastic resins and four polyurethane precursors", *https://plastics.americanchemistry.com/LifeCycle-Inventory-of-9-Plastics-Resins-and-4-Polyurethane-Precursors-Rpt-Only/*, accessed January 20, 2021

[22] Martens, H., Goldmann, D.: *Recyclingtechnik. Fachbuch für Lehre und Praxis*, 2. Auflage. Springer Vieweg, Wiesbaden, 2016

[23] Erema: "VACUREMA® Bottle-to-Bottle", *https://www.erema.com/assets/media_center/folder/VACUNITE_VACUREMA_B2B_2019_09_DE.pdf*

[24] Veolia: "Recycling nach dem URRC-Verfahren", *https://www.veolia.de/leistungen/entsorgung/abfallsortierung-und-verwertung/pet-recycling-pfandflaschen/veolia-pet*, accessed January 26, 2021

[25] Ingenieur.de: "PET-Recycling", *https://www.ingenieur.de/technik/fachbereiche/werkstoffe/pet-recycling-einfach-aetzend/*, accessed January 26, 2021

[26] Krones: "Krones Recycling-Lösungen für PET", *https://www.krones.com/media/downloads/recycling-loesungen_pet_de.pdf*, accessed January 26, 2021

[27] EFSA: "Scientific Opinion on the safety assessment of the processes "Roxane Nord" and "Stute", based on KRONES® technology used to recycle postconsumer poly(ethylene terephthalate) (PET) into food contact materials", *https://efsa.onlinelibrary.wiley.com/doi/pdf/10.2903/j.efsa.2014.3886*, accessed January 26, 2021

[28] Ragaert, K., Hubo, S., Delva, L., Veelaert, L., Du Bois, E.: Upcycling of contaminated post-industrial polypropylene waste: A design from recycling case study. *Polymer Engineering & Science*, 58 (2018) 4, pp. 528–534, *https://doi.org/10.1002/pen.24764*

8.4 Mechanical Recycling of Styrenics

8.4.1 Mechanical Recycling of Polystyrene

Herman Van Roost

8.4.1.1 Introduction – Quality-Driven Circularity

Between 2018 and 2021, mechanical polystyrene (PS) recycling has undergone an impressive transformation, which makes it a textbook example illustrating the critical importance of quality for the recycling profession – not just for PS but for all polymers. Therefore, the central theme of this section on mechanical recycling of styrenics is *recyclate quality*, with all its consequences for a polymer's circularity, illustrated here by the special case of PS.

In the world of recyclates, the term "quality" is used in two very distinct ways:

- **"Acceptable quality"** for use in so-called "less demanding" applications, which do not require typical virgin-purity characteristics or narrow polymer properties; in such cases, the recyclate is often used because of its low cost compared to virgin resins. This describes the historical situation of most current PO recyclates (e. g. non-pressure pipes) and PS recyclates (e. g. coat hangers).
- **"Near-virgin quality"**, allowing a recyclate to be used in most applications which are generally served by virgin resin ("prime segments"); as a special case they can often return to their original application, so-called "closed loop" recycling. The best-known example is the one of PET bottles.

This observation suggests that the definition of recyclate quality is only relevant in function of its targeted second-life use. The definition and quantification of this "adequate quality" is developed further in this chapter.

Over time in the process of maturing circularity, a polymer tends to establish its second-life applications first in less demanding applications before gradually evolving into prime segments, driven by increasing recycling capabilities, and in the search for better margins - or simply to escape from saturated "less demanding" segments. The recent transformation "jump" of mechanical PS recycling was spectacular due to the huge gap that was bridged between the acceptable quality for the less demanding applications (90–95% PS polymer) and the food-contact purity required to access most of the PS packaging applications. In particular, the industry's ambition to close the loop "from yoghurt pot to yoghurt pot" allowed it to suddenly occupy, next to the PET bottle, the leading position amongst all circular plastic packaging applications.

This immediately affects the circular value chain: the high economic value of rPS as recycled content in food packaging provides the economic justification for the construction of new recycling plants, creating in turn a "pull" effect on the collection and sorting of PS waste packages, as required for ensuring the long-term supply of circular feedstock. In order to maximize the benefit of its new-found circularity, the value chain will start optimizing the packaging design, with beneficial results on sorting and recycling. This illustrates the "virtuous circle" of circularity at work.

The increased availability of high-quality recyclate from packaging also serves cross-sector PS circularity, and even include XPS (extruded PS) and EPS (expanded PS) in the same "circular galaxy", confirming the position and recognition of PS as a fully circular polymer.

This section on the mechanical recycling of PS integrates the basic thoughts and concepts that underpin the circular transformation of not only PS but of any polymer: an adequate definition of quality, the critical role of food contact, the mechanical recycling process for food contact rPS, cross-sector circularity, circular design, and how to put all this in industrial practice.

8.4.1.2 Adequate Recyclate Quality

8.4.1.2.1 Practical Quality Standard

Today's most important basis for the trustworthy use of recyclates is largely based on one *de facto* standard (which in practice is very efficient): the previous life application. The simple information that a recyclate comes from milk bottles, pots and trays, yoghurt pots, electrical and electronic equipment (EEE), automotive, or window frames, etc., largely describe the polymer's expected properties, the potential additives and legacy chemicals, and the potential residual contamination from the use phase. The previous-life application also determines the waste streams and processes that generate the recycling feedstock. From there, potential second-life uses are logically defined, and market values vs. recycling economics drive the resulting recyclate quality.

This practical quality standard leads to some interesting consequences:

- The short shelf life of packaging, which leads to no concern regarding legacy chemicals, puts this application in the position of most attractive "standard" feedstock for recyclate - even for non-packaging applications. Therefore, the presence of a material in packaging is a crucial asset for its circularity in general.
- This practical standard provides the basis of "closed-loop recycling": by returning to the same application, many technical and contamination concerns become irrelevant, and the single focus becomes the capability of the waste and recycling process to reproduce a recyclate that exactly matches the initial virgin polymer, by sorting and removal of impurities.
- In combination with "design for recycling" measures - as applied for many years in the PET bottle world - this leads already to a quite strong and effective standardization of recyclate quality: for example, clear rPET flakes or pellets, fit for manufacturing drink bottles again.

Inspired by this example, the approach to standardize the combination "previous application + capable process", potentially extended with "design for recycling" measures, is a good way of indirectly defining recyclate quality - just like for virgin plastics where defining the monomer, the polymerization process, and the process conditions will lead automatically to a defined quality for the virgin resin.

The analytical method described below is not at all intended to replace this current practical standard but is needed to provide a classification methodology and technical communication tool about recyclate quality.

8.4.1.2.2 Virgin Equivalency

Ideally, recyclates should be of such quality that they are *equivalent* to virgin resins in their use for the same applications. Therefore, recyclate quality can be judged by its "distance from virgin quality" or "Quality Distance (QD)", expressed in multiple dimensions relevant for their second-life application.

If, for a given quality dimension, the virgin quality is defined as 100%, the "Virgin Equivalency (VE)" of a recyclate will be VE = 100 – QD. A recyclate can then be defined by its "Profile of Virgin Equivalencies (PVE)", combining its quality dimensions.

Since the relevance/importance of quality dimensions depends on the targeted second-life application, standardizing them should enable the acceptability of a given recyclate for the chosen application to be judged. Hence, in a circular economy each plastic application should be described with its list of relevant quality dimensions and their corresponding "minimum required VE" (VE_{min}), opening up (or not) the use of a given recyclate for a particular application depending whether all its VE values are greater than the corresponding VE_{min}.

Starting from an observed VE deficiency (equating to an excess QD), its origin can be identified upstream in the circular value chain and drive a further improvement and standardization at the level of this origin (sorting, washing, etc.). Alternatively, the recyclate can be structurally diverted to an application for which the PVE is acceptable (e. g. transparent green or blue PET bottles are usually diverted to textiles).

A set of "standard diversions" results in a cascade of subsequent life cycles. The same practice is observed in all mature circular industries, such as the paper and cardboard industry, starting from 57 standard waste qualities that are a function of the product's origin, which are directed to specific second-life uses.

Figure 8.35 visualizes the suitability of two recyclates for use in an application based on its relevant quality dimensions and minimum required Virgin Equivalency.

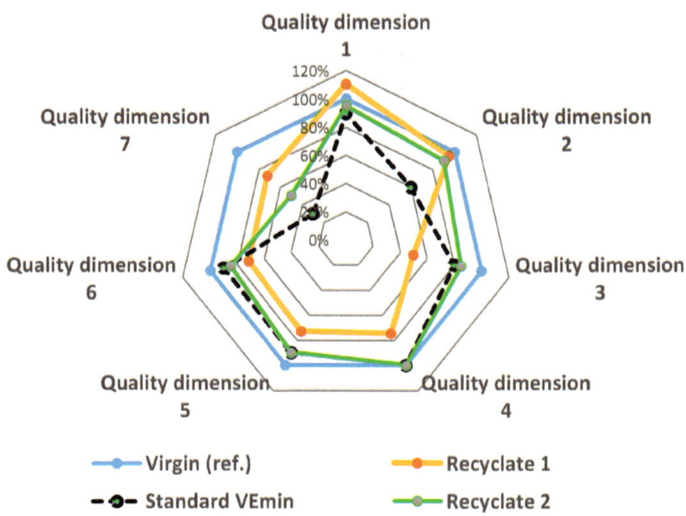

Figure 8.35 Virgin Equivalency of two recyclates compared to a standard for a given application. In this graph, Recyclate 1 does not pass the criteria for this application, despite excess quality in three dimensions, of which one even exceeds the quality level of the typical virgin. Recyclate 2 meets all minimum quality requirements except for a small shortage in quality dimension 6; it seems worthwhile to work on the origin of that shortage, or consider a small adaptation of the standard

8.4.1.2.3 Purity vs. Polymer Properties

Quality dimensions of plastic recyclate can be grouped into two distinct types: "purity" and "polymer properties", with limited interference between them.

Purity

The purity dimensions consist of every "non-purity" aspect of a recyclate compared to virgin polymer, with an extra negative score for each dimension in the event of inconsistency:

- **Smell:** Gradation from none, to weak, to acceptable (detergent type), to strong, to repellent (burnt or chemical type)
- **Black dots:** Gradation in terms of quantity and size
- **Non-target polymer content:** Gradation from 100% target polymer to binary or even multipolymer mixtures
- **Non-food/skin contact:** Gradation from certified food contact, to skin contact, to potential contamination in the waste and recycling process, to previous non-food application, to presence of non-food contact additives (and even presence of currently forbidden legacy chemicals)
- **Color:** Gradation from "natural" to translucent and sorted color, to light and dark gray
- **Additives and fillers:** Gradation from basic antioxidant formulation, to inert fillers, to undefined compounds with property enhancing characteristics, with or without compatibilizers

Figure 8.36 shows the purity dimensions of three PS recyclates as virgin equivalency.

The origin of these "purity-type" QD issues lies partially in the lack of capability of the applied waste and recycling process and partially in the initial article design, which may make it more difficult (or impossible) to properly recycle an object for a certain application.

Apart from comparing and classifying recyclates this way and structurally diverting part of them to a "less demanding" second life (to the extent that these applications are not yet saturated), structural improvement can only be expected by evolving towards:

- A high standard for the recycling process that applies the full range of available technologies, as explained in Section 8.4.1.4: deep sorting, hot washing, flake sorting, super-cleaning, fine melt filtration, etc.
- A high standard of article design: reduced use of pigments, minimum use of fillers, no combination with paper that creates smell and black dots, fewer polymers per article, and guaranteed food-contact compatibility even for non-food-contact applications.

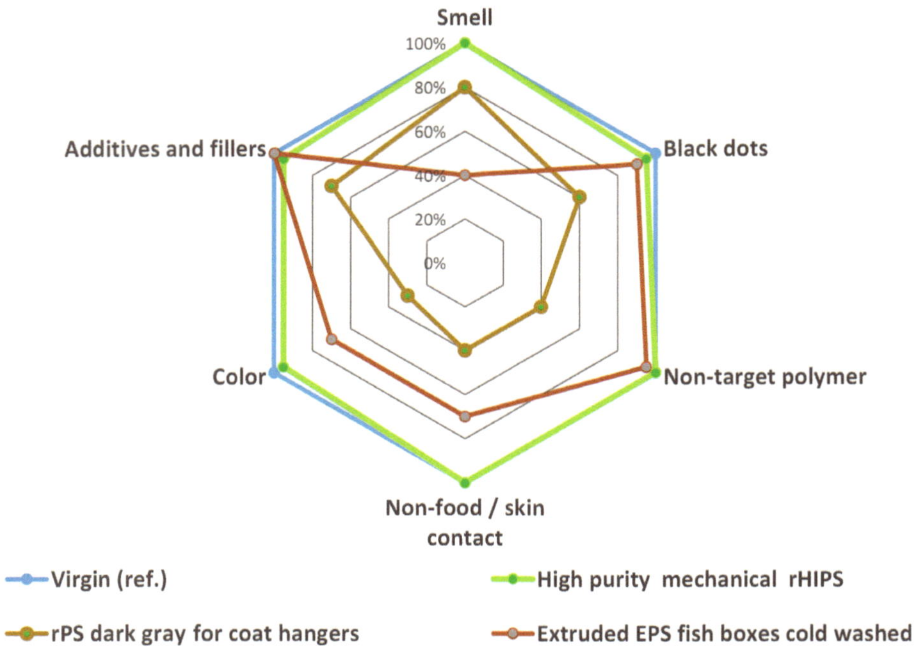

Figure 8.36 Virgin equivalency of three PS recyclates: Purity dimensions (virgin reference = 100%). The high-purity mechanical rHIPS from the recycling process described later in this chapter and designed for food contact is clearly superior in all purity dimensions. Its high virgin equivalency (VE) in all purity dimensions will allow it to easily penetrate "virgin segments". The cold-washed and extruded EPS fish boxes are quite good in several purity dimensions, but tend to have a limitation due to smell. The gray rPS recyclate is mainly fit for thick injected pieces like coat hangers

Polymer Properties

The polymer-property dimensions determine the technical/mechanical/processing suitability of the recyclate for a dedicated application relative to the virgin resins typically used for them, but independently from the above-mentioned purity dimensions. The waste and recycling process has a limited impact on these: their optimum target is to reproduce 100% of the pure polymer that is in their feedstock (apart from some fine-tuning with compounding at the end of the recycling).

Some of these polymer properties are quite generic and include basic information such as:

- Density
- Melt flow index (MFI)

Many others are specific requirements for an application – for example, for PE shrink film:

- Shrinkage in the machining and transverse directions
- Elongation at break

- Tensile strength at break
- Dart drop
- Slow puncture resistance
- Tear resistance

Figure 8.37 describes the polymer properties of a PE recyclate as virgin equivalency for the application “shrink hood for industrial or commercial palletized goods”.

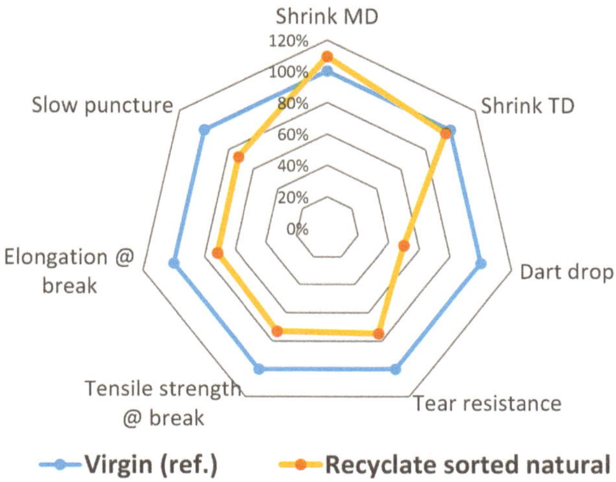

Figure 8.37 Technical properties of a PE recyclate for the application “industrial/commercial shrink hood”. It appears that a dedicated recyclate originated from pallet shrink hoods still shows the required shrink properties, but it no longer shows the same mechanical properties as the initial virgin shrink hood polymer, as would be expected for the standard industrial use

Observed QD issues can have different origins:

1. Simply the difference in requirements between initial and secondary application:
   - e.g. packaging waste used in automotive or EEE applications
2. Polymer degradation caused during the use phase or by the recycling process itself:
   - e.g. certain agricultural PE films that were exposed to UV for a whole season
   - e.g. the IV (intrinsic viscosity) reduction of PET bottle flakes due to exposure to water, especially during the hot caustic wash
3. Market mix effect of a wide variety of different sub-applications that are served with strongly different grades of the same polymer, coming out of the same waste collection and sorted and recycled in a single fraction:
   - e.g. HDPE bottles for milk vs. shampoo vs. detergent vs. aggressive chemicals
   - e.g. PE shrink film, which is usually collected together with stretch film and recycled together in one common fraction “natural film”

- e.g. PP packages produced by injection (high MFI) or by extrusion thermoforming (low MFI), sorted in and recycled from one common PP fraction
- e.g. PS packages from HIPS, GPPS, XPS, and EPS sorted in and recycled from one common PS fraction

Although scientific literature focuses a lot on Cause (2), polymer degradation, this phenomenon appears in practice to be quite limited; the above given PET and PE examples are rather an exception to that rule. Whereas it is obviously possible to damage and destroy the molecules by harsh treatment (excess shear and temperature), and whereas some polymers like PS are more insensitive to degradation than others, *polymers are in general very well (over)protected against normal exposure conditions during their production and use phase including the recycling process, and show very little molecular degradation at the start of their second life cycle.*

This leads to the conclusion that for a given recyclate from a given waste stream (e.g. PE packaging), the observed QD issues are mainly generated by Cause (3), mix effects, and that the QD inconsistencies correspond mainly to varying portions of different sub-applications in the waste stream. This conclusion is widely confirmed by observed seasonal variations of recyclate, not only in terms of color but also properties. For example, during the "garden season" rPP color darkness AND mechanical properties are influenced by an increased portion of plant pots and trays in the separate waste collection for plastics (the "yellow bin" in Germany and Austria).

This finding has several interesting consequences:

- A standardization potential for recyclate can be achieved by standardization of the virgin grades by polymer application! Indeed, if the wide range of different grades in a sorted polymer fraction could be reduced to a single grade of that polymer serving all sub-applications, the resulting recyclate – if properly recycled – would show a very low QD and be automatically fit for starting a new life in the same application.

  It is worthwhile to remark that the PS industry is relatively mature in terms of standardized virgin resins: instead of applying a wide range of dedicated polymer grades for each application, as is the case for PE and PP, PS convertors mostly use their own mixtures of two virgin standards – GPPS, and HIPS with high rubber content. The resulting waste is a mixture in a quite narrow range for each sub-application.

  Also, foamed PS articles (respectively XPS and EPS) are quite "standardized" in terms of polymer properties, and consistent: because they are pure GPPS with MFI adapted to a dedicated process, all virgin resins are within a narrow range of each other, and consequently so is their waste.
- It stresses the importance of a deep sorting phase as the first recycling step – not just separating colors, but also separating molecularly distinct forms (such as

homopolymers from copolymers, and high-MFI (injection) from low-MFI (extrusion) polymers), depending on the technological sorting capabilities. Not only will the properties of each sorted fraction be better adapted to dedicated applications in their second life, but they will be each much more consistent, because the varying mix portions at the input stage are transformed into varying volumes per fraction, not varying properties of the output mix.

An interesting phenomenon is that sorting objects by color indirectly implies that specific packages of a limited number of brands are put together in one fraction, producing a narrower spread of polymer properties that is closer to the requirements of these same brands.

An extra advantage of sorting by color is that it adds pigment circularity on top of polymer circularity. Mixed colors lead to an inconsistent "grayish" hue, which often must be made darker by adding carbon black, placing an additional cost on top of its already limited applicability. Sorted colors, however, can be re-used as such, or if necessary they can be fine-tuned with a reduced amount of pigment. This can result in a considerable cost saving, and provide a recyclate with higher economic value.

- A lack of properties (QD) due to a residual mix effect can easily be compensated by on-purpose mixing/compounding with either:
  - A virgin-grade ("booster") resin with over-performance in the relevant QD dimension. Current polymer technology has a lot of unused potential to produce over-quality molecules for which no pure virgin market applications exist, but which can be tapped for the purpose of boosting recyclate
  - A chosen recyclate that shows such over-performance (often from a different application)

Virtually all demanding applications in terms of mechanical properties can be served by a standardized compound (formulation) that combines a chosen recyclate (e.g. from packaging) with booster virgin or with a different recyclate – on condition that the recyclate purity corresponds to the minimum standard of the application. The biggest hurdle to the generalized use of this "boosted recyclate" is ignorance and psychological resistance: each industry sector has developed strict, high standards, and cannot imagine that these can be met with recycled content – especially not from other sectors. Overcoming this resistance cannot be done with a "push" approach, so it is imperative that the end user is committed to use recyclate, without compromising on quality.

Figure 8.38 shows how the polymer properties of a shrink hood recyclate can be restored by boosting with virgin material that has over-performance in the relative quality dimensions for the application. It was developed as a project subsidized by the Flemish region, by a consortium of Valipac (EPR organization), Wienerberger (end customer), Morssinkhof (recycler), Total (formulation and booster), and FEMA (federation of construction materials). In 2019 it won the first packaging award of the Federation of the European construction industry UFEMAT. See the website *https://www.wienerberger.com/en/press/press-releases/20191111-wienerberger-belgium-honored-with-ufemat-award.html*

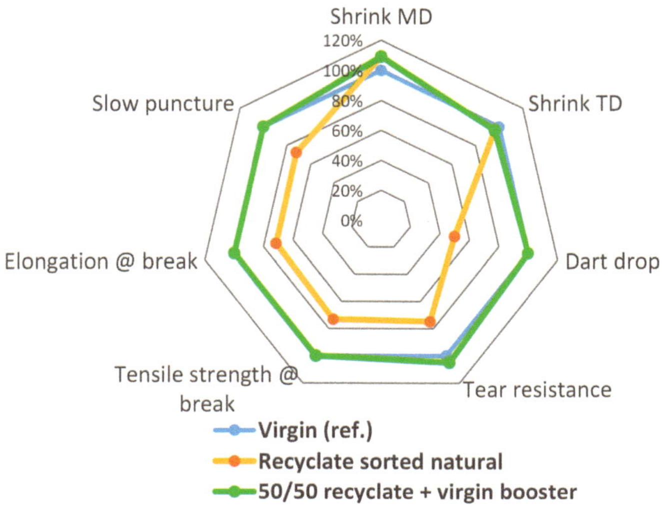

Figure 8.38 Technical equivalency of virgin resin and a polymer with 50% recycled content, for the application of a circular shrink hood. By a formulation of 50% recyclate and 50% of a virgin booster resin with over-quality, the virgin equivalency in the relevant dimensions for the shrink hood application is entirely restored

This opens up interesting potential:

- For more recycled content in the packaging sector, e.g. for UN packaging: a standard for dangerous products requiring high polymer performance
- For cross-sector circularity (see Section 8.4.1.6.2): reinforcing the construction sector's role to uptake recycled content from other sectors. This puts particularly the packaging sector in the central position as responsible provider of recycled content, not only to ensure its own circularity, but for the circularity of plastics in general.
- It places the virgin industry and technology in a key role to develop high-performance resins to "boost" recyclates for use in high-end market segments; in return, the renewed focus on technology will generate a new dynamic and attractiveness to the plastics industry.
- Finally, it creates a strategic interest for the entire plastics industry to preserve and care for circular design in all plastics sectors: whether it be for ensuring outlets for its own recyclate or for securing feedstock for its own recycled content, both are required for the recognition of plastics as the circular material which it intrinsically is.

8.4.1.3 The Importance of Food Contact Quality for Recyclates

Contrary to common belief, the quality dimension "food contact for recyclates" is not something for niche markets; it is, together with absence of in-mass coloration, the main entry gate to virtually all plastics segments that are today served by virgin resins.

It is therefore important for the circularity of all plastics to compare the relative position of recyclates to virgin resins in terms of the food contact aspect. Understanding the food contact property of virgin resins and its consequences for the value chain is part of this exercise.

The food contact property is also much less "binary" than commonly perceived; it exists in several gradations, and even a non-regulatory accepted level of food contact or equivalent purity has huge advantages for the penetration of a recyclate into "virgin segments".

A particular challenge for PS is indeed that most of its packaging applications are for food contact - just like the PET beverage bottle. Since the PET recycling industry has pioneered the food contact recyclate dimension, it is also important to understand the drivers of this "miracle".

8.4.1.3.1 Food Contact in the Virgin Industry Context

In essence, the petrochemical industry transforms a black, stinking, sticky, and toxic feedstock into a clean and transparent polymer, suitable for contact with food. The nature of the feedstock has never been in the way of acceptance for food-contact applications, due to the extreme sensitivity of any polymerization process to tiny levels of feedstock impurities, which imposes a strong impetus for upstream purification.

The initial plastics application in the 1950s was toys, and so there was little consideration about food contact. As the use for food developed, dedicated food-contact grades were produced on dedicated lines and with dedicated additive formulations.

However, over time, the virgin-plastics industry has evolved towards supplying virtually 100% food-contact grades to the market, even for products that are clearly intended for non-food markets. (Exceptions exist obviously where at the end a specific masterbatch is added for specific purpose). The main reason was *supply chain simplification.*

The required quality effort is impressive. It goes even beyond the parts of the process that are in contact with the product, and includes, for example, more expensive food-contact lubricants used for machines in the maintenance shop. Non-food-grade lubricants and liquids are even generally banned from the warehouse, simply in order to avoid potential non-food-grade contamination by human error during urgent maintenance to the polymerization process.

The primary reward was absolutely worth it: manufacturing simplicity, product range simplicity, inventory reduction, human error reduction, transition cost reduction, all leading to operational excellence. However, the unexpected secondary effect was even stronger and has deeply affected the entire plastics value chain:

- *The entire converter industry* benefits from the same simplicity:
  - Producers of food-contact articles can focus on the operational aspects of their profession, without concern that their production is accidentally contaminated by a resin error;
  - Producers of non-food-contact articles benefit from the overall upstream efficiency gain of their supplier; and they can focus on the management of their own use of non-food elements adapted to their downstream market, knowing that nothing else related to the polymer is in their way;
  - Converters of both food and non-food articles can produce these (from the same or from different polymer grades) on the same production lines without any transition cost, without silo cleaning, and with simplified feedstock management and no risk of food-relevant contamination from the operation;
  - Manufacturing plants don't risk workers health or social issues from employees exposed to the polymer;
  - Feedstock purchasers receive the widest possible choice between resins from the entire virgin market, focusing on "value for cost" instead of applicability for their specific purpose.
- *The brand owners* benefit from the polymer's food contact quality as an "insurance" guaranteed by large petrochemical corporations:
  - Corresponding to an obvious need for food-contact items; this segment is much larger than packaging, and includes (for example) water pipes, as well as fridge liners and internals;
  - But also non-food-related applications like skin-contact products (shampoo, soap, sun protection, cosmetics, personal care, etc.). Here, the food-contact quality guarantees "over-quality", protecting the brand owners from unexpected customer claims against unknown substances upstream from their own product design. These markets might never have been conquered without food-contact feedstock;
  - The same is true for toys, for which the historical polymer market became increasingly sensitive to the material and surface purity for children;
  - Finally even for non-food-contact articles, the high food-contact standard of the virgin plastic is interesting: this "over-quality" provides a simplification that avoids the need to investigate the physical nature of a lack of food-contact quality (in other words, if something is *not* food contact, it can be for many reasons and it is necessary to know which, for judging the risk).

In conclusion, it is clear that for recyclates, entering this historical food-contact-based cost and market optimization has many implications and challenges beyond the simple binary question “does the application require food contact or not?”. It is also clear that over time it will be in the overall interest of plastics circularity that the quality of recyclates in general should come as close as possible to a technical food-contact standard, whether effectively approved by a food contact regulation (for the reasons set forth in Section 8.4.1.3.3) or not.

8.4.1.3.2 The PET Bottle Recycling Reference

The development of closed-loop PET bottle recycling during the last 25 years has generated a wealth of technologies, stretched by an extreme quality target: reaching food contact while starting from post-consumer waste. It includes “deep” NIR sorting, hot caustic washing, flake sorting, and super-cleaning, with melt filtration as a final but less critical step due to the high upstream quality of the flakes.

Indeed, PET as the youngest mass-polymer had the perfect properties to be used for beverage bottles, but could not do this with a linear “throw-away” concept: recycling was the only way to consumer acceptability.

In comparison, the historical development of PO and PS recycling did not aim for this food contact quality, because these mature polymers had enough less demanding applications available as outlets, for which recyclate could serve as opportunistic substitute for virgin resin, based on lower price. As a result, industrial development was oriented towards cost efficiency: the lowest possible cost for an acceptable quality. These processes work with no or limited pre-sorting, cold wash, and density-based flake separation instead of flake sorting; the flakes of mixed color (“jazz”) are then extruded together into a gray product, and melt-filtered at the end to remove residual impurities that were not sufficiently removed in the earlier stages.

For any observer, the difference between typical rPET and typical rPO/rPS is striking: it seems to demonstrate that PET is *the* recyclable polymer (purity >99.9%, transparent, food contact) whereas recycling PO and PS seem doomed to be recycled into less demanding applications (purity ~95%, gray, often smelly, black dots). However, *this is not true*! The difference is entirely the consequence of the different primary technological and operational focus of rPET vs. rPO/rPS recycling: food contact quality vs. cost efficiency orientation.

Indeed, if the same stretched recycling technology approach of PET recycling (deep sorting, hot caustic wash, flake sorting, degassing) is applied to PO and PS recycling, the same high purity levels are obtained as with rPET – with narrow specifications, natural or bright sorted colors, and absence of odors and black dots. If, in addition, the waste feedstock is of the right quality and the used processing fluids are food-grade, then the obtained purity of these recycled materials will make them *technically* fit for contact with food.

The industrial references of "high-purity mechanical recycling" are growing: after an early start (1990) in the US with a development by virgin resin producer Union Carbide (since 2001 Envision Plastics), WRAP developed in 2008 closed-loop milk bottle recycling in the UK with Biffa and Closed Loop Recycling (now Veolia UK). More recently, the technology was expanded to all polyolefins (rigid and flexible) in a pioneer project known as CPO (Circular PolyOlefins), part of the so-called "New Plastics Economy" promoted by the Ellen MacArthur Foundation. This project was led by Tomra Sorting, and resulted in a series of industrial plants and projects in the Netherlands, Germany, and Bulgaria, amongst others.

The application of this technology on PE and PP packaging feedstock with proven food contact origin has led to several FDA-approved food-contact applications. However, in the European context they could never pass the EFSA requirement of low diffusion which is typical for PET. As explained in Section 8.4.1.4.4, polyolefins have a "high diffusion" property, meaning that they easily absorb substances from the (waste) environment which could then be passed on into the food.

This European PET monopoly on food-contact recyclate was broken in 2020 when Styrenics Circular Solutions together with Tomra Sorting extended the high-purity mechanical process to rigid and expanded PS packaging, leading to food-contact petitions towards the European Food Safety Authority (EFSA) – indeed, PS shares the "low diffusion" property with PET.

8.4.1.3.3 Many Ways for a Recyclate to be Non-Food-Contact

As explained in more detail in Section 8.4.1.4.4, regulatory food-contact status is not only based on applying a powerful recycling process, but also on managing the risk that unwanted substances would be present anyway, and potentially migrate from the recyclate into the food. These substances can have several origins:

- Originating from the recycling process:
  - By use of non-food-grade chemicals, lubricants, or additives etc., in direct or indirect contact with the waste or recyclate
- Originating from the waste feedstock:
  - From residual non-food content that was contained in the package's previous life cycle
  - By contamination of feedstock articles by misuse
  - By contamination of feedstock articles in the collection stage
  - Additives or materials voluntarily added in the design of a non-food contact article in the feedstock

The potential migration of substances into food is judged based on:

- The targeted food application:
  - Contact surface vs. content, intensity of the contact (e.g. liquid or fatty vs. fresh fruit or vegetable)

- Quantity of expected consumption
- Conditions of the contact: shelf life, temperature, etc.

- The polymer properties:
  - Polymers with low diffusion characteristics (PET, PS) will absorb smaller quantities of contaminants and better prevent the migration of any substances into food than others (PE, PP).

This results in a *gradation* of food contact statuses: e.g. a recyclate can be food-contact-approved for long shelf life, but only for non-fatty products; or only for short shelf life and cold storage conditions; or with the exception of microwave application; or only for non-intense contact with dry fruit and vegetables; or at maximum 50% recycled content; or passing FDA criteria but not EFSA criteria, etc.

Conversely, the requirement to pass multiple parallel criteria generates for recyclates a wide range of "non-food-contact typologies", based on not passing one or more food-contact criteria; e.g. from a minor to major non-food character:

- Because the targeted use is outside the regulatory accepted scope:
  - Other food, other temperature or contact duration, other % of recycled content, etc.
- Because the polymer type is too diffusive: even if the recyclate is proven to be very pure and technically acceptable for contact with food, there is a residual risk that non-intended substances are unintendedly absorbed and will end up migrating into the food:
  - This criterion is very limiting for PE and PP under EFSA rules, but less problematic for obtaining a favorable FDA opinion, all other aspects being acceptable. Apart from food-contact applications outside Europe, such favorable FDA opinion provides in Europe the "over-quality insurance" needed to access a wide range of skin-contact and other purity-oriented applications, as explained in Section 8.4.1.3.1.
- Because of a low-performance recycling process, from one of more of the following:
  - Not effectively removing all contamination from the previous use and waste stage, which often results in odors and visual aspect deficiencies
  - Using non-food-grade chemicals (surfactants, antifoam, additives, etc.) in the recycling process
  - Having an unadapted configuration (open belts collecting dust from the environment, etc.) or non-certified operation
- Because of unadapted feedstock, from one of more of the following factors:
  - Related to insufficient sorting
  - Related to the individual design of food-contact articles. This can include components not in contact with the food, such as glued labels and inks, which contaminate the plastic during the recycling

- Related to the individual design of non-food-contact articles. The use of (allowed) additives or colorants can prevent future recycling into food-contact materials, regardless of the capability of the recycling process. Obviously the presence of legacy chemicals, typical for long-life-cycle articles, is an even stronger reason for exclusion from any food-contact applications.

In conclusion, the closer a recyclate comes to food-contact status, the more simplicity it brings to its user – just like virgin resins. Reversely, "non-food-contact" is a complex recyclate property which has to be described in detail before the user can accept the recyclate's suitability for the target second-life application.

It is clear that the strongest barriers against approaching technical or regulatory food contact are at the article design stage. This is because they destroy upfront the potential economic value and investment justification into available technology to lift recyclate quality to the level where they can enter "virgin segments".

8.4.1.3.4 A Virtuous Circle towards Plastics Circularity

In order to penetrate "virgin segments", plastics circularity has to massively break out of the "vicious circle" currently defined by the following rationale, of which it is victim today:

- Food contact is believed to be a binary concept and only relevant to the food sector, which is a niche market. Other sectors don't need that level of purity, and anyway it is unrealistic for non-PET recyclates to target such segments.
- Therefore, there is no economic justification to invest in and operate high-purity recycling technology, which can produce technical food-contact purity.
- In return, there is ample freedom to design articles with (cheaper) non-food components, additives, colorants etc.
- And since the recyclate will not be food contact for that reason, it may as well not be food contact for recycling process reasons: allowing open-channel configurations, non-food-grade chemicals etc.

As a result, access to "virgin segments" remains indeed a niche for a very limited number of recyclates: a typical example of a "self-fulfilling prophecy"!

Instead, the circular plastics industry should be inspired by the historical (and radical) choice made by the virgin resin industry to *simplify* the attainment of top-quality resins, and aim to "plant the same flag on the moon" by reversing their historical dynamic into a virtuous circle, defined by the following rationale:

- Facing reality: Converters and brand owners of most "virgin segments" have conquered their markets and economically optimized their operation on the basis of a food-contact polymer feedstock, even where this is not strictly needed, thanks to the "over-quality insurance" provided by this food-contact property. Their new need for recycled content will open up some formal flexibilities but will *never reverse* this deep operational excellence concept. Recycled content in these "virgin segments" will be reserved to (near-)food-contact or similar high-purity recyclates.

- Fully exploit the proven technologies to produce top-quality recyclates, as close to food-contact quality as possible: High-purity mechanical recycling technology is today capable of producing recyclates with the technical purity levels needed for technical food contact, opening many new and lucrative markets. The level of investment is like that for the PET bottle recycling industry - considerable but within reach.

This was the route to circularity chosen by the PS industry.

8.4.1.4 Food Contact Mechanical Recycling Process of PS

The mechanical recycling process developed by Styrenics Circular Solutions and Tomra Sorting for closing the material loop from yoghurt pot to yoghurt pot is based on the same stretched quality target as PET beverage bottle recycling. Conceptually, it applies the same main process steps: deep sorting, hot washing, flake sorting and super-cleaning. However, each of these processes had to be enhanced in order to cope with the specific challenges of the yoghurt pot.

8.4.1.4.1 Deep Sorting

When looking at the post-consumer PS waste from packaging, it is striking how much of it has been used for food applications: whether used for dairy, meat, fish, sweets, eggs, vegetables, or drinks, they are all excellent feedstocks for food contact recyclate! Figure 8.39 shows just a small selection.

A second observation is that the PS packages can either be rigid or foamed, white, transparent or colored. White yoghurt pots are the largest fraction, making them the perfect target for closed-loop recycling back into yoghurt pots.

Figure 8.39 Variety of PS food-contact packaging (Image sources: Winai Tepsuttinun/Garsya/AlenKadr/Sergiy Kuzmin/Lushchikov Valeriy/Shutterstock.com)

At first sight, this wide variety of PS types and forms, having each different polymer properties, seems to be a concern: indeed, until 2018 the common industry belief was that PS objects could be perfectly distinguished from other polymers by NIR sorting, but that the different types and forms of PS could not be sorted from each other.

This assumption proved to be completely wrong when Tomra Sorting started the first tests to separate different styrenic polymers. To the general surprise of the industry, all the types and physical forms of PS could easily be recognized and separated by NIR and optical sorting technology: rigid from foamed PS, GPPS from HIPS, XPS from EPS. As shown in Figure 8.40, it even appeared feasible to separate HIPS fractions by their rubber content, and to distinguish them by the type of rubber used.

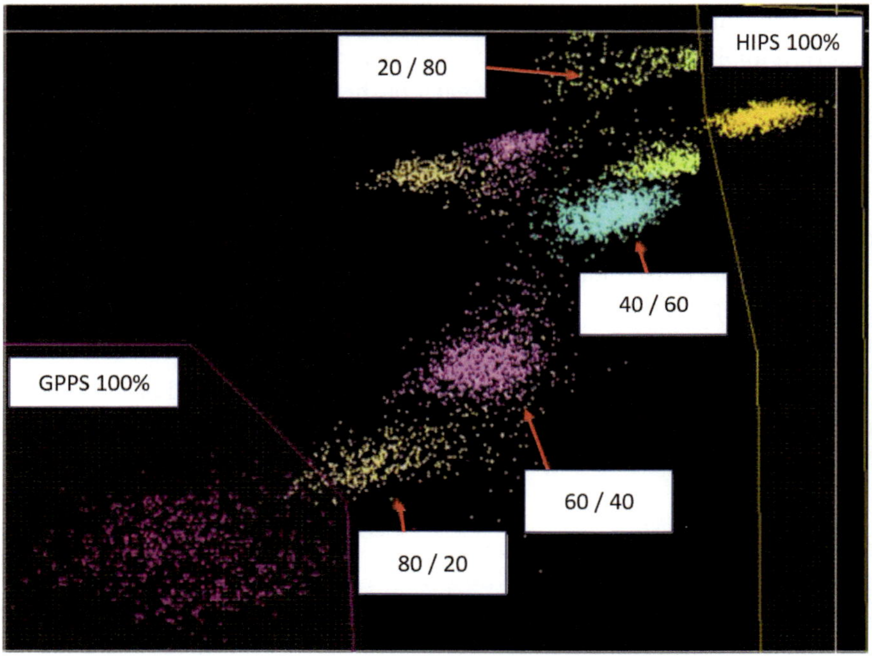

Figure 8.40 These NIR sorting spectra correspond to different mixtures of pure GPPS (lower left) and pure HIPS (upper right)

Since then, these new sorting possibilities are fully available to be combined with existing technologies such as optical (color) sorting, intelligent (package form) sorting, removal of objects based on unwanted substances, etc. Depending on the initial composition of sorted bales containing styrenic objects, it becomes possible to produce PS fractions with very precise composition. This deep sorting step is of critical importance for the quality of the future recyclate.

To produce a feedstock that is well adapted to be recycled into yoghurt pots, the logical choice is to use a two-step sorting process generating a "White HIPS" fraction, as shown in Figure 8.41. To ensure high precision, the objects smaller than 20 mm are first removed by a "fines screen". Then a two-step positive sorting is done, first on HIPS and second on white.

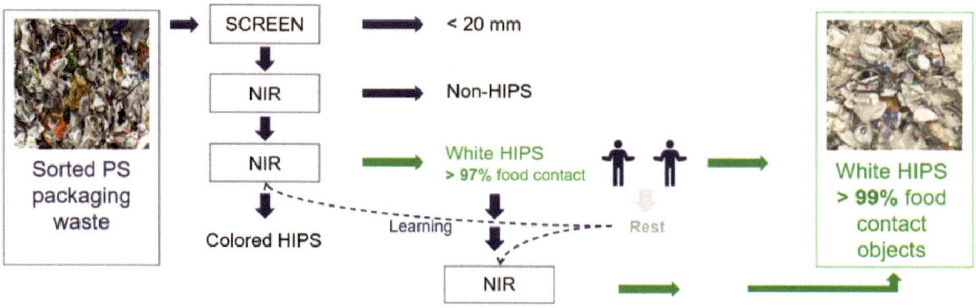

Figure 8.41 Production of PS feedstock for food contact by deep sorting

After a manual check, this fraction appears to contain already more than 97% food-contact items. Feeding the information about the non-food-contact objects back to the sorting machines for "learning" proved to result in an excellent feedstock composition of more than 99% food-contact items: an outstanding basis for the next process step.

8.4.1.4.2 Hot Washing

After a wet grinding step, the white HIPS objects are transformed into flakes but they still contain a lot of dirt, so they have to be cleaned. The dirt sticking to HIPS flakes from yoghurt pots is a combination of (mostly dairy) food residue and non-PS materials from lids and labels: paper, aluminum, aluminized multilayer film, and also glues and inks. All of these must be thoroughly removed in the washing step, which is the heart of the high-purity recycling process.

By definition, cleaning is moving dirt: the dirt needs to be moved from point A (where it doesn't belong) to point B (where it belongs). To do this, the dirt must be loosened from the surface, made transportable and transported.

The principles underpinning the washing process are described by "Sinner's Circle", invented in 1959 by Dr. Herbert Sinner, a chemical engineer and Head of Detergent Development at Henkel. Initially developed for the washing of cloths, Sinner's Circle is nowadays used much more widely, in pretty much everything that needs to be cleaned. It consists of four functional process variables, which have to be applied in a certain ratio in order to clean an object optimally with water. These four variables are shown in Figure 8.42:

- Mechanics (actually mechanical labor, in the form of movement or friction)
- Temperature
- Chemistry (e. g. detergent)
- Time

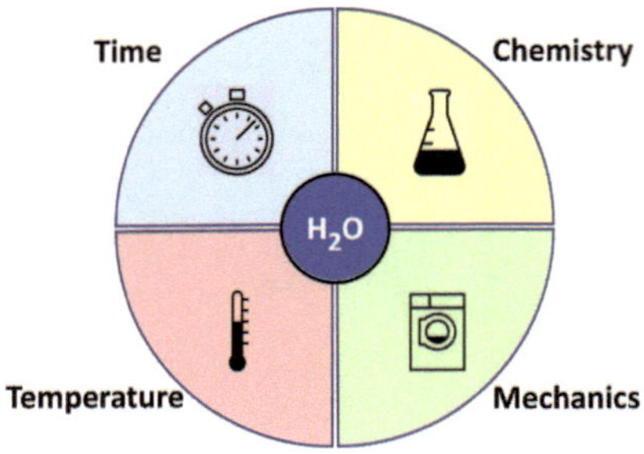

Figure 8.42
The four process variables of Sinner's Circle

At constant cleaning performance, these four factors all affect each other. Increasing the size of one factor automatically means that the total of the other three factors becomes smaller, and *vice versa*. The optimal combination of factors is different for every washing purpose, since each type of "dirt" will be more sensitive to a specific factor instead of others.

The main role of the water is to be the medium that transports the other factors. It is needed to dissolve chemicals, transfer heat to the object, and transport dirt away from the object – and it can also can play an important role in applying mechanical force.

The commercial hot wash processes on the market for PET bottles do not all apply the four functional variables in the same way; this allows them to differentiate themselves in terms of water or energy consumption while still doing the required cleaning job for PET bottles. However, when applied to PS flakes, this leads to quite different washing results – most of which are simply below the requirement for the return to food contact.

Testing the performance of a washing process can be done by applying on the washed flakes an even more severe set of functional variables (e. g. higher temperature, more or different detergents, longer time, and more friction), and subsequently analyzing the quality of the washing water. If the initial washing was perfect, everything was already removed and no more "dirt" will be removed, despite all efforts. Poor washing, however will result in contaminated water, which is often even visible with the naked eye, as shown in Figure 8.43.

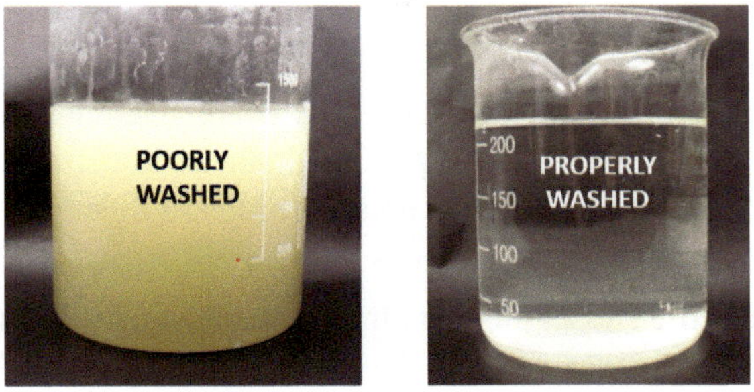

Figure 8.43
Testing the washing quality by means of the washing water from an extra more severe washing

When comparing the PS washing challenge to the one of PET bottles, it is clear that PS is more difficult: not only due to the dairy residue but also because the PET industry has over the years developed strict guidelines for the "design for recycling" of their bottles, avoiding materials and chemicals that are difficult to remove. Therefore, the current washing processes as available for the PET bottle industry cannot simply be copied, but must be enhanced for the purpose of cleaning PS packaging waste.

For the proper washing of PS post-consumer flakes, the best results are obtained with the following combination of Sinner's Circle washing variables, visualized in Figure 8.44:

- Low friction: This avoids a situation where PS flakes are broken into small pieces or pulverized, reducing the recycling yield and the effectiveness of the subsequent flake sorting step;
- Temperature of 85 °C: High enough for thorough cleaning, but still sufficiently below temperatures where PS and other polymers would become soft or sticky;
- 3% Caustic soda: This concentration allows the cleaning mixture to deal with the organic residues, paper, glues and inks and even to dissolve the aluminum lids;
- 15 minutes residence time.

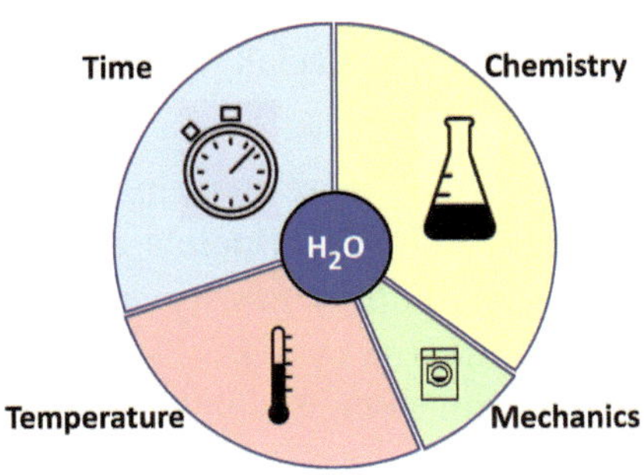

Figure 8.44
Sinner's circle for PS washing

Tests have revealed that it is of key importance to keep the caustic soda concentration at a constant level during the entire 15 minutes of the hot wash step, by a control loop with continuous caustic soda make-up, in order to compensate for its consumption by the organics and aluminum. And it goes without saying that to achieve a food-contact-grade product, the caustic soda must be of food-grade quality itself.

It also appeared to be important that the subsequent rinsing step, following the hot caustic wash, is also performed with hot water at 85 °C: otherwise the flakes will again accumulate residual dissolved dirt on their surfaces. After the washing step, the flakes are dried with hot air and passed through an air classifier to remove flexibles originating from plastic lids and labels.

At the end of this washing process, the PS flakes contain more than 96% pure HIPS polymer. Residues, aluminum, paper and glues are absent, and the remaining non-HIPS material entirely comprises other polymers (PET, PO).

8.4.1.4.3 Flake Sorting

In most PET wash processes, the wet stage is used as an opportunity to apply a flake density separation in a float-sink tank, which allows separation of the sinking PET bottle flakes from the floating HDPE cap flakes. This practice is not applied for PS because it causes a 10–20% loss of yield, because some PS yoghurt pots apply a light foaming, making these flakes float, together with the unwanted polyolefin fraction.

Instead, the production of high-purity PS by separation of all non-HIPS and non-white plastics is entirely based on optical and NIR flake sorting technology.

Successful application of flake sorting technology is dependent on the size of the flakes: excellent sorting results are obtained on flakes of 4 mm or larger, but below 2 mm the sorting is not effective. Therefore, the flakes have to pass a sieving step: removing first all fines below 2 mm, and subsequently separating the main target fraction >4 mm from a secondary fraction between 2 and 4 mm. The latter has still some potential as HIPS recyclate with good purity, especially because flake sorting technology is constantly improving.

The >4 mm fraction is then passed through one or more steps in which the flakes are sorted on the same criterion of "white HIPS". The principle of such a sorting machine is shown in Figure 8.45: the flakes slide by gravity at constant speed down an inclined plate and over a detector, followed by a precise pneumatic blowing mechanism. For full effectiveness, this requires double-sided detection capability, because several yoghurt pot brands are coextruded with a white layer on the inside but colored on the outside. After the flake sorting step, the resulting recyclate flakes have a PS purity of >99.9%.

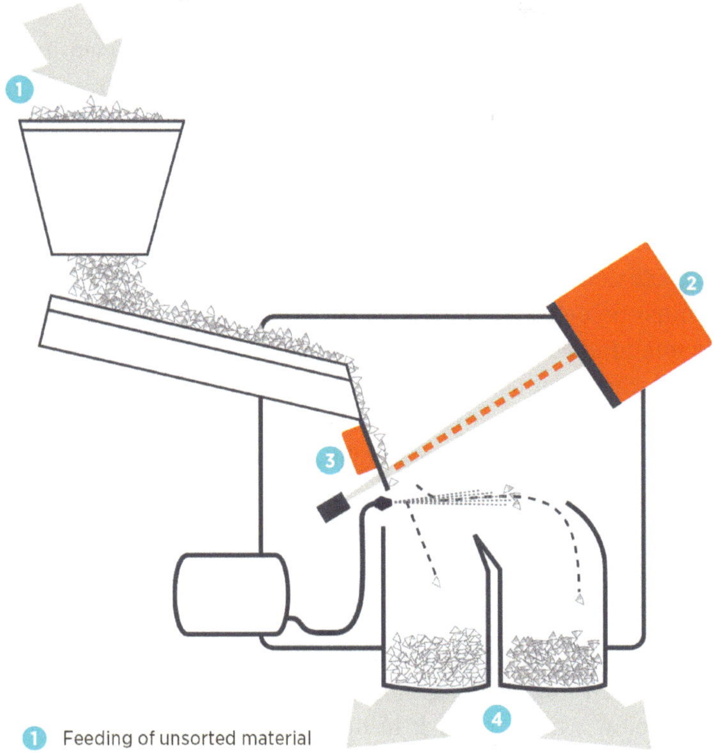

1 Feeding of unsorted material
2 Scannerbox for material and color detection
3 Electromagnetic sensor (optional)
4 Separation chamber

Figure 8.45 Functioning principle of a flake sorter

Application for foamed PS:

The above-mentioned washing and flake sorting process also appears to be perfectly applicable to foamed PS packages (e. g. XPS trays), and this is despite the often-present LDPE film cover and the inner textile for absorption of juice. The main difference at the end is that the flakes contain up to 50% of their weight of rinsing water (lab tests have shown that there is no residual caustic soda).

The solution to this is not to remove the water with thermal drying (which would be energy-intensive, and generate dust and electrostatically charged flakes), but instead to use the extra weight from the water content to facilitate the subsequent flake sorting step. Afterwards, an agglomerator can be easily applied to remove the water from the foamed flakes prior to the extrusion step.

8.4.1.4.4 Super-Cleaning

A process for producing recycled PS that is intended to be used in contact with food must not only be capable of producing very pure product from a *regular* waste stream of food contact packages, but also to remove concentrations of substances which may be present by possible *misuse*, down to a level which is considered safe by EFSA. Such removal is done by an extra process step called "super-cleaning" or "decontamination".

The cleaning efficiency (capability to remove substances) of the super-cleaning technology must be determined by means of a so-called "challenge test": feeding purposely contaminated recyclate into the super-cleaning device and comparing the output with the input concentration of substances. To be representative of all possible classes of contaminants of concern, the contaminants are a cocktail of substances with different molecular weights and polarities.

This measured cleaning efficiency is then applied to a maximum reference contamination from misuse of post-consumer PS packages, resulting in a maximum residual contamination level c_{res}, which must be below the maximum allowed modelled concentration c_{mod} that is evaluated safe by EFSA.

This c_{mod} is calculated using generally recognized conservative migration models, such that the related migration cannot give rise to a dietary exposure exceeding the threshold below which the risk to human health would be negligible. Maximum daily consumption data are available in the EFSA database for each type of food (e.g. yoghurt), leading to a maximum concentration of substances in the food, $c_{food,max}$, which should not be exceeded at the moment of consumption.

The concentration of substances in food c_{food} due to migration during time t out of a polymer with initial concentration of substances $c_{P,0}$ is modelled by the scheme shown in Figure 8.46.

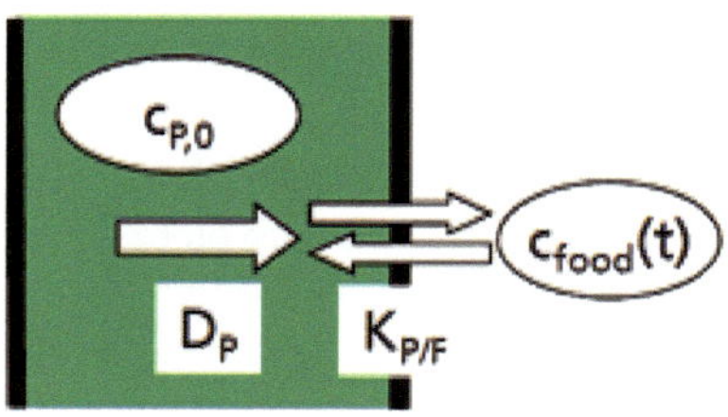

Figure 8.46
Migration process of substances from a polymer into food

c_{food} depends on:

- The packaging geometry: Surface/volume ratio (small size is worse) and material thickness (thick material = larger reservoir of substances)
- Temperature: Higher temperatures lead to higher polymer diffusion coefficient D_P

- Time t: Longer shelf life results in higher diffusion until the equilibrium polymer-food is reached
- Distribution coefficient $K_{P/F}$ of the substance between food and polymer at equilibrium
- The polymer diffusion coefficient D_P: This is a highly important factor due to the large differences between polymers, resulting in a division into "high-diffusive polymers" (PE, PP) and "low-diffusive polymers" (PET, PS). As an example, the diffusion coefficient of PS is 100,000 times lower than the one of PE. This "low-diffusive" property prevents surrounding contaminants from easily entering the polymer, and prevents any substances present in the polymer from migrating easily into the food: a double benefit.

The overall evaluation procedure is summarized in Figure 8.47.

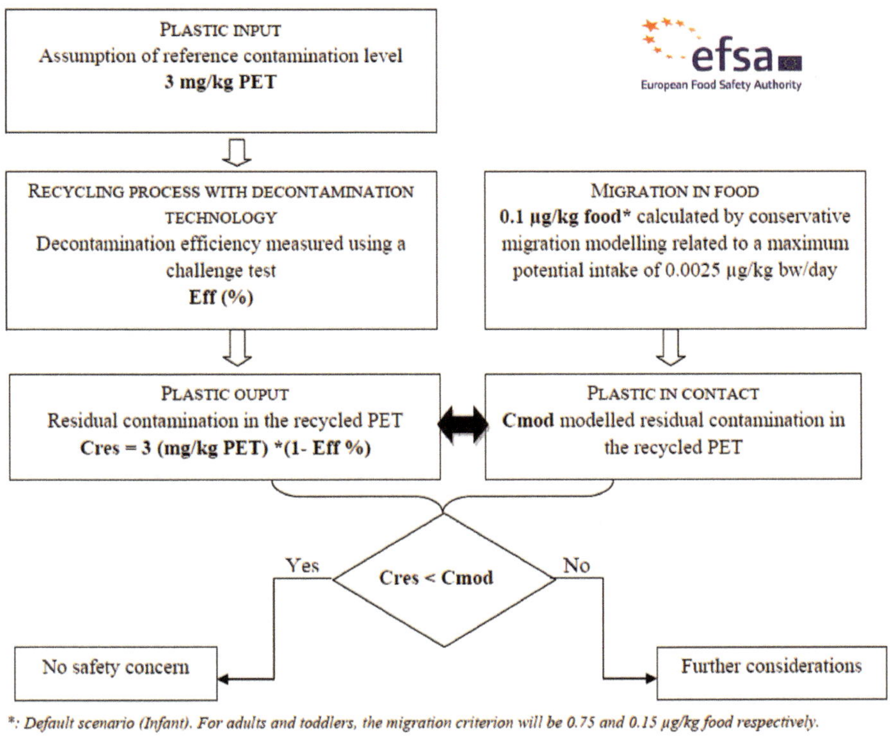

Figure 8.47 EFSA evaluation process as published in its scientific opinion on recycled PET

Two families of super-cleaning technologies are typically in use by the PET recycling industry and can be considered candidates for super-cleaning PS. Physically they all combine a high temperature with vacuum during a certain residence time.

Solid-Phase Super-Cleaning

This technology consists of applying a high temperature and vacuum to solid flakes or pellets for a considerable residence time (2–3 h), sometimes with active stirring or circulation. The high temperature forces the flakes or pellets to "sweat out" the contaminating substances by diffusion. Once arrived at the surface of the flake or pellet, the substances are removed by the vacuum. Obviously the "low diffusion" property of PS works here as a resisting force against the super-cleaning objective. The process works best for flakes and foamed flakes, because of the smaller diffusion distance before reaching the surface than is the case for pellets.

For PET, the applied temperature is 160–180 °C, but PS has to be limited to ~100 °C to stay in the solid phase. In order to achieve the same cleaning effect, residence times have to be significantly increased (e.g. 8–12 h), to the point where it becomes costly to industrialize.

Typical technologies are those used by Starlinger, Krones, and Erema. The main differences relate to the use of flakes or pellets, the pre-heating method, and the use of gravitational plug flow or the use of stirring or circulation.

Figure 8.48 Krones' solid-phase super-cleaning technology

Melt-Phase Super-Cleaning

Here the temperature is increased above the polymer melt temperature, typically 240 °C, which is no problem for PS. Then, depending on the technology, a large surface of molten recyclate is exposed to a deep vacuum, typically 1–10 mbar, allowing substances to be driven out of the polymer.

The first industrial application of this kind for PS is operational in Japan, where since 2010 PS food trays have been recycled back into food-contact material, by means of the Gneuss MRS (multi-rotation section) extrusion technology. As shown in Figure 8.49, this consists of an initial single screw to melt the ingoing flakes, followed by a multi-rotor screw section that expands the melt surface by a factor of 100 while constantly renewing it and exposing it to the high vacuum.

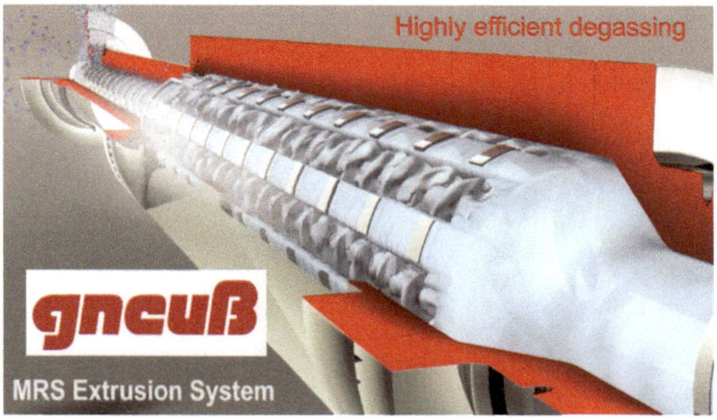

Figure 8.49
View of multi-rotor extruder screw of the Gneuss MRS extruder

The second application of melt-phase technology to achieve the production of food-contact PS was realized by the EPS-SURE project, which transformed used EPS fish boxes into food-contact PS for use in yoghurt pots and fish trays. The project obtained a grant from the European Commission through the European LIFE+ Program. It started in 2017 under a Spanish consortium coordinated by Cicloplast, an association that integrates all the companies of the plastics sector in Spain, in which Anape, Coexpan, El Corte Inglés and Total Petrochemicals Iberica took part. In June 2020, the project announced that it had achieved the technical EFSA requirements for food contact, on the basis of a proprietary melt-phase super cleaning technology realized by Total Petrochemicals Iberica.

Another commercial melt-phase technology, illustrated in Figure 8.50, was developed by the Austrian company NGR, whereby molten polymer is pumped through a horizontal die plate with a large number of fine holes, creating liquid strands that fall into a vacuum. The combined parallel strands have a very large surface from which the applied vacuum removes most of the contaminating substances before the polymer reaches the liquid surface underneath. The horizontal part increases the residence time to ~25 min, which is important for increasing the IV of recycled PET, but which has little added value for recycled PS.

Figure 8.50 Scheme of NGR's super-cleaning equipment P-REACT

Both Gneuss and NGR technologies have been successfully challenge-tested with PS recyclate by Styrenics Circular Solutions, each showing largely sufficient cleaning efficiencies and leading to an EFSA petition for use at 100% in food-contact applications.

8.4.1.4.5 Extrusion and Melt Filtration

For removing the very last imperfections from the high-purity white rHIPS, which may have a visual or mechanical effect in dairy packaging applications, it is important to apply a melt filtration of 50 µm. Such very fine filtration systems have become increasingly widely available.

This role as final correction of a near-perfect recyclate is different from melt filtration in the regular mechanical PS or PO recycling, where it is actively used to remove paper and non-target polymer, compensating for low washing performance and for the absence of flake sorting. It is also different from melt filtration in virgin polymer extrusion, where the melt filter serves mainly as a guard to avoid metal pieces contaminating the polymer.

The position of this melt filtration step in the high-purity recycling process for food contact depends on the super-cleaning technology. If the super-cleaning is done on solid flakes then the extrusion and melt filtration comes at the end, but if the super-cleaning is done in the melt phase or on pellets, then the melt filtration happens directly before the super-cleaning.

To allow the melt filter to properly remove unmolten particles, the polymer should be molten without much shear, because high shear would break the particles into smaller pieces, which would then pass through the filter more easily. Therefore, a single-screw extruder is usually used, often complemented with a preheating step before the flakes enter the extruder. A twin-screw extruder with high mixing performance should be avoided before the melt filtration.

After the melt filtration and prior to pelletizing or further compounding, the molten polymer should pass a degassing stage, to remove potential gases and odors developed by particles in the melt filtration step.

Melt filtration devices come in several types, each with advantages and disadvantages. An ideal melt filtration system must be adapted to the specific feedstock quality and to the sensitivity of the downstream application. For high-purity mechanically recycled PS, the quantity of unmolten particles is low and their size is small, but downstream dairy applications are very demanding.

The best melt filtration system will be characterized by:

- The technical capability to remove both hard and soft unmolten particles of small size at constant low pressure drop over the filter and with low % loss of filtrate
- Minimal disruption of the steady extrusion process, and minimal need for operator intervention by automatic renewal of the filtration surface or elements
- Absence of dead spots where material can accumulate and degrade into internally generated contaminants
- Low cost of consumables and maintenance, and simplicity for operator interventions

Some well-known manufacturers of melt filtration systems are Ettlinger, Mas, Pall, Gneuss, Erema, and Starlinger.

8.4.1.4.6 Industrial Blueprint for a High-Purity PS Recycling Plant

The powerful process sequence "deep sorting + hot caustic wash + flake sorting + super-cleaning + fine melt filtration" as described above is not only suitable for PS and XPS, but for any type of rigid plastic. When applied to PE or PP, the recyclate will be much purer than the usual gray recyclates produced with low cost-oriented recycling processes, and capable to access sensitive markets like skin contact or cosmetics; and when applied to PS or PET, the recyclate will even be suitable for food-contact applications.

This characteristic has a double advantage for investors:

- It allows immediate building of a fully competitive unit of economic size: typically 50 kt/a mixed feedstock.
- It allows rPS production to be ramped up, in accordance with the growing feedstock availability and market.

The plant configuration (Figure 8.51) starts with a common deep sorting line of 10 t/h input capacity, running 5 days per week and producing different deep-sorted fractions which are temporarily kept in inventory. When a fraction is available in sufficient amount, it is recycled as a batch in one of the two identical lines of 4 t/h input capacity with hot washing, flake sorting, super-cleaning and melt filtration, running in continuous operation. The batch operation allows optimization of the internal inventory in a trade-off with the batch change frequency.

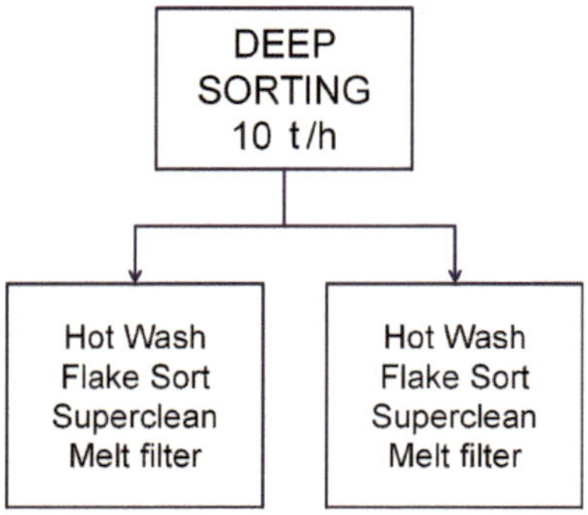

Figure 8.51
Basic configuration of a PS recycling plant

The feedstock can be pre-sorted PS bales or mixed bales containing PS, or a combination. For sufficient rPS production, a part of the feedstock should be pre-sorted PS. The complexity of the deep sorting line (expressed as number of sorted fractions) depends on the desired feedstock flexibility and on the requirement for color sorting.

The output of the plant is 33 kt/a of high-purity mechanical recyclate, comprising >99.9% PS (or other polymer), which is odorless, color-sorted, melt-filtered at 50 μm. Starting with feedstock from typical bales with 95% "object purity" and 70% target polymer content, this corresponds to 66% overall yield, however with an excellent target polymer yield close to 95%: see the material balance table in Figure 8.52. The plant occupies a surface of about 30,000 m²; its overall investment cost is estimated about €40 million.

High Purity Mech. Recycling - main flow	Process		PS polymer content		% of PS input
	Ton	Yield %	%	Ton	
Baled PS objects feedstock	1,000	100%	70,0%	0,700	100%
PS objects in bales feedstock	0,950	95,0%	73,7%	0,700	100%
Deep sorted PS objects	0,922	92,2%	73,7%	0,679	97,0%
Hot washed flakes	0,710	71,0%	95,5%	0,678	96,8%
Sorted flakes	0,663	66,3%	99,9%	0,663	94,7%
Super cleaned flakes	0,663	66,3%	99,9%	0,663	94,7%
Melt filtered pellets	**0,654**	**65,4%**	**99,95%**	**0,653**	**93,3%**
+ Recycled PS by-product	**0,010**	**1,0%**	**96,6%**	**0,010**	**1,4%**
= Total output	**0,663**	**66,3%**	**99,9%**	**66,3%**	**94,7%**

High Purity Mech. Rec. - other flows	Process		PS polymer content		% of PS input
	Ton	Yield %	%	Ton	
Sorting residue	0,079	7,9%	26,8%	0,021	3,0%
Sludge	0,212	21,2%	0,6%	0,001	0,2%
Flake sorting residue	0,046	4,6%	32,1%	0,015	2,1%
Total material loss	0,337	33,7%	11,0%	0,037	5,3%
Melt filtrate	0,010	1,0%	96,6%	0,010	1,4%

Figure 8.52 Material balance of the high-purity mechanical recycling process, producing a >99.9% pure PS recyclate from bales of pre-sorted PS packaging

In a conservatively estimated feedstock environment, the full production cost of the recyclate allows an outstanding competitiveness with virgin PS.

Figure 8.53 Typical view of a high-purity mechanical recycling plant

8.4.1.5 Waste Feedstock for Circular PS

8.4.1.5.1 PS Packaging in EU Waste Streams

Current Situation

Guaranteed availability of recycling feedstock is a key requirement for any investor in a high-purity recycling plant. A feedstock study for Styrenics Circular Solutions (SCS), complemented with field sorting tests by SCS and Tomra Sorting, has shown that the total amount of PS used for packaging in the EU is close to 800 kt/a. Due to the short shelf life of packaging, these materials certainly become waste during the same year. Since PS packaging waste contains about 25% non-PS polymer in the form of food residue, paper, aluminum and other plastics, these returning packages represent *about 1 Mt of waste per year* - on average 2.3 kg per inhabitant per year.

For the whole EU27+, only 30% of the PS packaging waste is collected in a separate collection (in Germany and Austria known as the “yellow bin”), where it occupies a share of 4% relative to the total amount of plastics in that waste stream. The amount of PS packages in the residual waste or “gray bin” appears to be 2.5 times higher - however, this is not only the case for PS but also for PP, PE, and PET!

Figure 8.54 summarizes the findings of pre-incinerator plastics extraction tests by Tomra Sorting: PS represents 3.2% of all plastics, corresponding to 4.9% of the rigid plastics fraction.

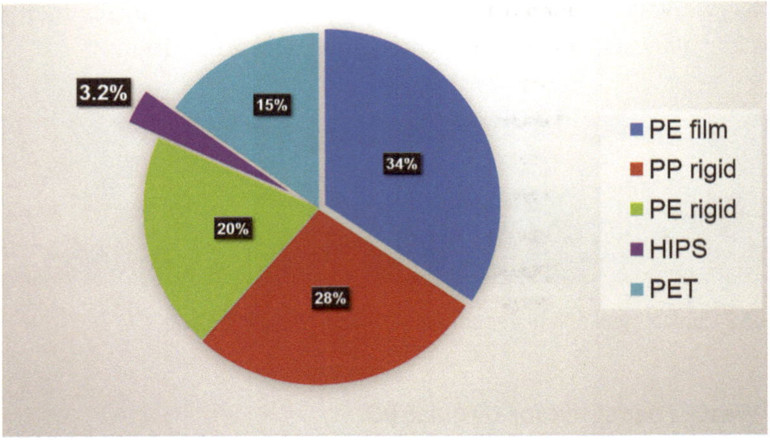

Figure 8.54 Typical composition of pre-incinerator extracted plastics fraction in Germany. PS represents 3.2% of all plastics incinerated

PS packages end up in the incinerator feed by several mechanisms:

- Initial sorting by the consumer in the "gray bin" instead of the "yellow bin", for various reasons (e.g. reluctance to keep it inside due to the low collection frequency of the yellow bin, or confusion about the instructions), or by error even in the compostable waste bin.
- A lot of yoghurt is consumed "out of home" where no "yellow bin" collection is organized, for example, in offices, schools, hospitals, and industry.
- Immediate diversion of small yoghurt pots as "fines" at the entrance in sorting centers, by having an inlet sieve mesh that is too large. Once in the residue stream, these packages never reach the sorting equipment.
- Lack of incentive for the sorter to improve PS sorting efficiency, due to insufficient demand for sorted PS, causing it to have a negative market value similar to incineration fee.

Several of the identified reasons are in a positive improvement dynamic: separate collection of domestic packages is being organized for "out-of-home" locations, and new sorting centers are equipped with smaller inlet sieves or with extra sorting capability on their fines stream.

An important factor driving these improvements will be the economic motivation of EPR actors and sorters to produce more sorted PS because of its interesting downstream value as feedstock for high-purity recycling plants.

Future Potential

However, when looking at the massive amount of perfectly sortable and recyclable PP at the entrance of incinerators in Germany, an additional reflection for the longer-term supply of PS is unavoidable: indeed, sorted PP *does* have a strong demand today, but still the majority of PP packages don't reach the sorting stage. The positive conclusion is that there is a huge extra potential recycling feedstock to be tapped for all plastics. This explains the important gap between known quantities of plastics going to the packaging market and reported quantities of waste coming back for recycling.

At the same time it forces us to temper the ambition on the separate collection. Indeed, if this is the situation in a "mature collection country" like Germany, after 20+ years of "all plastics" collected in the "yellow bin" by a relatively disciplined population, by which mysterious influence would this further improve? And why would other countries in 5 years from now do better than Germany today? The most logical conclusion is that the current situation in Germany is reasonably to be considered a *ceiling for the future separate collection potential in the EU.*

8.4.1.5.2 The Potential of Pre-Incinerator Extraction

Residual Waste Plastics as Circular Feedstock

In our thinking about waste and recycling, a strong belief is that a plastic object in the residual waste ("gray bin") is lost as recycling feedstock, due to the heavy contamination with organics and other materials. However, this conviction dates from 25 years ago, when sorting was done manually. The smell and concern for workers' health is understandably the cause of the still-surviving psychological rejection and push-back against this stream as a circular feedstock.

However, today's sorting technology is perfectly capable of extracting and sorting plastics from the waste stream entering incinerators. Many incinerators have for a long time removed PVC because of their operational permit (dioxins in combustion gases), and this extraction can easily be extended to "all plastics". This is common practice in part of Poland and several other East European countries, and is increasingly done in the Netherlands and Scandinavia, with ongoing projects in Germany and Austria.

Just like the objects form the separate collection, these extracted plastic objects can be sorted in modern sorting operations; called "post-sorting". In Poland, some sorting centers work one shift on feedstock from separate collection and one shift on pre-incinerator extracted plastics. Some years ago the post-sorting produced bales of inferior quality, but there has been such a focus on improving this, that today the outgoing bale quality (expressed as "object purity") is the same or sometimes better than bales from separate collection. Only the smell is stronger (obviously), causing a slightly lower market value than similar sorted fractions from the separate collection.

The higher contamination level can be an issue for the low-cost oriented mechanical recycling plants. However, this is not at all the case for the high-purity mechanical recycling process, as explained in Section 8.4.1.4. Application of deep sorting, hot caustic washing, flake sorting and super-cleaning yields a recyclate that is transparent or brightly colored, odorless and free of black spots; it can hardly be distinguished, visually or analytically, from high-purity recyclate from a separate collection. Despite its "gray bin" origin, it is really "near-virgin" and obviously superior to the regular mechanical recyclates from low-cost-oriented recycling processes, even if those recycle feedstock from a separate collection.

The above-mentioned result was demonstrated and documented in the previously-mentioned CPO project, with Tomra Sorting as project leader and key expertise generator. As a member of Styrenics Circular Solutions, Tomra Sorting helped to replicate all CPO observations and conclusions for styrenic plastics, to the genuine surprise of the PS industry and of the external world.

Economics of Pre-Incinerator Extraction

Contrary to common belief, there is a strong economic incentive for incinerators to extract plastics from their feedstock.

Since the major debates dating from before the EU Plastics Strategy, the plastics industry has developed an internal conviction that its waste is serving an essential purpose at municipal waste incinerators ("waste-to-energy") by contributing its high calorific content to the waste, allowing the overall mass to burn and be transformed into energy. It is generally believed that extracting the plastics beforehand will require this energy source to be replaced with gas, to support the incineration. The reality is completely different, as explained below.

For the vast majority of municipal waste incinerators, the waste feedstock is much too rich in energy (13–15 GJ/t), compared to their design (10–11 GJ/t). And the situation is gradually worsening, because of the increasing diversion of organic waste to composting, which removes organics and water content from the incineration feedstock. (Comment: This is not the case in specific installations using a prepared stream of high calorific solid fuel; like cement kilns. But these are a small minority compared to the municipal waste incinerators.) As a result, their incineration capacity is bottlenecked by energy content. Every single ton of plastics (25–35 GJ/t) in the waste reduces the waste incineration capacity by ~2–3 t, forcing them to operate below their design and economic optimum, as shown schematically in Figure 8.55. So the extraction of one ton of plastics opens the way for ~€200–300 extra incineration revenue (gate fee) at constant energy production, guaranteeing a rapid return on the extraction investment.

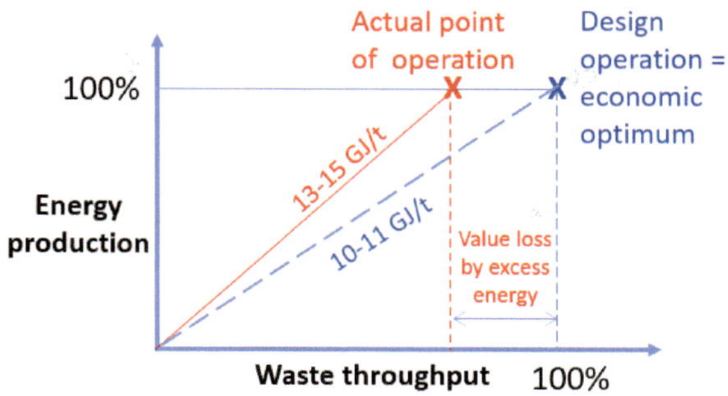

Figure 8.55 Economic optimization of a typical municipal waste incinerator by extraction of plastics from the feedstock

The EU societal impact of the pre-incinerator extraction of plastics is three-fold:

- Elimination of the need to build hundreds of incinerators, because the currently installed incinerator capacity will cover a large part of the future needs.
- Reduction of waste incineration tax for the population: indeed, increasing this tax is the current "business model" used by most municipalities to compensate for decreasing capacity of their incineration plants.
- Generation of massive extra feedstock for circular plastics, including PS.

8.4.1.6 Circularity of Polymers – The Unique Position of PS

8.4.1.6.1 Open Loop vs. Closed Loop: Is that Really the Question?

The styrenics industry has recently demonstrated the mechanical recycling of PS packaging waste into high-purity recyclate, fit for use in food-contact applications thanks to the low diffusive property of PS. The EFSA petitions underway will open the use of this recyclate into the same or similar food packages as their previous life cycle.

Effectively closing the loop from yoghurt pot to yoghurt pot is a high-visibility "flag on the moon", driving the PS value-chain actors to cooperate, not only for this common development but also to continue enhancing the design of the yoghurt pot, for ensuring that more and better PS recyclate continues to be available to become yoghurt pots, over and over again.

In the case of PET bottles, closing the loop back into the same packaging application offers the highest circular value – but that should not be generalized. Is there a way to judge the circular value of packaging vs. other plastic applications? Is food-contact purity, although an impressive and almost symbolic achievement, a sufficient criterion? Compliance with food-contact requirements is a key asset not only for packaging but also others like (for example) fridge internals or water pipes: are these applications, with much longer useful life, to be considered "downcycling" vs. the yoghurt pot or the PET bottle? On all these questions the answer is of course NO.

Apart from serving as recycled content in yoghurt pots, high-purity mechanical PS recyclate can replace virgin PS in almost any "prime segment" application, wherever its highest economic and societal value can be contributed. Conversely, all regular "prime segments" will positively benefit from the availability of recyclate on the condition that it has near-virgin purity. As a consequence, the value of a second-life destination of a recyclate is not represented by the "open loop vs. closed loop" opposition but only *by the quality level of the recycled material.*

8.4.1.6.2 A Material's Circularity and Circularity Matrix

A society's circularity is defined by the extent to which its material demand is met with recycled molecules. The Figure 8.56, from the EC document "Monitoring framework for the circular economy" published January 2018, shows the circularity of resources in the EU in 2014:

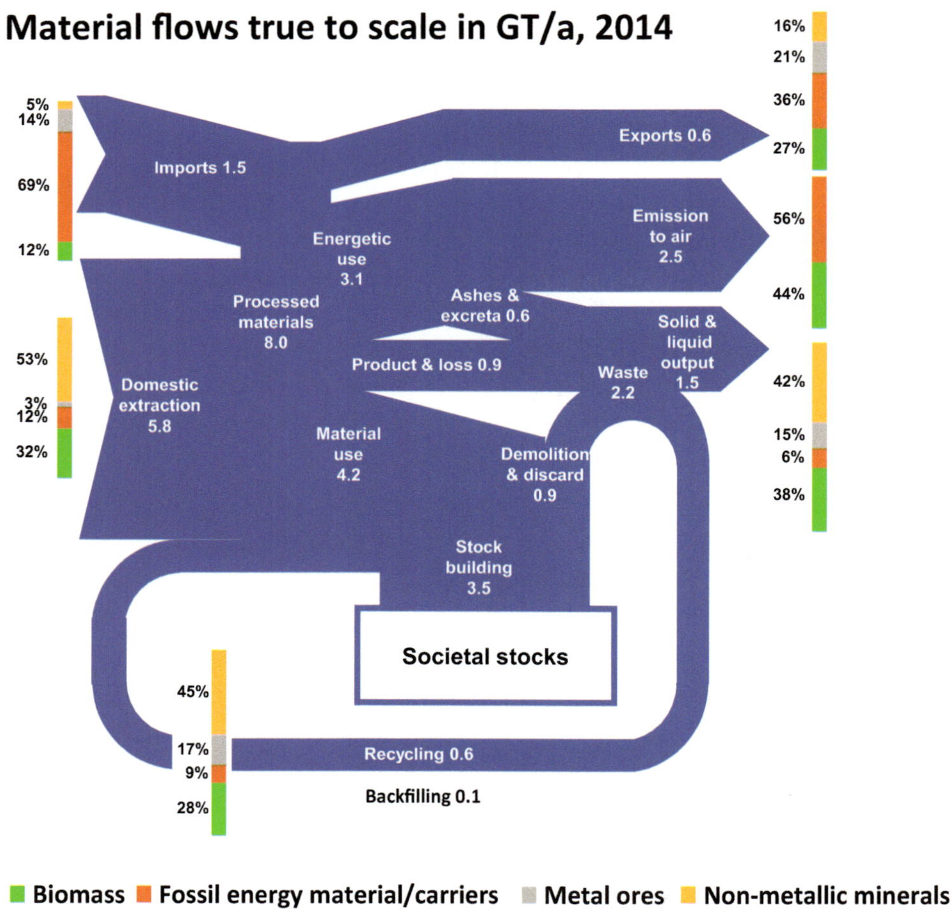

Figure 8.56 EU material balance 2014

The total internal material demand was 4.2 Gt, of which 0.6 Gt was met with recycled materials, leading to a 14% circularity rate. The graph also shows that 3.5 Gt or 83% of the material input in 2014 remained in use beyond that year ("societal stock building"), and that the 0.6 Gt recycled material was generated from 0.9 Gt of waste, produced from the historical societal stocks.

Of course, not all sectors have the same contribution to this circularity result, depending on the length of the products' useful life cycle in each sector. The packaging sector, with a typical life cycle shorter than one year, produces most of its waste during the same year, whereas the materials used by the construction sector tend to remain in "societal stock" for many decades before becoming waste.

Conversely, every sector will have typical requirements related to its produced objects, causing restrictions for the use of recycled materials depending on their previous life sector. The longer the initial life cycle, the more the material will tend to be "worn out" or "outdated" - or even considered risky because of their potential to contain legacy substances that are now regulated.

This "stock building" phenomenon in each sector applies to all used materials in that sector, and also to plastics. Therefore, the transition of plastic molecules after their production, from used objects to waste and circular feedstock, is spread over time, as schematically shown by Figure 8.57.

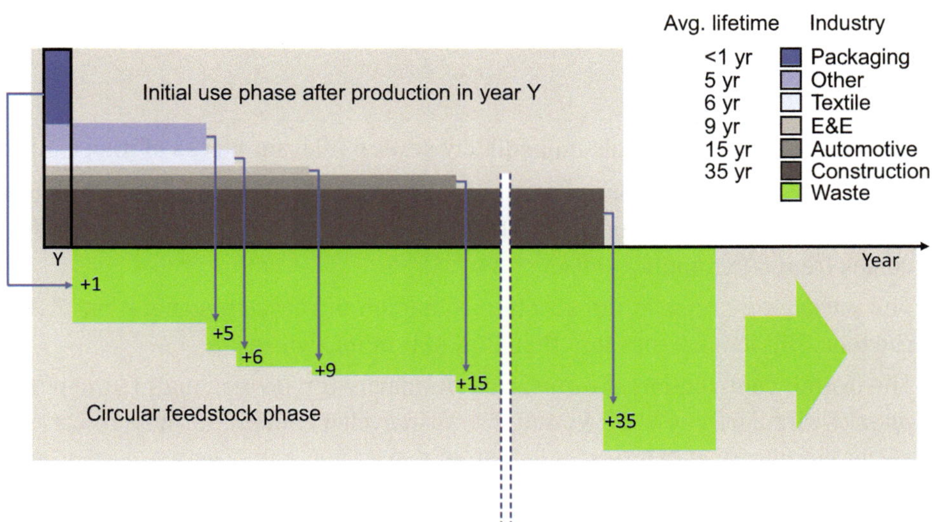

Figure 8.57 The transition of virgin polymer to circular feedstock

So, when looking at the plastic waste generated in a particular year Y, its total volume is composed of the consumed packaging plastics of that same year (because their lifetime is shorter than one year) plus waste volumes V_i of applications A_i with respective lifetimes T_i, which were produced in the past, during years $Y - T_i$ to Y.

The best and most effective representation of a material's circularity is its "circularity matrix", showing for a dedicated year its recyclates generated and consumed by application, or aggregated by sector. The matrix in Figure 8.58, used by the Circular Plastics Alliance (CPA), provides the best current view on current EU plastics circularity, showing all plastic recyclates produced and consumed in each sector; it is used as a starting point for managing the industrial ambition to consume 10 Mt of recyclate in Europe by 2025.

All polymers		Second life	Current flows 2014 - 2018							
			Agriculture	Packaging	EEE	Construction	Automotive	Other	Total	% of recyclate to own sector
Initial application	Demand >>>		1 741	20 429	3 174	10 138	5 069	10 650	51 200	
		Recyc. uptake	122	620	46	981	177	2 051	3 997	
	Waste	Recyc.prod.								
Agriculture	638	302	122	69	-	56	-	55	302	40%
Packaging	14 228	2 932	-	544	38	545	105	1 699	2 932	19%
EEE	630	72	-	-	8	23	23	19	72	11%
Construction	1 009	362	-	7	-	333	-	22	362	92%
Automotive	590	328	-	-	-	24	48	256	328	15%
Total	17 095	3 997	122	620	46	981	177	2 051	**3 997**	**1 056**
		% recycled content	7,0%	3,0%	1,4%	9,7%	3,5%	19,3%	7,8%	
% of recycled content from own sector			100%	88%	17%	34%	27%			26%

Figure 8.58 Circularity matrix of all plastics in the EU – yearly figures for the period 2014–2018

The circularity matrix reveals immediately several relevant pieces of information:

- The construction sector appears to be leading both in terms of using its own recyclate (92%) and in terms of overall recycled content (10%), of which two-thirds comes from other applications;
- The packaging sector is the absolute champion of recyclate generation: 75% of the total. But it consumes less than 20% of it in its own sector;
- The overall contribution of own-sector circularity (green diagonal) to the overall plastics circularity is 1056 kt out of 4 Mt recycled content, or only 26%. Cross-sector circularity (74%) constitutes the vast majority of the overall plastics circularity.

Although all closed-loop circularity happens within the same sector, it is important to realize that own-sector circularity is not all a closed loop: for example, most recyclate produced from PET bottles is consumed by PET trays! As a matter of fact, closed-loop circularity is an important symbolic material quality, drives cooperation and technology development (a "flag on the moon"), and is a trigger for responsible behavior of the producer through circular design; but it should not be confused with a material's effective circularity.

Conceptually, sectors or applications can either contribute to a material's circularity by generating a lot of recyclate with interesting qualities (packaging), or by their capability to use recyclate as recycled content (construction), or in both ways. This cross-sector circularity is the basis on which all circular materials thrive: a donor vs. receiver segmentation based on the natural characteristics of each segment.

By its nature, plastic packaging is in the role of a "universal donor" for all polymers, with positive destiny to become the circular feedstock of choice for all other plastic sectors. It is not only a large source of circular feedstock with predictable availability, but its purity in terms of absence of legacy chemicals is structurally guaranteed by its short shelf life.

This is similar to the situation with blood donation: blood antigens place limits on the ability to transfuse red blood cells between people belonging to different blood groups, as shown in Figure 8.59. The following circularity principles can be recognized:

- Every blood type is donor and receiver of itself;
- Blood type 0– is a universal donor;
- Blood type AB+ is a universal receiver;
- Other blood types are partial donors and partial receivers.

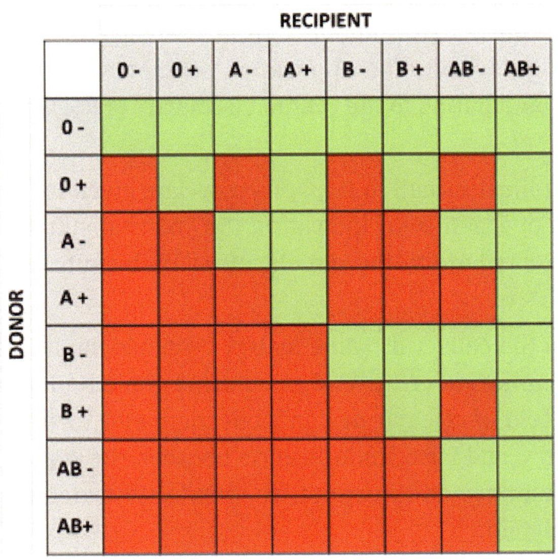

Figure 8.59
Blood donor and receiver types

The optimal management of blood transfers from a given stock of blood is strongly based on the optimized use of this donor–receiver concept.

8.4.1.6.3 Circular Unification of Rigid and Foamed PS

Rigid PS and its foamed versions (EPS, XPS) are basically the same molecules: their differences are in the same range as those between HDPE homopolymer (milk bottle) and copolymer (shampoo flacon, detergent bottle). They follow different processing routes to become packages with very distinct properties, but as a circular feedstock they belong together in a single category, serving the PS circularity with similar constraints as other plastics: the need for collection and sorting, sensitivity to design for recycling factors, etc.

Large parts of the PS industry have reached a unique level of maturity, where convertors make their own optimized mix out of "standard" GPPS and HIPS grades, in contrast to the numerous differentiated grades of other polymers, with each optimized for their application. This practice of "convertor mixing" generates a wide range of possibilities to use recycled content, by simply adapting the type of virgin resin in the mixture to the properties of the recyclate (whether it comes from rigid GPPS or HIPS or from foamed packages) and the target end properties of the second-life application.

For the above-mentioned reasons, rigid and foamed PS must be considered together when it comes to circularity. To do otherwise would be like separating milk bottles from shampoo bottles and detergent bottles, providing unnecessary constraints on HDPE circularity.

This circular unification leads to one single combined circularity matrix for rigid and foamed PS, as shown in Figure 8.60. The application segments are chosen based on the relevance of their contribution to the overall circularity of PS, whether it is as a donor or receiver segment.

Every cell represents the second-life destination (top) of a recyclate given its initial application (left), with the level of green color indicating the "natural suitability" for a recyclate to successfully contribute to the circularity of PS by substituting virgin PS resin.

By applying quantitative data to the cells, this same matrix becomes the management tool *par excellence* for planning, steering, and monitoring the circularity of PS by dedicated actions in the field of circular design, collection, and sorting, for pledging as an integrated industry, and for communication between the stakeholders of the value chain. Such qualitative and quantitative circularity matrixes can easily be composed for every polymer.

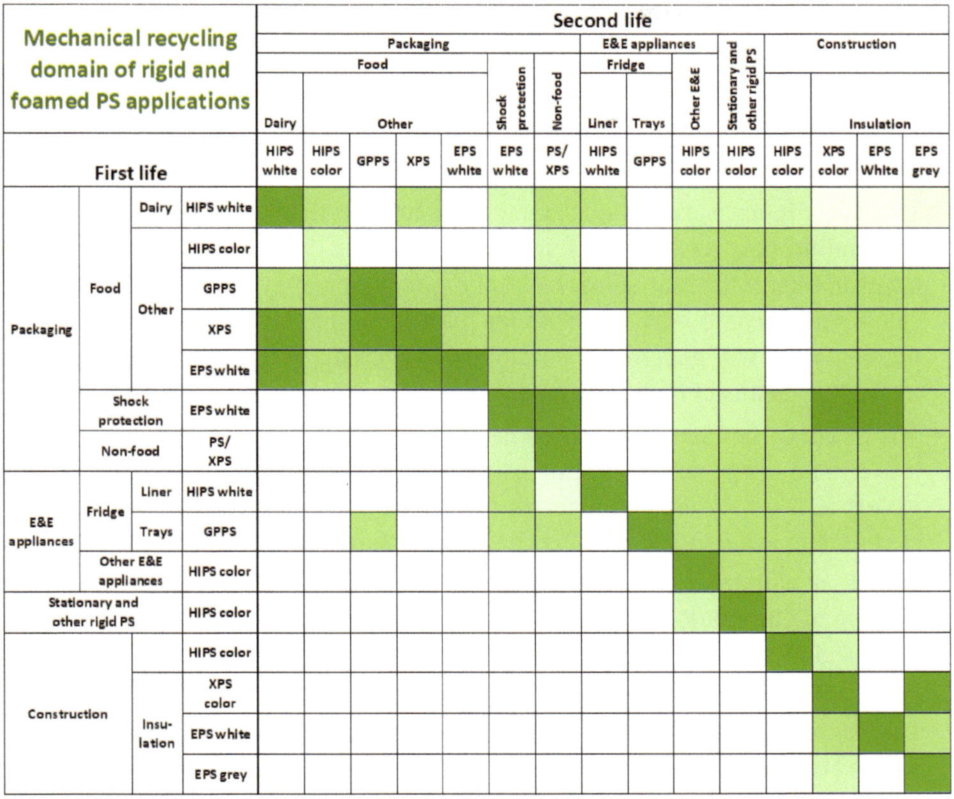

Figure 8.60 Unified circularity matrix for rigid and foamed PS. The shade of green indicates the "natural suitability" of a recyclate to successfully contribute to the circularity of PS

8.4.1.7 Circular Design of PS Packages

In essence, this section has been about "design for recycling", but the name "circular design" emphasizes the importance, when initially designing an object, to look beyond the first life cycle towards high value second-life applications. The targeted second-life segment can be the initial one (closed loop) but also another high-value segment of the circularity matrix, wherever it may add a high economic and societal value by replacing virgin PS. The recommended design measures may be different in function of the chosen second-life application, but should be the result of a voluntary process and not be used as an excuse for not taking any design measures.

8.4.1.7.1 Design for Increased Circular Yield

In general, current designs of PS packages are not problematic for their circularity, compared to some other packaging types like multilayer flexibles. Nevertheless, packaging design is a dynamic subject, and a future focus on circularity will be needed to avoid deviation to linearity. The example of the PET bottle industry serves as inspiration to the PS value chain, not only for technical solutions but also for the required structural cooperation between the stakeholders.

The design aspects that impact the quantity and quality of high-purity mechanical PS recyclate are the same for rigid and foamed PS packages, and similar to packages of other plastics:

- The use of in-mass coloration cannot be undone. In the best case these packages will end up in a color sorted fraction, and become colored recyclate for a dedicated market with that color; most often, however, they are mixed with other colors and become gray recyclate for markets which can accept this. Avoiding pigmentation will ensure the possibility to access a wide range of "virgin segments" by means of high-purity recycling.
- The use of carbon black makes a package invisible to sorting machines, causing it to end up in the residue stream going to incineration. The simple replacement of carbon black by a black pigment solves this without changing the visual aspect of the packaging, increasing the yield of the sorting process.
- The use of aluminum lids causes a part of the yoghurt pots to be deviated into the fraction of aluminum packages, because the sorting step for aluminum is at the very beginning of the sorting process, and the attractive force of the "eddy current" sorters is quite powerful.
- The presence of paper labels and the glue to attach them to the packages increases the variable cost of the high-purity mechanical recycling process, and in less powerful recycling processes they are the reason for residual smell and black dots in the recyclate (the result of being burnt in the extruder). The alternative is to replace paper by plastic and minimize the use of glue.

Besides these measures, an interesting development with potential to improve the circular yield is in the PS industry's development pipeline: "the mono-material PS yoghurt pot" with cup, label, and lid in PS, maximizing the PS polymer content in the recycling feedstock – which is sometimes confused with "yield of the recycling process".

This confusion deserves an explanation: when we mention for a regular mechanical PS recycling process "a recycling yield of 75%" then this is the result of a target polymer yield of ~95% and a recyclate purity of ~90%. The rest of the "lost" mass is simply other material present in the feedstock, correctly removed by the recycling process:

- ~5% of the bales are sorting errors (because the typical bale specification is 95% "object purity").
- 20–25% are other materials in (or on) the correctly sorted objects: organic residue, paper labels, aluminum, other polymer components. The sorting process *cannot* remove these; it is the mission of the recycling to do that as much as possible with as little as possible loss of the target polymer.

So this “75% recycling yield” reflects mainly the ~70% target polymer content of the feedstock, taking into account a good sorting quality (95% “object purity”) and an outstanding target polymer recycling yield (95%), which is the only relevant yield figure describing the performance of a recycling process: “how much of the target polymer in the feedstock is transformed into recyclate”. Indeed, nobody expects the recycling process to transform residue or paper into PS! On the contrary, attempts to increase the “recycling yield” without increasing the polymer content in the feedstock will mathematically result in a quality degradation of the recyclate.

This is not specific to PS recycling of course – for example, the overall yield of LDPE recycling of agriculture film is barely 50% – simply because there is 50% earth/dirt attached to the feedstock. Any feedstock preparation which can increase its polymer content, by reducing the amount of dirt, will result in an increased yield of the recycling process. Reversely, running the same recycling process and feedstock at 5% higher yield would mean that the recyclate will contain 10% more dirt.

Figure 8.61 shows the mass balance of such a “regular” mechanical PS recycling process – representative of most current PS or PO recycling, with the quality suitable for less demanding applications or to serve as a typical feedstock for a chemical recycling process.

Mechanical recycling - main flow	Process		PS polymer content		% of PS input
	Ton	Yield %	%	Ton	
Baled PS objects feedstock	1,000	100%	70,0%	0,700	100%
PS objects in bales feedstock	0,950	95,0%	73,7%	0,700	100%
Objects after non-plastic removal	0,940	94,0%	73,7%	0,693	98,9%
Cold washed flakes	0,740	74,0%	90,0%	0,666	95,1%

Mechanical Recycling - other flows	Process		PS polymer content		% of PS input
	Ton	Yield %	%	Ton	
Sorting residue	0,060	6%	12,3%	0,007	1,1%
Sludge	0,200	20%	13,3%	0,027	3,8%
Total material loss	0,260	26%	25,6%	0,034	4,9%

Figure 8.61 Mass balance of a regular PS or PO recycling process

In comparison, the high-purity recycling process (99.9% purity) in Figure 8.35 also has ~95% target polymer yield, but through more steps (CapEx) and higher intensity (CapEx, OpEx) its overall recycling yield is “only” ~65%. However, the difference is 10% higher purity, meaning that 10% non-target polymer was further removed.

8.4.1.7.2 Generalized Packaging Design for Food Contact

In Section 8.4.1.6 on circularity, we showed that plastic packaging is in the role of a “universal donor” for all polymers, with strong potential to become the circular feedstock of choice for all other plastic sectors.

In Section 8.4.1.3 on food contact, we concluded that the strongest barriers against approaching technical or regulatory food contact are at the article design stage, because they destroy upfront the potential economic value and investment justification of available technology to lift recyclate quality to the level where they can enter “virgin segments”.

However, some virgin materials and components without food contact compliance are used in packages - inks for example, or simply non-food-contact materials and components that are acceptable for a non-food first life cycle. The presence of such materials and components in the packaging waste creates unnecessary contamination of this main circular feedstock stream, preventing it from fully playing its role as a universal circular feedstock. (In the analogy with the blood types, it is as if the universal donor of blood is contaminated with a virus, creating a barrier against its use despite its attractive characteristics).

Therefore it is the author's opinion that the packaging sector should adopt a generalized "circular material excellence standard", prescribing that *all virgin materials and components used in packages should be food-contact-compliant: the design of a plastic article should never prevent a future potential (near-)food contact of its recyclate, even if its initial life cycle would allow it.* So inks, paper, pigments, glues, additives, etc. would no longer be a source of risk for the downstream recyclate user.

For obvious reasons, recyclates themselves should be exempt from the rule: in any case, the uptake of non-food-contact recyclate in packaging is still very small, so its disturbance to the waste quality will be limited.

Thanks to the short shelf life of packaging, this standard would clean up the entire packaging waste stream in a very short time. The resulting quality guarantee will remove a big concern about the main recycling feedstock, as it did with virgin resins. It will not automatically generate food-contact-quality recyclate, but it will make the "non-food-contact" property much less complex and risky: from now on, its purity will depend mainly on the performance of the recycling process.

In such an environment, recyclers will feel encouraged to invest in high-purity recycling technology with the capability to technically remove non-food residues below the levels of concern. Recyclers who prefer to continue serving less-demanding segments will also benefit because of an improved balance between offer and demand.

The high-purity recyclates will spontaneously become attractive to serve as recycled content back into new packaging. Even without regulatory food contact status, their technical composition will be so pure that this "near-food contact" or "technical food contact" property deserves to receive a dedicated recognition as a standard as well. This high-purity standard, subject to quality audits, can then become a sufficient criterion to formally open access to skin contact and similar applications.

In addition, other sectors may be interested to voluntarily join this "circular material excellence standard", driving future circularity with visible branding by means of a dedicated label and stimulating further uptake by making it a criterion in circular procurement policies, complementary to recycled content.

The technical exceptions to this standard will drive innovation projects which currently find no justification, or they can remain exceptions for which innovative recycling methods will be required: *the "virtuous circle" of circularity at work!*

8.4.2 Mechanical Recycling of Styrene Copolymers

Eike Jahnke

8.4.2.1 Introduction

The class of styrenic polymers (or styrenics for short) is a diverse family of materials [1]. From a volume perspective, the homopolymer, polystyrene (PS) is the largest member of the family with 15 Mt/a produced worldwide when its impact-modified version, HIPS, is included. Acrylonitrile butadiene styrene (ABS) follows as the second largest member, which is produced on the order of 10 Mt/a, and rising [2]. In addition to these two main products, various other styrene copolymers are available commercially, with the most prominent examples being ASA, SBCs, SAN, SMMA, SMA, AMSAN, and several types of blends with polycarbonate and polyamide (Figure 8.62) [3]. However, in comparison to PS and ABS, those styrenic specialties can be considered niche products in the context of the circularity discussion, as their overall annual volume is at least one order of magnitude smaller. Consequently, their relative shares in the various waste streams are considered to be relatively small. This article will focus on the (mechanical) recyclability of the two large-scale polymers, PS and ABS.

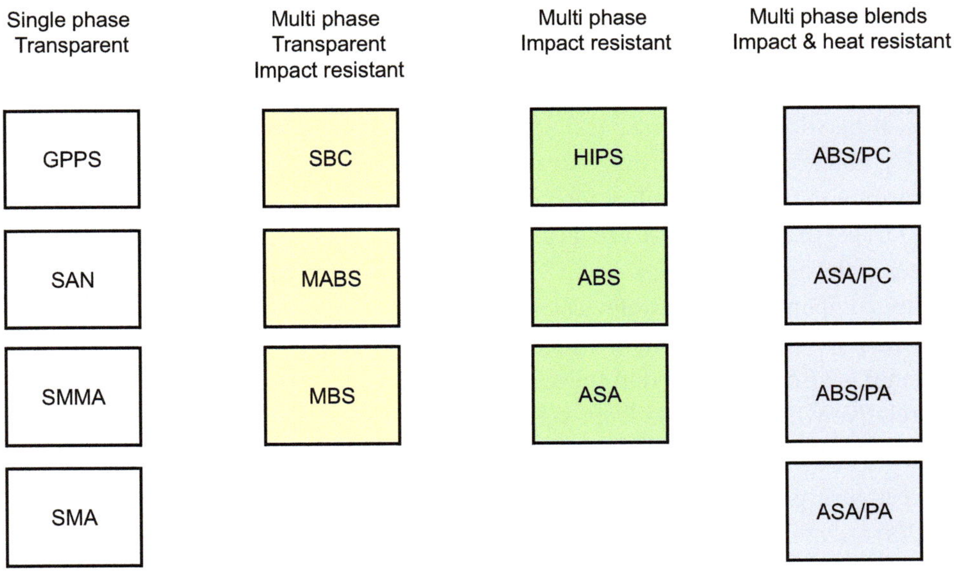

Figure 8.62 Classification of most commercially available styrenics (according to the INEOS Styrolution portfolio in EMEA for 2021). PS/HIPS and SAN/ABS contribute the majority volume share, while the other copolymers and blends are sold in significantly lower volumes annually

PS and ABS are both used in a multitude of industries and applications. PS is abundantly used in packaging applications, helping to extend the shelf life of many foods or keeping goods safe in the form of EPS packaging. Other large application fields are in XPS foams for energy-saving insulation in homes, electrical and electronic equipment (EEE), and household applications. The latter is also one of the main application sectors for ABS, while its combination of easy processing and balanced property profile has also led to a plethora of applications in the automotive, toys, construction, and medical industries. ABS is therefore generally considered the largest-volume member of the engineering plastics group.

While applications produced with "virgin" (i.e., conventionally produced) styrenics already have a significantly positive impact on a sustainable society during their lifetime (insulating houses, supporting lighter automobile designs, extending the shelf life of foods), it is clear that end-of-life solutions need to be developed and brought to scale for these valuable plastics as well. Because of the ever-growing annual production of new plastics including styrenics driven by a growing world population and increasing shares of middle-class households, the amount of plastic waste is also increasing [4]. This is also "supported" by increasingly short use times of certain items [5].

The wide distribution of applications also causes styrenics waste to return back through multiple waste streams. For example, PS as well as ABS make up major portions of the WEEE (waste electrical and electronic equipment) stream, which due to regulations [6] has proven to be an attractive source for recyclate in Europe, with other regions pursuing similar concepts. HIPS and ABS represent together 55% of plastics used in electrical and electronic equipment according to Maris et al. [7], and even 80% according to a report by Stenvall et al. [8]. WEEE is segmented according to the exact application category from which it originates. To name only the largest examples, these are large household appliances, small household appliances, telecommunications and IT equipment, as well as consumer equipment. Many components of such devices like housings, fridge in-liners or other mainly visible parts are predominantly made from either HIPS or ABS, while PP (polypropylene) is also widely used in these applications. Hence, styrenics can be retrieved in relatively high amounts from WEEE waste streams. If a proper sorting methodology is applied, very favorable purities and properties are achievable.

ABS is also present to a significant extent in cars, rendering end-of-life vehicles (ELVs) an interesting source for post-consumer recycling (PCR) of ABS. This is likewise a regulated area [9], and similarly qualified for retrieval of ABS waste, although the material quality is on the lower side of the spectrum (see below).

Another pool of potential waste retrieval, especially for PS, is household waste, with collection schemes running in various countries due to the abundance of PS in food packaging (for example, yoghurt pots). This leads to a relatively high share of 4–7% of PS in the household waste stream that is worthwhile extracting.

For ELV and WEEE waste streams, it has been proven that float-sink separations give rise to a relatively pure mixed fraction of styrenics, as they share a very similar density profile (1.04–1.11 g/cm^3 depending on the pigmentation level). Thus they can be separated from polyolefins (densities <1 g/cm^3) and other polymer families with higher densities, such as PET, PVC, PC, PA, and flame-retarded versions of PS and ABS (1.15 g/cm^3 and higher). Further methodologies such as electrostatic techniques, NIR identifications, and many others are used (or can be used) to increase the purity level. The separation of PS from household waste predominantly relies on NIR identification, which has proven to be a highly efficient methodology.

As laid out in several other chapters of this book, and in a recent overview article [10], several recycling technologies are available or are under development to treat plastic waste with the aim of recycling. This of course also applies for styrenics either as a component of the mixed plastics waste or for sorted fractions of individual styrenics materials. These technologies include chemical or feedstock recycling like depolymerization of PS, which is explained in detail in Section 9.1. In addition, dissolution (Chapter 10) or gasification (Chapter 11) are potential routes to treat styrenics-rich waste. While advanced recycling approaches are currently in development and scale-up phases with implementation time scales stretching into the mid- and late 2020s, mechanical recycling is already commercially applied and widely available.

Generally speaking, greenhouse gas considerations mean that it is preferable to use mechanical recycling, as it keeps the polymer chains intact and uses significantly less energy for conversion into new products [11, 12] – see also Section 9.1. For example, in a life cycle analysis investigating the recycling of electronics waste to recycled polymers, Wäger et al. demonstrated the combined energy demand (and thus the greenhouse gas footprint) for such processes is six to ten times lower than using conventional plastics from fossil sources [12]. The quality requirements for the input material for mechanical recycling approaches are considered the highest in terms of purity and retained mechanical and rheological performance. The reason is the fact that the materials are only brought in the molten state and are re-extruded but otherwise used as they are when re-processed and prepared for the next life cycle. The presence of impurities is of especially high concern [13–15] and needs to be avoided by very laborious sorting and cleaning procedures depending on the intended end application. However, new sorting technologies are currently in development and should contribute to tackle such challenges on the way to a fully circular plastics economy [16, 17].

8.4.2.2 The Chemistry of PS and ABS during Production, Processing and During Their Life Cycles

In order to assess the mechanical recyclability of ABS and the potential limitations thereof, one needs to understand the chemistry of virgin PS and ABS as a starting point. Furthermore, a detailed look at the effects that occur in the melt during processing and reprocessing, as well as the ageing mechanisms that occur during the (in some cases) very long life times of the applications have to be analyzed. Based on such analysis, successful recycling processes can be designed, leading to high-performing materials. Nothing less is required to guarantee acceptance by a large range of customers for a hopefully circular approach, i.e., a real “drop-in” solution that will not be considered a downcycling solution. Hence, performance on par with the initial applications the material was recovered from needs to be the final goal of mechanical recycling offerings.

The base materials for PS and ABS are the styrene homopolymer PS (also known as general-purpose PS, GPPS) as well as styrene–acrylonitrile copolymer (SAN). Both are amorphous transparent materials. GPPS is colorless while SAN has a slightly yellowish base color stemming from neighboring acrylonitrile repeat units in the main chain polymer which can form cyclic, conjugated chromophores. Both polymers are produced in large quantities (multi megatonne scale), and by far the most popular synthetic route is free-radical polymerization.

The applications of GPPS and SAN can be widened significantly when these very stiff but brittle materials are impact-modified [18]. The most commonly used method involves the introduction of polybutadiene-based rubber phases. In case of PS, one typically uses the mass-polymerization approach in which PBR or SBR rubbers are dissolved in a styrene solution, with radical styrene polymerization leading to a graft reaction combining both phases, and producing the opaque, high-impact PS (HIPS). The rubber content is typically in the single-digit percentage range [19]. Similarly, one can also synthesize “mass-ABS” by carrying out the graft reaction on the PBR or SBR in the presence of a styrene–acrylonitrile mixture. Figure 8.63 (right) shows an example of such a material. It can be clearly seen that the dark PBR particles (due to staining in the TEM imaging process) also contain SAN (bright material) on the inside.

Notwithstanding the above, the process that is by far more frequently applied for ABS synthesis is so-called emulsion polymerization. Here, the PBR particles are prepared in a separate batch emulsion process. The nearly perfectly spherical particles obtained are grafted (core–shell structure) with SAN in order to make them compatible with the SAN phase, which is added by extrusion during the compounding of the emulsion ABS. Due to the refractive index mismatch of the rubber and the SAN phase, ABS is also an opaque material. The rubber phase contents within commercially available ABS grades is relatively broad, ranging from 5% to 30%

[20]. As a rule of thumb, ca. 20 wt% of ABS is made up of polybutadiene rubber (industry average). In order to tune the properties of ABS, the SAN constitution can be varied. The average molecular weight, as well as the relative ratio of SM and AN, have a significant influence on the performance of the ABS blend. While the former characteristic governs mainly the rheological behavior, the latter provides superior mechanical performance and chemical resistance if a higher AN content is chosen. However, due to the higher costs and the effects from a more yellowish base color, the AN contents are typically kept as low as the desired property profile allows. In a similar way to butadiene, the AN content in a typical ABS grade can be considered ca. 20% by weight, while the range in commercially available grades spans from ca. 15% to 30%. For any commercial ABS grade. The co-monomer ratios are thus chosen for a variety of reasons targeted at specific applications.

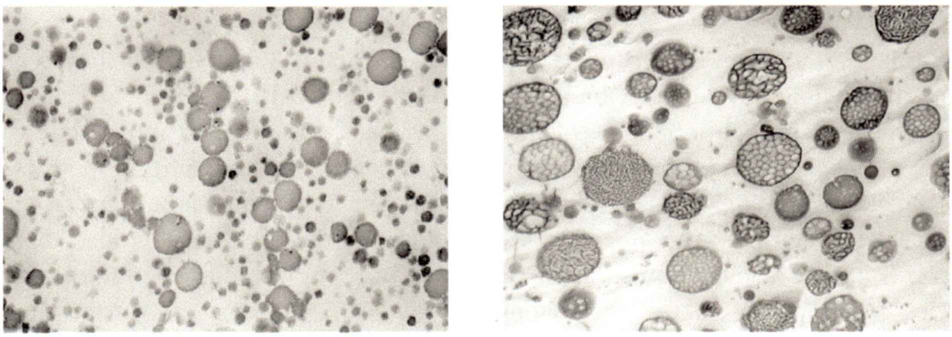

Figure 8.63 Left: TEM image of an emulsion-polymerized ABS. The bimodal size distribution of the spherical polybutadiene particles are visible as darker objects (due to staining). An SAN graft shell is applied to the particles in order to ensure good adhesion to the SAN hard phase (bright areas). Right: TEM image of an ABS produced using the "mass-polymerization" approach. Here, polybutadiene rubber is added to a mix of SM and AN and grafting occurs also inside the PBR particles

Based on the above-described composition ranges of the GPPS/HIPS and SAN/ABS couples, chemical intuition would lead to the suspicion that the polybutadiene moieties with their very reactive double-bonds would be the starting point for (photo) oxidative degradation. To a lesser extent (but inevitably becoming more pronounced over long time intervals) the PS and SAN polymer chains would also be expected to become subject to photooxidation, since the aromatic structure in the styrene repeat units absorb light in the UV range, with the expected consequences of reaction initiation. But beyond this intuition, numerous detailed investigations have been published in the literature. Several excellent summaries of these findings have recently been provided by Perrin and Signoret et al. [21], Fiorio et al. [22] and Quiles-Carillo and Balart et al. [23] in the introductory sections of their very interesting papers on the recyclability of styrenics. Although these findings are summarized below, it is recommended to study the full range of original citations given in those three publications.

As one would suspect, it is indeed the PB phase which is attacked during the initial stages of degradation [24–27]. Shimada et al. investigated the thermooxidative reaction in ABS at elevated temperatures (70–90 °C) [26] by IR spectroscopy. While the absorption bands associated with SAN did not show any change, the formation of hydroxyl and carbonyl groups was observed at the expense of the bands stemming from PB. Thus, it was concluded that the reaction involves the creation of free radicals and the formation of hydroperoxides. This process is then followed by the decay of those peroxides, triggering the formation of a variety of oxidation products and finally also cross-linking of the rubber phase. In a follow-up investigation, a comparable result was obtained for UV-induced oxidation: the PB phase suffers comparable oxidation while the SAN phase was largely unchanged under the given conditions [27].

Gardette et al. have carried out a similar investigation [28], in which they used relatively long-wavelength irradiation (300–400 nm) on HIPS and identified the species formed by titration. Next to the species expected from the rubber degradation, they also found conjugated, aromatic moieties which are mainly responsible for color formation and indicate that chain scission phenomena are occurring. They widened their work to ABS and SAN, and found that the attack on the rubber phase is the starting point for the degradation in HIPS and ABS. The radical species formed in the rubber degradation process also lead to matrix damages in a subsequent step (Figure 8.64). In pure SAN and PS the photooxidation starts more slowly, but this ultimately leads to chain scissions and conjugated (colored) species. In summary, thermal- or UV-induced stress causes molecular weight reduction and color formation in PS and SAN, leading to lower viscosities and slightly reduced mechanical properties. In combination with polybutadiene rubber, i. e., in HIPS and ABS, the rubber phase is attacked even more rapidly, causing embrittlement of the overall material and triggering an even greater acceleration and more pronounced decay of the matrix materials.

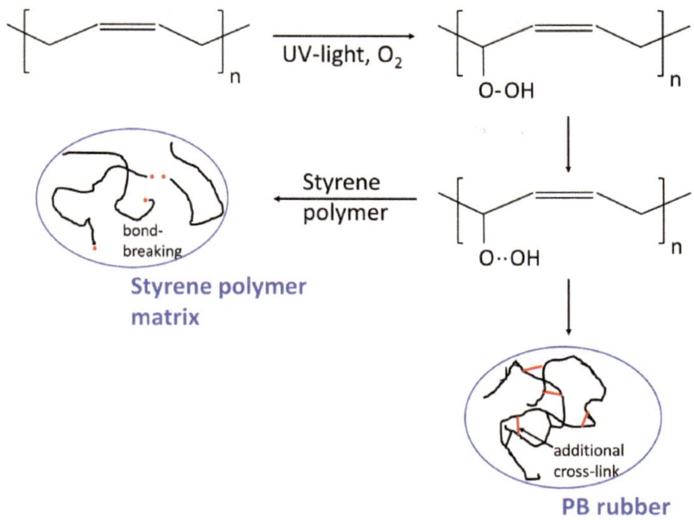

Figure 8.64 Initial attack on polybutadiene by oxygen leading to formation of oxidation products, rubber cross-linking, and subsequent attack on the PS or SAN matrix including color formation [21]

With this in mind, several authors have investigated if and how these effects of ageing can be influenced. One straightforward possibility is the pigmentation of the material: Davis and Gordon showed as early as in 1974 that the most damage is caused by irradiation below 340 nm while any wavelength above 450 nm is not very effective in triggering the decay mechanisms [29]. Hence, the effective absorption of irradiation by dyes or pigments absorbing below 450 nm will clearly stabilize styrenic materials - one reason why black material shows much lower signs of degradation, and partly the reason why so many WEEE applications are colored in dark tones. While the effect of purposely coloring polymeric materials is rather obvious, there are also reports in the literature investigating the effect of the discoloration observed during ageing itself. For instance, it was found during the investigation of design objects from the 1960s [30] that photooxidation only happens in the top 70 µm layer (approximately) due to a passivation effect. This can either be caused by subdued oxygen migration efficiency through the degraded layer, or the effective absorption of UV irradiation by the yellow layer on the surface caused by the ageing mechanisms themselves [31].

Several investigations have recently been published on the reprocessing of styrenics that were partially combined with ageing tests [32–37]. Karahaliou et al. found that multiple re-extrusions lead to distinct discoloration while the mechanical properties of ABS are relatively stable upon such "recycling" [33]. Rahimi et al. reported similar findings after five re-processing cycles, i.e., they described a slight decrease in Charpy impact performance and a reduced shrinkage. They attributed both effects to shortening of polymer chain lengths as well as crosslinking of the ABS rubber [36]. Hamarat et al. extended the methodology to 12 re-extrusion cycles, including the mixing of virgin material into already re-processed ABS. Overall, the results are well in line with the previous findings: the melt volume rate (MVR) increases, while the impact strength suffers (initially quickly, then more slowly with an increasing number of re-extrusions). The reasons remain chain scission in the matrix and cross-linking, as well as oxidation of the rubber phase [37]. Perez et al. also subjected ABS to multiple, in their case ten, re-extrusion cycles [34]. They added an investigation of accelerated UV ageing for the samples after one, five, and ten re-processing cycles. Their results seem to indicate a synergistic effect of the damage, as the effect of UV-ageing was more severe when the re-processing had already initiated the detrimental mechanism. This could hint at a depletion of stabilizers that are part of commercial ABS resins. An interesting combination of re-processing and heat ageing was reported by Salari et al. [32]. They re-extruded ABS five times, and observed a profound color change (yellowing), an increase in MVR (attributed to chain scission under the heat and shear stress in the extrusion process), as well as a significant drop in impact properties. Interestingly, the lowest impact loss came from the extrusions themselves. However, when Charpy specimens were heat-aged before testing, the drop was more significant, with the worst results being obtained when the Charpy bars were notched before ageing. This clearly supports the hypothesis that thermal ageing has a pronounced surface effect, as the Charpy test is highly notch-sensitive. Also, Boldizar et al. has reported on such a combination of multiple re-extrusions and heat ageing of specimens with overall very similar results [38].

Very recently, Perrin and Signoret et al. went to the extent of investigating re-processing as well as UV-ageing effects of PS, SAN, HIPS, and ABS in an excellent report [21]. They established that the butadiene rubber indeed initiates photooxidation, because differently photodegraded specimens of PS and SAN (varying in duration of photodegradation, and by the use of natural and artificial ageing) were significantly more stable than their impact-modified counterparts (HIPS and ABS, respectively). They also established that ABS is more stable than HIPS. During multiple re-extrusion cycles of HIPS and ABS, the authors found a significantly more pronounced coloration in the bulk, especially for previously photodegraded materials in both natural and artificial conditions. The mechanical performance clearly suffered from reprocessing but significantly more so for aged samples, which indicates that the oxidized species, which interestingly also differed between naturally and artificially aged samples, cause more harm during extrusion. They seem to act synergistically in the degradation processes. Especially in naturally aged samples, the discolorations and mechanical property loss was more pronounced. The authors speculate that this could be linked to a stronger but shallower oxidation within artificially aged materials due to the rapid formation of a protective layer. Also, the absence of longer wavelengths in the accelerated ageing could lead to less severe or different chemical alteration compared to natural conditions.

A different topic has recently been addressed by Noeli et al.: the recyclability of chrome-plated ABS [39, 40]. As will become evident below, one major challenge for a full applicability of recycled ABS is the color freedom. The various dyes and pigments used in colorful applications end up in the plastic waste, and for mechanical recycling a mixture of these colors is inevitable leading to gray tones on average. An interesting alternative to obtain non-colored or "naturally-colored" ABS would be chrome-plated materials, which represent a high-volume application for ABS. The authors successfully removed the metal layer by nitric acid, but the strong oxidative power of the acid also caused damage to the polymer, which upon reprocessing showed very strong discoloration (Figure 8.65). When switching the acid to lactic acid, the effect was strongly reduced but overall the color change was still very pronounced. The mechanical properties were impaired only slightly, which emphasizes that discoloration becomes evident even at very small ratios of affected material.

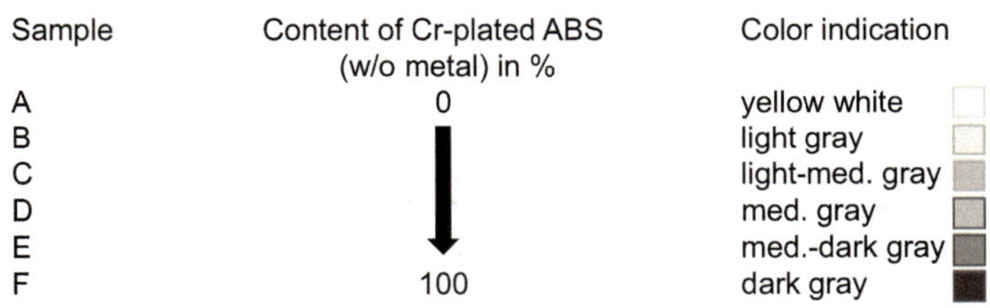

Figure 8.65 Specimen after reinjection of de-metallized ABS materials with increasing amount of acid-treated material. From (a) zero to (f) 100% [39]

In summary, both HIPS and ABS are polybutadiene-modified styrenic materials. Thermo- and photooxidative attack is initiated in the more susceptible rubber phase, where it leads to oxidation and cross-linking. As a result, the impact performance and ductility of the materials suffers over time. The PS matrix is less stable than the SAN matrix but, triggered by the rubber oxidation side-products, the matrix polymers are also damaged. The effects are discoloration and chain scission, which further reduces the mechanical performance of the materials. Depending on the life cycle of the applications and the conditions of use, a more or less pronounced property decay is to be expected. The considerations need to be taken into account when post-consumer mechanical recycling of HIPS and ABS are targeted.

8.4.2.3 Prospects, Prerequisites and Limitations of the Mechanical Recycling of Styrenics

The previous section showed that the underlying chemistry of the processes involved in the thermal ageing and photooxidation of styrenics, and the effects of re-processing them, are well-understood. It is clear that any pre-existing oxidative defect can catalyze further degradation and impair properties if no countermeasures are taken.

While the above-mentioned investigations were carried out on pure styrenics in most cases, it is clear that under industrial recycling conditions using real waste, one needs to consider much worse starting conditions. Inhomogeneous polymer mixtures that might result from imperfect sorting can be considered a significant issue due to the incompatibility of the polymer classes. Hence, proper sorting is the prerequisite to improve the starting point of the mechanical recycling task. One relatively minor issue in this context is the mixing or averaging of nearly all commercially available ABS or HIPS grades within the waste stream. The fact that further impurities and additives like flame-retardants, pigments, or fillers will inevitably be present in the waste stream makes successful mechanical recycling even more challenging. As was recently presented by Wagner et al. [13, 14], it will take several years until problematic additives that have long been banned from use in virgin materials will be depleted in the waste stream. This is simply a function of the long life time of some articles in the waste stream. In the following section, an attempt will be made to demonstrate which quality levels of mechanically recycled styrenic products can be achieved, in ascending order – mainly depending on the quality of the sorting step at the beginning of the recycling process.

Mechanical recycling of mixed plastic waste has existed for a long time. But applications for the more or less completely unsorted mixed plastic waste do not go much beyond the often-cited park benches. The incompatibilities of the different polymer families based on different polarities or rheologies make attempts to compatibilize such differences a very complex task, limiting the application range of such products significantly. Hence, the performance of such mixtures is far below the ones expected from any of the pure constituents. However, such efforts to make use of less well-sorted mixed plastic waste are clearly preferable compared to landfilling or incineration. Therefore, whenever no further sorting is available or commercially viable, such solutions are highly welcome.

When looking at the next level of sorting, i.e., the stage at which styrenics in general have been removed from the mixed waste fraction, a clear performance improvement becomes possible. Several reports have been published about the optimization of the ABS/HIPS system obtained mainly from WEEE [41–44]. The repeated investigation of this material pair is because both have a high abundance in the WEEE stream [7, 8]. As they have very similar densities, they can be separated from other plastics relatively easily by flow-sink techniques. ABS/HIPS blends show complex phase morphology dependency, as briefly illustrated in Figure 8.66. Both matrices (SAN for ABS, and PS for HIPS) can become the matrix or dispersed phase for the material pair ABS/HIPS, depending on the content of each polymer in the blend. It has also been reported that the shape and size of dispersed phases also depend on the system composition. Therefore, properties such as the Young's modulus (*E*) and other mechanical properties of the material are directly affected by composition of the polymer mixture. And because the proportion can vary with the incoming waste composition, the properties of such mechanically recycled styrenic blends can significantly vary over time.

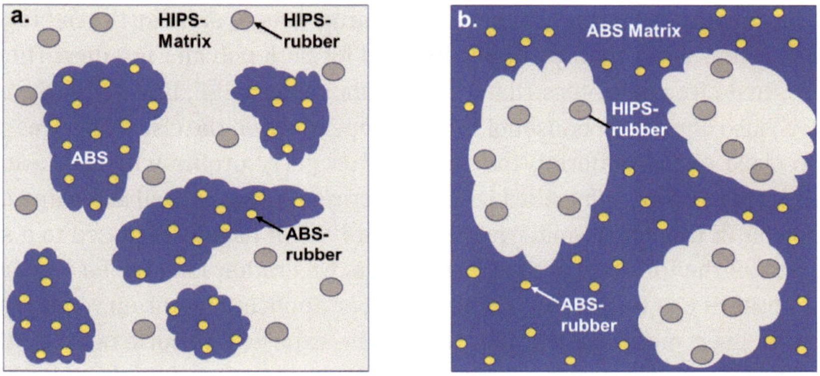

Figure 8.66 Differences in morphology of ABS/HIPS blends: (a) HIPS-rich blend; (b) ABS-rich blend [43]

It has also been attempted to add further polymers like SAN or SBS rubber, as reported by Vazquez et al. [41, 42]. The most significant improvement was obtained when low amounts of SBS was added to HIPS or when virgin SAN was added to ABS-rich mixtures. While these investigations prove that more-or-less acceptable performance can be obtained from PCR/HIPS or PCR/ABS blends, the list of applications for such materials remains relatively short. Nevertheless, these attempts are valid in order to make some use of the waste stream and to enable recycling of a larger segment of the styrenics waste, in the event that only (or primarily) manual sorting techniques are available.

From a technical perspective, the pairing of ABS and PVC is less challenging, as these polymers are almost homogeneously miscible. Furthermore, as PVC often appears in high quantities in certain WEEE streams, Jaidev et al. looked at such blends as a way to treat parts of the WEEE stream [45]. The authors reported that ABS/PVC blends showed decent mechanical performance, whereas HIPS/PVC mixtures were inferior due to immiscibility. The authors claim that blending ABS and PVC is industrially applicable, but the idea of creating such a blend system seems at least questionable for future circularity efforts. Such a blend would pollute the waste stream after the next life cycle, and is thus not considered advisable. As an intermediate conclusion, one could state that blending HIPS and ABS and potentially improving the blend performance by the addition of another styrenics-based compatibilizers like SBCs could be regarded as somewhat superior to the mixed plastic approach. Moreover, if applications for such blends do exist, there are only a limited number of arguments against the pursuit of such a route. One potential counter-argument could be that the already nearly pure material pairing should rather be separated a final time to enable an even better use. In any case, a large-scale, industry-wide commercial break-through is not expected for this approach.

Clearly more promising are applications of pure sorted PCR/PS and PCR/ABS fractions. Several recycling companies in Europe, Asia, and North America have achieved a very good quality and purity level for both polymer families. They typically originate from styrenics-rich waste streams like WEEE, ELV, or (in the case of PS/HIPS) also often from household waste. Depending on the circumstances during the life cycle of the material, the resulting property profile will vary a lot. The mechanisms of rubber phase oxidation and cross-linking as well as chain scission phenomena in the matrix materials SAN and PS will have progressed to a significant extent if the life span of the plastics parts was rather long (as for ELV) and/or if the exposure conditions to UV light (outdoor applications) or elevated temperatures have been more challenging. Hence, the expected performance profile will depend heavily on the waste source. From the experience of working with many sorters and recyclers in recent years, the current author (at INEOS Styrolution) can state that the performance of ABS stemming from ELV waste is relatively poor. The impact strength of ABS coming from the ELV waste stream can be as low as 3 kJ/m^2 and often does not exceed 5 kJ/m^2, which is merely above the performance of a virgin SAN. Hence, one can assume that the capability of the rubber phase to modify the impact properties of the system is nearly completely depleted due to oxidation and cross-linking. On the other hand, the quick exchange rate of modern electronic devices enables the separation of relatively high-performing ABS samples from WEEE, which can feature Charpy impact values of 8 kJ/m^2, and in some cases even double-digit readings after proper sorting. Such products have been actively marketed by the same sorting/recycling companies for a couple of years. The property profile of such post-consumer waste ABS allows for a good application field, and they clearly offer the best GHG footprint improvement [12]. Ford et al. have reported a case study proving that products with good performance can be designed using such commercially available 100% PCR ABS grades [46]. Many more commercially viable examples have been developed in recent years, with the number of players in the sorting/plastics waste recycling field increasing, and OEMs becoming more and more interested to apply such materials in mass production [47, 48].

Naturally, the same sorters and recyclers investigate how they can further improve the performance of such grades by the addition of relatively small amounts of compatibilizers, performance boosters, virgin ABS materials, or restabilization packages. There are a couple of interesting examples in the literature on how to best approach such challenges. For instance, Scaffaro et al. investigated mixtures of virgin and post-consumer recycled (PCR) ABS containing 0, 30, 50, 70, and 100% of the latter, and presented the full range of rheological and mechanical data. While the rheology is largely unaffected (despite the generally higher MVR of the PCR ABS due to chain scission effects caused by ageing), the mechanical properties, as expected, are lower for the recyclate mixtures. Interestingly, the mechanical properties depend relatively linearly on the PCR ABS content after the initial blending/compounding step. However, the authors also investigated two further re-processing cycles, and found that the performance drops significantly for all mixtures (Figure 8.67).

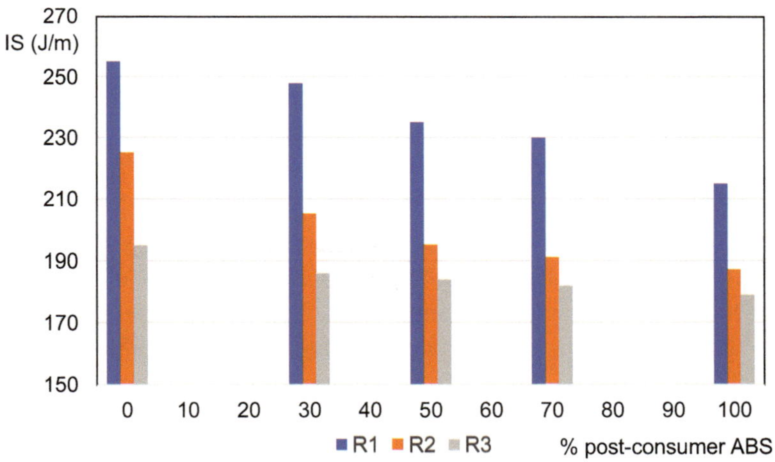

Figure 8.67 Impact strength of virgin/PCR ABS blends after one, two and three re-processing cycles [49]

In another report, Shu et al. have investigated the use of SBS rubber as a performance booster, but found this to be not very effective [50]. This result is in line with a previous report, which showed limited usefulness of SBS modification in ABS, although it did show improvements for HIPS [42]. However, when the SBS was partially epoxidized, the compatibilization performance increased significantly due to covalent bond formation during the extrusion process. It is well-known that such reactive extrusion effects, aiming at compatibilization, can be achieved for several polymer classes using (for instance) copolymers including maleic anhydride [51].

Yet another interesting study on the effect of adding various stabilizer packages to ABS was carried out by Fiorio et al. [22]. They investigated multiple re-processing and artificial ageing in the presence of various commercially available stabilizers, and showed that such stabilizers have clear positive effects on the recyclability and oxidation resistance of ABS. A similarly interesting investigation on the effect of re-processing on the stabilizers within ABS was published by Bai et al. [52], who concluded that the correct selection of stabilizers has a decisive influence on the material's performance level.

Several recycling companies specializing in sorting and providing styrenics have added such "performance-boosted" materials, which come relatively close to the specification ranges of commercially available ABS resins. However, it remains to be proven whether the knowledge about the optimum strategy (including the addition of virgin ABS, appropriate re-stabilization, or modification by reactive extrusion) will become fully available to recycling companies. Additionally, the resulting recipe costs might come close to or even exceed those of virgin ABS producers, as the backward integration is missing (see below).

The highest achievable performance level of a mechanically recycled ABS can only be a full recovery of the initial specification range of conventionally produced ABS. As has been shown by the various examples of investigations of multiple re-processing runs, ABS is fairly stable towards such stress, especially when the conditions are very gentle [35]. In that sense, the market for post-industrial recycled (PIR) ABS has already been established for a long time. Many injection-molding companies collect their material residue, shred it, and reuse it as part of their normal operation. A very prominent example of such permanent reuse comes from the dairy food packaging industry, who produce yoghurt pots by stamping HIPS sheets. After the first stamping run, ca. 70% of the material is converted to yoghurt pots, while the remaining 30% is sheet residue, which is reground and reused in sheet extrusion and stamping, with the 30% residue being returned to the cycle again. Pots made from such a process can therefore be considered to be "industrially recycled", and properly stabilized styrenics will show hardly any effect from such multiple cycles.

While not all plastics converters can directly reuse their processing scraps, they tend to at least collect and sell it to specialized companies who then offer post-industrial regranulates made from those. The performance of such materials should be nearly on par with virgin materials, as they only went through one or a few extrusion and/or injection-molding cycles with hardly any further heat or UV ageing exposure. It is clearly advisable to optimize this kind of recycling stream, as it makes the best use of the valuable material and prevents the formation of more waste. However, it will not solve the general post-consumer waste issue.

This being said, a real commercial breakthrough of PCR ABS seems to require a similar performance level to PIR or virgin ABS. Given the ageing characteristics of the butadiene rubber and the long life spans of ABS applications in electronic devices or even cars, a PCR ABS grade consisting fully of recyclate seems not to be a viable solution. However, a real "drop-in" solution seems desirable, as this could directly replace conventional ABS - at least from a performance point of view. Hence, it seemed advisable to develop PCR ABS grades that do not compromise on mechanical or rheological properties relative to its virgin equivalent, and can be used as such a "drop-in" solution. On the back of this paradigm, INEOS Styrolution developed the Terluran® ECO products with the aim to maximize the PCR ABS content while also providing a wide specification range overlap with standard conventional ABS grades. Following this aim, grades featuring 50 and 70% recycled content have been developed, which come close to the performance characteristics of Terluran® GP-22 (one of the most widely used ABS grades globally). Consequently, these grades are called Terluran® ECO GP-22 MR50 and Terluran® ECO GP-22 MR70. Table 8.6 shows that the MVR is slightly above that for Terluran® GP-22, while the Charpy impact strength of MR50 is only slightly below 20 kJ/m^2, as a production average over the last 1.5 years. In addition, the Vicat B50 and the Young's modulus (E) values are close to the virgin counterpart.

Table 8.6 Example Properties of Terluran® GP-22 ECO and its Conventional Counterpart

Grade (PCR share)	Young's modulus, E [MPa]	Notched Charpy [kJ/m^2]	MVR 220/10 [cm^3/10 min]	Vicat B/50 [°C]
Terluran® GP-22 (0%)	2300	22	19	96
Terluran® ECO GP-22 MR50 (50%)	2000	20	22	92
Terluran® ECO GP-22 MR70 (70%)	2000	19	20	92

The challenge to constantly achieve such performance levels lies in all components of the recipe, high-quality sorting of the PCR ABS input, as well as a well-designed compounding procedure. A combination of a very efficient ABS rubber concentrate and further additives and stabilizers ensures that the defects and decomposition products present in the PCR ABS do not impair the final material performance. Of course, these non-standard recipes come at a certain cost, but the experience of the last two years has proven that the Terluran® ECO grades can be offered at a price that is very close to the conventionally produced Terluran® in Europe. However, backward integration into the most decisive recipe components has clearly turned out to be necessary to render this project commercially successful.

Terluran® ECO GP-22 MR50 and MR70 are currently offered in standard black. However, more products with even higher performance profiles are currently being added to the Novodur® ECO specialty product line. These will include Novodur® P2H-AT (enhanced flow, general-purpose), Novodur® H801 (high heat tolerance, high impact strength), and Novodur® HH-106 (high heat tolerance, balanced property profile), which will also become available in a range of custom-colored variations. The efforts needed to reach such higher performance levels are of course more demanding, and require more sophisticated sorting and additivations. However, in these cases the projections also show that the products will reach prices comparable to those that customers are used to from the Novodur® product line. It seems feasible to achieve significant PCR ABS content levels of at least 50% while the target of offering absolute "drop-in" solutions remains unchanged. Once customers can reliably exchange PCR ABS for conventional materials and achieve significant greenhouse gas savings at very comparable costs, the activation energy to pursue such projects is at the same level as for standard projects aiming to qualify new materials for certain applications. With this in mind, one can be more optimistic about the economic success, giving rise to the expectation that an increasing share of the ABS waste can be extracted from suitable waste streams, and that a circular economy is a big step closer for ABS.

When considering the PS/HIPS material family, the situation is comparable to ABS if the WEEE stream is the source of the post-consumer waste. The properties of a PS fraction will have suffered significantly even if high purity is reached after sorting. Therefore, it seems a somewhat unrealistic goal to aim to fully recover the properties of conventional PS/HIPS. Also in this area, industry players are working on offerings that are expected to be commercialized in the near future. The challenge for PS mechanical recycling becomes even larger as a considerable fraction of PS applications will require food contact approval. Certain fractions of the household waste stream offer very valuable resources in this respect. For instance, yoghurt pots, which constitute a significant application of HIPS/PS mixtures, form a sizeable share of separately collected household plastic waste. These PS items can be sorted from household waste very effectively, and even today, purities >99% can be achieved in an economical manner applying modern NIR sorting technologies. Further enhancements of such sorting processes could employ artificial intelligence, which identifies and sorts out food contact items made from PS in a first step before the fast NIR technology then distinguishes between individual food contact items (if required) for improving the quality or color sorting. As the lifetime of a yoghurt pot is very short (a couple of weeks), and it is stored in the dark and under refrigerated conditions for most of the time, the mechanical and chemical properties of PS/HIPS used in this application will have hardly changed. In fact, reprocessed yoghurt pots are indistinguishable in their behavior compared to a commonly used conventional HIPS/PS 50:50 mix. Based on a "super-clean" technology similar to that established in the PET recycling industry in recent years, perfectly clean PS that even qualifies for new food-contact applications has been achieved by Styrenics Circular Solutions. As most yoghurt pots are (naturally) white, the recyclates are also available perfectly white. Colored yoghurt pots are sorted out and will be used for different food-contact applications. Other closed-loop recycling streams have similar high probabilities of success, and are currently in development by INEOS Styrolution. These examples prove that sorting purities of 99.9% are clearly feasible if the conditions are well-chosen, and that properties of recycled grades on par with the virgin materials are essential for commercial success.

8.4.2.4 Mechanically Recycled PCR ABS – An Economic Case Study

The advantage of mechanical recycling in terms of greenhouse gas impact compared to other recycling technologies is a clear driver to maximize its application. While this is true for all polymers that can be recovered from existing waste streams with justifiable efforts in terms of energy and cost, it is clearly particularly valid for ABS. The abundance of WEEE and its projected volume growth paired with the relatively high share of ABS in this particular waste stream have already led to several sorting companies engaging in the sorting of styrenics-rich fractions, and even pure PS/HIPS and ABS subfractions. The sorting costs depend on the applied sorting techniques and the process yield. The amount of sorting residue (the mixed waste fraction that is left after all separable PS/HIPS, ABS, and PP has been removed) is currently adding to the process costs, as it requires incineration to be disposed of. As a rough estimate for the costs, it should be fair to assume a similar price for a roughly sorted ABS fraction as for "contaminated PS waste" destined for microwave-assisted depolymerization, i.e., approx. €200/t. The costs of such recycled material are expected to come down in future with increasing economy of scale, improved sorting technology, and a potential future use of the residual fraction of mixed waste plastic as a valuable input material for pyrolysis or gasification, once these techniques are operational on an industrial scale. The latter cost factor could then be turned into a positive contributor of extra revenue for sorters/recyclers.

As described above, the production of high-quality mechanically recycled ABS materials – either in the 100% PCR content range or in materials aiming to compete with the properties of conventionally produced ABS (like Terluran® ECO) – cannot be compounded using "contaminated flakes". The required quality level for those applications reaches >96%, and in the best cases 98–99%. Hence, more intensive sorting efforts need to be performed to improve the purity, which will certainly add significant costs. But even if costs increase significantly, the economics are still reasonable when looked at from various angles, as described below.

Bottom-up view

The composite costs of conventionally produced ABS (60% styrene monomer, 20% butadiene, and 20% acrylonitrile) ranged between €750/t and €1310/t in the representative time interval of January 2019 to December 2020 (the data shown in Figure 8.68 were compiled using monthly contract prices of all three raw materials). The low point in May 2020 was reached at the peak of the Covid-19 crisis in Europe, when crude oil and base chemicals traded at long-term lows. ABS could therefore be produced in world-scale plants at very competitive costs, and producers were adding the required margin to arrive at the ABS market price (also shown in Figure 8.68). In times of balanced supply and demand, those curves run relatively equidistantly. Standard conventional ABS is a "primary product", i.e., it is extracted from the industrial process in a single step. Specialty ABS that is equipped with customer-specific additive packages or pigmentation is considerably more expensive due to the required "secondary extrusion step", the additivation costs as well as typically slightly higher margin expectations due to the customer-specific solution that is offered. In total, such cost components will add a relatively higher cost component, compared to the difference between a standard ABS price and the composite costs included in the latter.

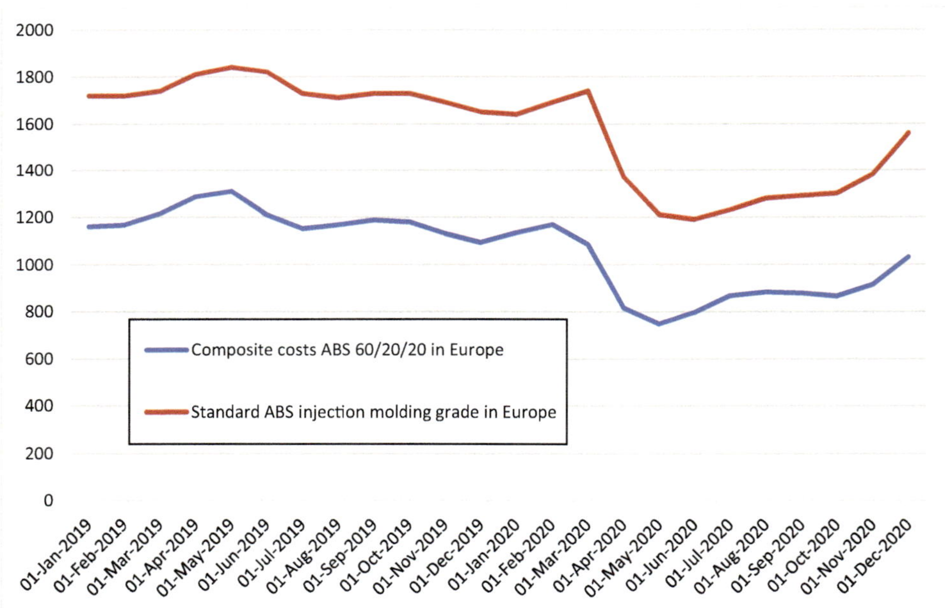

Figure 8.68 Composite costs for ABS and mean standard ABS sales prices in Europe from January 2019 to December 2020

When PCR ABS grades are produced, they require a compounding step comparable to a "secondary extrusion step" to get from the stage of PCR flakes to the finished compound using a twin-screw extruder or comparable equipment, plus the usual peripherals. Depending on the compound recipe, additives like stabilizers, processing aids, performance boosters, or virgin ABS must be added. While these recipe components might be relatively expensive for small- to medium-sized recyclers/ sorters who engage in the marketing of PCR ABS grades, it is fair to assume that backward-integrated, global ABS suppliers would add those recipe components at cost levels comparable to the composite cost mix of virgin ABS and those of typical additivation packages.

Even if somewhat higher-performing virgin ABS components and additives or higher relative amounts thereof would be required to boost the performance level of the PCR ABS flakes within the compound to conventional level, the total additional recipe costs should not increase in a prohibitively high manner. Consequently, one would need to assume that the acceptable price level for sufficiently purified PCR ABS flakes needs to be below the composite cost of conventional ABS if an equal sales price is targeted. The cost difference would need to be on the level of the cost difference between "primary" and "secondary" ABS operations which - cost of recipe components left aside - can be considered to be several hundred euros per tonne [51]. When taking the composite cost average for 2019 and 2020 of ca. €1060/t into account, a PCR ABS flake price clearly above a value of €200/t seems acceptable. For recyclers, who offer grades with inferior performance compared to conventionally produced ABS, the room to maneuver might be somewhat smaller, as they will most likely have to sell their products at a relative discount (although products with a completely comparable property profile should not face such constraints).

At the extreme low point of the composite costs around May 2020, when the sales prices of standard ABS were also at historically low levels, the acceptable PCR ABS flake price must have been significantly below the (at that time) valid level of €750/t. It has indeed been the case that under these circumstances, as well as in the following months of low conventional ABS sales prices, sales of PCR ABS materials have not been profitable. However, under normal, average market conditions, sorters and recyclers can run a profitable PCR ABS flake business. In addition, companies marketing mechanically recycled PCR ABS compounds at acceptable quality and performance level - or even levels comparable to virgin grades - should also find profitable conditions when engaging in this market.

Top-down view

If a standard ABS is considered the point of comparison for PCR ABS products, a look at the sales price range of those materials should be a valid starting point. As shown in Figure 8.64, the price of a standard injection-molding ABS grade moved between €1840/t in May 2019 and €1190/t in June 2020, considering the same period as above. The price average for this clearly relatively volatile episode is calculated at ca. €1575/t. Depending on (i) the costs of running a "secondary compounding step" for each market participant and (ii) the performance level of the offered PCR ABS relative to a standard virgin ABS product, the acceptable price level for PCR ABS flakes with a high enough purity can be extrapolated. For both product categories (PCR ABS with inferior properties as well as real "drop-in" solutions), the circularity aspect of a significantly lower greenhouse gas footprint can be assumed a real value for the customers. Hence, the "drop-in" solutions are usually offered with a slight increase in price compared to conventional materials. Only at times when the composite costs of virgin material are so low that the fossil-driven process is much more advantageous does the price increase for PCR ABS need to be significant to cover at least the "secondary costs". For PCR ABS grades with inferior properties, it is the performance difference and the applicability for certain end-products that will govern the achievable sales price. However, an average sales price for a standard ABS of ca. €1575/t does indicate that a very cost-efficient secondary operation will be required and a high-performing product will be needed. If both are given, a PCR ABS flake price significantly above the €200/t level for "contaminated PS" (or ABS in this case) will be supported by the economics. In volatile times when prices hit long-term lows, it will be essential to have a very favorable cost structure to run a PCR ABS product that is also profitable in times of crisis.

In conclusion, the costs required to convert mixed WEEE into first a styrenics-rich fraction and then into individual post-consumer PS and ABS fractions of high purity can be covered by well-equipped sorters and recyclers. Such businesses can be run in a profitable manner, and investments to increase the available volumes are foreseeable. However, it seems advisable to close the performance gap to conventional standard ABS in order to be able to charge prices on at least this level. Large ABS suppliers who have the benefit of size, backward-integration into recipe components, as well as a well-established channel to market seem equipped to offer attractive products at competitive prices while still staying profitable. The latter could be the case, even without virgin-plastic taxes or CO_2 certificates being brought into the equation to promote circular products, as could happen in the coming years. Well-designed mechanically recycled PCR ABS grades are available already today, and poised to gain significant market share in the coming years.

8.4.2.5 Conclusion

In conclusion, it is expected that mechanical recycling solutions will gain importance and market share for both PS and ABS in the 2020s. They have come and are here to stay [53]. Overall, mechanical recycling materials have the most favorable greenhouse gas footprint of all recycling technologies foreseen for styrenics. And very importantly, this technology is readily available now, and attractive product solutions are commercially available. The achievable qualities by sorting certain waste streams are suitable for a wide range of applications, and more (and better) qualities as well as application fields are going to be developed in the coming years. The economic foundation is already present, even without the implementation of virgin-material taxes or the introduction of CO_2 pricing mechanisms. The circularity is already proven from waste collection, sorting to achieve the required PCR flake qualities, and PCR PS and PCR ABS product performance.

However, also a lot still needs to be achieved. For instance, higher shares of the waste streams like WEEE, ELV, and household waste need to be properly sorted, by investment in sorting capacity and quality. A more favorable economy of scale at the sorting step would be required to help increase the profitability of the waste treatment step and to bring down prices of high-quality sorted plastic flakes. Of course, the compounders and ABS producers also need to become more innovative and make even more attractive offers accessible to plastic converters and OEMs, who at their end need to look deeper into the applicability of PCR ABS or PCR PS solutions based on mechanical recycling.

In summary, this value chain needs to become stronger by joint innovation approaches and intensified collaboration across the whole value chain, to make the best use of the available high-quality plastic waste. Then, the fraction of materials well-suited for mechanical recycling will lower the waste pile in the most effective manner, as it is fed back in a truly circular economy while the remaining, significantly smaller part of the waste will be subjected to more advanced recycling technologies that are described elsewhere in this book. Overall, styrenics are fit for mechanical recycling, and can contribute to a circular economy not only in their first life, but in many subsequent iterations of the lifecycle.

References for Section 8.4.2

[1] Nießner, N., Wagner, D.: *Practical Guide to Structures, Properties and Applications of Styrenic Polymers*, Smithers Rapra Technology, 2012

[2] Nießner, N., Jahnke, E.: *Kunststoffe*, (2020) 8, pp. 11–16

[3] Nießner, N., Weinkoetz, P., Villegas, M.: *Kunststoffe*, (2020) 8, pp. 24–28

[4] Robinson, B.H.: *Science of The Total Environment*, (2009) 408, pp. 183–191

[5] Bovea, M.D., Ibáñez-Forés, V., Pérez-Belis, V., Quemades-Beltrán, P.: *Waste Management*, (2016) 53, pp. 204–217

[6] Directive 2002/96/EC of the European Parliament and of the Council of January 27, 2003 on waste electrical and electronic equipment (WEEE)

[7] Maris, E., Botané, P., Wavrer, P., Froelich, D.: *Minerals Engineering*, (2015) 76, pp. 28–37

[8] Stenvall, E., Tostar, S., Boldizar, A.: *Waste Management*, (2013) 33, pp. 915–922

[9] Directive 2000/53/EC of the European Parliament and of the Council of September 18, 2000 on end-of life vehicles

[10] Vollmer, I., Jenks, M.J.F., Roelands, M.C.P., White, R.J., van Harmelen, T., de Wild, P., van der Laan, G.P., Meirer, F., Keurentjes, J.T.F., Weckhuysen, B.M.: *Angewandte Chemie, International Edition*, (2020) 59, pp. 15402–15423

[11] Schyns, Z.O.G., Shaver, M.P.: *Macromolecular Rapid Communications*, (2020), 2000415

[12] Wäger, P.A., Hischier, R.: *Science of the Total Environment*, (2015) 529, pp. 158–167

[13] Wagner, F., Peeters, J., De Keyzer, J., Duflou, J.,Dewulf, W.: *Polymer Engineering & Science*, (2018) 58, pp. 485–492

[14] Wagner, F., Peeters, J.R., De Keyzer, J.,Janssens, K., Duflou, J.R., Dewulf, W.: *Waste Management*, (2019) 96, pp. 206–214

[15] De Tandt, E., Demuytere, C., Van Asbroeck, E., Moerman, H., Mys, N., Vyncke, G., Delva, L., Vermeulen, A., Ragaert, P., De Meester, S., Ragaert, K.: *Waste Management*, (2021) 119, pp. 315–329

[16] Mastellone, M. L.: *Science of the Total Environment*, (2020) 718, pp. 137233

[17] Li, J., Li, C., Liao, Q., Xu, Z.: *Journal of Cleaner Production*, (2019) 213, pp. 838–844

[18] Han, Y., Lach, R., Grellmann, W.: *Die Angewandte Makromolekulare Chemie*, (1999) 270, pp. 13–21

[19] Alfarraj, A., Nauman, E. B.: *Polymer*, (2004) 45, pp. 8435–8442

[20] Wypych, G.: ABS poly(acrylonitrile-co-butadiene-co-styrene), in: *Handbook of Polymers*, Elsevier, 2016, pp. 5–11

[21] Signoret, C., Edo, M., Lafon, D., Caro-Bretelle, A.-S., Lopez-Cuesta, J.-M.: *Journal of Polymers and the Environment*, (2020) 28, pp. 2055–2077

[22] Fiorio, R., Villanueva, S., Diez, A. S., D'hooge, D. R., Cardon, L.: *Materials*, (2020) 13, p. 212

[23] Balart, R., Garcia-Sanoguera, D., Quiles-Carrillo, L., Montanes, N., Torres-Giner, S.: *Polymers*, (2019) 11, p. 281

[24] Adeniyi, J. B.: *European Polymer Journal*, (1984) 20, p. 291

[25] Adeniyi, J. B., Kolawole, E. G.: *European Polymer Journal*, (1984) 20, p.: 43

[26] Shimada, J., Kabuki, K.: *Journal of Applied Polymer Science*, (1968) 12, p. 655

[27] Shimada, J., Kabuki, K.: *Journal of Applied Polymer Science*, (1968) 12, p. 671

[28] Jouan, X., Gardette, J.-L.: *Journal of Polymer Science, Part A*, (1991) 29, pp. 685–696

[29] Davis, A., Gordon, D.: *Journal of Applied Polymer Science*, (1974) 18, pp. 1173–1179

[30] Saviello, D., Pouyet, E., Toniolo, L.: *Analytica Chimica Acta*, (2014) 843, pp. 59–72

[31] Mailhot, B., Gardette, J.-L.: *Polymer Degradation and Stability*, (1994) 44, pp. 237–247

[32] Salari, D., Ranjbar, H.: *Iranian Polymer Journal*, (2008) 17, pp. 599–610

[33] Karahaliou, E.-K., Tarantili, P. A.: *Polymer Engineering & Science*, (2009) 49, pp. 2269–2275

[34] Perez, J. M., Vilas, J. L., Laza, J. M., Arnaiz, S., Mijangos, F., Bilbao, E., Leon, L. M.: *Journal of Polymers and the Environment*, (2010) 18, pp. 71–78

[35] Peydro, M. A., Parres, F., Crespo, J. E., Navarro, R.: *Journal of Materials Processing Technology*, (20103) 213, pp. 1268–1283

[36] Rahimi, M., Esfahanianb, M., Moradi, M.: *Journal of Materials Processing Technology*, (2014) 214, pp. 2359–2365

[37] Hamarat, I., Kuram, E., Ozcelik, B.: *Proceedings of the Institution of Mechanical Engineers, Part E: Journal of Process Mechanical Engineering*, (2018) 232, pp. 449–458

[38] Boldizar, A., Möller, K.: *Polymer Degradation and Stability*, (2003) 81, pp. 359–366

[39] Kurek, A. P., Ranzan, T., Cardoso, I. N., Dotto, M. E. R., Balzer, P. S., Sellin, N.: *Journal of Applied Polymer Science*, (2020) 137, p. 48372

[40] Cardoso, I. N., Ranzan, T., Kurek, A. P., Sellin, N.: *Journal of Polymers and the Environment*, (2020) 28, pp. 826–833

[41] Vazquez, Y. V., Barbosa, S. E.: *Journal of Polymers and the Environment*, (2017) 25, pp. 903–912

[42] Vazquez, Y. V., Barbosa, S. E.: *Journal of Environmental Management*, (2018) 217, pp. 381–390

[43] Hirayama, D., Saron, C.: *Polymer*, (2018) 135, pp. 271–278

[44] Brennan, L. B., Isaac, D. H., Arnold, J. C.: *Journal of Applied Polymer Science*, (2002) 86, pp. 572–578

[45] Jaidev, K., Suresh, S. S., Gohatre, O. K., Biswal, M., Mohanty, S., Nayak, S. K.: *Waste Management & Research*, (2020) 38, pp. 312–321

[46] Ford, P., Fisher, J.: *Journal of Cleaner Production*, (2019) 236, p. 117490

[47] Laird, K.: *Sustainable Plastics*, 2020, *https://www.sustainableplastics.com/news/electrolux-takes-floor-100-circular-vacuum-cleaner-prototype*, accessed December 2021

[48] MGG Polymers, 2020, *https://mgg-polymers.com/de/news/blog/nachhaltig-gruener-kunststoff-in-neuen-elektro-und-elektronikgeraeten*, accessed December 2021

[49] Scaffaro, R., Botta, L., Di Benedetto, G.: *European Polymer Journal*, (2012) 48, pp. 637–648

[50] Shu, L., Li, Y., Li, J., Wang, W., Vupputuri, S., Yang, D., Dong, B., Cao, N., Pan, D., Das, R., Guo, Z.: *Macromolecular Materials and Engineering*, (2020) 305, p. 2000284

[51] Internal knowledge, INEOS Styrolution

[52] Bai, X., Stein, B.K., Smith,K., Isaac, D.H.: *Progress in Rubber, Plastics and Recycling Technology*, (2012) 28, pp. 1–14

[53] Bucknall, D.G.: *Philosophical Transactions of the Royal Society A*, (2019) 378, p. 0268

8.4.3 Mechanical Recycling of Expandable Polystyrene (EPS)

Norbert Nießner

EPS (expandable polystyrene) has been known for almost 80 years. While foamed polystyrene was first reported in 1935 by Munters and Tandberg [1], Fritz Stastny of BASF developed a polystyrene bead foam (tradename Styropor®), which is the foundation of today's EPS [2]. It was introduced at the Düsseldorf K-Fair in 1952 [2]. EPS is based on polystyrene beads that contain approximately 5% of a low boiling hydrocarbon, such as pentane. The pentane-containing beads are typically produced by aqueous suspension polymerization, or by extrusion [2]. In a subsequent step, the beads are subjected to heat, typically in a "steam-chest mold" where the pentane evaporates and at the same time expands the beads [3]. Applications are, for example, insulation panels for building and construction, often containing halogenated flame retardants for safety reasons [3]. Other applications for EPS range from fish boxes to moisture-protecting "chips" for packaging to additives for concrete [4].

EPS is a lightweight insulation material – it saves fuel during transportation and therefore reduces climate-relevant CO_2 emissions. Due to its high insulation performance, EPS makes a valuable contribution towards energy saving, especially for refrigerated containers or even for whole buildings [4].

EPS can be recycled several times without significant deterioration of its polymer properties and can be reused in a large variety of applications. As such, EPS helps us to protect our climate.

Typical ways of recycling are mechanical recycling to PS and EPS as well as energy recovery. Mechanical recycling usually starts with the grinding of the EPS waste. Several alternatives for further valorizing of the ground waste are available afterwards:[4].

- *Recycling as EPS*: The waste material can be used again for typical EPS applications. The most common EPS recycling pathways are:
  - the use of ground EPS for the production of new packaging for household goods and electric/electronic applications. A special case of EPS mechanical recycling are fish boxes [9]: Fish boxes, consisting of 100% EPS are widely recycled at scale all across Europe. They are designed for recycling and easy to collect and sort at big collection sites (fish processors, markets, harbors), and are then mechanically recycled, the preferred treatment option, because it has the lowest environmental impact. Norway, Denmark, the Netherlands, and Greece have already achieved fish box recycling rates of 90%, while a number of other countries are not far behind, including Ireland, Portugal (75%), and the UK (70%).
  - the use as insulation improvers for construction materials like bricks, concrete, cement), and the use as material for soil improvement (e.g. drainage, substrate for plants).
- *Recycling as PS*: in this process, the molten or ground EPS is converted to compact granules of polystyrene. This recycled product can be injection molded or extruded to finished or semi-finished goods, or it can be used (after extrusion and degassing) to produce EPS once again.

In 1992 EPS organizations from Asia (AMEPS), Europe (EUMEPS), and North America (EPS-IA) formed the International EPS Alliance (INEPSA) to enhance global EPS recycling and environmental stewardship. Thirty-one countries around the world have signed the landmark International Recycling Agreement, pledging their commitment to global EPS recycling [7].

To summarize: EPS can be easily recycled according to different pathways. Nevertheless new routes are under investigation to further boost the EPS (and XPS) recycling quota. Two examples are given below:

- The *EPS-Sure Project* is an industry consortium project, with the aim to demonstrate the feasibility of manufacturing recycled polystyrene (rPS) with high quality to satisfy the demanding requirements for applications in food contact [5]. The project aims to:
  - define a sustainable model to collect, transport, and recycle EPS fish boxes, so they can be converted into PS pellets for the manufacture of food contact packaging,
  - implement procedures and good practices for EPS fish boxes waste management: collection logistics, storage, pretreatment (compaction, briquetting, washing), and recycling,
  - set up a demonstrative pilot plant in order to produce PS pellets from recycled EPS fish boxes waste and to produce PS sheet containing recycled PS.

- The PolyStyreneLoop project involves the setting up of a large-scale demo plant as a closed-loop solution for the recycling of polystyrene (PS) insulation foam waste. The demonstration plant in Terneuzen, Netherlands, has a capacity to recycle 3300 tons per year of PS-foam demolition waste containing HBCD. The plant is supposed to produce approximately 3000 tons of Loop-PS recyclate to be used for the production of new insulation material. The HBCD removed during the solvent-based purification process is treated by the Bromine Recovery Unit (BRU) of a third party located next to the PolyStyreneLoop plant [6]. The owner Polystyrene Loop B.V. fell into bankruptcy on March 8th, 2022. After a consortium of German companies had purchased the Dutch plant from the bankruptcy, it was re-opened on July 22nd, 2022.

References for Section 8.4.3

[1] Munters, C. G., Tandberg, J. G.: US-B 2,023,204, 1935

[2] Scheirs, J., Priddy, D. (Eds.) in: *Modern Styrenic Polymers*, John Wiley & Sons, 2001, p. 13 ff.

[3] Gausepohl, H., Gellert, R. (Eds.) in: *Kunststoff Handbuch 4 – Polystyrol*, Carl Hanser Verlag, Munich, 1996, p. 567 ff.

[4] *https://eumeps-powerparts.eu/*, accessed April 5, 2022

[5] *http://www.life-eps-sure.com/en/the-project*, accessed November 19, 2021

[6] *https://polystyreneloop.eu/*, accessed April 5, 2022

[7] *https://epsrecycling.org/*, accessed April 5, 2022

[8] Brandrup, J., Bittner, M., Menges, G., Michaeli, W. in: *Recycling and Recovery of Plastics*, Carl Hanser Verlag, Munich, 1995, p. 635 ff.

[9] *https://eumeps-powerparts.eu/recycling/fishboxes*, accessed April 5, 2022

8.5 Mechanical Recycling of Engineering Thermoplastics

Hannah Mangold, Caroline Beyer

Mechanical recycling is a polymer-to-polymer recycling, while chemical recycling converts polymers to the respective monomers or lower molecular weight products. Mechanical recycling is preferred if the starting material has a high purity and is a mono-material, as the energy input is usually lower compared to chemical recycling pathways. There are several possibilities for mechanical recycling that range from melting of the polymer to obtain a recycled material with similar quality as a virgin polymer, to downcycling to another application with lower requirements [1].

Engineering thermoplastics are often not suitable for mechanical recycling due to the very high requirements on their quality, e.g. in terms of mechanical strength and temperature resistance, and the degradation of the polymers which occurs in further processing steps. Also, engineering plastics often contain many additives to improve their properties and are tailored for specific applications. Moreover, mixed or contaminated plastic waste is a huge challenge. For many waste streams, such as end-of-life vehicles (ELV) or waste from electrical and electronic equipment (WEEE), all materials are shredded together and besides valuable metals, plastics is often seen as an undesired by-product. Therefore, dedicated sorting technologies for the polymeric materials are needed, as well as recycling processes that can deal with a certain amount of contamination.

Due to the above-mentioned restrictions, a lot of techniques for mechanical recycling take advantage of the solubility of one material in a solvent to separate the polymer of interest from additives or other admixed polymers. This allows for the separation of materials on the polymer level in more complex mixtures.

Further techniques that are important in the mechanical recycling of engineering plastics relate to the use of new additives in the recycled products that compensate for the negative effect of the recycling process. Examples for these additives are stabilizers against further oxidative or photo-oxidative damage, compatibilizers or adhesion promoters that improve the mechanical properties in contaminated or mixed materials, and chain extenders or other cross-linking agents that counterbalance the change in mechanical properties due to the reduction in molecular weight during the recycling process.

8.5.1 Market Size and Recycling Challenge

End-of-life vehicles are an important source of plastic waste, as about 25% of a car consists of plastics. Several engineering plastics are part of these 25%: polyamides (PAs) are found, for example, in the hub cap and the motor cover, polyurethanes (PUs) are used in the seats, polycarbonates (PCs) are found in outer parts of the car body and headlamps, polybutylene terephthalate (PBT) is found in electronic housing, plugs, and bumper parts, and polyoxymethylene (POM) is found in bearing components and connectors. All polymers are shredded and are therefore mixed in the automotive shredder residue/fraction (ASR). An additional challenge for obtaining automotive waste material lies in the long use time and prevalent export of cars – for example only about one-sixth of the cars produced in Germany remain in the country, while the rest is exported to other countries.

A quickly growing waste stream is waste from electrical and electronic equipment (WEEE). The EU has imposed a directive of collecting this waste stream, with a collection target of 65% in 2019 of the total weight of EEE put on the market. The plastics used in EEE comprises about 13% of the total consumption of plastics in Germany [2]. The level of plastics content in WEEE is quite diverse, ranging from 25% in a computer to 64% in a coffee machine [3]. Moreover, the types of polymers that are used are also highly mixed, with polypropylene (PP), high-impact PS (HIPS), acrylonitrile-butadiene-styrene (ABS), polyurethane (PU), and polycarbonate (PC) taking up the largest volumes [4]. An additional problem is caused by the use of brominated flame retardants, which make mechanical recycling difficult.

About 18 Mt of polyurethanes are produced every year world-wide, which means it is the 6th most used polymer [5]. In Germany alone, 930 kt were produced in 2019 [2]. The main use is for buildings, vehicles, and furniture. Most people spend a significant time of their day actually on polyurethane, as mattresses and car seats are made using polyurethane foam. The European Bedding Association states that up to 30 million mattresses are disposed of every year, and they estimate that 60% end up in landfill and 40% are incinerated [6]. This shows a great potential for recycling options.

Recycling of polyurethanes from applications in buildings is partly hampered by their long use time. Plastics use in buildings has increased significantly over the last 40 years, but with a usage time of (for example) about 40 years for windows and 80 years for plastic piping, the production of new material and the availability of waste material diverge significantly.

The application of polyamides is also diverse. Worldwide, about 8 Mt is produced annually, of which about 354 kt is produced in Germany [2]. The volume share of PA6 is slightly higher than for PA66, while polyphthalamides (PPAs) are for more specialized applications and their volume is significantly smaller. For both PA6 and PA66 the most important applications are textile fibers, including carpets, industrial fibers, and automotive parts, e.g. glass-fiber reinforced PA66. PA6 is also used in multi-layer film packaging, together with other plastics such as PE, PP, or PET.

In general, sorting of plastic waste is crucial before mechanical recycling can take place. This can be done, for example, by NIR spectroscopy [7]. TrinamiX has developed a spectrometer that can not only distinguish common plastic household waste such as PE, PP, and PET, but also engineering plastics such as PA6 and PA66 [8]. However, further development is necessary as (for example) black plastics cannot be sorted with NIR technology.

Recent initiatives like the "Digital Watermarks HolyGrail 2.0" project [9] (of which more than 120 companies and organizations across the value chain of the packaging industry are members) aims to simplify sorting of end-of-life consumer goods packaging, and hence contribute to higher recycling quotas. The digital watermarks have the format of a postage stamp, are invisible to the human eye, and can include a variety of information such as manufacturer, material composition, and food or non-food application. This information can be analyzed by a high-resolution camera during sorting and contribute to more accurately sorted waste streams that simplify recycling. The initiative is planning an industrial pilot project for feasibility studies.

8.5.2 Major Applications and Recycling Process

8.5.2.1 Polyamides

Mechanical recycling opportunities of polyamides are manifold. The choice of an appropriate recycling strategy depends on the composition of the waste source, the associated materials (e.g. other polymers, fine particles) and hence the difficulty of separation, as well as the target re-use application. Major techniques include extraction in an appropriate solvent and separation of associated materials without PA degradation or melt extrusion. The following two sub-sections describe example procedures for the major polyamide 6 and polyamide 66 fields of application. Some described processes are suitable for both PA6 and PA66, which is why reading both sub-sections is recommended.

8.5.2.1.1 Polyamide 6

Multilayer Films

APK AG commercialized their solvent based Newcycling® process for the recycling of mixed plastic wastes stemming from packaging applications. Their process allows the re-use of recycled materials in the original application. While the use of post-industrial waste is well understood, in 2021 APK started to investigate the suitability of post-consumer wastes [10]. In their process, the plastic waste is first shredded, and associated materials removed. Solvent treatment dissolves one of the compounds, so that the polymers can be separated in a multi-step procedure and further processed to granules [11]. The optimum solvent for polyethylene dissolution is a mixture of alkanes, isooctane, and cycloalkanes [12]. This procedure allows the recycling of both PA6 (Mersamid® PA-6 NCY) and LDPE (Mersalen® LDPE NCY) from multilayer films [13].

Recently, BASF SE demonstrated the regranulation of PA6/PE multilayer films to monolayer films. Multilayer films with a PA content <10% were intermixable with PE without any additive, while the treatment of films with higher PA concentrations was demonstrated with the help of modifiers and the application of appropriate converting conditions [14].

Textiles

Spandex is a common additive in textiles to improve elasticity, giving the well-known stretching effect. It is a polyurethane containing poly(tetramethylene ether) glycol (PTMEG or polyTHF). For recycling of textiles in which PA6 or PA66 is the major compound, the admixture with spandex limits re-use applications. Aquafil, an Italian synthetic fiber producer, has described a method of removing spandex from PA6 or PA66 that takes advantage of spandex's thermal instability. Their process involves thermal treatment at 150–220 °C (for PA6) or 150–260 °C (for PA66) under inert gas in an autoclave to degrade spandex. This is followed by several washing steps with polar solvents such as ethanol to wash off spandex-related compounds. Solvent removal gives the purified polyamide compound, which can be further recycled. The used solvent can also be recycled [15].

The recycling of PA containing textiles mixed with cotton has been described by the Research Institute for Textile Chemistry and Physics at the University of Innsbruck. The application of a mixture of water, ethanol, and calcium chloride allows selective dissolution and hence separation of polyamide, which is received as a powder for further re-use, e. g. in spinning processes to form new fibers. The process is currently being evaluated in the lab phase, and an expansion to a pilot scale is planned within the next four years [16].

Carpets

The separation of PA6 or PA66 in carpets from polyolefins like polypropylene is possible by application of a polyamide-selective cyclic ester solvent at temperatures above 165 °C. The solvent must be capable of dissolving or suspending the polyamide at temperatures above the melting temperature of the polyolefin to form two immiscible phases that allow polymer separation. The process is usually operated at ambient pressure, so that high-boiling solvents such as ethylene carbonate, propylene carbonate, or butylene carbonate (or mixtures thereof) are preferred solvents. The resulting polyamide is claimed to be suitable for any further application [17, 18].

Another polyamide-specific solubility strategy involves the application of a caprolactam–water mixture. In a temperature-dependent manner, PA6 is dissolved from post-industrial or post-consumer carpet mixtures at temperatures between 120 and 180 °C, while higher temperatures of 180 to 240 °C enable PA66 dissolution. Evaporation of caprolactam and water or the precipitation of the polymer give the solidified polyamide, which can be used in molding applications or in carpet fiber manufacturing [19].

Engineering Plastics

The re-use of extruded PA6 parts (e. g. gears, bars) in carbon fiber composites has been demonstrated by Romanian researchers, who compared the physical properties of recycled and virgin PA6 applied in layers on carbon fiber composites. After dissolving PA6 in formic acid, the material was applied to the carbon fiber composites. The mechanical properties of recycled and virgin PA6 showed insignificant differences, therefore the application of recycled PA6 seems reasonable [20].

German researchers have demonstrated the regranulation of cast polyamide waste. The choice of additive (lubricant, oxidant) modifies physical properties of the regranulate, which extends the possible applications. Research is ongoing to further explore these opportunities [21].

Fibers

DuPont has described a general procedure for the recovery of aliphatic polyamides (PA6, PA66) from admixed materials by dissolving the polyamide at elevated temperatures in polar solvents such as anhydrous polyols (e.g. ethylene or propylene glycol) or aliphatic carboxylic acids (e.g. acetic or propionic acid). After dissolving at elevated temperatures, insoluble materials are filtered off. The polyamide is then precipitated by the addition of another aliquot of the same solvent at lower temperatures, which gives a mixing temperature that is below the solubility temperature of the polyamide. Taking advantage of differing solubilities of PA6 and PA66, their separation from a polyamide mixture is possible by careful adjustment of the quenching temperature. The use of the same solvent for dissolving and quenching is an elegant way to simplify solvent recycling at low costs. However, careful handling is required, as these solvents are reactive and can therefore alter the polymer's molecular structure. The procedure is described to work for polyamide separation from carpet backings, reinforced fibers (glass fibers) and other associated polymers [22].

8.5.2.1.2 Polyamide 66

Textile Fibers Including Carpets

For the recycling of polyamide materials doped with fine particles such as TiO_2, DuPont's solvent-based procedure described above [22] is not applicable, as the polyamide's viscosity is too high to allow the filtration that would enable TiO_2 removal. In order to get access to recycled polyamide from such post-industrial or post-consumer waste streams, DuPont uses a solvent to dissolve and partially depolymerize the polyamide material in the solvent. They aim at lowering the average molecular weight to less than 90% of its initial value, in order to decrease viscosity to an extent that filtration and hence removal of fine, insoluble particles becomes possible. After the purification, polyamide is recovered from the solvent, and to illustrate the potential of this approach, repolymerization to a higher viscosity is also described. An example of a starting material is a composition of 92% polyamide 66 with a mixture of latex and polypropylene, as found in carpets [23].

Described examples for re-use of PA66 material stemming from waste carpets are various vehicle parts, where the polyamide material is processed with glass fibers, impact modifiers, and further additives to meet the required properties [24, 25].

Industrial Fibers

Airbags are a typical PA66 application with high quality requirements. Their production comes along with significant amounts of waste that is highly attractive for mechanical recycling.

Hyundai Motors has described the recycling of silicon-rubber-coated airbag waste. They first swell the silicon rubber by treatment with aqueous fluoroboric acid and then treat it with propanediol to separate the silicon rubber from the PA66. The leftover PA66 is then melted and extruded to give high-quality PA66 [26].

Airbus has described the re-processing of vacuum bags (used, for example, in aircraft part production) into filaments for fused filament fabrication, or powders for 3D printing processes with application in aircraft part production. The procedure involves simple grinding and extrusion of material that passed quality control [27].

Rhodia Operations uses air bag wastes as an ingredient together with polyamide, reinforcing or bulking fillers, and further typical additives (e.g. stabilizers, antioxidants, dyes, and pigments) as a feedstock for molding or extrusion processes, with products including technical plastics, (mono-)filaments, yarns, and fibers [28]. The patent was acquired by Polytechnyl (Solvay), and now belongs to Domo Chemicals. They promote these glass-fiber reinforced recyclates under the brand name Technyl® 4earth® [29], which is for example applied by the German high-pressure cleaning equipment manufacturer Kärcher in the production of spray lances for their high-pressure cleaners [30].

Engineering Plastics

As well as the mechanical recycling of airbag wastes, Hyundai Motors has also described the mechanical recycling of selected waste vehicle parts (radiator fan, radiator, end tank, and engine cover) by grinding, washing with a microbial detergent aqueous solution, and mixing of the parts in a defined ratio, followed by chain extension reactions to ensure mechanical properties close to virgin PA66 [31]. The latter steps are conducted in the melt of recycled polymer, followed by extrusion molding to give the final product. Hyundai Motors further described the use of the above-mentioned recycled PA66 in radiator shrouds to meet the requirements of the EU's End-of-Life Vehicle Directive [32].

Recently, Italian researchers showed that mechanical properties of glass-fiber-reinforced end-of-life automotive parts deteriorate as the glass fiber length is reduced. Even though a reuse in the same application is therefore not possible, the recyclate could be used in other suitable applications in the automotive sector, in order to comply with the EU Directive [33].

8.5.2.2 Polyurethanes

For the mechanical recycling of polyurethanes, as a first step sorting and subsequently crushing, shredding, or grinding of PU waste is necessary. This creates a feedstock that can be used for further processing steps, such as rebonding, molding, or regrinding, which are necessary to obtain a material that can be re-used. For pure PU waste streams mechanical recycling can have the advantage of low cost and relatively simple realization, but the mechanical performance of the recovered products is often lower than the virgin material [34]. Therefore, either new applications need to be identified for the recovered products, or virgin and recycled polymers need to be mixed to get the desired properties.

8.5.2.2.1 Rebonding with Adhesives

In the rebonding process the PU is first shredded into small flakes and then mixed with an adhesive, usually an isocyanate in the range of 5–10 wt% with respect to the shredded PU. This mixture is subjected to a high-temperature water stream to melt the PU, which is then compressed. PU foam from vehicles or buildings can be used in this process, but flame retardants might be a problem [35]. Dow Chemical published a patent on that process as early as 1992, in which they describe the recycling of a flexible foam by curing polymer shreds with a binder, with improved tear resistance being obtained in the recycled product [36].

8.5.2.2.2 Molding and Extrusion

Molding and extrusion processes are based on the thermoplastic nature of polyurethanes. Here, a suitable temperature and pressure, depending on the structure of the PU, is needed to melt the waste material. Without using an adhesive, this waste material can then bond together in its molten state. Often the properties of the recycled materials are inferior, and therefore closed-loop recycling is not possible [37]. Melt extrusion with a twin screw extruder at 230 °C was used by Kunchimon et al., who recycled a mix of PA6 and PU from textiles to give hybrid fibers [38]. The fiber strength was decreased with respect to PA6 fibers due to the formation of pores in the structure, but increased compared to PU.

The influence of hot-pressing and melt-mixing on the chemical and physical properties of polyurethanes was studied in detail by Lee et al. [39]. With an increasing number of melt mixing cycles the number average molecular weight (M_n) decreased while the polydispersity index (PDI) increased. This led to a decrease in tensile strength and elongation at break.

A detailed study of the effect of recycling by extrusion and regranulation on the property changes of thermoplastic PU was carried out by Wölfel et al. [40]. They observed an exponential decrease in molar mass in the initial recycling runs until it reached about 40 kg/mol, resulting in the material becoming brittle. They also used the material to melt-blow nonwoven fabrics. The properties were degraded compared to the virgin material, but by using a mix of recycled and virgin material, it might be possible to obtain a material with tailored properties.

8.5.2.2.3 Regrinding and Powdering

Regrinding and powdering can be used for PU foams if the waste material is ground into a fine powder with a particle size of (usually) less than 200 µm. Various processes can be used for the grinding, such as two-roll milling, cryogenic grinding, pellet milling, or precision knife cutting [41]. This rigid-foam powder can then be mixed with a polyether polyol and an isocyanate to make new polyurethanes that can be re-used (for example) in car seat foams [42]. The powder can also be used as a filler in new PU products, as its chemical structure is very similar and therefore mixing is easily possible. Up to 20% of filler can be used without negatively affecting mechanical properties. It acts as a damping agent and can be used for absorbing energy or sound insulation [34].

8.5.2.2.4 Solvent-Based Recycling

The process of dissolution and precipitation could also be used for polyurethanes, although there is no commercial process using this method. A recent patent by Hebei University describes the subsequent steps of crushing and then dissolving the PU in a low-boiling solvent that can be recovered and reused at low energy costs. Example solvents are chloroform, acetone, ethyl acetate, and acetic acid, with varying ratios of solvent to PU of 0.5–100 [43].

8.5.2.3 Other Engineering Plastics

The recyclability of a fiber-reinforced poly(butylene terephthalate) (PBT) was shown by researchers from Ford Motors in 1996. They added a silane to the reground composites and could afterwards use different molding techniques to obtain a composite with mechanical properties as good as virgin short-fiber composites [44]. An extensive review of the recycling of fiber-reinforced thermoplastics was carried out by Pegoretti, who describes the techniques used for the mechanical recycling of composites and the effect on the fibers; examples for PBT, various polyamides, and poly(ether ether ketone) (PEEK) are given [45].

The reprocessability of PEEK was studied by McLauchlin et al. [46]. They found that PEEK can be re-processed by an injection molding process after a grinding step. At least three cycles of molding, then regrinding, then re-molding were possible without deterioration of the tensile strength, as the degree of crystallinity of the polymer (studied by DSC measurements) remained intact. However, chain scission was a problem, and led to a reduction in elongation at break after three cycles.

Yu et al. investigated the mechanical properties of recycled polysulfone (PSU) plastics [47]. Recycling was done by melting and extrusion of wasted nonwoven PSU material. They showed that the tensile properties remained similar, while the impact strength was inferior in the recycled product. The thermal stability and long-term durability of the recycled material decreased, as the molecular structure is slightly degraded during the recycling process. Similar results were found by Li et al. [48], who investigated the effect of hydrothermal aging on recycled post-industrial PSU. Recycling was done by washing, drying, melting, and subsequent extrusion and compounding. They found that stress relaxation and creep resistance were inferior in the recycled product due to shorter molecular chain lengths, while the static tensile properties remained similar.

Repeated grinding and injection molding was used for recycling of a ternary blend of PBT, PC, and ABS in a publication by Kuram et al. [49]. They studied the impact of recycling on mechanical, chemical, and rheological properties of the materials with and without glass fibers, and found that impact strength increased with increasing recycling number, while tensile strength decreased.

8.5.3 Major Applications of Recycled Engineering Plastics

As described above, mechanical recycling in a closed-loop system is often difficult for engineering plastics, as they usually contain a manifold of additives and are designed for specialized applications. The recycled product often exhibits diminished physical properties and cannot be used in high-performance or food-contact applications. However, the recycled parts can be used in different applications where (for example) quality requirements are lower, or where they can be mixed with virgin polymers to give new plastics with the desired properties. An example of this is Nypel®, where about 25% of recycled PA material is mixed with virgin polymer to give a PA resin for injection molding and extrusion compounds [50]. Waste directives, such as the EU's End-of-Life Vehicle Directive and the extended producer responsibility (EPR) give incentives to make the end-of-use treatment financially profitable.

References for Section 8.5

[1] Plastics Europe: "Recycling und energetische Verwendung" (2021), *https://www.plasticseurope.org/de/focus-areas/circular-economy/zero-plastics-landfill/recycling-and-energy-recovery*, accessed May 4, 2021

[2] *Stoffstrombild Kunststoffe in Deutschland 2019*, Conversio 2020

[3] Meinander, M., Mroueh, U.-M. (Eds.): *Directions of future developments in waste recycling. VTT Technology*, (2013) 60

[4] Buekens, A., Yang, J.: Recycling of WEEE plastics: A review. *Journal of Material Cycles and Waste Management*, (2014), pp. 415–434

[5] Simon, D.: *Waste Management*, (2018), pp. 147–171

[6] Recycling Special: Europe: *http://www.europeanbedding.eu/wp-content/uploads/2014/08/Budapest-Beds.pdf*, accessed March 21, 2021

[7] Brunner, S., Fomin, P., Kargel, Ch.: Automated sorting of polymer flakes: Fluorescence labeling and development of a measurement system prototype. *Waste Management*, (2015) 38, pp. 49–60

[8] Trinamix: "Plastic sorting made easy", *https://trinamixsensing.com/plastics* , accessed May 14, 2021

[9] AIM. *About Digital Watermarks*, (2019), *https://www.aim.be/priorities/digital-watermarks/*, accessed May 4, 2021

[10] Fachverband Kunststoffrecycling. "APK AG zielt mit Newcycling® auf neue Geschäftsfelder für Kunststoffrezyklat", (2021), *https://www.bvse.de/gut-informiert-kunststoffrecycling/nachrichten-recycling/6846-apk-ag-zielt-mit-newcycling-auf-neue-geschaeftsfelder-fuer-kunststoffrezyklat.html*, accessed April23, 2021

[11] APK AG: "Newcycling® - wie aus Verpackungsabfällenhochwertige Kunststoffgranulate werden" (2019), *https://www.apk-ag.de/newcycling/prozess/*, accessed April 23, 2021

[12] Wohnig, K., et al.: *Solvent and method of separating a plastic from a solid within a suspension*, WO2018/114047, PCT, 2018

[13] APK AG (2019): *https://www.apk-ag.de/produkte/newcycling-granulat/*, accessed April 23, 2021

[14] BASF SE. "Mechanisches Recycling von Polyamiden in Mehrschichtfolien" (2020), *https://chemicals.basf.com/global/de/Monomers/polyamide_intermediates/mechanical-recycling.html*, accessed May 4, 2021

[15] Robello, R., et al.: *Method of polyamide fiber recycling from elastomeric fabrics*, WO2013/032408, PCT, 2013

[16] Huntinger, H.: "Durchbruch im Recycling von Kleidung mit Polyamid" (2020), *https://innovationorigins.com/de/neue-methode-ermoglicht-das-recycling-von-kleidung-mit-polyamid-erstmals-im-geschlossenen-kreislauf/*, accessed May 4, 2021

[17] Mauldin, L. B., Cook, J. A.: Separation of polyolefins from nylons. US2005/0272913, United States, 2005

[18] *Separation of polyolefins from polyamides*, WO2005/118691, PCT (2005)

[19] Mohajer, Y., et al.: Selective polyamide recovery from multi-component materials. US5889142, United States, 1999

[20] Pelin, C.-E., et al.: Recycling and resusing polyamide 6 extruded waste products to manufacture carbon fiber based composites. *Annals of the Academy of Romanian Scientists, Series on Physics and Chemistry Sciences*, (2017) 2, pp. 91–105

[21] Formisano, B., Göttermann, S., Bonten, C.: Recycling of cast polyamide waste on a twinscrew. *AIP Conference Proceedings*, (2016) 1779, 140002

[22] Subramanian, P. M.: Recovery of polyamide using a solution process. US5430068, United States, 1995

[23] McKinnon, M. S.: *Solvent-based recovery and recycle of polyamide material*, WO01/94457 PCT, 2001

[24] Chen, Y., et al.: Composite material vehicle front-end module, an automobile engine fan, oil bottom shell, radiator water chamber, comprises PA66 recycled waste carpet, toughening agent, and alcoholysis-resistant glass fiber. CN106751804, China, 2017

[25] Cai, Q., et al.: Composite material useful in periphery of automobile engine, comprises recycled waste carpet raw materials, toughener, glass fiber, lubricant, additional toughener and antioxidant. CN108359237, China, 2018

[26] Lee, D. J., Yang, J. H.: Recycling method of airbag fabric waste involves separating silicone rubber from airbag fabric waste using 1,2-propanediol, and obtaining nylon by melting resultant airbag, extrusion-molding, drying and pelletizing. KR2012005255, South Korea, 2012

[27] Guinaldo Fernandez, E., et al.: Method of recycling plastic nylon 6,6 from vacuum bags to obtain filaments for powder for 3D printing processes. EP3718724, EU, 2020

[28] Basire, C.: *Recycling polyamide airbags*. US2017/0292020, United States, 2017

[29] Domo: "Technyl 4Earth" (2021), *https://www.technyl.com/product/technyl-4earth/*, accessed May 10, 2021

[30] "Recycled PA6.6 meets Kärcher's performance and environmental requirements", *Plastics Today*, (2018), *https://www.plasticstoday.com/recycling/recycled-pa66-meets-krchers-performance-and-environmental-requirements*, accessed May 10, 2021

[31] Jung, D.S., Lee, D.J., Yang, J.H.: Manufacture of recycled polyamide 66 resin involves pulverizing lungs radiator shroud, lungs radiator fan, lungs radiator end tank and lungs engine cover, washing with microorganism detergent aqueous solution, molding and palletizing. KR2012005254, South Korea, 2012

[32] Lee, D.J., et al.: Polyamide resin composition comprising fiber reinforced polyamide pellet and molded article thereof. US2017/0275458, United States, 2017

[33] Pietroluongo, M., et al.: Mechanical recycling of an end-of-life automotive composite component. *Sustainable Materials and Technologies*, (2020) 23, e00143

[34] Datta, J., Włoch, M.: "Recycling of Polyurethanes", in: *Polyurethane Polymers*, Elsevier (2017), pp. 323–358

[35] Zia, K.M., BhattiIja, H.N., Bhatti, A.: Methods for polyurethane and polyurethane composites, recycling and recovery: A review. *Reactive and Functional Polymers*, (2007) 67, pp. 675–692

[36] Nafziger, J.L., Lowenkron, S.B., Koehler, C.E., Stevens, B.N.: Flexible polyurethane rebond foam having improved tear resistance and method for the preparation thereof. US 5312888, US, 1994)

[37] Yang, W., Dong, Q., Liu, S., Xie, H., Liu, L., Li, J.: Recycling and disposal methods for polyurethane foam wastes. *Procedia Environmental Sciences*, (2012) 16, pp. 167–175

[38] Kunchimon, S.Z., Tausif, M., Goswami, P., Cheung, V.: Polyamide 6 and thermoplastic polyurethane recycled hybrid fibers via twin-screw melt extrusion. *Journal of Polymer Research*, (2019) 26, p. 162

[39] Lee, Y.-H, Kang, B.-K., Kim, H.-D., Yoo, H.-J., Kim, J.-S., Huh, J.-H., Jung, Y.-J., Lee, D.-L.: Effect of hot pressing/melt mixing on the properties of thermoplastic polyurethanes. *Macromolecular Research*, (2009) 17, pp. 616–622

[40] Wölfel, B., Seefried, A., Allen, V., Kaschta, J., Holmes, C., Schubert, D.W.: Recycling and reprocessing of thermoplastic polyurethane materials towards nonwoven processing. *Polymers*, (2020) 12, p. 1917

[41] Zia, K.M., Bhatti, H.N., Bhatti, I.A.: Methods for polyurethane and polyurethane composites, recycling and recovery: A review. *Reactive & Functional Polymers*, (2007) 67, pp. 675–692

[42] Lin, Y.F.: Waste rigid polyurethane foam plastics recycling method. CN101096426 (2008)

[43] Xionglei, W.: Method for recycling waste polyurethane. CN202011158031 (ed.) Hebei University of Engineering, 2020

[44] Chu, J., Sullivan, J.L.: Recyclability of a glass fiber poly(butylene terephthalate) composite. *Polymer Composites*, (1996) 17, pp. 523–531

[45] Pegoretti, A.: Towards sustainable structural composites: A review on the recycling of continuous-fiber-reinforced thermoplastics. *Advanced Industrial and Engineering Polymer Research*, (2021) 4, pp. 105–115

[46] McLauchlin, A.R., Ghita, O.R., Savage, L.: Studies on the reprocessability of poly(ether ether ketone) (PEEK). *Journal of Materials Processing Technology*, (2014) 214, pp. 75–80

[47] Yu, L., Zhao, D., Wang, W.: Mechanical properties and long-term durability of recycled polysulfone plastic. *Waste Management*, (2018) 84, pp. 402–412

[48] Li, Z., Yu, L., Yan, Z.: Effect of hydrothermal aging on mechanical properties and long-term durability of recycled polysulfone. *Journal of Polymers and the Environment*, (2021), DOI: 10.1007/s10924-021-02182-7

[49] Kuram, E., Ozcelik, B., Yilmaz, F., Timur, G., Sahin, Z.M.: The effect of recycling number on the mechanical, chemical, thermal and rheological properties of PBT/PC/ABS ternary blends with and without glass-fiber. *Polymer Composites*, (2014) 35, pp. 2074–2084

[50] BASF SE: “Nypel® Polyamide Resins”, *https://plastics-rubber.basf.com/northamerica/en/performance_polymers/products/nypel.html*, accessed May 15, 2021

8.6 Mechanical Recycling of PVC

Jason Leadbitter

8.6.1 Introduction

Poly(vinyl chloride) or simply PVC is unique among all other commodity thermoplastics. Its backbone comprises a hydrocarbon component with a repeat chlorine atom (–CH_2–CHCl–), and it is precisely because of this chlorine atom that uniqueness is created. The rather large size of the chlorine atom, compared to its neighboring carbon and hydrogen atoms, is responsible for PVC’s unique characteristics compared to most other polymers. For example, it imparts a significant tolerance to an extraordinary large range of formulated additives that can be incorporated into PVC, including minerals, fillers, pigments, modifiers, stabilizers and, in particular, plasticizers. And these can be added, in some cases, at exceptionally high levels and, within reason, without the final product falling apart.

This tolerance means that PVC products are not one formulation but many. Rather like an artist’s color palette, where the pigments are replaced with a diverse range of additives, this allows the highest possible level of creativity from the very soft touchy-feel of a highly plasticized (flexible) PVC synthetic leather, almost resembling the texture of jelly, to the hard, durable rigidity in unplasticized (rigid) PVC-U window frames. Without doubt, this is the reason why PVC has found so many uses in society across an incredibly high number of diverse applications – from cable sheathing to blood bags and everything in between. Yet it is this level of creativity that is also responsible for the recycling challenges of what to do with discarded PVC products at their end-of-life. The challenge is further exacerbated if we then consider not just the diversity of the composition of the PVC formulation itself, but the opportunities created by the production of composites from PVC, such as the combination of a PVC with polyester fibers for use in flexible applications such as tensile roofing membranes, or in combination with aluminum foil for its use in pharmaceutical blister packaging.

Fortunately for most PVC applications, such composites are not required to be disassembled that frequently. Indeed, PVC has symbolically become known as *the* construction polymer. Best estimates demonstrate that approximately two-thirds of

PVC applications on the market today have long to very long lifespans, in some cases often outliving humans (as for PVC pipes, which can last more than 100 years). At the same time, short-lived products, such as PVC packaging, are in decline, often being replaced by other thermoplastics (although the unique combination of properties that PVC brings to certain applications, such as disposable medical devices and pharmaceutical blister packs, cannot be easily replicated by alternatives).

8.6.2 How is PVC Manufactured?

Prior to discussing the recycling opportunities for PVC, it is first instructive to understand how PVC is manufactured. Like most other commodity thermoplastics, PVC's starting position is predominantly from fossil feedstocks, including either natural gas or naphtha supplied via the oil and gas extraction industries, and which are fed directly into a steam cracker. The feedstocks produced from the crackers include ethene gas ($CH_2=CH_2$), commonly referred to as ethylene, as well as several other hydrocarbon feedstocks.

The other component of PVC, chlorine, is derived from the chlor-alkali industry. This is achieved by the electrolysis of salt (sodium chloride) which is dissolved in water to form brine, the electrolysis of which generates chlorine, sodium hydroxide and hydrogen. Through the chemical combination of ethene (43%) and chlorine gases (57%), an intermediate product is produced, known as 1,2-dichloroethane (commonly, ethylene dichloride, EDC). This product is further "cracked" to produce the PVC monomer, known as vinyl chloride monomer (VCM) and hydrogen chloride. VCM is a highly hazardous chemical and a gas at room temperature. It can be safely transported in the form of a liquified gas under pressure, or converted on an integrated site with facilities to process the upstream raw materials directly into PVC resin through a polymerization process, which enables PVC resins to be produced with levels of residual VCM satisfying strict regulatory limits.

There are two main polymerization processes yielding either suspension PVC (S-PVC) or emulsion PVC (E-PVC). Suspension PVC is produced by far in the highest volumes, whilst emulsion PVC is used in more specialized applications. Both processes use water as a polymerization medium, with the mode of polymerization differing. For each polymerization process, minor formulation ingredients as well as temperature control are used to vary the polymer chain length and particle size. Although the products are chemically the same, this leads to PVC resins being produced over a range of molecular weights. Within the PVC industry, the nomenclature that is used to describe the chain length is known as the K value, and this can range from relatively low molecular weight (K50), typically for use in injection molding applications, to very high molecular weight (K100) for flexible articles with rubber-like properties typically processed by extrusion, and used in applications such as dynamic window gaskets. The properties of PVC are influenced by its average molecular weight and its molecular weight distribution.

8.6.3 PVC Compounding

On its own PVC is completely useless – just like the analogy of baking of a cake where PVC is the "flour", it requires the careful selection of other formulation ingredients. Perhaps the most important additive to accompany PVC is a heat stabilizer: addition levels vary but typically range from around 1–3% by weight. Their main function is to prevent thermal decomposition of the PVC to hydrogen chloride due to the high temperatures and shear during processing. Other additives include plasticizers, which are used for modifying the polymer to create flexible products: these can be added at significantly higher levels, even as much as 40% by weight. There are many other additives, all with their unique contribution towards the final application. For example, a wide range of fillers are used in many PVC applications, principally to reduce cost, although some specialized fillers can be used to enhance the physical properties – such as to improve the resistivity for cables, or to extend the fire-retardant properties of the final product. Other more commonly used additives include impact modifiers, process aids, blowing agents, smoke and flame suppressants, colorants, antioxidants, antistatics, and viscosity modifiers.

The process that is required to convert the PVC resin and its associated additives into processable products is known as PVC compounding and requires two stages: mixing and melt compounding (for suspension PVC), or mixing into a plastisol or paste (for emulsion- or paste-forming PVC polymers). The process of manufacturing PVC products requires an additional stage, although in some processes during the compounding stage shaping can occur to form either an intermediate or final product. Frequently, the processing of PVC involves three stages: mixing, compounding, and shaping, and may frequently overlap into one or more other phases.

8.6.4 PVC Applications in Use

PVC has come to be known as *the* construction polymer, principally because more than two-thirds of PVC across the globe is used in construction applications. Figure 8.69 outlines the main sectors for PVC applications in the EU, the two largest being profiles including window frames, as well as pipes and fittings. Other significant construction applications where PVC is used include cables, flooring and roofing membranes. Whilst other PVC markets include automotive exterior, interior, and seat coverings, there are some niche packaging and medical device applications.

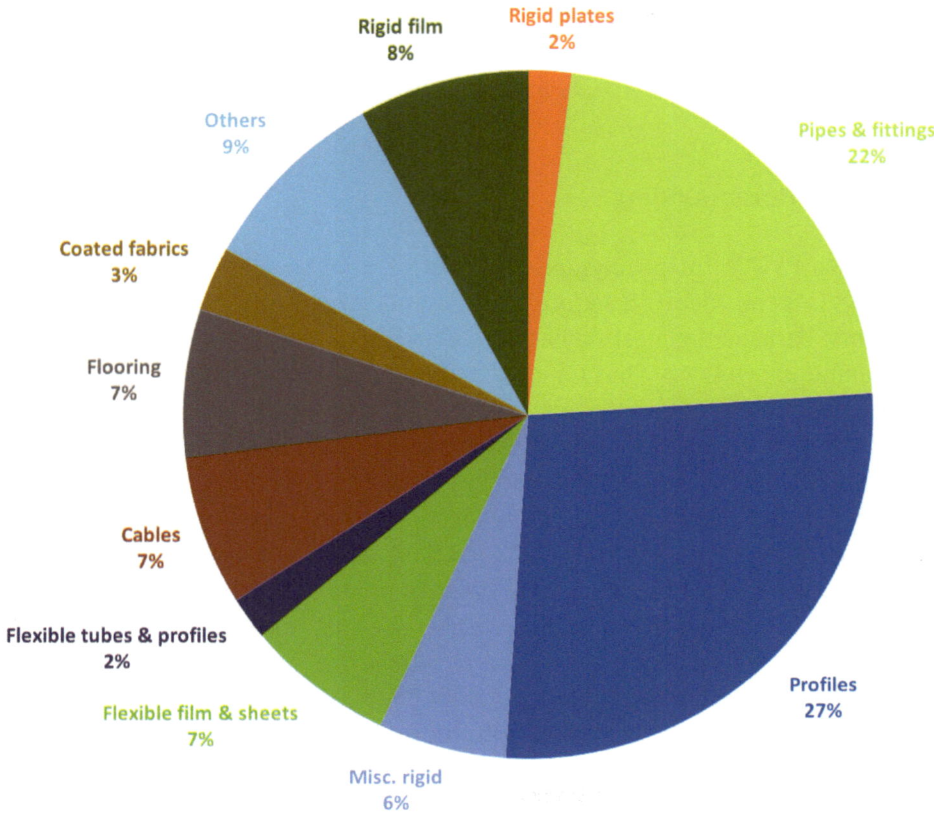

Figure 8.69 PVC applications in the EU [1]

PVC is a versatile material that offers many possible applications. As well as the applications listed above, it has also found widespread use in fashion and footware, credit cards, vinyl records, and synthetic leather, as well as stationery and protective veneers for hardware products.

8.6.5 Waste Hierarchy and End-of-Life Options

When a product has reached the end of its natural life, or remains functional but its life has been prematurely terminated, there are several options for disposal. Before reviewing these, it is first instructive to look at a typical waste hierarchy triangle, which usually starts by considering product design, i. e. before the product has ever been produced (Figure 8.70).

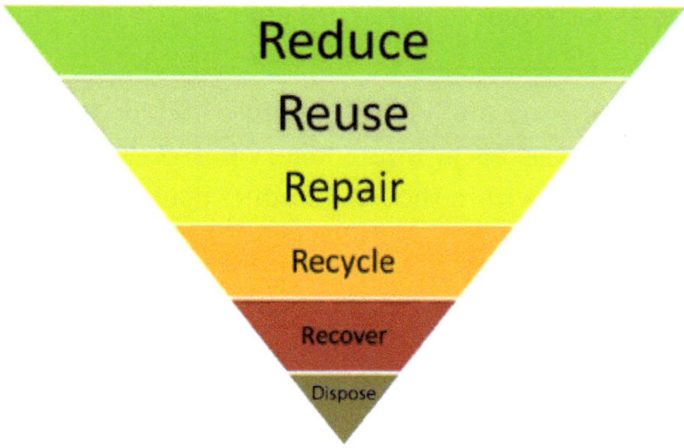

Figure 8.70
Waste hierarchy

Reduce

This is often overlooked, but serves as a reminder that reductionism is an important consideration. There are numerous examples of reductionism within the PVC industry, including reduced wall thickness in PVC pipes, while achieving the same strength as thicker pipes through a process known as biorientation. Other examples include the extensive use of cellular PVC for applications in housing construction, e.g. fascias and soffit boards, where a chemical foaming process caused by the addition of additives during the extrusion and forming processes creates expansion gases. This produces a cellular structure that significantly lightens the product, thereby substantially reducing the amount of PVC consumed per functional unit of product.

Reuse

Whilst examples of reuse in PVC are uncommon, there are examples of older PVC window frames being repurposed for use in other housing, and numerous textile products such as tarpaulins being repurposed into bags, etc. However, most discarded PVC products fall into the remaining three categories: recycling, incineration with energy recovery, or disposal via landfill.

Repair

Our predecessors were renowned for their ability to make repairs. In today's mass-consumer market this has been sadly overlooked. When most items become broken, our modern society has a tendency towards disposal and replacement. This *modus operandi* is rarely questioned, but there are signs from the European Commission for change, and indeed a resolution was adopted in the European Parliament [2] in November 2020 to grant consumers a "right to repair" by making repairs more appealing, systematic, and cost-efficient, whether by extending guarantees, providing guarantees for replaced parts, or better access to information on repair and maintenance.

For several PVC products, repairs are often feasible. These include welding of vinyl roofs, as well as repairing failed gaskets or ironmongery in PVC-U window profiles. Often it is usually the components within the PVC product itself that require repair rather than any failure of the PVC material itself.

Recycle

For almost all discarded PVC products that are sent for recycling, the most common route is mechanical recycling, but other techniques are emerging. There are several mechanical recycling techniques available, and are described in more detail later in the section.

Recover (Incineration with Energy Recovery)

PVC has a thermal heat value of approximately 19 MJ/kg, which is similar to coal, and higher than the average heat value of municipal waste (11 MJ/kg) used to generate electricity. Therefore, it can make a useful contribution as a fuel for power generation through waste incineration. Historically, there were numerous accusations regarding PVC's contribution to dioxin formation in municipal incinerators, which were largely associated with much older incinerators that did not conform to EU standards.

In July 2000, the European Commission published a Green Paper on the environmental issues of PVC [3]. The Commission stated that: "It has been suggested that the reduction of the chlorine content in the waste can contribute to the reduction of dioxin formation, even though the actual mechanism is not fully understood. The influence on the reduction is also expected to be a second or third order relationship. It is most likely that the main incineration parameters, such as the temperature and the oxygen concentration, have a major influence on the dioxin formation". It states further that "at the current levels of chlorine in municipal waste, there does not seem to be a direct quantitative relationship between chlorine content and dioxin formation". These views are based on many in-depth studies carried out in various parts of the world.

Vehlow and colleagues (at the Forschungszentrum Karlsruhe Technik und Umwelt Institut für Technische Chemie) investigated the behavior of a large range of chlorinated products in the feed of municipal solid waste incinerators [4]. His conclusion was: "The increased Cl and Br levels (in the feed) caused no significant increase of the concentration of PCDDs or PCDFs in the raw gas". These findings are in line with many studies concerning the possible impact of PVC materials in the feed of municipal waste incinerators.

The most extensive study was performed in the United States [5], where the impact of the waste feed chlorine content on PCDDs and PCDFs emissions from 155 facilities was analyzed. The conclusion was: "The hypothesis that the amount or type of chlorine in the waste feed to combustion units is directly related to PCDDs/PCDFs concentrations measured at the combustion outlet is not supported by the preponderance of the data examined during this study".

A study [6] included chlorine feed concentrations from less than 0.1% to 80% (1900 test results from 169 facilities, including municipal solid waste incinerator, hazardous waste incinerators, hazardous waste fired boilers, cement kilns, biomass combustors, and laboratory-, bench- and pilot-scale combustors). The study showed no statistically significant relationship between chlorine input and PCDD/F stack concentration. A study later performed at the Umeå University [7] showed that the chlorine source and level are unimportant for formation of chlorinated organic pollutants.

During the thermal treatment process that takes place in municipal solid waste incinerators, PVC along with other chlorinated wastes will contribute toward the production of hydrogen chloride (HCl) gas, which, along with other gas emissions such as sulfur dioxide (SO_2) require treatment such as neutralization, to avoid emissions to the atmosphere. Flue gas cleaning systems include a number of options such as flue gas scrubbing with sodium bicarbonate or lime. Further details have been published by Johnsen [8].

There are several other processes for creating value from the flue gas treatment processes, including SolVair® [9]. In this process the acid gas is neutralized by sodium bicarbonate, and the resulting sodium chloride is purified before being used in the soda ash process for glass production.

Disposal (via Landfill)

Disposal via landfill is the least attractive option, given that it represents a significant loss in value of resources. Whilst the long-term effects of disposal of PVC have been significantly researched and the process deemed safe, there are numerous regulatory initiatives that are driving more sustainable options. For example, several countries have already banned the landfilling of untreated organic waste, including Germany, Denmark, and Sweden, and this trend will continue.

8.6.6 How to Go About Recycling of PVC?

On paper, the recycling of thermoplastics should be relatively straightforward. For example, any polymer chemist worth their salt will state that by their very nature thermoplastics are melt-reprocessable (i. e., they become soft when heated in order that they can be processed, and that they reharden upon cooling, without appreciable change in physical properties) – and that this process can be repeated. This is what defines a thermoplastic and makes it so different to a thermoset. In contrast, the chemical bonds of thermosets don't just exist throughout the backbone of the polymer, but the spaghetti-like strands of the polymer backbone are chemically tied to one another, forming a three-dimensional lattice; once chemically formed they retain their shape and cannot be reprocessed through simply melting and reshaping. Since PVC is a thermoplastic, there are usually no such chemical cross-links, and so in principle PVC should be reasonably tolerant of the recycling process.

However, before any such reprocessing can take place there is the significant dilemma of ensuring that any discarded PVC product is disposed of in a way that ensures some form of collection. This is not an easy step, and indeed it has taken decades of hard work to ensure that such infrastructure exists for the key applications. Indeed, this also varies from one application to another. Perhaps one of the most well-understood demonstrations of plastics collection is within the packaging arena. Most European countries now operate various curbside collection points for household plastics packaging waste, and very high recycling rates are now achievable, particularly for high-volume products such as HDPE or PET bottles. However, for PVC its use in packaging has drastically reduced, and it remains less than 2% of the European plastics packaging market. In such applications PVC can often be "seen" as a contaminant, as its critical mass makes it less attractive for separate collection and sorting.

In contrast, if we turn to construction applications, which is a much more significant sector for the PVC industry, a whole new recycling industry has been set up to ensure that at end-of-life, discarded PVC products can be given a new lease of life via mechanical recycling. This is essential for the circular economy and to ensure that the polymer is kept in circulation for as long as possible.

8.6.7 The Dawn of a New Industry

During the late 1990s and early 2000s, PVC came under severe political pressure; it was singled out by the pressure group Greenpeace as unrecyclable, and further political pressures mounted at the European Commission level, leading to a green paper on PVC [3]. It was clear that if the European PVC industry wanted a future, it would have to develop the infrastructure to ensure that the industry could transition from a linear model of mine–use–dispose to a circular model, and in addition give up on some of its more environmentally damaging additives such as heavy-metal stabilizers and a number of plasticizers that were increasingly being reclassified as toxic. Through a unique process, the whole value chain of the European

PVC industry formed a voluntary commitment, known as Vinyl 2010 [10] with the specific challenges of developing a circular economy and phasing out harmful additives.

Vinyl 2010 had four main contributors: The European Council of Vinyl Manufacturers (ECVM), The European Council of Plasticisers and Intermediates (ECPI), The European Stabiliser Producers Association (ESPA), and the European Plastics Converters (EuPC). One of the key commitments of Vinyl 2010 was to achieve 200 kt PVC recycling per annum across Europe, over and above what was already recycled within regulated waste streams, this was further increased to 250 kt as the EU expanded. Several years into the commitment it was clear that to ensure the recyclability of PVC there needed to be an organized collection and sorting infrastructure, and this was achieved by the formation of Recovinyl® [11]. After the successful completion of the first 10-year voluntary commitment, the industry developed an even more ambitious voluntary commitment known as VinylPlus® [12]

8.6.7.1 VinylPlus®

The official launch of VinylPlus took place in June 2011. Building on the foundations of Vinyl 2010, the initiative took the original commitment a significant step further, and was aligned to a sustainability framework known as The Natural Step [13]. The commitment was embedded in five key sustainability challenges. The first four challenges were technical in nature, whilst the fifth challenge consisted in promoting sustainability awareness across the entire value chain. Challenge No 1 (controlled-loop management of PVC) identified the need to build on the recycling infrastructure of Vinyl 2010, and provided a "stretch target" of achieving 800 kt/a of PVC recycling by the end of 2020. This represented more than a three-fold increase on the recycling rates achieved at the end of the Vinyl 2010 initiative.

8.6.7.2 Recovinyl®

The Recovinyl® organization was established in 2003 to both develop and encourage the recycling of PVC in Europe. It has been the "glue" to bring both the recyclers and converters together and thereby the start of the transition from a linear to circular economy (Figure 8.71). In its early days of development, the organization provided a financial stimulus from the Vinyl 2010 sponsors directly to the recyclers, to assist them in developing the necessary infrastructure throughout the whole process, from collection and sorting to investments in new recycling technologies, and finally to ensuring that the recycled PVC compounds found their way back to the converters for use in new PVC applications. This infrastructure should not be underestimated, and has steadily grown from the early 2000s to be a significant presence in today's market.

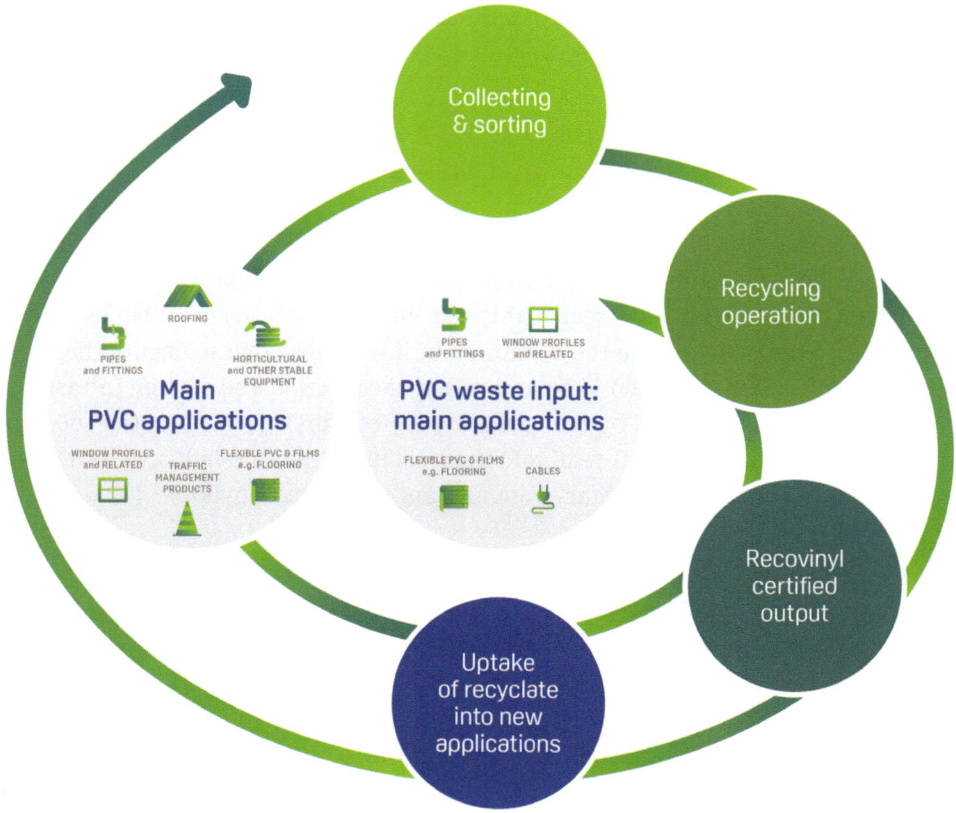

Figure 8.71 Schematic representation of Recovinyl operations (Courtesy of Recovinyl)

Today, Recovinyl continues to monitor and verify the recycling of PVC waste, thereby ensuring the uptake of PVC recyclate. It is by far the largest contributor of recycled volumes within the VinylPlus program. The number of PVC recyclers has steadily grown, with an established European network of more than 175 members. It now operates in over 16 member states as well as the UK.

The latest development of Recovinyl has been the launch in early 2021 of RecoTrace™ [14]. This is an enhanced system of data collection to monitor, verify, and report PVC recycling and uptake across Europe. RecoTrace has also been developed to align with the ambition of the newly launched Circular Plastics Alliance [15] and EU policy initiatives that aim to encourage the recovery, recycling, and reuse of all plastics. This follows the European Commission's call for pledges [16] for the whole of the European plastics industry to boost recycling. The commitment from VinylPlus is to recycle at least 900 kt/a of PVC into new products by 2025.

The key developments of RecoTrace™ are to ensure that recyclers and converters can register their data in a format that is straightforward, as well as remaining safe and secure. In this way there will be full accountability and traceability of PVC recycling across Europe.

8.6.7.3 Successes of the VinylPlus Initiative – European Recycled PVC Market Size

Figure 8.72 highlights the tremendously successful growth of PVC recycling across Europe from the early 2000s until the present. This growth is considerably larger than the almost stagnant market growth for virgin PVC sales across Europe over the same timescale, and demonstrates the success of the Recovinyl process. By far the largest market for recycled PVC products has been achieved via the window profile market, and the associated rigid profiles market. The sector making the largest contribution to this market is EPPA [17], the European Trade Association of PVC Window System Suppliers.

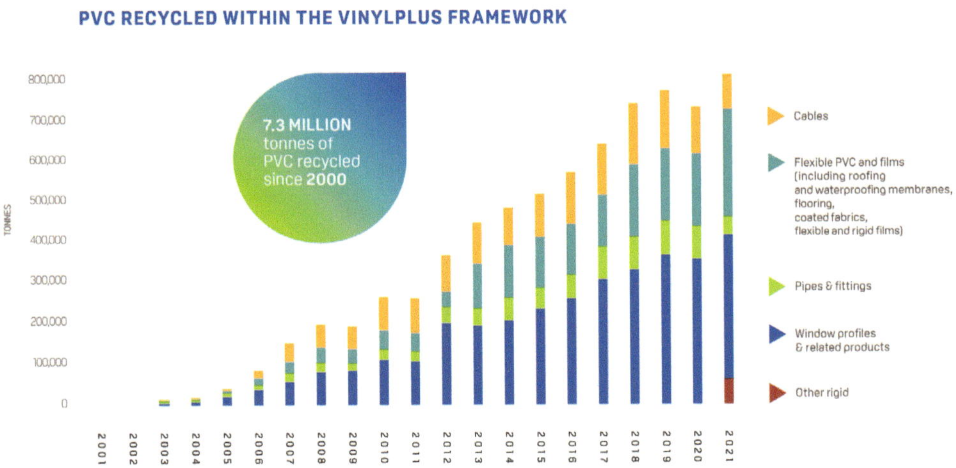

Figure 8.72 Audited PVC recycling volumes achieved via VinylPlus (Courtesy of VinylPlus)

The second largest recycling segment relates mainly to flexible PVC applications, including flooring, roofing, and waterproofing membranes.

Whilst the third largest is the recycling of cables, this remains a significant volume and is more than 100 kt/a. Unlike the other applications, the cables are not recycled back into new cabling, due to the potential for even small amounts of metal residues (such as copper wire) to be present, which would pose an unnecessary risk for re-use in cabling by causing a short circuit.

8.6.8 European Markets for Recycled PVC

The Recovinyl initiative has been tremendously successful at ensuring that recycled PVC products remain in controlled loops, often being recycled into the same or similar applications from the discarded product.

The main markets for recycled PVC are highlighted in Figure 8.73. By far the largest sector for recycled PVC includes windows and profiles, whilst for flexible PVC a significant volume of recycled cable is used in road traffic management systems. Other important sectors include pipe applications and floor coverings, as well as roofing and horticultural uses. The PVC industry can demonstrate controlled-loop recycling whereby discarded PVC from building and construction PVC is recycled back into either the same or similar construction products. Instead, the vast majority of recycled cable products are used in road traffic management systems.

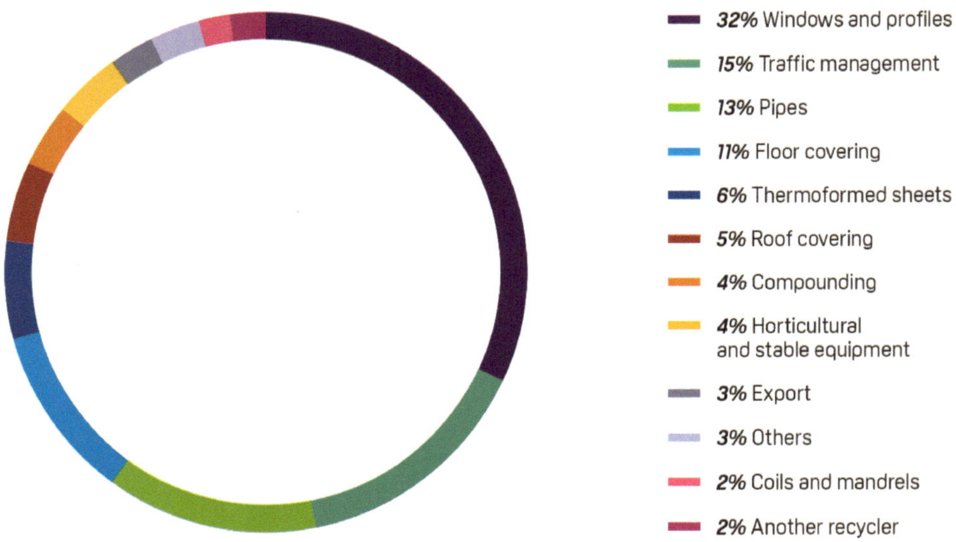

Figure 8.73 Results of the RecoVinyl traceability survey 2019, relating to the use of controlled-loop recycled PVC in 2018 (Courtesy of Recovinyl)

8.6.9 The Recyclability of PVC

Even though PVC formulations can be both complex and hugely diverse, the recyclability of PVC products can be quite remarkable, provided there has been some degree of collection and pre-sorting. For example, the formulation of PVC window frames will vary from one window profile manufacturer to another, yet the recyclability of PVC-U window frames from a wide range of manufacturers that have reached their end-of-life will produce recyclate fit for re-use either into new window frames or for use in an alternative suitable construction application. Ironically, for some PVC formulations that contain impact modifiers, there is evidence that some physical properties can improve after being recycled compared to the virgin product.

One of the first publications in this area [18] demonstrated that PVC products could pass through up to 10 thermal history processes without any significant loss in their physical properties, even if recycled at 100%. This paper has become the industry benchmark, and is often quoted in the media. Interestingly, the stabilizer used to protect the PVC during its original manufacture will come back to life during the thermal recycling process, even after remaining largely dormant during the use phase of the product. The paper claimed that even after the fourth-generation regrind (10 thermal histories in total) there was still a reasonably comfortable margin of stabilization.

In this paper, two PVC formulations were evaluated, both via an injection molding process, which tends to have harsher shear rates compared to extrusion. Another observation was that one of the formulations contained an impact modifier and this product's impact properties marginally improved after the 1st-generation regrind, suggesting optimum dispersion of the impact modifier occurred only after the 1st recycling process. Generally speaking, all other physical properties remained the same even after the 4th-generation regrind. Color changes were observed in one of the formulations, but this was considered only an aesthetic issue, whilst a lubrication issue was also detected in one of the injection-molded grades (smaller injection molding shot size), but these were deemed to be minor issues given that the physical integrity of the recyclate remained completely intact without the need for additional formulation additives.

In summary, it can be concluded that PVC is highly tolerant from a recycling perspective, and does not require additional stabilizers or other additives, or virgin PVC resins to compensate for any changes in melt rheology. Indeed, because of the insignificant changes in physical properties reported in the above-mentioned paper (even after several recycling passes), it could be concluded that the main polymer backbone remains intact. During processing, the labile chlorine from the PVC backbone can detach and react with a hydrogen atom – a mechanism that is commonly known and that leads to a process of conjugated unsaturation. A PVC formulation without stabilizer would first rapidly show signs of color change, typically from white to yellow, followed by a rapid progression to stronger colors and ultimately black. These color changes are due to increasing levels of unsaturation (double bonds) within the main polymer chain. Hence the main purpose of the stabilizer is to initially react with the hydrochloric acid that is initially generated during the decomposition process, thereby protecting the PVC during thermal treatment processes associated with compounding, molding, and/or extrusion.

8.6.10 Legacy Additives and Regulatory Issues

A major challenge for recycled PVC is the issue of legacy additives. These are additives that were permitted for use at the time of the product's manufacture but are now subject to strict regulation. Hence, they can be present in older PVC formulations, and pose significant challenges when these products have reached their end-of-life and considered for recycling. Several such additives are either prohibited or restricted by legislation, and two of the most significant are discussed below.

8.6.10.1 Lead Stabilizers

Lead compounds were widely used as the main stabilizer system in PVC formulations in many countries up until the 21st century. However, from 2001 onwards the European PVC industry has progressively phased them out through its subsequent Voluntary Commitments. By the end of 2015, lead stabilizers were no longer used in any new PVC formulations. However, there remains a substantial volume of PVC products in use containing this stabilizer, and the industry remains committed to ensuring that such products continue to be recycled safely, as well as exploring novel technologies that may lead to removal during the recycling process.

In November 2019, the REACH Committee accepted the European Chemical Agency's proposal for revising the limits of lead content in articles containing recycled PVC, with a derogation period of 15 years, to be reviewed after 7½ years. The draft regulation was sent for scrutiny to the European Parliament and Council. On 21 January 2020, the European Parliament ENVI Committee adopted a motion for resolution objecting to the draft EU Commission regulation, considering it incompatible with the aim and content of the REACH Regulation (which has placed a restriction on the use of lead compounds in new manufacture). On 12 February 2020, a plenary session of the European Parliament voted in favor of this resolution objecting to the draft Commission regulation. The draft proposal was then returned to the Commission for review.

ECHA's proposal for lead content limits requires that recycled PVC containing lead be completely encapsulated within a layer of virgin PVC in soft applications and in specific rigid applications. Indeed, a large proportion of recycled PVC containing lead stabilizers is already encapsulated within products, such as three-layered pipes in which the inner and outer layers are based on virgin PVC and the core layer contains PVC recyclate, both in pipes and window profiles. VinylPlus has started a project on the development of soft PVC applications with a barrier layer. The project will investigate materials suitable for such barriers, including virgin soft PVC and rigid PVC films. It will look at the available converting technologies and at application's durability, recyclability, and costs. A dedicated consortium will also be set up for the development of new technologies.

8.6.10.2 Low Molecular Weight Orthophthalate Plasticizers

A number of orthophthalate plasticizers have also come under increasing scrutiny. The main one is di-2-ethylhexyl phthalate (DEHP or DOP). This plasticizer was extensively used in a wide range of flexible PVC applications but due to increasing concerns regarding its toxicity, specifically toxicity to reproduction, it has been identified as a substance of very high concern (SVHC) requiring authorization for permitted uses under REACH Regulations. For most applications, the PVC value chain has substituted DEHP with a range of alternative safer plasticizers. Given its widespread use in applications such as cables, it must be assumed that DEHP is a legacy additive in older cables, typically those manufactured prior to 1999 when the revised toxicity classification was made. Consequently, the industry has been working on various risk assessments to demonstrate that recycled products containing DEHP are safe for use in some applications, and authorizations under REACH for such products have been recommended by the European Chemicals Agency.

8.6.11 Benefits of PVC Recycling from a Life Cycle Perspective

According to the latest progress report from VinylPlus [19], over 6.5 Mt of PVC has been recycled since 2000. This has saved an estimated 13 Mt of CO_2 because of not producing the equivalent virgin PVC. In addition, the amount of primary energy saving can be anything up to 90% lower compared to virgin PVC.

References for Section 8.6

[1] PVC Applications [Webpage]: ECVM, (2017), *https://pvc.org/pvc-applications/*

[2] Höglund, I.: Parliament wants to grant EU consumers a "right to repair" [Press release], EU Parliament, November 25, 2020, *https://www.europarl.europa.eu/news/en/press-room/20201120IPR92118/parliament-wants-to-grant-eu-consumers-a-right-to-repair*

[3] Environmental issues of PVC, EU Commission, (2020), *https://eur-lex.europa.eu/legal-content/SL/TXT/?uri=CELEX:52000DC0469*

[4] Vehlow, J., Bergfeldt, B., Hunsinger, H.: PCDD/F and related compounds in solid residues from municipal solid waste incineration – A literature review, *Waste Management & Research*, (2006) 24, pp. 404–420

[5] Rigo, H. G., Chandler, A. J., Lanier, W. S.: *The Relationship Between Chlorine in Waste Streams and Dioxin Emissions from Waste Combustor Stacks*, ASME, 1995

[6] Rigo, H. G., Chandler, A. J., Lanier, W. S.: The relationship between chlorine in waste streams and dioxin emissions from waste combustor stacks [ASME Research Report, CRTD Vol. 36], ASME Committee on Industrial and Municipal Wastes, 1995

[7] Wikstrøm, E.: *The role of Chlorine during Waste Combustion*, [Ph. D. Thesis], Umeå University, 1999

[8] Johnsen, T.: Understanding the basics: How acid gases from PVC energy recovery are neutralized [White Paper], VinylPlus, (2015), *https://pvc.dk/wp-content/uploads/2016/02/Understanding_the_basics_hcl.pdf*

[9] Clean air solutions for higher performance [Webpage], SOLVAir®, *https://www.solvairsolutions.com/en/global-presentation*, accessed December 2021

[10] Vinyl 2010 Voluntary Commitment [Webpage], British Plastics Federation, *https://www.bpf.co.uk/members/Vinyl_2010_Voluntary_Commitment.aspx*, accessed December 2021

[11] Recovinyl [Webpage], *https://www.recovinyl.com/*, accessed December 2021

[12] VinylPlus [Webpage], *https://vinylplus.eu/*, accessed December 2021

[13] The Natural Step [Webpage], *https://thenaturalstep.org/*, accessed December 2021

[14] Recovinyl Launches RecoTrace – An advanced data system for circular PVC [Press release], VinylPlus, January 3, 2021, *https://www.vinylplus.eu/news/recovinyl-launches-recotrace-an-advanced-data-system-for-circular-pvc/*

[15] Circular Plastics Alliance [Webpage], *https://circular-plastics-alliance.com/en/welcome-to-the-circular-plastics-alliance/*, accessed December 2021

[16] European Circular Economy Stakeholder Platform [Webpage], European Union, *https://circulareconomy.europa.eu/platform/en/commitments/pledges#:~:text=In%20January%202018%2C%20the%20European,in%20the%20EU%20by%202025*, accessed December 2021

[17] European Trade Association of PVC Window System Suppliers [Webpage], *https://www.eppa-profiles.eu/*, accessed December 2021

[18] Leadbitter, J., Bradley, J.: Closed-loop recycling opportunities for PVC, *Current Trends in PVC Technology Conference*, Institute of Polymer Technology and Materials Engineering, Loughborough University, November 1997

[19] NEW! VinylPlus Progress Report 2021 [Press release], VinylPlus, May 17, 2021, *https://www.vinylplus.eu/resources/new-vinylplus-progress-report-2021/*

8.7 Compatibilization of Blends for Mechanical Recycling

Norbert Nießner, Bianca Wilhelmus

8.7.1 Classification of Polymer Blends

Blending of polymers is a convenient way to obtain materials with improved properties or with a better performance-to-cost ratio. The properties of more than one polymer may be combined, which is a usually cheaper and less time-consuming strategy than to develop entirely new polymeric materials. Blending is typically done in standard equipment such as twin-screw extruders [1, 2].

Polymer blends can be divided into three categories on the basis of their phase behavior: miscible (either fully or partially), immiscible, or alloy. Miscible polymers form a homogenous blend; the domain size is comparable to the dimension of a macromolecular statistical segment [3]. Immiscible polymers show a heterogeneous structure, with separated domains of each polymer. The phase morphology is coarse, the interface between the domains is sharp, and the adhesion between the phases is poor, resulting in inferior properties [2]. The size of the domains increases with increasing concentration of the dispersed phase, due to coalescence [4]. The interface and morphology of immiscible blends can be improved by compatibilization; the immiscible polymer blend then becomes an alloy. Compatibilization suppresses coalescence and decreases interfacial tension, thus enhancing interfacial adhesion [1].

There are two ways to achieve compatibilization. Firstly, a compatibilizer can be added as a third polymer. Common compatibilizers are either graft- or block copolymers. Secondly, a copolymer can be formed "in situ" by reactions of the different polymers during the blend preparation, also called "reactive blending" or "reactive extrusion". Compatibilized blends have gained significant importance as high-performance alloys. In recent years, more research was conducted in further enhancing their properties and finding alternative compatibilizers for otherwise immiscible blends [3, 4]. An overview of the different types of polymer blends is given in Figure 8.74, to provide a fundamental understanding of their basic properties.

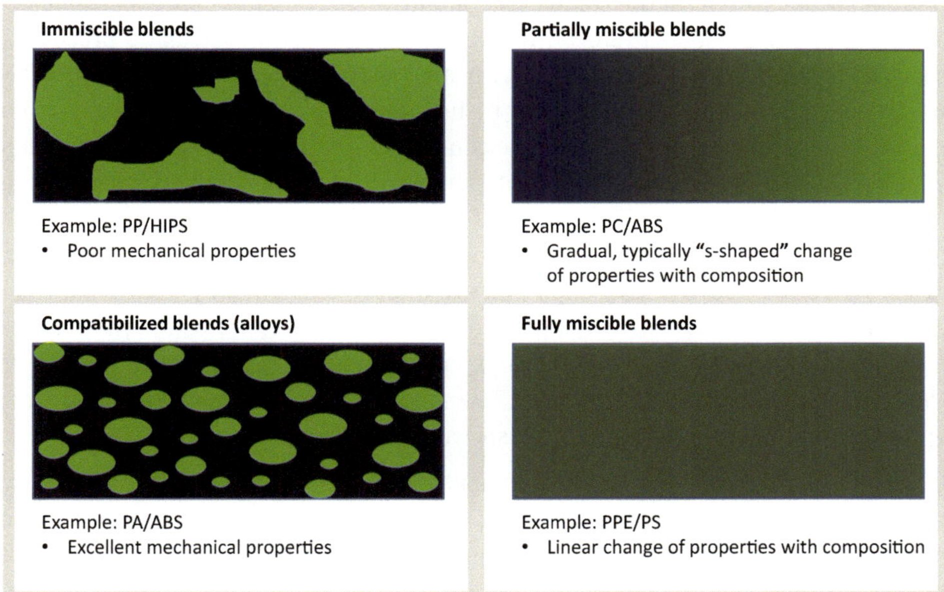

Figure 8.74 Different types of polymer blends

8.7.2 General Principle of Compatibilizers

Compatiblizers often are macromolecular species exhibiting interfacial activities with both blend components. In case of block copolymers, for instance, one constitutive block would be miscible with one blend component, and the other constitutive block would be miscible with the other blend component [2]. Figure 8.75 provides a schematic representation how different block copolymers or graft copolymers are positioned at the interface of two otherwise immiscible polymers. These compatibilizers reduce the interfacial tension, prevent coalescence of the domains, and thus generate and stabilize a finer morphology. If the compatibilizer penetrates both polymer phases deeply enough to entangle with the constitutive polymer chains, the interfacial adhesion is improved. This is necessary to provide efficient stress transfer from one polymer phase to the other and for preventing crack propagation, resulting in better mechanical properties [2]. Block and graft copolymers can be tailored to fit the blend components, to ensure the best achievable compatibilization effect.

It is also possible to form the compatibilizer during the blending process instead of adding a pre-made compatibilizer. In this so-called reactive extrusion process, a reactive compound that is miscible with one blend component, and reactive towards functional groups of the other blend component, is added to the reaction mixture. The reactive compound can be generated by free-radical copolymerization or melt-grafting of reactive groups onto chemically inert polymer chains. During reactive compounding, by the reaction of those groups with the functional groups of the second polymer, block or grafted copolymers are formed directly at the interfaces. Therefore, reactive extrusion often is a more efficient way to achieve compatibilization than the addition of pre-made compatibilizers, provided the reactivity of the reactive and functional groups is sufficient, particularly with regard to the short mixing time in the extruder. Examples of reactive groups are anhydride, epoxide, oxazoline, carbodimide, and isocyanate groups [2].

Especially in reactive extrusion, the compatibilization efficiency is also affected by processing conditions such as the temperature, feed rate, and screw speed of the extruder [5]. However, there are still many open questions regarding the effect of mixing procedures or sequence of adding reactants on the final product. It is believed that the addition of reactants at different stages of the reaction can affect the morphology development and chemical conversion of the whole reaction. Changing the sequence of adding reactants could alter factors such as residence time, component viscosity, mixing, initial distribution of the functionalized polymer, etc., affecting the final compatibilization. Further research in this area could provide a better understanding of the processes and therefore enable optimization and control of the compatibilization process.

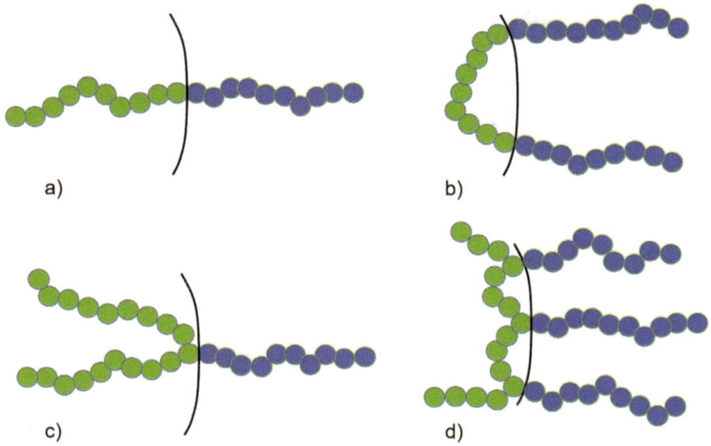

Figure 8.75 Schematic representation of the compatibilization by a) diblock, b) triblock, c) single-graft, and d) multi-graft copolymers at the interface of a heterogeneous polymer blend

8.7.3 Selected Examples of Compatibilized Blends

Important commercial examples for compatibilized blends are ABS/PA6 alloys. Due to the immiscibility of ABS and PA, compatibilizers such as styrene-acrylonitrile-maleic anhydride (SANMA) are used [6]. Polyamides are attractive polymers for reactive compatibilization, as they offer reactive functionalities through their amine and carboxyl end-groups [7]. Much research is also conducted to study the effectiveness of adding reactive polymers to the SAN phase of ABS to control the morphology and produce extremely tough blends [7]. The amount of anhydride present in compatibilizers strongly affects the reactivity and morphology in ternary polyamide blends. Thus, when the compatibilizer used is miscible with the SAN phase and has a maximum amount of anhydride functionality, the largest reduction in the dispersed-phase domain size can be achieved. Figure 8.76 (top) shows the morphology of a non-compatibilized ABS/PA6 blend, with PA6 as dispersed phase in the ABS matrix. Figure 8.76 (bottom) depicts how the morphology of a SAN/PA6 blend changes due to compatibilization: the formerly coarse domains are transformed into a much finer structure.

Besides SANMA, other types of compatibilizers are also suitable for compatibilization of ABS and PA, including methyl methacrylate-glycidyl methacrylate (MMA-GMA), methyl methacrylate-maleic anhydride (MMA-MA) copolymers and imidized acrylic polymers [8–10].

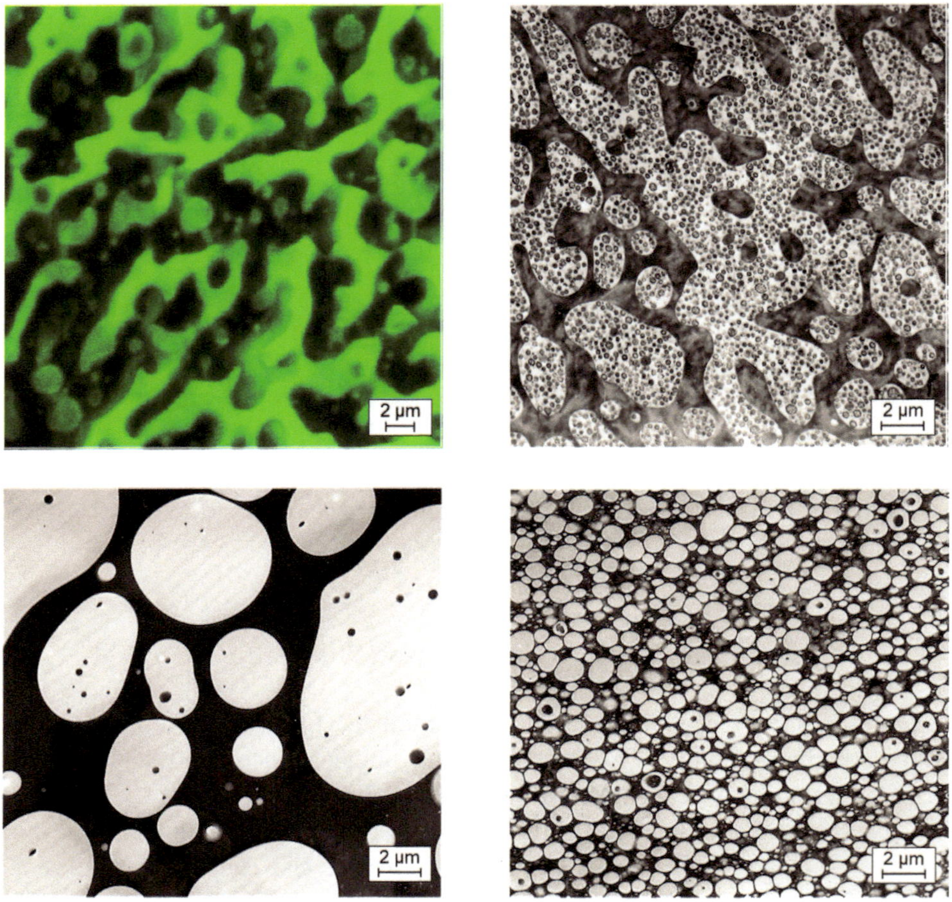

Figure 8.76 Top: Morphology of uncompatibilized ABS/PA6 60/40, stained with OsO_4 and phosphotungstic acid (courtesy of INEOS Styrolution). Bottom: Morphology of uncompatibilized (left) and compatibilized (right) SAN/PA6 (60/40), stained with phosphotungstic acid (Courtesy of INEOS Styrolution)

Another example of reactive compatibilization is the use of an amine-functional styrene-acrylonitrile (SAN-amine) polymer for a bisphenol A polycarbonate/acrylonitrile-butadiene-styrene (PC/ABS) blend. SAN-amine is miscible with the SAN matrix in the ABS, and the pendant secondary amine groups then react with PC at the carbonate bond, forming a SAN-g-PC copolymer at the interface [11]. This reaction utilizes very well the general principle of selecting a compatibilizer according to suitable reactive groups and appropriate blocks, respectively.

Styrene-butadiene block copolymers are well suited to compatibilize blends of PP and PS. Figure 8.77 shows how the microstructure changes by adding SBC to a blend of PP and HIPS, significantly improving mechanical properties such as tensile strength and elongation at break.

In addition to block or graft polymers, sometimes also inorganic fillers are applied, either as the sole compatibilizer or in combination with block or graft copolymers. Examples are talc, kaolin, organoclays such as montmorillonite, and carbon nanotubes [1, 2, 12–14].

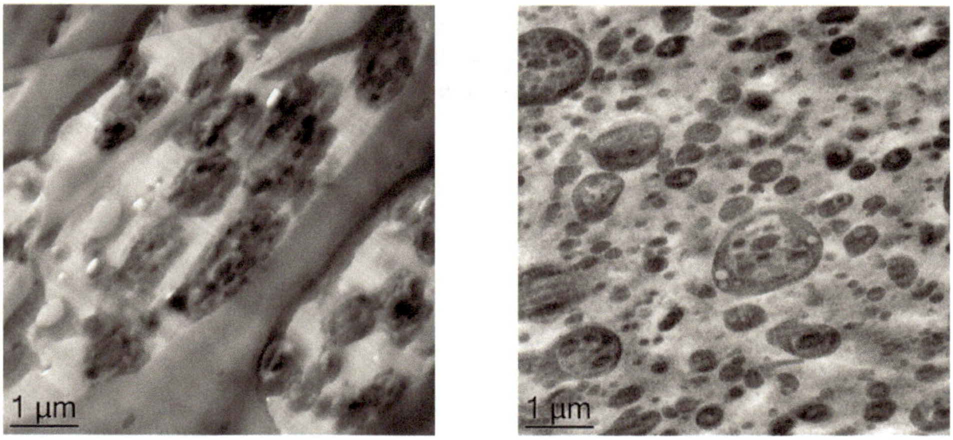

Figure 8.77 Left: Uncompatibilized PP/HIPS blend. Right: PP/HIPS blend compatibilized with SBC (Courtesy of INEOS Styrolution)

8.7.4 Compatibilization in Mechanical Recycling

In addition to the industrial examples given above, the investigation of blend systems has a very practical use when it comes to mechanical recycling of end-of-life plastics. This feedstock often contains impurities in the form of other polymers, even if it is (pre-)sorted. Therefore, it is desired to have options for the mechanical recycling of this kind of feedstock as well, without loss of properties. Another important aspect is the degradation of plastics, both prior to recycling and during recycling procedures [1]. In order to achieve acceptable product quality, this has to be counterbalanced by stabilizers, additives, or reactants, such as chain extenders, if applicable. The processing temperature of a compatibilization step needs to be chosen so that it is high enough to melt all blend components, but as low as possible to avoid further degradation of the polymer chains.

The composition of the end-of-life plastics strongly varies. The three most important sources for plastic waste are packaging, waste electric and electronic equipment (WEEE) and end-of-life vehicles (ELV). All three waste streams are handled differently. In countries with an established collection scheme, mixed packaging waste is usually collected at the curbside and later sorted into separate fractions. Often, there is a separate collection for PET bottles, sometimes even encouraged by a deposit scheme. The remainder of packaging materials mainly consists of polyolefins (PP, LDPE), PS and PET. Even after sorting, some challenges for mechanical recycling remain, for example, the recycling of multilayer packaging (including PET/PE, PET/PE/EVOH and PA/PE) and the recycling of fractions that contain a few percent of other polymers [15].

The composition of WEEE plastics depends a lot on the type of equipment that is to be recycled. For instance, large household appliances such as refrigerators mainly consist of ABS and PS, whereas in other items such as printers, computers and monitors PC/ABS and PP can also be found [16]. However, such items are often presorted during or immediately after collection, so that special fractions are generated and marketed – for example, a PS fraction from fridge recycling. Typical challenges for WEEE would be the recycling of PP/PS mixtures from small appliances, or PS/ABS mixtures.

End-of-life vehicles are usually shredded, resulting in light and heavy fractions. As plastics are used for many different parts in a car with very diverse property requirements, the polymers in the shredder fraction are also numerous: PP, ABS, PMMA, PC, PA, POM, and PUR, to name just a few. Mostly, they end up in the light fraction [17]. The huge variety of polymers, as well as some residual metals in the shredder fraction, provide a special challenge to sorting and mechanical recycling.

As outlined above, each waste stream requires a dedicated treatment, but, as a general strategy, compatibilization might help to solve certain problems in mechanical recycling, especially when it comes to feedstock that is still contaminationed by other polymers. Table 8.7 summarizes a few typical polymeric systems that occur during recycling of end-of-life plastics, and possible compatibilizers that can be used to improve the mechanical properties of such systems.

Table 8.7 Suitable Compatibilizers for Different Polymeric Systems [1, 2, 15, 18]

Polymeric system	Compatibilizer
PP/PE	EVA, EPDM, EPM, SEBS
PP/PS	SBS, SEBS, EVA
PP/PET	PP-g-AA, PP-g-MA, SEBS, SEBS-g-MA, SEBS-g-GMA, EVA, EVA-g-MA
LDPE/HDPE	CPE, EVA
PE/PET	PE-g-AA, PE-g-AN, PE-g-GMA, SEBS, SEBS-g-GMA, SEBS-g-MA, EGMA
PP/ABS	PP-g-AA, PP-g-HEMA, PP-g-SAN
ABS/PC	ABS-g-MA, PP-g-MA, PMMA
PC/PS	PC-g-PS, PS-PAr, SEBS-g-PC
PP/PA	SEBS-g-MA, PP-g-MA, PP-g-AA
ABS/PA	SEBS-g-MA, SAN-MA

While blending of polymers offers possibilities for the creation of materials with enhanced properties for special applications, compatibilization of blends often is a necessity and is able to boost properties of mechanically recycled end-of-life plastics.

References for Section 8.7

[1] Utracki, L. A., Wilkie, C. A. (Eds.): *Polymer Handbook*, 2nd Edition (2014), Springer, New York

[2] Koning, C., Van Duin, M., Pagnoulle, C., Jerome, R.: *Progress in Polymer Science*, (1998) 23, pp. 707–757

[3] Utracki, L. A.: *Polymer Engineering & Science*, (1995) 35, pp. 2–17

[4] Kitayama, N., Keskkula, H., Paul, D. R.: *Polymer*, (2000) 41, pp. 8053–8060

[5] Özkoç, G., Bayram, G., Quaedflieg, M.: *Journal of Applied Polymer Science*, (2008) 107, pp. 3058–3070

[6] Jafari, S. H., Pötschke, P., Stephan, M., Warth, H., Alberts, H.: *Polymer*, (2002) 43, pp. 6985–6992

[7] Majumdar, B., Keskkula, H., Paul, D. R., Harvey, N. G.: *Polymer*, (1994) 35, pp. 4263–4279

[8] Araujo, E. M. in: *Polymeric Materials: New Research*, Caruta, B. M. (Ed.) (2005), Nova Science Publishers, New York, pp. 85–105

[9] Lai, S. M., Li, H. C., Liao, Y. C.: *European Polymer Journal*, (2007) 43, pp. 1660–1671

[10] Kudva, R. A., Keskkula, H., Paul, D. R.: *Polymer*, (2000) 41, p. 335–349

[11] Wildes, G. S., Harada, T., Keskkula, H., Paul, D. R., Janarthanan, V., Padwa, A. R.: *Polymer*, (1999) 40, pp. 3069–3082

[12] Sung, Y. T., Fasulo, P. D., Rogers, W. R., Yoo, Y. T., Yoo, Y., Paul, D. R.: *J. Appl. Polym. Sci.*, (2012) 124, pp. 1020–1030

[13] Mazzucco, M. L. C., et al.: *Journal of Composite Materials*, (2016) 50, pp. 771–782

[14] Salzano De Luna, M., Filippone, G.: *European Polymer Journal*, (2016) 79, pp. 198–218

[15] Ragaert, K., Delva, L., Van Geem, K.: *Waste Management*, (2017) 69, pp. 24–58

[16] Chauhan, V., Varis, J., Kärki, T.: *Procedia Manufacturing*, (2019) 39, pp. 502–508

[17] Buekens, A., Letcher, T. M.: In: T. M. Letcher (Ed.), *Plastic Waste and Recycling*, Elsevier Academic Press, London, 2020, pp. 401–414

[18] Kum, C. K., Sung, Y.-T., Kim, Y. S., Lee, H. G., Kim, W. N., Lee, H. S., Yoon, H. G.: *Macromolecular Research*, (2007) 15, pp. 308–314

9 Chemical Recycling

9.1 Basic Principles

9.1.1 An Overview of Major Recycling Technologies

Bianca Wilhelmus

9.1.1.1 Introduction

This chapter introduces the chemical recycling options for various kinds of feedstock, along with an overview of the existing technologies and the past and current industrial uses of these technologies. After a short overview of reactor types and technologies, thermolytic processes are discussed, from the early stages to most recent applications. This is followed by a review on solvolytic processes for selected polymer types.

There are various ways to recycle plastics after their disposal, both mechanical and chemical. However, there is no "one-size-fits-all" approach – for every special waste problem, a specific decision is required. The choice of the most suitable recycling method depends on the quality and composition of the waste, and it must consider environmental as well as economic criteria. Chemical recycling plays a key role, especially for soiled waste plastics, for mixed plastics or multilayered packaging products, and for plastics that up to now have not been recyclable [1, 2]. The importance of chemical recycling is also illustrated by IUPAC (the International Union of Pure and Applied Chemistry) who selected "Turning Plastics to Monomers" (another way of describing chemical recycling) as one of the top ten emerging technologies in 2019 [3].

Chemical recycling processes aim to convert plastic waste into low-molecular-weight, reusable building blocks like monomers, or to harvest fuels and feedstock for subsequent processes [4]. The idea behind all these processes is to keep the molecules in the loop, without any downcycling.

Whereas the terms "chemical recycling" and "feedstock recycling" are frequently used interchangeably, some authors distinguish between chemical recycling (which produces monomers and other chemicals that can be used for polymer production again) and feedstock recycling (which produces chemicals that can be used for other purposes, such as fuels, syngas or naphtha) [5, 6]. The line between chemical and feedstock recycling is not always clear, but often drawn arbitrarily, as most technologies produce monomers, building blocks, and other chemicals at the same time [6].

Chemical recycling comprises several methods, including thermolysis (pyrolysis or gasification) and solvolysis.

Several factors influence the product distribution resulting from **thermolysis** of plastics. One of the most important factors is the chemical composition of the resins, as the primary reaction products are directly related to the chemical structure and composition of the resin, as well as to its decomposition mechanisms. Furthermore, the thermolysis temperature, the heating rate, and the residence time play important roles: higher operating times and high heating rates facilitate bond breaking, whereas a longer residence time favors secondary conversion processes, with a tendency to higher coke and tar production. Higher temperatures, the use of vacuum, or dilution of the educt are correlated to the formation of smaller (gaseous) molecules; lower temperatures and increased pressure result in more liquid products and secondary product formation [7].

Those technologies employing higher temperatures are commonly referred to as **gasification**. This is usually operated at temperatures higher than 550 °C in an air-lean or oxygen-deficient environment, producing fuel gases with high heating value, such as CO, H_2 and CH_4, and a solid residue [8, 9].

Pyrolysis is commonly done at temperatures below 800 °C, often below 550 °C, with the complete exclusion of oxygen, and sometimes with a catalyst. The products are gaseous or liquid, along with a solid residue. In most cases, the liquid pyrolysis oil is the major - and desired - product [2, 9, 10].

Some of the basic principles of thermal decomposition of thermoplastics are [7, 11]:

- Decomposition into monomer units, also referred to as "unzipping": prevalent in PMMA, PTFE and PS
- Random fragmentation of the polymer chain into fragments with variable length: mainly in polyolefins such as PP and PE
- Elimination of simple, stable molecules: seen in PVC with the formation of HCl

In detail, the decomposition itself can be a combination of initiation, propagation, chain transfer and termination reactions [11–13]. The modes of decomposition differ depending on the type of polymer. Table 9.1 lists the some of the most common thermoplastic polymers and their principal thermal decomposition products [7]. A more detailed description of the mechanisms in the depolymerization of PS, PP, PE, and PET will follow in the next sections of this chapter.

Table 9.1 Major Thermoplastic Polymers and Their Most Important Thermal Decomposition Products [7, 11, 14–19]

Polymer	Mode of thermal decomposition	Main low-temperature products	Main high-temperature products
PE	Random chain rupture	Waxes, paraffin oils, α-olefins	Gases and light oils (e. g. methane, ethene, benzene)
PP	Random chain rupture	Petroleum jelly, olefins	Gases and light oils
PVC	Elimination of HCl from the chain, chain dehydrogenation, and cyclization	Hydrochloric acid, benzene	Toluene
PS	Combination of unzipping and some chain rupture	Styrene and its oligomers	Styrene and its oligomers
PMMA	Unzipping	Methyl methacrylate	Methyl methacrylate and gases (e. g. carbon monoxide, carbon dioxide, propene)
PTFE	Unzipping	Tetrafluoroethylene	Tetrafluoroethylene, hexafluoropropene, cyclo-perfluorobutane
PET	β-Hydrogen transfer, rearrangement and decarboxylation	Benzoic acid, terephthalic acid, monomethyl terephthalate, carbon oxides	Carbon dioxide
PA-6	Unzipping	Caprolactam	Caprolactam, cyclopentanone, hydrogen cyanide

In addition to the thermolytic decomposition of polymers, there is a second, potentially more selective way to recover monomers and building blocks from polymers: **solvolysis**. Usually, the applied temperatures are much lower than those in thermolysis. Solvolysis processes such as hydrolysis, alcoholysis, or ammonolysis cleave specific bonds, e. g. ether, ester, or acid amide bonds, and thus are limited to polymers with these bonds (condensation polymers), such as PET, PU, PA, and PC [2, 20]. In contrast to solvolysis, the term "dissolution" refers to cleaning a polymer by dissolving it and subsequently precipitating it again. No bonds are cleaved, and therefore this method is not considered as a chemical recycling process within the context of this chapter.

Solvolytic processes can be carried out in a wide range of solvents and co-solvents, with or without catalysts, by applying different temperatures and pressures, even under supercritical conditions [21]. Solvolysis is also an option for the recycling of composite materials such as fiber-reinforced polymers, since it allows separation of fibers from the polymer building blocks [22].

9.1.1.2 Technologies

Several technologies are suitable for chemical recycling processes. The primary reactors used in these processes are based on only a few design types, briefly described in this section [7, 22].

Screw reactors such as **extruders** are widely available, have capacities of more than 1 t/h, and are commonly used in the polymer industry for mixing and melting. Operated at elevated temperatures, they can decompose the polymer feed, but a major drawback is their limited operating temperature. Extruder pyrolysis processes have been used to depolymerize PS or PMMA into their monomers or to produce waxes from polyolefins, as done by Japan Steel Works or Union Carbide [7, 23, 24]. If depolymerization of PS or PMMA is intended, the monomers are collected via the degassing domes of the extruder [25].

Tubular or shaft reactors provide a simple setup, but lack the mixing and plasticizing effect of screw extruders [7]. Depending on the design and specific purposes, they are operated horizontally or vertically [26, 27].

In **rotary kilns**, the tumbling action facilitates mixing of the waste, but the sealing of the kiln is a difficult task and may cause safety issues [7]. Rotary kilns are well-established in the treatment of municipal solid waste, but by adjusting the residence time, the spectrum of products can be optimized. Some examples of rotary kiln pyrolysis processes are the Landgard® process for syngas production, the NKK PVC dechlorination process, and the Kobe Steel process for tire pyrolysis [28–31].

Fixed-bed reactors provide a simple solution for lab-scale reactions, but they have relatively low heating rates due to their low heat transfer coefficient [32]. Gases and vapors can be discharged from the reactor during pyrolysis, but solid feedstock and solid reaction products do not move and have to be taken out after the reaction [32]. When larger quantities of solid materials are handled, the non-uniform temperature distribution inside the reactor causes inefficiencies [10]. Fixed bed reactors are sometimes used in catalytic conversions [7].

In **fluidized-bed reactors**, the fluidized bed acts as a heat transfer agent [7]. The heating rate is high, enabling fast pyrolysis processes and an efficient cracking of polymers despite their low thermal conductivity and high viscosity [32]. However, there are special demands on the feedstock size, as the particles need to float within the reactor. Separation of chars from the bed material is also challenging [32]. Scale-up from lab-scale to large industrial scale is possible, but special attention has to be paid to the column design and hydrodynamics to ensure transferability [33].

Gas–solid vortex reactors utilize a rotating bed in a static geometry, with centrifugal forces replacing the gravitational force [34]. High centrifugal acceleration ensures intense heat and mass transfer [22]. The operational reliability of these reactors suffers from limitations, mainly caused by vibrational issues [34].

Stirred-tank reactors are common if a liquid phase is involved. The can be operated discontinuously (batch mode) or continuously. The stirring improves the heat transfer. Special care has to be taken because sometimes reaction products can accumulate in the downstream ducting, and so clog the equipment [7]. It is possible to operate stirred reactors in a cascade, in order to employ a series of different reaction temperatures. This is especially useful if the feed consists of a mixture of polymers with different decomposition temperatures [35].

Autoclaves are closed systems, and reactants or products are neither added nor removed during the reaction. Autoclaves are usually operated in a batch mode, providing the possibility to run reactions under pressure [7, 32].

In a **salt or lead bath reactor**, the polymer waste floats on the surface of the molten bath, with the bath acting as a heat transfer agent [36]. The process has some drawbacks, such as the need for periodic shutdowns to remove residues that accumulate on top of the layer [7], contamination of the reaction products, and high energy consumption [36].

9.1.1.3 Historical Overview of Thermolytic Processes

Industrial application of waste pyrolysis and gasification evolved in the 1970s and 1980s, not only for the treatment of household waste, but sometimes also for decontamination, e.g. removal of polychlorinated dioxins and furans [37]. In many cases, the original driving force behind developing these technologies was finding a solution to the increasing quantities of waste, as an alternative to landfill. Initially, the waste was incinerated to generate energy, but later, the idea of utilizing waste to create chemical building blocks arose, especially with growing amounts of plastic in household waste, particularly from packaging.

Technologies for thermal plastic recycling often evolved from waste incineration technologies, with additional steps regarding input sorting and preparation or purification of the resulting gases, for example. Many of these approaches were not economically viable, so that they stopped after a certain period. Generally speaking, technologies for household waste incineration have had a higher chance of surviving in the long run than technologies for recycling of pre-sorted plastics. Sometimes the technologies never grew out of a pilot or demo scale, while in other cases initiatives stopped after subsidies were phased out.

Although it is impossible to mention all global developments and initiatives for pyrolysis and gasification, the most prominent ones will be described in this section (see Table 9.2 for a summary). Several processes are based on a combination of pyrolysis followed by gasification, while others involve only pyrolysis.

Table 9.2 Examples of Industrial Pyrolysis and Gasification Plants

Process	Technology	Plant owner	Location	Duration
PYROPLEQ®	Rotary kiln	Babcock Krauss-Maffei	Burgau, Germany	1984–2015
ConTherm®	Rotary kiln	RWE	Hamm, Germany	2001–2010
Landgard®	Rotary kiln	City of Baltimore	Baltimore, USA	1974–1979
Schwel-Brenn-Verfahren	Rotary kiln	City of Fürth	Fürth, Germany	1997–1999
Mitsui R21	Rotary kiln	Local communities	Japan (several locations)	Operational
Oxygen Refuse Converter Purox	Shaft reactor	Union Carbide	Charleston, USA	1974–1979
Nippon Steel Direct Melting	Shaft reactor	Local communities	Japan and Korea (several plants)	Operational
Thermoselect	Shaft reactor	EnBW	Karlsruhe, Germany	1999–2004 [a]
BASF Conversion Process	Tubular reactor	BASF	Ludwigshafen, Germany	1994–1996
Veba Oil Combi Cracking Process	Liquid-phase reactor and fixed-bed hydrotreater	Kohleöl-Anlage Bottrop/Veba Oil	Bottrop, Germany	1992–1999
BGL (British Gas Lurgi)	Fixed-bed reactor	SVZ Schwarze Pumpe GmbH	Spreetal, Germany	2000–2007
GSP/Noell	Entrained flow reactor	Local communities	Several countries	Operational
BP Chemicals Polymer Cracking	Fluidized-bed reactor	BP Chemicals	Grangemouth, UK	1994–1998
Ebara TwinRec	Fluidized-bed reactor	Local communities	Japan (several locations)	Operational
Ebara Ube	Fluidized-bed reactor	Showa Denko	Kawasaki, Japan	Operational
Hochtemperatur Winkler (HTW®)	Fluidized-bed reactor	Rheinbraun AG	Berrenrath, Germany	1986–1992
Enerkem	Fluidized-bed reactor	City of Edmonton	Edmonton, Canada	Operational

[a] Adapted process operational in several plants in Japan.

One important thermolysis technology in the last quarter of the 20th century was the **PYROPLEQ®** process, incorporated in several plants in Austria, Germany, Italy, Switzerland and Korea [38]. This process, developed by PLEQ and later, after acquisition, by Mannesmann Demag Energie- und Umwelttechnik, involves an externally heated rotary kiln and process temperatures from 450 to 500 °C, mostly followed by gas combustion at 1200 °C, to produce heat or power [39, 40]. The PYROPLEQ® process was used in the Babcock Krauss-Maffei plant in Burgau/Günzburg, Germany, was established 1984, and closed in 2015 [32, 41]. The plant consisted of two lines with a capacity of 3 t/h each. The pyrolytic gases were incinerated in a subsequent gasifier [41].

With the **ConTherm®** process, RWE carried out a comparable process in their Westfalen power plant in Hamm-Uentrop, Germany, with an annual capacity of 100 kt, starting in 2001 [41]. The plant was closed in 2010 due to economic reasons [41]. The main disadvantage of the process was a significant amount of solid residues and fly ash discards that had to be disposed in landfill, turning it into an environmentally less attractive process compared to other technologies [40].

The **Landgard®** process, used by Monsanto's Enviro-Chem Systems, is another example of pyrolysis in a rotary kiln [42, 43]. An industrial plant in Baltimore, USA, with a capacity of 450 t/d, operated from 1974 until its closure in 1979 due to technical difficulties [41]. Shredded mixed waste was heated up to 1000 °C, and resulting hot gases were passed through a boiler to generate steam [44]. The process was assumed to have advantages over conventional incineration by reducing the volume of waste, but several technical problems occurred, including waste storage, excessive wear of the kiln lining, slag formation, and pollution [44, 45].

The **Schwel-Brenn-Verfahren""** (smolder burn technology), combining rotary kiln pyrolysis at 450 °C with a high temperature gasification, was developed by Karl Kiener [46] and improved by Siemens [47, 48]. A plant with the nominal capacity of 100 kt/a was operated in Fürth, Germany, from 1997, but closed in 1999 due to technical problems [48]. The technology was licensed to Mitsui Engineering & Shipbuilding Co., Ltd and to Takuma Co., Ltd, who started several demonstration and industrial plants in Japan [39]. The technology was further modified to the **Mitsui R21** process and is now considered as fully commercialized in Japan, the first plant in Yame Seibu (close to Fukuoka) starting in 2000 [40, 49].

Union Carbide developed the **Oxygen Refuse Converter Purox** process, a high-temperature thermolysis in a vertical shaft reactor with the addition of oxygen [44, 50, 51]. The waste was fed from the top of the reactor, first passing a drying zone. In the middle of the reactor, thermolysis took place, while oxygen was injected from the bottom of the reactor, where also primary combustion and melting of the slag occurred [29]. A demo plant with 90 t/d capacity for syngas production from waste operated sporadically from 1974 to 1977 in Charleston, USA, but the technology was not brought to larger scale [41, 44].

The **Nippon Steel Direct Melting** process also uses a vertical shaft reactor [52]. It is one of the few fully commercialized advanced processes for thermal waste treatment running since the early 1980s, with a number of plants in Japan and Korea [9, 40]. Pre-crushed waste is thermolyzed at 1000 °C, while oxygen is introduced at the bottom of the reactor. The resulting syngas is combusted with air in a second reactor, to finally generate steam [9, 40]. At a later stage, Nippon Steel also investigated the liquefaction of plastic waste into oils [53].

The **Thermoselect** process, developed by Thermoselect SA, Switzerland, also combines pyrolysis in a shaft reactor with a high-temperature gasification process, but in contrast to several other approaches, the pyrolysis vapor is cleaned [54, 55]. This offers some flexibility, as the gas cannot only be used as fuel, but also as feedstock for other purposes [40]. The first commercial plant applying the Thermoselect process was built in Karlsruhe, Germany, and operated from 1999 to 2004 [40, 41]. The nominal capacity of the plant was 225 kt/a [56]. This process, together with the BGL process, in considered as one of the most advanced gasification processes [41]. The waste feedstock is first compressed and then pyrolyzed at increasing temperatures up to 800 °C. After a residence time of 1–2 h, the waste is completely decomposed to vapor and solid residues, and the vapor is gasified for a few seconds at 1200 °C in a second reactor, while the solid residues are ignited at 1600–2000 °C [40]. The Thermoselect process was also adapted in several plants in Japan [41]. Between 1999 and 2006, Kawasaki Steel (later JFE Steel Corporation, after a merger with NNK Steel) commissioned seven plants with a total of 16 Thermoselect gasifiers [57].

In Japan, **Sanyo Electric** used microwave-heated tubular reactors to melt plastic waste, followed by pyrolysis in a second reactor [31, 58, 59]. A test plant with a capacity of 1 t/d was constructed in 1971, and a pilot plant with a capacity of 3 t/d started in 1972 at the Gifu site [60].

A three-step mixed plastic waste conversion process was investigated at **BASF** in Ludwigshafen, Germany, with the waste stemming from the German packaging collection system. A pilot plant with a nominal capacity of 15 kt/a operated from 1994 to 1996 [7]. The first process step was liquefying the waste input in an agitated reactor. To absorb the hydrochloric acid evolved from any PVC present in the feed a water washer was used in this initial step. In the second process step, the liquid oil was heated up to 400 °C in a tubular reactor, forming gases, naphtha, aromatic fractions, and high-boiling oils, which were separated in the third step [7, 61]. Most of the resulting products could be used in the local Verbund site, e. g. the naphtha in the steam cracker, and some monomers directly in the production of polyolefins. Although the pilot plant demonstrated the suitability of the process, it was not transferred to a larger scale, because the quantities of packaging waste needed for economical operations were not available at that time [7].

In the **Veba Oil Combi Cracking** (VCC) process, depolymerization of mixed plastic waste at 350–400 °C was followed by two downstream process steps. The gaseous product of the depolymerization was partially condensed and fed into a hydrotreater where the hydrochloric acid was removed. The resulting chlorine-free condensate and gas were mixed with the depolymerizate and hydrogenated in a liquid-phase reactor at 400–450 °C under pressures of up to 100 bar. The final product was a synthetic crude oil, free of chlorine and poor in oxygen and nitrogen [6, 62–64]. A plant at the former Kohleöl-Anlage Bottrop (KAB) site, Germany, originally started in 1981 as a hydrogenation plant for producing naphtha and gas from coal, but was modified later to accept waste as input. In 1992/1993, the depolymerization unit was added [6], with a nominal capacity of 80 kt/a. Due to economic reasons, it stopped in 1999 [6]. A fruitless attempt to sell the pilot plant via internet auction in 2000 raised some media furor [65]. In 2002, the entire refinery was sold to BP. The technology received an upswing when an increasing market demand and higher oil prices caused BP later to construct a new pilot plant based on the VCC process in its Naperville facility in the US. The technology was also licensed to NAIF-NK in Tartastan and Yangchang Petro in China [66].

The **BGL (British Gas Lurgi)** slagging bed gasifier process used high-temperature gasification of waste to produce syngas and a vitrified slag [67]. The fixed bed Lurgi gasifier was originally developed to gasify lignite and coals [68]. From 1958 to 1965, British Gas modified the Lurgi gasifier and ran a demo plant in the UK with a capacity of 100 t/d [69]. The first commercial BGL gasifier operated from 2000 to 2007 in Spreetal, Germany, at SVZ (Sekundärrohstoff-Verwertungszentrum Schwarze Pumpe GmbH) [69]. Gasification took place at temperatures of up to 1600 °C and pressures of 25 bar. Organic compounds were converted to gas while inorganic compounds were bound in the slag. The slag viscosity was adjusted by adding limestone. The nominal capacity of the BGL gasifier at SVZ was 312 kt/a [48]. The plant produced syngas for methanol production and power generation from household waste and coal, but finally closed due to economic reasons [41].

As well as the BGL gasifier, SVZ also used an **entrained flow gasifier**, developed by Deutsches Brennstoffinstitut (DBI) in Freiberg, Germany, in 1975 under the name **GSP** (Gaskombinat "Schwarze Pumpe") [69, 70]. The technology was acquired by Noell Group in 1991 (later Babcock and Future Energy), before being taken over by Siemens in 2006. This gasifier was a top-fed, top-fired reactor with a water wall screen in the gasifier able to handle liquid or solid feedstock, and produced syngas [69]. A comparable gasifier for utilization of "nitrogen-organic compounds" was installed at the BASF Seal Sands facility in the UK, but was decommissioned in 2009 [71, 72]. Siemens gasifiers operating with this technology have since been installed in a number of other countries such as Czech Republic, China, and the USA [73].

Another example of entrained-flow gasification is the **Texaco** process [74]. This process is based on the liquefaction of pre-shredded waste, followed by gasification in an entrained-flow gasifier at 1200–1500 °C with a mixture of oxygen and steam as the gasification agent [5]. After purification, clean and dry syngas is obtained [22]. Any chlorine present in the plastic waste input is captured in the washing step by adding ammonia and thus forming NH_4Cl [64]. A pilot plant with a capacity of 10 t/d was built in California in the 1970s [64, 75], and since then several other projects have been realized globally [76]. Texaco and Chevron merged to Chevron-Texaco in 2001 and sold the gasifier technology to GE Energy in 2004.

The fluidized-bed technology is yet another suitable method for thermolysis of plastic waste. Fluidized-bed reactors offer several advantages, for instance scalability, feedstock flexibility, no moving mechanical parts in the hot reactor regions, and a very homogenous temperature distribution, permitting a consistent product spectrum. The University of Hamburg, with research originally led by Sinn and Kaminsky, has pioneered investigations into fluidized-bed pyrolysis of plastic waste since 1970 [5, 31, 77, 78]. In contrast to many other fluidized-bed processes, predominantly developed in Japan, the **Hamburg** process was intended for the production of organic chemicals with a high aromatic content [77]. The University of Hamburg operated a pilot plant with a capacity of 20 kg/h and a demo plant with 120 kg/h throughput. The fluidized bed consisted of 0.2–0.5 mm quartz sand particles, and was indirectly heated up to 600–900 °C. Product gases were condensed and distilled [77]. Sinn, Kaminsky and their co-workers investigated the pyrolysis behavior and product spectra of various feedstocks such as polyolefins, polystyrene, mixed plastic waste, and scrap tires, and provided valuable insights for subsequent generations of researchers.

Based on the fluidized-bed process developed by Sinn and Kaminsky, the company **D.R.P.** (Deutsche Reifen- und Kunststoffpyrolyse GmbH), an affiliate of the German company Eckelmann, started a pilot plant in Ebenhausen, Germany, in 1984, which was taken over by Asea Brown Bovary (ABB) in 1985 [79]. In two units with a capacity of 1 t/h each, tires and waste polyolefins were pyrolyzed at 700–800 °C. Although a stable operation was achieved, the facility was closed due to economic reasons [41].

A comparable concept, also based on fluidized-bed pyrolysis, was the **BP Chemicals Polymer Cracking Process** (PCP), started with research on a laboratory scale by BP Chemicals [7, 80]. A consortium of companies (BP Chemicals, Elf Atochem, Enichem, DSM, and CREED) worked on the improvement of a pilot plant demonstration unit. A pilot plant with a nominal capacity of 400 t/a was operated from 1994 to 1998 at the BP site in Grangemouth, UK, followed by a feasibility study for the installation of a 25 kt/a plant (EU project "POLSCO") [38, 81]. Plastic waste was pyrolyzed in the absence of air at 500 °C, which is lower than in the Hamburg process. Chlorine in the feedstock was converted to HCl and subsequently reacted with lime [6]. The pyrolysis of mixed plastic waste resulted in the formation of ethylene, propylene and a wax-like hydrocarbon mixture. In ratios up to 20%, this could be used with naphtha as feed for the conventional steam crackers on-site [5]. Although the process was shown to produce a stable hydrocarbon source as a product from mixed plastic waste, the economics of a full-scale plant were not favorable enough, due to the dependence on the price of virgin mineral oil [82].

An important example of a fluidized-bed process combined with a gasification step is the **Ebara TwinRec** process [83]. First, the waste is pyrolyzed at 500–600 °C in a revolving fluidized-bed gasifier with a silica sand bed. The pyrolysis gases are further reacted at temperatures of 1350–1450 °C in the second step, while slagging the solid residues [40, 48, 84]. A pilot plant with a capacity of 7 t/d was operated in Sodegaura, Japan, from 1995 to 1997, followed by a 20 t/d plant in Fujisawa from 1997 to 2001 [48, 85]. Commercial operation of a plant with two lines of 225 t/d each started in 2000 in Aomori, Japan, followed by other commercial facilities in Japan [85, 86].

In another development by Ebara Corp., in this case jointly with Ube Industries Ltd, plastic waste is gasified at 10 bar in a two-step process [87]. In contrast to the TwinRec process, the **Ebara Ube** process does not incinerate the syngas but uses this as raw material or fuel [48]. In 2000, a first industrial plant based on the Ebara Ube process with a capacity of 30 kt/a was established close to Ube, Japan [48, 88]. The Ebara Ube process was also commercially applied at a scale of 70 kt/a in the Kawasaki plant of Showa Denko in 2003, which is still in operation today. Recently, JCG Holdings Corp. announced the conclusion of a licensing contract with Ebara Environmental Plant Co., Ltd, Ube Industries Ltd, and Showa Denko K. K. to cooperate on engineering, procurement, and construction for plastic waste gasification facilities based on the Ebara Ube process [89].

Another example of high-temperature fluidized-bed gasification is the **Hochtemperatur Winkler (HTW®)** process, an evolution of the early steam/oxygen coal gasifier patented by Winkler in 1922 [90, 91]. From 1986 to 1992, Rheinbraun AG ran a demo plant with a capacity of 25 t/h in Berrenrath, Germany, to carry out combined gasification of waste and coal at 10 bar. The syngas was further processed to methanol by DEA [92]. **ThyssenKrupp** acquired the HTW® technology from RWE in 2010, and still offers it today [93, 94].

One other commercially available example of fluidized-bed gasification is the **Enerkem** process [95]. In a plant in Edmonton, Canada, Enerkem produces syngas (and subsequently methanol and ethanol) from municipal waste and biomass [91]. Founded in 2000, Enerkem initially used biomass as feedstock, but later also used municipal and plastic waste [96]. The waste feedstock is first treated to remove chlorine and sulfur. The second step is gasification at 600–700 °C to a crude syngas, which is then further cleaned and processed [97]. The industrial-scale plant with a capacity of 100 kt/a started in 2014 [98, 99]. Enerkem is part of a consortium planning a waste-to-chemicals plant in Tarragona, Spain, operational in 2025, with a capacity of 400 kt/a [100]. Further expansion to other locations in Canada was announced [100].

Some of the new approaches for pyrolysis and gasification will be described in Section 9.1.1.4.

In contrast to many of the thermolytic processes described here, technologies for solvolysis are still in the development phase [98]. A few recent developments will be covered on the following pages.

9.1.1.4 Recent Approaches for Pyrolysis and Gasification

Although the basic technologies for plastic recycling are known and a lot of attempts have been made in the last 50 years to recover valuable chemicals from plastic waste, many of the processes never grew beyond the pilot scale. Apart from the Enerkem gasification, very few other processes (amongst them the Mitsui R21 process, the Ebara TwinRec process, and the Ebara Ube process) are considered to be fully commercialized. It might not be a coincidence that these technologies originated from Japan, because the environmental policy of the Japanese Government has traditionally promoted plastics recycling. In 1997, the Japanese government established a containers and packaging recycling law for PET bottles and in the year 2000 for other plastics. This law created attractive markets and tremendously increased plastic recovery [7, 88]. Nevertheless, the main driver still seemed to be waste elimination and not the recovery of building blocks or the utilization of the recycled materials as starting materials for the production of new goods. However, once waste collection, pyrolysis and/or gasification are established, it needs only one additional step to upgrade the existing plants to recover the reaction products for further use in the chemical industry.

Globally, a trend to increasingly stringent standards and legislation for reducing the impact of waste plastic in the environment is apparent [101]. There is a need for improved waste management, including the use of waste as a valuable resource for the circular economy [102]. In the light of increasing amounts of plastic waste generated each year, the European Union has set a clear target that by 2030, all plastic packaging is either reusable or can be recycled, and that more than half of plastic waste will actually be recycled. The implementation of this strategy is also supported by EU research funding, such as the Horizon 2020 program [103].

Like Japan, China and South Korea have adopted national "top-down" strategies to enable the circular economy [104]. China has recently begun to reject plastic waste from abroad, and it is expected that other countries like Malaysia will follow [2]. Brazil, India, and the USA are taking the "bottom-up" approach [104]. In the USA, due to the absence of a federal recycling law, state and local governments are responsible for their own recycling legislation [105].

The lack of suitable waste streams, logistics, or commercial viability are often named as reasons for the discontinuation of demonstration or pilot plants [2]. The availability of (pre-)sorted plastic waste with a consistent minimum quality is also a precondition for a stable, economical operation of a recycling plant. Compared to the 1990s, there are indications that the general conditions for successful chemical recycling have changed in recent years. Legislation and regulation are strong drivers towards a circular economy, but increasing consumer awareness of environmental concerns, better collection schemes and easier access to recycling, higher fees for landfill, and improved waste sorting play an important role in increasing the availability of suitable feedstock as well. Chemical recycling processes benefit from an integrated flow of material and energy within the framework of a chemical industry site, helping to ensure the utilization of all products and residues in an economical way [1].

Additionally, the self-commitment of the chemical industry should not be neglected. To increase recycling rates and foster innovative technologies, numerous industry initiatives have been founded, some of them related to general recycling of all polymer types, others dedicated to only one polymer type, such as Styrenic Circular Solutions (SCS) for polystyrene [106], Petcore Europe for polyethylene terephthalate [107], Vinyl Circular Solutions for polyvinyl chloride [108], and the Polyolefin Circular Economy Platform (PCEP) for polyolefins [109], to name only a few.

Many of the aforementioned processes were intended to generate energy (e.g. in the form of steam), or oils as fuel replacement, or syngas. The recovery of monomers and other building blocks from plastic waste has gained more interest lately, shifting from creating fuels to creating raw materials for the chemical and plastics industry [110]. Several new approaches were started, and many major plastic manufacturers have announced own efforts or collaborative projects to utilize the output of recycling processes in their products [110, 111]. The number of companies currently active in the field of chemical recycling is astonishingly large, especially for fuel or syngas production; therefore the following selection of companies can only highlight a few of the recent developments, concentrating on the conversion of plastic waste to building blocks, monomers, or naphtha. Table 9.3 lists the current most prominent examples.

Table 9.3 Examples of New Polymer Recycling Initiatives Based on Pyrolysis and Gasification

Company	Country	Input	Main products
Agilyx	USA	Mixed plastics	Synthetic crude oil
		Polystyrene	Styrene
BASF (ChemCycling®)	Germany	Mixed plastics	Synthetic crude oil
Recycling Technologies	UK	Mixed plastics	Synthetic crude oil, wax
		Polystyrene	Styrene
GreenMantra Technologies	Canada	Mixed plastics	Waxes, additives
		Polystyrene	Styrene
Pyrowave, Inc.	Canada	Polystyrene	Styrene
Enval Ltd	UK	Aluminum-plastic laminate packaging	Synthetic crude oil, aluminum
Plastic Energy Ltd	UK	Mixed plastics	Synthetic crude oil
New Hope Energy	USA	Mixed plastics	Synthetic crude oil
Licella Pty Ltd/ReNew ELP	Australia/UK	Mixed plastics or other feedstock	Synthetic crude oil
OMV	Austria	Mixed plastics	Synthetic crude oil

Table 9.3 Examples of New Polymer Recycling Initiatives Based on Pyrolysis and Gasification *(continued)*

Company	Country	Input	Main products
Indaver NV	Belgium	Polyolefins	Waxes
		Polystyrene	Styrene
Brightmark LLC/RES Polyflow	USA	Mixed plastics	Synthetic crude oil
Dyneon GmbH	Germany	Polytetrafluoroethylene	Tetrafluoroethylene

Agilyx, founded in 2004 under the name Plas2Fuel, now parented by Norwegian-based Agilyx AS, pyrolyzes waste plastic. First, the technology was used to produce fuel from mixed plastics, but in 2015, due to declining oil prices, the company moved to the production of synthetic crude oil, and later also to the depolymerization of polystyrene into styrene monomer [112]. The process starts with melting the plastic waste in an extruder before pyrolyzing it in a continuous dual-screw reactor [97, 113]. A pilot plant for mixed waste pyrolysis in Oregon, USA, with a nominal capacity of 50 t/d has been operated since 2014, followed by a PS recycling plant in 2018 [67, 112]. New recycling plants based on the Agilyx technology are planned in Chiba, Japan, by the polymer manufacturer Toyo Styrene Co., Ltd and with other partners in Europe and the US [112].

Late in 2018, **BASF** announced a successful pilot production run with oils from their **ChemCycling®** plant. The process was developed in cooperation with Recenso GmbH (formerly Alphakat Engineering GmbH) and its daughter company Carboliq GmbH in Ennigerloh, Germany [114, 115]. Mixed plastic waste is liquefied at 400 °C by the combined application of thermal, catalytic and mechanochemical (tribochemical) mechanisms, applying friction as the only source of energy. The process is named **Catalytic Tribochemical Conversion** (CTC), also known as the Dieselwest process [115]. The resulting crude oil was fed into the steam cracker at the BASF Verbund site in Ludwigshafen, Germany, to finally produce olefins. The first commercial products from the process (allocated by mass balance) were launched in 2020 [114, 116]. Carboliq employs the CTC process in a modular concept, with a capacity of 600–1000 kg/h input for each module [115].

The company **Recycling Technologies**, UK, founded in 2011, has developed a technology to transform mixed plastic waste into shorter-chain hydrocarbons, marketed under the name Plaxx® and intended for the use as naphtha or crude oil replacement or as refinery wax [117]. The process is pyrolysis at 400–600 °C in a fluidized-bed reactor [118]. A demonstration plant in Swindon, UK, was inaugurated in 2018. A first commercial-scale unit is planned in Scotland [119]. The company has commercialized modular units ("RT7000") with a capacity of 1 t/h, equating to about 7 kt/a [117]. Together with INEOS Styrolution and Trinseo, two plants with a capacity of 15 kt/a each are planned in France and Belgium [119].

GreenMantra Technologies, founded in 2010, started a commercial-scale catalytic depolymerization plant for the conversion of mixed plastics to waxes and additives in 2014 in Brantford, Canada. The process is carried out by feeding the molten plastic waste into a high-pressure stirred-tank reactor, where catalytic depolymerization takes place at 300–600 °C [120]. Later, the technology was also adopted to the depolymerization of polystyrene [121, 122]. A partnership with Sun Chemical was established in 2017 to use polymers from recycled PS waste in ink formulations [111, 122].

Pyrowave Inc., another Canadian startup, is active in the use of microwave radiation for polystyrene depolymerization [123]. A catalyst is used to absorb microwaves, transfer the heat to the waste, and thus initialize the depolymerization [124]. The resulting reaction mixture is separated in two sequential condensers in order to split styrene from other reaction products [97]. In 2020, Pyrowave signed a joint development agreement with Michelin Group to develop an industrial demonstration plant in the next three years [125].

Enval Ltd, UK, has specialized in the microwave-assisted pyrolysis of aluminum-plastic laminate packaging. Both pyrolysis oils and aluminum are recovered [126, 127]. The process runs continuously at 500–600 °C in a reactor with two interconnected, stirred chambers that contain a bed of carbon black [97, 128]. The offered equipment is based on a modular concept with 2 kt/a capacity each.

Plastic Energy Ltd, UK, is a supplier of modular pyrolysis plants, converting mixed plastics into a pyrolysis oil branded **TACOIL** (thermal anaerobic conversion oil). Thermal degradation of the waste takes place in a stirred-tank reactor [129]. Two plants in Spain have been operational since 2014 and 2017, respectively [130, 131]. Several polymer manufacturers (including SABIC Innovative Plastic, Total, and INEOS) have tested the oil as cracker feedstock and are planning to start recycling facilities based on this technology in the next years. Further recycling plants are foreseen to be constructed globally [130, 132, 133].

New Hope Energy, USA, builds pyrolysis plants and operates a commercial scale recycling facility with a capacity of 53 kt/a in Texas since 2019 [134]. The continuous process runs with the use of molten salt as heat exchange medium [135]. A cooperation agreement with Chevron Phillips was announced, and a capacity expansion to more than 300 kt/a is planned [134].

The companies **Licella Pty Ltd**, Australia, (previously Ignite Energy Resources Ltd) and **ReNew ELP** ("End-of-Life-Plastics"), UK, use a catalytic hydrothermal reactor (**Cat-HTR**®) [136–138]. The reaction takes place in supercritical water and produces naphtha, gas oils, and wax residues [139, 140]. Besides plastic waste, biomass and other feedstock can also be processed to a crude oil. A pilot plant with a capacity of 10 kt/a was commissioned in New South Wales, Australia, in 2012 [140, 141]. ReNew ELP plans the construction of an 80 kt/a industrial plant in Teeside, UK, with the help of **Mura Technology**, a joint venture of Licella and Armstrong Capital, using the Licella technology [142].

The **Re-Oil** process developed by **OMV** aims for the depolymerization of polyolefins and polystyrene. Since 2018, a pilot plant with a capacity of 800 t/a has been operational in the OMV refinery in Schwedt, Austria, and further scale-up of the process to 200 kt/a is planned [143]. The polymer waste is molten in an extruder and then mixed with a solvent carrier medium. A liquid, high-boiling refinery side-product acts as a solvent to improve the viscosity and thermal conductivity of the polymer melt. The mixture is heated to 400 °C in a tubular reactor, where the depolymerization takes place. Light reaction products are removed and distilled, and unconverted plastic and heavier products are circulated to the reactor once more. The main product is a synthetic crude oil [136, 143, 144].

In collaboration with Ghent University and the University of Antwerp, the company **Indaver NV**, Belgium, has started the project **Plastics2Chemicals** for the depolymerization of dedicated streams of polyolefins and of polystyrene to base chemicals [145]. A 15 kt/a demo plant in Antwerp is planned, and a collaboration with INEOS Styrolution is announced to align the output of the depolymerization process with feedstock specifications of the polymerization process [146].

The company **Brightmark LLC** (controlling owner of **RES Polyflow** since 2018), USA, is constructing a pyrolysis plant with a capacity of 100 kt/a in Indiana [147]. In a continuous process, plastic waste is moved through multiple reactor zones by either a conveyor belt or an auger with a fixed speed. In every reactor zone, the pyrolysis gases are removed and condensed to a crude oil [97, 148].

A special case is the depolymerization of PTFE by **Dyneon GmbH**, a daughter company of 3M, in their Burgkirchen/Gendorf facility, Germany. Since 2015, the company has been operating a pilot plant for monomer recovery from PTFE waste, with a nominal capacity of 500 t/a [149]. The process was developed in a public-funded project in cooperation with the University of Bayreuth and the research institute InVerTec [150].The depolymerization takes place at 550–700 °C in a fluidized-bed reactor heated by microwaves [151].

A unique process that does not really belong in the same category as the other thermolytic processes mentioned so far, is the **Accelerated Thermal Oxidative Decomposition®** employed by the startup company **Novoloop**, USA (formerly **BioCellection Inc.**). HDPE and LDPE are treated with an oxidizing acid, for example nitric acid, at 60–200 °C in a stirred-tank reactor and thereby broken down into oxygenated organic compounds such as succinic, glutaric, adipic, pimelic, suberic, and azelaic acid [97, 152, 153].

9.1.1.5 Solvolytic Processes

The thermolytic processes mentioned above are all more or less suitable for the depolymerization of addition polymers such as polyolefins or polystyrene; some processes are intended for general application to mixed waste, whereas some are for presorted feedstock or for only a single polymer type. In contrast to these, for the recycling of condensation polymers, another possibility exists: solvolysis. Solvolysis is a potentially more selective way to recover monomers from polymers containing ether, ester, and acid amide bonds, and has gained significantly more interest in R&D and industrial applications in recent years. Most of the initiatives focus on PET or PU, and to a lesser extent on PA, PC, and PLA [2].

Processes and technologies for solvolysis of the most important condensation polymers will be discussed below. In contrast to the thermolytic methods discussed in Section 9.1.1.4, solvolytic recycling methods are mostly dedicated to a specific polymer type; therefore, the section is organized by polymer type rather than by technology.

9.1.1.5.1 Polyethylene Terephthalate

Although polyethylene terephthalate (PET) is often recycled mechanically, degradation of the polymer chains after a few cycles means that there is a need for a supplementary chemical recycling route. There are several possibilities for the solvolysis of PET, and depending on the process, different products (such as terephthalic acid (TPA), bis(2-hydroxyethyl) terephthalate (BHET), and dimethyl terephthalate (DMT)) are created [2, 154]. Table 9.4 summarizes the principal methods. After purification, BHET and DMT can be used directly for the re-polymerization to PET, but TPA must be converted to BHET first [2].

Table 9.4 Solvolysis Processes for the Chemical Recycling of Polyethylene Terephthalate

Process	Primary reaction products [a]	Company examples
Glycolysis	BHET and oligomers, EG	▪ CuRE Technology B. V. ▪ Eastman Chemical Co. ▪ Garbo SRL ▪ IFP Energies nouvelles ▪ Ioniqa Technologies B. V. ▪ JEPLAN Inc. ▪ Nan Ya Plastics ▪ PerPETual Global Technologies
Hydrolysis	TPA, EG	▪ BP Chemicals ▪ Carbios ▪ Gr3n ▪ Rittec Umwelttechnik

Table 9.4 Solvolysis Processes for the Chemical Recycling of Polyethylene Terephthalate *(continued)*

Process	Primary reaction products [a]	Company examples
Methanolysis	DMT, EG	▪ Eastman Chemical Co. ▪ ITOCHU Co. ▪ Loop Industries, Inc.
Ammonolysis	TPA diamide, EG	Only on lab scale
Aminolysis	BHETA, EG	Only on lab scale

[a] BHET: bis(2-hydroxyethyl) terephthalate; BHETA: bis(2-hydroxyethylene) terephthalamide; DMT: dimethyl terephthalate; EG: ethylene glycol; TPA diamide: terephthaldiamide; TPA: terephthalic acid.

GLYCOLYSIS is the oldest method for PET depolymerization. Besides the formation of monomers, glycolysis also creates other products and low-molecular-weight oligomers that are useful for the chemical industry, but it is best suited for high-quality, pure feedstock. Copolymers, colorants or dyes are rather difficult to remove and require special procedures [22, 155]. Glycolysis works by catalytic transesterification of PET with an excess of a glycol at 110–270 °C. Ethylene glycol (EG), diethylene glycol (DEG), propylene glycol (PG), polyethylene glycol (PEG), 1,4-butanediol or hexylene glycol can be used [22, 155].

CuRE Technology B. V. is operating a pilot plant in Emmen, Netherlands, for the glycolysis of PET to oligomers, targeted on colored polyesters, including carpets and textiles. A commercial-scale facility with a capacity of 25 kt/a is planned [156].

Eastman Chemical Co. stated the construction of a polyester recycling plant with a capacity of 100 kt/a in Tennessee, USA, to be completed in 2022. Eastman named their polyester solvolysis processes “polyester renewal technology”, comprising both glycolysis and methanolysis [157–159].

In 2019, **Garbo SRL**, Italy, announced a cooperation with Plastipak to implement a recycling plant for PET glycolysis [160]. The process, named ChemPET, was developed in 2017 [161]. After glycolysis with ethylene glycol, five purification steps are used in their process to remove solids, other polymers, salts, and other contaminants for creation of pure BHET crystals [162].

In 2020, **IFP Energies nouvelles**, its subsidiary **Axens** and **JEPLAN Inc.** signed a joint development agreement to commercialize the “Rewind® PET” process, in which PET glycolysis will be combined with crystallization and distillation steps for purification. JEPLAN has acquired the industrial PET bottles chemical recycling plant from PET Refine Technology in Japan, which runs with a capacity of 22 kt/a. The technology was originally developed by Aies Co., Ltd [163, 164]. Since 2018, JEPLAN has operated a 2 kt/a PET textile demo plant, which should now be brought to a larger scale within this collaboration [165].

Ioniqa Technologies B. V., a spinoff from Eindhoven University, Netherlands, applies magnetic ionic liquids as the solvent and/or catalyst for the glycolysis of PET to BHET. After the reaction, water is added and the magnetic fluid can be separated in a magnetic decanter from the aqueous solution of BHET [162, 166]. A 10 kt/a pilot plant was commissioned in 2019 [167].

Nan Ya Plastics, Taiwan, has commercialized PET glycolysis in a 216 kt/a plant. Under the brand ECOGREEN, PET bottle flakes are processed to yarn and fibers [168, 169].

The company **perPETual Global Technologies** utilizes ethylene glycol for glycolysis. Since 2010, the process has been used on an industrial scale in Nashik, India, where 2 million plastic bottles per day are converted into filament yarn [170, 171]. In the process, PET is not completely depolymerized to monomers, but to low molecular weight oligomers instead [2].

Teijin Fibers Ltd also operated a glycolysis plant in Japan from 2003 to 2018 with a capacity of 62 kt/a ("ECOCIRCLE"), but withdrew from production due to economic reasons and export constraints to China [164, 172].

HYDROLYSIS is another industrially important process for PET recycling. This can take place in acidic, alkaline, or neutral aqueous media, often at elevated temperatures and pressure to break the polymer chain into terephthalic acid and ethylene glycol [173].

BP Chemicals started to construct a hydrolysis pilot plant in their facility in Naperville, USA, to apply their Infinia® process [174]. Early in 2021, the business was sold to **INEOS**, who continued to develop the technology and pilot plant [175].

Carbios, France, applies enzymatic hydrolysis of PET [176, 177]. Selective depolymerase enzymes even enable the recycling of blends, without requiring high temperatures [2, 162]. The hydrolase enzyme was chosen as a compromise between productivity and enzyme cost. It is estimated that the cost for the enzyme needed for recycling represents 4% of the price for virgin PET [178]. Funds for the construction of a pilot plant with a planned capacity of 50–100 kt/a have been acquired [169, 172].

Gr3n, Switzerland, uses a caustic ethylene glycol solution for alkaline hydrolysis in a microwave-heated reactor [179]. Ethylene glycol is recovered by distillation, and terephthalic acid salts are precipitated by hydrochloric acid. An electrolysis process is present for HCl and NaOH recovery [162]. An initial demonstration plant was started in 2021 in Italy, a 30 kt/a industrial plant will follow [180].

Germany-based **Rittec Umwelttechnik** has also developed a process for PET hydrolysis. The continuous alkaline hydrolysis process aims especially for the depolymerization of multilayer or colored packaging waste, as an extruder is used to apply shear force to the material [181, 182].

Other than glycolysis or hydrolysis, there are only a few technologies using **METHANOLYSIS** that have progressed beyond the laboratory scale. One possible reason could be the production of other alcohols and polymer derivatives during methanolysis, which provide difficulties for product purification [169].

One of the few examples is **Loop Industries, Inc.**, Canada. The company uses methanolysis and a purification process [169, 183]. Together with Suez, Loop Industry plans to build an industrial-scale recycling plant in France with a capacity of around 70 kt/a, to be commissioned in 2025 [169, 184].

The recycling of waste PET fibers and fabrics from pre- and post-consumer textiles is targeted by the RENU project, initiated by **ITOCHU Co.**, Japan [185]. A commercial plant with a capacity of 30 kt/a is already operational [169].

Polyethylene terephthalate ammonolysis and aminolysis are reported only on the laboratory scale; therefore, these methods are not discussed here in detail.

9.1.1.5.2 Polyurethanes

Another class of polymers in need of innovative recycling procedures are polyurethanes. A large variety of isocyanates and polyols can be used to create polyurethanes, so PU rather stands for a plastics industry segment than for a single, well-defined polymer resin [186]. While, in principle, mechanical recycling (other than regrinding and the use as filler) is possible for thermoplastic polyurethanes (TPE-E), this is not possible for the PU foams due to the cross-linked structure of these thermosets [187]. In Europe, flexible foams and rigid PU foams together represent about 68% of total polyurethane production [188].

Pyrolysis is not the optimal choice for PU recycling, because due to the nitrogen present in the polymer, HCN is formed, so that the pyrolysis gas must be monitored for toxicity [22, 32]. A viable option for the chemical recycling of polyurethane is hydrolysis to polyols and amine-based products. Recycling of PU waste by hydrolysis started as early as the 1970s in the USA [186]. Although different hydrolysis processes are possible, the most important method of recovering monomers from PU today is glycolysis [2]. Aminolysis and phosphorolysis processes are still in the stage of preliminary investigation at a laboratory scale [187].

GLYCOLYSIS implies the conversion of the ground PU waste in high-boiling-point glycols, usually diethylene glycol (DEG), with a catalyst and diethanolamine (DEA) as a co-reagent at temperatures of 180–220 °C [186]. The main limitation of glycolysis is the difference in process parameters for flexible and rigid foams, requiring separation of the used waste prior to recycling [188]. Various polyurethane processors have evaluated glycolysis technologies on pilot scales up to 1 t/h, but plans for larger-scale, commercial plants (with capacities of ca. 50 kt/a) have not materialized yet, mostly due to the lack of sufficiently homogeneous, good-quality end-of-life PU feedstock, as well as long-term contractual feed supply. In Europe, several small-scale commercial glycolysis plants exist, e.g. in Italy, Austria, Germany, and the UK [189].

One of the recent approaches is the RENUVA® project, initiated by **Dow Polyurethanes**, Switzerland, for the glycolysis of PU mattresses. In the frame of this project, Dow inaugurated an industrial-scale plant in Semoy, France, in 2021 [190].

Flexible polyurethane foams are recycled at **Ikano Industry**, Poland, with glycolysis technology developed by H&S Anlagentechnik GmbH [191–193].

RAMPF Eco Solutions GmbH & Co. KG, Germany, offers plants for polyol production from PU and PET waste by solvolysis (glycolysis, acidolysis, and polyolysis). Plants have already been installed in several countries [161, 194, 195].

9.1.1.5.3 Polyamides

Compared to other plastics, polyamides provide a greater challenge for depolymerization due to the stability of the backbone amide bond [196]. The two commercially most important polyamides are poly(hexamethylene adipamide) (PA-6,6) and polycaprolactam (PA-6). PA-6 consists of only one monomer, caprolactam, and depolymerization is possible (for instance) with superheated steam in the presence of a catalyst. PA-6,6 results from polymerizing two monomers, adipic acid and hexamethylene diamine, adding more complexity to the depolymerization process [197]. For both PA-6 and PA-6,6, depolymerization by hydrolysis, glycolysis, or ammonolysis have also been reported, but mostly only on a laboratory scale [116, 198, 199].

Aquafil Group, Italy, has operated a commercial plant with a capacity of 40 kt/a for the depolymerization of PA-6 in Slovenia since 2011. In 2018, a 16 kt/a plant dedicated to carpet recycling in Phoenix, USA, was started. The recycled polymer is branded ECONYL® [169, 200, 201].

9.1.1.5.4 Polycarbonate

The most widely used polycarbonate is poly(bisphenol A carbonate) [202]. Various methods for the chemical recycling of PC to recover raw materials have been reported, including thermal pyrolysis, alcoholysis, aminolysis, and hydrolysis [198]. However, solvolytic processes are complicated due to the rigidity of the macromolecular backbone that protects the carbonate bond and provides some resistance, therefore dissolving (or at least swelling) of the PC surface is necessary [202]. Despite several patented processes from industrial companies, viability on the industrial scale seems to be lacking [203].

9.1.1.6 Outlook

Chemical recycling of polymers has been a topic in academic and industrial research for decades. Many proposals have been launched over the years, whether in pyrolysis, gasification, or solvolysis. Despite a typically enthusiastic beginning, very often projects have been stopped, and demonstration or pilot plants closed.

Why is it that chemical recycling initiatives, even very promising ones, so seldom thrive? There are numerous possible reasons, and how these combine to culminate in closure in individual cases is usually a matter for speculation.

- **Economic reasons:** As far as industrial companies are involved, processes need to be economically viable. All costs for the recycling process, including feedstock preparation, and treatment and purification of products, must be compared with the cost of virgin material. Even considering that a small-scale recycling process can hardly compete with processes in efficient, optimized world-scale polymer plants, the cost of recycled products needs to be at least within certain limits to be accepted by the market. However, the market situation is also influenced by government regulations and by public perception. If regulations require, for instance, a certain amount of recycled material within a product, or a quota of products to be recycled at the end of their use phase, this might change the picture and enable profitable recycling processes.
- **Feedstock availability:** Quite a few industrial recycling initiatives have been stopped due to lack of feedstock. To ensure the smooth operation of a recycling facility, continuous availability of suitable feedstock is required. In the past, this was sometimes not certain: recycling processes and pilot plants were in place, but feedstock supply was lacking. However, modern collection schemes and consumer awareness of recycling are now helping to provide enough plastic feedstock.
- **Feedstock quality:** Plastic waste is not a homogenous polymer fraction. Post-consumer waste in particular is a mixture of various polymer types, containing additives, colorants, impurities and contaminants, moisture, and dirt. Not every recycling process can cope with every kind of feedstock, and often a costly pre-treatment or sorting step is required. Sometimes the heteroatoms introduced by the contaminants (oxygen, nitrogen, sulfur, chlorine and the like) cause trouble. Not every process is sensitive to the same contaminants or shows similar robustness to variations in quality, so the feedstock should be chosen accordingly to ensure a smooth recycling process.

Although there are still some challenges to be overcome, chemical recycling of plastic waste can provide many benefits. Contaminants such as halogenated flame-retardants can be removed very effectively, and it is possible to create virgin-like polymers from waste polymers, which enables a versatile use in all applications, even in food contact or medical purposes. Chemical recycling as a supplement for mechanical recycling can increase overall recycling rates, because it utilizes a wide range of feedstock that otherwise would be incinerated. Numerous initiatives both in academic research and in industry in recent years show that there is a high demand for efficient chemical recycling processes, and there is now new momentum to tackle and overcome the challenges that lie ahead.

References for Section 9.1.1

[1] Sasse, F., Emig, G.: *Chemical Engineering & Technology*, (1998) 21(10), pp. 777-789

[2] Vollmer, I., Jenks, M.J.F., et al.: *Angewandte Chemie, International Edition*, (2020) 59(36), pp. 15402-15423

[3] Gomollón-Bel, F.: *Chemistry International*, (2019) (April-June), pp. 12-17

[4] Rickert, J., Cerdas, F., Herrmann, C.: *Procedia CIRP*, (2020) 90, pp. 426-431

[5] Thunman, H., Berdugo Vilches, T., Seemann, M., et al.: *Sustainable Materials and Technologies*, (2019) 22, p. e00124

[6] Tukker, A.: *Plastics Waste: Feedstock Recycling, Chemical Recycling and Incineration*, (2002), Rapra Technology Limited, Shrewsbury, UK

[7] Buekens, A.: In *Feedstock Recycling and Pyrolysis of Waste Plastics: Converting Waste Plastics into Diesel and Other Fuels*, Scheirs, J., Kaminsky, W. (Eds.) (2006), John Wiley & Sons, Ltd, Chichester, pp. 1-41

[8] Brems, A., Dewil, R., Baeyens, J., Zhang, R.: *Natural Science*, (2013) 5(6), pp. 695-704

[9] Arena, U.: *Waste Management*, (2012) 32, pp. 625-639

[10] Chen, D., Yin, L., Wang, H., He, P.: *Waste Management*, (2015) 37, pp. 116-136

[11] Peterson, J.D., Vyazovkin, S., Wight, C.A.: *Macromolecular Chemistry and Physics*, (2001) 202(6), pp. 775-784

[12] Westerhout, R.W.J., Waanders, J., et al.: *Industrial & Engineering Chemistry Research*, (1997) 36(6), pp. 1955-1964

[13] Kruse, T.M., Sang Woo, O., Broadbelt, L.J.: *Chemical Engineering Science*, (2001) 56(3), pp. 971-979

[14] Simon, C., Kaminsky, W.: *Polymer Degradation and Stability*, (1998) 62(1), pp. 1-7

[15] Kaminsky, W., Franck, J.: *Journal of Analytical and Applied Pyrolysis*, (1991) 19, pp. 311-318

[16] Yoshioka, T., Grause, G., Eger, C. et al.: *Polymer Degradation and Stability*, (2004) 86(3), pp. 499-504

[17] Martín-Gullón, I., Esperanza, M., Font, R.: *Journal of Analytical and Applied Pyrolysis*, (2001) 58-59, pp. 635-650

[18] Bockhorn, H., Donner, S., Gernsbeck, M., et al.: *Journal of Analytical and Applied Pyrolysis*, (2001) 58-59, pp. 79-94

[19] Leichtnam, J.N., Schwartz, D., Gadiou, R.: *Journal of Analytical and Applied Pyrolysis*, (2000) 55(2), pp. 255-268

[20] Grause, G., Tsukada, N., Hall, W.J., et al.: *Polymer Journal*, (2010) 42(6), pp. 438-442

[21] Oliveux, G., Dandy, L.O., Leeke, G.A.: *Progress in Materials Science*, (2015) 72, pp. 61-99

[22] Ragaert, K., Delva, L., Van Geem, K.: *Waste Management*, (2017) 69, pp. 24-58

[23] Japan Steel Works Ltd, DE Patent 2310463 (1973)

[24] Sperber, R.J., Rosen, S.L.: *Polymer-Plastics Technology and Engineering*, (1974) 3(2), pp. 215-239

[25] Schäfer, P., Hopmann, C., Facklam, M., et al.: In *Advances in Polymer Processing*, Hopmann C., Dahlmann, R. (Eds) (2020), Springer Vieweg, Berlin, Heidelberg, pp. 37-49

[26] Walendziewski, J.: *Fuel Processing Technology*, (2005) 86(12-13), pp. 1265-1278

[27] Saad, J.M., Williams, P.T.: *Energy and Fuels*, (2016) 30(4), pp. 3198-3204

[28] Li, S.-Q., Yan, J.-H., Li, R.-D., et al.: *Experimental Powder Technology*, (2002) 126, pp. 217-227

[29] Boegly, W.J., Mixon, W.R., Dean, C., Lizdas, D.J.: *Solid Waste Utilization: Pyrolysis*, Report No. ANL/CES/TE-77-15 (1977), Oak Ridge National Lab., Tenn. (USA)

[30] Asanuma, M., Ariyama, T., Iemoto, M.: *Development of PVC Dechlorination Process*, (2001), NKK Techn. Rep. Jap. Ed., pp. 12-15

[31] Kaminsky, W., Menzel, J., Sinn, H.: *Conservation & Recycling*, (1976) 1, pp. 91-110

[32] Czajczyńska, D., Nannou, T., Anguilano, L., et al.: *Thermal Science and Engineering Progress*, (2017) 3, pp. 171–197

[33] Horio, M., Nonaka, A., Sawa, Y., Muchi, I.: *AIChE Journal*, (1986) 32(9), pp. 1466–1482

[34] Kovacevic, J.Z., Pantzali, M.N., Heynderickx, G., et al.: *Chemical Engineering Science*, (2014) 106, pp. 293–303

[35] Bockhorn, H., Hentschel, J., Hornung, A., et al.: *Chemical Engineering Science*, (1999) 54(15–16), pp. 3043–3051

[36] Smolders, K., Baeyens, J.: *Waste Management*, (2004) 24(8), pp. 849–857

[37] Gleis, M., In *Waste Management*, (Vol. 3). Thomé-Kozmiensky, K.J., Thiel, S. (Eds.) (2012), TK Verlag, Neuruppin, pp. 403–413

[38] Al-Salem, S.M., Lettieri, P., Baeyens, J.: *Waste Management*, (2009) 29(10), pp. 2625–2643

[39] Malkow, T.: *Waste Management*, (2004) 24(1), pp. 53–79

[40] Livingston, W.R.: *Technical and Economic Assessment of Energy Conversion Technologies for MSW*, (2002), DTI Sustainable Energy Programmes Report 02/1347

[41] Gleis, M., In *Energie aus Abfall, Band 8*, Thomé-Kozmiensky, K. J, Beckmann, M. (Eds.) 2011), TK Verlag, Neuruppin, pp. 437–465

[42] Monsanto Enviro-Chem Systems, Inc., US Patent 3 862 887 (1973)

[43] Sperber, R.J., Rosen, S.L.: *Polymer-Plastics Technology and Engineering*, (1974) 3(2), pp. 215–239

[44] 121–135

[45] Rensfelt, E., Östman, A.: *IEA biomass agreement Task X. Biomass utilization Biomass thermal gasification activity Sub-task 6 – Gasification of waste Summary and conclusions of twenty-five years of development*, (2000), Energimyndigheten Report FBT 00/11

[46] Kiener, K., US Patent 4 142 867 (1979)

[47] Siemens AG, DE Patent Application 197 29 597 (1997)

[48] Quicker, P., Neuerburg, F., Noël, Y., et al.: *Sachstand zu den alternativen Verfahren für die thermische Entsorgung von Abfällen*, (2017), Umweltbundesamt Texte 17/2017

[49] Mitsui Eng. & Shipbuild Co. Ltd, JP Patent 3686152 (1996)

[50] Union Carbide Corp., US Patent 3 801 082 (1972)

[51] Union Carbide Corp., US Patent 4 017 273 (1976)

[52] Nippon Steel Corp., JP Patent 1985011766 (1978)

[53] Mikata, N., Takeuchi, T., Hashimoto, S., et al.: *Thermal Cracking Properties of Waste Plastics*, (1996), Nippon Steel Technical Report

[54] Thermoselect AG, EP Patent 0 443 596 (1991)

[55] Thermoselect AG, EP Patent 0 742 039 (1996)

[56] Kaiser, B., et al.: Paper presented at VDI-Wissensforum "BAT- und preisorientierte Dioxin-/Rauchgasreinigungstechniken für Verbrennungs- und Feuerungsanlagen", München, September (2002)

[57] Themelis, N.J.: *Waste Management World*, (2007) July, Article 304395

[58] Sanyo Electric Co. Ltd, JP Patent 1983056396 (1976)

[59] Sanyo Electric Co. Ltd, JP Patent 1983029833 (1975)

[60] Takihara, T.: *Kankyo Sozo*, (1973) 3(1), pp. 18–26

[61] BASF AG, EP Patent 0713906 (1995)

[62] John Brown Deutsche Engineering GmbH and Veba Oil Technology, US Patent 6 251 148 (1996)

[63] Rahman, F., Siddiqui, M.N., Redhwi, H.H., et al.: *Energy Sources, Part A: Recovery, Utilization and Environmental Effects*, (2006) 28(15), pp. 1353–1363

[64] Fatima, S.: *American Journal of Engineering Research*, (2014) 3(11), pp. 93–108

[65] Jung, A.: *Der Spiegel*, (2000) (16), pp. 84–87

[66] Ramirez-Corredores, M. M.: *The Science and Technology of Unconventional Oils: Finding Refining Opportunities*, (2017), Elsevier Academic Press, London

[67] Oasmaa, A., Laatikainen-Luntama, J., et al.: *Thermal conversion of plastic- containing waste: A review*, (2017), CLIC Innovation Oy Report D4.1-22

[68] Sharman, R. B., Lacey, J. A., Scott, J. E.: *The British Gas slagging gasifier: A springboard into synfuels*, (1981), British Gas International Consultancy Service Report

[69] Breault, R. W.: *Energies*, (2010) 3, pp. 216–240

[70] Greil, C., Hirschfelder, H., Turna, O., et al.: *Proceedings of 5th European Gasification Conference*, Noordwijk, The Netherlands (2002), pp. 20028–20037

[71] Paper presented at the Gasification Technologies Conference San Francisco, CA (2001)

[72] *https://www.constructionnews.co.uk/archive/so-long-to-seal-sands-03-03-2009/*

[73] Schingnitz, M., Mehlhose, F.: Paper presented at the International Freiberg Conference on IGCC & XtL Technologies (2005)

[74] Texaco Inc., US Patent 3 544 291 (1968)

[75] Perry, H.: *Evaluating the Status of the Texaco Gasifier*, (1984), Resources for the Future Report DOE/ET/16056-T1

[76] Preston, W. E.: Paper presented at the Gasification Technologies Conference San Francisco; CA (2000)

[77] Kaminsky, W.: *Journal of Analytical and Applied Pyrolysis*, (1985) 8, pp. 439–448

[78] Carl Robert Eckelmann AG, US Patent 4 203 804 (1977)]

[79] Frede, M., Ruckdeschel, W.: *Zeitschrift für Umweltchemie und Ökotoxikologie*, (1996) 8(4), pp. 207–212

[80] BP Chemicals Ltd, US Patent 5 481 052 (1993)

[81] Tukker, A., Groot, H. de, Simons, I., et al.: *Chemical Recycling of Plastics Waste (PVC and Other Resins)*, (1999), TNO Report

[82] McCarthy, B. E. C.: *Developments in the End-of-Life Management of Plastics*, (2005), Master thesis, University of Lund

[83] Ebara Corp., JP Patent 3714438 (1997)

[84] Buekens, A., In *Encyclopedia of Sustainability Science and Technology*, R. A. Meyers (Ed.) (2012), Springer, New York, pp. 5235–5296

[85] Pigneri, A.: *Gasification Technologies Review*, (2014), Talent With Energy Pty Ldt. Report

[86] Quicker, P., Neuerburg, F., Noel, Y., et al.: *Status of Alternative Techniques for Thermal Waste Treatment*, (2015), Expert Report for the Federal Ministry for the Environment, Nature Conservation, Building and Nuclear Safety, Report no. 29217

[87] Ebara Corp. and Ube Industries, Ltd, US Patent 6 902 711 (2000)

[88] Yoshioka, T.: Paper presented at the 7th International Symposium on Feedstock Recycling of Polymeric Materials, New Delhi, India, October (2013)

[89] *https://www.jgc.com/en/news/2020/20201006.html*

[90] I. G. Farbenindustrie AG, DE Patent 437970 (1922)

[91] Iannello, S., Morrin, S., Materazzi, M.: *KONA Powder and Particle Journal*, (2020), 37, pp. 114–131

[92] Renzenbrink, W., Wischnewski, R., Engelhard, J., et al.: Paper presented at the Gasification and Syngas Technologies Council Conference (1998)

[93] Ullrich, N.: Paper presented at the 6th International Freiberg Conference on IGCC & XtL (2014)

[94] *https://www.thyssenkrupp-industrial-solutions.com/de*

[95] Enerkem Inc., CA Patent 2 789 790 (2000)

[96] Biello, D.: *Scientific American*, (2011), August

[97] Smith, R.C.: *Journal of Polymer Science*, (2020) 58(10), pp. 1347–1364

[98] Mamani-Soliz, P., Seidl, L.G., Keller, F., et al., In *Recycling- und Sekundärrohstoffe, Band 13*, Holm, O., Thomé-Kozmiensky, E., Goldmann, D., Friedrich, B. (Eds.) (2020), TK Verlag, Neuruppin, pp. 268–284

[99] Solis, M., Silveira, S.: *Waste Management*, (2020) 105, pp. 128–138

[100] *https://enerkem.com/newsroom/releases/*

[101] Aguado, J., Serrano, D.P., San Miguel, G.: *Global NEST Journal*, (2007) 9(1), pp. 12–19

[102] Drzyzga, O., Prieto, A.: *Microbial Biotechnology*, (2019) 12(1), pp. 66–68

[103] European Commission, *A European Strategy for Plastics in a Circular Economy. Communication from the Commission to the European Parliament, the Council, the European Economic and Social Committee and the Committee of the Regions*, (2018)

[104] Geng, Y., Sarkis, J., Bleischwitz, R.: *Nature Comment*, (2019) 565, pp. 153–155

[105] *https://www.ncsl.org/research/environment-and-natural-resources/state-and-federal-efforts-to-revitalize-recycling.aspx*

[106] *http://styrenics-circular-solutions.com/*

[107] *https://www.petcore-europe.org/*

[108] *https://vinylplus.eu/*

[109] *https://pcep.eu/*

[110] Billiet, S., Trenor, S.R.: *ACS Macro Letters*, (2020) 9(9), pp. 1376–1390

[111] Maisels, A., Hiller, A., Simon, F.G., *Chemie-Ingenieur-Technik*, (2021) 93(11), pp. 1742–1750

[112] *https://www.agilyx.com/*

[113] Agylix Corp., US Patent 10 731 080 (2019)

[114] *https://www.basf.com/global/en/who-we-are/sustainability/we-drive-sustainable-solutions/circular-economy/mass-balance-approach/chemcycling.html*

[115] *https://carboliq.com/pdf/1910_CARBOLIQ-flyer-K-2019.pdf*

[116] Pohjakallio, M., Vuorinen, T., In *Plastic Waste and Recycling*, Letcher, T.M. (Ed.) (2020), Elsevier Academic Press, London, pp. 359–384

[117] Gear, M., Sadhukhan, J., Thorpe, R., et al.: *Journal of Cleaner Production*, (2018) 180, pp. 735–747

[118] Recycling Technologies Ltd, US Patent 10 760 003 (2018)

[119] *https://recyclingtechnologies.co.uk*

[120] Greenmantra Recycling Technologies Ltd, US Patent 8 664 458 (2010)

[121] Greenmantra Recycling Technologies Ltd, US Patent Application 2019/0194375 (2019)

[122] *https://greenmantra.com/*

[123] *https://www.closedlooppartners.com/research/advancing-circular-systems-for-plastics/*

[124] Pyrowave Inc., US Patent 9 487 708 (2012)

[125] *https://www.pyrowave.com/en/*

[126] Horodytska, O., Valdés, F.J., Fullana, A.: *Waste Management*, (2018) 77, pp. 413–425

[127] *https://www.enval.com/*

[128] Ludlow-Palafox, C., Chase, H.A.: US Patent 7 951 270 (2004)

[129] Plastic Energy Ltd, WO Patent Application 2020/065316 (2019)

[130] *https://plasticenergy.com/*

[131] Antelava, A., Damilos, S., Hafeez, S., et al.: *Environmental Management*, (2019) 64, pp. 230–244

[132] *https://www.sabic.com/en/news*

[133] *https://www.ineos.com/*

[134] *https://newhopeenergy.com/*

[135] Combs, J. D.: US Patent 9 714 391 (2015)

[136] Lechleitner, A., Schwabl, D., Schubert, T., et al.: *Österreichische Wasser- und Abfallwirtschaft*, (2020) 72(1–2), pp. 47–60

[137] Ignite Energy Resources Pty Ltd and Licella Pty Ltd, US Patent 8 579 996 (2008)

[138] Licella Pty Ltd, Ignite Energy Resources Pty Ltd, Licella Fibre Fuels Pty Ltd, US Patent 9 944 858 (2011)

[139] *https://muratechnology.com/*

[140] Maschmeyer, T.: Paper presented at DBU Conference "Neue Entwicklungen und Tendenzen in der Depolymerisation von werkstofflich schwer verwertbaren Altkunststoffen", Osnabrück, Germany, November (2019)

[141] *https://www.licella.com.au/*

[142] *https://renewelp.co.uk/*

[143] Hofer, W.: Paper presented at DBU Conference "Neue Entwicklungen und Tendenzen in der Depolymerisation von werkstofflich schwer verwertbaren Altkunststoffen", Osnabrück, Germany, November (2019)

[144] OMV Refining and Marketing GmbH, US Patent 9 920 255 (2012)

[145] Pinkel, M., Deditz, J., Sarc, R.: *Study about Plastic Sorting and Recycling*, (2019), Report for COWI A/S, Denmark

[146] *https://www.indaver.com/be-en/home/*

[147] *https://www.brightmark.com*

[148] RES Polyflow LLC, US Patent 10 711 202 (2017)

[149] *https://www.3mdeutschland.de/3M/de_DE/presse-de/pressemeldungen/ganzemeldung/?storyid=f29c372a-a0ce-4ba7-96b3-758490c120d4*

[150] Universität Bayreuth, *Rückgewinnung fluorierter Monomere aus Reststoffen*, (2010), Deutsche Bundesstiftung Umwelt, Abschlussbericht 25198-31

[151] 3M Innovative Properties Co., US Patent 8 344 190 (2009)

[152] Biocellection Inc., US Patent Application 2020/0199325 (2019)

[153] *https://www.novoloop.com/*

[154] Al-Sabagh, A. M., Yehia, F. Z., Eshaq, G., et al.: *Egyptian Journal of Petroleum*, (2016) 25(1), pp. 53–64

[155] Geyer, B., Lorenz, G., Kandelbauer, A.: *Express Polymer Letters*, (2016) 10(7), pp. 559–586

[156] *https://curetechnology.com/*

[157] *https://www.eastman.com/Company/Circular-Economy/Solutions/Pages/Polyester-Renewal.aspx*

[158] Eastman Chemical Co., US Patent 6 191 177 (2000

[159] Eastman Chemical Co., US Patent 6 410 607 (2000)

[160] *http://www.garbosrl.net/*

[161] Ügdüler, S., Van Geem, K. M., Denolf, R., et al.: *Green Chemistry*, (2020) 22(16), pp. 5376–5394

[162] Hoenderdaal, W. G.: Paper presented at Petcore Europe Workshop, September (2017)

[163] Aies Company Ltd, US Patent 6 630 601 (1999)

[164] Plastic Waste Management Institute: *An Introduction to Plastic Recycling*, (2019), Plastic Waste Management Institute, Tokyo

[165] *https://www.ifpenergiesnouvelles.com/article/recycling-pet-axens-ifpen-and-jeplan-will-demonstrate-and-commercialize-innovative-process*

[166] Ioniqa Technologies B. V., US Patent 10 703 703 (2018)

[167] *https://ioniqa.com/*

[168] *https://www.npc.com.tw/j2npc/enus/home*

[169] Hann, S., Connock, T.: *Chemical Recycling: State of Play*, (2020), Eunomia Report for CHEM Trust

[170] *https://www.perpetual-global.com/*

[171] Petrecycle Ltd, US Patent 7 214 723 (2004)

[172] Kawai, F., Kawabata, T., Oda, M.: *Applied Microbiology & Biotechnology*, (2019) 103(11), pp. 4253–4268

[173] Raheem, A. B., Noor, Z. Z., Hassan, A., et al.: *Journal of Cleaner Production*, (2019) 225, pp. 1052–1064

[174] *https://www.bp.com/*

[175] *https://www.ineos.com/businesses/ineos-aromatics/ineos-infinia/*

[176] *https://carbios.fr/en/*

[177] Carbios, US Patent 10 385 183 (2015)

[178] Tournier, V., Topham, C. M., Gilles, A., et al.: *Nature*, (2020) 580(7802), pp. 216–219

[179] GR3N SAGL, European Patent EP 2 736 968 (2012)

[180] *http://gr3n-recycling.com/*

[181] Rittec Umwelttechnik GmbH, WO Patent Application 2020/053051 (2019)

[182] Biermann, L., Brepohl, E., Eichert, C., et al.: *Green Processing and Synthesis, (2021) 10(1), pp. 361–373*

[183] Loop Industries, Inc., US Patent 10 640 442 (2018)

[184] *https://www.loopindustries.com/en/*

[185] *https://renu-project.com/*

[186] Nikje, M. M. A., Garmarudi, A. B., Idris, A. B.: *Designed Monomer & Polymers*, (2011) 14(5), pp. 395–421

[187] Simón, D., Borreguero, A. M., de Lucas, A., et al.: *Waste Management*, (2018) 76, pp. 147–171

[188] Kemona, A., Piotrowska, M.: *Polymers*, (2020) 12(8), p. 1752

[189] *http://www.isopa.org/media/2605/chemical-and-feedstock-recycling.pdf*

[190] *https://www.dow.com/en-us/* ~

[191] *https://ikanoindustry.pl/index*

[192] *https://www.hs-anlagentechnik.de/de/recyclingreaktoren-fuer-post-consumer-matratzen.html*

[193] H & S Anlagentechnik GmbH, US Patent Application 2021/0017354 (2017)

[194] *https://www.rampf-group.com/de/*

[195] Rampf Holding GmbH & Co. KG, EP Patent Application 3 590 999 (2019)

[196] Worch, J. C., Dove, A. P.: *ACS Macro Letters*, (2020) 9, pp. 1494–1506

[197] Wang, Y.: *Waste and Biomass Valorization*, (2010) 1(1), pp. 135–143

[198] Achilias, D. S., Andriotis, L., Koutsidis, I. A., et al., In *Material Recycling – Trends and Perspectives*, Achilias, D. S. (Ed.) (2012), Intech, pp. 3–64

[199] Češarek, U., Pahovnik, D., Žagar, E.: *ACS Sustainable Chemistry & Engineering*, (2020) 8(43), pp. 16274–16282

[200] *https://www.aquafil.com/*

[201] Aquafil Engineering GmbH, EP Patent 2 917 273 (2013)

[202] Grause, G., Kärrbrant, R., Kameda, T., et al.: *Industrial & Engineering Chemistry Research*, (2014) 53, pp. 4215–4223

[203] Bozzano, G., Dente, M., Del, R., In *Material Recycling – Trends and Perspectives*, Achilias, D. S. (Ed.) (2012), Intech, pp. 115–132

9.1.2 Petrochemical Feedstock Recycling

Dieter Stapf

Against the backdrop of increasing production of packaging waste, an increased emphasis on the problem of microplastics in the marine environment, and the need for action in dealing with residual waste locally as a result of Chinese and other Asian import bans on waste, the EU Commission published the first EU plastics strategy in January 2018 [1]. In this strategy, 55% of plastic waste will have to be recycled in 2030, which will require the deployment of additional recycling capacity of ca. 10 Mt/a. The individual measures to be taken by the EU-27 countries are currently focussed on packaging waste [2]. A particular challenge is that the recycling rate of plastics will need to increase significantly to meet these targets. As a minimum prerequisite, suitable waste collection as well as sorting systems have to be implemented, and depending on the impact of these, additional technology pathways may be needed.

Clearly, waste recycling must be a pillar of the "circular economy" concept. A circular economy for plastics is hampered mainly by the nature of waste. Affordability is an issue, too. Plastic waste typically does not appear as a pure polymer. Recycling processes are either physical (meaning that materials are separated from each other but the molecular structure is not changed) or chemical (meaning that the molecular structures are changed and new materials are made from waste). Waste recycling requires the rigorous implementation of waste collection in general, but can be an effective means of recovering materials that can be re-used and recycled using standard, well-established processes. Not all materials can be recycled easily and at low cost, however, and in the absence of energy recovery, these are inevitably sent to landfill.

There are numerous barriers to the physical (mechanical) recycling of plastic waste, resulting in waste recycling rates that are below the targets mentioned above. These include: composite materials that cannot be separated mechanically, constraints associated with sorting technologies, non-melting thermosets and elastomers that cannot be re-granulated for second use, end-of-life products in complex mixtures and the associated high costs of separation, and ensuring the quality of secondary raw materials to maximize recycling. This results in the majority of plastic waste not being recycled; these include packaging waste sorting residues, shredder residues, thermal insulation composite systems, and waste from electric and electronic equipment. The recycling rate is higher for production waste than for post-consumer waste due often being single origin and having low levels of contamination. To date, post-consumer plastics waste in developed countries consists of equal proportions of packaging materials and engineering plastics. The size of the gap between current recycling rates and the targets varies from country to country, in the same way as waste collection systems and application of (physical) sorting as a prerequisite for recycling do.

It is also difficult to remove the need for carbon entirely in many industrial processes. In a future independent of the use of fossil hydrocarbons, carbon will be available from biomass, anthropogenic wastes and residue streams, and from unavoidable CO_2-streams. Thus, the feedstocks utilized by the chemical industries in particular will change dramatically. Continuously increasing manufacturing of chemical products as well as high energy-density compounds (e.g. aviation fuels) will require these carbon feedstocks, and in order to meet climate protection targets. However, fuels are not covered in this chapter, as waste-based fuels are not regarded as waste recycling.

The aforementioned challenges and opportunities have increased interest in the potential of chemical recycling (also referred to as 'feedstock recycling' and 'waste-to-chemicals') to contribute to the recycling rate as an extension to waste sorting and as an alternative to waste incineration. Moreover, by reducing the fossil feedstock share in chemical manufacturing, chemical recycling of waste providing suitable chemical feedstock can reduce the carbon footprint of the manufacturing industry. A recent technology trend is a focus on chemical recycling of packaging waste to petrochemical feedstock, especially as a "drop-in" naphtha surrogate for multi-product steam cracking processes, accompanied by the "mass-balance approach" to the value chain, in order to recognize the contribution of waste-based virgin plastic to achieving the required recycling rates for plastics.

This section describes the basic principle of petrochemical feedstock recycling. It starts with an overview of plastics recycling processes according to their physical or chemical nature. Then, waste-to-petrochemicals process chains are discussed, linking plastic waste changes with the chemical feedstock requirements. Finally, key aspects are outlined.

Later in this book, Chapter 11 classifies thermochemical recycling processes and discusses the readiness, limits, and opportunities of the most important technologies for the case of mixed plastic waste, and gives guidance on process selection.

9.1.2.1 Waste & Recycling

Approximately 44% of global plastics production of ca. 390 Mt/a (including ca. 30 Mt/a of recyclates) is consumed by the packaging sector [3]. Worldwide, plastic waste generated amounts to ca. 250 Mt/a, of which nearly 57% is packaging plastics. The gap between production and waste and the high importance of packaging plastic waste is due to the short life-cycle of packaging versus the long life of engineering plastics products.

The situation is different for some developed countries with high gross national product that export technical products. Here, packaging amounts for only 25% of plastics production, followed by the construction sector, automotive, and electrical and electronic products [4]. In these countries, packaging accounts for only about 40% of plastic waste, and often plastics recycling is already emphasized.

In both situations, a large amount of plastic is "stored" in plastic products in use, and this huge quantity, which has not been designed for recycling, grows continuously. This shows that (i) current focus on packaging waste recycling tackles the most important problem and can be a first step into a circular economy for plastics, and (ii) plastic products other than packaging, which are more difficult to recycle, will be of higher importance in the future.

Today, most plastic recyclates are sourced from separate collection of products consisting of relatively pure resins such as plastic bottles or manufacturing residues, or from sorting of collected post-consumer packaging waste by separating polyolefinic packaging products by highly developed sensor-based sorting technologies [4, 5]. The overwhelming amount of plastic waste that is not recycled is mixed plastic waste. Figure 9.1 illustrates the composition of mixed plastic waste for the example of a sorting residue from collected packaging waste. Besides different packaging plastics, this waste sample includes other packaging materials (e.g. metals and engineering plastics, but also biomass and mineral contamination). Many products included are composite materials, and all plastics included are functionalized, meaning that besides the pure resins they contain other plastic and non-plastic components, depending on the specific application they were used for. Achieving higher recycling rates means recycling this mixed plastic waste.

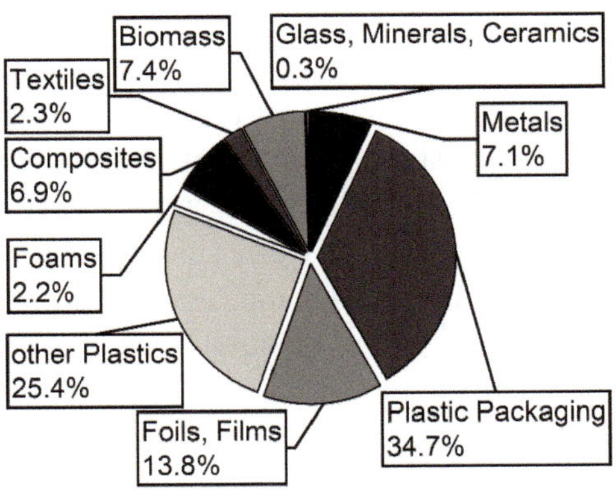

Figure 9.1 Macroscopic composition of a random sample of the lightweight packaging sorting residue as received, determined by a manual sorting analysis [6]

This section mainly emphasizes processes and technologies capable of processing mixed wastes at high temperature. The base thermal waste treatment process is waste incineration ("waste-to-energy", WtE), i.e. combustion of waste with sufficient oxidant to allow full conversion of the hydrocarbon content to CO_2 and water.

For alternative thermal treatment (thermochemical conversion) which recycles waste to new feedstock, we distinguish between sub-stoichiometric oxidant availability (gasification) and processes with no additional oxidant feed (pyrolysis).

Gasification is an established process for production of chemicals and liquid fuels from coal; however, its use to manage waste streams is much less developed. The desired gasification product is syngas (primarily CO and H_2) of a specific composition depending on the downstream utilization process. Autothermal gasification in general delivers the necessary temperature level for a high syngas yield by direct heating, i.e. partial combustion of the feed, resulting in the byproducts CO_2 and H_2O. Like the one-step incineration process, gasification processes can be applied to generate energy in combined heat and power (CHP) plants. However, in contrast to incineration, gasification can also be set up as a two-stage combustion process, in which the feed is gasified first and the syngas is burned afterwards to generate heat. It is also possible to utilize the syngas as a secondary feedstock, which can be converted in downstream chemical synthesis to either "green fuels" or to basic chemical products – which demonstrates the flexibility of gasification pathways for waste management.

The definition of **pyrolysis** includes purely thermal cracking of the feedstock, catalytic cracking, and/or hydrocracking. In case of real mixed waste, it typically includes chemically bound oxygen, resulting in oxygen containing products such as water. Due to the complex degradation mechanism of plastics as well as biopolymers yielding a broad molecular product distribution, any pyrolysis process produces at least three multicomponent product phases, namely permanent gas, condensable vapours, and charcoal. Pyrolysis is an endothermic process that requires reaction heat supply through indirect heating of the reactor. It can be provided by means of combustion of part of the pyrolysis products. The purpose of pyrolysis is to convert biomass or mixed waste into secondary feedstocks for the production of fuels or chemicals. Although also being called "liquefaction", due to the nature of the pyrolysis process in some cases, it cannot liquefy more than approximately 50 mass% of the feed, depending on its composition and the pyrolysis conditions. Furthermore, the multi-phase nature of the product stream can result in a need for additional downstream processes.

The scheme in Figure 9.2 organizes recycling processes according to their physical or chemical nature. Recovered components maintain their chemical nature during physical recycling processes (e.g. material recovery of a specific polymer), whereas chemical processes, in general, break up the initial material (e.g. recovery of the monomer as building block for new material).

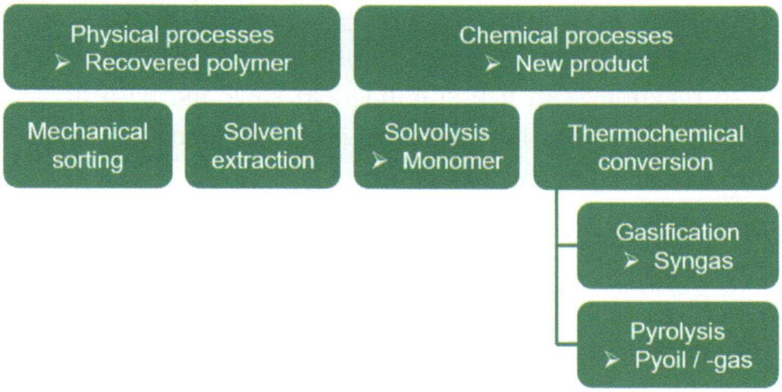

Figure 9.2 Recycling processes for plastic waste and respective main products

Physical separation processes (mechanical treatment, see Chapter 8) are well-established technologies to separate valuables from waste mixtures. Higher plastic recycling rates can be achieved by broad application of sensor-based sorting technologies, or in some cases by applying polymer-type-specific solvents. However, these technologies are characterized by high specific cost compared to basic mechanical waste pretreatment (such as the crushing, screening, etc. used for metals and minerals). The long-term increase of eco-products ("design for recycling") will also lead to a higher rate of recovery in the packaging sector. Here the focus lies on physically separable thermoplastics. The polymer is recovered as so called recyclate and can be processed into recycled plastic products. An obstacle is the lower quality of these mechanically recycled products compared to new virgin products. In general, re-granulates cannot achieve the quality of virgin material, because of (a) chemically bound ingredients such as ultra-fine particles (which can affect color) and catalyst residues, and (b) different mechanical properties resulting from the reduction in polymer chain length during post-processing.

Solvent extraction of one or more polymer types from plastic products also falls into the category of mechanical recycling. If applicable to a given mixture, this can result in clean polymer, thus overcoming the issue of quality. Initial applications on separately collected products are currently being run on a demonstration scale.

Solvolytic **chemical separation processes** are mainly applied to pure plastic waste streams available from plastic manufacturers or from plastic processing industries. Recently, solvolysis has focussed on suitable (pure resin) post-consumer products that are separately collected, such as plastic bottles and matrasses (see Section 9.1.1). This depolymerization through chemolysis is mostly possible with polycondensates. The monomers are recovered from the mixture and can be reprocessed to high-quality virgin material. Key obstacles to increased adoption of this processing technology include contamination of the feedstock and the high process-specific separation effort.

Non-selective solvent liquefaction applying a hot, pressurized solvent aims to break down the waste polymer structures primarily into a “slurry” of liquid components. These reactions can use supercritical solvents, with an example being hydrothermal liquefaction of biomass. Like gasification and pyrolysis, it falls under the category of thermochemical conversion.

In general, **thermochemical processes** (thermal conversion, thermolysis) of waste decomposes mixtures of substances at high temperatures into the basic organic “building blocks” of chemistry, from which new products of high quality are produced. During processing, metallic and mineral impurities are separated and hazardous organic substances are destroyed. In principle, thermochemical processes are also suitable for those waste fractions that cannot be sorted by the above-mentioned processes, for example the non-pure fractions resulting from separate collections of packaging waste, such as sorting residues and shredder fractions. In general, mixed composite materials as well as plastics contaminated with catalyst residues or additives such as flame retardants can be recycled thermochemically.

In the following sections, chemical recycling of mixed plastic waste is considered. Herein, consecutive mechanical and chemical recycling go hand-in-hand in a waste-to-chemicals process chain, in which thermochemical recycling by pyrolysis or gasification avoids incineration of mechanical recycling sorting residues.

9.1.2.2 The Waste-to-Chemicals Process Chain

Taking the example of pyrolysis, Figure 9.3 shows the major feedstocks and intermediate products and the transforming process steps in the waste-to-chemicals process chain. Collected mixed plastic waste is first recycled mechanically. These processes are described in detail in Chapter 8. The products of these initial sorting steps are high-quality standard thermoplastics (which can be used as recyclates after washing and re-granulation), as well as residues. These residues typically include valuable materials like ferrous and non-ferrous metals, and also residues to be disposed of or treated, such as mineral fractions or waste water. Depending on the input waste stream and the mechanical pretreatment processes, a large combustible fraction remains as waste sorting residues. Its volume can be as high as 60–80% of the initial mixed waste [5]. Because modern political and legal boundary conditions do not allow landfilling of this fraction (see Chapter 5), the usage options are either energy recovery or chemical recycling.

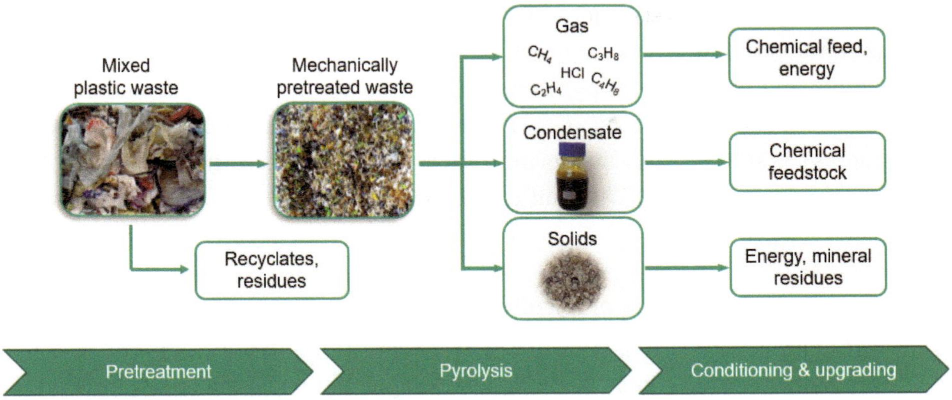

Figure 9.3 The waste-to-petrochemicals process chain: combined mechanical and chemical recycling using the example of pyrolysis

Energy recovery requires further conditioning of waste sorting residues into waste-derived fuels according to the requirements of the type of downstream energy recovery plant involved. These fuels can be specified as solid recovered fuel (SRF) if they meet the classification and specifications given in EN 15359 "SRF – Specification and classes". An international standard for SRF, which is a more strictly defined sorted waste, is under development (ISO/TC 300 "Solid recovered fuels"). The most common options for using these fuels are in waste incineration plants for municipal solid waste, specific SRF plants, cement plants, and co-combustion in sewage sludge incinerators or coal fired boilers. Conditioning of SRF for this application includes mixing of mechanically pretreated waste fractions and adjustment of the fuel particle size distribution.

For recycling of waste sorting residues, similar conditioning to that for energy recovery would be necessary [7]. Here, we consider chemical recycling of this mechanically pretreated waste by pyrolysis. Processing of mixed waste by pyrolysis is described in more detail in Chapter 11.

Depending on waste feedstock, the pyrolysis technology applied, and desired downstream use, pyrolysis products require specific conditioning and upgrading:

- Pyrolysis solids include all minerals and metals separated from the waste (or added to the process to remove contaminants), as well as the pyrolysis char. This fraction typically needs to undergo thermal treatment, e.g. in a combustion step before further processing, and by this can contribute to deliver the pyrolysis process energy.
- Condensable pyrolysis products of feedstock derived from mixed plastic waste include a small water phase and a large oil phase. Pyrolysis water is to be treated for disposal, either through co-combustion with other pyrolysis product fractions, or through regular waste-water treatment. The pyrolysis oil itself, as the main product, is subsequently upgraded.

Non-condensable pyrolysis gas includes valuable gas species such as olefins as well as combustion products such as CO_2. In particular, for pyrolysis of polyolefinic waste, the yield and value of pyrolysis gas can be so high that it becomes worthwhile to use it as feedstock, too. Otherwise, due to its relatively high heating value this fraction is incinerated, delivering process heat to fulfill the requirements of the pyrolysis process itself, as well as for supplying external heat or generating electrical energy.

9.1.2.3 Upgrading and Use of Pyrolysis Oil as Petrochemical Feedstock

The degradation of waste plastics into petrochemical feedstock paves the way to a closed plastics recycling loop, because when replacing fossil feedstock in polymer value chains, virgin-quality polymer product is generated from waste-based feedstock. By overcoming product quality issues of mechanical recycling and by replacing plastic waste incineration, chemical recycling can be a crucial step in increasing recycling rates, if the whole "circular economy" value chain performs better than fossil-feedstock-based polymer products in key areas such as climate impact and energy demand. The determinants of this are techno-economic feasibility and affordability. Recent studies have addressed the advantages of chemical recycling of packaging waste plastics by pyrolysis, including the impacts of avoided incineration [8–10].

By focusing on providing plastic-waste-based pyrolysis oils, large-scale existing chemical manufacturing infrastructures and value chains can be used, and extra investment is avoided. The waste-to-chemicals technical challenge is shifted to the plastic waste pyrolysis part of the value chain, and to the generation of suitable pyrolysis oils according to the specifications of downstream petrochemical processing units. The generation of waste-based pyrolysis oils and their composition are discussed in Chapter 11. The basic options for using pyrolysis oil feedstock are given in Figure 9.4. Here, key petrochemical value chains are ranked by their technology-specific feedstock quality requirements in terms of acceptable impurities:

- On the bottom level, partial oxidation of heavier hydrocarbon fuels by means of entrained flow gasification is a common process to provide synthesis gas for production of C1 chemical intermediates with an H/C ratio of 1. As a large-scale process, it is located in some petrol refineries as well as on large chemical production sites. In this application, waste-based pyrolysis oil replaces heavy fuel oil. Therefore, some sulfur and heavy metal content as well as higher viscosity (along with higher molecular weight) is more acceptable than for any other petrochemical value chain. Even tolerances of halogen species can be higher because of the implemented upstream syngas cleaning steps before use for chemical synthesis.
- The next quality level is fluidized catalytic cracking (FCC) of gasoil, which is the largest chemical reaction step in petrol refining. FCC is the worldwide most im-

portant source for C_4 olefins, and is nearly as important as a propylene source as naphtha steam cracking [11]. Pyrolysis oil as a surrogate for gasoil must be free of heteroatoms to a large extent. In contrast, hydrocarbon content can vary widely, including aliphatic species (and their isomers) and aromatic species. Molecular weight is significantly more limited than for heavy fuel oil. Also, olefin content is restricted. The hydrotreatment step in some petrol refineries is not considered here as a pyrolysis oil outlet, because it instead produces fuel components than chemical synthesis feedstock, and thus does not contribute to recycling.

- Naphtha steam cracking where plastic-waste-based pyrolysis oil replaces part of the fossil feedstock is the most direct way to close the plastic loop by chemical recycling. Naphtha steam cracking is the major source of small olefins. Similar restrictions on feedstock specification apply as for the FCC process. In particular, olefin content must be very low. In addition, boiling behavior, and more specifically start and end points of the boiling curve of waste pyrolysis oil, must be close to the feedstock design values for the specific steam cracker plant. Because pyrolysis oils are process-dependent multi-component mixtures, chemical pretreatment through adapted hydroprocessing is necessary to modifiy their product composition and properties before they can be blended with (or added to) light or heavy naphtha feedstock.
- Thermal separation of high-value chemicals from waste-derived pyrolysis oils can be an option, if the content of specific olefinic or aromatic components in the multi-species mixture is exceptionally high. Therefore, the waste feedstock content of polymers that tend to form these intermediate species during pyrolysis must have been high (see Table 9.2) and pyrolysis parameters and technology must be adapted. The feasibility of thermal depolymerization with high monomer selectivity is addressed in Chapter 11.
- In general, waste-based as well as biomass-based pyrolysis oil cannot fulfill all specifications of crude fossil-based oils. Due to its nature, the chemical composition and physical properties vary widely and cannot be adjusted except by hydrotreatment processes, if a narrow parameter variation and small deviations from petrochemical value chain feedstock specifications is required. This “heteroatom challenge” includes removal of halogens, removal or conversion of char- or water-forming precursors, and removal of components that can deactivate or poison catalysts in subsequent processes of the value chain. This challenge is partly overcome by blending pyrolysis oil with standard feedstock in these large-scale processes, as well as by choosing the petrochemical value chain that provides the best fit to the affordable pyrolysis gas quality. Due to the large size of the petrochemical value chains, it is reasonable to assume that blending capacity will be sufficient in the future, even with increasing volumes of recycled liquid feedstocks.

- A key issue for "climate-neutral" waste- and biomass-based feedstocks is that they have a lower hydrogen content than fossil feedstock (although plastic-waste-based feedstock is better than biomass-based feedstock, and both are far better than CO_2). Therefore, as the proportion of these feedstocks used in chemical production is increased, so does the amount of (renewable) hydrogen required. The highest chemical recycling efficiency will be achieved when plastic-waste-based feedstocks are converted to products with a similar C/H ratio.

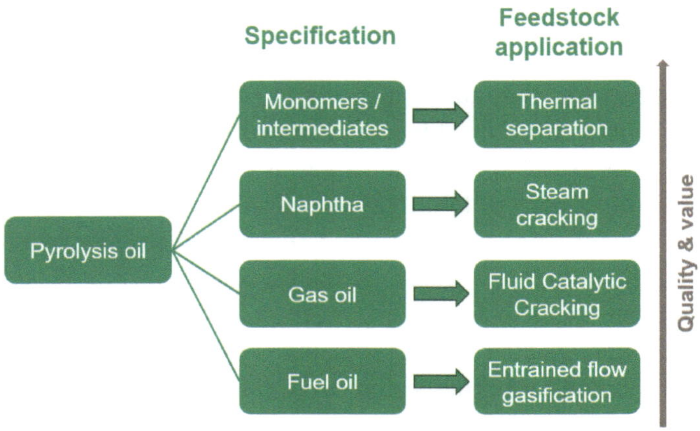

Figure 9.4 Petrochemical feedstock recycling options through pyrolysis oil – specifications to be achieved by the entire pyrolysis process according to feedstock application

9.1.2.4 Decentralized Versus Centralized Chemical Recycling

Chemical recycling plants can be decentralized, i. e. located at the primary (mechanical) waste treatment plant where collected waste is made available for further processing, or centralized, i. e. at the petrochemical production site where the waste-based feedstock is processed.

The overall efficiency of the waste-to-chemical process chain can depend strongly on this choice in terms of both the environment and the economy. Two key aspects influence this decision – transport effort and yield of the chemical recycling product. Whilst waste is typically lower in density and energy content than a liquid chemical recycling product that is determined as feedstock, it is nevertheless typically collected and transported over long distances. Specifically in urban and industrial regions, transport effort will thus not be a major consideration on decentralized versus centralized plant location. This is not the case if plastic foam is being recycled, when transport effort can be a deciding factor on the location.

If a key product of a chemical recycling process is gaseous, decentralized recycling is not favored. At decentralized locations, gaseous products cannot typically be utilized as feedstock and their value must be recovered energetically. Combustion of the gases brings little economic and environmental value. Thus, gasification processes, as well as pyrolysis processes with a high yield of gaseous products compared to pyrolysis oil, benefit from a central location and integration into petrochemical sites [12]. Pyrolysis of mixed packaging plastic waste is one such case. Here, on the one hand, the oil yield of thermal pyrolysis can be limited due to the heterogeneous waste mixture, because it depends highly on the cleanness and preselection of the waste feedstock, i.e. a high content of polyolefins. If catalytic pyrolysis is applied to such a feedstock, the yield can dramatically shift from the oil to the gas phase. Furthermore, the quality of the pyrolysis gas can be drastically increased. Therefore, central pyrolysis at a steam cracker site would be preferable, where gases as well as oils are utilized as feedstock, and where upgrading of pyrolysis oil can be done centrally.

9.1.2.5 The Status of Petrochemical Feedstock Recycling

The present trend towards chemical recycling is steam cracker feedstock recycling [13] of packaging plastic waste by means of pyrolysis. Packaging waste is the largest and most recognized waste stream, pyrolysis of which is clearly easier than converting engineering plastics to chemical feedstock, while also allowing design of demonstration plants and subsequent developments with a higher technology readiness level (TRL). Focusing on decentralized provision of pyrolysis oils initially avoids larger investments and integrates into existing petrochemical value chains with a focus on plastics-to-olefins recycling. These technologies and their current applications are given in Section 9.1.1.

By leading to new products, thermochemical recycling of mixed plastic waste has the potential to overcome mechanical recycling issues, as well as problems from the relative purity of waste feedstocks when used in solvent-based processes. Therefore, it is worth significantly increasing the technology readiness level of pyrolysis and gasification processes by improving feedstock flexibility, with a focus on removal of contaminants and heteroatoms from liquid products. Furthermore, these processes address recycling of composite materials and non-melting plastics. They can therefore complement mechanical and chemical recycling process chains towards the goal of achieving higher recycling rates and economic feasibility, helping chemical recycling compete with waste incineration.

Thermochemical processing of mixed plastic waste is the scope of R&D efforts. Because so far there are only a few large-scale applications for selected packaging waste only, data availability on chemical recycling processes that allow for process evaluation is very limited. Laboratory data cannot be scaled-up because mixed feedstocks as well as applied technology have a critical impact on product yields and compositions. Chapter 11 will address mass and energy balances, yields, and application examples of pyrolysis and gasification technologies.

References for Section 9.1.2

[1] *A European Strategy for Plastics in a Circular Economy*, [Communication from the Commission to the European Parliament, the Council, the European Economic and Social Committee and the Committee of the Regions], Brussels, 2018. *https://eur-lex.europa.eu/legal-content/EN/TXT/?qid=1516265440535&uri=COM:2018:28:FIN*

[2] Witts, H., Hopfensack, L., Azak, G.: Europäische Abfallvermeidungspolitik. *Müll und Abfall*, (2018) 8, pp. 388–393

[3] Lindner, C., et al.: *Global Plastics Flow 2018*, Conversio Market & Strategy GmbH, Mainaschaff, 2019

[4] Lindner, C., Schmitt, J.: *Stoffstrombild Kunststoffe in Deutschland 2017*, Conversio Market & Strategy GmbH, Mainaschaff, 2018

[5] Christiani, J., Beckamp, S.: Was können die mechanische Aufbereitung von Kunststoffen und das werkstoffliche Recycling leisten? *Energie aus Abfall*, 17, TK-Verlag, Neuruppin, 2020, pp. 139–152. *https://www.vivis.de/2019/01/energie-aus-abfall-band-16/148/#more-148*

[6] Zeller, M., et al.: Chemical recycling of mixed plastic wastes by pyrolysis – Pilot-scale investigations. Submitted for publication.

[7] Stapf, D., et al.: Pretreatment of municipal solid waste (MSW) for gasification (Biomass pre-treatment for bioenergy – Case study 3). IEA Bioenergy report, 2019. *http://task36.ieabioenergy.com/publications/biomass-pre-treatment-for-bioenergy-case-study-3-msw-pre-treatment-for-gasification/*

[8] Broeren, M., Lindgreen, E.R., Bergsma, G.: Verkenning chemische recycling – Update 2019. Report, CE Delft, Delft, 2019

[9] Russ, M., et al.: Evaluation of pyrolysis with LCA – 3 case studies. Report, Sphera solution GmbH on behalf of BASF SE, 2020

[10] Volk, R., et al.: Techno-economic assessment and comparison of different plastic recycling pathways – A German case study, *Journal of Industrial Ecology*, (2021) 1–20. *https://doi.org/10.1111/jiec.13145*

[11] Bender, M.: An overview of industrial processes for the production of olefins – C_4 Hydrocarbons, *ChemBioEng Reviews*, (2014) 1 (4), pp. 136–147

[12] Thermal processes for feedstock recycling of plastics waste. Study conducted by KIT and Conversio Market & Strategy GmbH on behalf of BKV GmbH and Plastics Europe AISBL, 2019. *https://www.bkv-gmbh.de/en/info-zone/studies.html*

[13] Stapf, D., Ciceri, G., Johansson, I.: Trends and drivers in alternative thermal conversion of waste. IEA Bioenergy Report, 2020. *https://doi.org/10.13140/RG.2.2.15439.10409*

9.1.3 Plastic Recycling by Depolymerization

Achim Schmidt-Rodenkirchen, Stephan Aschauer, Thorsten Gerdes, Norbert Nießner

9.1.3.1 Introduction

If recycling of a polymer via a solvent is not possible, one option is pyrolysis or depolymerization, with the aim of recovering the monomers, instead of incineration or gasification. While pyrolysis may involve nonspecific product compositions, depolymerization is focused on the most selective recovery of monomers possible. Depolymerization is thus a special form of pyrolysis.

It has to be noted that there are other pyrolysis and gasification processes for chemical recycling, as well as other reactor types - a good overview is available [1]. In this section, the focus is on depolymerization by pyrolysis in a fluidized-bed reactor.

Because most bulk plastics have a low cost/performance ratio, losses by byproducts and energy demand must be avoided. Particularly for high-performance polymers such as PTFE, PS, or PMMA, a depolymerization reaction has the potential to be economically useful. It is these tight economic constraints in particular that lead to the reactor selection being a complex and challenging task. This section introduces fluidized-bed technology for the chemical recycling process design.

The thermal decomposition of technical polymers directly into the monomer is in most cases a direct conversion from a solid to a gas or liquid. The basis is the conversion from a high-molecular-weight polymer to the monomer via chain breaking. Fluidized-bed reactors have a lot of advantages for solid–gas reactions. In general, these pyrolytic reactions are fast and endothermal because the reverse reaction, polymerization, is exothermic. In a polymerization process one of the main challenges is transferring heat out of the reactor. In the case of depolymerization, the challenge is the energy supply into the reactor and into the feedstock.

In chemical recycling technologies, it must be considered that, in addition to the purity of the feed, the process conditions in the reactor also define the yield of the monomer, and hence the success of the process.

In addition to avoiding side-reactions and decomposition of the monomer product, the process temperature and residence time must be optimized within tight constraints, because:

- High temperatures lead to decomposition of the monomer;
- High residence times lead to side reactions and complex waste products.

In combination with a strongly endothermal process, a reactor design should be chosen that provides the following properties:

- Quick conveying of the polymer into the reactor
- Flash pyrolysis by intensive heat transfer
- Quick transport of the raw gas out of the reactor

The fluidized-bed reactor (FBR) combines all these attributes, and W. Kaminsky carried out a lot of research on what later came to be known as the "Hamburger Verfahren" (or "Hamburg procedure") [2]. There are now many reports of successful depolymerization processes in fluidized beds [3–5].

Additionally, the FBR technology enables instant separation of the monomers from the polymer. The gas-phase monomer molecules are diluted in the fluidization gas after being formed from the polymer chain. Hence the local concentration of the monomer is reduced, and side-reactions are suppressed. This is a fundamental difference compared to, for example, extruders, stirring vessels, and rotary kilns.

9.1.3.2 Design Issues for Fluidized-Bed Reactors

9.1.3.2.1 Thermodynamics

For the design of a FBR, the optimization of the heat transfer is one important challenge, in addition to fluid dynamics. A badly designed reactor can lead to cold zones inside the reactor and eventual bed breakdown while plugging the reactor with molten polymer. For this reason, the scalability of the reactor is limited to the heat transfer.

In addition to the heat capacity of the feed to be applied, the enthalpy of the depolymerization reaction must of course be provided. To furnish the monomers, the polymer chain must be broken. Therefore, a good basis for the calculation of the energy demand of a depolymerization reactor is to use the data from the exothermal reverse reaction, the polymerization. However, this approach does not describe the entire problem, since incomplete reactions to oligomers, the presence of impurities in the feed (such as copolymers and fillers), and the enthalpy of evaporating monomers are not considered. However, it still usually gives a sufficient estimate for necessary energy to be transported/transferred. Table 9.5 provides thermodynamic data for the (exothermal) polymerization reactions of some important monomers.

Table 9.5 Energy Demands of Polymerization: Heat of Reaction and Adiabatic Temperature Rise of Monomers [6]

Chemical name	$-\Delta H_R$ [kJ/mol]	$-\Delta H_R$ [kJ/kg]	$\Delta T_{adiabatic}$ [K]
Acrylonitrile	77	1423	720
Acrylic acid	67	930	465
Methyl acrylate	80	930	465
Butadiene	73	1350	675
ε-Caprolactam	16	15	70
Chloroprene	68	770	385
1,1-Dichloroethylene	76	780	390
Ethylene	101	3620	1810
Ethylene oxide	95	2160	1075
Isobutylene	48	855	430
Isoprene	75	1100	550
Methyl methacrylate	59	590	295
α-Methylstyrene	35	295	150
Propylene	84	2000	1000
Vinyl propionoate	86	860	430
Propylene oxide	76	1310	650
Styrene	70	670	335
Vinyl acetate	88	1020	510
Vinyl chloride	71	1135	550

9.1.3.2.2 Kinetics

In general, the FBR technology is suitable for many purposes. For example, other suitable industrial applications are the hydration of nitrobenzene to aniline [7] and the most applied process in the world - the fluid catalytic cracking (FCC) to form gasoline. The effect of well-mixed particles in a fluidized bed (with gaseous or liquid feed) can be used for various chemical reactions:

- Decomposition of parts of the gas phase: Chemical vapor deposition from a gaseous precursor onto seed crystals (i.e. solid particles of the bed), while the gaseous precursor is consumed (e.g., silicon production from mono-silane (SiH_4)) [8]
- Decomposition of the solid phase: Combustion of particles in which the particle is consumed by the reaction between the solid and gaseous phase (e.g., municipal waste combustion),
- Two-phase solid particles: A stationary phase is inert to the reaction but used for heat transfer, while the second phase is decomposed by the supplied heat (e.g., pyrolysis of PTFE in a SiC bed),
- Regeneration of catalysts: Coked catalyst is burned with oxygen for regeneration.

For the explanation of physical correlations, it is assumed in the subsequent section that, in addition to the bed material, the polymer is present in the form of particles in the reactor and the polymer particle shrinks during pyrolysis.

The kinetics of the depolymerization process is as important for the design of the reactor dimensions as for the suppression of unwanted side-reactions. These side-reactions are determined by the residence time of the monomers in the reaction zone. If the reaction is fast, equilibrium in the reactor is reached at short residence times. Hence, the residence time can be limited approximately to the required reaction time derived from kinetics. A good design based on known kinetics enables a high throughput and reduces the risk of side-reactions. Pyrolysis processes are reactions of first order. An overview is given in Figure 9.5.

Compared to PE, the depolymerization of PS is faster at the same pyrolysis temperature. In addition, PS is one of the few polymers decomposing with high selectivity into its monomer, when heated above the ceiling temperature T_c. Compared to biopolymers like wood or straw, the temperature dependence of the pyrolysis reaction of synthetic polymers is more pronounced. Coal combustion is one of the most common fluidized-bed processes. The reverse conclusion is that depolymerization of polymers should be well-suited for FBR applications.

Unlike most polyolefins, PS tends to form less coke during depolymerization in an FBR [3]. The problem of coke formation of PS is due to elongated residence times [10]. For this reason, short, limited residence times as typically realized in FBR are advantageous.

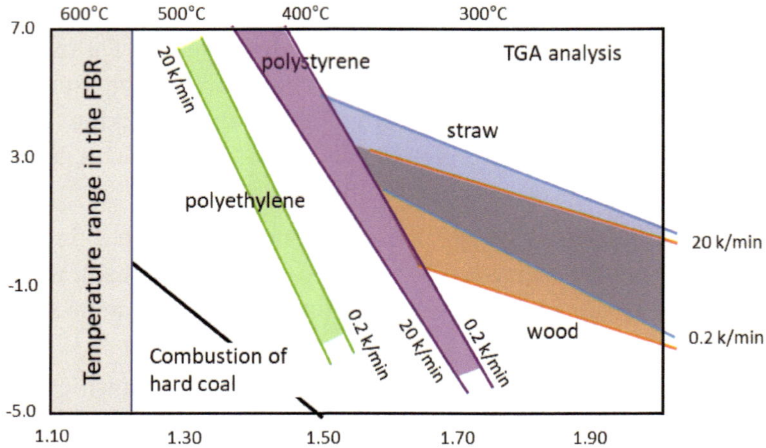

Figure 9.5 The kinetics of depolymerization: Arrhenius plots for biopolymers, PE, and PS compared to combustion of black coal [9]

To limit the residence time, the particle size of the feed material in FBR is also important. The decomposition of the particle can be described by the shrinking core model, which is shown in Figure 9.6. The heat for depolymerization is transported by the fluidization gas or by a particle of the stationary phase (particles of the fluidized bed) to the reaction zone, where monomers are generated by decomposing the polymer chain. The monomer is then transported through the outer core to the particle surface, and finally released to the bulk phase. The assumption of a shrinking core model is only valid if the particles are conveyed by pneumatic transport, or the feeding screw is cooled. If particles melt in the feeder, other decomposition models must be considered.

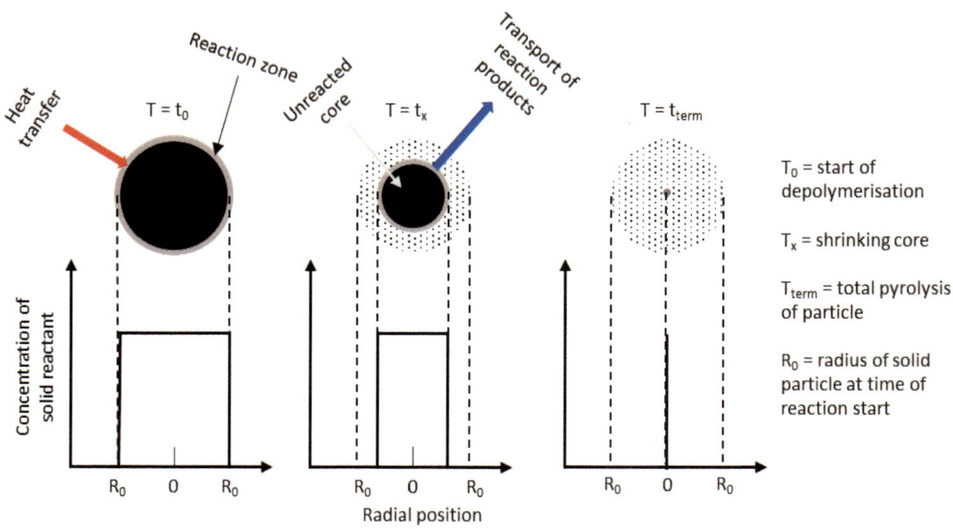

Figure 9.6 Shrinking core model: theoretical decomposition of a polymer particle [11]

In the case that the polymer feed is molten before decomposition starts, the monomer is diluted in the melt. From there it must diffuse to the surface of the molten particle and then evaporate into the surrounding gas phase. The resulting profile of the monomer concentration with respect to the residence time in the reactor is shown in Figure 9.7, and defines the complete residence time that is responsible for consecutive reactions.

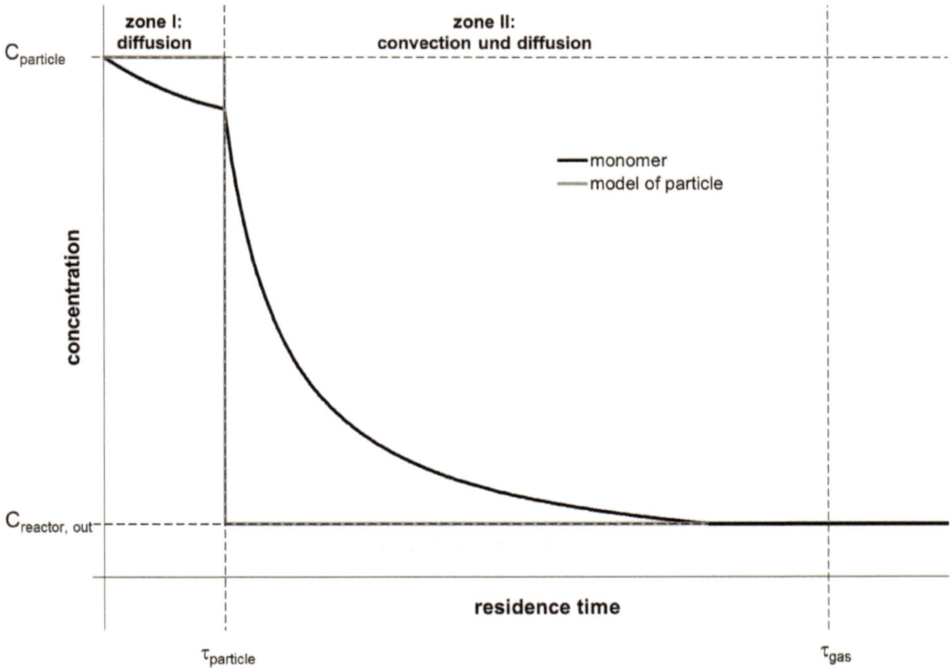

Figure 9.7 Scheme of the monomer concentration depending on residence time for depolymerization [12]

9.1.3.3 Fluidized-Bed Reactor Technology

The advantage of FBR reactors over other reactors (like plug-flow reactors or stirrer vessel reactors) is the high rate of mass and heat transfer between the solid and the fluid phase.

In this section, only reactors with a solid–gas arrangement are considered. Solid–liquid FBR reactors are not considered, because they are of minor importance for recycling. Because transfer limitations are low, the reactor type is appropriate for fast reactions and short residence times of the feed material. These properties predispose this technology for use in fast (flash) pyrolysis processes with residence times below 2 s. For a simple overview of the classification of the reaction velocity, a rough classification from the field of biomass pyrolysis is given in Table 9.6.

Table 9.6 Overview and Comparison of Different Types of Biomass Pyrolysis [13, 14]

Pyrolysis type	Temperature (°C)	Residence time	Heating rate	Proportion of solid products [%]	Proportion of liquid products [%]	Proportion of gaseous products [%]
Fast	~500	<2–3 s	High	~12	~70	~13
Moderate	~500	10–30 s	Medium to high	~25	~50	~25
Slow (charring)	~400	Hours to days	Low	~35	~30	~35
Slow (torrefaction)	~250	10–60 min	Low	~80	~5	~20

To adapt the residence time of the solid in the reactor to the characteristics of flash pyrolysis, an upper particle size to which the feedstock is ground should be determined via the shrinking core model and the kinetics of depolymerization.

9.1.3.3.1 Physical Basics

The main principle of the FBR technology is the movement of particles within a fluid inside a vertical pipe. If the pressure drop on the particles caused by the upward passing fluid is higher than gravity and adhesion forces of the particles, particles overcome the adhesion forces and are ultimately lifted ("transcended"). Upon achieving a certain gas velocity, the "loosening point", all particles start to move stochastically and chaotically. The interaction of forces on one particle in a tube is illustrated in Figure 9.8 in order to visualize the fundamental phenomenon of fluidization.

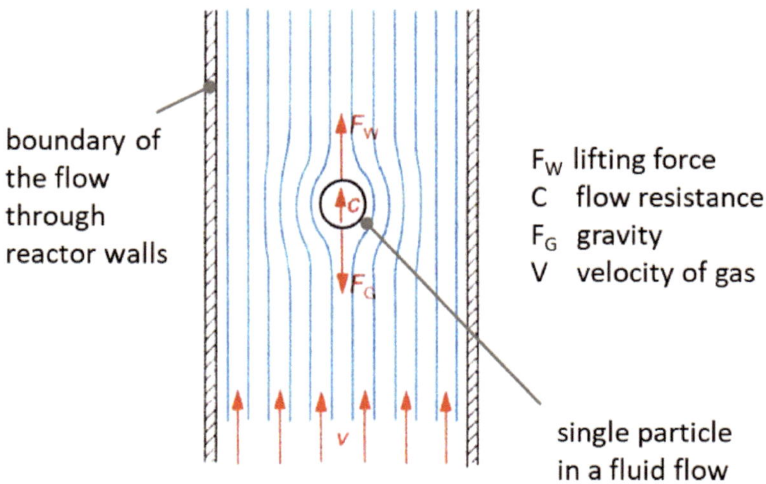

Figure 9.8 The equilibrium of forces that operates on one particle in a fluidized bed: gravity acts against the lifting force and flow resistance [15]

Within a fluidized bed, the particle-particle, particle-wall, and fluid interactions result in a complicated system. Photographs of the phenomena taking place during the fluidization of particles are shown in Figure 9.9.

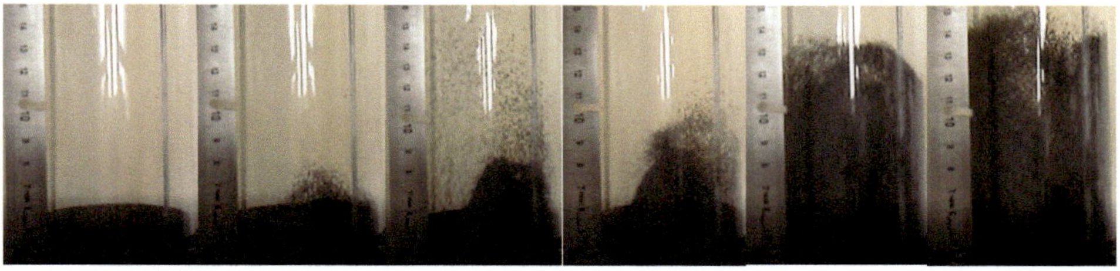

Figure 9.9 Fluidization of rounded SiC particles (d_{50} = 500 µm; $\phi_s \approx 0.9$) in air

The way in which the particles behave in a fluidized bed is determined by the properties of the particles as well as those of the fluid. The most important properties of the fluid for the fluidization are density and viscosity.

While the buoyancy force in gas-solid systems is usually negligible, density and viscosity influence the fluidization and may be described by the Reynolds number and the properties of the bed material. The most important material properties of the particles are density, size, and shape. These properties influence the equilibrium of forces and hence the amount of fluidization gas that is needed to surpass the minimum superficial fluidization velocity (loosening point). Flat particles, or particles with many snagging points, are difficult to fluidize. While size and density are clearly-defined quantities, to describe the influence of shape, the dimensionless quantity sphericity (ϕ_s) has been introduced [11], defined as:

$$\phi_s = \frac{\text{Surface area of sphere}}{\text{Surface area of particle}}\left(\text{of the same volume}\right) \qquad \text{Equation 9.1}$$
$$= 1 \text{ for spheres; } < 1 \text{ for all others}$$

Finally, the fluidization properties of the particles have been summed up into Geldart classifications. The relationship between particle and fluid property itself is represented in a Geldart diagram (Figure 9.10).

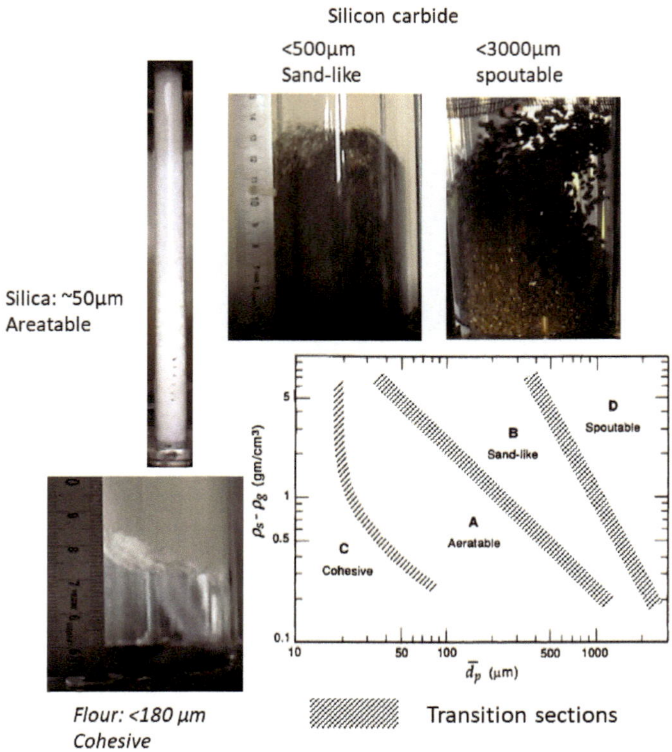

Figure 9.10 Geldart diagram [16] showing the dependency of the particle and fluid properties with examples:
(A) Aeratable: Small particle size with low density; smooth fluidization; low gas velocities.
(B) Sand-like: <500 μm with ρ_p (density of the particle) <4 g/cm^3; good fluidization with vigorous bubbles.
(C) Cohesive: Very fine powder; interparticle forces greater than gas drop.
(D) Spoutable: Large and/or dense particles; difficult to fluidize; large exploding bubbles; spouted beds; uneven pressure distribution

While the description using the Geldart classification is phenomenological, the loosening point can be determined by measuring the pressure drop as a function of the superficial velocity. An example is provided in Figure 9.11. The pressure drop over the bed increases linearly until the transition between fixed-bed and fluid-bed is reached. At this point, a local maximum of the differential pressure is reached. This phenomenon is due to the adhesive forces between the particles. Beyond this pressure maximum (the loosening point), the pressure drop is nearly constant until the velocity of the gas gets so high that particles start to leave the bed. The range over which the pressure drop is constant indicates a homogeneously fluidized bubbling bed.

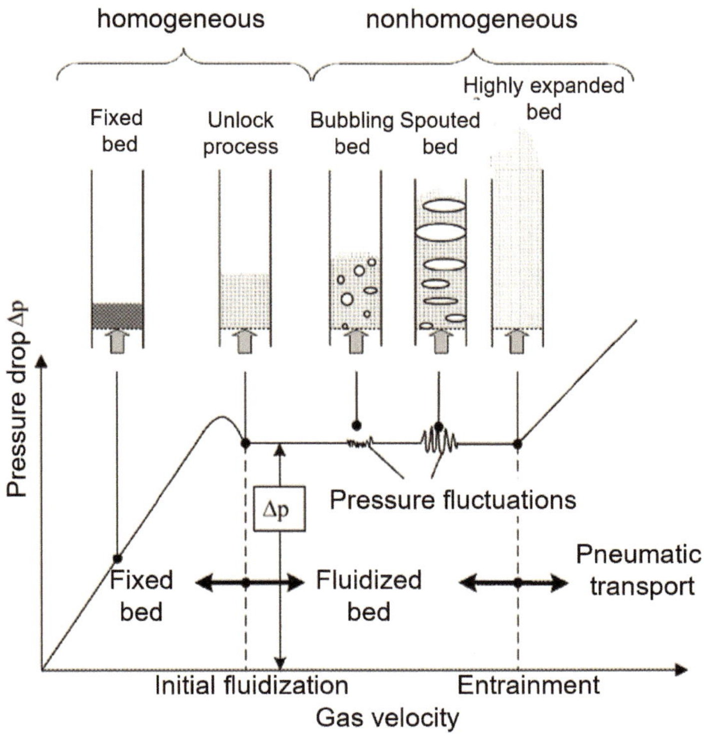

Figure 9.11 Relationship between the pressure drop over a bed of particles and the superficial velocity [17]

In the case of gaseous decomposition products, the reactor design must take into consideration the increase of the gas volume due to evaporation. Depending on the polymer and the process conditions, the amount of monomer in the gas stream can exceed 50%. This effect influences the fluid dynamics in the FBR, which in turn must be considered by controlling the amount of fluidization gas.

The Reh diagram (Figure 9.12) combines all the parameters that determine the fluidization of particles: the superficial gas velocity (u or u_0), as well as the properties of particles and gas (particle diameter (d_p), particle density (ρ_p), fluid/gas density (ρ_f or ρ_g), dynamic viscosity (v), kinematic viscosity (η), dynamic viscosity of the fluid (η_f), and acceleration due to gravity (g). To account for the complex interactions, dimensionless numbers are used again:

$$\text{Reynolds number, } \mathrm{Re} = \frac{\text{inertial force}}{\text{viscosity}} = \frac{u_0 d_p \rho_f}{\eta_f} \qquad \text{Equation 9.2}$$

$$\text{Archimedes number, } \mathrm{Ar} = \frac{\text{lifting force}}{\text{inertial force}} = \frac{d_p^3 g}{v^2} \frac{\rho_p - \rho_g}{\rho_g} \qquad \text{Equation 9.3}$$

$$\text{Froude number, Fr} = \frac{\text{inertial force}}{\text{gravitational force}} = \frac{3}{4}\frac{u^2}{gd_p}\frac{\rho_g}{\rho_p - \rho_g} \quad \text{Equation 9.4}$$

$$\text{Omega number, } \Omega = \frac{u^3 \rho_g}{\nu g\left(\rho_s - \rho_g\right)} \quad \text{Equation 9.5}$$

By plotting the characteristics of the fluid against those of the particle, all regimes of pneumatic transport during fluidization can be represented graphically. The Archimedes number in dependency of the Omega Number also combines the properties of the fluid and the particles, and helps to narrow down the fluidization regime with characteristic lines. Figure 9.12 describes the regimes of fixed-bed, fluidized-bed, and pneumatic transport. A stable, well-fluidized particle bed has a void fraction (ε) of about 0.45–0.65. The yellow area indicates the limits for economic process parameters.

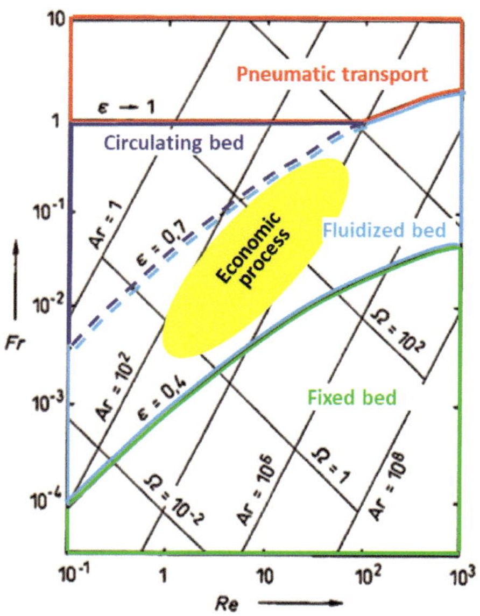

Figure 9.12
Reh diagram for the classification of the fluidization regimes, showing the regimes of fixed bed ($\varepsilon < 0.4$), fluidized bed ($\varepsilon > 0.4$), and pneumatic transport ($\varepsilon > 1$) [18]

The high degree of particle mixing in fluidized beds ensures excellent mass and heat transport. The processes in the fluidized bed are completely chaotic, which makes it difficult to determine the heat transfer coefficients. Hence, the description of the heat transfer can again be achieved by using dimensionless numbers. One preferred method is the application of the Nusselt number (Nu) and the Reynolds number (Re). The Nusselt number relates the heat conductivity to the total heat transfer rate, and enables the calculation of the heat transfer coefficient (α) with the known heat conductivity of the fluid (λ_f).

$$\alpha_{p,f} = Nu \frac{\lambda_f}{d_p} \qquad \text{Equation 9.6}$$

Figure 9.13 shows the interaction of the Nusselt and Reynolds numbers in a fluid bed (shown by the shaded area). At low Reynolds numbers the ratio of heat transfer to heat conductivity of the fluid is low. This compilation of experimental data enables the prediction of the heat transfer into the bed by calculating the Reynolds number. According to Kunii et al. [11], the formation of gas channels or big bubbles in the fluid bed reduces the heat transfer due to reduced particle contacts.

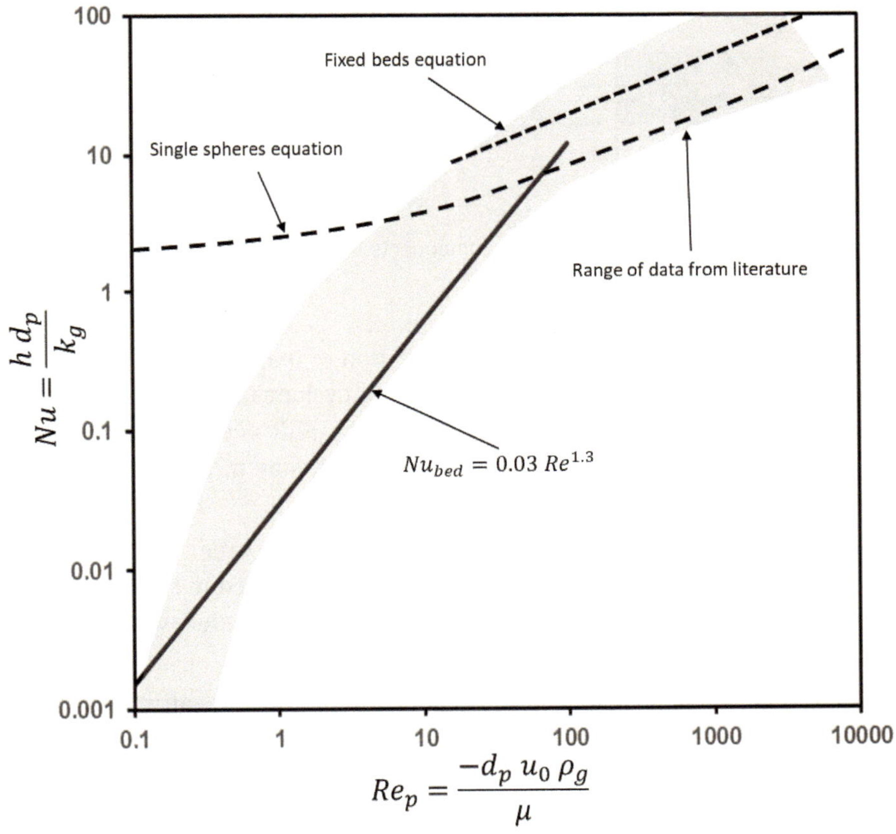

Figure 9.13 Dependency of the number of Nusselt number (Nu) on the Reynolds number (Re) [11]

9.1.3.3.2 Technical Design

Figure 9.14 provides an overview of the most important terms and components of an FBR. Two variants are usually distinguished:

- The stationary bed (with a void fraction between 0.4 and 0.7 according to the Reh diagram), in which the bed particles are kept as stationary phase in the reactor.

- The circulating bed (with a voids fraction between 0.7 and 1), in which the particles are continuously discharged from the reactor together with the fluid. Downstream, the particles are precipitated via a cyclone and recycled into the reactor.

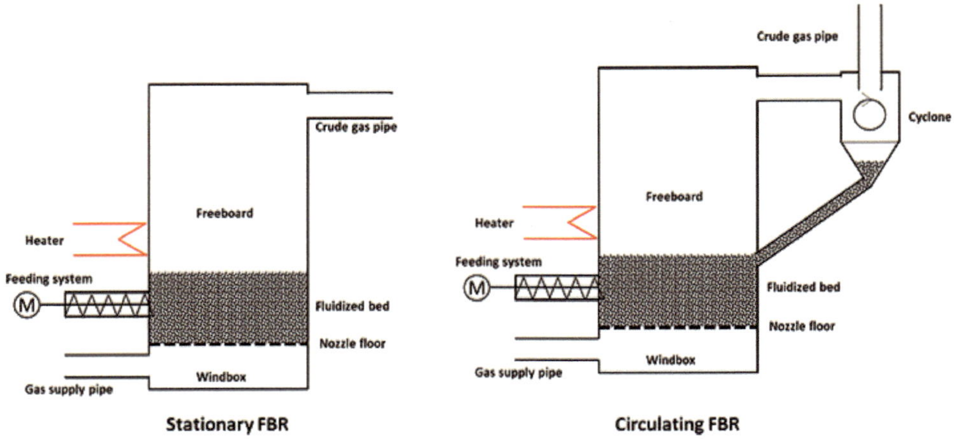

Figure 9.14 Schematic showing the main components of (left) a stationary FBR and (right) a circulating FBR

One issue in recycling polymers is the separation of inert fillers. They can be entrained and then disposed from the raw gas via cyclones if they are fine enough compared to the stationary phase, so long as the FBR does not have a settling chamber in the freeboard. Otherwise, the fillers will stay in the bed and must be run off by an additional outlet.

If, however, all solids are to be retained in the reactor (in order to be drawn off at the bottom, if necessary), a "calming zone" must be installed in the free space. As a rule of thumb, this is a widening in the freeboard, which reduces the gas velocity below pneumatic transport conditions.

Another requirement is that for uniform fluidization over the entire cross-section. If uniform gassing is not ensured, there is a risk of inhomogeneous mixing. As a consequence, in regions of poor fluidization the heat transfer can be significantly decreased, and the bed can collapse, which may lead to a blocked reactor. As a countermeasure, careful design of the nozzles is necessary. In conclusion, a larger pressure drop should exist in the nozzle than in the bed. Local inhomogeneities in the bed then do not directly lead to channeling inside the bed. Figure 9.15 shows possibilities for appropriate nozzle design. Blocking of the nozzles, high abrasion rates of the nozzles, and the slippage of particles from the bed into the windbox may occur due to improper design of the nozzles.

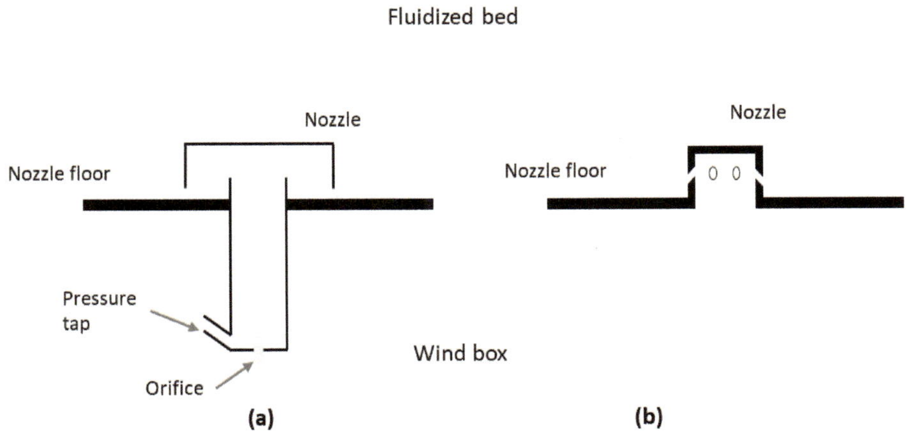

Figure 9.15 Optional nozzle design to ensure robust pressure distribution over the cross-section of the reactor [11]

Since in polymer recycling the possibility of direct firing of the fluidized bed with support burners is not possible, the energy input via reactor wall and gas may not be sufficient. This problem occurs especially when the reactor is scaled-up. In these cases, options include heat pipes, dielectric heating (e.g., by microwaves) or the usage of a regenerator as a second-stage reactor, often applicable in large-scale applications. The temperature distribution of a microwave-heated fluidized bed is shown in Figure 9.16.

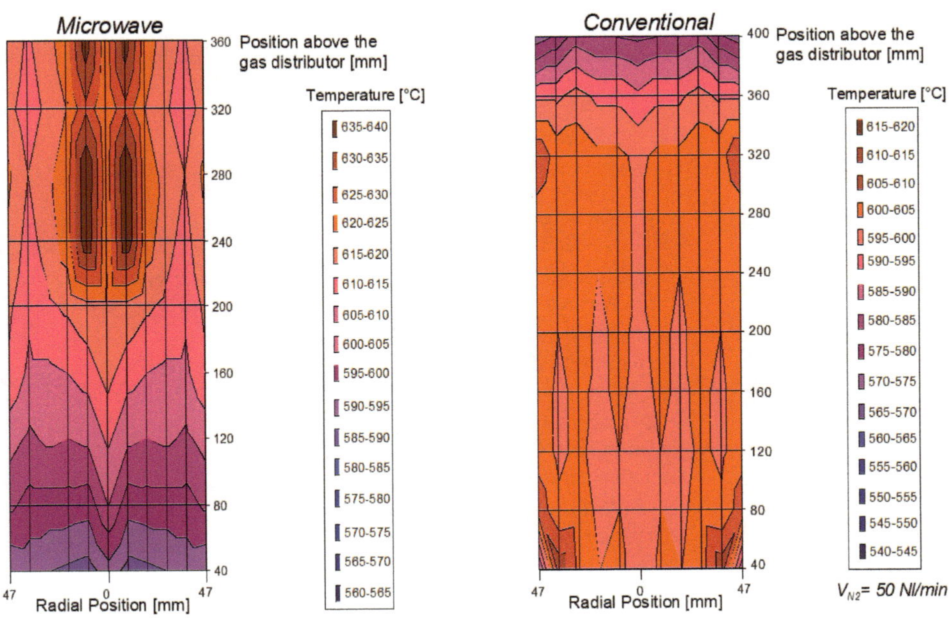

Figure 9.16 Temperature distribution in a microwave-heated bed. The microwave source was located at the top of the reactor [19]

If energy input becomes a problem during scale-up, the choice of stationary phase material can be a key to the solution. If microwave heating is an option, the dielectric loss of the material is an important factor, whereas if a second stage is used, the heat capacity of the bed material limits the heat transfer to the pyrolysis reactor. On the other hand, the density determines the consumption of the fluidization gas. Silicon carbide has very good thermal properties combined with low density. The thermal properties of some chemically inert materials are compiled in Table 9.7.

Table 9.7 Thermal Properties and Density of Some Chemically Inert Materials

	SiC	ZrO_2	Al_2O_3	AlN	Si_3N_4	Stainless steel	SiO_2
Density [$g\ cm^{-3}$]	2.8	6	3.9	3.3	3.4	7.9	2.2–2.7
Heat conductivity [$W\ m^{-1}\ K^{-1}$]	85–120	2–3	35.6–39	70–285	15–25	20	0.3
Heat capacity [$J\ kg^{-1}\ K^{-1}$]	1125	456	990	738	500–900	500	835

9.1.3.3.3 Raw Gas Treatment

Downstream of the reactor, the second critical process step takes place. The design of the raw gas treatment unit depends on several parameters concerning the fluidization gas, the reactivity of the monomers, and the impurities in the feed.

One of the most important parameters is the selection of the fluidizing gas. Condensable gases, inert gases, or non-condensable gas by-products from the raw gas are possible candidates. The advantages and disadvantages are listed in Table 9.8.

Table 9.8 Brief Discussion of the Downstream Behavior of the Fluidization Gas

Fluidizing gas	Advantage	Disadvantage
Condensable (steam)	▪ High heat capacity may improve heat transfer into the particle ▪ Concentration of the monomer downstream the quench ▪ In situ elimination of inert fillers from compounds and halogen impurities ▪ High process safety due to water in the quenching system together with the monomer ▪ High process safety due to dilution by steam in the reactor	▪ Conditioning of the water for recycling ▪ High enthalpy stream ▪ Steam must be produced

Fluidizing gas	Advantage	Disadvantage
Inert gas (N_2)	▪ Low energy consumption due to low heat capacity ▪ High process safety due to dilution ▪ Chemically inert	▪ Losses of the enthalpy of vaporization ▪ Higher amount of expensive inert gas needed ▪ Higher losses of product in the exhaust gas due to the vapor pressure of the monomer ▪ Large process units needed to treat the inert gas ▪ Risk of accumulation of unwanted components by recycling N2
Non-condensable gases (NCGs) from the process (e. g. methane, ethane, hydrogen, ...)	▪ Low energy consumption due to low heat capacity ▪ Gas from the process, no need for external supply	▪ Chemically not inert ▪ Higher losses of product in the exhaust gas due to the vapor pressure of the monomer ▪ Large process units needed to treat the NCG

Behind a FBR, cooling of the raw gas is normally required. The process unit downstream of the reactor can be designed as a classical tube-bundle heat exchanger. The disadvantage of this option is the risk of scaling effects due to precipitated particles on the large surfaces of the heat exchanger.

A quenching design with a direct contact between the cooling media and the raw gas is also possible. No heat transfer areas in direct contact with the raw gas are required, and thus scaling effects are avoided. In flash pyrolysis, it can be advantageous to quench the raw gas quickly to suppress side-reactions. As well as the advantages listed in Table 9.8, some authors have demonstrated [2, 3, 12] that steam stabilizes the monomers in the gas phase and prevents them from re-polymerizing. This effect is also used in a well-known industrial process: steam cracking. The direct quenching system can work with water, or with oil with a high heat exchange efficiency. The short distance from the reactor to a water or oil quench has the advantage of reducing the exposure time of the raw gas to high temperatures.

If fillers from the composites in the feed materials are fine enough, the FBR can have a "classifying" effect, by which the fines are discharged from the reactor and are precipitated in the quenching media. The advantage is the immediate separation of the particles from the gas flow without the risk of blocking subsequent process units. On the other hand, separation of fines from water or oil can be a challenging issue.

Using a water quench, corrosive impurities such as halogen-containing compounds can be neutralized in situ. The opposite concept to this is a dry process chain, with precipitation of fines in, for example, a hot cyclone or a filter, or a combination of cyclone, filter, and heat exchanger. The major disadvantages are long residence times on high temperature levels, and blocking of the system, for example by fillers.

The combination of flash pyrolysis, fast quenching with water, and short paths is shown in Figure 9.17, in a concept developed by the R&D institute InVerTec. The fluidization with steam and short residence times achieve styrene yields up to 80 wt% with PS as feed material.

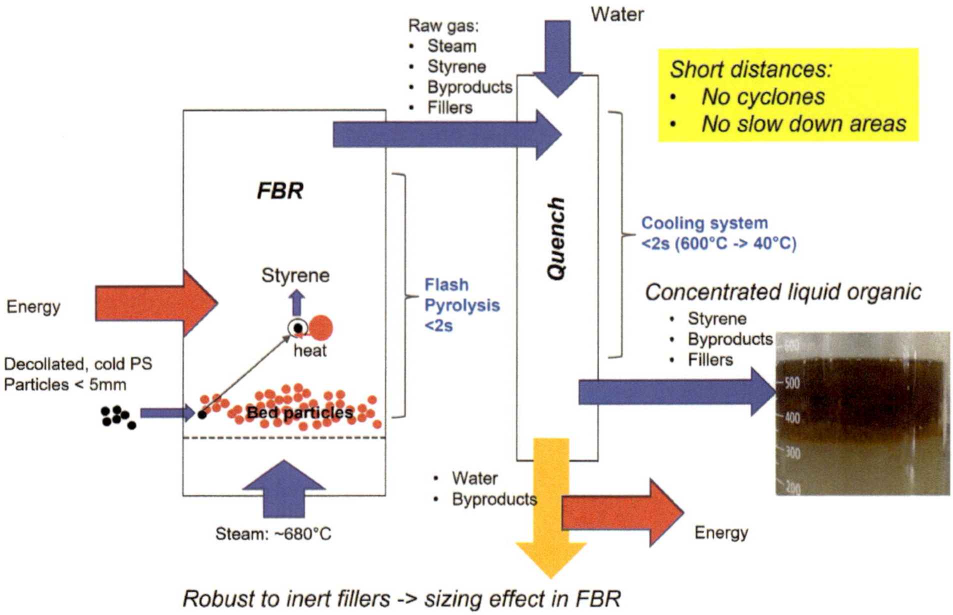

Figure 9.17 Concept of the FBR-based InVerTec miniplant for PS depolymerization

Kaminsky et al. [2, 3] have successfully investigated various plastics, including specialties like PTFE or styrene butadiene rubber (SB), in their test depolymerization plant with a capacity of 30 kg/h. The main part of the plant, shown in Figure 9.18, consists of a 450 mm FBR reactor with a quartz sand particle bed. The FBR is indirectly heated by incineration of process gas. The raw gas from the FBR is first liberated from dust in a cyclone, and subsequently cooled in a quencher with product oil. Afterwards, the condensable parts of the gas are condensed in two columns and the rest is purified in an electrostatic precipitator. This purified gas is used for combustion and heat generation, and has a heat content of about 50 MJ/m^3.

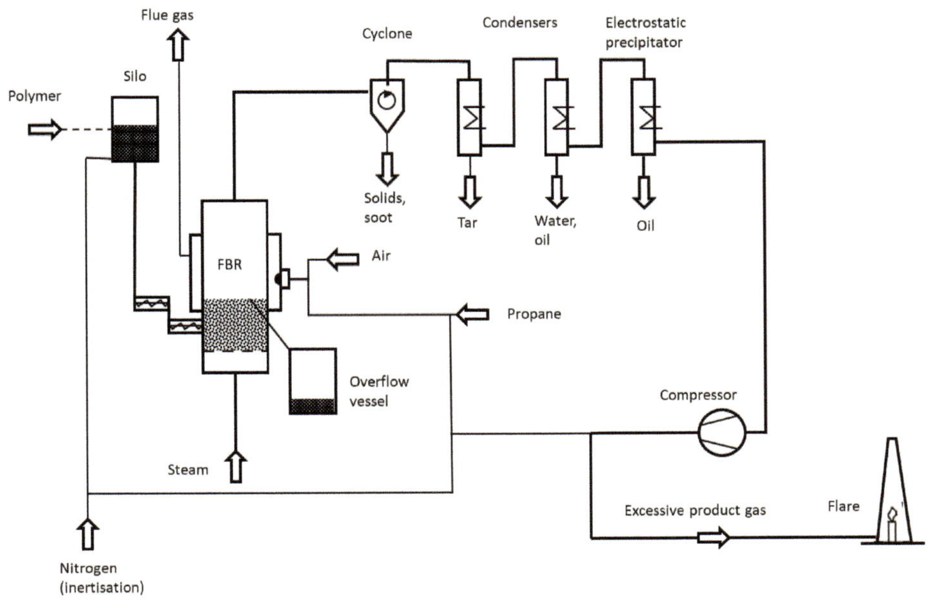

Figure 9.18 The design of the test depolymerization plant known as the "Hamburger Verfahren" (or "Hamburg procedure"), developed by Kaminsky et al. [3]

This process provides high yields of monomer for the three polymers considered suitable for thermal depolymerization, as shown in Table 9.9. PMMA gives the best performance, with 97.2 wt% of MMA, followed by PTFE with 76.0 wt% of TFE. PS gives 64.9 wt% styrene at 580 °C and 76.8 wt% styrene at 520 °C. As expected, bulk polymers made of olefins show much poorer yields of monomers, of less than 30%.

Table 9.9 Results of Pyrolysis Trials According to the "Hamburger Verfahren" of the Kaminksy Group

Polymer	Temperature [°C]	Monomer [wt%]	Gas [wt%]	Oil [wt%]	Residue [wt%]
Polystyrene	580	64.9	9.9	24.6	0.6
PTFE	555	76	18.8	5.2	0.3
PMMA	450	97.2	1.25	1.4	0.15
PE	740	25.6 [a]	30.2	42.4	1.8
PP	760	17.6 [a]	32	48.8	1.6

[a] Sum of ethylene and propylene.

Kaminsky et al. have even shown that mixtures of plastics (like those in municipal waste) are in general suitable for pyrolysis. However, with respect to the optimization of monomer yield, individual pyrolysis should be considered due to the differences in optimum pyrolysis temperature. However, steam is not completely inert, because it shifts the product mixture slightly towards olefins [17].

9.1.3.3.4 Scale-Up

Due to the complex subprocesses that depend on turbulent fluid flow conditions and fluid-particle interactions in an FBR, dimensionless numbers are the basis for classical scale-up regarding fluid dynamics and heat transfer. These correlations were briefly discussed in Section 9.1.3.3.1.

The optimization of FBR technology for the treatment of non-homogeneous and constantly-changing feed materials is a technical challenge. In order to achieve good product yields, tailored solutions are usually required, and a variety of technologies have been developed to achieve this. For example, to increase the required energy input in the endothermic processes, options include:

- Heat pipes
- Superheated fluidization gas
- Using material with high heat capacity for the stationary phase
- Dielectric heating of the FBR particle bed by microwaves
- Design of an autothermal process by partial combustion
- Using a second reactor as a regenerator for coal combustion to provide the energy

One consequence is that concepts developed for one scenario are usually only partially transferable to another, and therefore a plant developments is usually unique. As in most plastic converting processes, the heat transfer is the most crucial aspect, and insufficient heat transfer leads to blockages or stickiness of the material, which is of major concern at any scale. Because of these difficulties, fluidized-bed reactors are mostly scaled by only a factor of 10–200 [11, 25] in capacity, whereas in contrast, stirred-tank reactors can be scaled by a factor of 1000–10,000 [26]. For well-known FBR processes like the fluid catalytic cracking with a good data-base also factors of 1,000 are feasible [27].

Of course, these days simulation programs can support the design and scaling of these reactors. Table 9.10 gives a short overview of the most-used mathematical models used for FBR simulation. Special, commercially available programs like ERGUN enable very detailed description of the fluid dynamics. But even general-purpose tools like Aspen plus® offer possibilities to simulate the behavior of an FBR, embedded into a complex flowsheet.

Table 9.10 Classification of Modelling Approaches for Gas–Solid Systems [21]

Name	Gas phase	Solid phase	Phase coupling	Scale
Discrete bubble model	Lagrange	Euler	Resistance function for gas bubbles	Industrial (10^1 m)
Two-fluid model	Euler	Euler	Gas-solid resistance function	Pilot plant (10^0 m)
Unresolved discrete particle model	Euler (unresolved)	Lagrange	Gas-particle resistance function	Laboratory (10^{-1} m)

Name	Gas phase	Solid phase	Phase coupling	Scale
Resolved discrete particle model	Euler (resolved)	Lagrange	Boundary conditions on particle surface	Laboratory (10^{-2} m)
Molecular dynamics	Lagrange	Lagrange	Elastic collisions on particle surface	Mesoscopic (<10^{-3} m)

In addition to the theoretical challenges, there are practical issues to consider in order to achieve a robust design, especially with a complex technology like FBR. Many of these problems in the reactor design are only revealed during operation – for example, the effects of halogen impurities or of solids and encrustations ("scaling") due to aluminum input.

Besides the reactor itself, the downstream processing steps are of course also crucial for a successful plant design. The treatment of the raw gas or raw oil must be robust with respect to changing gas compositions and load of inert solid particles. These process units must be designed considering the quality requirements of subsequent processing steps and the equipment interfaces to polymer production.

While pyrolysis and incineration is already established at a world scale, as discussed in Section 9.1.3.3.5, depolymerization has so far been developed up to the demonstration scale only.

Table 9.11 gives an overview of three depolymerization processes that have successfully achieved a TRL of 6–9 (or are still in development).

Table 9.11 Depolymerization Processes Currently Under Development

Designer	Recycling Technologies	Recycling Technologies	3M Dyneon/InVerTec e.V.
Owner	Recycling Technologies	Ineos Styrolution	3M Dyneon
Plant place	Swindon	Wingles	Burgkirchen
Country	UK	France	Germany
Stage/TRL	7	Under development	7
In-feed material	MSW	Polystyrene	PTFE
Running since	2018	Planned for 2023	2015
Current capacity [t/a]	7000	15,000	500
Temperature [°C]	400–600	600	600
Pressure [bar]	1	1	1
Product	Hydrocarbons	Styrene	Tetrafluoroethylene Hexafluoropropene

9.1.3.3.5 Industrial Applications of Fluidized Beds in Plastic Recycling

There are only a few industrial applications of processes for converting plastic to something useful at a world scale. This section does not cover just depolymerization, but broadens the scope to include technical processes such as gasification and incineration, since world-scale applications of depolymerization processes based on fluidized beds are in development right now (see Table 9.11). Besides, fluidized-bed reactors are currently the only reactor type at world scale that allow chemical re-utilization of plastic waste. Table 9.12 shows the three industrial applications of fluidized beds for plastic recycling operating at a world scale.

Table 9.12 Comparison of the Processes for MSW Recycling in World-Scale Fluidized Reactors

Designer	Ebara-Ube	Ebara-Ube	Enerkem [a]
Owner	Showa Denko	Infraserv Hoechst	City of Edmonton
Plant place	Kawasaki City	Frankfurt Hoechst	Edmonton
Country	Japan	Germany	Canada
Process	Gasification	Incineration	Gasification
In-feed material	MSW	MSW	MSW + Biomass
Running since	2004	2003	2012
Capacity [t/a]	64,000	700,000	110,000
Temperature [°C]	1300	850	600
Pressure [bar]	16	1	1
Plant size [m²]	7400	–	–
Product	CO_2: 430 t/d Ammonia: 175 t/d	Steam: 6360 t/d Electric power: 70 MW	Ethanol: 100 t/d
Investment [€ million]	–	300	200
Designer	Ebara-Ube	Ebara-Ube	Enerkem

[a] At the time of writing (2021), Enerkem plants are no longer running at a full commercial scale.

Most important is the Ebara-Ube fluidized-bed PTIFG (pressurized twin internally circulating fluidized-bed gasifier) process. This process is applied at Showa Denko plant in Kawasaki City to produce ammonia (175 t/d) and dry ice (CO_2, 430 t/d) from municipal waste at a capacity of 64 kt annually since 2004 (see Figure 9.19). The process has run uninterrupted since then, and there is a special "fluidization" fluid dynamics design, which makes the processing of different household wastes possible. In Figure 9.20 the principle of this double rolling-circulation fluid dynamics concept is shown. The process consists of a "low-temperature" gasifying stage at 600–800 °C and a post-O_2-burner stage at 1300 °C to completely convert the gases to syngas (CO–H_2 mixture). There is a water quench to cool the process gas and to define the reactor pressure by the liquid water height.

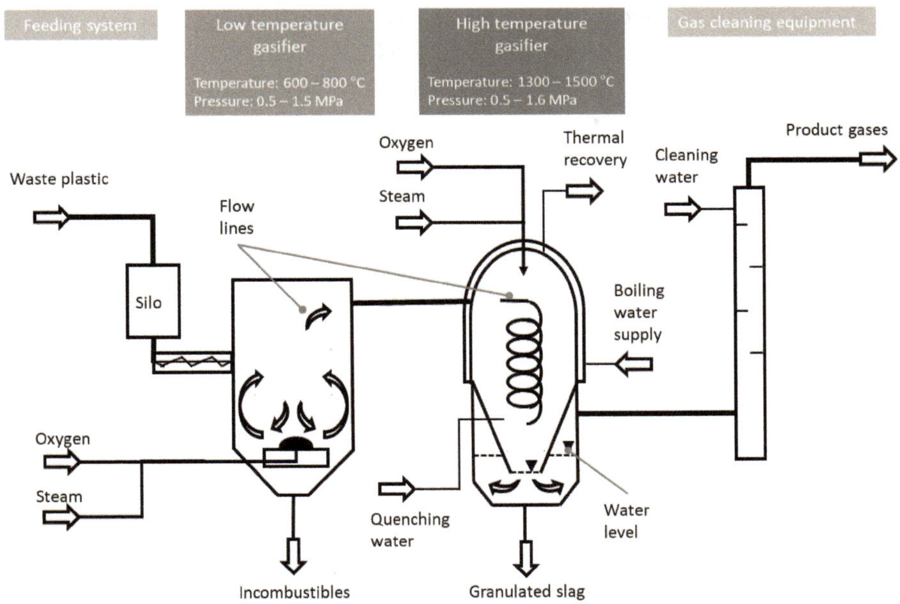

Figure 9.19 Process flow of the Ebaras PTIFG process at Showa Denko (*https://www.eep.ebara.com/en/products/gas.html*)

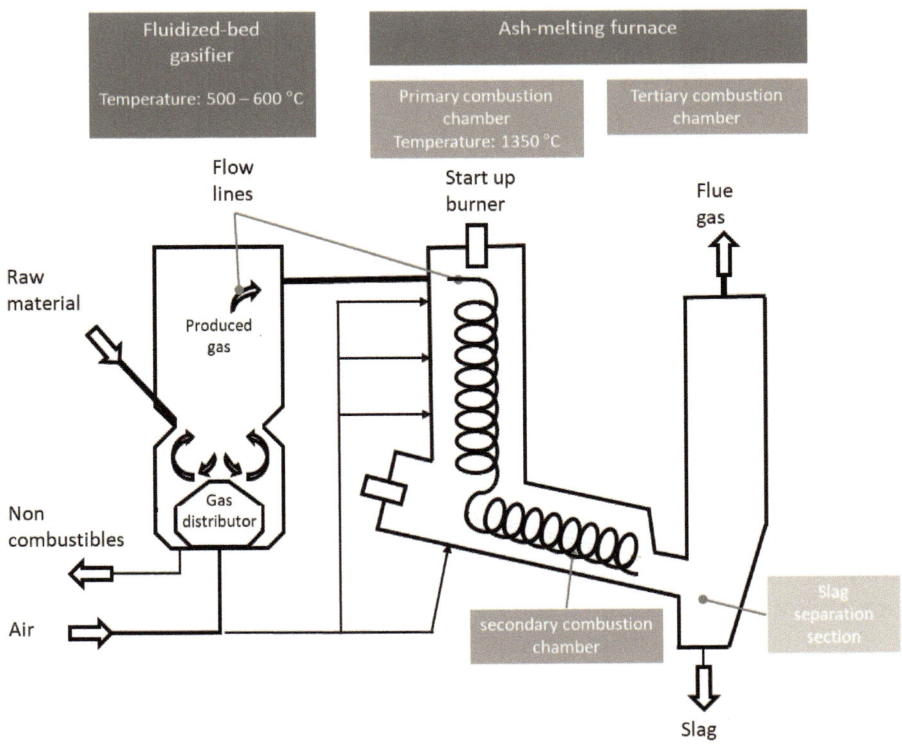

Figure 9.20 Detailed drawing of the two gasifying stages of the Ebaras PTIFG process at Showa Denko [22]

The (re)use of plastic according to the process is particularly valuable in Japan, because steam reforming of natural gas to produce syngas is uneconomical, due to the fact that natural gas must be transported by ship rather than by pipeline (a consequence of the country being an island subject to regular earthquakes). The government's strategy is therefore to eliminate waste as completely as possible, and to achieve this it is raising the gate fees for landfilling up to €400/t.

The Ebera process has also been used at the chemical park in Frankfurt/Hoechst, Germany, to provide steam and electricity to the industrial location. Here, 700 kt/a of municipal solid waste (MSW) is converted to 6360 t/d steam and 70 MW electricity. This is useful for supplying the 16 bar steam needed for distillation processes at the site, which would otherwise be difficult because of the lack of exothermic processes employed there. Figure 9.21 shows the process in Frankfurt/Hoechst.

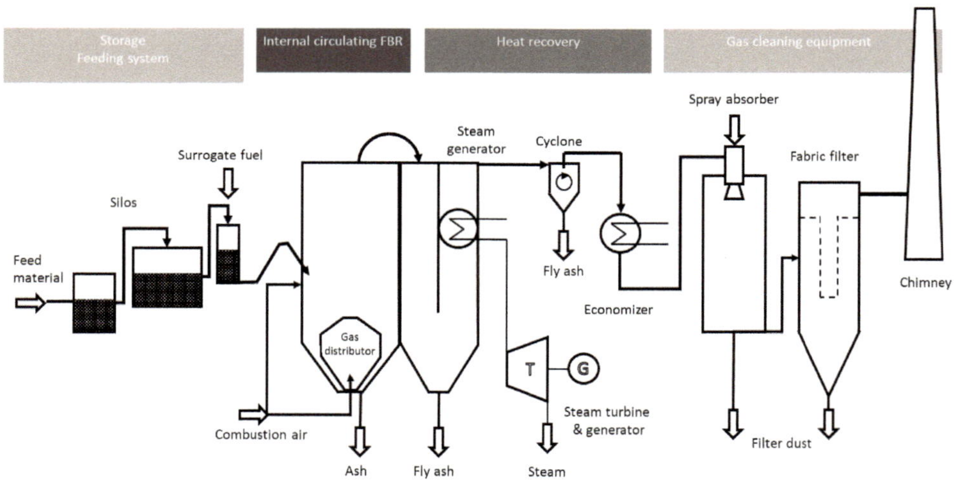

Figure 9.21 Ebaras process applied at Frankfurt/Hoechst to produce steam and electricity from MSW (*https://www.infraserv.com/media/content/leistungen/EVA_dt_rz_web.pdf*)

The third industrial application is at Enerkem in Edmonton, Canada, where a fluidized bed with a capacity of 100 kt/a is applied to produce methanol or ethanol from MSW and biomass (Figure 9.22). The process is comparable with the Ebara process, with the exception that the high-temperature stage is not present, and that the fluid dynamics is less complex. Enerkem is planning to build a plant at the Port of Rotterdam with a capacity of 200 kt/a. The plant in Edmonton is not at full commercial scale, so the next development step will be in the Port of Rotterdam.

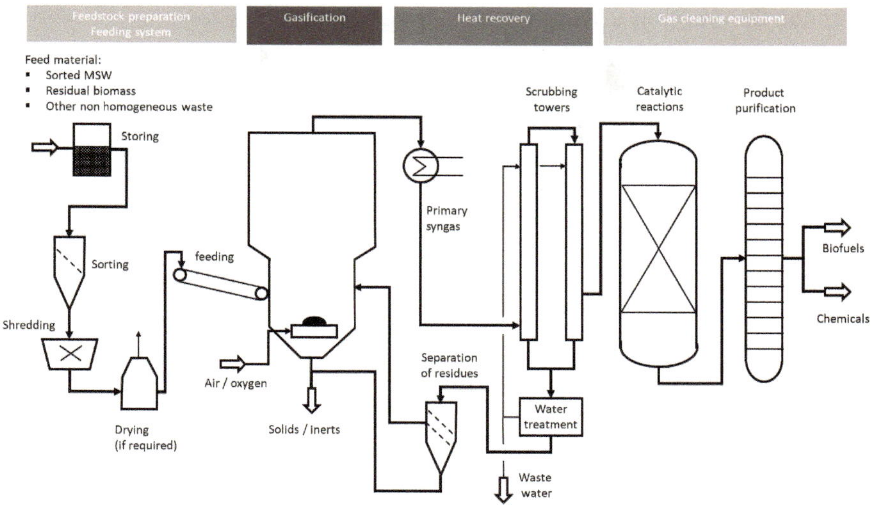

Figure 9.22 Enerkem's fluidized-bed process for the production of methanol and ethanol [23]

As well as Enerkem, there is a process to generate a valuable liquid oil from MSW, run by Recycling Technologies. These plants are based on studies at the University of Warwick, UK, and have a proposed plant size of 7 kt/a. An overview of the process is shown in Figure 9.23. The reactor concept can be considered as a fluidized bed on the commercial scale, but more along the lines of "numbering-up" rather than classical scale-up.

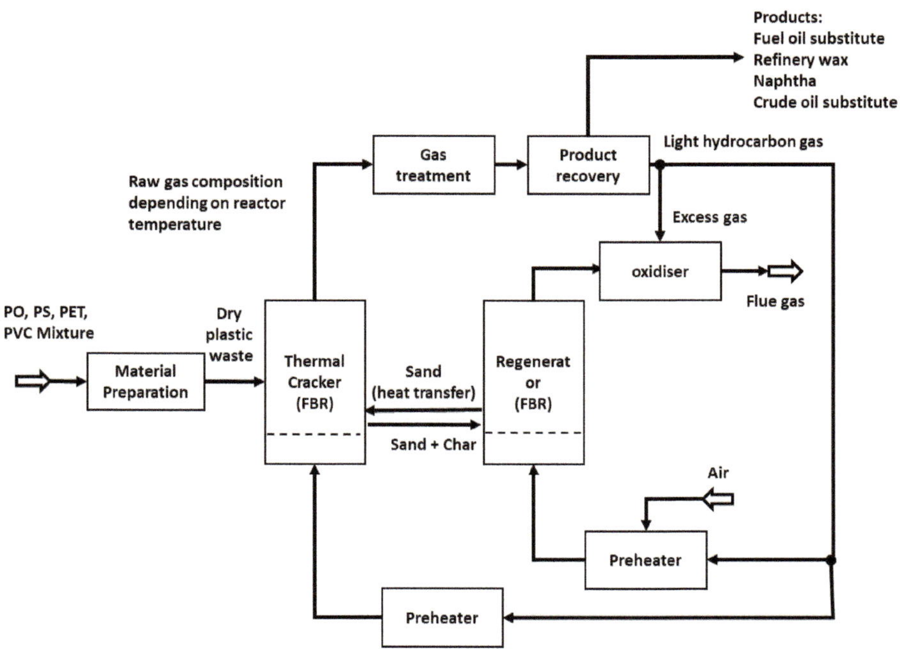

Figure 9.23 Process flow used by Recycling Technologies [24]

References for Section 9.1.3

[1] Mamani-Soliz, P., Seidl, G. L., Keller, F., Lee, R. P., Meyer, B. in: *Recycling und Sekundärrohstoffe, Band 13*, (Recycling und Rohstoffe) (Eds: Thomé-Kozmiensky, E., Holm, O., Friedrich, B., Goldmann D.), Thomé-Kozmiensky Verlag GmbH, Nietwerder, 2020

[2] *Feedstock Recycling and Pyrolysis of Waste Plastics*, (Eds: Scheirs, J., Kaminsky, W.), Wiley-Interscience, New York, 2006

[3] Simon, C. M., Kaminsky, W.: Chemical recycling of polytetrafluoroethylene by pyrolysis, *Polymer Degradation and Stability*, (1998) 62, pp. 1-7

[4] Kaminsky, W., Franck, J.: Monomer recovery by pyrolysis of poly(methyl methacrylate)(PMMA), *Journal of Analytical and Applied Pyrolysis*, (1991) 19, pp. 311-318

[5] Eger, C.: *Chemisches Recycling von PMMA im Wirbelschichtreaktor*, 1st ed., Shaker, 1999

[6] *Polyreaktionen und Polymerisations Fähige Systeme Polyreaktionen*, Berufsgenossenschaft Rohstoff und Chemische Industrie, 2015, *https://downloadcenter.bgrci.de/resource/downloadcenter/downloads/R008_Gesamtdokument.pdf*

[7] Kahl, T., Schröder, K.-W., Lawrence, F. R., Marshall, W. J., Höke, H., Jäckh, R. in: *Ullmann's Encyclopedia of Industrial Chemistry*, Wiley-VCH Verlag GmbH & Co. KGaA, Weinheim, Germany, 2000

[8] Filtvedt, W. O., Mongstad, T., Holt, A., Melaaen, M., Klette, H.: Production of silicon from SiH_4 in a fluidized bed, operation and results, *International Journal of Chemical Reactor Engineering*, (2013) 11, pp. 57-68

[9] Rumpel, S.: *Die Autotherme Wirbelschichtpyrolyse zur Erzeugung Heizwertreicher Stützbrennstoffe*, Karlsruhe, 2000

[10] Kodera, et al.: Reactor design and operation of small-size liquefaction plants for waste plastics; *Feedstock Recycling of Plastics - Selected Papers presented at the Third International symposium on Feedstock Recycling of Plastics*, Karlsruhe, Germany, September 2005

[11] Kunii, D., Levenspiel, O., Brenner, H.: *Fluidization Engineering*, 2nd ed., Elsevier Science, Burlington, 2013

[12] Aschauer, S.: *Untersuchung zur Zersetzung von PTFE und PTFE-Co-Polymeren in der Wirbelschicht, Diplomarbeit*, Univeristy of Bayreuth 2008

[13] Bridgwater, T.: Biomass for Energy, *Journal of the Science of Food and Agriculture*, (2006) 86, pp. 1755-1768

[14] Kaltschmitt, M., Hartmann, H., Hofbauer, H.: *Energie aus Biomasse: Grundlagen, Techniken und Verfahren*, 3rd ed., Springer Vieweg, s. l. 2016

[15] Siegel, W.: *Pneumatische Förderung. Grundlagen, Auslegung, Anlagenbau, Betrieb*, Vogel, 1991

[16] Geldart, D.: Types of gas fluidization, *Powder Technology*, (1973) 7, pp. 285-292

[17] Howard, J. R.: *Fluidized Bed Technology: Principles and Applications*, Hilger, Bristol, 1989

[18] *Lehrbuch der technischen Chemie*, (Eds: Gmehling, J., Brehm, A., Baerns, M.), Thieme, Stuttgart, 1996

[19] Gerdes, T., Tap, R., Bahke, P., Willert-Porada, M. in: *Advances in Microwave and Radio Frequency Processing*, (Eds: M. Willert-Porada), Springer Berlin Heidelberg. Berlin, Heidelberg, 2006

[20] Kaminsky, W., Schlesselmann, B., Simon, C.: Olefins from polyolefins and mixed plastics by pyrolysis, *Journal of Analytical and Applied Pyrolysis*, (1995) 32, pp. 19-27

[21] van der Hoef, M. A., van Sint Annaland, M., Deen, N. G., Kuipers, J. A. M.: Numerical simulation of dense gas-solid fluidized beds: A multiscale modeling strategy, *Annual Review of Fluid Mechanics*, (2008) 40, pp. 47-70

[22] Themelis, N. J. in: *16th Annual North American Waste-to-Energy Conference*, ASMEDC 2008

[23] Iaquaniello, G., Centi, G., Salladini, A., Palo, E. in: *Methanol*, Elsevier, 2018

[24] Gear, M., Sadhukhan, J., Thorpe, R., Clift, R., Seville, J., Keast, M.: A life cycle assessment data analysis toolkit for the design of novel processes – A case study for a thermal cracking process for mixed plastic waste, *Journal of Cleaner Production*, (2018) 180, pp. 735–747

[25] Carlsmith, L. E., Johnson, F. B.: *Ind. Eng. Chem.*, (1945) 37, p. 451

[26] Zlokarnik, M.: Scale-up: Modellübertragung in der Verfahrenstechnik, Wiley-VCH, 2000

[27] Katsuma, T., Dozono, T.: AIChE Symp. Ser., 83(255), 86 (1987)

9.2 Chemical Recycling of Polyolefins

Adrian Griffiths, Alejandro Sánchez, Stephanie Loo

9.2.1 Introduction

Polyethylene and polypropylene, the most common polyolefins, account for nearly 50% of the total plastic demand in Europe [1]. They are used in a large number of applications since their value-to-cost ratios are typically hard to beat. This scale and diversity of use combined with low production cost presents a challenge when it comes to the stewardship of this resource at the end of its life.

The humble plastic bag in some ways epitomizes the problem of polyolefin waste with ghostlike images of bags drifting in our oceans, begging an environmental question that must be answered. This same image also illustrates the attributes of polyolefins – they perform far better than many alternative materials, are ubiquitous, are unbelievably cheap, and are so easily deemed valueless after use. Consequently, to ensure that polyolefins are recycled at scale, the solution must be extremely cost-effective so that it can be made available all around the world. In Europe, legislation and fiscal incentives will make many recycling approaches economical, but to really move the needle on the recycling of polyolefins, the solution must valorize this material after use, regardless of where it is used.

We must also be clear about the definition of recycling, as "recyclable" does not equate to "sustainable". Turning used plastic into another product that cannot itself be recycled merely prolongs its disposal. The material must be circular, meaning used material can be returned to virgin-quality material time after time. However, even being circular does not necessarily make it sustainable. If driving the material around the circular economy creates carbon emissions, the quest remains incomplete.

Chemical recycling is not a panacea; it does not on its own make polyolefins sustainable. However, chemical recycling in combination with several other resource management tools could play a significant role in a vision where polyolefins are recyclable, circular, and sustainable.

9.2.2 Basic Technical Principle

Chemical recycling promises to add to the toolbox of existing methods for recycling plastics. Types of chemical recycling typically include purification, monomer recycling (i.e., depolymerization), and feedstock recycling (Figure 9.24). Purification is considered in more detail in other sections, as is monomer recycling, which has relatively limited application to polyolefins unless the plastic is being turned into a gas rather than oil. In this section, the specific role of feedstock recycling is in view - a pathway by which used plastic is converted back to oil or gas form, to displace the use of virgin feedstocks.

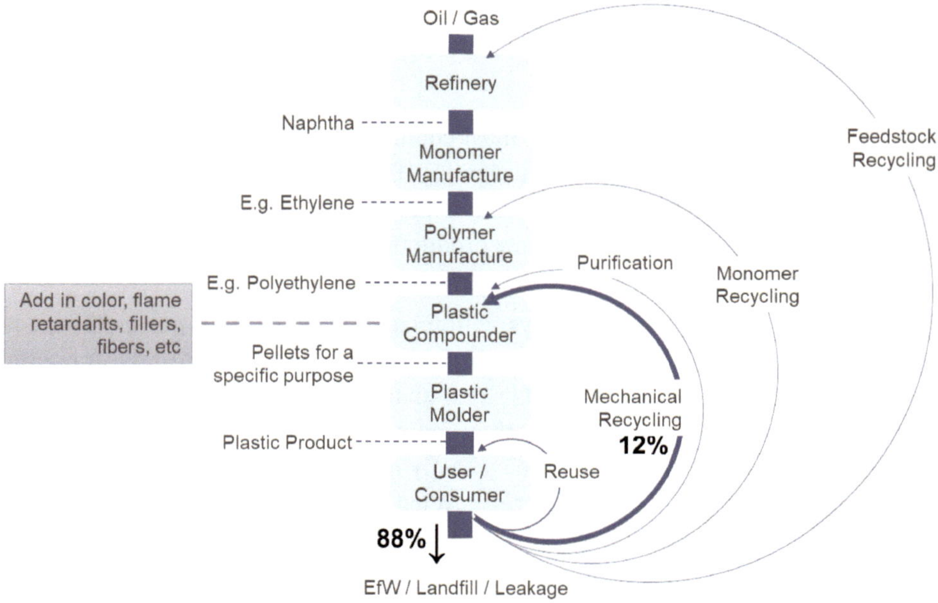

Figure 9.24 Plastic production and recycling today

By recreating a form of crude oil to produce naphtha and polymers using the same processes and equipment that were used to make the original plastic pellets, the optimal plastic for each use can be produced with no compromise in quality. This potential for feedstock recycling to help close the loop on plastics is promising, but as shown in Figure 9.24, the capacity for this recycling option remains negligible, relative to the scale of polyolefin production. Feedstock recycling cannot be achieved by any one company; it requires the whole value chain to work together (Figure 9.25).

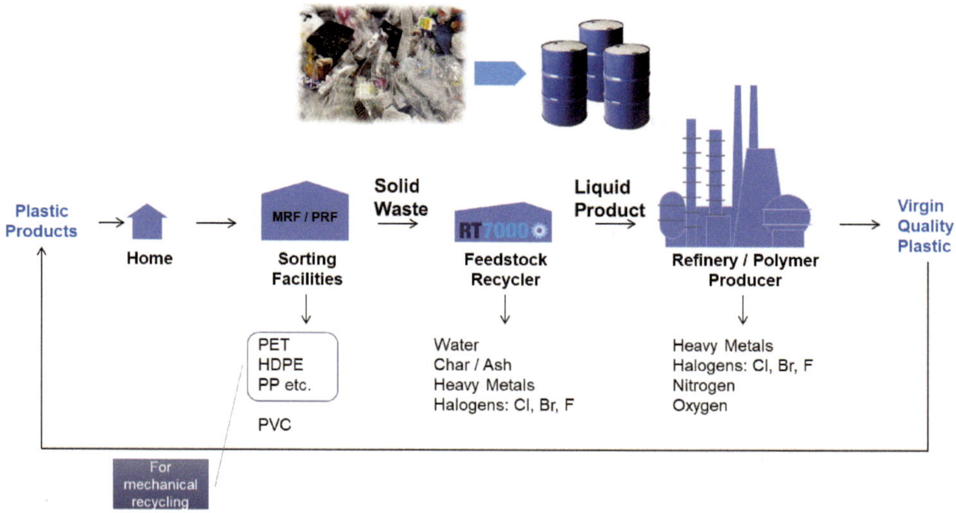

Figure 9.25 Plastics recycling value chain

The resource and waste management industry plays a vital role in collecting materials and routing them to the optimal recycling pathway: mechanical recycling, purification, monomer recycling, or in this case, feedstock recycling. The routing decision is discussed in greater length in Section 9.2.6.3.

The role of the feedstock recycler is to convert a solid waste (plastic) into a liquid product (oil). This could be done by the resource management industry, by specialist companies, or by petrochemical companies. Each of these business models are being tried in various locations and at differing scales. Where the conversion technology is best located and at what scale will be considered later. However, in all cases the output from the feedstock recycler, usually a liquid oil, is reintroduced into the petrochemical industry as a feedstock with the aim of producing more virgin-quality polyolefins.

9.2.3 Examples

1. **Input material:** This is often referred to as residual plastic waste (RPW), which means post-consumer plastic from which material suitable for mechanical recycling, purification, or monomer recycling has been removed, typically using near infra-red (NIR) sorting systems. The NIR system should also remove as much PVC and PTFE from the input material as possible, as the chlorine and fluorine is not wanted in the feedstock oil. RPW is typically post-consumer waste material in the form of films, laminated packaging (crisp packets, food pouches etc.), or contaminated material (with labels, food residues etc.), though other

plastics such as PS and PA (within laminated structures), and low levels of PVC that escape the NIR sorting, can also be present.

2. **Processing:**

   a) **Material preparation:** Material is normally shredded in order to open plastic bags etc. and allow their contents to be removed. Post-consumer plastic is often wet, so the shredded RPW must be dried to below 5% moisture content to avoid lowering the efficiency of the process. This drying can be done with waste heat from the process itself. Drying the material in a trommel arrangement also has the advantage of knocking off food dust residues and other contamination which adheres to plastic, especially wet plastics. Following the drying, the material is passed through a ballistic separator to remove grit, glass, and metal objects that were trapped in the plastic waste.

   b) **Cracking:** Initially this was achieved using batch reactors, where plastic was put into a chamber, the air removed and replaced with an inert atmosphere, then the temperature raised to around 450–500 °C where the long polymer molecules "crack" into a range of shorter molecules such as waxes, oils, and gases. After the plastic has reacted, batch reactors must be cooled, cleaned, and reloaded. This is not the optimal approach from an efficiency or cost perspective, and so for low-value olefins (where these considerations are essential), more efficient continuous reactors have been developed.

   In most pyrolysis processes, a range of by-products are created, such as char (which is frequently contaminated with heavy metals from the plastics), water, and halogens (chlorine, fluorine, and bromine). These typically combine with hydrogen to create strong acids that must be neutralized, otherwise the equipment itself will corrode very rapidly.

   Continuous reactors take many forms: heated screw reactors, rotating kilns, and fluidized-bed reactors (FBRs). Each have their advantages and disadvantages, but for post-consumer polyolefin material where contamination levels can be high and char is a persistent feature (from polyester-type material mixed in with the polyolefin, as well as paper and other biocontamination), the self-cleaning nature of the FBR is a distinct advantage. FBRs are also flexible from a size perspective: the same approach can be deployed from less than 1 t/h to hundreds of tonnes per hour.

3. **Refining:** Raw pyrolysis oil is not normally suitable for use directly in steam crackers, though there is a trend for chemicals, including plastics, to be produced straight from the crude, so crackers in such complexes can take raw pyrolysis oils. However, to be used as the feedstock in older steam crackers, some refining of the raw pyrolysis oil is required to further remove heavy metals, halogens, nitrogen, and oxygen molecules (from the polyesters, PA, ABS etc. that are present, if not a pure polyolefin waste stream). This usually requires some

form of hydrotreating, which is sometimes done by the specialist feedstock recycler, but can be done by the petrochemical industry, as it does when refining crude.

Given the very small capacity for plastic-to-oil conversion, the outputs were, until relatively recently, considered too insignificant by the petrochemical industry to be used as a feedstock. Consequently, the anticipated use of the pyrolysis oil was as a fuel. The pressure from consumers to ensure that plastic is recycled has changed this mindset, and the petrochemical industry is now aligning with consumers, resource managers, and chemical recyclers to ensure that the whole value chain works together to make sure the output quality from feedstock recycling is progressively similar to virgin-quality plastic.

9.2.4 Products: Pyrolysis Oils

Thermally cracking polyethylene molecules (which may have 500,000 carbon atoms in HDPE, although UHMWPE can have up to 6,000,000 carbon atoms) [2] into shorter molecular units using pyrolysis alone does not typically produce one type of hydrocarbon [3]. Instead, it produces an array of molecules including chars, waxes (35+ carbon atoms per molecule), liquids, and gases (≤5 carbon atoms), all with varying molecular structures, including paraffins, isoparaffins, olefins, naphthenes (ringed paraffins), and aromatics (Figure 9.26).

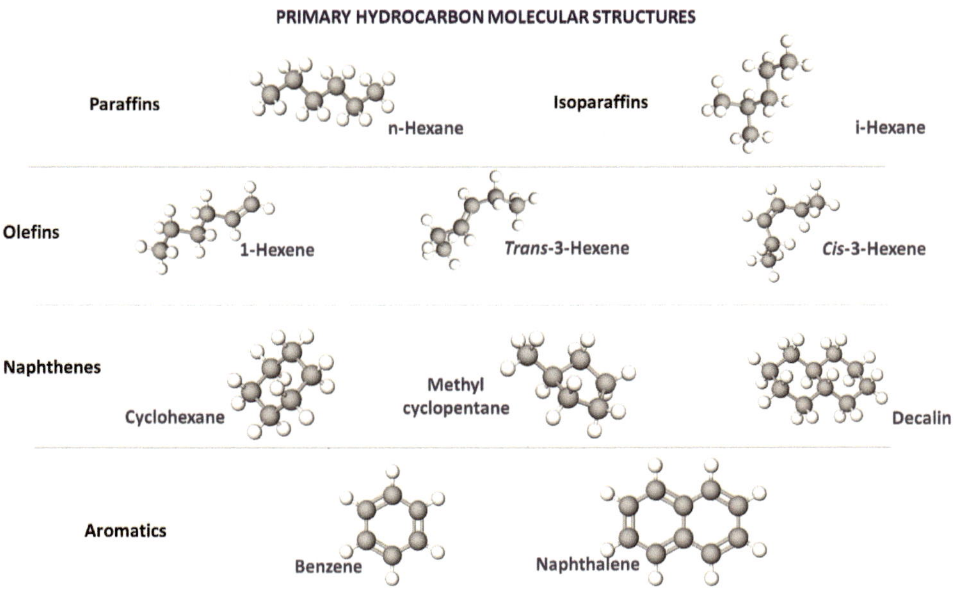

Figure 9.26 Molecular structures of primary hydrocarbons [4]

This complexity is further compounded because polyethylene, like any other polyolefin (i. e., polypropylene, polybutene, and mixtures of these) entering a pyrolysis reactor is not only made up of carbon and hydrogen. To create plastic products with desirable properties, additives are added into the plastic [5]. The types of additives commonly used can be broadly categorized into:

- Functional additives (stabilizers, antistatic agents, flame retardants, plasticizers, lubricants, slip agents, curing agents, foaming agents, biocides, etc.)
- Colorants
- Fillers (mica, talc, kaolin, clay, calcium carbonate, barium sulfate)
- Reinforcements (glass fibers, carbon fibers)

These compounds must be removed from the resulting oil, either during the feedstock recycling process or at a refinery. The complexity does not stop there. Polyolefin waste (POW) is often mixed with food residues, surfactants, other plastics, and other unexpected substances such as mercury (Hg), the source of which is sometimes unknown.

So, what does pyrolysis oil look like and how does it compare to crude oil? In general terms, the oil produced from polyolefins has a high ratio of aliphatics over aromatics (>75%), even when the input plastic is relatively dirty, wet, and contaminated [3]. This is good because high concentrations of paraffins are appreciated by refineries and petrochemical complexes due to their high conversion and efficiency (lower energy consumption). Crude oil desirability, and thus price, is determined by its chemical and physical properties [4]. In general, heavier oils with a higher content of elements other than carbon and hydrogen are less valuable, as more processing is required when compared with lighter oils such as Brent [6]. More specifically, the value of a crude oil is influenced by the American Petroleum Institute (API) grade, total acidity number (TAN), and the content of nitrogen (N), oxygen (O), sulfur (S), and heavy metals [6]. Table 9.13 shows these attributes for a range of crudes compared to the output oil (Plaxx®) from a feedstock recycling pyrolysis plant developed by Recycling Technologies Ltd when processing polyolefins. Table 9.14 shows representative content of metals.

Table 9.13 Physico-Chemical Properties for a Range of Crudes and a Pyrolysis Oil [7–11]

Property	Pyrolysis oil (Plaxx®)	Fossil crude		
		Light (North Sea)	Medium (Iraq)	Heavy (Venezuela)
API	32–58	38.9	24.3	16.98
Density [kg/m^3]	740–870	830.5	908	953
Acidity [mg$_{KOH}$/g]	0.1–2.5	0.05	0.3	2.7

Property	Pyrolysis oil (Plaxx®)	Fossil crude		
		Light (North Sea)	Medium (Iraq)	Heavy (Venezuela)
Elemental analysis	**ASTM D5291**			
Carbon mass [%]	82–87	86.6	88.6	84.4
Hydrogen mass [%]	12.2–14.5	12.2	7.4	7.9
Nitrogen mass [%]	1 (max.)	0.93	1.8	4.9
Oxygen mass [%]	0.2–2.5	0.27	2.5	2.8
Sulfur	**ASTM D2622**			
Sulfur [mg/kg]	11.1–25	3900	20,000	47,000
Chlorine [ppm]	42–150	200	570	1500
SARA analysis	**IP 469**			
Saturates mass [%]	75–98	65.70	52.37	25.53
Aromatics mass [%]	0.2–10	11.30	15.62	28.45
Resins mass [%]	1.5–12	22.30	27.49	31.40
Asphaltenes mass [%]	0.2–15	0.70	4.52	14.62
Boiling point cut	**D2887/D7169**	**D86/D1160**		
IBP [°C]	60–450	36.1	38	85
FBP [°C]	160–750	585 Distillates >585 Residues (11.45%)	585 Distillates >585 Residues (30.05%)	585 Distillates >585 Residues (11.45%)

Table 9.14 Metal Content (ppm) for a Range of Crudes and a Pyrolysis Oil [12–16]

Metal	Pyrolysis oil (Plaxx®)	Fossil crude		
		Light (Saudi Arabia)	Medium (Saudi Arabia)	Heavy (Venezuela)
API	**32–58**	**38.5**	**24.4**	**16.98**
Al	90.1–250	1.1	1.1	11
V	3.5	2.2	52	350
Cr	13.6	0.52	<1	13
Mn	<2.0	<0.05	<1	15
Co	<2.0	–	2	4.5
Ni	<2.0	0.55	25	150
Cu	6.9	0.15	2	46
Zn	<2.0	0.12	<1	56
As	<2.0	–	–	39
Fe	120–200	0.41	4	146
Cd	<2.0	0.52	–	73
Sb	<2.0	–	–	32
Pb	10.9	0.25	<2	12

Table 9.14 Metal Content for a Range of Crudes and a Pyrolysis Oil [12–16] *(continued)*

Metal	Pyrolysis oil (Plaxx®)	Fossil crude		
		Light (Saudi Arabia)	Medium (Saudi Arabia)	Heavy (Venezuela)
Ca	1200–3000	0.4	<2	1000–10,000
K	20.6	0.06	<2	
Mg	225–750	0.25	<1	
Na	15.4	2.2	2	
P	14.9	–	<2	
Si	415–1000	0.4	5	

In Plaxx made from PO waste, the API grade, and the levels of nitrogen, sulfur, and aromatics are typically lower, and therefore the product is more valuable than a crude oil. Levels of resins, which are composed of fused aromatic rings with branched paraffin and polar compounds, are also lower. This may indicate a high feasibility of Plaxx being used in petrochemical upgrading applications, as the steric effects hindering the production of olefins may be reduced. On the other hand, the asphaltene levels are higher in Plaxx, as is the acidity and oxygen content. This may negatively affect the integrity of the plants in operation as well as the stability of the production, due to the corrosion and flow assurance issues [14, 17–20].

The levels of chlorine in Plaxx are noticeably lower than in crude. However, it is important to note that the form of chlorine found in both differs: in Plaxx it is mainly organo-chlorinated compounds, while in crude oil it is predominantly inorganic [21]. Inorganic chlorine compounds found in crude oil can be reduced using a conventional desalting process prior entering to the distillation column [4]. However, this process will not reduce the levels of the organo-chlorinated compounds found in Plaxx. This is a challenge that will require concerted effort to overcome.

Looking at metal contamination, crude oil contains K, Na, Ca, Cu, Pb, Fe, Mg, Na, Mo, Zn, Cd, Ti, Mn, Cr, Co, Sb, U, Al, Sn, Ba, Ga, Ag, and As. In general, these elements are present in the crude oils as organic salts (mainly as the chloride and sulfate of K, Mg, Na, and Ca), and as organometallic compounds of Ca, Cu, Cr, Mg, Fe, Ni, Ti, V, and Zn [20]. Some of them are present only at trace levels (ppm) and others may reach up to 1–2%. The concentration of trace metals in any crude oil is found to be inversely proportional to the API gravity of the oil. The type and quantity of metals in Plaxx is comparable to crude oils. Therefore, existing refining techniques and capacity can be used for their removal [4, 21–23].

Given the similarities between the oil produced by cracking polyolefin waste and crude, it is reasonable to conclude that current refining techniques will be able to refine this oil for petrochemical uses. The question then is where the refining should be done. The refining processes can be incorporated into the feedstock recycling system right after the pyrolysis reactor. This approach makes the plant more expensive, but also increases the value of the output oil and the number of petrochemical facilities that can utilize it. Alternatively, the refining stages can be added by the downstream petrochemical company in front of a steam cracker. This adds capital cost at this stage but provides the option of being larger in scale and integrated into the petrochemical complex. Another approach is that the feedstock recycling pyrolysis outputs could go to existing refineries where the refining technology and capacity already exists. However, not every refinery has the necessary processes, as it depends on the nature of the crude that the refinery is equipped to deal with.

Refineries are commonly referred to as simple or complex (Figure 9.27). Each key process is assigned a complexity factor or index, and therefore a correlation between refinery classes and complexity ranges is established. Refineries with cracking and coking capacity are generally referred to as "cracking" and "coking" refineries, respectively, or as "complex" or "very complex" refineries, to distinguish them from "simple" refineries, which do not have such units. In Europe and elsewhere, it is common to refer to complex refineries as "conversion" or "deep conversion" refineries [24].

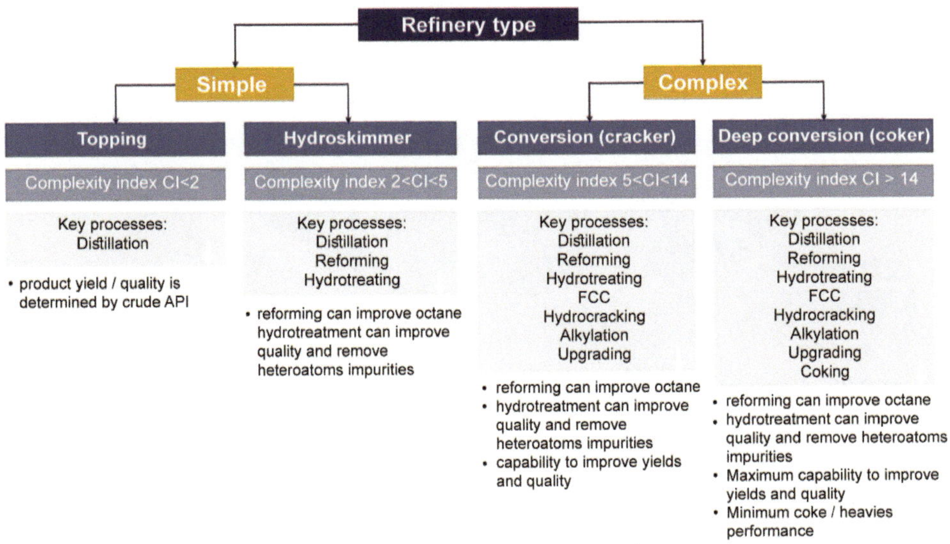

Figure 9.27 Refinery types and complexities [24]

Conversion processes in "complex refineries" carry out chemical reactions that crack large, high-boiling-point hydrocarbon molecules into smaller, lighter molecules suitable for gasoline, jet fuel, diesel fuel, petrochemical feedstocks, and other high-value products. The primary conversion processes that perform these operations are fluid catalytic cracking (FCC), hydrocracking, and coking. Complex refineries typically run heavier and more sour crude oils and produce more higher-quality products [24–27]. Such refineries can easily deal with the oil from the pyrolysis of a wide range of plastics, and certainly the output from polyolefins.

In Asia, the Middle East, and South America - areas that are experiencing rapid growth for gasoline demand and light products for petrochemicals - almost all new refinery construction can be classed as conversion or deep conversion, as shown in Figure 9.28 (complex refineries). In Europe, Japan, and Russia, hydro-skimming refineries are common. For pyrolysis oils to be accepted in these simple refineries, they may need to be blended into crude whilst volumes are low. Some of these refineries may then require further enhancement, or the oil from the plastic-to-oil conversion stage may need to be further enhanced, before significant volumes of pyrolysis oils can enter such refineries [25].

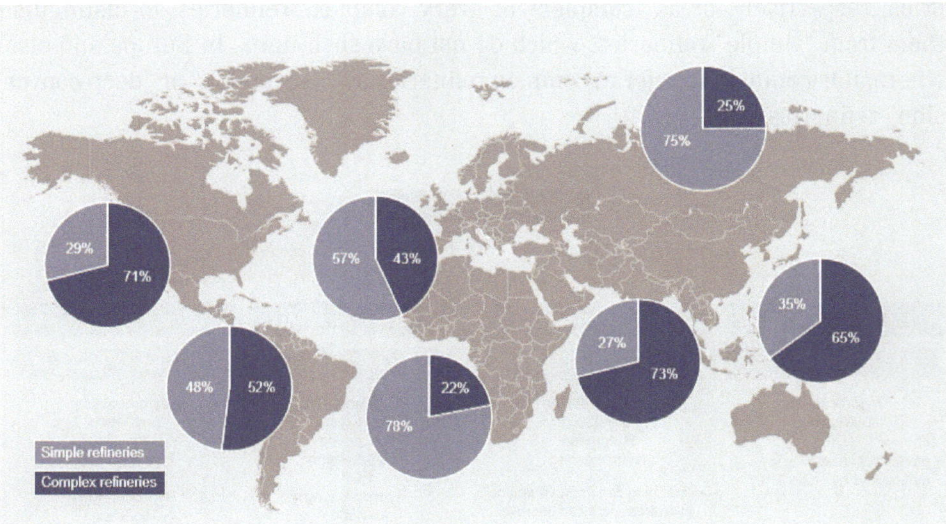

Figure 9.28 Simple and complex refineries by region [28]

Wherever possible, the most logical choice is to use existing refining capacity rather than constructing new infrastructure. However, the vast majority of the output of refineries today is petroleum fuels, so if pyrolysis oils are blended into the crude stream, it stands to reason that most of the molecules will end up a fuel, not naphtha for steam crackers, so requiring a "mass balance and allocation" approach to account for this material. This short-term loss of material into the fuel pool causes concern in some circles, it being seen as a less environmentally sound approach than the chemical recycler adding a refining stage and then supplying their output to a steam cracker. However, this argument misses two points:

1. Every barrel of oil coming from waste plastic means one less barrel of oil is taken out of the ground, which fully achieves the desired displacement of virgin feedstocks by leaving that carbon in the ground.
2. The refining industry is changing fast. While its main output is currently fuels, the days of burning carbon-based materials as fuel are numbered (see Chapter 6). With electric cars already outselling diesel cars in Europe, and China's petrol and diesel engine ban visible on the horizon, the fall-off of fuel demand will leave a lot of refining capacity available to be re-tasked to optimize their outputs to base chemicals and naphtha. Oil companies have already started this change, with new refining facilities being built that do not envisage fuel use at all, the so-called "crude oil-to-chemicals" (OTC) trend (Figure 9.29) [29, 30].

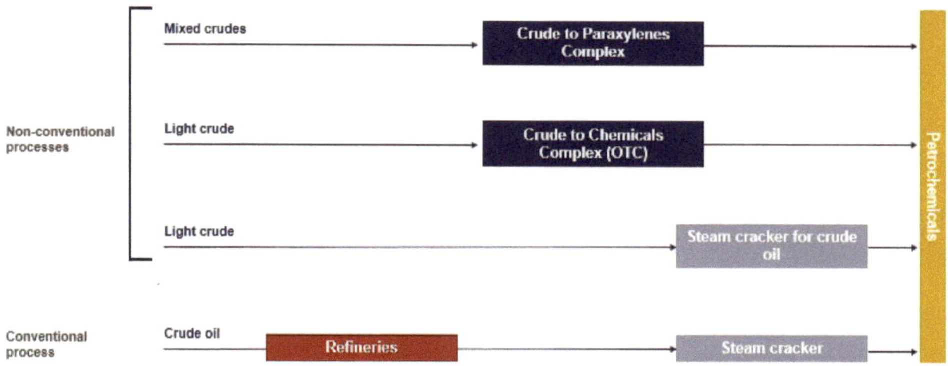

Figure 9.29 The crude oil-to-chemicals process [29]

So whilst mass balance and allocation will be needed in the short term to account for the pyrolysis oil entering a refinery, in the longer term refineries are moving away from fuels into chemicals, and towards circular economy inputs such as pyrolysis oils. For example, the major French oil and chemicals producer Total has converted their 101,000 bbl/d oil refinery (representing 7% of Total's European refining capacity) into a biorefinery for transport fuels, naphtha, and bioplastics, plus a plastics recycling plant [31].

As the capacity of feedstock recycling increases and the use of crude wanes, dilution will cease to be an option. Different routes integrating pyrolysis oils into refineries, traditional petrochemical facilities, or OTC are shown in Figure 9.30. This is a moving situation, and at best the basis for future discussions.

Completing the circular journey from the output of a pyrolysis reactor to virgin-quality plastic is technically feasible, and the ingenuity of the oil industry to remove unwanted materials and turn chemical chaos into order is, we are sure, capable of the challenge. Clearly there are mismatches in capabilities around the world, but the main technology building blocks exist. Silicon and chlorine, to single out two compounds, remain a challenge, but these too will be overcome.

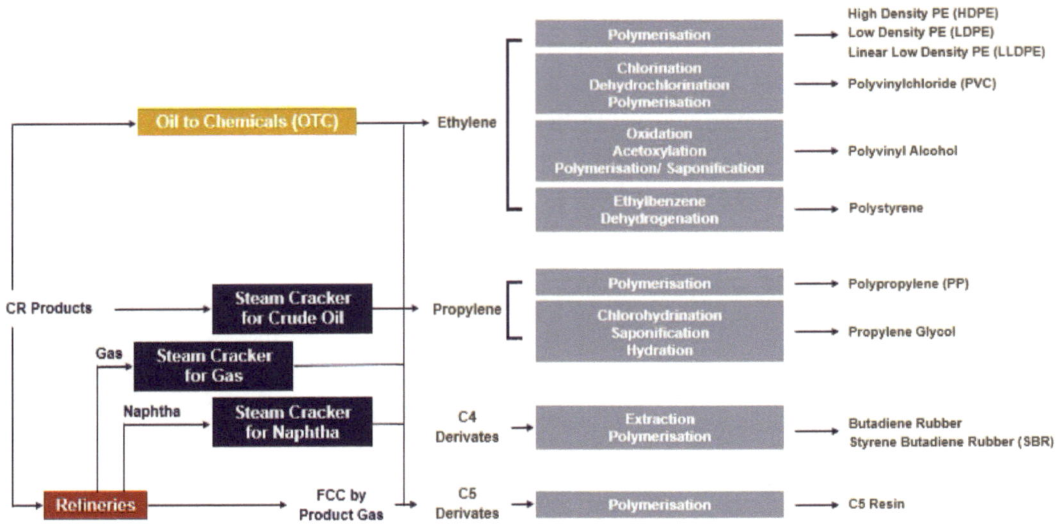

Figure 9.30 Routes for chemically recycled products [3, 17]

9.2.5 Applications

The use of feedstock recycling for pure polyolefins is relatively straightforward, but most pure polyolefins can be recycled mechanically. The real benefit of feedstock recycling is its suitability for processing contaminated materials, films, and laminated structures, including foil-backed materials. In focusing on such materials, feedstock recycling complements mechanical and monomer recycling. NIR sorters can separate the material for mechanical recycling, purification, monomer recycling, leaving residual plastic waste (RPW), which, following the removal of PVC and HDPE, forms the input stream. Applying feedstock recycling to RPW ensures that more than 90% of all domestic plastic waste is recyclable. However, it also requires a broadly polyolefin stream with appreciable levels of PS, PET, PA, and biomaterials, as discussed above.

The conversion of RPW into oil can be done on a small scale (<25 kt/a), a medium scale (25–200 kt/a), or a larger scale (>200 kt/a). This means that the process is applicable to:

- Resource and waste management sectors wishing to upgrade the material in their own facilities from a waste to a product

- Specialist chemical recyclers that establish themselves in strategic locations to provide a plastic-to-oil service
- Petrochemical companies wanting to switch away from virgin fossil feedstocks into recycled materials

Conventional petrochemical wisdom would say the bigger the better. But economies of scale can be achieved in different ways by building very large facilities or many small facilities. In cities and regions where the population density is high and waste plastic is aggregated in large quantities, building a recycling plant that approaches the scale of a petrochemical facility is feasible. Singapore, for example, has 900 kt of plastic waste going through a single waste facility, and Tokyo and some of China's super-sized cities will also have such quantities in single locations. However, for most of the world, getting anywhere near 1 Mt of plastic per annum into one location is a significant challenge.

To illustrate this challenge, consider the situation from the perspective of a trucking operation. It typically takes 1.3 t of contaminated (water, metal, grit, biomass) polyolefin waste to get 1 t of polyolefin suitable to start the feedstock recycling process. If each truck carries 20 t of baled plastics, then a 1 Mt/a plant receiving 1.3 Mt of material requires a convoy of 180 trucks per day (assuming 360 days of shipping per year). This is a considerable logistical challenge, and ignores the ash, metal, and glass contamination found in the waste plastic that must subsequently be disposed of.

The alternative is for the resource management sector to undertake the plastic-to-oil conversion and to aggregate the oil. If we consider the output of our illustrative 1 Mt/a feedstock recycling facility with an 85% yield (which is typical for polyolefins when the gases produced are used to power the process) this equates to 850 kt of feedstock. To transport this oil from waste facilities to the same point, at 27 t per truck, would require 87 trucks – less than half the 180 trucks needed to transport the plastic.

A criticism of this smaller-scale approach is that the capital expenditure (CAPEX) per tonne of processing capacity at small scale will never compete with the large scale. But what if machines were to be manufactured at volume? Putting cost to one side for a moment (as this will be considered in more detail in Section 9.2.6.2) one thing is certain: mass production allows many machines to be produced quickly. Millions of tonnes of processing capacity can be produced in assembly factories each year and sold to existing waste sites around the world.

The reality is that chemical recycling facilities at all scales have merit, and therefore its application at different positions along the value chain is valid. It is our view that a combination of both large- and small-scale feedstock recycling plants will enable the current gulf between plastic production and recycling capacities to be closed.

9.2.6 Market

In considering the market for feedstock recycling of polyolefins, it is important to consider several aspects:

- **Capacity:** Can sufficient recycling capacity be added fast enough to allow recycling to catch up with polymer production capacity?
- **Cost:** Is it feasible that recycling can eventually compete with virgin polyolefin production, eliminating the need for fiscal incentives?
- **Collaborators:** Will the options for recycling work together? Can the resource management industry select the optimal pathways?
- **Circularity:** Material is lost around the system – some plastics are durable in use, recycling has yield losses, and the plastic market is growing. Where will all this material come from?
- **Carbon:** Reduce, reuse, recycle is right. But recycling is not necessarily sustainable – can it be?

9.2.6.1 Capacity

The sheer scale of polyolefin production means that recycling capacity must be added at a phenomenal pace to meet the scale of production. Companies such as QuantaFuel and ReNew ELP are building 20–100 kt/a feedstock recycling plants, and there are up to 400 kt/a facilities in the pipeline. However, this rate of build will have to increase significantly to catch up with existing polyolefin manufacture.

Polyolefin production capacity is not standing still either, and when the polyolefin industry adds plants it does at a grand scale, such as Shell's new 1.6 Mt/a polyethylene plant in Pennsylvania. McKinsey suggests that global plastic production could exceed 1 Bt by 2050 (Figure 9.31). This implies a production capacity increase of 640 Mt over a 30-year period from the 2020 capacity of ca. 368 Mt.

If this growth is linear, 21 Mt/a of new plastics production capacity will be added each year. Figure 9.31 also suggests that CR could account for around 30% of plastic production by 2050, which would be 300 Mt of plastic output. If, for example, the yield from plastic to oil and back to plastic were 90%[1], 333 Mt of plastic recycling capacity will be needed by 2050. If added over 30 years on a straight-line basis, this would be 11.1 Mt/a of new chemical recycling capacity. But is this feasible?

[1] This assumes that pyrolysis is driven by zero-carbon electricity and all the outputs from pyrolysis are reused within the petrochemical system as products (see Section 9.2.6.4), with the return to plastic being determined by mass balance and allocation.

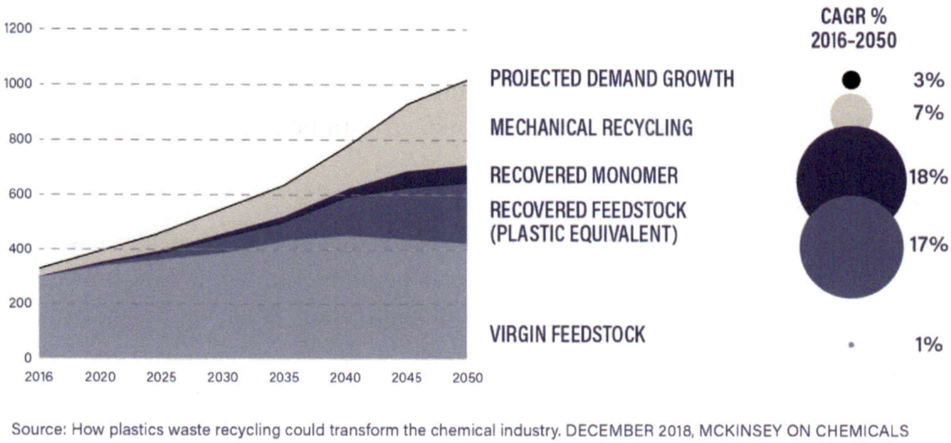

Figure 9.31 Global polymer demand and how it could be covered [32] © McKinsey (source: How plastics waste recycling cloud transform the chemical industry. DECEMBER 2018, MCKINSEY ON CHEMICALS)

The initial approach to chemical recycling has been for small companies to form, which then develop a process, secure a suitable site, contract with a company to supply the waste plastic and a petrochemical company to take the oil, raise the necessary investment, engage an engineering, procurement and construction (EPC) partner to construct the facility, and then operate it. This approach is valid, but it is hard for relatively small companies to achieve all this quickly enough to really move the needle on chemical recycling capacity. As a result, it is hard to see 11.1 Mt/a being added only in this way.

There are, however, two changes that are currently taking place that will make a significant difference - innovation and investment. On the innovation side, there are petrochemical giants beginning to build very large chemical recycling plants. Other companies, like Recycling Technologies Ltd, are looking to mass-produce smaller-scale plants. Both innovations will complement the medium-scale plants referred to earlier and boost the rate at which physical recycling capacity can be added. The other consideration is investment. Until now, investment in CR has been seen as optional and at "toe in the water" scale. By contrast, over the last 10 years plastic production has grown from 250 Mt/a to 368 Mt/a, around 11.8 Mt/a of new capacity [1]. It is worth noting that if "oil/gas to plastic" capacity can be added at ~11 Mt/a, with the right will there is no reason that plastic-to-oil capacity cannot be added as quickly. The question of "will" is really one of investment.

It is estimated that the growth of new oil/gas-to-plastic capacity has been funded by investments in facilities at ca. \$5–7 billion per annum per 1 Mt of new capacity, or \$60–80 billion per annum. Looking forward, as the growth in plastic demand is supplied, not from oil and gas (see Section 9.2.6.4), but from mechanical and chemical recycling (where the input material is waste plastic whose value is, in the case of polyolefins, close to zero and the output material arguably more valuable than plastic from fossil sources), this scale of investment will flow into chemical recycling, allowing the recycling capacity to catch up with production.

9.2.6.2 Cost

Given the low cost of production of virgin polyolefins and the typically low-value items made from them, discarded polyolefin products are frequently considered to have no value and are discarded without thought. There is no doubt that, with patience and perseverance, the chemistry barriers to feedstock recycling can be overcome. The combination of large-scale, centralized facilities, and medium- and small-scale mass-produced, distributed facilities can bridge the capacity gap. The question that remains is, can it be done economically? Currently the answer is probably not, at least not without the oil produced from plastic attracting premium pricing.

Premiums do exist, within Europe in particular. There are incentives for increasing recycling rates, and various forms of extended producer responsibility (EPR) are under development, which could prove pivotal in making feedstock recycling economically viable now, allowing it to grow and thereby address the underlying costs.

However, as the feedstock recycling approach matures, fiscal incentives should become superfluous, as costs will reduce enough that the value of the oil produced will provide support for the collection of a valueless material, create a profitable return on the investment in the conversion technology, and compete with the price of crude oil (Figure 9.32).

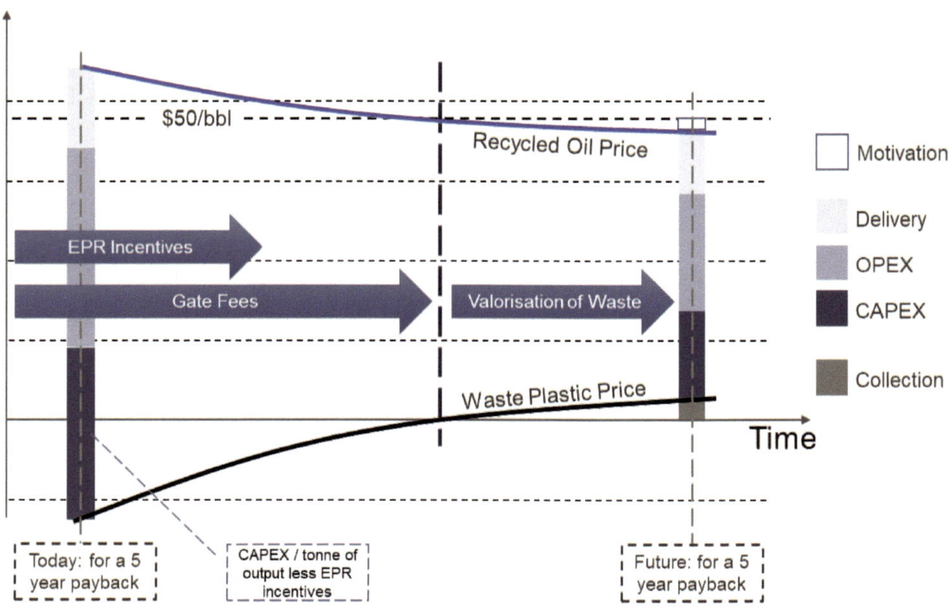

Figure 9.32 Price of recycled oil and waste plastic over time

This requirement for policy and fiscal incentives is very similar to the electricity generation industry. For example, when offshore wind was first considered, it was extremely expensive compared to installing alternatives such as a combined gas cycle power station. For offshore wind to compete with the incumbent technology, feed-in tariffs and contracts for difference were developed, enabling a fledgling sector to gain scale and reach the costs that are only possible with economies of scale. Now wind power and solar are the lowest-cost methods for producing electricity (Figure 9.33).

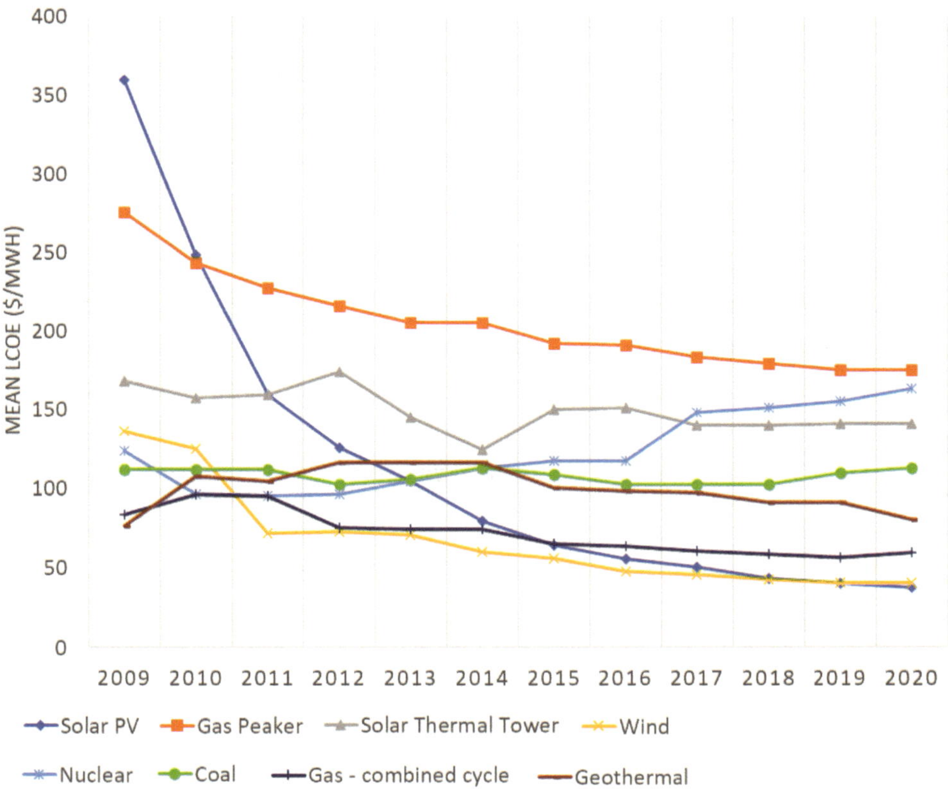

Figure 9.33 Levelized cost of electricity production from different technologies over time [33]

Capital Expenditure (CAPEX)

Economies of scale within the chemical sector have traditionally meant bigger reactors and higher-throughput systems. However, with wind and solar generation, economies of scale have come predominantly from the number of units installed rather than the size of each unit. It is true that wind turbines have become colossal (Siemens Gamesa's SG 14-222 DD wind turbine rotor is 222 m in diameter with a peak generation capacity of 14 MW [34]), but their output per unit remains tiny compared to the 6.7 GW capacity of the Tuoketuo Power Station, the biggest coal-fired power station on earth, albeit made up of a series of ~600 MW units. The secret to wind and solar success is mass-production; with increasing numbers of units comes reducing costs. In 1977, electricity from a photovoltaic cell cost $76/W. By 2013 it was $0.74/W [35]. Mass production, continual design improvements, technology evolvement, and experience have achieved enormous cost efficiencies. More recently, the levelized cost of electricity from offshore wind was $255/MWh in 2012, falling to $83/MWh by 2020, and forecast to fall further to $58/MWh by 2025 [36].

Mass-production of machines to convert plastic into oil for the global waste industry will involve thousands of machines, providing the opportunity for chemical recycling to gain the associated cost advantages. Therefore, as some facilities grow to achieve lower CAPEX per tonne of feedstock recycling capacity, others will achieve this by being mass-produced. Both routes lead to substantially lower CAPEX per tonne costs than exist currently.

Operating Expense (OPEX)

The mantra "bigger is better" has not just been driven by CAPEX costs, but also energy and labor costs. The more efficient a petrochemical complex is, the lower the energy cost per tonne. The bigger the throughput per hour, the more tonnes of product to distribute across the cost of operations and maintenance. In a petrochemical facility the input material usually comes via a pipeline from a refinery, so the input is available in vast quantities. However, when waste plastic is the feedstock it is normally located in a very distributed manner, so in feedstock recycling the cost of logistics must also be considered in the OPEX.

- **Energy** has been expensive, so any marginal improvement in efficiency has been welcome (hence bigger reactors), but this is changing quickly. Electrical energy for processing is becoming cheaper with the roll-out of wind and solar capacity, and many chemical processes are seeking to utilize this change. This shift to electricity will also apply to feedstock recycling, diminishing the advantage of large scale.
- **Labor** is optimally spread across many tonnes of throughput – an argument for larger facilities. This is a disadvantage for small plants, though tempered when co-located with waste sorting facilities where labor can be "shared" across functions. Over time, feedstock recycling plants will become easier to operate and the labor cost per tonne will diminish, but this will remain a significant component of the OPEX.
- **Logistics** must be carefully considered in the unit economics of CR. For a centralized CR facility, plastic will require drying (before or after transport), baling, transporting, and de-baling at its destination. Any non-plastic material will then

need to be separated and disposed of, requiring more transportation to yet another facility. Alternatively, at a decentralized CR plant, all sorting, drying, processing, and disposal of undesirable material will take place at the same location, with only the valuable oil product being transported to the petrochemical facility.

Whilst CAPEX and energy costs will come down over time and to a lesser extent so will labor, the cost of logistics is hard to overcome. Nevertheless, the opportunity for cost reductions, as exemplified by the experience of many other technologies, points to a future where feedstock recycling creates an oil at lower cost than that which is extracted from the earth.

9.2.6.3 Collaborators

Creating multiple pathways for dealing with a plastic product at the end of its life opens the possibilities of much higher levels of recycling. However, with this choice comes the challenge of selection: which route is optimal for each plastic item in the waste stream – mechanical, purification, monomer, or feedstock recycling (Figure 9.24)? This choice is particularly tricky with respect to polyolefins, which make up the lion's share of household waste compared to engineered plastics used in the automotive or electronic industry, where end-of-life pathways are more regimented.

Currently, near-infrared (NIR) sorting targets plastics such as PET and HDPE, removing them from the waste stream. But polymer-type separation is not the only concern. Color is also important, and the polymer streams must be further divided into clear and natural (for PET and HDPE respectively) or jazz (multicolor) prior to mechanical recycling.

Advancements in mechanical recycling as well as packaging design will mean increasing quantities of used plastics can be recycled this way, at least once. However, the mechanical recycling process tends to degrade the quality of the polymer, and the resulting material cannot always be used in, for example, food contact applications. Material that has been mechanically recycled as many times as possible and can't be usefully recycled again can be sent for feedstock recycling so that the hydrocarbons can be used to recreate virgin-quality plastics again. The challenge for the resource management industry is how to identify not only the polymer type and color, but also the number of times it has already been mechanically recycled to route the highest-grade material to mechanical recycling and previously recycled materials to feedstock recycling.

When purification and monomer recycling are also considered, the choices become even more complex. It cannot be the responsibility of the resource management companies to make these routing judgments. It is incumbent on the organization designing and producing products containing recycled materials to select the optimal route along which the product should be directed after its life. Tracer molecules could be embedded into the product that can be identified within the resource management infrastructure, allowing the product to be sent to the recycling route that the designer specified [37]. Whilst this is eminently possible from a technical perspective, it will need unprecedented levels of cooperation and probably legislation to become a reality.

9.2.6.4 Circularity

In theory, if every bit if plastic produced and used each year could be collected and recycled, then plastic could circulate indefinitely in a "utopian loop". Of the plastic produced each year, some is used in long-term applications (e.g., water pipes, building products, cars etc.) The durability of plastic makes it a go-to material in many long-term applications, and it will therefore not arrive back at the recycler's door in the same year it was made. Sadly, it is also a fact that some material never makes it back to a recycler at all, with people choosing to burn it (with or without energy recovery), an act of carbon release that has little justification (not even for its energy), except perhaps in a medical context. The losses due to durable use, incineration, and system leakage are exacerbated when one considers the yield from recycling itself. Each form of recycling will have a yield; no system is perfect. The conversion of polyolefin waste-to-oil has a yield of 80–85% depending on the mix and levels of contamination. If other plastics are combined with the waste, then the yield is further reduced. On top of the losses each year, we need more plastic. In the last 10 years (2010–2020) plastic production has risen from 250 Mt to 368 Mt, an increase of 11.8 Mt on average per year [1]. So, for a circular economy to work, the durable use, leakage, yield, and the augmentation of the inventory of the plastic circular economy has to come from somewhere.

McKinsey helpfully has estimated and represented this diagrammatically (Figure 9.31). Here the assumption is that the losses and the system augmentation will be derived from virgin feedstocks, which may be the case, meaning that whilst recycling will increase it will not be a circular material for a long time.

But there are alternatives to this scenario that could make plastic much more circular. The refining capacity that has been developed to make biodiesels and other biofuels from waste biomass may be part of the answer. From a carbon perspective, this bio-derived oil would be better used to top-up the plastic circular economy than using it as a fuel. Whether this means that such biowaste, following gasification, undergoes a Fischer–Tropsch process to produce oil or whether the gas goes directly into a cracker, it would be a much better use of it than simply burning it to produce electricity. There are also options such as the extraction of CO_2 from the atmosphere followed by synthesis with hydrogen produced by electrolysis – ways that add hydrocarbons into the plastics circular economy in a sustainable way, dramatically reducing the need for virgin oil and gas feedstocks, perhaps even to zero? See Figure 9.34.

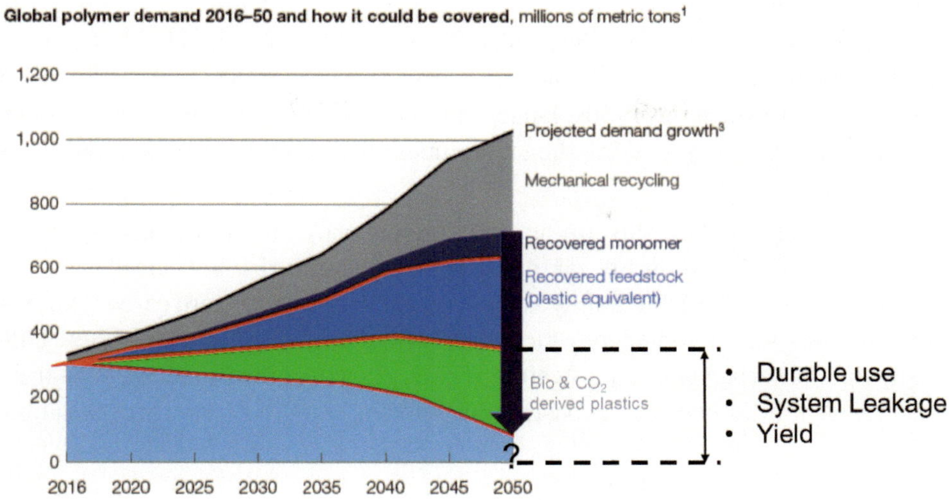

Figure 9.34 McKinsey's view with a suggestion for a more sustainable outcome

This all adds up to a system with hydrocarbons going around a virtuous circle (Figure 9.35), but is this also carbon-absorbing and hence sustainable? That depends on the whether the carbon absorbed by the biomaterial is more or less than the carbon emitted by the inefficiencies of the recycling mechanism.

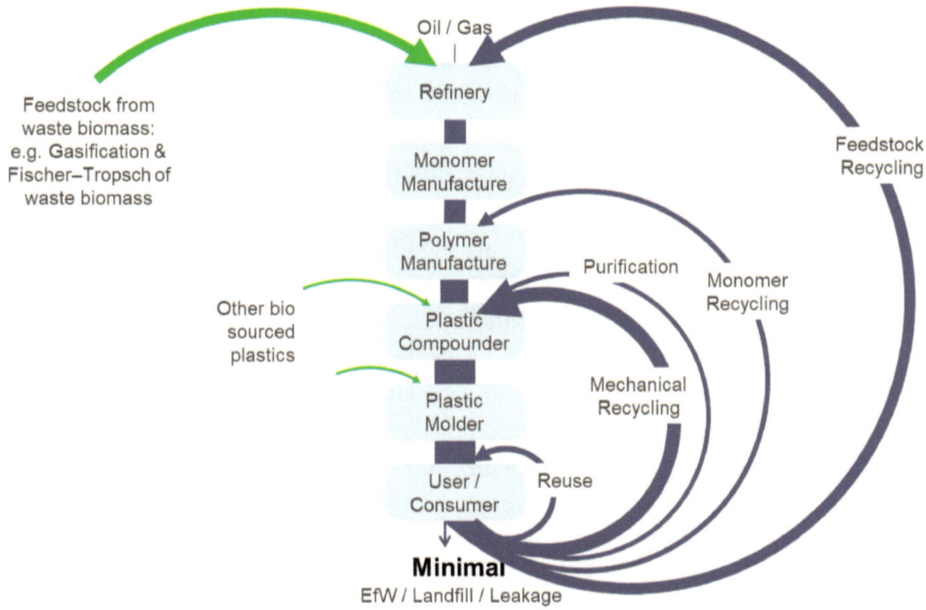

Figure 9.35 Plastic in the circular economy

9.2.6.5 Carbon Reduction

The Mauna Loa Observatory is located on the north flank of Mauna Loa Volcano, on the Big Island of Hawaii. There they have been measuring CO_2 concentrations in the atmosphere since the 1950s, longer than anywhere else on the planet. The increases seem relentless, taking the environment to a condition that humanity has never experienced before.

The consequences of all this are not fully understood, but without doubt, policymakers around the world are taking action to slow the rise and ideally put the system into reverse. Sustainability is often associated with net zero carbon, but net zero conjures up a knife edge, which when something subsequently shifts could easily tip into carbon emissions again. For plastic, or any other material, to be truly sustainable it must become net positive by reducing the CO_2 currently in the atmosphere.

This focus on removal of carbon from the atmosphere means the burning of anything with carbon in it for energy must stop, unless equipped with an effective and economical carbon capture device. As the magnitude of the carbon crisis permeates strategic thinking, the use of biomaterials as fuels will also have to stop. The thinking that biomaterial is only releasing the carbon that it has absorbed during the growing cycle and is therefore sustainable, is only valid if CO_2 concentrations in the atmosphere are within the normal sustainable range, but evidently they are not. So, if a living plant provides the service of extracting CO_2 from the environment, it is lunacy to subsequently burn that biomass and put it back into the atmosphere. This aversion to burning anything with carbon in it may seem like a dream. After all, in the real world we need energy for everyday life. We want to get on planes and visit exotic destinations. How can this occur without carbon release?

Necessity, it is often said, is the mother of invention, and concern over the impact on human health is one of the most powerful drivers of change. The speed of producing a vaccine for the Covid-19 pandemic attests to this principle, with the generally accepted time of vaccine development being shrunk from 10 years to around 10 months. The Great Horse-Manure Crisis of 1894 is another wonderful example. In 1894 the Times newspaper is said to have run an article that predicted "in 50 years, every street in London will be buried under nine feet of manure". The prediction was extrapolating from the dire situation they found themselves in, with horses being the main form of transport and the emissions from horses piling up in the streets creating a serious health hazard. Of course, the prediction never came true, as in 1908 Henry Ford started mass-producing the motor vehicle, and within a remarkably short time, the horses of London and other big cities around the world had essentially vanished. Ironically, it is the car itself that is now blamed for the air pollution that is injurious to the health of many city dwellers today. In the UK, the first person to have air pollution on their death certificate as a cause of death has just made the news [38]. China has linked air pollution and the use of cars, leading to a government announcement in September 2017 that they would ban the sale of cars powered by petrol and diesel [39]. They did not say by when, but this signal was enough to put change into motion. In Norway, electric vehicles already outsell petrol, diesel, and hybrids combined [40].

The rate of change is undeniable. It is not hard to see that by 2035 there could be virtually no new cars powered by petrol or diesel. Once something is linked to human health, change happens very fast. This brings us back to the point that burning anything with carbon in it simply for its energy is lunacy while the CO_2 concentrations in the atmosphere are as high as they are. The potential for global warming to have a catastrophic impact on human health is increasingly accepted. Thankfully, people have been thinking about the need to decarbonize fuel for some time. Whether the fuel used in the ships and aircraft of the future is hydrogen (from electrolysis), ammonia, or something else is not the subject of this section, but just as the car replaced the horse, and electric cars are replacing petrol and diesel versions, the days of using transport fuels containing carbon are numbered. Even the International Energy Agency (IEA) have now called for an end to the funding of new exploration for oil, gas, and coal [41].

The focus on carbon reduction is going to drive structural change in the petrochemical world, but it is not just someone else's problem, feedstock recycling must decarbonize too. In many of the processes currently making their debut, the thermal energy needed to crack the polymer comes from the input plastic. The pyrolysis operation produces a range of hydrocarbons, solids, liquids, and non-condensable gases such as methane, ethane, and ethylene. These light hydrocarbon gases are frequently used as the fuel to generate the heat needed to crack the plastic. In doing this, by combusting these gases, CO_2 is produced, though of course in far smaller quantities than if all the material was combusted in an EfW facility. Whilst this improvement is good news for carbon reduction, it is not sufficient to make feedstock recycling and plastic as a whole a sustainable material. But the feedstock recycling industry is still a nascent one. There are many techniques that are in development that will see the decarbonization of the plastic-to-oil conversion, boosting the mass yields to >90%, and so reducing the losses. Pyrolysis powered by electricity derived from zero-carbon generation is likely to feature soon, with the process gas from the pyrolysis of polyolefins also seen as a product rather than a fuel.

A circular economy for plastic has an immense amount to offer, and although there are many obstacles that will take considerable human energy and perseverance to work through, the sustainable vision is becoming clearer, and it is achievable.

9.2.7 Conclusion

Plastic in the environment has reached a level where to do nothing is no longer an option. Continuing to invest billions of dollars in new plastic production with the assumption that there is an ever-expanding marketplace for plastic from fossil material is now a folly. It is equally clear that recycled plastic is a terrific material from an LCA perspective, as is plastic made from CO_2 and bio-sources.

Whilst some recycling approaches, like mechanical recycling, are well-established, many of the techniques needed to ensure plastic is a sustainable material for the future are nascent. However, all the pieces of the puzzle exist, and if the level of investment that plastic from fossil sources has received is now directed at plastic from plastic, then the system can be created profitably, and plastic will remain the "miracle material" for many years to come.

References for Section 9.2

[1] Packaging Europe: HolyGrail 2.0: Ground-breaking recycling project enters its next stage with cross-industry backing. *Packaging Europe*, 2020. *https://packagingeurope.com/holy-grail-enters-next-stage-with-cross-industry-backing/[PLASTICSEUROPE2020]*, accessed November 19, 2021

[2] Kissin, Y.: *Polyethylene: End-Use Properties and their Physical Meaning*, Hanser Publications, Cincinatti, Ohio, 2013, pp. 1, 2

[3] Al-Ali AlMa'adeed, M., Krupa, I.: *Polyolefin Compounds and Materials: Fundamentals and Industrial Applications*, Springer International Publishing, Switzerland, 2016, pp. 327–333

[4] Gary, J., Handwerk, G., Kaiser, M.: *Petroleum Refining Technologies and Economics*, 5th Edition. CRC Press Taylor & Francis Group, Boca Raton, Florida, 2007, pp. 16–21

[5] Hansen, Erik; *Hazardous Substances in Plastic Materials*, COWI: Vejle, Denmark, 2013, pp. 11–12

[6] Chinenyeze, M., Ekene, U.: Physical and chemical properties of crude oils and their geolithic significances. *International Journal of Science and Research*, (2017) 6, pp. 1514–1521

[7] A full list of our crude assays, British Petroleum, 2020, *https://www.bp.com/en/global/bp-global-energy-trading/features-and-updates/technical-downloads/crudes-assays.html*, accessed November 21, 2021

[8] McKenna, A.: *Detailed characterization of heavy crude oils and asphaltenes by ultrahigh resolution Fourier transform ion cyclotron resonance mass spectrometry*, Florida State University Libraries – Electronic Theses, Treatises and Dissertations, 2009; *http://purl.flvc.org/fsu/fd/FSU_migr_etd-2534. https://www.bp.com/en/global/bp-global-energy-trading/features-and-updates/technical-downloads/crudes-assays.html*, accessed November 21, 2021

[9] Santos, J., Vetere, A., Wisniewski, A., Eberlin, M., Schrader, W.: Comparing crude oils with different API gravities on a molecular level using mass spectrometric analysis. Part 2: Resins and asphaltenes. *Energies*, (2018) 11, 10, pp. 571–590

[10] Santos, R., Loh, W., Bannwart, A., Trevisan, O.: An overview of heavy oil properties and its recovery and transportation methods. *Brazilian Journal of Chemical Engineering*, (2014) 31, 3, pp. 1–14

[11] Yassin, A., Mohamed, B., Morsi, M., Osman, M., Mohamed, M.: *Petroleum Refinery Process Design & Naphtha Methaforming – Simulation Using Aspen HYSYS*, Alexandria University Specialized Scientific Programs Gas & Petrochemicals Program, 2019, pp. 2–22

[12] Eqal, A. Correlation of metals with hydrocarbons of crude oil in Maysan Province Southern Iraq. *International Journal of Engineering Research & Technology (IJERT)*, (2015) 4, 9, pp. 424–428

[13] Young, A., Venditti, S., Burrueco, C., Yang, M., Waters, A., Davies, H., Hill, S., Millan, M., Crittenden, B.: Characterisation of crude oils and their fouling deposits using a batch stirred cell system, *Heat Transfer Engineering*, (2009) 32, pp. 216–227

[14] Wilberforce, O.: Profile of heavy metals in crude oil commonly consumed for medicinal purposes in Abakaliki. *Journal of Pharmacy and Biological Sciences*, (2016) 11, 3, pp. 43–44

[15] All, M., Bukhari, A., Saleem, M.: Trace metals in crude oils from Saudi Arabia. *Industrial & Engineering Chemistry Product Research and Development*, (1983) 22, 4, pp. 691–694

[16] Barros, H., Sajo-Bohus, L., Abril, J. M., Greaves, E. D.: Radioactivity concentration and heavy metal content in fuel oil and oil-ashes in Venezuela. *Radioprotection*, (2005) 40, 1, pp. S183-S189

[17] Matar, S., Hatch, L.: *Chemistry of Petrochemical Processes*, (2000) 2, Gulf Publishing Company, Houston, Texas, pp. 49–111

[18] Saintilan, N., Spangenberg, J., Chiaradia, M., Chelle-Michou, C., Stephens, M., Fontbote, L.: Petroleum as source and carrier of metals in epigenetic sediment-hosted mineralization. *Scientific Reports*, (2019) 9, 8283, pp. 1–7

[19] Pasbana, A., Nonahala, B.: The investigation of precision of analytical methods for determination of salt content in Iranian crude oils. *Analytical Methods in Environmental Chemistry Journal*, (2018) 1, pp. 23–28

[20] Zhao, X., Shi, Q., Gray, M., Xu, C.: New vanadium compounds in Venezuela heavy crude oil detected by positive-ion electrospray ionization Fourier transform ion cyclotron resonance mass spectrometry. *Scientific Reports*, (2014) 4, 5373, pp. 1–6

[21] Kaminski, M.: Comments on identification and determination of volatile chloroorganic compounds in crude oil at trace level. *Chemia Analityczna*, (2009) 54, 4, pp. 1–15

[22] Castaneda, L., Munoz, J., Ancheyta, J.: Current situation of emerging technologies for upgrading of heavy oils. *Catalysis Today*, (2014) 220–222, pp. 248–273

[23] Mandal, P., Abdullah, A., Rahmanc, M.: Total acid number reduction of 2,6-naphthalenedicarboxylic acid using subcritical methanol for reducing acidity of heavy oil: A kinetic study. *Procedia Engineering*, (2016) 148, pp. 1213–1219

[24] Kaiser, M.: A review of refinery complexity applications. *Petroleum Science*, (2017) 14, pp. 167–194

[25] Alabdullah, M., Gomez, A., Vittenet, J., Bendjeriou-Sedjerari, A., Xu, W., Abba, I., Gascon, J.: A viewpoint on the refinery of the future: Catalyst and process challenge. *ACS Catalysis*, (2020) 10, pp. 8131–8140

[26] Ancheytaa, J., Ranaa, M., Furimsky, E.: Hydroprocessing of heavy petroleum feeds. *Catalysis Today*, (2005) 109, pp. 3–15

[27] Sawarkar, A., Pandit, A., Samant, S., Joshi, J.: Petroleum residue upgrading via delayed coking: A review. *The Canadian Journal of Chemical Engineering*, 2007

[28] *Oil & Gas Review: World Oil Review 2020, https://www.eni.com/assets/documents/eng/scenari-energetici/WORLD-OIL-REVIEW-2020-vol1.pdf*, accessed November 19, 2021

[29] Charlesworth, R.: Crude oil-to-chemicals (COTC): An industry gamechanger? *The Annual GPCA Forum*, Dubai, UAE, 2019

[30] Tullo, A.: Why the future of oil is in chemicals, not fuels. *Petrochemicals*, 2019. *https://cen.acs.org/business/petrochemicals/future-oil-chemicals-fuels/97/i8*, accessed November 19, 2021

[31] Beacham, W.: Total, LyondellBasell moves highlight growing momentum of circular economy shift. ICIS, 2020. *https://www.icis.com/explore/resources/news/2020/09/29/10558015/total-lyondellbasell-moves-highlight-growing-momentum-of-circular-economy-shifthttps://www.bp.com/en/global/bp-global-energy-trading/features-and-updates/technical-downloads/crudes-assays.html*, accessed November 19, 2021

[32] Hundertmark, T., Mayer, M., McNally, C., Simons, T. J., Witte, C.: How plastics waste recycling could transform the chemical industry, 2018. *https://www.mckinsey.com/industries/chemicals/*

our-insights/how-plastics-waste-recycling-could-transform-the-chemical-industry, accessed November 19, 2021

[33] Lazard's levelized cost of energy analysis (Version 14.0), 2020. *https://www.lazard.com/media/451419/lazards-levelized-cost-of-energy-version-140.pdf*; *https://www.bp.com/en/global/bp-global-energy-trading/features-and-updates/technical-downloads/crudes-assays.html*, accessed November 19, 2021

[34] Bairstow, J.: Siemens Gamesa reveals 'world's biggest wind turbine'. *Energy Live News*, 2020. *https://www.energylivenews.com/2020/05/20/siemens-gamesa-reveals-worlds-biggest-wind-turbine/*, accessed November 19, 2021[35] Sunny uplands. *The Economist*, 2012. *https://www.economist.com/news/2012/11/21/sunny-uplands*, accessed November 19, 2021

[36] *Global Offshore Wind Report 2020*, Global Wind Energy Council, 2020. *https://gwec.net/global-offshore-wind-report-2020/*, accessed November 19, 2021

[37] *Plastics Europe: Plastics – The Facts 2020*, *https://www.plasticseurope.org/en/resources/publica tions/4312-plastics-facts-2020*, accessed November 19, 2021

[38] Air pollution: Coroner calls for law change after Ella Adoo-Kissi-Debrah's death. *BBC News*, 2021. *https://www.bbc.co.uk/news/uk-england-london-56801794*, accessed November 19, 2021

[39] Ebnesajjad, S.: Electric cars in China: An ambitious stab at reducing pollution. *New Materials & Applications*, 2017. *lsevier.com/new-materials-applications/electric-cars-china-ambitious-stab-reducing-pollution*, accessed November 19, 2021

[40] Dawson, B.: Norway becomes first country to sell more electric cars than petrol vehicles. *The Independent*, 2021. *https://www.independent.co.uk/news/world/europe/norway-electric-cars-b1782706.html*, accessed November 19, 2021

[41] *Net Zero by 2050: A Roadmap for the Global Energy Sector*, International Energy Agency, 2021. *https://www.iea.org/reports/net-zero-by-2050*, accessed November 19, 2021

9.3 Chemical Recycling of PET

Carsten Eichert, Vitalij Salikov, Stephan Scholl

9.3.1 Introduction

Polymers have become an important part of our daily life and economy. They provide cost-effective material solutions for many applications, such as, for example, lightweight materials in cars or aircraft, which save fuel and reduce CO_2 emissions, or as insulation materials, which saves energy [1]. However, the largest application of polymers in Europe is as packaging materials (39.6%), followed by building and construction (20.4%) and the automotive industry (9.6%) [2].

Packaging protects products from environmental influences or environment against hazardous products. For food products, it extends their life and thus contributes significantly to the reduction of food waste, and the lowering of food prices. Compared to common alternatives used for packaging of sensitive products, such as glass or metals, polymers, due to their low specific weight and low melting points, require less energy to be produced or transported. However, all the benefits of plastics only count if they are used efficiently and sustainably, and do not endanger consumer health and environment. Approx. 75% of plastic packaging consists of the so-called "big five" polymers: low-density polyethylene (LDPE), high-density polyethylene (HDPE), polypropylene (PP), polystyrene (PS), and polyethylene terephthalate (PET) [3].

PET has a very favorable combination of properties, justifying its widespread application [4]:

- Good dimensional stability and low shrinkage
- High strength, stiffness, hardness, and good toughness (also at low temperatures)
- Physiological harmlessness
- Low water absorption
- Good resistance to chemicals
- Favorable creep behavior and good resistance to environmental stress cracking

Sustainable handling of PET throughout the whole life cycle is crucial for the environment and protection of resources. The global production quantity of PET is huge. Besides packaging, polyethylene terephthalate is very widely applied in the textile and fashion sector, where PET is the most commonly used representative of the polyester family. Polyester accounts for more than half of all produced textile fibers [5]. The Ellen MacArthur Foundation reported the worldwide PET production of 78 Mt in 2017 [6], of which 98% was derived from non-renewable resources (so-called virgin PET). After being used, about 40% of produced PET is landfilled, 14% incinerated or used for energy recovery, and about 32% finds its way into the environment. The remaining 14% is collected for recycling, whereas only 2% of this amount is recycled in the true sense of a circular economy [6]. Under environmental conditions, PET is quite stable, and hence carries a high risk of persisting, for example as microplastic particles.

9.3.2 Overview: Options for Recycling of Feed Materials

The recycling of polymers is usually split into two categories: mechanical (or physical) and chemical approaches. There is some inconsistency in the literature in this regard – for example compare references [7] and [8], where solvent-based methods are related to mechanical or chemical recycling, respectively. Although use of chemicals is indispensable for the basic process step of polymer dissolution, in this contribution chemical methods are characterized by a change in the chemical structure of the polymer during the recycling procedure. Mechanical recycling methods usually do not introduce any changes to the chemical structure of the polymers, at least not intentionally. Through mechanical reworking of the polymer, contaminants are separated, and the polymer is converted to granules, which can be remelted and reshaped to new products. The use of chemicals in mechanical recycling processes is minimal, and serves special purposes like surface cleaning, suppression of discoloration, or degradation of polymer structure and properties [9–11].

Polymers consist of long macromolecules, which themselves are formed from primary building blocks (monomers, or co-monomers), covalently bonded to each other and forming a molecular chain. The polymers are usually subdivided into three basic types: thermoplastics, elastomers, and thermosets [4]. The macromolecules of elastomers and thermosets are irreversibly interconnected (crosslinked) to each other by strong chemical bonds, forming a three-dimensional loosely or tightly meshed network, respectively. After exceeding a temperature specific to the material, these polymers undergo decomposition without melting, and in addition they cannot be physically dissolved (i. e., without breaking the chemical structure). In thermoplastics, the chemically bonded macromolecules are held in an assembly only by relatively weak physical forces and entanglements. These polymers can be melted or dissolved in appropriate solvents, with the macromolecules mainly remaining intact. For the sake of completeness, so-called thermoplastic elastomers can be mentioned, which combine elastomeric properties with the melting and dissolution behavior of thermoplastics [4].

PET is a thermoplastic polymer and can therefore be recycled using both mechanical and chemical methods. In the following discussion, the typical mechanical PET recycling process, materials that are not amenable to mechanical recycling, as well as the toolkit of available chemical methods will be introduced.

PET has the recycling code "1" [12], and PET bottles are one of the most recycled plastic products. So far, only mechanical processes have been established for high-volume recycling of PET. A mechanical recycling process typically consists of:

1. Comminution of PET packaging
2. Separation of polyolefins and other impurity materials by swim-sink, air classifying, and other techniques
3. Cleaning/washing
4. Drying

This results in flakes [9, 13], which can be remelted and reshaped to new products (as-produced or mixed with virgin PET). Depending on the specifications of downstream users and the regulatory background of the intended application, process steps such as cleaning of the flake surface by etching with an alkaline solution, melting in an extruder, decontamination by degassing at elevated temperatures, and melt filtration are applied to produce a PET-granulate and to remove volatile and some non-volatile contaminants. However, the separation of non-volatile contaminants, pigments, and additives by filtration of the viscous polymer melt has a lower mesh-size limit in the double-digit μm range [14]. Thus, non-volatile contaminants may accumulate with an increasing number of recycling cycles [15].

During the recycling procedure, PET degrades to some extent due to chain scissions. This affects the processing and mechanical properties and can cause discoloration. The chain scissions can be induced either mechanically or chemically. Excessive shear stress during the melting (plastification) in an extruder can result in a significant reduction of the molecular weight [4]. In the presence of oxygen, PET is susceptible to thermo-oxidative degradation at elevated temperatures. Furthermore, PET polycondensation, as an equilibrium reaction, can be reverted to some extent in the presence of moisture within the polymer during processing at elevated temperatures. The hydrolysis proceeds much faster than the thermo-oxidative degradation, so that it cannot be prevented even by short processing times [5]. For restoration of the polymer molecular weight and further decreasing of the content of volatile contaminants, mechanical recycling can be extended by a solid-state post-condensation (SSP).

Some examples of technologies available for mechanical recycling of PET on the market are the VACUREMA® process by EREMA [9], the UnPET process by United Resource Recovery Corporation [10], the recoSTAR process by Starlinger [11], the OHL process by ProTec Polymer Processing [12], and the MetaPure process by Krones [10, 16–19].

Mechanical methods require quite homogeneous mono-material input. Therefore, they are mainly used for recycling of clear or slightly colored PET bottles, carefully sorted or obtained from the deposit-refund systems (so-called bottle-to-bottle applications). PET-containing waste fractions, which cannot be mechanically recycled, are usually landfilled, or used for energy recovery by incineration (waste-to-energy approach). The reasons for this practice are manifold and include, amongst others [20]:

- Limitations of the available methods for waste sorting
- Market-driven standards and regulatory requirements that cannot be fulfilled by means of the available sorting and mechanical recycling technologies
- High amounts of PET waste exceeding processing capacities of the installed recycling plants

As the shortest and lowest-energy way of closing the loop, mechanical recycling should be applied for PET recycling wherever technologically and economically appropriate. Since chemical methods are usually connected to an increased effort compared to mechanical methods, their primary applications should address treatment of materials not accessible for any available mechanical approach. Equivalently, the application of mechanical recycling should avoid unnecessary downcycling of high-quality products to low-quality recyclates if an economically and energetically justified chemical method is available. This approach brings together mechanical and chemical recycling methods as complementary technologies, allowing for the establishment of a comprehensive sustainable strategy in the recycling of polymers.

The most important PET waste categories that are not amenable for direct mechanical recycling are *multilayer packaging* and *polyester textiles*. It is also worth mentioning that chemical methods can process quite large amounts of off-spec materials generated during mechanical recycling, for instance undersized PET fractions. The large surface area of small particles is even more advantageous for a use in a chemical recycling process.

Multilayer Packaging

Food packaging must meet strict specifications in terms of functional and mechanical properties, toxicological safety, and price. Certain property profiles are technologically difficult or economically inefficient to achieve using a single polymer, e.g., providing a gas barrier by increasing the thickness. Consequently, some packaging consists of different coupled layers to combine specific advantageous properties of different materials. Typical examples for PET-containing multilayer materials are some thermoforms, flexible films, and bottles for sensitive food and beverage products where PET is usually combined with polyolefins (predominantly PE) or materials with high barrier properties, such as ethylene vinyl alcohol, polyamide, polyvinylidene chloride, aluminum, etc. According to Petcore Europe, PET thermoformed packaging amounts approximately to 1.1 Mt/a on the European market, of which about 40% consists of monolayer PET, 30% of PET/PE multilayer, and 30% of multilayer PET combined with high barrier materials [21]. Additionally, some packaging is heavily colored. Despite the large production quantity, these materials are usually incinerated for energy recovery or landfilled [22].

Polyester Textiles

The fashion industry is increasingly faced with the challenge of assuming its responsibility for discarded clothing and other textile articles. Several global companies have already committed to increase the use of recycled material in their products [5, 23]. Polyester is the most used material in textile manufacturing [5]. However, mechanical recycling of PET bottles or other food packaging material for subsequent use in the production of textile fibers is associated with a typical downcycling of the PET-containing feed. Valuable input material readily processable to recyclates appropriate for food contact materials is converted into textiles, which are typically heavily colored and combined with diverse chemicals and other materials. Up to now, no mechanical processes capable of producing food contact materials from discarded textiles is available, and the corresponding chemical methods are still under development, or have yet to meet regulatory challenges. Recycling of textiles is an even more challenging task than recycling of multilayer packaging, due to the diversity of chemicals used for dyeing and finishing, and the wide range of materials combined with polyester, such as cotton (cellulose), spandex (mainly polyurethane), and nylon (polyamide), to name a few. In Europe, less than 20% of discarded clothing is reused domestically or abroad as second-hand wear or downcycled into low-quality products, such as cheap blankets, industrial wiping rags, nonwovens, and thermal insulation materials. The rest is incinerated for energy recovery or landfilled [24].

9.3.3 Chemical Recycling and its Relevance for PET

As stated earlier, chemical methods are characterized by a change in the chemical structure of the polymer during the recycling procedure. In the following discussion, basic principles of chemical recycling approaches and their relevance for PET recycling are introduced.

9.3.3.1 Pyrolysis

By thermal decomposition approaches, such as pyrolysis of plastic waste, the polymer structure (macromolecules and intermolecular linkages, if any present) is broken down into components of lower molecular weight by means of a thermal treatment. The pyrolysis is carried out in the absence of oxygen. The polymeric feedstock is decomposed to pyrolysis oils, solid char, and pyrolysis gases. The thermal decomposition of polymers usually involves scission of the macromolecules at random positions. Therefore, a pyrolysis oil usually consists of a variety of hydrocarbons that can be fed back into a petroleum refining processes, e.g. in a steam cracker, replacing naphtha [8]. Controlled depolymerization to monomers by means of pyrolysis is only possible for few polymers, e.g. poly(methyl methacrylate) [25]. In contrast to these polymers and to polyolefins, PET is not the typical target polymer for methods based on pyrolysis, since the pyrolysis of PET usually yields comparatively small amounts of pyrolysis oil of unfavorable composition, and with high amounts of gaseous products [26]–[28].

9.3.3.2 Chemical Depolymerization

The term "chemical depolymerization" refers to all processes where the structure of the target polymer is broken up using chemicals. Compared to pyrolysis, the decomposition of macromolecules requires lower temperatures and occurs in a more controlled process, usually leading to monomers, which can be directly reused for polymerization of new polymer. Therefore, methods of chemical depolymerization are increasingly termed "monomer recycling" [29]. Since the monomers are substances of comparatively low molecular weight, a variety of well-established process engineering methods are available for their purification, such as filtration, adsorption, distillation, extraction, and crystallization. As a very promising method for recycling of mechanically non-recyclable PET-containing fractions, monomer recycling enjoys much attention, and is the focus of numerous research and development activities (see for example references [3, 30–42]). The method is described in more detail in a separate section below.

Other Applications of Chemistry or Chemicals for Recycling of PET

Additionally, some mechanical recycling processes utilize chemical methods or chemicals. Mechanical recycling of mixed plastic waste is in principle possible if the polymers are compatible (miscible). Within common polymers, only polycarbonate is compatible with PET [4]. Coextruded mixtures of immiscible polymers usually show coarse morphology and poor mechanical properties. *Compatibilization* is employed for processing mixed fractions or composite materials into usable polymer blends by addition of so-called compatibilizers, i.e., molecules improving the adhesion between single phases. Compatibilization allows the polymers involved to be used for one additional life cycle as a blend. The number of life cycles could potentially be increased as technologies become available to sort out the compatibilized blends. As compatibilizers, block or graft copolymers are often used, in which the first structural component (block or side chain) is compatible with one polymer of the blend, and the second component with the other polymer.

For instance, mixtures of PET with polyolefins PE or PP can be compatibilized by molecules of these polyolefins grafted with maleic anhydride (PE-g-MA or PP-g-MA) or by ionomeric compatibilizers [22].

The *delamination* method is based on the breaking of adhesive bonds between individual layers of multilayer materials using physical or chemical effects. In some approaches selective solvents, depolymerization aids, or surface-active substances (surfactants) are applied [22]. Before treatment, multilayer materials are usually crushed into small pieces to increase the contact surface available for attack by separation agents. The separation of individual layers is usually carried out at elevated temperatures under agitation. The delaminated components of the multilayer materials are further separated and processed by mechanical methods. The Saperatec process is typical delamination technology [43]. Currently, Saperatec is engineering an industrial-scale plant for treatment of post-production packaging films consisting of, among others, PET/Al and PET/Al/PE multilayers (where Al is aluminum). A good overview on available delamination methods is available [22].

The *selective dissolution* approach takes advantage of the different solubilities of polymers in different solvents. In a selective dissolution process, the polymer is dissolved by a selective solvent (sometimes developed especially for the target polymer) while the other components of the mixture such as other polymers, additives, etc. remain inert. During the dissolution, the macromolecules of the polymer are not broken down, and retain their structural integrity. The undissolved components can be separated by means of filtration or other suitable solid–liquid separation methods. Subsequently, the target polymer and the solvent are separated either by solvent evaporation or precipitation of the polymer by addition of an anti-solvent. Alternatively, the impurity components can be dissolved, and the target polymer remains intact and can be separated from solution, for example by filtration.

Like chemical depolymerization, selective dissolution is also a very promising approach for recycling of polyester textiles. For instance, Haslinger et al. developed a method of recycling polycotton textiles (polyester/cellulose blend), where the cellulose is selectively dissolved using an ionic liquid and the polyester can be separated by filtration [44]. Process conditions with minimum degradation of PET were identified. Grause et al. used tetrahydrofuran, cyclohexanone, or cyclopentanone to dissolve and separate PVC from PVC-coated polyester fabric [45]. Worn Again developed a process by which polycotton can be recycled by selective dissolution of PET and cellulose using solvent systems consisting either of dimethyl sulfoxide or 1,3-dimethyl-2-imidazolidinone and several additives [46]. The solubility of the components can be adjusted with the concentration of the additives in the base solvent.

Concerning recycling of PET packaging, Poulakis and Papaspyrides [47] and Achilias et al. [48] used solvent/non-solvent systems consisting of *N*-methyl-2-pyrrolidone/n-octane and benzyl alcohol/methanol, respectively, for non-selective recycling of PET packaging materials. Rensselaer Polytechnic Institute developed a process whereby the common packaging polymers PS, LDPE, HDPE, PVC, and PET are consecutively recovered from a mixture by selective dissolution using xylene at different temperatures [49], leaving PET as the last remaining undissolved fraction. Alternatively, PET could also be dissolved at 190 °C using tetrahydrofuran. The Fraunhofer Institute for Process Engineering and Packaging developed a process for PET recovery from mixed PET waste obtained from the curb-side collection system in Germany by means of selective dissolution [50].

As stated earlier, mechanical and chemical recycling methods can complement each other, helping to provide a comprehensive plastics recycling strategy within a circular economy. Figure 9.36 depicts an integrated view on the synergistic interplay of different recycling pathways in the circular economy, and the inclusion of their products in the value chain of plastics.

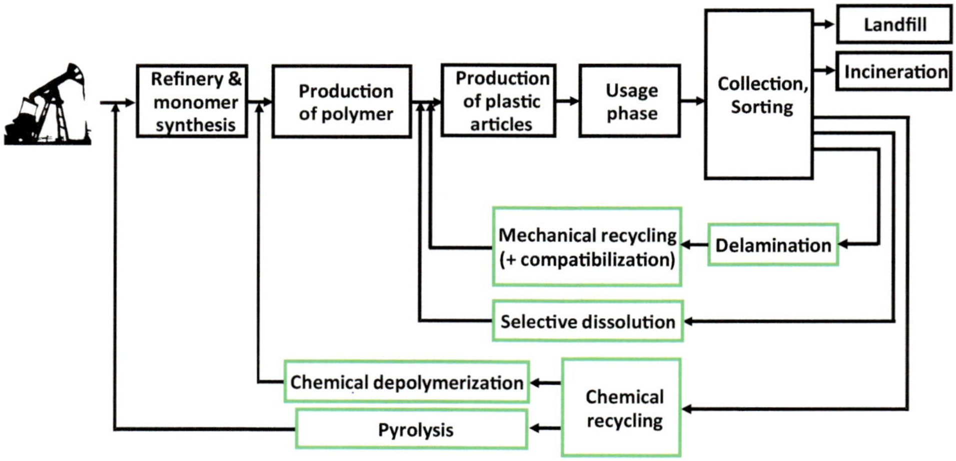

Figure 9.36 Combination of recycling methods to achieve a circular economy

9.3.4 PET Monomers

Figure 9.37a shows the structural formula of PET. Industrial production of PET primarily utilizes polycondensation of monoethylene glycol (MEG, Figure 9.37b) and terephthalic acid (TA, Figure 9.37c). An older industrial process for PET polymerization uses dimethyl terephthalate (DMT, Figure 9.37d), which is polymerized with MEG in a two-step process, consisting of transesterification and polycondensation. This approach was initially necessary, as no industrial method was available to purify the terephthalic acid. The widespread use of TA began in 1965 with the market entry of a high-purity and affordable product [5]. Currently, about 90% of PET is produced via the TA route [4].

On an industrial scale, terephthalic acid is produced by a catalyzed oxidation of *p*-xylene in acetic acid. The oxidation of methyl groups of *p*-xylene occurs via two intermediates, *p*-toluic acid and 4-carboxybenzaldehyde (4-CBA). Due to its structural similarity, 4-CBA co-crystallizes with terephthalic acid and becomes inaccessible for further oxidation [52]. The resulting product, termed crude terephthalic acid (CTA), is not suitable as a feedstock for the PET production, since monofunctional 4-CBA can cap the polymerization and limits molecular weight build-up. Additionally, crude TA is contaminated with catalyst residues and yellow impurities. For purification, CTA is dissolved in water at high temperatures and pressures, and the solution is fed into a reactor where hydrogen is added. In this reactor, the 4-CBA is converted into the more soluble *p*-toluic acid, which after TA crystallization remains predominantly in the mother liquor (among other impurities). The resulting product, referred to as purified terephthalic acid (PTA), is one of the purest high-volume chemicals sold by industry [52]. To reduce the plant and production costs of TA, several processes have been developed, where the purification is carried out using a second oxidation step. The resulting acid is often termed as medium-quality terephthalic acid (MTA) or qualified terephthalic acid (QTA) [53].

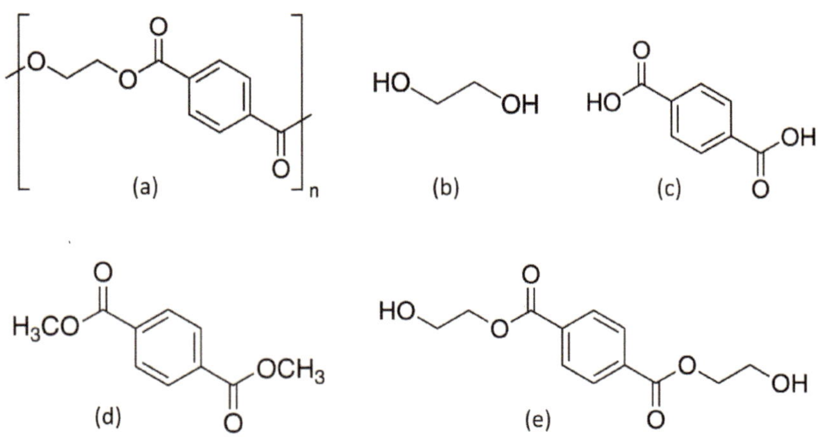

Figure 9.37 Structural formulas of (a) PET, (b) monoethylene glycol, (c) terephthalic acid, (d) dimethyl terephthalate, (e) bis(2-hydroxyethyl) terephthalate

For production of MEG, two routes are usually used. Initially, ethylene oxide is produced by the catalyzed oxidation of ethylene in presence of oxygen. Subsequently, ethylene oxide is either hydrolyzed to MEG or first converted to ethylene carbonate using CO_2. After that, the ethylene carbonate is hydrolyzed to MEG and CO_2. The newer second route is more efficient and should avoid the co-production of higher diols such as diethylene glycol, triethylene glycol etc.

The PET polycondensation is usually carried out as a melt condensation. Initially, PTA and MEG are mixed at slight MEG excess, forming a pumpable pasty product. When PTA and MEG are heated together, the first reaction product is mainly bis(2-hydroxyethyl) terephthalate (BHET, Figure 9.37e), which is further condensed to PET. The most common catalysts used in the PET melt condensation are antimony(III) oxide or acetate. The typical setup for polycondensation consists mainly of a series of continuous stirred-tank and disk-ring reactors (or wiped-film evaporators operated at temperatures exceeding the PET melting point), and vacuum. These conditions (agitated melt and vacuum) allow volatilization of the by-products of PET polycondensation, so that the molecular chains can grow. PET produced by melt polycondensation alone has a molecular weight which is sufficiently high for production of polyester fibers or films. For production of bottles, thermoforming sheets, and high-performance fibers, the molecular weight of PET is further increased by means of a solid-state post-condensation (SSP). In a SSP process, the polymer is heat-treated in the absence of oxygen at temperatures between the glass transition and melting points. SSP is carried out in vacuum or with purging with a dry inert gas to further drive off the by-products formed during the melt condensation and SSP itself, such as MEG and acetaldehyde. A good overview of SSP is available [51].

9.3.5 Monomer Recycling of PET

Monomer recycling is a chemical recycling approach, where the plastics are chemically depolymerized to their constitutional monomers. Subsequent isolation and purification of the monomers aim at achieving monomer specifications enabling a drop-in replacement of virgin monomers in existing industrial polymerization plants. The methods of PET monomer recycling are currently at different stages of technological development and commercialization, and have not yet reached a world-scale industrial application. Since the depolymerization methods for PET has been extensively reviewed elsewhere e.g., in references [32, 33], here they will only be briefly introduced, along with some technologies close to industrial application as well as some recent developments.

Chemical depolymerization of PET can be carried out (for example) by hydrolysis, methanolysis, and glycolysis reactions. Depending on the depolymerization approach, terephthalic acid (TA), monoethylene glycol (MEG), dimethyl terephthalate (DMT), and bis(2-hydroxyethyl) terephthalate (BHET) are the target products.

PET polymerization is a reversible process. Mixing the polymer with the by-products previously released and removed during the esterification and polycondensation forces these equilibrium reactions back to the side of the former reactants. In the methanolysis, glycolysis, and hydrolysis methods, these by-products are methanol, monoethylene glycol, and water, respectively. Figure 9.38 shows the possible points for including recyclates obtained in different PET monomer recycling processes back into the existing PET production chain.

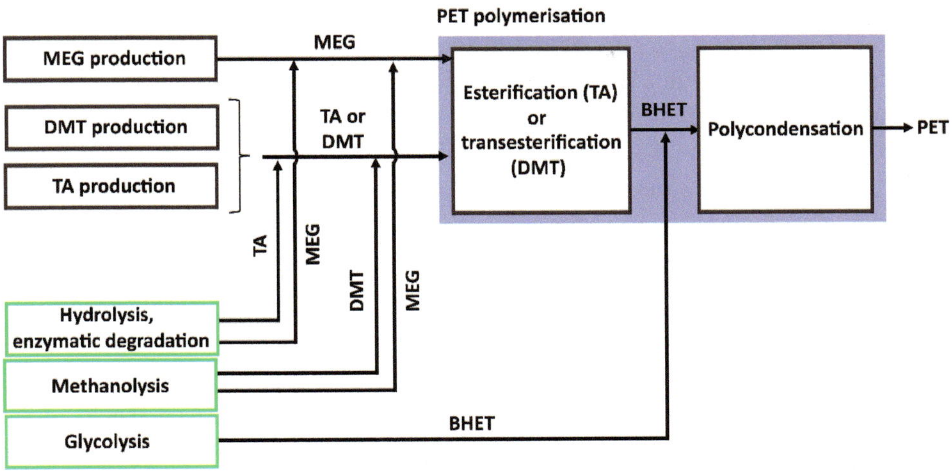

Figure 9.38 Integration of monomeric recyclates in the PET manufacturing

9.3.5.1 Methanolysis

In this approach, methanol is used as depolymerization agent. Prior to depolymerization, the PET input is usually mechanically comminuted to pieces in the mm-range to increase the specific surface area, which is a common procedure for all chemical depolymerization methods. The products of the methanolysis reaction are DMT and MEG (Figure 9.37b, d). The methanol can be applied as liquid, vapor or supercritical fluid [31]. The developed recycling methods using methanolysis for chemical depolymerization of PET were recently extensively reviewed [32]. DMT can be used as feedstock for the existing PET polymerization plants using the DMT route or, as proposed by Eastman Chemical Company, further converted into TA by means of hydrolysis [35].

9.3.5.2 Glycolysis

In course of the glycolysis reaction, the macromolecules are first broken down to oligomers and finally to BHET (Figure 9.37e). MEG is the most used glycol, whereas diethylene glycol, propylene glycol, and dipropylene glycol can also be used. Typical depolymerization temperatures range from 180 °C to 250 °C, and residence times are usually from 30 min to 8 h [33].

To achieve nearly complete depolymerization in an appropriate time, the use of a catalyst, such as zinc acetate, is unavoidable [32]. Ioniqa Technologies developed a novel catalyst complex consisting of a catalytic group (e.g. butylimidazolium iron tetrachloride) grafted onto agglomerated magnetic nanoparticles using an organosilane linker [54]. The reusable catalyst complex also helps to discharge hydrophobic contaminants, such as oligomers and some colorants, and can be separated by application of a magnetic field or other solid-liquid separation methods. In their method, PET glycolysis is carried out at temperatures of 180–210 °C with a reaction time of 160–180 min. The method has been successfully tested for recycling of various materials, including polyester textiles and multilayer PET trays with PE or PP layers.

To use recycled BHET for PET production, either a complete process including polycondensation equipment is required, or the existing infrastructure needs to be upgraded to use the recyclate together with fresh BHET (Figure 9.38). Jeplan has developed a method for recycling of PET bottles and polyester clothing based on glycolysis (BRING Technology™). Crystallization and distillation are used for purification. The technology developers claim that the chemical recycling units can be combined with the existing polymerization plants [55]. The glycolysis reaction does not necessarily have to be carried out until depolymerization is complete. Cumapol developed a process whereby partial glycolysis is carried out no further than is necessary for effective purification (CuRe process). After purification of the produced oligomers, the molecular weight is increased again by polycondensation within the same process [37]. DuPont Teijin Films uses glycolysis to recycle the post-consumer PET packaging (LuxCR™ process). The recycled BHET is repolymerized and used for production of biaxially-oriented PET films. The decontamination is performed by monomer and polymer filtrations, as well as by vacuum extraction for several hours at 270–300 °C [41]. Teijin Fibers developed a combined method for recycling of dyed polyester fibers using glycolysis for depolymerization and transesterification to DMT. The method consists of dye extraction steps using *p*-xylene and MEG, glycolysis, and transesterification to DMT [34].

9.3.5.3 Hydrolysis

The PET hydrolysis can be carried out under neutral, acid, or alkaline conditions with or without catalysts. In neutral hydrolysis, water or steam are usually applied at 200–300 °C and pressures of 10–40 bar [33]. Acid hydrolysis is predominantly carried out using sulfuric acid, although other mineral acids have also been employed. For PET depolymerization, either concentrated acids or high temperatures and long reaction times are required [33]. Liu et al. used ionic liquids for PET depolymerization, which could be recycled up to eight times [36]. At a depolymerization temperature of 170 °C, a TA yield exceeding 88% could be achieved within 4.5 h. In neutral and acid hydrolyzes, PET is directly depolymerized to TA and MEG. However, the poor solubility of TA at depolymerization conditions results in challenging equipment design and operational issues.

The alkaline hydrolysis of PET is usually carried out in the presence of an aqueous alkali metal hydroxide, such as NaOH or KOH. If NaOH is used in depolymerization reaction, the salt disodium terephthalate and MEG are produced. MEG can be recovered either from the depolymerization product by evaporation, or from the mother liquor remaining after the TA precipitation. After dissolution of the water-soluble depolymerization product, the solution can be purified from soluble and insoluble contaminants using standard methods. Subsequently, an acid, usually sulfuric or hydrochloric, is added to precipitate terephthalic acid. The virtually insoluble TA crystals can be separated by filtration or other solid-liquid separation methods. Together with TA, the sodium salt of the applied acid is formed; it can be isolated, purified, and offered for further use.

As in other monomer recycling methods based on hydrolysis, the depolymerization step of the alkaline approach poses a challenge, since it is usually carried out in a batch process and requires either long reaction times or high temperature and pressure, which interfere with process economics. An improvement was achieved by Benzaria et al., who developed a continuous process using PET with solid NaOH (or KOH or Na_2CO_3) in a reactive extrusion, where the depolymerization is initiated at 150–160 °C and proceeds at lowered temperatures within 2–6 min to a PET conversion of 40–60% [30]. After leaving the extruder, the mixture passes through a heating tunnel or kneader, where depolymerization proceeds for further 30 min at 80–150 °C to achieve a PET conversion to monomers exceeding 95–97%. The second apparatus is kept under low pressure or purged with nitrogen to separate MEG released during the reaction. Gr3n also designed a continuous process and an apparatus in which alkaline hydrolysis is intensified by heating with microwave irradiation (DEMETO approach) [39].

RITTEC, in collaboration with the Institute for Chemical and Thermal Process Engineering at the Technische Universität Braunschweig, developed a depolymerization approach using a reactive extrusion by adding NaOH (revolPET® approach) [42, 56]. In the process, the depolymerization takes less than 2 min and is carried out at moderate temperatures of 130–160 °C with conversion of PET to monomers of 92–97%. The method is applicable to the recycling of multilayer materials consisting of PET in combination with LDPE, HDPE, or PP. Since polyolefins are not chemically affected by the applied depolymerization procedure, they can be easily recovered from the depolymerization product solution. The revolPET® recycling process is shown schematically in Figure 9.39. Figure 9.40 shows the setup used for the process development at the pilot scale.

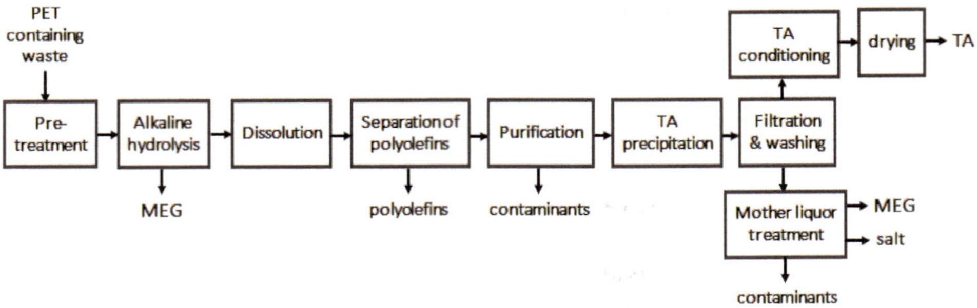

Figure 9.39 Block flow diagram of the revolPET® process

Figure 9.40 revolPET® pilot plant

Palme et al. used alkaline hydrolysis for selective depolymerization of polyester from polycotton textiles [38]. Using the phase transfer catalyst benzyltributylammonium chloride, PET hydrolysis could be completed within 40 min in 10% NaOH solution at 90 °C, without severe degradation of the cellulose.

Enzymatic degradation is a further promising low-energy approach for PET monomer recycling. In this method, enzymes catalyze the PET hydrolysis. A recent overview on research on PET-hydrolyzing enzymes is available [57]. At present, the method developed by Tournier et al. appears to be the technology most close to industrial realization (Carbios approach) [58]. They engineered an efficient PET depolymerase by introducing a disulfide bond in the enzyme LCC (leaf and branch compost cutinase), and applied the modified enzyme for depolymerization of amorphized and micronized post-consumer PET. A conversion to monomers exceeding 90% was achieved within 10 h at 72 °C. Quartinello et al. developed a combined method for depolymerization of polyester fibers [59]. The textiles are initially subjected to neutral hydrolysis at 250 °C and pressure of 40 bar, resulting in a conversion of PET into TA of 85% purity, plus oligomers. Subsequently, the intermediate product is hydrolyzed using HiC enzyme (*Humicola insolens* cutinase) to TA of 97% purity.

In addition to PET production, the recyclates recovered in monomer recycling of PET can potentially be used for production of other polymers. These include, for example, polybutylene terephthalate, some polyphthalamides, unsaturated polyesters, and alkyd resins. The applications of MEG are also numerous, and include its use as an antifreeze, de-icing agent, lubricant, and disinfectant. The possible applications for by-products of the monomer recycling based on alkaline hydrolysis, for instance sodium sulfate, can be found (among others) in paper and glass industries.

Table 9.15 provides a (non-exhaustive) overview of recycling processes based on chemical depolymerization.

Table 9.15 An Overview on Recycling Technologies Based on Chemical Depolymerization

Company or organization	Depolymerization method and other characteristic features	References
BP	Ethanolysis with subsequent oxidation	[60]
	Methanolysis and/or hydrolysis	[61]
Carbios	Enzymatic hydrolysis	[58, 62–65]
Chalmers University of Technology	Selective alkaline hydrolysis of polyester from polycotton textiles	[38]
Cumapol	(Partial) glycolysis	[37]
DePoly	Photocatalyzed alkaline hydrolysis	[66]
DuPont Teijin Films	Industrial scale process based on glycolysis	[41]
Eastman Chemical Company	Glycolysis	[67]
	Methanolysis with subsequent conversion to TA	[35]
	Ammoniolysis with subsequent conversion to TA	[68]
	Acetolysis with subsequent conversion to DMT	[69]
Garbo	Glycolysis	[70]

Table 9.15 An Overview on Recycling Technologies Based on Chemical Depolymerization *(continued)*

Company or organization	Depolymerization method and other characteristic features	References
Gr3n	Microwave-assisted continuous alkaline hydrolysis	[71, 72]
GREEN UNION	Glycolysis	[73]
IFPEN	Continuous alkaline hydrolysis in an extruder	[30]
Ioniqa	Catalyzed glycolysis. Use of magnetic nanoparticles or their aggregates as catalyst carrier	[54, 74]
Jeplan	Decoloration and glycolysis of polyester textiles. Conversion of BHET to further products	[75–77]
Loop Industries	Alkaline hydrolysis	[78]
	Catalyzed methanolysis	[79]
Nan Ya Plastics	Glycolysis. Industrial scale “bottle-to-fiber” process	[8, 80]
PerPETual Global Technologies	Glycolysis. Industrial scale “bottle-to-fiber” process	[8, 81]
Poseidon Plastics	Catalyzed glycolysis	[82]
Premirr Plastics	Microwave-assisted catalyzed glycolysis	[83]
RITTEC Umwelttechnik	Continuous alkaline hydrolysis in an extruder	[42, 56]
SABIC	Catalyzed glycolysis. Purification of BHET by means of continuous multi-column liquid chromatography	[84, 85]
Tejin Fibers	Decoloring of polyester fibers, glycolysis, and subsequent conversion to DMT	[34]

References for Section 9.3

[1] European Commission, *A European Strategy for Plastics in a Circular Economy*, 2018. *https://ec.europa.eu/environment/circular-economy/pdf/plastics-strategy-brochure.pdf*

[2] Plastic Europe, *Plastics – The Facts 2020: An Analysis of European Plastics Production, Demand and Waste Data*, 2020

[3] Aguado, A., et al.: Chemical depolymerisation of PET complex waste: Hydrolysis vs. glycolysis, *J. Mater. Cycles Waste Manag.*, (2014) 16, pp. 201–210

[4] Domininghaus, H.: *Die Kunststoffe und ihre Eigenschaften*, 6th ed. Springer-Verlag Berlin Heidelberg, 2005

[5] Lehmann, M. et al.: *Pulse of the Fashion Industry 2018*. Global Fashion Agenda and The Boston Consulting Group, 2018

[6] Boon, E.: A global momentum to rethink the plastics system, in *Petcore Europe Conference*, 2018

[7] Vogel, J., Krüger, F.: Chemical Recycling, *German Environment Agency*, 2020. *https://www.umweltbundesamt.de/sites/default/files/medien/5750/publikationen/hgp_chemischesrecycling_englisch_bf.pdf*

[8] Hann, S., Connock, T.: Chemical Recycling: State of Play, *Eunomia*, 2020

[9] Awaja, F., Pavel, D.: Recycling of PET, *Eur. Polym. J.*, vol. 41, no. 7, 2005, pp. 1453–1477

[10] United Resource Recovery Corporation, UnPET process, 2021. *http://www.urrc.net/new/pages/process.html*

[11] Sankar, S. V., Kumar, S. A.: Common additives used in recycling of polymers, in *Recycling of Polymers: Methods, Characterization and Applications*, 2016

[12] EU Commision, 97/129/EC: Commission Decision of 28 January 1997 establishing the identification system for packaging materials pursuant to European Parliament and Council Directive 94/62/EC on packaging and packaging waste, *Off. J. Eur. Union*, 1997

[13] Ragaert, K., Delva, L., Van Geem, K.: Mechanical and chemical recycling of solid plastic waste, *Waste Manag.*, vol. 69, 2017, pp. 24–58

[14] Ettlinger, Recycling of PET: Ultra-fine microperforation in melt filtration, *Recycl. Mag.*, 2019

[15] Brouwer, M. T., Alvarado Chacon, F., Thoden van Velzen, E. U.: Effect of recycled content and rPET quality on the properties of PET bottles, Part III: Modelling of repetitive recycling, *Packag. Technol. Sci.*, vol. 33, no. 9, 2020, pp. 373–383

[16] EREMA plastic recycling systems, VACUREMA bottle-to-bottle process, 2021. *https://www.erema.com/en/vacurema_bottle_to_bottle/*

[17] Starlinger, recoSTAR product line, 2021. *https://www.starlinger.com/en/recycling/recostar-product-line/recostar-pet-non-food/*

[18] ProTec Polymer Processing, Sustainable process engineering: Highest quality PET recycling, 2021. *https://www.sp-protec.com/en/materials-treatment/pet-recycling*

[19] Krones AG, Solutions for PET recycling MetaPure W and S, 2012. *https://www.krones.com/en/products/machines/solutions-for-pet-recycling-metapure-w-and-s.php*

[20] PlasticsEurope, Recycling and Energy Recovery, 2019. *www.plasticseurope.org/en/focus-areas/circular-economy/zero-plastics-landfill/recycling-and-energy-recovery*

[21] Hoenderdaal, W.: Introduction to Recycling PET Thermoforms, in *Petcore Europe Workshop September 2017*, 2017. *https://www.petcore-europe.org/images/news/pdf/1-recycling-pet-thermoforms-2017-introduction_wim-hoenderdaal.pdf*

[22] Kaiser, K., Schmid, M., Schlummer, M.: Recycling of polymer-based multilayer packaging: A review, *Recycling*, vol. 3, no. 1, 2018

[23] Global Fashion Agenda, 2020 Circular Fashion System Commitment, 2017. *https://www2.globalfashionagenda.com/commitment/#*

[24] Greenpeace, Fashion at the cross roads, 2017

[25] Kaminsky, W., Franck, J.: Monomer recovery by pyrolysis of poly(methyl methacrylate) (PMMA), *J. Anal. Appl. Pyrolysis*, vol. 19, 1991, pp. 311–318

[26] Anuar Sharuddin, S. D., Abnisa, F., Wan Daud, W. M. A., Aroua, M. K.: A review on pyrolysis of plastic wastes, *Energy Convers. Manag.*, vol. 115, 2016, pp. 308–326

[27] Yoshioka, T., Grause, G., Eger, C., Kaminsky, W., Okuwaki, A.: Pyrolysis of poly(ethylene terephthalate) in a fluidised bed plant, *Polym. Degrad. Stab.*, vol. 86, no. 3, 2004, pp. 499–504

[28] Park, C., et al.: Pyrolysis of polyethylene terephthalate over carbon-supported Pd catalyst, *Catalysts*, vol. 10, no. 5, 2020

[29] Petcore Europe, *PET Monomer Recycling. https://www.petmonomerrecycling.org/*

[30] Benzaria, J., Dawans, F., Durif-Varambon, B., Gaillard, J.-B.: Method for recovery of alkali metal or alkaline-earth metal terephthalate and of alkyleneglycol from polyethylene terephthalates, Patent application US 000005545746 A, 1995

[31] Sako, T., et al.: Depolymerization of polyethylene terephthalate to monomers with supercritical methanol, *J. Chem. Eng. Japan*, vol. 30, no. 2, 1997, pp. 342–349

[32] Shojaei, B., Abtahi, M., Najafi, M.: Chemical recycling of PET: A stepping-stone toward sustainability, *Polym. Adv. Technol.*, vol. 31, no. 12, 2020, pp. 2912–2938

[33] Sinha, V., Patel, M. R., Patel, J. V.: Pet waste management by chemical recycling: A review, *J. Polym. Environ.*, vol. 18, no. 1, 2010, pp. 8–25

[34] Teijin Fibers Limited, Method for recovery of valuable ingredient from dyed polyester fiber, Patent application EP 1914270A1, 2006

[35] Eastman Chemical Company, Process for recycling polyesters, Patent US 6472557B1, 1999

[36] Liu, F., Cui, X., Yu, S., Li, Z.: Hydrolysis reaction of poly(ethylene terephthalate) using ionic liquids as solvent and catalyst, *J. Appl. Polym. Sci.*, vol. 114, 2009, pp. 3561–3565

[37] Cumapol developing potentially groundbreaking polyester recycling technology, *Packaging Europe*, Issue 16.1, 2019. *https://packagingeurope.com/cumapol-developing-potentially-groundbreaking-polyester-recy/*

[38] Palme, A., Peterson, A., de la Motte, H., Theliander, H., Brelid, H.: Development of an efficient route for combined recycling of PET and cotton from mixed fabrics, *Text. Cloth. Sustain.*, vol. 3, no. 1, 2017

[39] Parravicini, M., Crippa, M., Bertele, M. V.: Method and apparatus for the recycling of polymeric materials via depolymerization process, Patent application WO2013014650A1, 2013

[40] Patterson, J. D.: Continuous depolymerization of poly(ethylene terephthalate) via reactive extrusion [Dissertation], North Carolina State University, 2007

[41] Petcore Europe, DuPont Teijin Films launches chemical recycling LuxCR™ depolymerisation process, 2019. *https://www.petcore-europe.org/news-events/227-dupont-teijin-films-luxcr.html*

[42] Rittec Umwelttechnik GmbH, Method, device and use for reprocessing substancially polyalkyleneterephthalate, Patent application WO 002020053051 A1, 2018

[43] Saperatec, Plant & Service, 2021. *https://www.saperatec.de/en/plant-service.html*

[44] Haslinger, S., Hummel, M., Anghelescu-Hakala, A., Määttänen, M., Sixta, H.: Upcycling of cotton polyester blended textile waste to new man-made cellulose fibers, *Waste Manag.*, vol. 97, 2019, pp. 88–96

[45] Grause, G., Hirahashi, S., Toyoda, H., Kameda, T., Yoshioka, T.:Solubility parameters for determining optimal solvents for separating PVC from PVC-coated PET fibers, *J. Mater. Cycles Waste Manag.*, vol. 19, no. 2, 2017, pp. 612–622

[46] Walker, A.: Dissolution process, Patent application GB2560726A, 2018

[47] Poulakis, J. G., Papaspyrides, C. D.: Dissolution/reprecipitation: A model process for PET bottle recycling, *J. Appl. Polym. Sci.*, vol. 81, no. 1, 2001, pp. 91–95

[48] Achilias, J. D., Giannoulis, A., Papageorgiou, G. Z.: Recycling of polymers from plastic packaging materials using the dissolution-reprecipitation technique, *Polym. Bull.*, vol. 63, no. 3, 2009, pp. 449–465

[49] Nauman, E. B., Lynch, J. C.: Polymer recycling by selective dissolution, Patent EP 0491836 B1, 1996

[50] Fraunhover IVV, PET-recyclate from post-consumer waste, 2021. *https://www.ivv.fraunhofer.de/en/recycling-environment/packaging-recycling/pet-recycling.html*

[51] Gantillon, B., Spitz, R., McKenna, T. F.: The solid state postcondensation of PET, 1: A review of the physical and chemical processes taking place in the solid state, *Macromol. Mater. Eng.*, vol. 289, no. 1, 2004, pp. 88–105

[52] Sheehan, R. J.: Terephthalic acid, dimethyl terephthalate, and isophthalic acid, in *Ullmann's Encyclopedia of Industrial Chemistry*, Wiley-VCH Verlag GmbH & Co. KGaA, Weinheim, 2012.

[53] ICIS, MTA: The key to PET cost advantage? 2005. *https://www.icis.com/explore/resources/news/2005/01/28/647563/mta-the-key-to-pet-cost-advantage*

[54] Ioniqa Technologies B. V., Improved catalyst complex and method of degradation of a polymer material, Patent EP 3394158 B1, 2016

[55] Jeplan, Technology, 2021. *https://www.jeplan.co.jp/en/technology/*

[56] Biermann, L., Brepohl, E., Eichert, C., Paschetag, M., Watts, M., Scholl, S.: Development of a continuous PET depolymerization process as a basis for a back-to-monomer recycling method, *Green Process. Synth.*, vol. 10, 2021, pp. 1–13

[57] Kawai, F.: The current state of research on PET hydrolyzing enzymes available for biorecycling, *Catalysts*, vol. 11, no. 2, 2021, pp. 1–10

[58] Tournier, V., et al.: An engineered PET depolymerase to break down and recycle plastic bottles, *Nature*, vol. 580, no. 7802, 2020, pp. 216–219

[59] Quartinello, F., et al.: Synergistic chemo-enzymatic hydrolysis of poly(ethylene terephthalate) from textile waste, *Microb. Biotechnol.*, vol. 10, no. 6, 2017, pp. 1376–1383

[60] BP Corporation North America Inc., Ethanolysis of PET to form DET and oxidation thereof, WO 2007/076384 A2, 2006

[61] BP Corporation North America Inc., Method for forming an aromatic diacid and/or an aromatic diacid precursor from a polyester-containing feedstock, WO 2015/103178 A1, 2014

[62] Carbios, Method for recycling plastic products, EP 2 922 906 B1, 2013

[63] Carbios, New polypeptides having a polyester degrading activity and uses thereof, EP 3517 608 A1, 2018

[64] Carbios, Process of recycling mixed PET plastic articles, WO 2015/173265 A1, 2015

[65] Carbios, A process for degrading plastic products, WO 2017/198786 A1, 2017

[66] École polytechnique fédérale de Lausanne, Degradation of plastic materials into terephthalic acid (TPA), ethylene glycol and/or other monomers that form the plastic materials, WO 2020/173961 A1, 2020

[67] Eastman Chemical Company, Glycolysis process for recycling of postconsumer PET, US 6410607 B1, 2000

[68] Eastman Chemical Company, Production of terephthalic acid and ethylene glycol from polyethlene terephthalate by ammoniolysis, US 6723873 B1, 2000

[69] Eastman Chemical Company, Recycling of polyethylene terephthalate with regeneration of acetic acid, US 6545061 B1, 2000

[70] Garbo SRL, CHEMPET Project, 2021. *https://garbo.it/en/chempet/*

[71] Parravicini, M. Crippa, M., Bertele, M. V.: Method and apparatus for the recycling of polymeric materials via depolymerization process, no. 12, 2013

[72] GR3N, Apparatus and method for depolymerization of polymers, WO 2021/094949 A1, 2020

[73] Green Union SRL, Treatment plant and method for PET glycolysis, EP 3638720 B1, 2017

[74] Ioniqa Technologies B. V., Method and reactor system for depolymerising a terephthalate-polymer into reusable raw material, WO 2021/032821 A1, 2020

[75] Jeplan Inc., Production method for bis(2-hydroxyethyl) 1,4-cyclohexanedicarboxylate, EP 3533 778 A1, 2017

[76] Jeplan Inc., Production method for trans-bis(2-hydroxyalkyl) cyclohexanedicarboxylate, and bis(2-hydroxyalkyl) cyclohexanedicarboxylate, EP 3533779 A1, 2017

[77] Jeplan Inc., Method for producing decolored polyester, decolored polyester, and decoloring agent, WO 2020/213032 A1, 2019

[78] Loop Industries, Polyethylene terephthalate depolymerization, W O 2017/007965 A1, 2016

[79] Loop Industries, Revolutionary, sustainable technology, 2021. *https://www.loopindustries.com/technology*

[80] Nan Ya Plastics, Recycle polyester filament, 2021. *https://www.npc.com.tw/j2npc/enus/prod/Fibers/Draw-Textured-Yarn(DTY)/Recycle Polyester Filament*

[81] Perpetual, Sustainable filament yarn, 2021. *https://www.perpetual-global.com/*
[82] Poseidon Plastics Ltd, Polymer recycling, WO 2021/028695 A1, 2020
[83] Premirr Plastics LLC, Process and system for depolymerizing plastics, WO 2020/061521 A1, 2019
[84] SABIC Global Technologies, Purification of monomer from recycle polyesters, WO 2016/010587 A1, 2015
[85] SABIC Global Technologies, Sustainable process for the recycling of polyethylene phthalate, WO 2017/151144 A1, 2016

9.4 Chemical Recycling of Polystyrene

Norbert Nießner, Achim Schmidt-Rodenkirchen

9.4.1 Introduction

Polystyrene is used as pure "general-purpose polystyrene" or as impact-modified "high-impact polystyrene". Major applications are food packaging (e.g. for dairy products like yoghurt), refrigerator inliner, housings for kitchen appliances, and electric/electronic devices.

Polystyrene has a unique property: the ceiling temperature T_c, at which the equilibrium between polymer and monomer is completely on the side of the monomer, lies in a "benign" range of approximately 390 °C [1]. In order to overcome kinetic hindrance and for depolymerization to proceed quickly, temperatures between 400 and 700 °C are needed, at least on the laboratory scale [2].

The beauty of PS depolymerization recycling is the formation of monomers, allowing purification and re-polymerization to polymers without loss in quality. Thus PS depolymerization is one of the few recycling processes that fulfils the criteria of "fully circular" recycling. It can – admittedly with technical yields well below 100% – be repeated literally indefinitely. A further benefit is the use of "fully circular", recycled styrene as a co-monomer in other polymers, such as ABS (acrylonitrile-butadiene-styrene) or SBC (styrene-butadiene copolymers).

9.4.2 Overview and Challenges

Several reactor concepts have been described for conducting depolymerization of PS. Major lab devices are batch reactors, due to their particularly simple setup. A long-established process for decomposing polymers into the corresponding monomers is depolymerization by twin-screw extruder, for which a patent application was filed as early as 1973 [3]. Figure 9.41 shows the schematic structure of a suitable extruder. The material is heated dissipatively by a high-shear-energy input, causing the material to decompose. The volatile monomers are collected by appropriate devices, while the melt-like residue is extruded out of the machine.

BASF developed extrusion depolymerization [4], and also operated a large pilot plant in 1994 to process mixed plastic waste from the "DSD" collecting scheme. In the early 1990s, PVC was a dominant plastic packaging material, hence the BASF concept comprised pre-treatment at elevated temperatures, in order to convert chlorine from PVC into HCl, which was subsequently removed. The reactor itself was reported to be a tubular cracker reactor, capable of cracking all kinds of commonly used plastics from packaging applications.

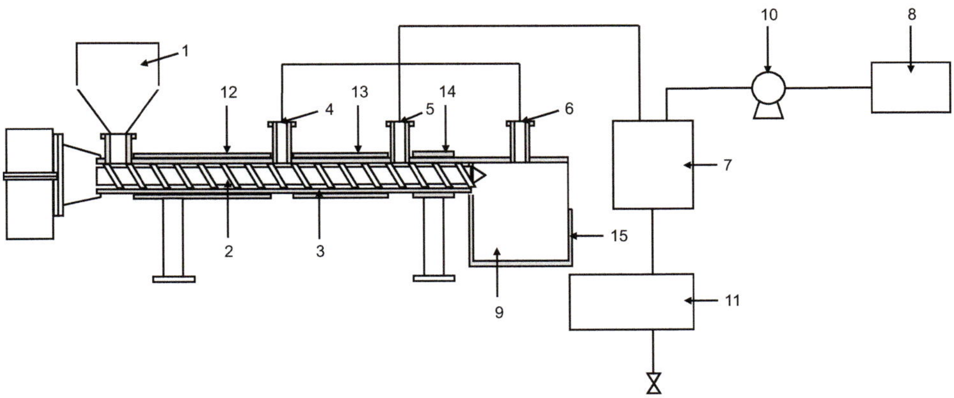

Figure 9.41 Extruder for depolymerization of PS [3].
Key: 1 filling funnel; 2 conveying screw; 3 extruder barrel; 4–6 drag out; 7 cooler; 8 tank; 9 tank for residues; 10 vacuum pump; 11 reservoir; 12–15 resistance heater

At about the same time, BP started to operate a fluidized-bed reactor at its site in Grangemouth, UK. Also here, the dominant packaging material (PVC) caused HCl formation, which was almost completely removed by washing with a caustic solution.

Within a fluidized-bed process, the polymer is fed into the reactor in the solid state. An auxiliary agent to build the fluidized bed (for example sand or silicon carbide) can be used. The composition of the depolymerization by-products in this process is strongly dependent on the residence time and temperature distribution. The fluidized-bed process allows precise process control and reaction temperatures are up to 700 °C, in order to maximize the yield of styrene monomer and oligomers [5–7].

As an alternative, the energy for thermal decomposition can be applied by microwave radiation. However, this requires the addition of a microwave-absorbing additive, such as carbon, since polystyrene itself is transparent to microwave radiation [8].

Table 9.16 summarizes the major features of different reactor concepts for PS depolymerization.

Table 9.16 Overview of PS Depolymerization Reactor Technologies

Reactor type	Example reference	Features
Batch	[9]	Simple, discontinuous process
Extruder	[4]	Robust, continuous process
Fluidized-bed reactor	[10]	Highest styrene yields reported in literature Low amounts of coke and gaseous by-products
Microwave-based pyrolysis	[11]	Efficient heating via microwave energy absorption
Vortex	[12]	Good heat transfer

These early depolymerization activities demonstrated the technical readiness of different thermal recycling processes. However, the availability of feedstock at a sufficiently high gate fee (required to economically run the plants) was a challenge in these early days. That is why – despite promising development and running pilot plants – this topic emerged again in the late 2010s and early 2020s.

In the meantime, companies including Agilyx [13] and Pyrowave [14] announced the successful depolymerization of PS on their pilot plants. A few notable points are:

- In November 2018, INEOS Styrolution and Agilyx announced a joint agreement for PS recycling in the USA [15].
- In 2019, AmSty and Agilyx announced the formation of a joint venture, Regenyx LLC, dedicated to fully recycling post-consumer PS materials back to new polystyrene products [16].
- In 2019, INEOS Styrolution won the ICIS prize "Innovation with best benefit to environment and sustainability" for the first demonstration of full recycling of PS via depolymerization [17].
- In June 2020, INEOS Styrolution and UK-based Recycling Technologies announced a joint agreement to develop depolymerization technology for post-consumer PS [18].
- In April 2021, INEOS Styrolution, Recycling Technologies, and Trinseo announced that they have reached a significant milestone in their plans to build commercial PS recycling plants in Europe [19].

If polystyrene is heated above 400 °C, decomposition products are formed. The main products are styrene monomer and oligomers in the form of dimers, trimers, and tetramers, but also benzene, toluene, ethylbenzene, and α-methylstyrene. The proportions of the individual constituents vary and depend on the experimental conditions, in particular on the temperature and the molecular weight of the polystyrene. As an example, consider the study by Achilias et al. in which the yield of styrene mono-, di-, and trimer in the liquid phase is over 80% (Table 9.17).

Table 9.17 Example of thermal degradation of PS [7]

Decomposition products	Structure	Yield
Styrene (monomer)	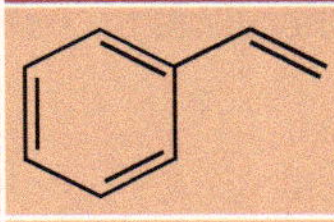	63.9%
2,4-Diphenyl-1-butene (dimer)	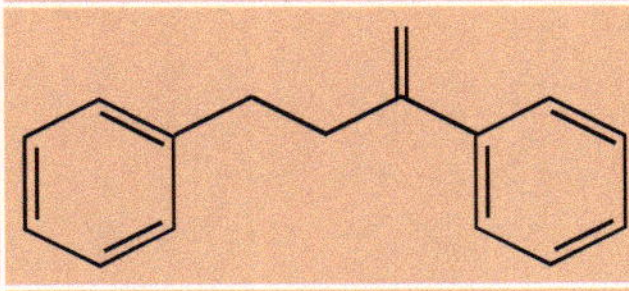	14.0%
2,4,6-Triphenyl-1-hexene (trimer)	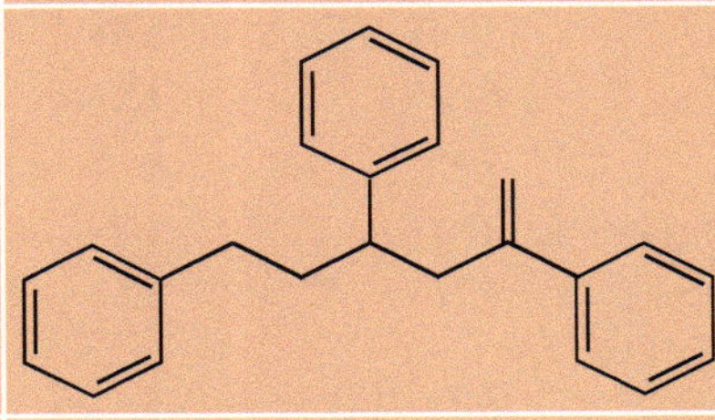	2.2%
Benzene & Toluene	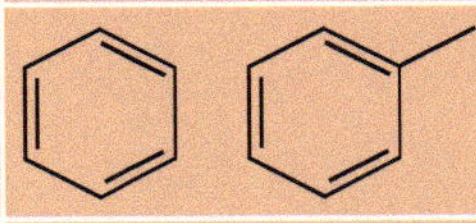	2.0%
α-Methylstyrene	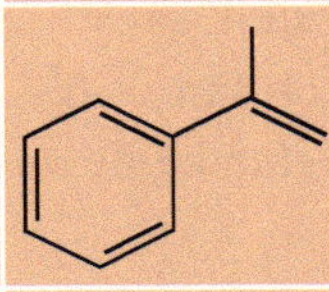	2.1%
Ethylbenzene	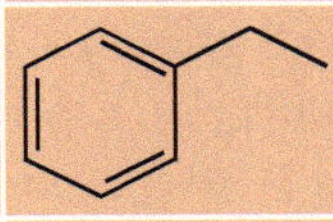	0.5%
Other aromatic products		15.3%

9.4.3 Laboratory Pre-Trials

To demonstrate the feasibility of the depolymerization process, experiments were carried out by the current authors using a glass bulb with a heating mantle under vacuum. For this purpose, the temperature on the outer wall of the glass flask and in the melt was monitored by means of temperature sensors in order to be able to simulate the process later in the reactor. We started with model substance "STYROLUTION PS_3".

The main products formed were verified by GC–MS. In addition to styrene, dimers and trimers as well as traces of other reaction products (including ethylbenzene and α-methylstyrene) were formed during depolymerization – due to the reaction mechanism (Table 9.18181818). We classified the unidentified reaction products into low- and high-boilers depending on their retention time.

Table 9.18 Analysis of PS Decomposition Products from Reactions in a Glass Vessel at 500 °C [20]

Decomposition products	"STYROLUTION PS_3"	"STYROLUTION PS_5"	Post-consumer recycled (PCR) sample from Duales System Deutschland (DSD) mixed waste fraction
Ethylbenzene	0.6%	3.4%	2.7%
Other low-boilers	<0.1%	6.4%	5.5%
Styrene	56.8%	43.2%	43.6%
α-Methylstyrene	0.4%	5.7%	4.0%
2,4-Diphenyl-1-butene (dimer)	14.2%	5.3%	4.6%
2,4,6-Triphenyl-1-hexene (trimer)	16.6%	8.6%	9.1%
Other high-boilers	6.7%	9.5%	9.6%
Char/losses	4.6%	17.8%	21.0%

"STYROLUTION PS_5" was investigated as another commercially available PS granulate. Due to a poly-butadiene content of 8%, this contains less polystyrene, which results in less styrene (43.2%) during depolymerization, but more low- and high-boilers (Table 9.18).

In addition to commercially available PS granules, a waste sample was also investigated in preliminary tests. For this purpose, a processed PS mixed fraction from the dual system was provided by the Department of Processing and Recycling of Solid Waste Material at RWTH Aachen University (I. A.R). The styrene yield here was 43.6%, based on the original quantity used. Compared to the depolymerization of pure PS granules, a higher proportion of low- and high-boilers also occurs here due to the presence of impurities (Table 9.18).

To carry out the experiments to investigate the influence of specifically changed process conditions and the influence of impurities, a customized laboratory reactor of the "versoclave" type on a 2 L scale from Büchi AG (Uster, Switzerland) was designed and set up for process control (Figure 9.42).

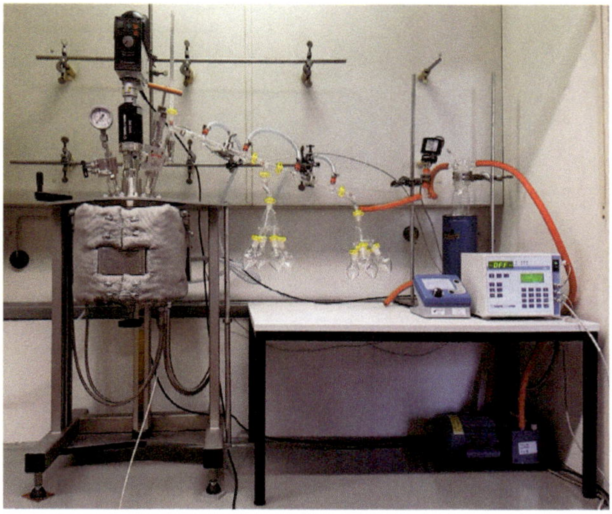

Figure 9.42 Büchi reactor for depolymerization of PS [20]

"STYROLUTION PS_3" depolymerization tests in the reactor showed comparable results to those carried out in the glass bulb. The composition of the condensates formed during depolymerization was analyzed by gas chromatography (Table 9.19).

Table 9.19 Comparison of PS Decomposition Products Depending on the Type of Vessel, at 500 °C [20]

Decomposition products	"STYROLUTION PS_3"	"STYROLUTION PS_5"
Ethylbenzene	0.6%	0.4%
Other low-boilers	<0.1%	<0.1%
Styrene	56.8%	54.0%
α-Methylstyrene	0.4%	0.2%
2,4-Diphenyl-1-butene (dimer)	14.2%	10.9%
2,4,6-Triphenyl-1-hexene (trimer)	16.6%	18.7%
Other high-boilers	6.7%	4.9%
Char/losses	4.6%	10.8%

9.4.4 Trials in the Fluid Bed Reactor (InVerTec)

Figure 9.43 shows the block diagram of the InVerTec styrene depolymerization process. The PS particles, which were 100% general-purpose PS (GPPS), 100% high-impact PS (HIPS), or a 50:50 blend, were conveyed via screw feeder (or pneumatically) into the bed of the FBR. The bed supplies heat into the reactor, and consists mainly of an inert material, such as sand or silicon carbide.

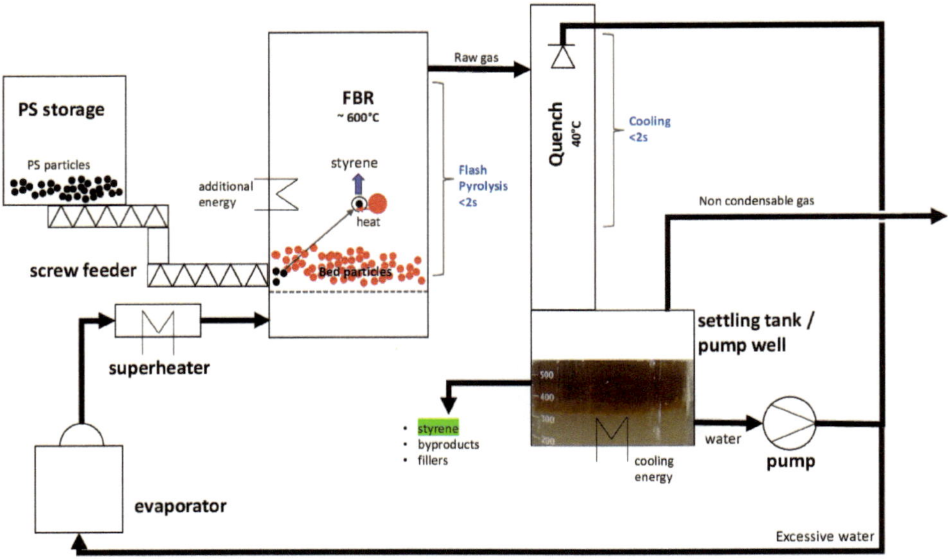

Figure 9.43 Block diagram of the InVerTec FBR depolymerization process (Courtesy by InVerTec)

The bed material is fluidized by steam or inert gas, which itself is preheated to reactor temperature. The PS collides with the hot bed material in the hot fluidization gas, and is immediately decomposed via flash pyrolysis into the monomers and by-products. The raw gas then enters the quenching system, where it is immediately cooled to prevent side-reactions. The quenching medium is water, which has a high heat capacity and enables the neutralization of acidic impurities. In the quench, the water is condensed, accompanied by condensable organics like styrene. The oil fraction separates from the water, and can be easily decanted. The water is cooled and then fed back into the process. Most of the solids were found in the crude oil fraction and must be filtered. The non condensable side-products leave the quenching tank at the top and can be used for energy production. A photograph of the miniplant is given in Figure 9.44, and this also shows further process units (such as coolers and exhaust gas burners) not shown in Figure 9.43.

Figure 9.44 Photograph of the equipment used for the InVerTec FBR depolymerization process (Courtesy by InVerTec)

The achieved compositions of the crude styrene oil are shown in Table 9.20. The highest styrene concentration was found at temperatures around 600 °C for all three cases. The largest portion of the condensed side-products has the dimer beside other high-boilers. In tests with HIPS, which contains butadiene as a co-monomer, the amount of styrene decreases with increasing HIPS fraction in the feed. The lower concentration of styrene also affects the yield. The content of high-boilers with dimer and trimer is higher than with pure GPPS.

Table 9.20 Comparison of PS Decomposition Oil

Decomposition products	100% GPPS ("KM05")	50% GPPS/ 50% HIPS ("KM13")	100% HIPS ("KM14/IN07-2")
Ethylbenzene	0.7%	0.8%	1.0% / 0.7%
Other low-boilers	0.2%	0.4%	0.3% / 0.1%
Styrene	81.1%	75.6%	80.0% / 68.0%
α-Methylstyrene	1.5%	1.4%	2.8% / 2.3%
2,4-Diphenyl-1-butene (dimer)	5.6%	5.9%	6.2% / 1.4%
2,4,6-Triphenyl-1-hexene (trimer)	0.5%	5.5%	3.0% / 0.14%
Other high-boilers	10.4%	10.4%	6.7% / 17.5%
Char/losses	<2%	<2%	<2%
Final yield of styrene	**80.7%**	**71.8%**	**(44.5%) / 52%**

The final yield of styrene is calculated from the measured mass balance as:

$$\text{Yield} = \frac{m_{\text{out,styrene}}}{m_{\text{in,styrene}}},$$

and this gives high styrene yields for GPPS and mixtures or GPPS and HIPS. During the trial with HIPS, there were repeated technical problems with the feeder, which is why no reliable steady state was achieved. However, a reduction in yield by HIPS seems likely.

9.4.5 Trials with PS and Impurities

To develop solutions for commercial-scale depolymerization technologies, it is of utmost importance that technologies show a certain robustness against impurities in post-consumer waste. In this section, we highlight the effect of impurities on the outcome of the PS depolymerization, in terms of yield, by-products, and kinetic effects.

In the first series, PE and PP were added to GPPS in amounts of 5–10%. The experimental setup was as mentioned in Section 9.4.2. In detail:

- PS supply: 3 g/min
- Temperature: 600 °C
- Residence time: 0.51 s
- Steam: 50 g/min

The result of the experiment is shown in Figure 9.45.

- Styrene concentrations are lower compared to GPPS depolymerization without impurities
- Within the margin of measurement error, the polyolefin contaminations do not retard the depolymerization of the GPPS portion.

In commercial-grade PS, we often find pigments and pigment combinations. Amongst them, titanium dioxide is the most frequent one, being used not only in the dominant (food) packaging applications, but also in white goods such as refrigeration inliner. The other frequently used impurity is carbon black.

Under same conditions as mentioned above, we depolymerized GPPS in presence of both pigments. The results are shown in Figure 9.46:

- The styrene amount is proportionally reduced when 5% TiO_2 is added. We cannot see a (significant) effect on yield of styrene.
- The addition of carbon black seems to have a yield-reducing effect; an addition of 2% results in approx. 10% reduction in styrene yield. However, such large amounts of carbon black are seldom used in consumer applications.

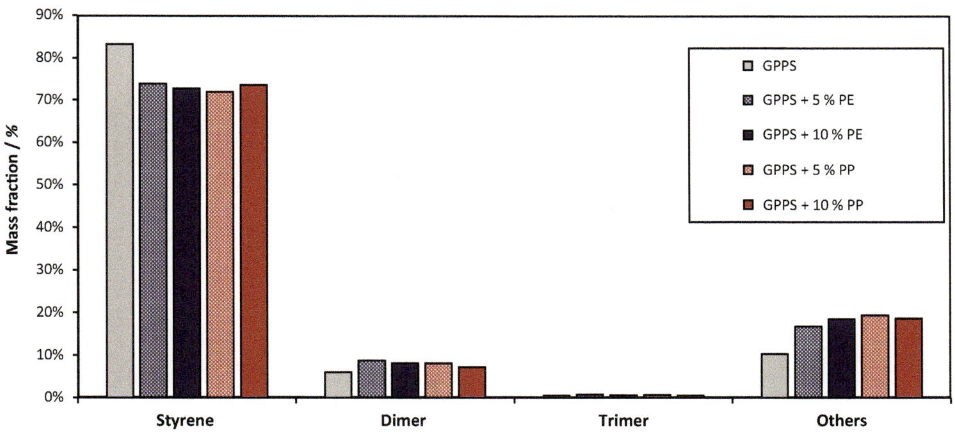

Figure 9.45 Depolymerization of GPPS in the presence of polyolefin impurities

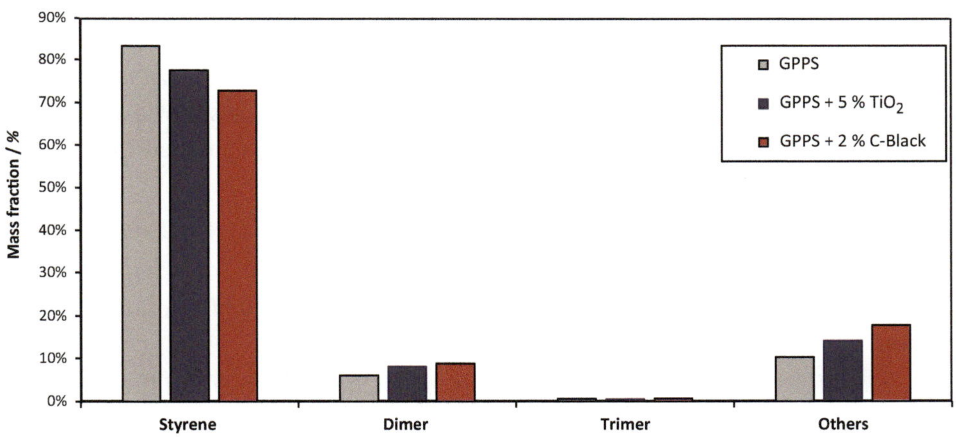

Figure 9.46 Depolymerization of GPPS in the presence of TiO_2 and carbon black

In our next experiment, we investigated the effect of adding ABS (acrylonitrile-butadiene-styrene) polymer to PS. In commercial sorting processes, it is quite easy to separate ABS from PS with existing detection methods, therefore it is likely that a pre-sorted, PS-rich waste will contain much less than 1% of ABS. Nevertheless, we investigated how 1% and 5% ABS influence the depolymerization of GPPS.

Under the same conditions as mentioned above, we depolymerized GPPS in the presence of ABS between 550 °C and 600 °C. The results are shown in Figure 9.47:

- 5% of ABS in GPPS inhibits the formation of styrene
- 1% of ABS in GPPS slightly reduces the yield of styrene

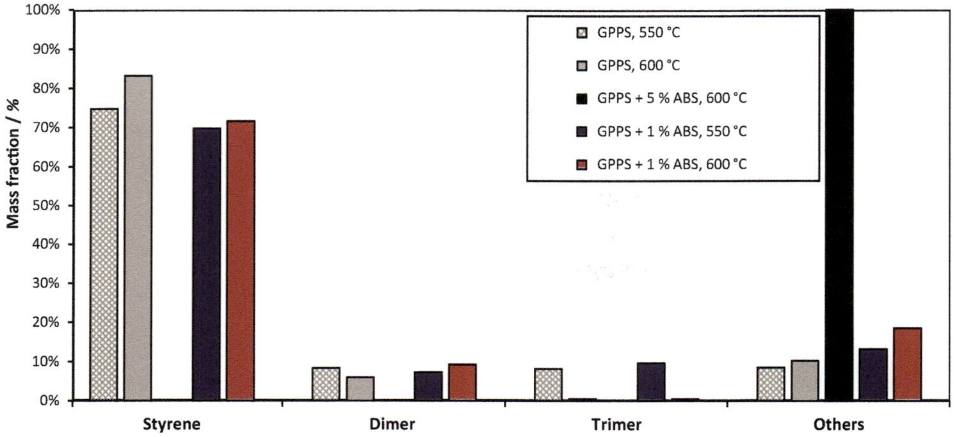

Figure 9.47 Depolymerization of GPPS in the presence of 1 and 5% ABS polymer

The final report of the public-funded project "ResolVe" [20] provides additional valuable insights into the robustness of the polystyrene depolymerization process. In a combination of discontinuous reactor (lab-scale) and twin-screw extruder, the depolymerization of GPPS and HIPS was systematically investigated in the presence of impurities. The following results were achieved:

- 10% of PMMA in GPPS: No negative effect on the yield of styrene from the GPPS portion
- 10% of PE in GPPS: No negative effect on the yield of styrene from the GPPS portion
- 10% of PP in GPPS: No negative effect on the yield of styrene from the GPPS portion
- 10% of PET in GPPS: Formation and deposit of terephthalic acid in the reactor
- 2% of carbon black in GPPS: Approx. 30% reduction of styrene yield after depolymerization
- 5% of TiO2 in GPPS: Only minor reduction of styrene yield observed
- 2% brominated flame retardant HBCD (hexabromocyclododecane) in GPPS: The amount of HBCD was depleted by almost 96% in the presence of BaO (a catalyst for the pyrolysis of PS [21]), while styrene yield was only slightly reduced. Small amounts of brominated organic products were detected:
  - 1-Bromo-1-phenylethane (0.3%)
  - 1-Bromo-2-phenylethane (<0.1%)
  - 1,2-Dibromo-1-phenylethane (<0.1%)
- 5% of dried algae as representative of organic impurities in GPPS: No negative effect on the yield of styrene from the GPPS portion

9.4.6 Conclusion

We can conclude that:

- The depolymerization of PS is a robust process that tolerates impurities such as PE, PP, and TiO_2, and that this tolerance is largely independent of the underlying process, whether discontinuous reactor, twin screw extrusion reactor, or fluidized-bed reactor.
- PMMA has an even lower ceiling temperature (T_c) than PS (290 °C compared to 390 °C). Therefore, PMMA depolymerizes at lower temperatures and obviously does not harm the depolymerization yield of styrene from PS.
- ABS in concentrations at or below 1% can be tolerated. At higher concentrations the PS depolymerization is significantly retarded.
- PET in concentrations at 5% is cleaved into its monomers; deposits of terephthalic acid were observed.
- The brominated flame retardant HBCD was almost completely depleted during PS depolymerization. The effect on styrene yield was minor.
- Organic model compounds did not affect the yield of styrene during PS depolymerization.

References for Section 9.4

[1] *https://www.polymerdatabase.com/*

[2] Maafa, I. M.: *Polymers*, (2021) 13, p. 225

[3] Tokushige, et al.: Verfahren zur kontinuierlichen thermischen Zersetzung von synthetischen makromolekularen Materialien, DE 2310463B2, Japan Steel Works Ltd

[4] Nortemann, A.: BASF AG; Process for the recovery of styrene from used polystyrene, EP-B 649,827 (Priority Oct 21, 1993)

[5] Bouster, C., Vermande, P., Veron, J.: Study of the pyrolysis of polystyrenes: I. Kinetics of thermal decomposition, *Journal of Analytical and Applied Pyrolysis*, (1980) 1, pp. 297–313

[6] Bouster, C., Vermande, P., Veron, J.: Evolution of the product yield with temperature and molecular weight in the pyrolysis of polystyrene, *Journal of Analytical and Applied Pyrolysis*, (1989) 15, pp. 249–259

[7] Achilias, D.S., Kanellopoulou, I., Megalokonomos, P., Antonakou, E., Lappas, A.: Chemical recycling of polystyrene by pyrolysis: Potential use of the liquid product for the reproduction of polymer, *Macromolecular Materials and Engineering*, (2007) 292, pp. 923–934

[8] Bartoli, M., Rosi, L., Frediani, M., Undri, A., Frediani, P.: Depolymerization of polystyrene at reduced pressure through a microwave assisted pyrolysis, *Journal of Analytical and Applied Pyrolysis*, (2015) 113, pp. 281–287

[9] Maafa, I. M.: Pyrolysis of polystyrene waste: A review, *Polymers*, (2021) 13, p. 225

[10] Scheirs, J., Kaminsky, W. (Ed): *Feedstock Recycling and Pyrolysis of Waste Plastics*, Wiley, 2006

[11] Undri, A., Frediani, M., Rosi, L., Frediani, P.: Reverse polymerization of waste polystyrene through microwave assisted pyrolysis. *Journal of Analytical and Applied Pyrolysis*, (2014) 105, pp. 35–42

[12] *https://www.lct.ugent.be/assets/vortex*

[13] *https://www.agilyx.com/*

[14] *https://www.pyrowave.com/en/*

[15] Press release April 26, 2018, *https://www.ineos.com/businesses/ineos-styrolution/news/agilyx-corporation-and-ineos-styrolution-sign-mou-to-advance-the-circularity-of-polystyrene/*

[16] Press release April 30, 2019, *https://www.businesswire.com/news/home/20190430005768/en/AmSty-and-Agilyx-Launch-Joint-Venture-Regenyx-LLC*

[17] *https://www.icis.com/subscriber/specialpublications/FileAsync?id=W5jYqy5gZvru%252bvykjuekc95XfGdJmC8WAI0alpNPUL9exFcf4OQDUjJ6YuNyAcij*

[18] Press release August 06, 2020, *https://recyclingportal.eu/Archive/57936*

[19] Press release April 20, 2021, *https://www.ineos.com/businesses/ineos-styrolution/news/ineos-styrolution-recycling-technologies-and-trinseo-progress-plans-for-the-first-polystyrene-recycling-plant-in-europe/*

[20] Niessner, N., Nosić, F., Wilhelmus, B., Nottbohm, J., Bittner, T., Schäfer, P., Kolb, T., Schedel, A., Feil, A., Rolf, K.: ResolVe - Recycling von Polystyrol mittels rohstofflicher Verwertung. Final report on the German Bundesministerium für Bildung und Forschung (BMBF) (No. 033R194C) (2021)

[21] Ukei, H., Hirose, T., Horikawa, S., Takai, Y.: Catalytic degradation of polystyrene into styrene and a design of recyclable polystyrene with dispersed catalysts, *Catalysis Today*, (2000) 62, pp. 67–75

9.5 Chemical Recycling of Engineering Thermoplastics

Hannah Mangold, Caroline Beyer

9.5.1 Introduction: Market Size and Recycling Challenge

Engineering thermoplastics are often not suitable for mechanical recycling due to the very high requirements on their quality and the degradation of the polymers that occurs in further processing steps. Moreover, mixed or contaminated plastic waste is a huge challenge. Therefore, processes are needed that can tolerate a certain amount of contamination.

One important recycling method for mixed plastic waste that cannot be mechanically recycled is thermochemical treatment. Inhomogeneous mixed plastic waste is for example used in the ChemCycling™ process, a thermochemical recycling process at high temperatures by which it is turned into pyrolysis oil. This process works very well for polyolefins, and is (for example) used for mixed "yellow bag" household waste or old tires [1].

If engineering plastics such as polyurethanes or polyamides are used, this pyrolysis oil contains heteroatoms, especially oxygen and nitrogen, which must be removed by catalytic treatment before feeding the oil into a steam cracker. Moreover, high aromatic content of the polymer is also not beneficial for the ChemCycling™

process, as a more elaborate hydrotreatment is necessary before the pyrolysis oil can be used in a steam cracker. This makes ChemCycling™ of other aromatic engineering plastics such as polybutylene terephthalate (PBT) less attractive.

Therefore, a less energy-intensive and more direct pathway to recycled engineering thermoplastics is the chemical recycling of polymers via their monomers. For this process, it is evidently more important to have pure waste streams, or processes that are robust to contamination and the presence of mixtures of materials. For certain applications, recycling is hindered due to highly mixed waste streams, where further progress in sorting is needed. Important waste streams are, for example, automotive shredder residue, or textiles. Both contain polyamides, but also other polymers that need to be separated before the re-monomerization process.

End-of-life vehicles are an important source of plastic waste, as about 25% of a car consists of plastics. Several engineering plastics can be found in this mixture: polyamides are found (for example) in the hub cap and the motor cover, polycarbonates are found in the outer parts of the car body and headlamps, PBT is found in electronic housing, plugs, and bumper parts, and polyoxymethylene (POM) is found in bearing components and connectors. After removal of the metal parts, all polymers are shredded and are therefore mixed in the automotive shredder residue. An additional challenge for obtaining automotive waste material lies in the long use time and prevalent export of cars; for example, only about one-sixth of the cars produced in Germany remain in the country, while the rest are exported to other countries.

A quickly growing waste stream is waste of electrical and electronic equipment (WEEE). The EU has imposed a directive relating to collection of this waste stream, with a collection target of 65% in 2019 of the total weight of the EEE put on the market. The plastics used in EEE comprise about 13% of the total consumption of plastics in Germany [2]. The plastics content in EEE is quite diverse, ranging from 25% in a computer to 64% in a coffee machine [3]. Moreover, the types of polymers that are used are also highly mixed, with polypropylene (PP), high-impact polystyrene (HIPS), acrylnitrile-butadiene-styrene (ABS), polyurethane (PU) and polycarbonate (PC) taking up the largest volumes [4].

About 18 Mt of polyurethanes are produced every year world-wide, which means it is the 6th most used polymer [5]. In Germany alone, 930 kt were produced in 2019 [2]. The main use is for building, vehicles, and furniture. Most people spend a significant time of their day in contact with polyurethane, as mattresses and car seats are made from polyurethane foam. The European Bedding Association states that up to 30 million mattresses are disposed of every year, and they estimate that 60% end up in landfill and 40% are incinerated [6]. This market therefore has great potential for recycling options.

Recycling of polyurethanes from applications in buildings is partly hampered by their long usage time. Plastic usage in buildings has increased significantly over the last 40 years, but with (for example) a usage time of about 40 years for windows and 80 years for plastic piping, the production of new material and the availability of waste material diverge significantly.

The application of polyamides is also diverse. Worldwide about 8 Mt is produced annually, of which about 354 kt is produced in Germany [2]. The volume share of PA6 is slightly higher than for PA66, while polyphthalamides (PPAs) are for more specialized applications and their volume is significantly smaller. For both PA6 and PA66 the most important applications are textile and industrial fibers and automotive parts, e. g. glass-fiber reinforced PA66. PA6 is also used in multi-layer film packaging, together with other plastics such as PE, PP, and PET.

9.5.2 Overview of Chemical Recycling Processes

9.5.2.1 Polyamides (PA)

The two major thermoplastic polyamides are PA6 and PA66. Various approaches for their chemical recycling are reported. This section focuses on the major technologies for depolymerization.

9.5.2.1.1 Polyamide 6 (PA6)

Chemical recycling of PA6 is highly attractive, as it is a polymer with a high carbon footprint (6.7 t CO_2 per tonne of polymer in Europe) [7]. Hence the production of its only monomer ε-caprolactamε (CPL) from fossil raw materials requires energy-intensive processes. Its molecular structure, being based on one monomer, simplifies chemical recycling and purification efforts, as there is only one valuable monomer that needs to be recovered. Going only one step back in the value chain from the polymer to the monomer is highly attractive, as it enables a circular economy on high level (Figure 9.48).

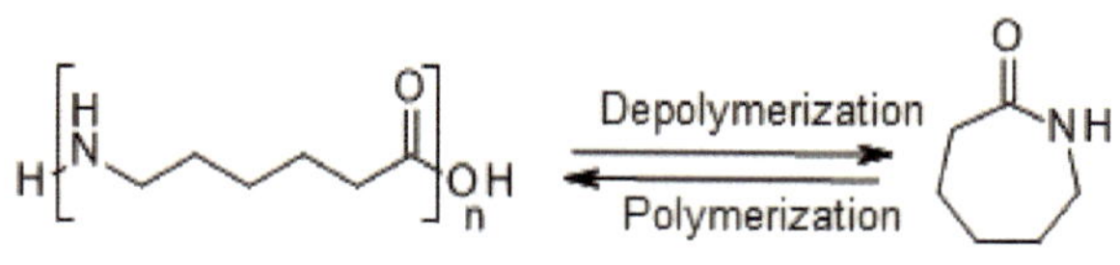

Figure 9.48 Depolymerization and polymerization of PA6

The three major techniques for PA6 depolymerization are hydrolytic, acidic, and alkaline depolymerization. All three depolymerization techniques have in common that caprolactam oligomers can be found as side-products. New approaches, such as the use of ionic liquids, are also described in the literature [8].

Hydrolytic Depolymerization

The hydrolytic and acidic depolymerization have a similar reaction mechanism, as they are the reverse reaction of the polycondensation. For PA6 polymerization, the water added to CPL acts as starter to enable the ring-opening of CPL to its open-chain form, aminocaproic acid. In addition, the water content is essential for the monomer–polymer equilibrium [9]. The hydrolytic depolymerization involves the hydrolysis of amide bonds without any catalyst, with water being the only reagent and solvent for products. Since the reactivity of amides in hydrolysis is sluggish, harsh reaction conditions such as high temperatures of up to 300 °C are required. As a consequence of the vapor pressure of water at such high temperatures (250 °C at 40 bar, 270 °C at 55 bar, 290 °C at 75 bar, 300 °C at 86 bar, and 310 °C at 99 bar) [10], the hydrolysis requires equipment that can handle high pressures of up to 100 bar. An excess of water ensures that the equilibrium is shifted towards the monomer. As the depolymerization can be done without any further additive apart from water, associated materials do not hamper PA6 depolymerization, as there is no catalytically active compound that could be deactivated. However, energy consumption is high, as an excess of water has to be heated up to 300 °C and evaporated after the hydrolysis.

Following the residence time of roughly 1 h, the pressure is reduced, so that some of the water evaporates. The now partly concentrated aqueous caprolactam solution is passed through a filter to separate any water-insoluble materials. Afterwards, the aqueous solution is further concentrated and the CPL produced can be purified by extraction or distillation [11].

In contrast to the acidic and alkaline process, CPL is kept in the vessel for the whole residence time in hydrolytic depolymerization. Hence a highly diluted aqueous solution is required to shift the equilibrium towards the monomer. The evaporation of CPL during depolymerization would remove large amounts of water from the aqueous phase, too, since the vapor pressure of water is higher than that of CPL under these conditions. This would result in a low water concentration and hence a low depolymerization rate. In the acidic process, the continuous steam supply compensates for this problem.

A high water-to-polymer ratio results in the mixture having a low viscosity, regardless of associated materials or side-products. Oily compounds are formed as a film on top of the aqueous layer and can be removed easily. At higher CPL concentrations, other undesired organic compounds can become soluble and hence complicate purification. The main side-products are caprolactam oligomers.

For glass-fiber-reinforced materials, the influence of water and temperature on caprolactam yield have been studied [11]. As CPL cannot be removed from the reaction vessel due to the above-mentioned reasons, the maximum caprolactam yield of approximately 78% at 250 °C was found to be determined by the water-to-polymer ratio, which needs to be at least 15:1. The caprolactam yield was further optimized by adjusting the temperature from 250 to 290 °C.

Acidic Depolymerization

The acidic depolymerization is a catalytic hydrolysis reaction at 230–325 °C that requires large amounts of catalyst (1–10 wt% H_3PO_4) and superheated steam at atmospheric pressure, which is passed through the mixture of polymer melt and catalyst [12]. The steam is required for two reasons: firstly, it enables chain scission, as it is required for the hydrolysis, and secondly, it is used as a carrier gas to remove CPL from the reactor, which further contributes to a shift of the equilibrium from the polymer towards the monomer. The addition of an acidic catalyst allows depolymerization to take place under moderate conditions compared to the hydrolytic depolymerization, as it facilitates the nucleophilic attack of water to the carbonyl group of the amide bond by protonation. The addition of a Lewis acid (e. g. $AlCl_3$, HNO_3, H_2SO_4 or aromatic sulfonic acids) has been shown to accelerate PA6 solubility [13].

The resulting aqueous solution of CPL needs to be further concentrated to furnish the valuable product. An oxidizing step using (for example) permanganate is used to remove impurities. Additional purification steps such as extraction, distillation [14], or crystallization [15] yield caprolactam in a quality that allows for re-polymerization. Overall, a moderate yield of roughly 56% can be achieved [14]. The process requires a two-stage evaporation of water: firstly, superheated steam is required for the reaction, and secondly, the resulting aqueous caprolactam solution needs to be concentrated to obtain the product of interest.

Although it benefits from moderate reaction conditions, acidic depolymerization has some downsides. The process is sensitive to feed impurities, such as glass fibers [11] or alkaline fillers ($CaCO_3$) used in carpets [12], because they might deactivate the acidic catalyst. These additives need to be removed in a pre-separation step to avoid limiting the recycling potential of the feedstock by acidic depolymerization. Hence, this technology is mainly suitable for relatively pure PA6 waste streams with a high PA6 content, e. g. production or fiber wastes [11], but not for post-consumer wastes. Also, the disposal of phosphorus-containing reaction residues is costly.

Alkaline Depolymerization

The alkaline depolymerization of PA6 is the reverse of anionic polymerization, and requires catalytic amounts of an alkaline catalyst to initiate caprolactam formation.

Wichterle et al. propose that alkaline depolymerization proceeds via catalytic deprotonation of an amide group at any point of the chain, followed by intramolecular cyclization to form CPL after elimination (Figure 9.49) [16]. This means that the polymer chain length is decreased from the middle of the chain. Hornung et al. support this proposal with kinetic data. Also, their comparison of kinetics of acidic, hydrolytic, and alkaline depolymerization showed that the alkaline depolymerization has the lowest energy of activation and hence gives the fastest reaction, followed by acidic depolymerization [17].

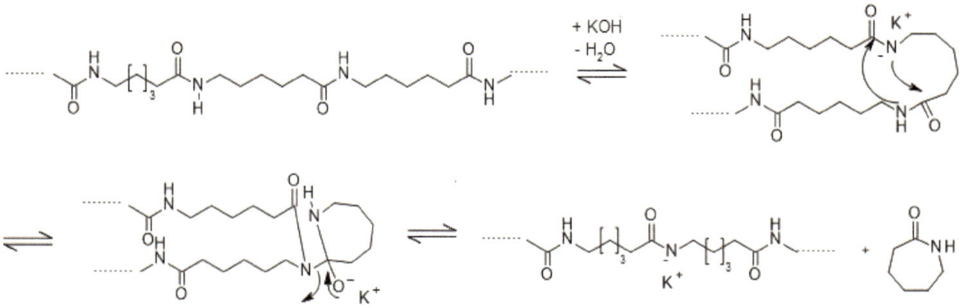

Figure 9.49 Reaction mechanism of alkaline depolymerization of PA6 according to kinetic data from Hornung et al. [17]

Typical catalysts, which are added up to 10 wt%, are alkali and alkaline earth hydroxides or carbonates, e.g. sodium or potassium hydroxide, and potassium carbonate. As there is no water that could act as carrier for CPL as in acidic depolymerization, alkaline depolymerization is typically operated under vacuum to trigger the equilibrium shift towards the monomer by continuous evaporation of CPL from the reaction mixture. As a consequence, CPL is received as a melt and not as an aqueous solution. The required temperatures are in the range 250–300 °C. As a consequence of the handling of molten CPL (which solidifies below 68 °C), smooth operation requires properly heated equipment to avoid cold spots that promote CPL deposits and hence blockages of the lines. Typical yields of raw CPL are >90% [18–20]. Additives such as di- or poly-carboxylic acid anhydrides have been reported to result in lower depolymerization temperatures, due to the formation of an imide co-catalyst that facilitates depolymerization and reduces melt viscosity. Hence energy savings are possible [21]. Other reported co-catalysts are iron(II) compounds such as iron ammonia sulfate, oxalate, or sulfate, which promote the formation of an amide anion in the polymer chain and hence lead to a reduced reaction time and increased CPL yield [22].

Alkaline depolymerization is an attractive route for the recycling of complex wastes, because inorganic or organic fillers (glass fibers, calcium carbonate, talc), dyes (titanium dioxide, iron oxides, carbon black, spinning dyes) and additives (UV and heat stabilizers, antistatic agents, flameproofing agents, antioxidants) are well-tolerated [18].

However, the alkaline catalyst does not just promote PA6 depolymerization to CPL, but also triggers base-catalyzed side-reaction of two CPL molecules to an imine or the corresponding diaminoketone in the presence of water (Figure 9.50) [23].

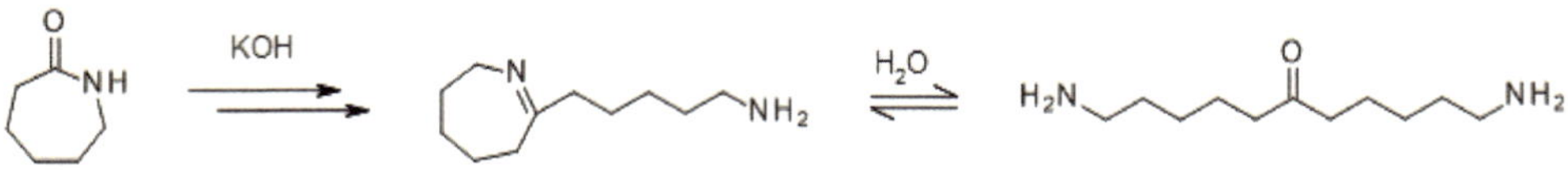

Figure 9.50 Base-catalyzed side-product formation of caprolactam

These side-products require additional purification steps, as they promote gelation of PA6 during polymerization [24] and have similar boiling points as CPL [25, 26]. Their removal is possible by precipitation or conversion to non-volatile compounds [27], for example, the transformation into a precipitating carbamate using CO_2 in combination with acetone [24]. Further purification steps are typically extraction [28] or distillation [18].

9.5.2.1.2 Polyamide 66 (PA66)

Polyamide 66 (PA66) consists of two monomers, adipic acid (AA) and hexamethylenediamine (HMD). There are various methods for chemical recycling of PA66. The most important ones are described below.

Hydrolysis

The hydrolysis of PA66 with water allows for a shift of the equilibrium between polymer and monomers towards the monomers and low-molecular-weight oligomers (Figure 9.51). The amount of water, temperature, and residence time determine the degree of depolymerization. However, full conversion is not achievable by neutral hydrolysis. Reported reaction conditions are (for example) 300 °C and up to 100 bar with a polymer-to-water ratio of 1:10, yielding 25% of AA after 45 min reaction time [29].

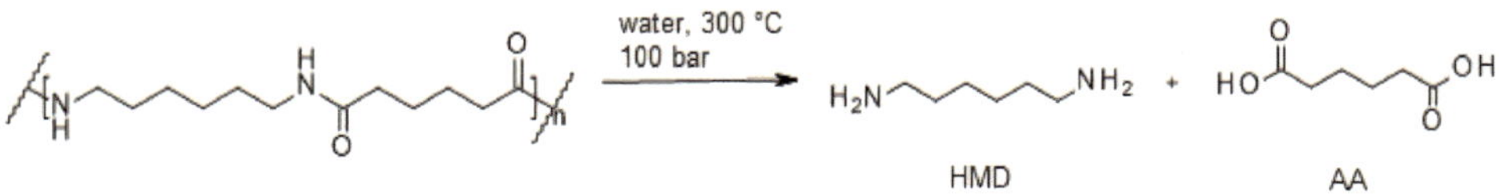

Figure 9.51 Neutral hydrolysis of PA66

Acidic Hydrolysis

Acidic hydrolysis uses strong mineral acids such as HCl [30] or H_2SO_4 [31, 32] to accelerate hydrolysis to AA and the corresponding salt of HMD. Using 30% hydrochloric acid, the reaction is carried out at roughly 110 °C for 5 h (Figure 9.52). The product mixture is separated by crystallization of crude AA, and after concentration, HMD hydrochloride crystallizes as well. Adipic acid is further purified by treatment with active carbon (for decolorization) and further crystallization. HMD hydrochloride can be purified by washing with ethanol. However, to receive HMD, equimolar amounts of a base, for example sodium hydroxide, are required, generating sodium chloride as a by-product [30]. After HMD release, the diamine is purified by distillation. The formation of an inorganic salt with organic impurities is a major disadvantage of acidic hydrolysis. Also, using corrosive HCl requires corrosion-resistant and hence expensive equipment. The application of microwave irradiation in acidic hydrolysis is also reported [33].

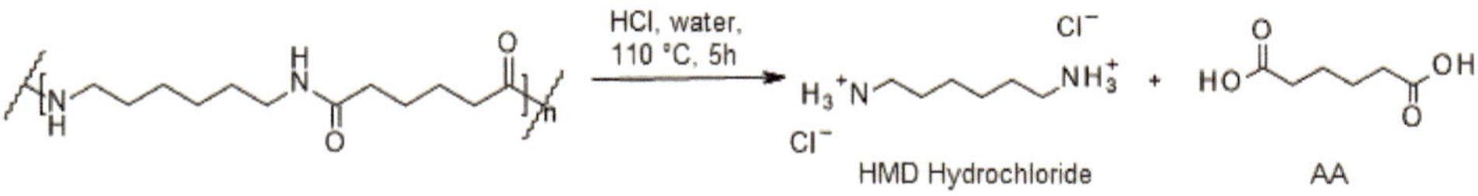

Figure 9.52 Acidic PA66 hydrolysis.

DuPont also described acidic hydrolysis using aliphatic mono-carboxylic acids, for example acetic acid, for the recycling of PA6 and PA66 mixtures at 110 °C and atmospheric pressure (Figure 9.53). This one-pot procedure circumvents the need for costly PA6 and PA66 pre-separation, and gives a mixture of *N,N'*-hexamethylene bisacetamide, AA (from PA66), 6-acetamidohexanoic acid and CPL (from PA6). The acetamide of HMD needs to be further reacted to give HMD [34].

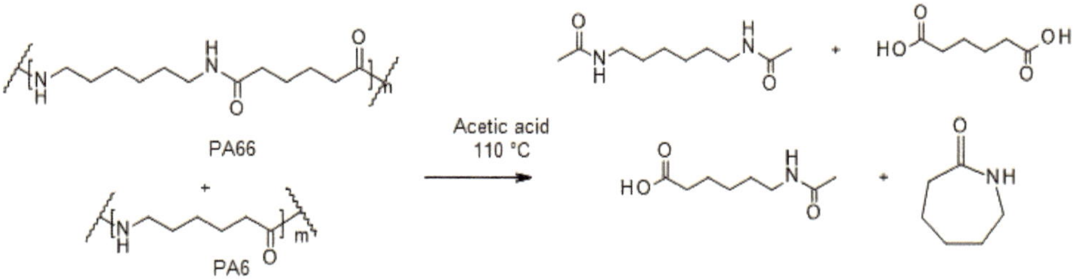

Figure 9.53 Acidic hydrolysis of PA66 and PA6 mixtures according to DuPont

In a further improvement, DuPont describes the conversion of 6-acetamidohexanoic acid (stemming from PA6) and *N,N'*-hexamethylene bisacetamide (stemming from PA66) to a mixture of adipic acid and alkylamide by means of oxidation using an oxidation catalyst or an electrochemical oxidation [35]. This procedure simplifies product purification and processing of PA6 and PA66 mixtures.

Rhone Poulenc describes an acidic hydrolysis of PA66 into diacids and the conversion of HMD into diacids using an oxidizing medium that includes nitrous groups e.g. from nitric acid, alkali metal and metal nitrites or nitrous vapors (NO_x). At temperatures below 100 °C, hydrolysis is possible in a time of 1–10 h. Cooling the reaction mixture to room temperature allows AA to be collected after crystallization.

Conducting the reaction in the presence of oxidation catalysts such as copper nitrate and ammonium vanadate enables the (partial) conversion of HMD into AA. Glutaric and succinic acid are also produced as minor compounds. The procedure is said to be suitable for PA6.10, PA6.12, poly-terephthalamides, polyisophthalamides, polyaramids and polyamides with an amino acid as building block (PA6, PA7, PA11, PA12). The process is said to tolerate inorganic fillers as well as inorganic and organic additives. Organic additives that are prone to the reaction conditions are either converted into carbon, CO, water, or other oxides, or do not undergo full oxidation and are separated from the product of value in the subsequent purification [36].

The use of nitric acid in acidic hydrolysis is also reported by DuPont. However, they describe the recovery of both AA and HMD while Rhone Poulenc did not isolate HMD but oxidized it to a dicarboxylic acid (adipic, glutaric, or succinic acid). After the acidic cleavage using nitric acid at temperatures of 70–110 °C, DuPont also separated the formed AA from the reaction mixture. The nitric acid was then catalytically hydrogenated to ammonia, so that HMD remained intact and could be separated from the reaction mixture without salt formation. Also, careful temperature control during hydrolysis to prevent nitrous acid formation is described as a powerful means to avoid formation of oxidation products. A yield of 90% for both AA and HMD is reported [37].

Another method for AA and HMD recycling is the treatment of PA66 with formic acid. Acidification with sulfuric acid after depolymerization enables AA crystallization. The addition of calcium hydroxide allows HMD to be collected by distillation [38].

Alkaline Hydrolysis

Complete re-monomerization can be achieved by alkali-catalyzed hydrolysis under increased pressure. In the presence of alkali or earth alkali hydroxides (e.g. NaOH), HMD and the sodium salt of adipic acid are formed (Figure 9.54). The reaction is described in various solvents, for example, water or alcohols, or alcohol-water mixtures at temperatures between 150 and 250 °C [39–41]. The reaction pressure depends on the solvent composition.

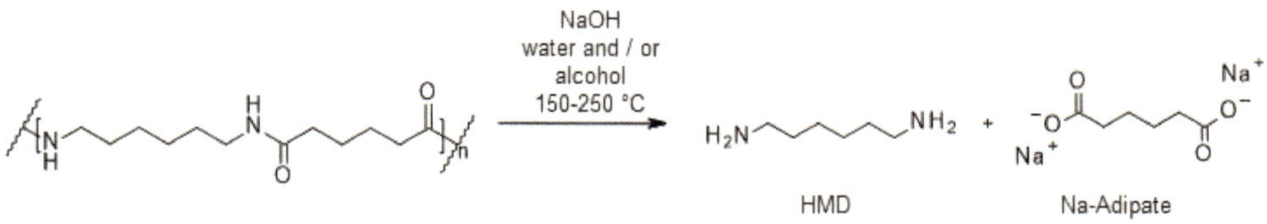

Figure 9.54 Alkaline PA66 hydrolysis

HMD can be isolated by extraction or distillation from the reaction mixture, while the conversion of Na-adipate to AA requires equimolar amounts of a strong acid, resulting in large amounts of inorganic salts with organic impurities. Also, the inorganic salt can complicate AA purification, which is usually done by crystallization. An elegant way to circumvent this salt accumulation is the conversion of Na-adipate to AA and re-usable NaOH by electrolysis, as described by BASF [42].

For mixtures of PA6 and PA66, the recovery of CPL from PA6 and HMD from PA66 without a pre-separation of the two polyamides is described by DuPont. The process requires an alkali or alkaline earth metal hydroxide (e.g. NaOH), and steam for stripping of CPL and HMD. An aqueous solution of CPL and HMD results. Typical reaction temperatures are between 250 and 300 °C [43]. A similar process has been described by DSM [44].

PA66 depolymerization in refluxing 50% NaOH solution in the presence of a phase-transfer catalyst is reported to give oligomers that were further depolymerized, giving a roughly 60% yield of adipic acid. HMD could not be isolated. The same process is reported to give high yields for PET depolymerization at room temperature [45].

Thermal Catalytic Depolymerization

Thermal catalytic depolymerization of PA66 catalyzed by carboxylic acid salts has been shown to happen at 280–330 °C via an anionic depolymerization mechanism, giving a mixture of 14- and 28-membered cyclic oligomers of adipic acid and HMD as the major products [46].

Reaction of PA46 and PA66 in a similar fashion has been described by DSM. However, they did not commercialize this recycling option, as pure cyclic diamides could not be achieved [47].

Ammonolysis and Aminolysis

Ammonolysis of PA66 using ammonia is described by DuPont and Invista, while Rhone Poulenc uses a primary or secondary amine (e.g. n-butylamine) in the aminolysis of PA66 (Figure 9.55).

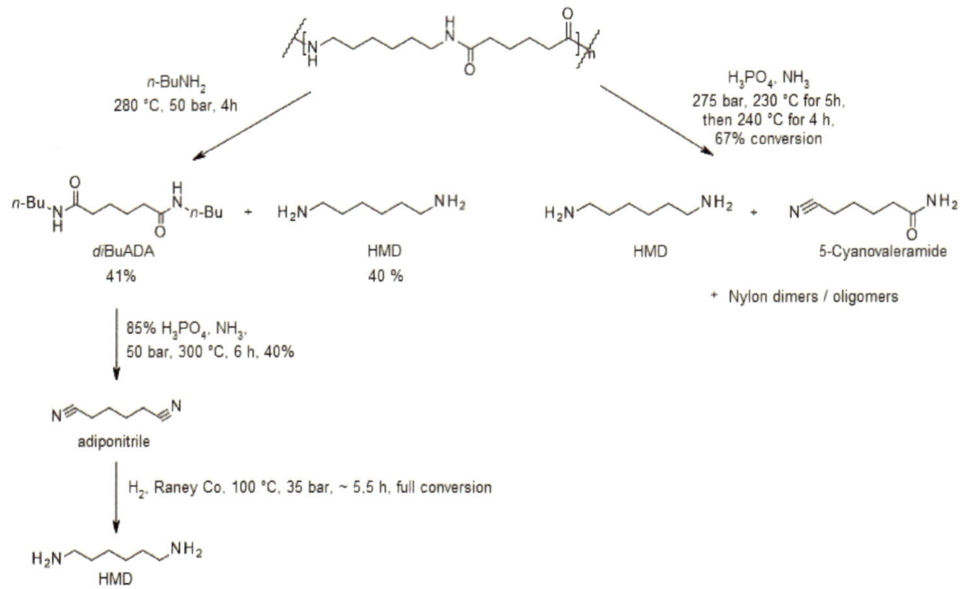

Figure 9.55 Aminolysis (left) and ammonolysis (right) of PA66

Ammonolysis requires high pressure (275 bar) and long reaction times (9 h) at elevated temperatures (230–240 °C) for the conversion of PA66 to a mixture of HMD, 5-valeroamide and dimers/oligomers of PA66, which can be purified by distillation [48–50].

The aminolysis of PA66 using an aliphatic amine enables better efficiency and milder reaction conditions (280 °C, 50 bar) compared to ammonolysis, because aliphatic amines such as n-butylamine are more nucleophilic than ammonia and volatize easily, which is beneficial for product purification [51]. Also, the yield of the monomeric products is higher.

The process described by Rhone Poulenc involves first aminolysis of PA66 giving HMD and the corresponding adipic acid amide (in the case of n-butylamine this is *N,N′*-dibutyl adipamide (diBuADA)) in yields of 41 and 40%, respectively. In a second step, diBuADA is further reacted with ammonia in an ammonolysis reaction to adiponitrile, which is the final product. However, the yield of adiponitrile is low, as reaction selectivity is poor [51, 52]. The hydrogenation of adiponitrile to HMD is possible with Raney cobalt [53].

Hydrogenation

Compared to the two-step procedure of ammonolysis (or aminolysis) and hydrogenation, the direct hydrogenation of polyamide 66 is less developed. Two approaches using promising ruthenium pincer catalysts were developed in parallel, which used polyamides and polyurethanes as starting materials. The first approach uses

DMSO as solvent for polyamides at higher temperatures. This approach has the drawback of catalyst deactivation by DMSO, and did not allow hydrogenative depolymerization of technical-grade materials. However, a PA66 conversion of 66% was achieved, giving hexanediol (HDO) and HMD in yields of 20% and 25% [54].

The second approach uses ruthenium catalysis in polar THF under reaction conditions of 200 bar and 200 °C (Figure 9.56). The solubility of the polymer material is crucial for reaction progress. Conversion of technical-grade polyamide 66 (Ultramid A27®) was demonstrated, even though the achieved yields of HMD (19%) and HDO (18%) need to be improved to become economically relevant. In addition, a deactivating effect of acidic end groups was observed, which currently limits applications [55]. The described procedures are also applicable for the hydrogenation of polyurethanes, giving the corresponding diamines and polyols.

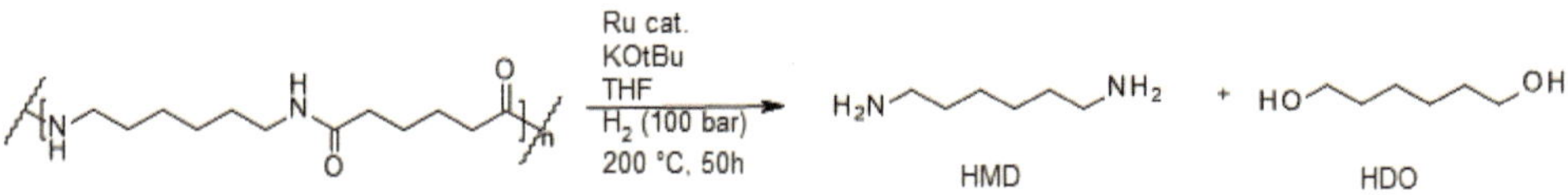

Figure 9.56 Hydrogenation of PA66

9.5.2.2 Polyurethanes (PUs)

Polyurethanes are produced by polyaddition reactions of an organic diisocyanate and a diol compound (Figure 9.57). They exist in many different forms with a wide range of properties - for example, from extremely flexible to extremely rigid. Usually the polyol component is varied to obtain the desired properties. Foamed polyurethanes are formed by adding water to the reaction mixture. The isocyanate groups react with the water to an amine, which in turn forms urea linkages with an isocyanate (-NH-C(=O)-NH-), at the same time as releasing carbon dioxide, which acts as a "blowing agent" during molding. Typically, the formulation of the system contains (beside the diisocyanate) various polyols, catalyst, stabilizer, and further additives or foaming agents. This mixture of components complicates recycling [56]. Flexible polyurethanes, such as flexible foams, are often produced with long-chain polyether polyols based on ethylene or propylene oxide; these are found in cushions in the automotive sector, or as mattresses and pillows. For rigid foams, short-chain polyols are used, which can be found in building insulation, coatings, and adhesives.

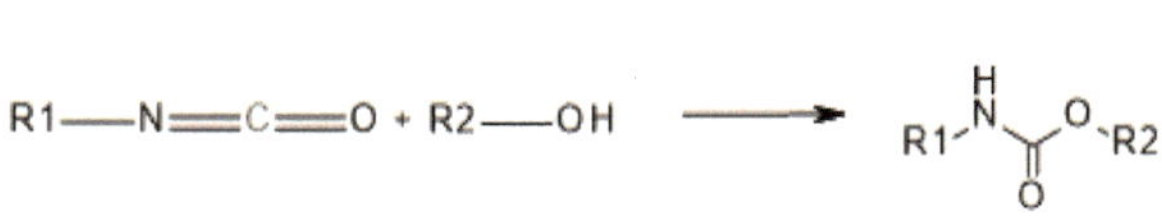

Figure 9.57 Synthesis of polyurethanes

The polymers can be decomposed by chemical reactions such as glycolysis, hydrolysis, or aminolysis, where a nucleophile with an active hydrogen atom can attack the polar group of the polyurethane bond. Thus, monomers or oligomers can be obtained. Purification methods such as distillation can be used to obtain the pure monomers, which in turn can be used to produce new "virgin-like" polymers.

Besides the chemical recycling methods that are described below, enzymatic recycling is also an option for obtaining monomers. On an industrial scale, enzymatic recycling has so far mainly used for PET (by Carbios), but enzymes such as cutinase, esterase and lipinase have been shown to depolymerize polyurethanes [57]. A huge advantage of enzymatic depolymerization is its selectivity towards a certain feedstock and the mild conditions that are used, which leads to a lower CO_2 footprint if highly effective enzymes are used [58].

Figure 9.58 shows a summary of possible recycling and disposal pathways for polyurethanes [59]. Of the chemical recycling pathways, so far only glycolysis has passed the research stage and made it to the pilot and production stages. However, further improvements are necessary to make the process profitable on a large scale.

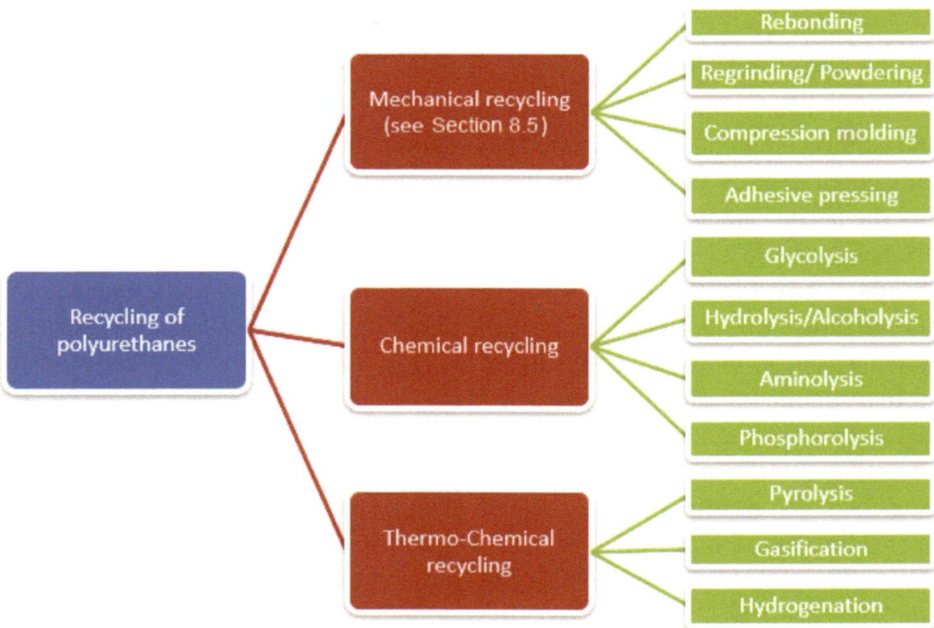

Figure 9.58 Recycling options for polyurethanes

9.5.2.2.1 Hydrolysis and Alcoholysis

Figure 9.59 shows that hydrolysis of a polyurethane produces amines, alcohols, and CO_2. This was the first chemical recycling process investigated for polyurethanes, by the Ford Motor Company in the 1960s [56]. Temperatures above 450 °C and a pressure of 50 bar was used. The process needs a high energy input, and the separation of the amines consumes additional energy; therefore the process was never implemented commercially.

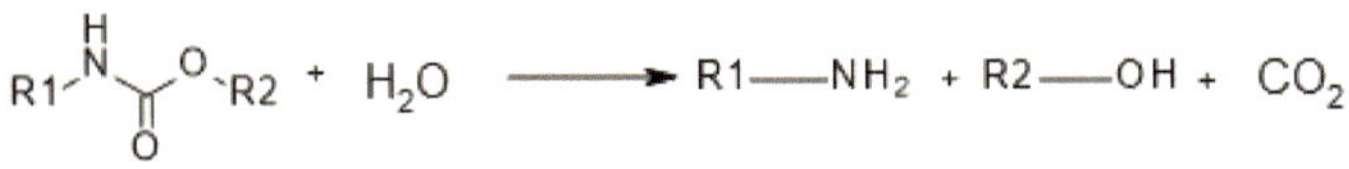

Figure 9.59 Hydrolysis of polyurethanes

If water and glycol are combined in a hydroglycolysis process, less drastic operating conditions can be used. This works for flexible foams, and pure polyols can be obtained [60].

Using alcohols (such as methanol or butanol) instead of water doesn't influence the reaction conditions drastically, but makes the separation of the amine and the polyol slightly easier. However, due to the high energy costs the process was also not implemented on commercial scale.

9.5.2.2.2 Aminolysis

Figure 9.60 shows the aminolysis of a polyurethane: Usually, alkanolamines such as monoethanolamine are used as cleavage agents, and lower reaction temperatures (80–160 °C) than in the glycolysis can be used. Basic catalysts (e. g. NaOH or Al_2O_3) accelerate the reaction [61]. For rigid foam scrap, diethylenetriamine (DETA) was used as a cleaving agent and NaOH as the catalyst at a reaction temperature of 180 °C. Methylene diphenylamine (MDA) was obtained as a product in addition to urethane oligomers. In this reaction, NaOH acts not only as a catalyst but can also react with the urethane linkage and form amines and alcohols [62]. However, similar to hydrolysis and alcoholysis, this process has never passed the research stage, and has not been scaled to industrial production.

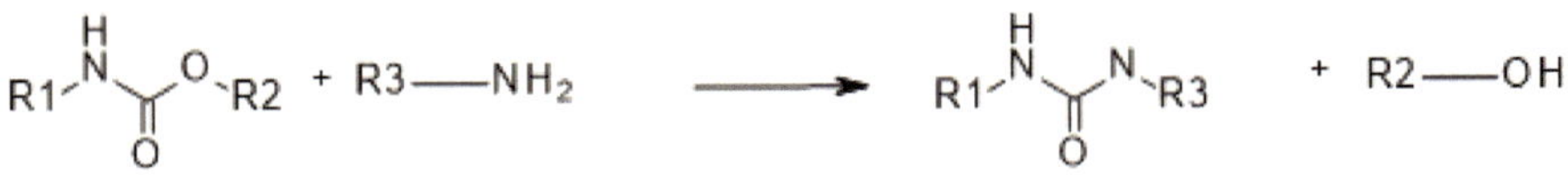

Figure 9.60 Aminolysis of polyurethanes

9.5.2.2.3 Phosphorolysis

The phosphorolysis reaction is valuable to obtain products which can be used in the synthesis of new polyurethanes with flame-retardant properties. In this reaction, esters of phosphonic or phosphoric acids are used as a cleavage agent, at reaction temperatures around 160–190 °C [63]. Similar to glycolysis, two phases can be obtained, where the upper phase contains the polyol with the phosphate functionality. This glycolyzate can then be used to synthesize polyurethanes or other polymers with flame-retardant properties [64].

9.5.2.2.4 Glycolysis

Glycolysis is the most widely used process for chemical recycling of polyurethanes. The reason for this is that glycols have a relatively high boiling point, and therefore no pressure needs to be applied for reactions at elevated temperatures (usually in the range 160–250 °C) [5]. The transesterification reaction that takes place is shown in Figure 9.61.

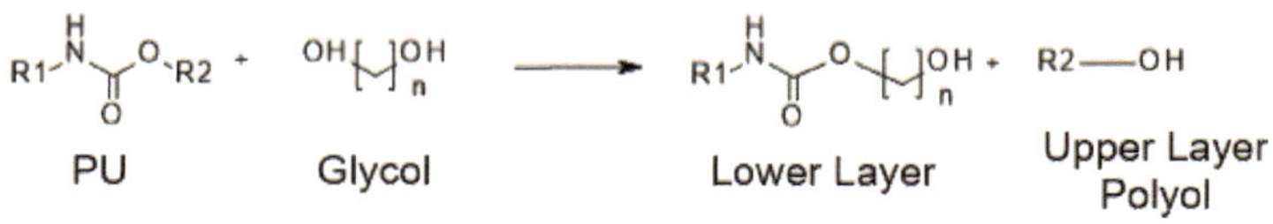

Figure 9.61 Glycolysis of polyurethanes

Different types of catalysts can be used, such as alkaline hydroxides, alkaline acetates, amines or organometallic compounds [5]. If a large excess of glycol is used, the reaction mixture splits into two phases, where the lower layer contains the glycol and the amines or carbamates, while the upper layer contains the recovered polyols. Thus, the polyols can be separated and worked up in the next step., for example with an extraction process. The lower phase needs some further reaction steps such as acidolysis to produce a graft polyol of minor quality, or hydrolysis to recover the diamine. Another option for the lower phase is further thermochemical treatment in gasification or pyrolysis. A detailed review of glycolysis processes has been published by Simon et al. [5] Recently, Recticel published a patent on the further improvement of the split-phase glycolysis process [65]. They use a glycol with a melting point below 200 °C (such as diethylene glycol (DEG), diglycerol, ethylene glycol or 2,4-bis(hydroxymethyl)-1,5-pentanediol). After glycolysis and phase separation, the upper phase containing the polyols is subjected to purification processes such as extraction and/or ion exchange. The lower phase is then subjected to further hydrolysis to obtain the diamine in high yields, notably for DEG (100% yield) and 2,4-bis(hydroxymethyl)-1,5-pentanediol (95.5% yield). It is further purified by an evaporation process.

Looking at all possible chemolysis processes for polyurethanes, glycolysis has the highest technological maturity level. Single-phase glycolysis plants have been established for shoe soles and rigid foams; however profitability could not be shown. One reason for this is the high price of the glycolysis agent that is used and the poor separability of the components. Therefore, development of split-phase glycolysis is crucial, and also the use of crude glycerol (a biodiesel byproduct) can help to improve profitability [66].

9.5.2.2.5 Thermochemical Recycling

The thermal degradation of polyurethanes has been studied in several scientific publications. A large variety of process conditions (and thus results) can be found, which is also due to the large variety of polyurethanes used, for example, various diisocyanates (e.g. MDI, TDI, HDI), polyols and chain extenders. These different monomers lead to different dissociation temperatures and resulting products. At about 200–300 °C isocyanates and alcohols are formed, while at 500 °C in nitrogen atmosphere a mixture of gases such as toluene, benzene, ethylene, and propylene is produced [59].

9.5.2.3 Polybutylene Terephthalate (PBT)

About 75% of PBT is found in industrial or automotive applications, while about 25% can be found in small electrical parts. Therefore, an elaborate sorting process is needed prior to chemical recycling processes. These processes need to be robust and contain further separation and cleaning steps, as PBT usually contains many additives (e.g. glass fibers) or blending components (e.g. PET).

The chemical recycling of PBT can be approached in similar ways to the much more widely used polyethylene terephthalate (PET), which is described in Section 9.3.

Due to its lower availability as a waste material, a smaller volume of literature and patents can be found on the recycling of PBT. The main methods include hydrolysis, alcoholysis, and glycolysis; reaction products and further cleaning steps differ depending on the method used.

Hydrolysis with a high yield is described in detail by Sato et al. [67]. In this publication, high-temperature liquid water around 500 K is used to decompose PBT into terephthalic acid (TPA) and tetrahydrofuran (THF). Yields of 98% were achieved for TPA recovery, and the depolymerization rate could be increased by adding additional TPA.

Alcoholysis reactions have been investigated using supercritical methanol [68] or ethanol [69], where 1,4-butanediol (1,4-BDO) and dimethyl terephthalate (DMT) or diethyl terephthalate, respectively, are obtained. Another possibility is glycolysis, for example by using an excess of BDO, which has the advantage of milder reaction conditions.

9.5.2.4 Other Engineering Plastics

Compared to the large-volume engineering plastics like PA and PU, there is a smaller volume of literature on the recycling of other engineering plastics, such as polyoxymethylenes (POMs) or polysulfones (PSUs), mainly due to their smaller overall volumes.

For POMs the annual world-wide production is 1.7 Mt [70]. As mentioned before, its application is (for example) in smaller parts in cars or in consumer electronics, as it is used for high-performance engineering components. Other applications include use in toothbrushes, in zippers, and in ski bindings.

For POMs chemical recycling via depolymerization is possible, but has not been extensively investigated. Several patents have shown that depolymerization of POMs into formaldehyde or trioxane is achievable by using a cationic catalyst or a metal oxide [71, 72]. Beydoun et al. recycled POM using a catalyzed reaction with 1,3-propanediol. They obtained cyclic acetals, which can be used in a variety of applications, such as fuel additives, solvents, or intermediates [70].

Polysulfones are engineering plastics that are widely used in medical applications, such as dialysis membranes, filtration membranes, and baby bottles owing to their mechanical properties, good thermal stability and also high stability toward chemicals and acids. Their world-wide production is around 70 kt annually. The aminolytic depolymerization of polysulfones has been described in the literature [73]. Here, ethylenediamine (EDA) reacts with the aryl ether bond in a nucleophilic substitution reaction. The reaction was carried out at 90 °C in *N,N*-dimethyl acetamide (DMAc). Depending on the reaction time, the degree of aminolysis could be changed, resulting in various oligomer lengths.

9.5.3 Applications of Major Recycled Engineering Plastics

The need for recycled plastics is especially high in applications close to the end-consumer. Here, a strong market driver exists due to the need for sustainable solutions. The recycled plastics that are obtained from re-polymerizing monomers obtained in the recycling process usually have identical properties to virgin plastics that are directly obtained from petrochemicals. Therefore, they can be used for any application where the original plastics are found. An additional benefit of the recycled polymers is their CO_2 abatement cost. Compared to the virgin polymer, emissions in the range of 1–3 t CO_2 per tonne of polymer can be avoided [74]. Life cycle analysis of the ChemCycling™ process has shown that a considerable amount of CO_2 emission can be saved when using pyrolysis oil instead of naphtha for the manufacturing of plastics. This is mainly due to the fact that pyrolysis of plastic waste emits 50% less CO_2 than incineration [75]. For each plastic waste stream a detailed life cycle analysis is necessary, and it has been shown that mechanical recycling is not necessarily the most environmentally friendly pathway, but chemical recycling has to be considered in terms of saving CO_2 equivalents [76].

Polyamides are currently chemically recycled primarily for applications in textiles. An example of this is ECONYL® made by Aquafil. They collect end-of-life textiles, carpets, and fishing nets made from PA6 and also pre-consumer waste, and use chemical recycling to turn it first into caprolactam and eventually into ECONYL® textile fibers [77].

Similarly, for polyurethanes, end-consumer applications such as mattresses and car seats are of greatest interest for recycling.

When engineering plastics are recycled via pyrolysis (e.g. using ChemCycling™), a "mass balance" approach is often used. In this way, materials with identical properties to virgin materials and with a better CO_2 footprint are obtained. The first customer products produced with ChemCycling™ include food packaging, refrigerator components, and insulation boxes [78].

Recently, BASF announced that outdoor clothes sold by VAUDE will be produced from old car tires using the pyrolysis/mass balance approach [79]. This product has its own brand name (Ultramid® Ccycled™), allowing end-consumers to directly see the circular economy approach for this material.

References for Section 9.5

[1] BASF SE. *https://www.basf.com/global/en/who-we-are/sustainability/we-drive-sustainable-solutions/circular-economy/mass-balance-approach/chemcycling.html*, accessed April 04, 2021

[2] *Stoffstrombild Kunststoffe in Deutschland 2019*, 2020

[3] Meinander, M. (Ed.), Mroueh U.-M. (Ed.), Bacher, J., Jermakka, J.:*Directions of Future Developments in Waste Recycling*, VTT Technology 60, 2013

[4] Buekens, A., Yang, J.: Recycling of WEEE plastics: A review. *Journal of Material Cycles and Waste Management*, (2014) 16, pp. 415–434

[5] Simón, D., Borreguero, A. M., de Lucas, A., Rodríguez, J. F.: Recycling of polyurethanes from laboratory to industry, a journey towards the sustainability. *Waste Management*, (2018) 76, pp. 147–171

[6] *http://www.europeanbedding.eu/wp-content/uploads/2014/08/Budapest-Beds.pdf*, accessed March 21, 2020

[7] PlasticsEurope. *Polyamide 6 (PA6)*, Brussels, Belgium. February 2014

[8] Kamimura, A., Yamamoto, S.: An efficient method to depolymerize polyamide plastics: A new use of ionic liquids. *Organic Letters*, (2007) 9, pp. 2533–2535

[9] Herzog, B., et al.: *Ullmann's Encyclopedia of industrial chemistry – Polyamides*, 2020. Weinheim, Wiley-VCH Verlag GmbH & Co. KGaA, 2020

[10] VDI-Gesellschaft Verfahrenstechnik und Chemieingenieurwesen (GVC), *VDI Heat Atlas*, Second Edition. Berlin, Heidelberg, Springer-Verlag, 2010

[11] D. Braun et al., Rohstoffliches Recycling von glasfaserverstärktem Polyamid-6. *Chemie Ingenieur Technik*, (2001) 73, pp. 183–190

[12] Corbin, T. F., Davis, E. A., Dellinger, J. A.: Reclaiming epsilon-caprolactam from nylon 6 carpet. US005169870A United States, 1991

[13] Fujita, K., Nakade, S.: Depolymerising nylon 6 using phosphoric acid and Lewis acid. JP06032775 Japan, 1994

[14] Corbin, T.F., Davis, E.A., Dellinger, J.A.: Reclaiming epsilon-caprolactam from nylon 6 carpet. US5169870 United States, 1992

[15] Born, C., et al.: Verfahren zur Herstellung von Caprolactam aus Polyamid-haltigen Abfällen. EP1371644 EU, 2003

[16] Wichterle, O., Sebenda, J.: Die polymerisation von ε-caprolactam mit alkalicarbonaten II. Über die kinetik und den mechanismus der alkalischen polymerisation des ε-caprolactams. *Collection of Czechoslovak Chemical Communications*, (1957) 22, pp. 1353–1367

[17] Bockhorn, H., et al.: Kinetic study on the non-catalysed and catalysed degradation of polyamide 6 with isothermal and dynamic methods. *Thermochimica Acta*, (1999) 337, pp. 97–110

[18] Kopietz, M., et al.: Obtaining caprolactam by cleavage of molten polycaprolactam. US5536831 United States, 1996

[19] Nielinger, W., et al.: Gewinning von Caprolactam durch thermische Spaltung von Polyamid 6. EP0529470 EU, 1992

[20] Mukherjee, A.K., Goel, D.K.: Depolymerization of poly-ε-caprolactam catalyzed by sodium hydroxide. *Journal of Applied Polymer Science*, (1978) 22, pp. 361–368

[21] Müller, W., Seyfarth, H.-E., Taeger, E.: Depolymerisation of polycaprolactam-by melt depolymerisation in presence of alkaline catalyst, with addn. of di- or poly-carboxylic acid anhydride to form imide co-catalyst. DE 4228496 Germany, 1994

[22] Depolymerisation of polycaprolactam-by heating in the melt in the presence of conventional alkaline catalysts, with iron-II cpds. as cocatalysts. DE 4311621 Germany, 1994

[23] Nawrath, G.: Diaminoketone aus Lactamen. *Angewandte Chemie*, (1960) 72, p. 1002

[24] Moens, L.: Purification of caprolactam from recycled nylon. US5919927 United States, 1999

[25] Denisenko, S.N., Pasch, E., Kaupp, G.: Die ersten tricyclischen Diaziridine – Ein einfacher Syntheseweg. *Angewandte Chemie*, (1989) 101, pp. 1397–1399

[26] Mastaglio, D.: Verfahren zur Herstellung von Diaminoketonen. DE1131697B, 1960

[27] Mastaglio, D.: Verfahren zur Herstellung von Diaminoketonen. DE1131697 Germany, 1962

[28] Müller, W., Taeger, E.: Verfahren zur Rückgewinnung von Caprolactam aus PA6-Teppichabfällen. DE10048824 Germany, 2002

[29] Fukusato, R., et al.: Recovery of raw material component from nylon 6,6, involves separating precipitate and raw material component content aqueous solution from hydrolyzed product under low temperature and/or low pressure. JP2001302597 Japan, 2001

[30] Niu, Y.: Prodn. of adipic acid and hexa:methylene-di:amine hydrochloride by depolymerising polyamide-66 at elevated temp. in presence of strong acid and water. CN87101541 China, 1988

[31] Zhang, Z.: Process for producing adipic acid and hexanediamine by depolymerization of nylon-66. CN1611477 China, 2005

[32] Monet, G.P.: Recovery of adipic acid and hexamethylendiamine from nylon. US3069465 United States, 1962

[33] Cesarek, U., Pahovnik, D., Zagar, E.: Chemical recycling of aliphatic polyamides by microwave-assisted hydrolysis for efficient monomer recovery. *ACS Sustainable Chemical Engineering*, (2020) 8, pp. 16274–16282

[34] Moran, Jr, E.F.: Conversion of nylon 6 and/or nylon 6,6 to monomers. US5310905 United States, 1994

[35] Moran, Jr, E.F., McKinney, R.J.: Conversion of nylon 6 and/or nylon 6,6 to adipic acid. US5468900 United States, 1995

[36] Dos Santos, E., Mettvier, P., Gubelmann, M.: Process for treating a material comprising a polymer by hydrolysis. US5905168 United States, 1999

[37] Davis, D., Wilhoit, E.: Nitric acid hydrolysis of polyamides. WO97/00846 PCT, 1997

[38] Guo, Z., et al.: Method for producing hexane diacid and hexanediamine by using nylon-66 disaggregation. CN101423478 China, 2009

[39] Miller, B. M.: Nylon hydrolysis. US2840606 United States, 1958

[40] Wiest, G., Stahl, H.: Verfahren zur Rückgewinnung von Dicarbonsäuren und Diaminen aus Polyamiden. DE1088063 Germany, 1961

[41] Chambert, F.-R., Joly, D.-J., Wersinger, J.-E.: Method for treating waste of polyamides. FR1070841 France, 1954

[42] Seeliger, U., et al.: Method for the simultaneous preparation of dicarboxylic acids and diamines by decomposing polyamides into their constituent monomers. WO9325513 PCT, 1993

[43] Moran, Jr, E. F.: Nylon component reclamation. US5266694 United States, 1993

[44] Thomissen, P. J. H.: Depolymerization of polyamides. EP0875504 EU, 1998

[45] Polk, M. B., et al.: Nylon 66, nylon 46 and PET phase-transfer-catalyzed alkaline depolymerization at atmospheric pressure. *Polymer-Plastics Technology and Engineering*, (1999) 38, pp. 459–470

[46] Langhammer, M., Lüderwald, I., Weingärtner, M.: Formation of large cyclic amides by catalytic depolymerization of polyamides. *Die Makromolecular Chemie*, (1986) 187, pp. 829–835

[47] Vulic, I., Jacobs, R. E. W. E.: Thermal catalytic depolymerization of polyamides. *Polymer Bulletin*, (1993) 31, pp. 55–60

[48] Gallagher, F.: Monomer recovery process for contaminated polymers. WO97/49652 PCT, 1997

[49] McKinney, R. J.: Ammonolysis of nylon. US5302756 United States, 1994

[50] Fergusson, S. B., Yan, Y.: Separation of the products of polyamide ammonolysis. WO01/70665 PCT, 2001

[51] Bodrero, S., Canivenc, E., Cansell, F.: Chemical recycling of polyamide 6,6 and polyamide 6 through a two step ami-/ammonolysis process. *Polymer Preprints*, (American Chemical Society, Division of Polymer Chemistry), (1998) 39, pp. 150–151

[52] Canivenc, E., Thierry, J.-F.: A process for the de-polymerisation and re-cycling of waste polyamide(s)-to form monomers contg. N, by stages of amminolysis and ammonolysis. WO1997003040 PCT, 1997

[53] Duch, M. W., Allgeier, A. M.: Deactivation of nitrile hydrogenation catalysts: New mechanistic insight from a nylon recycle process. *Applied Catalysis A: General*, (2007) 318, pp. 190–198

[54] Kumar, A.: Hydrogenative depolymerization of nylons. *Journal of the American Chemical Society*, (2020) 142, pp. 14267–14275

[55] Zhou, W., et al.: Depolymerization of technical-grade polyamide 66 and polyurethane materials through hydrogenation. *ChemSusChem*, (2021) 14, pp. 4176–4180

[56] Behrendt, G., Naber, B. W.: The chemical recycling of polyurethanes. *Journal of the University of Chemical Technology and Metallurgy*, 2009, pp. 2–23

[57] Magnin, A., Pollet, E., Perrin, R., Ullmann, C., Persillon, C., Phalip, V., Avérous, L.: Enzymatic recycling of thermoplastic polyurethanes: Synergistic effect of an esterase and an amidase and recovery of building blocks. *Waste Management*, (2019) 85, pp. 141–150

[58] Vollmer, I., Jenks, M. J. F., Roelands, M. C. P., White, R. J., van Harmelen, T., de Wild, P., van der Laan, G. P., Meirer, F., Keurentjes, J. T. F., Weckhuysen, B. M.: Beyond mechanical recycling: Giving new life to plastic waste. *Angewandte Chemie, International Edition*, (2020) 59, p. 15402

[59] Datta, J., Włoch, M.: Recycling of polyurethanes, in: *Polyurethane Polymers: Blends and Interpenetrating Polymer Networks*, Elsevier, 2017

[60] Gerlock, J., Braslaw, J., Zinbo, M.: Polyurethane waste recycling. 1. Glycolysis and hydroglycolysis of water-blown foams. *Industrial & Engineering Chemistry Process Design & Development*, (1984) 23, pp. 545–552

[61] Sheratte, M. B.: Process for converting the decomposition products of polyurethane. US 4110266 1978

[62] Chuayjuljit, S., Norakankorn, C., Pimpan, V.: Chemical recycling of rigid polyurethane foam scrap via base catalyzed aminolysis. *Journal of Applied Polymer Science*, (2002) 1, pp. 19–22

[63] Troev, K., Atanasov, V. I., Tsevi, R.: Chemical degradation of polyurethanes II. *Journal of Applied Polymer Science*, (2000) 76, pp. 886–893

[64] Grancharov, G., Mitova, V., Shenkov, S., Topliyska, A., Gitsov, I., Troev, K.: Smart polymer recycling: Synthesis of novel rigid polyurethanes using phosphorus-containing oligomers formed by controlled degradation of microporous polyurethane elastomer. *Journal of Applied Polymer Science*, (2007) 105, pp. 302–308

[65] Vanbergen, T., de Vos, D., Claes, L., Verlent, I., de Geeter, J.: Improved method of recycling polyurethane materials. PCT/EP2020/072323 2021

[66] Simón, D., Borreguero, A. M., de Lucas, A., Rodríguez, J. F.: Valorization of crude glycerol as a novel transesterification agent in the glycolysis of polyurethane foam waste. *Polymer Degradation and Stability*, (2015) 121, pp. 126–136

[67] Sato, O., Yamaguchi, A., Murakami, Y., Shirai, M.: Depolymerization of poly(butylene terephthalate) into terephthalic acid and tetrahydrofuran in high-temperature liquid water. *Chemistry Letters*, 2015, pp. 1312–1314

[68] Shibata, M., Masuda, T., Yosomiya, R., Meng, L.-H.: Depolymerization of poly(butylene terephthalate) using high-temperature and high-pressure methanol. *Journal of Applied Polymer Science*, 2000, pp. 3228–3233

[69] Pan, Z., Shi, Y., Liu, L., Jin, Z.: Depolymerization of poly(butylene terephthalate) in sub- and supercritical ethanol in a fused silica capillary reactor or autoclave reactor. *Polymer Degradation and Stability*, 2013, pp. 1287–1292

[70] Beydoun, K., Klankermayer, J.: Efficient plastic waste recycling to value-added products by integrated biomass processing, *ChemSusChem*, 2020, pp. 488–492

[71] Haubs, M., Lingnau, J., Kurz, K.: Process for recycling polyacetals. PCT/EP2012/073544 2012

[72] Hatsuhiko, H., Hiroaki, N.: Method for recovering formaldehyde from polyacetal resin. JP4447444 B2 [ed.] Daicel. 2006

[73] Wang, L., Cui, Y., Wang, N., Zhang, H., Zhu, B., Zhu, L., Xu, Y.: Aminolytic depolymerization of polyarylsulfones. *Polymer Degradation and Stability*, (2014) 103, pp. 69–74

[74] *https://cefic.org/app/uploads/2020/12/CEFIC_Quantis_report_final.pdf*, accessed April 05, 2021

[75] SE, BASF. *https://www.basf.com/global/en/who-we-are/sustainability/we-drive-sustainable-solutions/circular-economy/mass-balance-approach/chemcycling/lca-for-chemcycling.html*, accessed April 04, 2021

[76] Schwarz, A. E., Ligthart, T. N., Bizarro, D. G., De Wild, P., Vreugdenhil, B., van Harmelen, T.: Plastic recycling in a circular economy; determining environmental performance through an LCA matrix model approach, *Waste Management*, (2021) 121, pp. 331–342

[77] *https://www.econyl.com/de/*, accessed March 18, 2021

[78] *https://www.basf.com/global/en/who-we-are/sustainability/we-drive-sustainable-solutions/circular-economy/mass-balance-approach/chemcycling.html*, accessed March 18, 2021

[79] BASF SE. *https://www.basf.com/global/de/media/news-releases/2021/03/p-21-157.html*, accessed March 19, 2021

9.6 Chemical Recycling of PTFE (as a Model for Other Polymers)

Achim Schmidt-Rodenkirchen, Klaus Hintzer, Thorsten Gerdes

9.6.1 Introduction

PTFE (polytetrafluoroethylene) and other perfluorinated polymers like FEP (copolymers of tetrafluoroethylene (TFE) and hexafluoropropylene (HFP)) or PFA (a copolymer of TFE and perfluorovinylether) are niche polymers, with a total production in 2015 of 160 kt, compared to polymers overall of 322 Mt [1]. Of the perfluorinated polymers, PTFE has the highest share, at 52% [2].

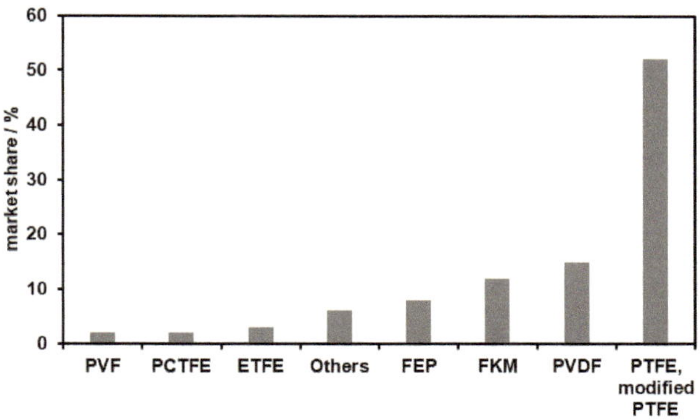

ETFE	Ethene-tetrafluoroethene copolymer
FEP	Fluorinated ethene-propene copolymer
FKM	Fluoroelastomers
PCTFE	Polychlorotrifluoroethylene
PFA	Perfluoroalkoxy/ propylfluroroalkoxy copolymer
PTFE	Polytetrafluoroethylene
PVDF	Poly(vinylidene fluoride)
PVF	Poly(vinyl fluoride)

Figure 9.62 About 60% (PTFE + FEP) of the total volume of produced perfluorinated polymers are suitable for chemical recycling [2]

Their unique properties, including thermal and chemical stability, low dielectric constant, high dielectric strength, and low surface energy (ideal for anti-adhesive applications), turns PTFE and other perfluorinated polymers into indispensable materials [3, 4]. Due to these outstanding material properties, the market is predicted to grow by approximately 5% per year [5].

The conventional production of PTFE is a very energy-intensive procedure with many process steps, and 220 MJ is needed for 1 kg of PTFE [6]. The production of Cl_2, which is needed for PTFE production among other applications, is one of the largest energy consumers worldwide [7]. Perfluorinated polymers are therefore valuable products. An overview of the energy demand for various polymers, including a ranking of PTFE, is shown in Table 9.21.

Table 9.21 Cumulative Energy Demand (CED) of Bulk Plastics and some Fluorinated Examples for Comparison of Energy Consumption [8]

Polymer	Cumulative energy demand (CED)		Ref.
	MJ per kg polymer	toe[a] per t polymer	
Polytetrafluoroethylene (PTFE)	220	5.2	[6]
Polyethylene terephthalate (PET)	54	1.3	[9]
Polyethylene (PE)	65	1.5	[10]
Polystyrene (PS)	71	1.7	[10]
Polyvinyl chloride (PVC)	53	1.3	[10]
Synthetic rubber	64	1.5	[10]
Nylon 6-6	163.0	3.9	[10]
Ethylene tetrafluoroethylene (ETFE)	150–170	3.6–4.0	[11]

[a] 1 tonne of oil equivalent (toe) = 42 GJ

The manufacture of TFE/HFP monomers consumes the lion's share of the energy demand in making perfluoropolymers (>99%). There have been many efforts to recover perfluorinated monomers from perfluoropolymers; all approaches were based on the finding that these polymers can be pyrolyzed at high temperatures into TFE/HFP.

The use of pyrolysis to convert perfluorinated polymers back to monomers is an example of a rare phenomenon in the materials industry; only a few other polymer classes (e.g. polystyrene, polyacrylates) can be converted into monomers upon heating.

Thus, perfluorinated polymers are ideal candidates for depolymerization, which provides the following benefits:

- High demand of resources = large savings potential
- Expensive material = high margin (US $20/kg) [12]
- Unique properties = high future potential
- Low volume of production = low scale for reaching industrial world scale
- Thermal decomposition behavior favors the formation of the monomer = high yields of monomer possible (over 90% yields of monomer/intermediate are obtainable under certain conditions)

Another aspect of recycling PTFE is due to the chemical composition of the polymer. PTFE and other perfluorinated polymers consist solely of carbon and fluorine. Efficient recycling of the material can close the fluorine loop and will reduce the consumption the mineral fluorspar, which is considered as a critical raw material for the European economy [13].

All these facts make PTFE an ideal candidate for upcycling, and although these benefits are well-known, to date the 3M company is the only one to have scaled-up the process, with the opening of a production plant at Gendorf, Germany, in 2015

[14] with a capacity of a few hundred tonnes. A picture of the building, which in addition to the pyrolysis reactors also contains the mechanical treatment and cleaning, is shown in Figure 9.63.

Figure 9.63
500 t/a Demonstration plant for chemical recycling of PTFE (courtesy of 3M/Dyneon) [15]

9.6.2 Manufacturing Process for PTFE and Other Perfluorinated Polymers

Five raw materials are needed for PTFE production, and the sequence is shown in Figure 9.64. Electrolysis of sodium chloride in water provides chlorine (Cl_2) which is subsequently converted by radical substitution with methane into chloroform ($CHCl_3$). In a parallel process, fluorspar (CaF_2) is converted with highly concentrated sulfuric acid into HF and gypsum. The key starting material ($CHClF_2$, known as R22) is then made from HF and chloroform. R22 is afterwards pyrolyzed at 900 °C into tetrafluoroethylene (TFE) and hexafluoropropylene (HFP) by splitting off one HCl molecule per R22 molecule. The polymerization of the monomers takes place in an aqueous suspension or emulsion.

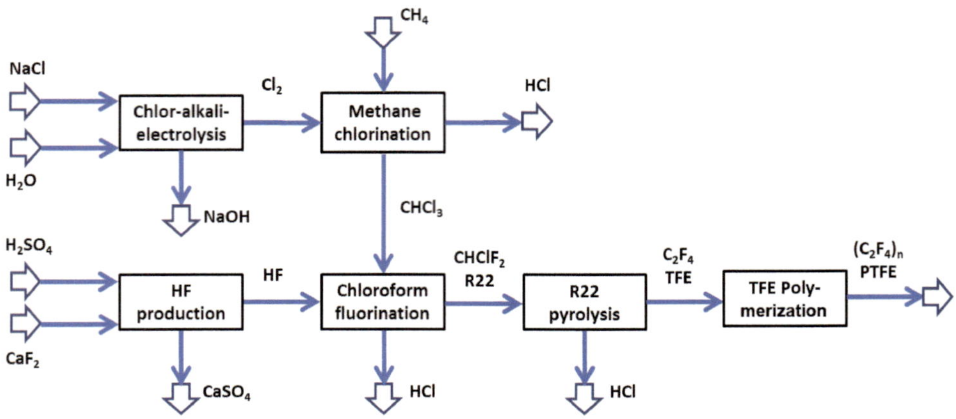

Figure 9.64 Process of conventional PTFE production with major precursors and by-products [6, 16]

The consumption of the raw materials is shown in Figure 9.65. Aside from the huge amount of raw materials needed for PTFE production, some coproducts like HCl are contaminated with HF from the various process steps, which makes further application difficult.

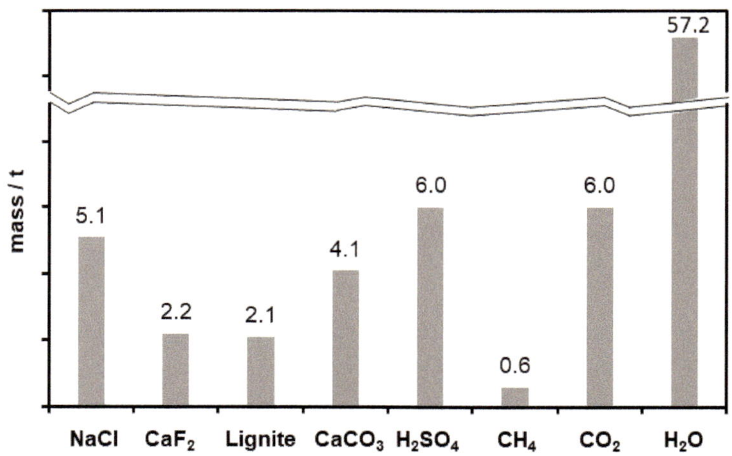

Figure 9.65 Raw materials required for the R22 route [17]

The energy used in producing PTFE from raw materials is shown in Figure 9.66. By far the largest amount is spent on producing TFE monomer, with less than 1% of the energy being used for the polymerization process. Consequently, if the polymer could be successfully recycled into the monomer, most of the energy can be saved.

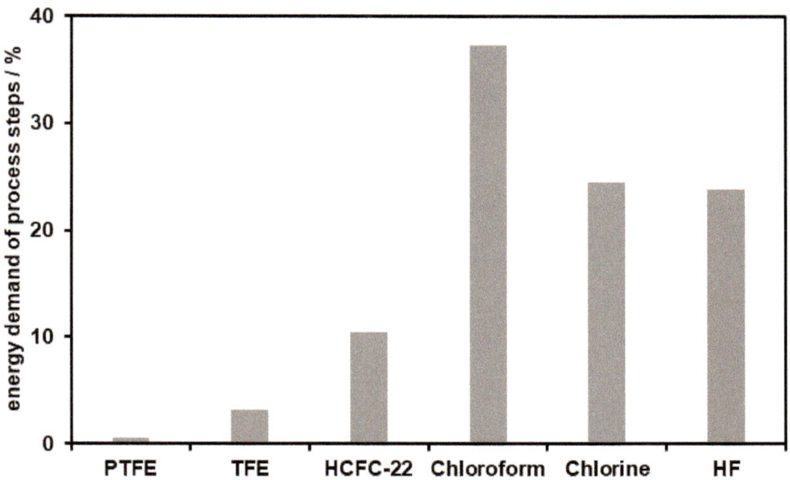

Figure 9.66 Ratio of energy demand of the major process steps on the total production process of PTFE in aqueous polymerization [6]

9.6.3 Mechanism of PTFE Depolymerization

The formation of the key monomer TFE is based on the recombination of two difluorocarbene (CF_2) molecules. In classical TFE production, the CF_2 molecule is obtained from the pyrolysis of R22.

By heat treatment/pyrolysis of PTFE, the polymer is degraded back to CF_2 [3]. The polymer breaks at temperatures above 500 °C statistically, by forming two radicals. The bond strength in the polymer chain increases from the radical along the chain (Figure 9.67). Consequently, decomposition starts at the radical end, and proceeds from there back along the chain [18].

$\cdot CF_2 - CF_2 - CF_2 - CF_2 - R_F$

184 326 356 359

Bonding strength in kJ
R_F = perfluorinated rest of polymer chain

Figure 9.67
C–C bond strength in perfluoroalkyl radicals

The dependence of the rate of the depolymerization reaction on temperature and the degree of degradation is given in Figure 9.68. The influence of the temperature approximately doubles with each 10 K, which is in accordance with the Arrhenius equation (Equation 9.7).

$$k = k_0 \exp\left(\frac{-E_a}{RT}\right) \qquad \text{Equation 9.7}$$

The influence of substances like fillers, in this case carbon, on the degradation behavior is also interesting, and illustrates the need to perform experimental investigation to understand the decomposition of complex mixtures of PTFE, perfluorinated polymers and end-of-life (EOL) materials.

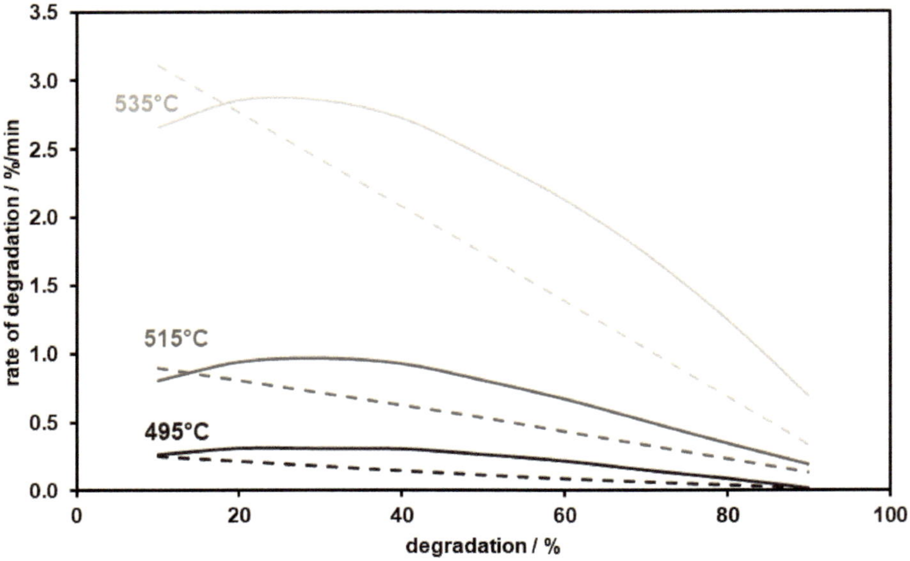

Figure 9.68 Thermal reaction rate vs. degree of decomposition for PTFE (-○-) and PTFE with 10% Carbon (-●-) [19]

Ultimately, difluorocarbene is obtained from the depolymerization reaction. Due to the strong C–F bonding the generated molecule is stable enough to recombine into a TFE molecule before further decay. This reaction is very fast, because the activation energy E_a for the reaction of two radicals is 0 kJ/mol (Equation 9.8).

$$2CF_2 \rightleftharpoons C_2F_4 \qquad \text{Equation 9.8}$$

Due to the reactivity of the CF_2 molecule, side-reactions become relevant to achieving optimum reaction control. In Figure 9.69 the formation of homologs is shown. The side-product hexafluoropropylene (HFP) is also a valuable product, used (for example) to produce the polymer FEP, and c-C_4F_8 can be reconverted into TFE or HFP. Higher homologs must be avoided, along with the formation of the toxic carbonyl fluoride (fluorophosgene) by the action of oxygen. The mechanism of the formation of higher homologs, such as HFP and c-C_4F_8 is described in the literature [20].

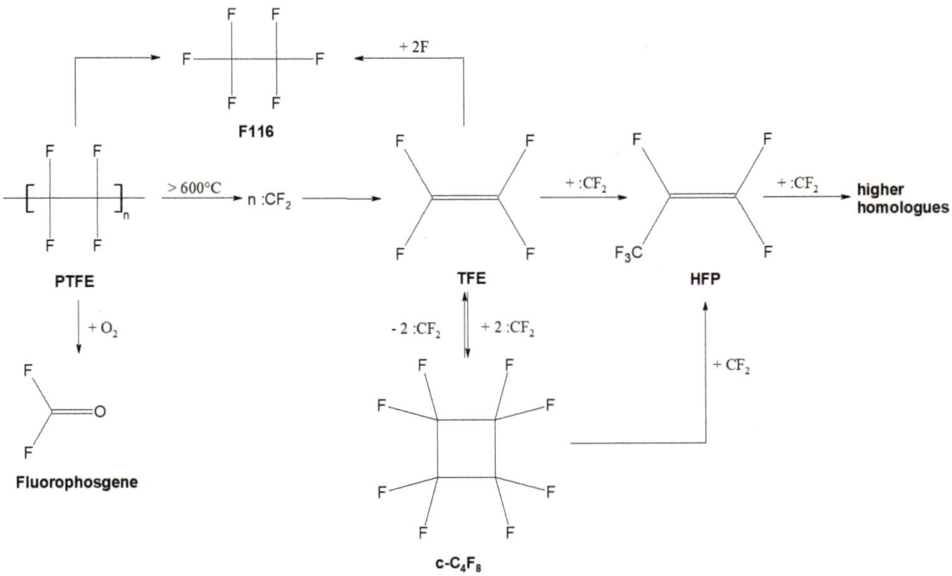

Figure 9.69 Potential side-products generated during the depolymerization of PTFE via CF_2 [8, 16, 20]

9.6.4 Recycling PTFE and Perfluorinated Polymer Materials

9.6.4.1 PTFE Production Waste

The amounts of scrap, wet waste materials or off-specification materials of unfilled PTFE from manufacturers are usually in the lower percentage range. In contrast, the quantity of scrap resin from processors (generated during the finishing process) or end-use articles is usually in the range 10–30%, and in some areas it is above 50%. These high quantities are the result of very specific processing technologies for PTFE (molding, sintering, machining, and cutting). For unfilled PTFE-scrap resins, there are three established recycling paths:

1. Sintered, unfilled PTFE-scrap resin is cleaned from all contaminants and milled into certain particle size classes, which can be reused, for example, in Ram extrusion applications. This so-called "repro-PTFE" can also be mixed with virgin PTFE to a certain content; such repro-PTFE materials get specific designations and are typically used for less demanding end-use applications.
2. Clean, unfilled PTFE can be thermally degraded into low-molecular-weight PTFE, and such processes are commercially established. The thermal degradation of high-molecular-weight scrap PTFE occurs at temperatures of about 500 °C in ovens, kneaders, or extruders [21], and the obtained low-molecular-weight PTFE is further milled into particles a few microns in size; such materials are often called PTFE micropowders.

3. Alternatively, clean, unfilled PTFE can be degraded by high-energy radiation such as X-rays, gamma rays, or electron beams. The degradation of high-molecular-weight PTFE by electron beam irradiation is widely used commercially, and in practice continuous processes are used to improve the economics. After irradiation, the material is milled to the desired smaller particle sizes. During the irradiation, significant amounts of perfluorinated carboxylic acids are formed, which are a major concern, and as a result this recycling path may be discontinued in the future.The thermo/radiation-degraded PTFE micropowders are mostly used as additives to plastics, inks, oils, lubricants, and coatings.

9.6.4.2 PTFE Composites Production Waste

In contrast to clean, unfilled PTFE, for which some recycling options are well-established, and the recycling rates reach significant levels, there no large-scale recycling technologies for PTFE mixtures. This is primarily due to the presence of a large variety of different fillers (e.g., glass fibres, graphites, carbons, metals, ceramics, organic fillers, and pigments) and to the variable amounts of fillers in the PTFE compounds. Because of the large amounts of scrap streams (at least 10–30%), and considering that landfilling is not sustainable and becoming increasing heavily regulated or even banned, scrap PTFE mixtures scrap are increasingly likely to be recycled into TFE/HFP.

9.6.4.3 Perfluorinated Thermoplasts Production Waste

Clean, unfilled, scrap materials from manufacturing and processing of perfluorinated polymers (e.g. FEP, PFA) are generated in the low percent range. Nearly all these materials are recycled back into the different processes; the end-use properties are almost unaffected.

9.6.4.4 End-of-Life (EOL) PTFE and Perfluorinated Materials

In some cases, used EOL PTFE and perfluorinated thermoplasts (e.g. FEP, PFA) are recycled by special cleaning processes and end up in the repro-PTFE or PTFE micropowder markets; PFA and FEP might be reused in applications where the quality requirements are much lower. Overall, most EOL materials end up in landfills, incineration plants, or blast furnaces. PTFE is often combined with lots of fillers like carbon black, graphite, bronze, MoS_2, short and long glass fibers, and colorants to achieve certain properties. Sealings contain often short glass fibers or carbon black. Typical EOL fluoropolymers are shown in Figure 9.70.

Figure 9.70 Examples for recycling of end-of-life (EOL) PTFE: Pipe sealings with glass fibers (left); semi-finished product with carbon black (middle); machine-cut tubes from heat exchangers (right)

Architectural materials (Figure 9.71, left) are commonly made of long fibers with a PTFE coating or laminate. Those fibers can cause problems in some process units like cyclones or reactors. Some materials contain of course unknown impurities, such as PVC in bulk material originating from the reprocessing of wires (Figure 9.71, right). In this case, the influence of chlorine on the chemical reactions and the position of elimination is yet to be investigated.

Figure 9.71 Architectural materials made from PTFE with long fibers (left), and polymer jacket of wires (right)

9.6.5 Recycling Concepts for Perfluorinated Materials

Fluoropolymers can generally be recycled according to three main categories: thermal, chemical, and mechanical. This is illustrated in Figure 9.72. The block schema for the fluorinated polymers shows the backward integration of the various recycling routes into the classic manufacturing route. The further back the integration reaches, the higher the energy required for recycling. Landfilling and thermal recycling should be avoided in future. The other recycling routes are discussed in the following sections.

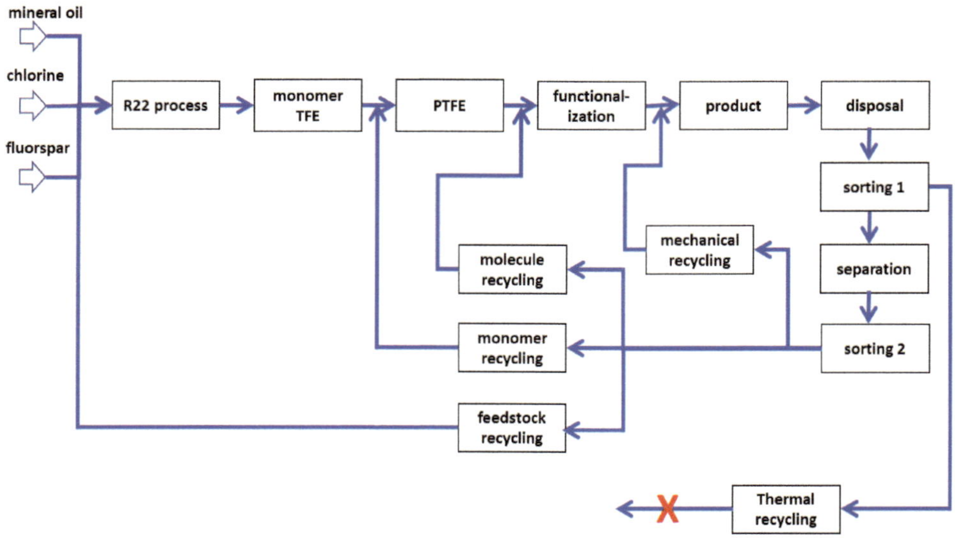

Figure 9.72 Typical classification of polymer recycling options used for PTFE [11]

9.6.5.1 Thermal Recycling

With a calorific value of 5400 kJ/kg [23], PTFE and perfluorinated polymers are not suitable for generating thermal or electrical energy because the value is below the required limit of 11,000 kJ/kg for application as a surrogate fuel [23].

However, these materials can be combusted in a typical temperature range from 900 °C to 1000 °C by a commercial waste incineration plant without forming toxic organic byproducts [24]. In such an incineration process, fluoropolymers release corrosive gases (e.g. HF). For processing limited amounts of fluorinated materials in municipal waste incineration plants, the plants must be equipped with appropriate corrosion protection [5] and off-gas treatment, which is commonly done by capturing the formed HF with (for example) CaO, to generate fluorspar (CaF_2) [25].

Because the calorific value of PTFE and perfluorinated polymers is below 6000 kJ/kg, and because they are chemically inert [4, 23], not biodegradable [26] and do not emit toxic components, landfilling is allowed in accordance with local regulations. The disadvantage of this practice is the accumulation of chemical inert perfluorinated polymers in the environment [8].

9.6.5.2 Mechanical Recycling

The major processing route for mechanical recycling consists of cleaning, milling, and reprocessing to repro-PTFE or even PTFE micropowder [3]. This is suitable for highly pure material without fillers.

The reprocessing is done under raised pressure and temperature, commonly in RAM extrusion [23]. The physical properties can deviate significantly from virgin

PTFE, and this has to be considered for the further applications [23]. To reduce the loss of properties, repro-PTFE is often blended with virgin material [5].

Mechanical recycling of PTFE makes sense if the available input materials are sufficiently pure and do not contain fillers. However, recycling such polymers multiple times results in diminished intrinsic properties and overall lower quality [8, 22].

However, in most cases mechanical recycling does not apply to EOL materials nor to most compounds outside the processor of the semi-finished product [23]. To avoid landfilling or combustion, chemical recycling is usually the only option.

9.6.5.3 Chemical Recycling of Perfluorinated Polymers

As shown in Figure 9.72, chemical recycling can be integrated into a common recycling process between combustion (thermal recycling) and mechanical recycling. Chemical recycling itself embraces three subcategories.

Here the following terms are used, which are general enough to cover possible sub-variants:

- Molecule recycling
- Monomer recycling
- Feedstock recycling

By using the solubility of the polymer molecules in certain solvents, the molecules can be separated in the solvent, the impurities precipitated, and finally the molecules reused. A process for dissolving PTFE polymer composite is not yet known.

In terms of fluorinated materials, the generic term "feedstock recycling" would include incineration processes for production of HF/CaF_2 and CO/CO_2. Because this process is very far back-integrated (i. e. recycled material streams are reintroduced into the production process far upstream) because all carbon–fluorine bonds are destroyed, it should only be considered for partially fluorinated materials when there is no other chemical recycling route available [27].

As discussed in Section 9.6.3, perfluorinated polymers can be depolymerized into the monomers TFE and HFP. This process is related to "monomer recycling". The thermal degradation of PTFE to the corresponding monomers was first described in the 1950s [28]. A continuous process for this approach was described in 1982 [29]. Further investigations were carried out [8, 16, 33] until high yields of over 90% were obtained using in a continuous fluidized-bed process.

Despite the high yield of valuable products, significant amounts of HF as a by-product are produced. In addition, the depolymerization reactor and the subsequent process steps has to cope with fillers from compounds and other possible impurities. The formation and presence of HF is a unique characteristic compared to other polymers.

The depolymerization process requires process units that are tolerant of formed impurities and able to stabilize the reaction mixture by carefully controlling the residence time of TFE and HFP. It is also advantageous if the crude monomer mixture can be fed into an existing TFE/HFP infrastructure, because this would reduce investment costs.

Figure 9.73 gives an insight into the process design that can meet all the following process requirements:

- Minimum residence time of generated monomers
- Quick elimination of HF from the product stream
- Accumulation of solids in one process unit

The preconditioned (milled) PTFE and EOL materials are buffered and subsequently transported into a fluidized-bed reactor (FBR) unit. In the FBR the fluoropolymer material, together with the bed material, is fluidized with superheated steam above 550 °C. The fluoropolymers are decomposed using flash pyrolysis, with smaller fillers (e.g. from polymer composites) being blown out of the reactor; the chemically inert bed material remains in the FBR.

Downstream of the reactor, the crude gas mixture is immediately quenched by water to reduce the temperature to below 50 °C. The quench water is neutralized by NaOH so that HF is converted to NaF. After separating the remaining moisture from the gas flow, the raw gas can be separated into the desired components using existing distillation columns.

The process steps to make the monomers consume 99.5% of the cumulative energy demand (CED) of a PTFE or FEP material [6]. Therefore, upcycling these monomers has the potential to save a lot of energy, and at the same time provide new polymers without any loss of performance.

The excess water from the quenching process is fed into a wastewater treatment plant after separation of the solids. The solids were collected in one process unit and can be filtered easily from the quenching water. This quenching procedure therefore offers the following advantages:

- “Freezing” the chemical equilibrium of the raw gas
- Elimination of corrosive components from the raw gas
- Precipitation of solid by the droplets of the quenching nozzles

Figure 9.74 gives an example of a GC analysis taken from a trial at the InVerTec miniplant running depolymerization of PTFE. The plant operates according to a patented process [30]. The chromatogram shows the valuable products TFE and HFP, along with TFE dimer, c-C_4F_8 (which can be used to make additional HFP). By-products in the low percentage range are C_2F_6, C_3F_6 and some C_4 species.

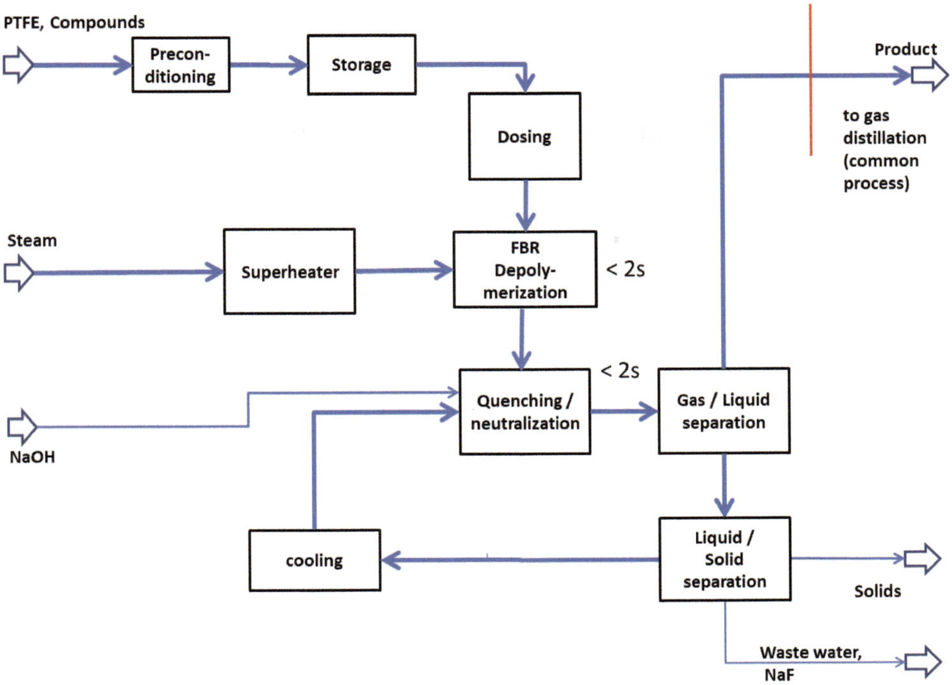

Figure 9.73 Block diagram of the recycling process via the depolymerization route

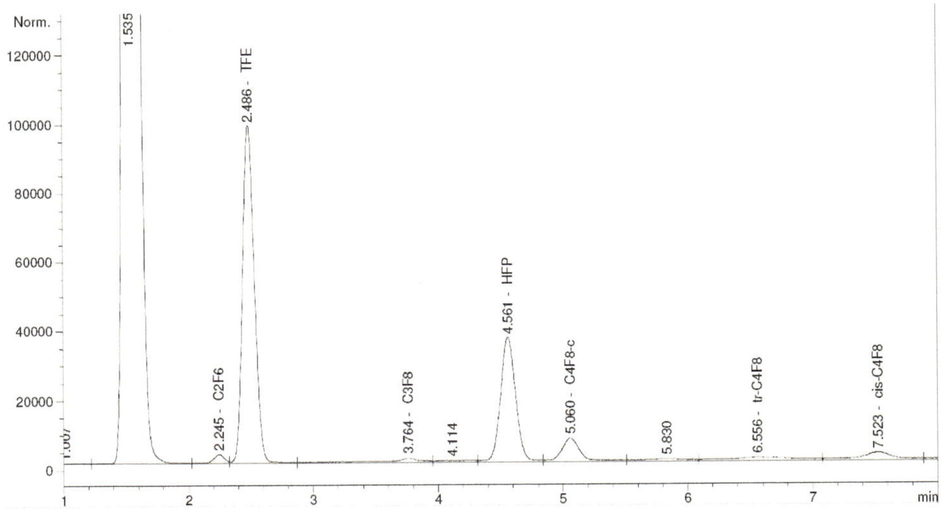

Figure 9.74 Example GC analysis showing the composition of the raw gas from a PTFE pyrolysis at the InVerTec miniplant [31]

The control room of the miniplant in shown in Figure 9.75. For safety reasons, the miniplant itself is located in a ventilated housing in the background.

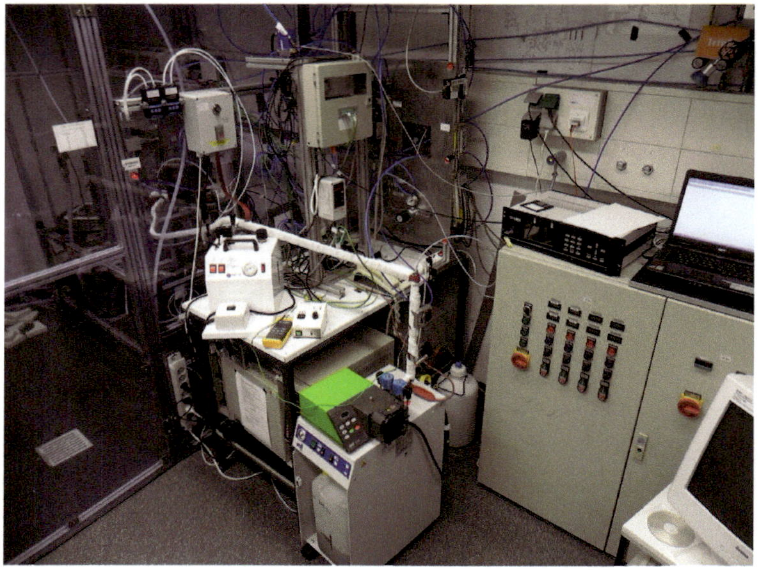

Figure 9.75 Control room for PTFE pyrolysis at the InVerTec miniplant, showing steam supply, control systems, and GC equipment (Courtesy of InVerTec)

If materials are processed that contain fillers that are very difficult to fluidize, other reactor concepts are needed to avoid accumulation of fillers and cementing (plugging) of the fluidized bed. For that purpose, the authors [31] have developed a forced stirred moving-bed (SB) reactor, which can be installed into the depolymerization process instead of the FBR unit [32]. The continuous-bed material (which remains in the SB reactor permanently) is mixed with a stirrer at the depolymerization temperature. Fillers, which are fed into the reactor together with the polymer, are dispersed in the stirred bed, broken by the force of the moving bed, and can then be removed from the quenching system in the same way as for the FBR reactor. A picture of the laboratory reactor is shown in Figure 9.76.

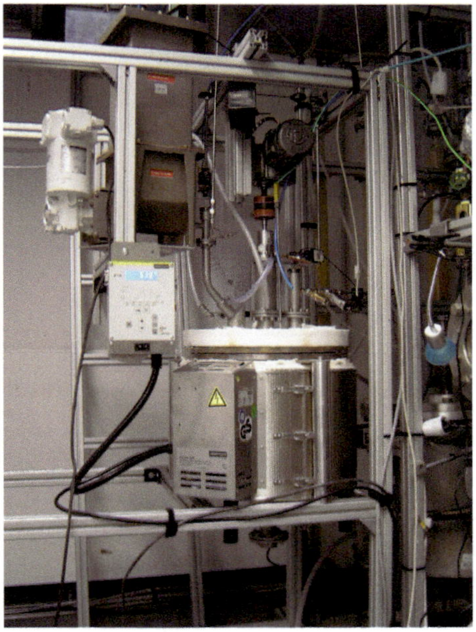

Figure 9.76
The stirred-bed reactor of the InVerTec miniplant Courtesy of InVerTec)

9.6.6 Environmental Aspects of the Presented Process

The environmental impact of the upcycling process described above was investigated by a life-cycle assessment, compared against conventional PTFE production. The main findings are summarized in Figure 9.77. Beside the significant reduction in the CO_2 footprint of the recycled material, the reduction in chlorine and hydrochloric acid is impressive. Thus, there is not only the possibility to prevent the landfilling of a chemically inert material, but also to make the polymer production much more environmentally friendly, and to close material cycles.

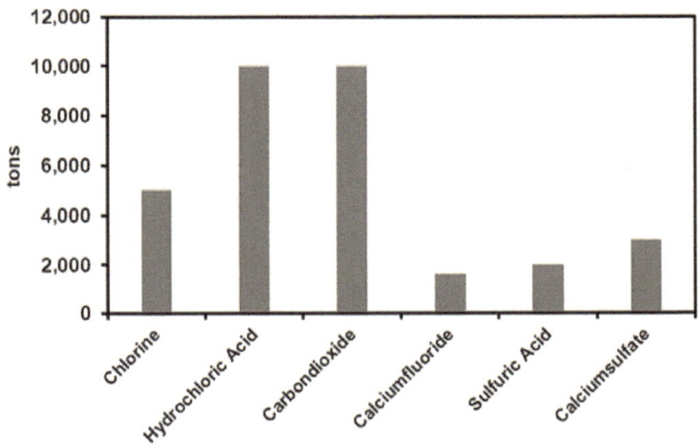

Figure 9.77 Environmental benefit from pyrolyzing perfluoropolymers back to TFE [3]

With all the new recycling options, the fluoropolymer industry can establish closed loops throughout the whole value chain with close to zero emissions for perfluorinated polymers. A closed fluorine cycle (see Figure 9.78) with optimized raw material streams, sound energy balances, and low environmental burdens will be also the target for partially fluorinated materials.

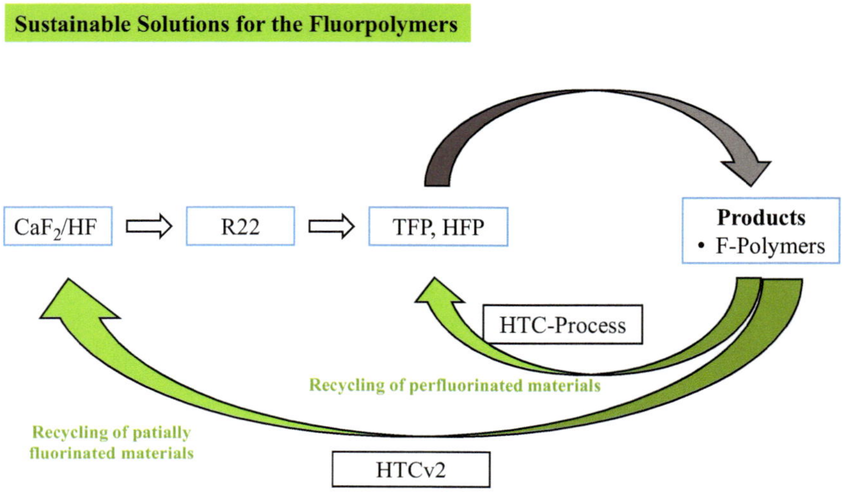

Figure 9.78 Sustainable solutions for fluoropolymers [3]. HTC-Process = PTFE depolymerization, HTCv2-Process = Conversion of partially fluorinated materials

References for Section 9.6

[1] Weltweite und europäische Kunststoffproduktion in den Jahren von 1950 bis 2019, *Statista, https://de.statista.com/statistik/daten/studie/167099/umfrage/weltproduktion-von-kunststoff-seit-1950/*

[2] Krämer, R., Schlipf, M.: Fluorpolymere: Gesellschaftliche Megatrends geben Fluorpolymeren Aufschwung, *Kunststoffe*, (October 2016), pp. 110–115, *https://multimedia.3m.com/mws/media/14309860/article-about-fluoropolymers-gesellschaftliche-megatrends-geben-fluorpolymeren-aufschwung.pdf*

[3] Dams, R., Hintzer. K.: Industrial aspects of fluorinated oligomers and polymers, in *Fluorinated Polymers, Volume 2: Applications*, (2016), RSC Polymer Chemistry Series No. 24, The Royal Society of Chemistry

[4] Carlson, P., Schmiegel, W.: Fluoropolymers, Organic, in *Ullmann's Encyclopedia of Industrial Chemistry*, 6th edn (2006), Bohnet, M. (Ed.), Wiley-VCH, Weinheim, Germany, pp. 10230–10269

[5] Drobny, J. G., *Technology of Fluoropolymers*, (2009). Taylor & Francis CRC, Press. Available online at *http://books.google.de/books?id=RXexW4aEYe8C.*

[6] Abraham, V.: Comparative life cycle assessment and environmental impacts of PTFE production and recycling, Masters thesis, University Bayreuth, Germany 2013

[7] Schmittinger, P., et al.: Chlorine, in *Ullmann's Encyclopedia of Industrial Chemistry*, 6th ed., Bohnet, M. (Ed.), Wiley-VCH, Weinheim, Germany, 2006, pp. 5362–5457

[8] Jess, A., Wasserscheid, P.: *Chemical Technology*, Wiley-VCH, 2013

[9] Shen, L., Nieuwlaar, E., Worrell, E., Patel, M.K.: Life cycle energy and GHG emissions of PET recycling: Change-oriented effects, *International Journal of Life Cycle Assessment*, (2011) 16, pp. 522–536

[10] Patel, M.: *KEA (Cumulative Energy Demand) für Produkte der Organischen Chemie*, Working paper of project no. 104 01 123 for the German Environment Agency (Bundesumweltamt): Erarbeitung von Basisdaten zum Energieaufwand und der Umweltbelastung von energieintensiven Produkten und Dienstleistungen für Öko-Bilanzen und Öko-Audits, Fraunhofer-Institut für Systemtechnik und Innovationsforschung (FhG-ISI), Karlsruhe, Germany, 1999

[11] Weible, S., Bauer, I.: *Background Report for the LCA. ETFE Construction Element. Environmental Product Declaration for ETFE construction element Project phase "LCA"*, 3M Dyneon, Nowofol, Vector Foiltec, 2011

[12] Ashby, M.F.: *Materials Selection in Mechanical Design*, Butterworth-Heinemann (Elsevier), Oxford, 2017

[13] *Critical Raw Materials for the EU*, Report of the ad-hoc working groups on defining critical raw materials, European Commission, 2010

[14] Hochleistungskunststoffe aus Burgkirchen, *3M Press release*, (2020) *https://www.3mdeutschland.de/3M/de_DE/presse-de/pressemeldungen/ganzemeldung/?storyid=f29c372a-a0ce-4ba7-96b3-758490c120d4*

[15] SGL und Dyneon werten Fluorpolymere auf, *Chemie Technik*, (2017) *https://www.chemietechnik.de/markt/sgl-und-dyneon-werten-fluorpolymere-auf.html*

[16] Aschauer, S.: Untersuchung zur Zersetzung von PTFE und PTFE-co-Polymeren in der Wirbelschicht, Diploma Thesis, University of Bayreuth, Germany, 2008

[17] *3M Dyneon Fluoropolymers: Up-cycling – Closing the Loop*, 3M, 2016, *https://multimedia.3m.com/mws/media/907323O/up-cycling-fluoropolymers-brochure.pdf?fn=Up-Cycling_Brochure_EN.pdf*

[18] Errede, L.A.: The application of simple equations for calculating bond dissociation energies to thermal degradation of fluorocarbons, *Journal of Organic Chemistry*, (1962) 27, pp. 3425–3430

[19] Fock, J.: On the influence of carbon black on the thermal degradation of polytetrafluoroethylene, *Journal of Polymer Science Part B: Polymer Letters*, (1968) 6, pp. 127–131

[20] Puts, G., Crouse, P., Ameduri, B.: *Handbook of Fluoropolymer Science and Technology*, John Wiley and Sons, 2014, Chapter 5

[21] Hintzer, K., Schwertfeger, W.: *Handbook of Fluoropolymer Science and Technology*, (2014), John Wiley and Sons, Chapter 21

[22] Koltzenburg, S., et al.: *Polymere – Synthese, Eigenschaften und Anwendungen*, Springer-Verlag, Berlin, Heidelberg, 2014

[23] *Recycling von Fluorpolymeren*, (Technisches Merkblatt 10) (2017), pro-K Fluoropolymergroup, *https://www.pro-kunststoff.de/assets/Merkbl%C3%A4tter%20und%20Co/FP%20TM-10-Recycling-von-Fluorkunststoffen.pdf*

[24] Henkel, R.: *Gore-Tex: Keine Gefahr beim Verbrennen von PTFE*, (2019), *https://www.ispo.com/maerkte/gore-tex-keine-gefahr-beim-verbrennen-von-ptfe*

[25] Stubenvoll, J., Bohmer, S., Szednyj, I.: *Stand der Technik bei Abfallverbrennungsanlagen*, Bundesministerium für Land- und Forstwirtschaft, Umwelt und Wasserwirtschaft, Wien, 2002

[26] Wang, L., Li, Q., Mei, T., Shi, L., Zhu, Y., Qian, Y.: A thermal reduction route to nanocrystalline transition metal carbides from waste polytetrafluoroethylene and metal oxides, *Materials Chemistry and Physics*, (2012) 137, pp. 1–4

[27] Grot, S., Grot, W.: Recycling of used perfluorosulfonic acid membranes, US7255798B2

[28] Madorsky, S.L., Hart, V.E., Straus, S., Sedlak, V.A.: Thermal degradation of tetrafluoroethylene and hydrofluoroethylene polymers in a vacuum, in: *Journal of Research of the National Bureau of Standards*, (1953) 51, pp. 327–333

[29] Merkel, J.: Pyrolyse von Polytetrafluorethylen, Polyamiden und Polyethanen in der Wirbelschicht-Massen und Stoffbilanzierungen, Dissertation, Universität Hamburg, 1982

[30] Hintzer, K., et al.: Process of making fluoroolefins by thermal decomposition of fluorinated materials (2010), 3M, WO 2010/039820

[31] Schmidt-Rodenkirchen, A., Aschauer, S., Gerdes, T., Hintzer, K., Zipplies, T., Willert-Porada, M., *Rückgewinnung Fluorierter Monomere aus Reststoffen*, (2010), final report 25198-31

[32] Hirschberg, M., et al.: Upcycling perfluoropolymers into fluorinated olefins (2021), 3M, WO 2021/165923

[33] BMBF-Abschlussbericht, *Rückgewinnung von Calciumfluorid als Sekundärrohstoff für Fluorpolymere*, (2015), Az. 003R080D

10 Dissolution

Andreas Mäurer, Martin Schlummer

10.1 Classification of Polymer Recycling by Dissolution

Plastic recycling by dissolution refers to recycling technologies that make use of a physical interaction of liquids and thermoplastic polymers, to allow dissolution without degrading the polymer chain to monomers or oligomers. In their dissolved state polymers, can be separated from foreign materials or foreign polymers easily and effectively. In a final stage, polymer and liquid are separated again, allowing circulation of the liquid within the process and producing a solid, clean recycled polymer.

In the literature and patents, dissolution has been termed "selective dissolution" [1], "selective dissolution and precipitation" (DSP) [2], or "solvent-based purification" [3], and is related to trade names like CreaSolv®, PureCycle Technologies℠, Newcycling®, and VinyLoop.

Despite its physical nature, these technologies have often (and falsely) been classified as chemical recycling [4]. In chemical recycling, degradation of polymers is targeted in order to provide feedstock chemicals that may be used for gasoline production, or to enter a polymer value chain from the very beginning. In contrast to dissolution, the product of a chemical recycling process is not a polymer that can directly be substituted for virgin polymers in plastic converters.

Mechanical recycling also applies physical principles in sorting and refining plastics, like spectroscopy [5], size reduction [6], magnetic, eddy current and washing pre-treatment processes for removal of foreign materials, and finally a remelting step, preferably including a melt filtration step [7]. In contrast to dissolution, pre-treatment measures have to be very effective, as the final re-melting/melt filtration step allows for a limited amount of non-target polymers only (<2%). The dissolution principle works with much higher loads of-non target components in the feed, but is most effective with a maximum of 40%.

Between classical mechanical and dissolution-based recycling, another field of technology is currently evolving, which mainly addresses multilayered polymer structures [8, 9]. Such delamination approaches attack adhesive layers in order to weaken the multilayer structure and to enable mechanical separation of the adjacent films. In these particular cases, solvents interact with the component that is not recycled, whereas the recycled materials are only subjected to a kind of surface cleaning.

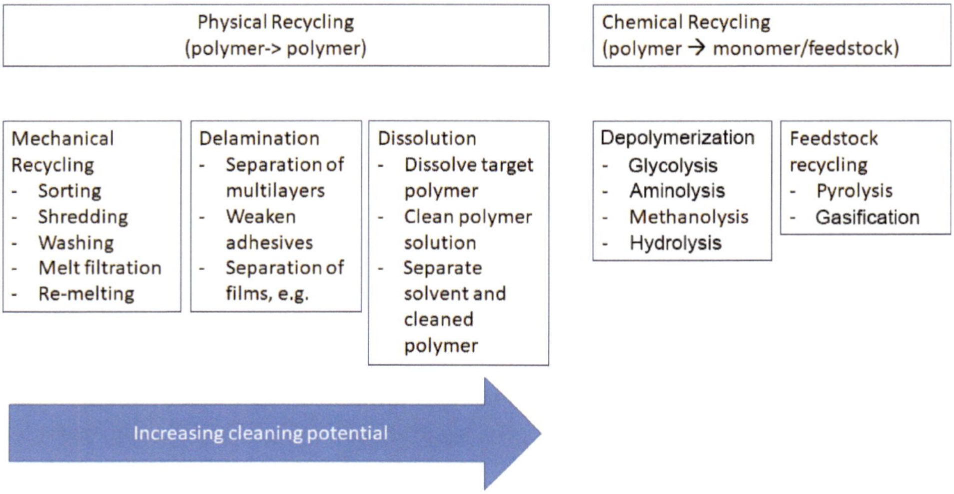

Figure 10.1 Differentiation of physical and chemical recycling processes

10.2 Solvent Selection

The understanding of interactions between specific polymers and solvents is key for dissolution-based polymer recycling. The basic principles of solvent selection have been recently reviewed by Zhao et al. [3]. First of all, the authors state, the closer the solvent and polymer are in polarity, the easier the polymer will dissolve in that specific solvent. In such a case, the interaction between the polymer and solvent prevails over the cohesion of the polymer macromolecules. To describe both polarity of solvents and polymers, several solubility parameter systems have been developed, with the Hansen solubility parameters (HSPs) being the most prominent ones for dissolution-based polymer recycling [10]. HSP values take into account three different types of interaction, namely dispersive van der Waals (δ_D), polar (δ_P) and H-bond forces (δ_H). A solubility parameter "distance" R_a between a polymer (pol) and the solvent (sol) can be calculated:

$$R_a^2 = 4\left(\delta_{D,pol} - \delta_{D,sol}\right)^2 + \left(\delta_{P,pol} - \delta_{P,sol}\right)^2 + \left(\delta_{H,pol} - \delta_{H,sol}\right)^2$$

The relative energy difference (RED) is defined as the ratio of R_a to R_0, where R_0 is the radius of the solubility sphere of a given polymer and R_a is the distance from a given solvent point to the center of the sphere. Thus, the RED value indicates if a polymer dissolves in a solvent, with RED <1 indicating dissolution, RED values around 1 hinting to partial solubility, and RED >1 showing a weak solvent-polymer interaction [3].

Thus, HSP may help identifying selective solvents for a given target polymer; however, the issue of selectivity is strongly influenced by the kind of polymers and the complexity of mixed plastics being present in a specific feedstock. Selectivity is crucial for mixtures of polymers with very similar polarities like ABS and HIPS, which are potentially present in plastics from electronic waste, or for different polyolefins in plastic packaging.

In addition to solvents, anti-solvents are often used in dissolution and precipitation processes. Thus, the selection of solvents also includes finding of liquids that do not interact with the polymer and lead to precipitation if they are added to a polymer solution. Zhao et al. [3] have collected typical solvents and anti-solvents for a list of standard polymers from the literature.

Besides selectivity, however, there are some other crucial factors for the selection of effective solvents, including toxicity, safety (particularly in industrial applications, where contaminated materials are present, heavy machinery is used, and where laboratory levels of cleanliness and safety are difficult or impossible to achieve), and potential for recovery. Toxicity is crucial, and not only for the working environment of the recycling plant itself. Low amounts of solvents will remain in the recycled polymer after the final drying step, typically well below 1000 ppm. If the selected solvent (or anti-solvent) is toxic or listed as a substance of very high concern (SvHC), such low levels may hinder a second-life application of the recycled polymer [11]. Safety issues reflect that most solvents used in dissolution processes are organic liquids and thus flammable. However, the higher the flash and self-ignition temperatures of specific solvents, the safer is their application in industrial-scale recycling facilities, and the lower the cost of explosion-proof equipment. The almost complete recovery of solvents (and anti-solvents) from the polymer solution as well as from solvent-containing side-streams of the process, however, is also important for economic closure of the solvent cycles. In this respect, high-boiling points (correlating with high - and therefore safe - flash points and ignition points) hamper most evaporation-based recovery processes. However, boiling points that are too low may also complicate recovery, as they increase the energy demand for complete solvent condensation [12]. Finding a good balance between selectivity, toxicity, safety, and recovery has become the business model of specialized solvent suppliers, such as the company CreaCycle, which offers specific solvents for the CreaSolv® process [13].

In addition to the above-mentioned technical selection criteria, economic and market aspects play a decisive role [14]. Solvents traded at low global volumes are in general more expensive than commodity products, and this might be especially relevant for upcoming groups of solvents like ionic liquids, solvents extracted from natural resources, or very specific fine chemicals [15]. Depending on the solvent-loss rate of a dissolution-based process (which itself is dependent on the choice of solvents), solvent cost may reach a substantial share of the total operating cost. Furthermore, availability not only implies cost of a solvent. The global/regional production capacities require a sufficient volume that can easily buffer the requested volumes of a dissolution process. Otherwise, missing investments in solvent supply capacities may then directly jeopardize an industrial scale dissolution process.

Last but not least, environmental factors may influence solvent selection. These include solvent parameters that may negatively affect LCAs (life cycle assessments) of dissolution processes, which take the environmentally adverse effects of solvent production into account. This gave rise to the employment of "green" solvents from natural sources like geraniol, limonene, cymene, menthol, eucalyptol, or similar compounds [14, 16]. However, in this respect, it must be stated that a natural origin does not directly imply low toxicity, safe handling, long-term stability, suitable availability, and a reasonable price range. All these factors need to be checked carefully.

From the above discussion of solvent selection, it will be evident that a clear-sighted choice of solvent and anti-solvents has a dramatic effect on the economic and technical success of a recycling process, and should take into account all considerations, even if they are partially indicating contradictive solvent properties.

■ 10.3 Unit Operations

Dissolution processes describe (in general) a sequence of unit operations starting with a dissolution step followed by cleaning and occasionally precipitation operations. The final steps include the separation of solvents from the recycled polymer as well as from any side-streams, in most cases at elevated temperatures.

A typical and universal dissolution scheme is sketched in Figure 10.2, depicting the CreaSolv® process, which has most recently been described as the most versatile dissolution and precipitation process [11].

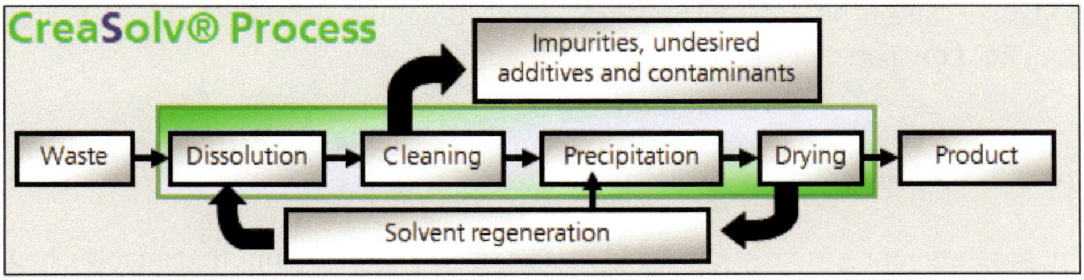

Figure 10.2 Typical sketch of unit operations of a dissolution process, exemplified by the CreaSolv® process

Each unit operation may involve even more than one technology, and requires specific technical equipment that allows safe treatment of polymers and solvents at different temperature and pressure levels.

10.3.1 Dissolution

Two thermodynamic transport processes are involved in polymer dissolution, namely solvent diffusion and chain disentanglement. Consequently, the dissolution time decreases with the decreasing particle size and increasing diffusivity of polymers, because of the increased mass transfer rate [17]. Due to the rise of diffusivity with temperature, dissolution time also decreases with temperature (e.g. [18, 19]).

The dissolution process of a polymer starts from the surface, whereas a sequence of different shells evolves comprising an infiltration layer, a swollen layer, a gel layer, and the dissolved polymer, each containing increasing shares of solvent [3]. To keep the gel layer as small as possible, dissolution is mostly supported by agitating the solvent–polymer mixture.

Batch dissolvers, tandem operation, and continuous dissolution aggregates have been discussed for technical-scale unit operation. However, in order to choose the most suitable types and geometries of dissolution aggregates, a number of aspects need to be considered, including procedures for safe and explosion-proof filling and operation, as well as the volume and physical state of the undissolved rejects.

It is important to understand that some pairs of solvents and polymers do not just interact physically, and may undergo chemical reactions that lead to degradation of the polymer chain. Knappich et al. [19] report on the dissolution of polybutylene terephthalate using two polar solvents. Whereas the polar aprotic solvent showed a good performance with negligible degradation effects in nitrogen atmosphere and minor degradation in oxygen atmosphere, significant polymer degradation was observed in both atmospheres when using the protic polar solvent. With oxygen, degradation started after 10 min of dissolution time and decreased the molecular weight of the polymer by a factor of more than 3.

10.3.2 Cleaning

In dissolution processes, cleaning is a substantial operation unit that enables such processes to provide purified recycled polymers. Cleaning separates the dissolved target polymer from undissolved solids or co-dissolved non-target compounds. Both kind of side-streams may also be further purified and valorised by downstream processing [19–21].

Separation of undissolved solids is generally performed by mechanical solid–liquid separation processes such as coarse (~500 µm) and fine (<100 µm) filtration making use of sieves, filters of different kinds, or by sedimentation with or without gravimetric support (e.g. [22–24]). Most scientific papers on dissolution processes report laboratory-scale filtration equipment (e.g. [25, 26]) operating between 20–100 µm. Technical-scale equipment is only rarely described, since the number of dissolution-based processes at pilot demo or even commercial scale is still small, and technical details are kept confidential. Thus, Naviroy et al. [27] described the use of a self-cleaning high-performance melt filter device that is tailor-made for polymer solutions. An older patent from 1998 describes the use of a series of filters (200 µm down to 5 µm) as well as the application of filter aids like activated carbon or diatomaceous earth to adsorb co-dissolved and undissolved matter before filtering them off by filters down to 0.01 µm. These two extreme kinds of filters indicate the potential of cleaning processes on the one hand, and show on the other hand the wide range of polymer solution viscosities these reports deal with.

Centrifugation was also proposed to clean polymer solutions on the laboratory and small technical scales [24, 28], and exhibited comparable performance to fine filtration [24]. Centrifugal decanting was also reported for the 10 kt/a commercial-scale PVC recycling process VinyLoop. Solid–liquid separation, however, fails if cleaning of co-dissolved contaminants or impurities is requested. In such cases, the dissolved state of polymers enables another cleaning principle: extraction. Extractive cleaning of dissolved polymers applies when a second liquid phase is introduced into the system. This may work by precipitation of the dissolved polymer, which builds a gel or particle phase in a liquid matrix solvent (and anti-solvent, see Section 10.3.3). Small molecules, like (for example) legacy additives, do not precipitate and can be eliminated with the liquid phase. This cleaning approach is also used for extraction of polymer additives in chemical analysis (e.g. [29]). In dissolution-based recycling it has been proposed to eliminate brominated flame retardants from plastics from waste electric and electronic equipment (WEEE) [30] or expanded polystyrene (EPS, XPS) [20]. Wagner and Schlummer [31] also proposed this approach for the extraction of plasticizers from waste soft-PVC, as currently explored in the European Horizon 2020 project "Circular Flooring".

10.3.3 Precipitation

Precipitation may assist extractive cleaning of dissolved polymer as described above. It may, however, also be used to facilitate or speed up solvent recovery from the dissolved polymer.

Precipitation is initiated by changing the dissolution condition, either by reducing the temperature (e.g. [32]) or by adding an anti-solvent [21, 33–35]. In the latter case, solvent recovery needs to separate solvent and anti-solvent sufficiently to reapply them in the process.

10.3.4 Drying and Solvent Recovery

Efficient and cost-effective separation of cleaned polymers from solvent (and anti-solvent) is key for dissolution processes, in order that marketable products can be provided. In these operations, temperature- and vacuum-assisted evaporation of the solvents applies in general. As many papers and patents on dissolution processes present laboratory-scale experiments, non-scalable laboratory equipment (such as vacuum-drying cabinets) is often used to perform solvent separation [24, 25, 34, 36].

Some papers propose (small-) technical scale equipment, one of these being Knappich et al. [36], who evaporated solvent from dissolved PBT in a non-agitated heated vacuum chamber, before completing the drying process in a state-of-the-art melt degassing extruder. Knappich et al. [22] removed the solvent from an ABS solution by a wall-heated stainless-steel vacuum dryer with 50 L capacity equipped with special agitator units for high-viscosity applications. Residual drying was performed by a melt degassing extruder. A 1994 patent of Nauman et al. [1] proposed flash-evaporation of overheated, pressurized polymer solutions. Cella et al. [37] investigated the drying of polystyrene dissolved in ethyl acetate using either direct degassing extrusion or a tubular evaporator. Drying in the tubular evaporator was incomplete, and although extrusion drying was almost complete, temperatures of up to 200 °C caused slight polymer degradation (yellowing).

■ 10.4 Dissolution Technologies in Circular Value Chains

In general, dissolution-based recycling may be considered if -mechanical recycling fails. This is the case for complex mixtures of polymers, composites, voluminous materials, and plastics that contain contaminants or so-called legacy additives, which are considered to be hazardous today, but were not when the plastic was made [31].

10.4.1 Waste Streams Suggested for Dissolution in the Literature

There are an increasing number of studies reporting dissolution-based polymer recycling approaches. They cover target polymers being present in complex mixtures of plastic and eventually other materials, as for example waste electric and electronic equipment (WEEE). Another group of reports targets composite materials, e.g. composites of two or more polymers (e.g. multilayer packaging films), composites of polymers and metals (e.g. metal-plating), or composites of polymer and other materials (e.g. fiber composites). Voluminous waste streams like foams (packaging or insulation) made from EPS or XPS make up a third field of operation. A fourth are waste plastics containing legacy additives or hazardous compounds or off-odors, such as polybrominated diphenyl ethers in WEEE, DEHP and HBCDD in building and construction waste (PVC flooring, external thermal insulation composite systems (ETICS)), or odorous post-consumer plastic packaging.

Depending on the type of waste and its origin (post-consumer vs post-industrial waste), dissolution has been proposed with different degrees of purification efforts, ranging from simple filtration to comprehensive multistep cleaning including filters, centrifuges, and extractive cleaning.

Based on the literature mentioned above, Table 10.1 summarizes examples of waste streams already treated by dissolution, the targeted polymers, and (if applicable) other target materials recovered by this process.

Table 10.1 Waste Streams Treated with Dissolution-Based Polymer Recycling Processes in the Literature

Waste stream	Target polymer	Other target materials	Purified from	Scale of operation	Reference
Packaging					
Multilayer packaging	LDPE, PP	Al	Adhesives	Lab	[14, 15]
Food packs	PE, PP, EPSP, PS, PET	–	–	Lab	[40]

Waste stream	Target polymer	Other target materials	Purified from	Scale of operation	Reference
Pharma packs	PE, PP, PVC	Al	–	Lab	[40]
Beverage cartons	LDPE	Al, board	–	Lab	[21]
Mix of LDPE, HDPE, and PP packaging	LDPE, HDPE, PP	–	–	–	[1, 41]
				Technical	[27]
Composites					
Mix of PA 6 and PA 6.6	PA 6, PA 6.6		–	Lab	[34]
Polymer metal composite	PBT	Cu	–	Small technical	[19]
Metal-plated ABS	ABS	Ni, Cu	–	Small technical	[36]
PVC-coated PET fibers	PVC	PET	–	–	[18]
Wind turbine blades	Elium®	Glass fiber	–	–	[39]
Pigmented ABS	ABS	–	Pigments	–	[24]
Waste PVC	PVC	–	–	–	[38]
Waste foams					
EPS	–	–	–	–	[12, 16]
GPPS, EPS	PS	–	–	–	[37]
Waste streams containing legacy additives or hazardous compounds					
EPS	PS	Br	HBCD	Lab	[20]
WEEE	ABS, PS, PC	Br	BFR	Lab	[28, 30]
WEEE	ABS, PS, PC	–	–	Lab	[40]
Flooring	PVC	Plasticizer	Phthalates	–	[31]

The waste streams presented describe the different polymer waste regimes, including packaging (PE, PP, EPS), WEEE (ABS, PS, PC/ABS), post-industrial automotive parts (PBT, ABS, PA), other composites (ABS, Elium®), as well as building and construction (PVC, EPS, XPS).

10.4.2 High-Level Applicability of Recycled Materials

The use of dissolution-based recycling is intended to generate high-purity recovered polymers, and enable production of high-quality recycled grades that exhibit qualities equal or very close to virgin polymers. Such quality levels will be key for transferring our current linear value chains into circular economies, since only they enable re-use in the same application and thus, closed-loop recycling strategies applicable in many different market sectors.

However, as apparent in Table 10.1, reports on laboratory-scale dissolution processes by far exceed reports on larger-scale treatments. This may due to a general cautious communication strategy for technical-scale dissolution equipment, in order to prevent disclosure of intellectual property rights. Further, it indicates the highly competitive situation for dissolution technologies during the last decade.

As a result of numerous European and international projects that covered dissolution as an alternative to mechanical recycling, initial reports on the high-level application of dissolution-based recyclates have been shared with the scientific community in the last few years. With respect to the European Horizon 2020 project Multicycle, Fraunhofer and Amcor reported the application of recycled polyolefins from German end-of-life packaging waste in the production of new non-food packaging films, thus indicating their high film-type quality level. In the European CloseWEEE project, PC/ABS was recovered from fractions of small domestic WEEE appliances and used test-wise for the body of electric toothbrushes. With respect to construction waste, recycled bromine-free polystyrene was physically recycled from flame-retarded ETICS, and could be re-applied in the production of new EPS insulation bricks (European Life project PolyStyreneLoop).

10.4.3 Industrial-Scale Dissolution Processes

The first large-scale installation for dissolution processes was built in Italy for recycling of PVC, when strong pressure was put on the vinyl industries in about the year 2000. The VinyLoop plant in Ferrara reached finally a capacity of 10 kt/a and provided clear proof of the technical and economic feasibility of dissolution [38], despite the competitiveness of the PVC commodity market. However, due to additional pressure on REACH-regulated phthalate-based plasticizers in waste plasticized PVC, VinyLoop stopped operation in 2018. Dissolution-based PVC recycling is currently continued under trade names Texyloop® and Polyloop, with the latter providing mobile equipment for production on the 500 t/a scale since 2021.

In 2011, APK AG in Germany, started a commercial scale dissolution process for PE from PE-based composite packaging materials. In the following years, different post-industrial PE feedstocks were tested, before the plant capacity was increased to 8 kt/a in 2018.

Another industrial dissolution process for packaging started operation in Indonesia in 2018, driven by the company Unilever, based on patents and developments by Fraunhofer IVV. Flexible multilayer packaging derived from Indonesian landfills was used as feedstock and processed to recycled polyolefins. Recycled polymers were successfully re-used in new flexible non-food packaging items (sachets).

Using pressurized dissolution, Procter & Gamble introduced the PureCycle technology for PP and announced the building of a 50 kt/a facility in Ohio to recover PP.

The PolyStyreneLoop project scaled up Fraunhofer IVV's CreaSolv® process, and erected a 3 kt/a plant in Terneuzen (Netherlands) for the recovery of PS from waste EPS insulation panels, including removal of the phased-out brominated flame retardant HBCDD (subsequently subjected to destruction and bromine recovery). The grand opening ceremony of the plant was celebrated in summer 2021.

The upcoming widespread availability of industrial-scale dissolution facilities indicates the increased interest of the market in its outstanding purification potential and a market request for purer and better-performing recycled polymers. After years of development, dissolution is now perceived as sufficiently mature for technical implementation in various value chains.

Developers and operators of such technologies have just started networking in order to jointly promote the benefits of physical recycling by dissolution, and to avoid misinterpretation of such technologies as "chemical recycling". Increased knowledge will be gathered during all these industrial-scale operations, and this will further enlarge knowhow on processes and equipment. In doing so, it will support dissolution as one of the most promising technologies enabling circularity of plastics.

References for Chapter 10

[1] Nauman, E. B., Lynch, J. C.: Polymer recycling by selective dissolution (Patent EP0491836B1), 1994

[2] Papaspyrides, C. D., Kartalis, C. N.: A model study for the recovery of polyamides using the dissolution/reprecipitation technique. *Polymer Engineering and Science*, 40 (2000) 4, pp. 979–984

[3] Zhao, Y. B., Lu, X. D., Ni, H. G.: Solvent-based separation and recycling of waste plastics: A review. *Chemosphere*, (2018) 209, pp. 707–720

[4] Schlummer, M., Fell, T., Mäurer, A., Altnau, G.: The role of chemistry in plastics recycling: A comparison of physical and chemical plastics recycling. *Kunststoffe International*, 5/2020, pp. 34–37

[5] Araujo-Andrade, C., Bugnicourt, E., Philippe, L., Rodriguez-Turienzo, L., Nettleton, D., Hoffmann, L., Schlummer, M.: Review on the photonic techniques suitable for automatic monitoring of the composition of multi-materials wastes in view of their posterior recycling. *Waste Management Research*, (2021) 39, pp. 631–651

[6] Maisel, F., Chancerel, P., Dimitrova, G., Emmerich, J., Nissen, N. F., Schneider-Ramelow, M.: Preparing WEEE plastics for recycling – How optimal particle sizes in pre-processing can improve the separation efficiency of high quality plastics. *Resources, Conservation and Recycling*, (2020) 154, 104619

[7] Ragaert, K., Delva, L., Van Geem, K.: Mechanical and chemical recycling of solid plastic waste. *Waste Management*, (2017) 69, pp. 24–58

[8] Kaiser, K.: Recycling of polymer-based multilayer packaging: A review. *Recycling*, 3 (2018) 1, p. 26

[9] Mumladze, T., Yousef, S., Tatariants, M., Kriukiene, R., Makarevicius, V., Lukosiute, S. I., Bendikiene, R., Denafas, G.: Sustainable approach to recycling of multilayer flexible packaging using switchable hydrophilicity solvents. *Green Chemistry*, (2018) 20, pp. 3604–3618

[10] Hansen, C.M., Just, L.: Prediction of environmental stress cracking in plastics with Hansen solubility parameters. *Industrial & Engineering Chemistry Research*, (2000) 40, pp. 21–25

[11] Vollmer, I., Jenks, M.J.F., Roelands, M.C.P., White, R.J., van Harmelen, T., de Wild, P., van der Laan, G.P., Meirer, F., Keurentjes, J.T.F., Weckhuysen, B.M.: Beyond mechanical recycling: Giving new life to plastic waste. *Angewandte Chemie, International Edition*, 59 (2020) 36, pp. 15402–15423

[12] Singhal, R., Ishita, I., Sow, P.K.: Integrated polymer dissolution and solution blow spinning coupled with solvent recovery for expanded polystyrene recycling. *Journal of Polymers and the Environment*, (2019) 27, pp. 1240–1251

[13] The CreaSolv® Process, CreaCycle, *www.creacycle.de/en/the-process*, accessed March 2022

[14] Gorre, R.C., Tumolva, T.P.: Solvent and non-solvent selection for the chemical recycling of waste polyethylene (PE) and polypropylene (PP) metallized film packaging materials. *IOP Conference Series: Earth and Environmental Science*, 463 (2020) 1

[15] Samorì, C., Cespi, D., Blair, P., Galletti, P., Malferrari, D., Passarini, F., Vassura, I., Tagliavini, E.: Application of switchable hydrophilicity solvents for recycling multilayer packaging materials. *Green Chemistry*, (2017) 19, pp. 1714–1720

[16] Garcia, M.T., Gracia, I., Duque, G., de Lucas, A., Rodriguez, J.F.: Study of the solubility and stability of polystyrene wastes in a dissolution recycling process. *Waste Management*, (2009) 29, pp. 1814–1818

[17] Miller-Chou, B.A., Koenig, J.L.: A review of polymer dissolution. *Progress in Polymer Science*, (2003) 28, pp. 1223–1270

[18] Grause, G., Hirahashi, S., Toyoda, H., Kameda, T., Yoshioka, T.: Solubility parameters for determining optimal solvents for separating PVC from PVC-coated PET fibers. *Journal of Material Cycles and Waste Management*, (2017) 19, pp. 612–622

[19] Knappich, F., Hartl, F., Schlummer, M., Mäurer, A.: Complete recycling of composite material comprising polybutylene terephthalate and copper. *Recycling*, 2 (2017) 2, p. 9

[20] Schlummer, M., Maurer, A., Wagner, S., Berrang, A., Fell, T., Knappich, F.: Recycling of flame retarded waste polystyrene foams (EPS and XPS) to PS granules free of hexabromocyclododecane (HBCDD). *Advances in Recycling & Waste Management*, (2017) 2

[21] Georgiopoulou, I., Pappa, G.D., Vouyiouka, S.N., Magoulas, K.: Recycling of post-consumer multilayer Tetra Pak® packaging with the selective dissolution-precipitation process. *Resources Conservation and Recycling*, (2021) 165, 105268

[22] Knappich, F., Schlummer, M., Mäurer, A., Prestel, H.: A new approach to metal- and polymer-recovery from metallized plastic waste using mechanical treatment and subcritical solvents. *Journal of Material Cycles and Waste Management*, (2018) 20, pp. 1541–1552

[23] Hoehn, A.F.: Polymer deformulation by solvent solution filtration. WO1998051717A1

[24] Arends, D., Schlummer, M., Maurer, A.: Removal of inorganic colour pigments from acrylonitrile butadiene styrene by dissolution-based recycling. *Journal of Material Cycles and Waste Management*, (2012) 14, pp. 85–93

[25] Hadi, A.J., Najmuldeen, G.F., Ahmed, I.: Potential solvent for reconditioning polyolefin waste materials. *Journal of Polymer Engineering*, (2012) 32, pp. 585–591

[26] Gutierrez, C., Garcia, M.T., Gracia, I., de Lucas, A., Rodriguez, J.F.: Recycling of extruded polystyrene wastes by dissolution and supercritical CO2 technology. *Journal of Material Cycles and Waste Management*, (2012) 14, pp. 308–316

[27] Naviroy, P., Treacy, J., Urffer, C.: Chemical recycling of plastics by dissolution. Senior Design Reports (CBE) (2019) 115. *https://repository.upenn.edu/cbe_sdr/115*

[28] Schlummer, M., Popp, L., Trautmann, F., Zimmermann, D., Maurer, A.: Recovery of bromine and antimony from WEEE plastics. *Electronics Goes Green*, (2016), pp. 1–5

[29] Ugduler, S., Van Geem, K.M., Roosen, M., Delbeke, E.I.P., De Meester, S.: Challenges and opportunities of solvent-based additive extraction methods for plastic recycling. *Waste Management*, (2020) 104, pp. 148–182

[30] Schlummer, M., Maurer, A., Leitner, T., Spruzina, W.: Report: Recycling of flame-retarded plastics from waste electric and electronic equipment (WEEE). *Waste Management Research*, (2006) 24, pp. 573–583

[31] Wagner, S., Schlummer, M.: Legacy additives in a circular economy of plastics: Current dilemma, policy analysis, and emerging countermeasures. *Resources, Conservation and Recycling*, (2020) 158, 104800

[32] Fischer, E., Schleich, R., Kaiser, T., Bungert, B., Thiele, A., Gorski, G.: Method for recovering mixed plastic matter (Patent WO2002048244A1). Der Grüne Punkt – Duales System Deutschland AG, 2002

[33] Achilias, D.S., Giannoulis, A., Papageorgiou, G.Z.: Recycling of polymers from plastic packaging materials using the dissolution-reprecipitation technique. *Polymer Bulletin*, (2009) 63, pp. 449–465

[34] Kartalis, C.N., Poulakis, J.G., Tsenoglou, C.J., Papaspyrides, C.D.: Pure component recovery from polyamide 6/6 6 mixtures by selective dissolution and reprecipitation. *Journal of Applied Polymer Science*, (2002) 86, pp. 1924–1930

[35] Hadi, A.J., Najmuldeen, G.F., Bin Yusoh, K.: Dissolution/reprecipitation technique for waste polyolefin recycling using new pure and blend organic solvents. *Journal of Polymer Engineering*, (2013) 33, pp. 471–481

[36] Knappich, F., Klotz, M., Schlummer, M., Wolling, J., Maurer, A.: Recycling process for carbon fiber reinforced plastics with polyamide 6, polyurethane and epoxy matrix by gentle solvent treatment. *Waste Management*, (2019) 85, pp. 73–81

[37] Cella, R.F., Mumbach, G.D., Andrade, K.L., Oliveira, P., Marangoni, C., Bolzan, A., Bernard, S., Machado, R.A.F.: Polystyrene recycling processes by dissolution in ethyl acetate. *Journal of Applied Polymer Science*, (2018) 135, 46208

[38] *PVC Recycling Technologies*, [Brochure], VinylPlus, 2015

[39] Cousins, D.S., Suzuki, Y., Murray, R.E., Samaniuk, J.R., Stebner, A.P.: Recycling glass fiber thermoplastic composites from wind turbine blades. *Journal of Cleaner Production*, (2019) 209, pp. 1252–1263

[40] Achilias, D.S., Antonakou, E.V., Koutsokosta, E., Lappas, A.A.: Chemical recycling of polymers from waste electric and electronic equipment. *Journal of Applied Polymer Science*, (2009) 114, pp. 212–221

[41] Hadi, A.J., Najmuldeen, G.F., Ahmed, I.: Polyolefins waste materials reconditioning using dissolution/reprecipitation method, *2nd International Conference on Chemistry and Chemical Process*, (Y. Dan (Ed.)), 2012, pp. 281–286

11 Thermochemical Processes

Dieter Stapf

A few polymer types preferentially decompose to their monomers in chemical processes. This so-called depolymerization is seen for polycondensates, as well as for PMMA, PS, and PTFE. Basically, pure polyesters like PET (just like polyamides and polyurethanes) can be depolymerized chemically by solvolysis. Thermal degradation of polystyrene yields styrene monomer and its derivatives when the heating rate and reactor residence time are moderate, and a similar reaction is observed for thermal decomposition of PMMA and PTFE. All other polymers decompose statistically, or initially form non-monomer products that are favored by the molecular structure, which results in a broad product spectrum. Also, process parameters such as temperature have a tremendous impact on the distribution of thermochemical degradation products because of chemical attack of intermediate species and subsequent reactions during radical-based decomposition. Key aspects of the degradation of pure polymers, and their products, are explained in more detail in Chapter 2 and Section 9.1.1).

When co-polymers, polymer mixtures, or polymers accompanied by other reactive components like biomass or halogens are to be recycled chemically, intermediate species interact during degradation. Due to modified degradation mechanisms, chemical recycling of mixtures results in a product spectrum that is different from that of pure polymers. Furthermore, plastic products themselves contain non-plastic additives, which can also be reactive to a large extent. Examples are fillers like lime, titanium white (and even carbon black), stabilizers, flame retardants, and polymerization catalyst residues. In addition, both process parameters and technology determine recycling product yields and quality.

As real waste rarely consists of pure substances, waste recycling processes striving for the pure polymer or the monomers have to deal with issues resulting from this fact. Waste composition and properties can hamper its processing, and in any case, mechanical recycling of plastics results in lower product quality than virgin polymer. Thermosets, elastomers, and composite materials cannot be recycled exclusively mechanically, but solvent extraction as well as solvolysis of suitable polymers requires well-defined waste feedstock. Thermochemical recycling basically overcomes these issues by transforming mixed waste into feedstock for new products with virgin quality. To achieve this, thermochemical processes must decompose polymers and non-plastic hydrocarbons to petrochemical feedstock as well as destroy contaminants and remove byproducts.

This chapter outlines thermochemical recycling processes for plastic waste. Key technologies are described and are integrated into full waste-to-chemicals process chains. Section 11.1 briefly covers the basic requirements of processing plastic waste, followed by Section 11.2 and Section 11.3 on the most important thermochemical recycling technologies and well-developed application examples. The aspect of mass and energy balance is emphasized, where this information is available, in order to address process-efficiency and provide data for sustainability evaluation. Section 11.4 gives an outlook on the application of technologies to the thermochemical recycling of mixed plastic waste. Parts of the technology subsections have been published elsewhere [1, 2].

11.1 Challenges and Approaches

11.1.1 Plastic Waste

The challenge of plastic waste recycling results from the nature of waste. It increases with the difficulty in collecting and sorting a specific polymer, and with the grade of functionalization of a specific plastic product. Mechanical recycling can be applied to thermoplastics that can be recovered as pure polymer, but any ingredient of the polymer product impacts the recyclate quality. Dissolution provides opportunities to physically recover polymers from higher functionalized plastic products with a defined (but not too complex) composition, including simple composite materials. The situation is similar for solvolysis, insofar as the polymer type is suitable to be depolymerized by a solvent. Thus, if applicable, solvolysis can recycle even specific thermosets, and as a chemical process it overcomes recyclate quality issues. However, solvent-based recycling processes require a constant waste quality, with each process being limited to a specific sortable product. By this, non-thermochemical plastic waste recycling is successful with post-consumer plastic products specifically from the packaging sector and with well-defined manufacturing residues. Opportunities to expand plastics recycling result from future product design-for-recycling, and from application of mechanical recycling and solvent-based technologies to well-defined post-consumer products that can be collected separately. Applications are given in the respective chapters of this book.

Most plastic waste is not easy to recycle. Many long-life products from the non-packaging sectors are highly functionalized, and post-consumer products appear as mixed waste with a broad and often unknown specification. In principle, thermochemical recycling can be applied to any such plastic waste, because composites can be thermally destroyed, organic and mineral/metal parts can be detached, and polymers and their ingredients can be decomposed to the level of small molecules that can be separated from each other. Major mixed-plastics waste streams are able to sort residues from lightweight packaging waste (LWP-SR), building demolition plastic wastes such as external thermal insulation composite systems (ETICS), commercial waste (e.g. from the construction sector (CA)), waste electronic and electrical equipment (WEEE), and shredder residues (e.g. from end-of-life vehicles (ASR)). In developed countries, by far the largest plastics waste fraction ends up in residual household waste (RHW). The major component of RHW is of biogenic origin. The plastics fraction in RHW is approximately 10–20%, depending on to what extent separate collection of packaging waste is in place. Table 11.1 gives an overview of the basic characteristics of representative random samples of these heterogeneous wastes.

Table 11.1 Examples of Post-Consumer Mixed Plastic Waste Types and Their Key Polymers

Waste	Main polymers and appearance	Contaminants	Lower heating value [MJ/kg] [2]
LWP-SR	PE, PP, PET, PVC, PS Fluff	See Figure 9.1	25
ETICS	Either EPS, or XPS, or PUR Composite foam	Mortar, plastering, polymer fiber, flame retardants	34
CW*	PE, PP, PVC Bulky, heterogeneous	Mineral and metal adherence, paper, cardboard	34
WEEE	Styrene copolymers, PVC, PP, PE, PA, PUR Mixed fine fraction	Fillers, color pigments, metal fragments, flame retardants	23
ASR	PP, PE, PA, PUR, PVC Mixed fine fraction	Fillers, color pigments, fiber, flame retardants	31
RHW	(all) Bulky, heterogeneous	Organics, sand, metals, glass, etc.	10

* CW = commercial waste.

Any of these wastes consist of more than just hydrocarbons, which are the desired products of petrochemical feedstock recycling. This causes the “heteroatom challenge” – meaning that species containing nitrogen, oxygen, sulfur, chlorine, bromine, and other heteroatoms have to be removed from the recycled products during processing.

11.1.2 Process Parameters

Product yields from chemical recycling are impacted by the feedstock, namely the amount of reactive substances present, its morphology and phase of matter, the process parameters, and the uniformity of reaction conditions, as decided by the respective technology. Briefly, the important impacts of process parameters are:

- **Stoichiometry:** Whilst pyrolysis is a thermal degradation process with indirect heating of the reactants without additional oxidizers, gasification processes are directly heated through partial combustion of the feedstocks by the oxidizer added to the reactor. The oxidizer content in the system must be sufficient to convert all carbon to its monoxide.
- **Minimum temperatures** are required during pyrolysis to allow for full thermal degradation of the specific polymers as well as other reactive components, typically between 350 and 450 °C. Thermodynamic equilibrium of gasification typically requires temperatures above 750 °C to assure full degradation of hydrocarbon species.
- **Minimum dwell times** in pyrolysis and gasification are determined by the reaction kinetics of degradation and partial oxidization, respectively. Reaction times strongly depend on temperature. At higher pyrolysis residence times, secondary reactions occur, which (i) cause further degradation of large molecules and (ii) promote formation of aromatics as well as char.

As mentioned above, the presence of catalytically active species or radical-forming reactive species, as well as added catalysts, can speed up reactions, such that smaller molecules are formed earlier and product distributions change. A similar situation can arise to form aromatics and char during pyrolysis.

Both pyrolysis and gasification are typically controlled by heat and mass transfer. Thus, mixing and back-mixing in the reaction system have a large impact on the reaction conditions and their distribution and uniformity. Typically, due to restrictions in heat transfer, residence times in pyrolysis reactors are longer than indicated by the end pyrolysis temperature. Gasifiers, on the other hand, require higher than kinetically indicated temperatures to overcome the limits of mass transfer. These non-idealities in most cases result in broad product distributions, and are strongly dependent on the specific technology.

11.1.3 Flexible Thermochemical Technologies and Technology Readiness

To be technically and economically viable, these thermochemical conversion technologies have to be integrated into waste processing chains. Thus, the full process chain (see Figure 11.1) usually consists of the (mechanical) pretreatment, the (thermochemical) conversion, and the cleaning (upgrading) of the main product. Mechanical and chemical recycling get combined and interlocked such that the waste-to-chemicals process chain gets optimized.

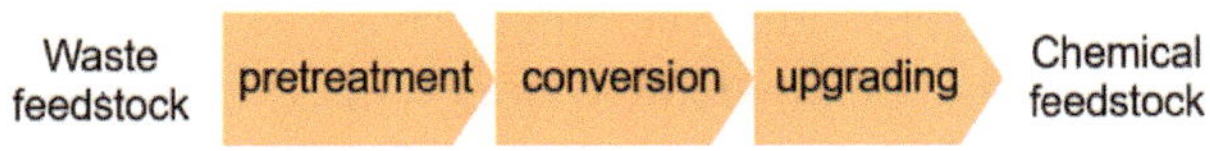

Figure 11.1 Process chain for the thermochemical recycling of wastes to secondary (chemical) feedstocks

Because of its chemical composition, mixed waste causes operational challenges for any thermal conversion system – for example, different types of corrosion. In addition, the physical nature of waste complicates mechanical feeding into such systems. In order for mixed plastic waste to be used in such systems, it usually needs pretreatment to remove non-combustible materials, homogenization to minimize operational variations, and ideally transformation so that its physical nature is compatible with mechanical feeding systems.

Mechanical pretreatment can allow waste to meet the physical and chemical specifications required of a specific thermochemical conversion technology. The pretreatment processes are relatively straightforward and involve several stages of sorting, separating, size reduction, and in some cases, biological treatment. Capital costs for the pretreatment systems are moderate and generally worth the benefit of making a low-cost, readily available feedstock stream available. The overall economic analysis is therefore favorable, but viability depends strongly on received gate/tipping fees associated with collecting the municipal waste.

Specific thermal treatment technologies have different feedstock quality requirements for the pretreatment step (mainly particle size, heating value, and chemical composition). This will be referred to in the sections with the specific technologies.

During the thermochemical degradation process, the heating rate, process temperature, stoichiometry, and residence time parameters must be optimized to maximize product yields and process reliability. Often, upgrading of the conversion products is necessary to meet the secondary feedstock specifications. Process chain byproducts can be valuable substances (e.g. metals) or energy carriers (e.g. heat), as well as material streams to be disposed of (e.g. ashes, wastewater).

As described in the literature, process suppliers or operators use a broad variety of methods and technologies for the feedstock recycling of waste at high process temperatures, which all fall under the categories of pyrolysis or gasification. An outline of today's high- and low-technology readiness level (TRL) chemical recycling technologies is given in Section 9.1.1.

So far, most technologies have not had their applicability proven on the industrial scale over a longer period, and so their TRLs are relatively low. Hereafter, we consider those thermochemical recycling technologies that have (a) reached a high TRL *and* size, and that have therefore been able to operate in a steady state over a longer period, by referring to industrial applications (TRL 7 – demonstration system, TRL 8 – first-of-a-kind commercial system, TRL 9 – full commercial application), or (b) only reached a lower TRL (TRL 4 – lab-scale validation, TRL 5 – large-scale prototype, TRL 6 – prototype system), as for current efforts to bring a residual part of the recycling process chain to an industrial scale. This is specifically the case for the important upgrading steps necessary to guarantee the qualities of the products for further use in the process industry.

11.2 Gasification for Chemical Recycling as Petrochemical Feedstock

Gasification in the context of waste recycling is done with the intention of converting biogenic energy resources and anthropogenic waste fractions (e.g. residual biomass, refuse-derived fuels, plastic wastes) into a high quality syngas containing the chemical building blocks H_2 and CO. Syngas is an industrial feedstock used globally for the synthesis of chemical intermediates and products for direct sale on the market. Use of syngas energy by combustion, or as feedstock of synthetic fuels is not regarded as "recycling" per se. Biogas, a mixture of CO_2 and CH_4 originating from biological treatment of organic waste, is not in the scope of this chapter.

Historically, gasification emerged as a sub-process of pyrometallurgy, whereby oxidized metals and coke typically react to form the desired raw metals and lean gases, which are used energetically. Dedicated gasification technologies became industrially relevant in the second quarter of the 20th century in order to avoid overdependence on petroleum imports. Both mobile wood-gas carburetors and industrial-scale coke-based syngas generators were developed up to high TRL. Although not being competitive on the world market, coal-based synthetic fuels were of specific relevance for closed economies such as in South Africa, and significant capacities were installed and operated over a long period (and are still in use). These developments form the basis for feedstock recycling of waste. The technological challenge of reliably producing clean syngas for fuel synthesis is similar to that of chemically recycling waste via gasification.

A second development that assisted gasification technologies came from the oil crisis in the 1970s. Then, large-scale, high-quality syngas production and high-temperature gas cleaning were developed to feed gas turbines in ultra-efficient power plants of the "integrated-gasification, combined-cycle" type. For reason of energy market prices, most of these large-scale demonstration plants have been closed. Since then, application of gasification to generate second-generation synthetic biofuels has been brought to the demonstration scale. Here, highly developed gasification technology is applied to heterogeneous, pretreated feedstock that is closer in nature to conventional waste. Furthermore, attempts to generate more bioenergy-based electricity has resulted in the installation of several hundred small-scale and easy-to-run decentralized gasifiers, mainly in Europe, during the past decade, but these have lower overall process efficiency. Hydrothermal treatment of biomass is not of significant relevance, so far.

Waste gasification on an industrial scale was developed for energy utilization as an alternative to state-of-the-art incineration technology. For more information, a detailed review of waste gasification is available [3].

11.2.1 Process and Technologies

Any combustible solid feedstock undergoes the consecutive physical and chemical conversion steps of drying, evaporation, char gasification and gas-phase oxidization of combustibles during a thermal process with presence of an oxidant. Depending on the zonal segmentation of the reactor interior, the intermediate product flows between the zones, and the conversion steps occurring, final syngas quality can vary significantly. In the overall gasification process, sub-stoichiometric addition of oxidant will (theoretically) be sufficient to convert all char. In high-temperature gasification, relatively-clean syngas can be generated, and thermodynamic equilibrium can be achieved by introducing additional oxidant; the price of this is the part of the feedstock needed for combustion to produce CO_2 and H_2O. If lower process temperatures are applied, the synthesis gas yield is higher, but it contains problematic non-equilibrium components such as trace-gas species, residual methane, and high-boiling hydrocarbons (tars).

There are many design variants for industrial-scale gasification. For the gasification of carbonaceous solid fuels, fixed-bed, fluidized-bed, and entrained-flow gasification processes have been developed; the latter is also particularly suitable for (viscous) liquid feedstocks. As a rule, fixed-bed gasifiers require coal as an additional feedstock when used on biomass or waste, as they have bed structural requirements similar to those of a blast furnace. With regard to the achievable synthesis gas quality (residual methane and tar contents), entrained-flow gasifiers are outstanding, fluidized-bed gasifiers (operated in either circulating or bubbling mode) follow thereafter, and fixed-bed gasifiers (especially in counter-current operation) deliver the lowest qualities. Conversely, the pretreatment cost for feedstocks of an entrained-flow gasifier are the highest, due to the fact that entrained-flow gasification requires either liquid or pulverized solid feedstock at a maximum grain size of approximately 0.1 mm. Fluidized-bed reactors can accept up to 50 mm grain size and fixed-bed gasifier's particle size distribution limits are even lower, but a significant part of the feed must show persistent physical and thermal stability to form a stable matrix inside the reactor. The fluidized-bed technology is described in more detail in Section 9.1.3.

All three gasifier technologies mentioned above are scalable up to 100 MW in size, but only entrained-flow gasifiers can achieve a 1000 MW size due their high reaction density, specifically when they are pressurized. Entrained-flow gasifiers can operate at up to 80 bar, thus avoiding additional costly syngas compression up to the pressure level necessary for thermal separation technologies, chemical synthesis, or gas turbine fuels, and supporting very high throughput of feed with a relatively small reactor volume. Usually, pressure levels of fluidized-bed and fixed-bed gasifiers are lower.

Today, industrially scalable entrained-flow gasifiers dominate the market for new installations on fossil feedstocks used, and the use of fixed-bed gasifiers is decreasing. Fluidized-bed gasifiers have little significance in industrial-scale syngas production, but a large number of small-scale units have been recently applied, mainly for the use of pure biomass in bioenergy.

Long-established metallurgical industry technologies are also used to gasify waste. These include the shaft furnace in the production of non-ferrous metals, and the blast furnace used in raw iron production. A special form is gasification in a limestone moving bed, derived from the lime industry. The particularly high process temperatures lead to the separation of the feed substances (separation of the molten metallic or mineral fraction). The process gas contains the hydrocarbons in partially oxidized form. Plastic waste in this case replaces part of the coke or the blow-coal, and acts as a reducing agent. The gas qualities are comparable to those of fixed-bed gasifiers. The use of waste as a predominant feedstock is carried out only on the medium-scale for vitrification of slags and for the energetic use of the synthesis gas, especially in Japan [4], and has reached a high TRL. The requirements for waste quality and permissible waste components are higher than those of the above-mentioned types of gasifiers, whereas the raw gas qualities are lower due to the pyrolysis gas components contained; resource and energy efficiency are also lower.

Even higher temperatures are reached when the energy is provided electrically, to generate a reactive plasma. Thermodynamically, there are only slight advantages in the gas composition over, for example, entrained-flow gasification. The external energy consumption is particularly high. The application of plasma gasification has so far mainly been carried out on hazardous waste at a small scale.

11.2.2 Syngas Upgrading – Cleaning of the Raw Synthesis Gas

Waste feedstock recycling by gasification requires an exceptionally high syngas quality due to the subsequent catalytic synthetic processes. In fact, the quality needs to be higher than for energetic utilization of syngas in turbines for power generation. Nitrogen dilution as well as hydrocarbon content of the syngas has to be avoided, and so the upstream gasifier has to be operated with oxygen or steam as gasification agent, similar to combined heat and power (CHP) applications. The raw synthesis gas coming from the gasifier must be purified and conditioned to meet the required specifications of, for example, methanol synthesis. In particular, the limit on concentrations of trace gases containing sulfur, halides, and nitrogen are much lower than for combustion exhaust gases, and a substantial proportion of particles must be removed. Small- and large-chain hydrocarbons must also be removed from the syngas. In addition, the H_2/CO ratio must be chemically adjusted depending on the downstream process of interest.

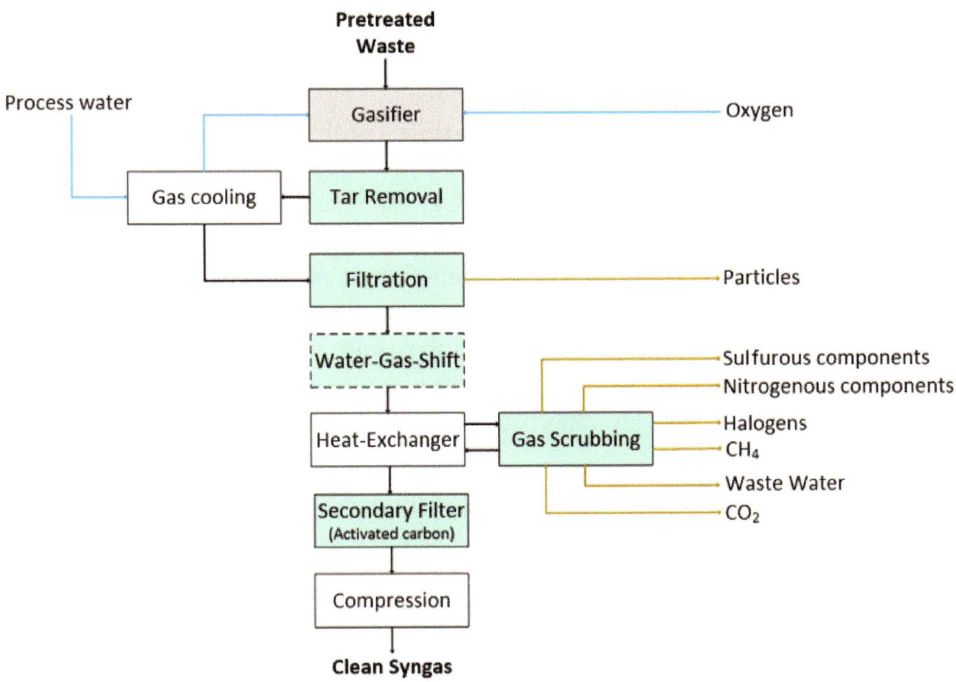

Figure 11.2 Major material flows and process steps in syngas production by gasification of pretreated waste

Figure 11.2 shows the necessary process steps in the synthesis gas treatment chain. It contains heat recovery steps to enable overall autothermal processing and high energy efficiency. Tar removal steps have to be integrated at specific condensation temperatures in order to manage plugging. This is the key technology issue that places limits on the feedstock, as well as limiting the load flexibility of the plant. Following removal of particulates by (for example) high-temperature filtration, a water-gas-shift reactor is included to adjust the H_2 and CO content over a suitable temperature range, followed by a multi-component gas scrubbing technology, placed upstream. Finally, compression is carried out to achieve the required methanol reactor pressure.

It has to be pointed out that this gas treatment and conditioning effort is significant in determining the processing cost of waste feedstock recycling. Furthermore, raw syngas quality critically depends on the gasification technology applied. Thus, the cost of raw gas treatment in the case of entrained-flow gasification would be much lower than in the case of fluidized-bed gasification, which ranks above fixed-bed gasification. On the other hand, the feedstock requirements of the gasifier and thus pretreatment cost vary with technology. Therefore, fluidized-bed gasification, having the lowest requirements, ranks above fixed-bed gasification and entrained-flow gasification. Furthermore, fixed-bed gasification requires a stable coke particle reacting bed, typically achieved by adding fossil coke to the waste feedstock.

11.2.3 Large-Scale, High-TRL Gasification Applications for Chemical Recycling

Gasification as part of a thermochemical recycling process aims at providing secondary feedstock from waste for C_1-chemicals such as methanol and C_1-based value chains such as methanol-to-olefins. For our purposes, Syngas-based fuel synthesis processes such as the Fischer–Tropsch synthesis are not considered as recycling; however, as they contribute significantly to greenhouse gas emissions when fed with natural gas, there is an opportunity for waste to contribute to emissions reduction in that sector, too.

Industrial gasification today mainly uses entrained-flow gasification, with many installations in the beginning of the 21st century for gasification of pulverized coal. All gasification developments of the 20th and the beginning of the 21st were driven by local availability of cheap carbon-rich feedstocks. Usually, these gasifiers are oxygen- and steam-blown, in order to avoid nitrogen dilution in downstream gas-processing plants and to increase process efficiency.

The first applications of gasification technologies for waste feedstock recycling were driven by attempts to recycle plastics in the late 20th century. Specifically in Germany, two industrial-scale gasification technologies on the 100 MW scale were operated over a longer period with different wastes (and mixtures of them) in order to produce methanol. In both cases, existing technologies for lignite feedstock developed to avoid overdependence on oil imports were adapted to process refuse-derived fuel (RDF) resulting from municipal solid waste (MSW) or even collected packaging plastic waste.

At the German Berrenrath plant, a high-temperature Winkler (HTW) gasifier was run on lignite, RDF and plastics using oxygen, steam, and CO_2 as gasification agents. This gasification technology is characterized by a higher freeboard temperature of approximately 950 °C, located above the bubbling fluidized bed at 10 bar. This results in a better syngas quality. With a heat load of 175 MW, it fed a 100 kt/a methanol plant. The plant was shut down after 10 years of successful operation in the mid-1990s, for economic reasons, mainly methanol prices.

At the Sekundärverwertungszentrum Schwarze Pumpe (SVZ) in Germany, which was already co-processing waste, sewage sludge, and lignite in revamped fixed-bed Lurgi gasifiers, for the first time a British Gas Lurgi slagging gasifier was installed in early 2000. Different types of waste were converted to syngas in the BGL gasifier at 25 bar, and fed a 100 kt/a methanol plant. The minor syngas quality issues of the fixed-bed gasifier were overcome by the extensive complementary plant structure of this production site, where significant tar quantities removed from the syngas served as feedstock for an entrained-flow gasifier. SVZ operations were terminated in 2005, mainly for two economic reasons: (i) the methanol production cost of this rather small dedicated production site were too high compared to world-scale methanol plants based on natural gas and, (ii) for mixed wastes and sorting residues, SVZ had to compete with cheaper waste incineration, because legal regulations at the time allowed low recycling rates for collected plastic waste material.

Both technologies – HTW as well as BGL gasifiers – could not be operated fully on RDF, but required significant co-processing of char-forming fossil feedstock (namely lignite) and waste. In addition, RDF had to be pelletized before being fed to the gasifiers.

This is not the case for the Showa Denko gasification plant at Kawasaki, Japan, which has been operated since 2004 with an increasing amount of pure collected and presorted polyolefinic packaging waste feedstock, which is pelletized on-site before being fed to the gasifier. Here, a two-stage gasification process is used at 10 bar, which consists of a steam/oxygen-blown bubbling fluidized-bed step at temperatures below 800 °C, followed by an steam/oxygen-blown slagging entrained-flow-like step to guarantee excellent syngas quality at ca. 1500 °C. The syngas is then converted in a water-gas-shift reactor, and hydrogen is removed and fed as part of the feedstock to a world-scale ammonia plant. Due to its specific nature (requiring a high process temperature and not making use of the feedstock's carbon content), the cold gas efficiency and economics of this process is significantly lower than for the aforementioned gasifiers.

At Edmonton, Canada, Enerkem has operated (since 2017) a solid recovered fuel (SRF)-to-methanol process, which is using pretreated residual household waste

(RHW) with a higher content of non-plastic waste of biogenic origin. It is based on its own proprietary technology that has been developed over a longer period and in several projects and plants with different sizes. Further projects have been announced in various countries. Again, this process uses bubbling fluidized-bed gasification technology and staged addition of oxidants. The fluidized bed is operated at below 700 °C due to the low SRF ash melting point, followed by a freeboard at higher temperatures to minimize tar formation. After flue gas treatment and recirculating of the tars to the fluidized-bed reactor, a proprietary methanol-to-ethanol process is applied at a smaller scale. Reactor over-pressure and capacity are smaller than for the processes mentioned above.

A more detailed description of these last three waste-to-methanol processes is available [4].

In addition to dedicated gasification processes, pyrometallurgical processes can be used to chemically recycle waste. Specifically, during processing of iron ore to raw iron, plastic waste is used to replace part of the reducing agent. Constraints here are that the waste feedstock ingredients must not disturb the production of raw iron or subsequent steel manufacture. This excludes many plastic waste types. In blast furnace processes, collected and sorted polyolefinic plastic waste is used to replace part of the coke but not the charge of the furnace. This application of waste feedstock is state-of-the-art and has been successfully applied at the Voest Alpine steel mill in Linz, Austria, for a long time. Feedstock utilization of the diluted syngas coming from these melting-type furnaces has not been applied so far due to the bad gas quality resulting in a high gas treatment and cleaning effort. Recent efforts aim at making chemical use of steel mill gases. Also, using lime shaft furnaces as waste gasifiers has been subject to R&D activities. The TRL of the latter processes is still low.

11.2.4 Balancing of a Gasification-Based Waste-to-Chemicals Process Chain

Gasifiers can be modeled as thermodynamic equilibrium reactors based on the specification of the feedstock composition and of process heat losses. By adding oxygen or steam to the waste feedstock, the necessary reactor temperature and syngas composition are reached with the highest possible theoretical cold gas efficiency. The cold gas efficiency describes the possible synthesis gas yield in terms of chemically bound energy of the feedstock, thus taking into account the effort to achieve the gasification temperature by burning part of the feedstock. Theoretical gasification temperatures can be as low as 750 °C, resulting in high cold gas efficiency. Due to real process limitations, higher temperatures have to be chosen, resulting in lower process efficiency. Nevertheless, for real gasification processes, process-dependent deviations from thermodynamic equilibrium have to be taken into account by adjusting theoretical syngas composition with respect to unreacted, non-equilibrium hydrocarbon species.

Here, we consider gasification of pretreated MSW in a fluidized-bed gasifier to provide syngas for methanol synthesis. This example case corresponds to a similar fully integrated large-scale process that is operated by Enerkem (see above). Because detailed mass and energy balances that allow process evaluation have not been published so far, the results shown here are based on process simulation that has been reported in detail elsewhere [5]. It includes mechanical pretreatment of the MSW collected by applying state-of-the-art SRF production technology, SRF gasification, and upgrading of the raw syngas to fulfill the demands of catalytic methanol synthesis. For an oxygen/steam-blown gasification of pretreated MSW considered here, the equilibrium balance was calculated at 950 °C. To consider the non-equilibrium syngas tar and methane contents, experience from the literature was taken into account.

Figure 11.3 shows the material flow diagram for this case. The overall process acts to separate the substances, and so recovering the metallic value products, separating off mineral products, and recirculating carbon from waste to syngas. The carbon mass fraction of the synthesis gas as desired recycled feedstock is significantly higher than the plastics carbon mass fraction of the waste. In the case of fluidized-bed gasification, about one third of the initial waste mass is converted into synthesis gas. Based on the carbon used, it is about 46%. All other waste carbon is converted to CO_2 during incineration of the fine mineral residue from mechanical treatment, to achieve the necessary gasification process temperature, and in particular to adjust the required hydrogen content of the synthesis gas via the water-gas-shift reaction during syngas upgrading. Primarily due to the lack of hydrogen in the waste feedstock, roughly half of the waste carbon is recovered when making methanol alone. In the energy balance (not shown here), the thermochemical process part enables energy-independent operation of the entire process chain due to the supply of high-quality process heat, which also takes into account the electricity requirement. The overall process energy efficiency of the chemical recycling process chain can be up to 90%, by making use of both gasification cold gas efficiency and high-value syngas sensible heat.

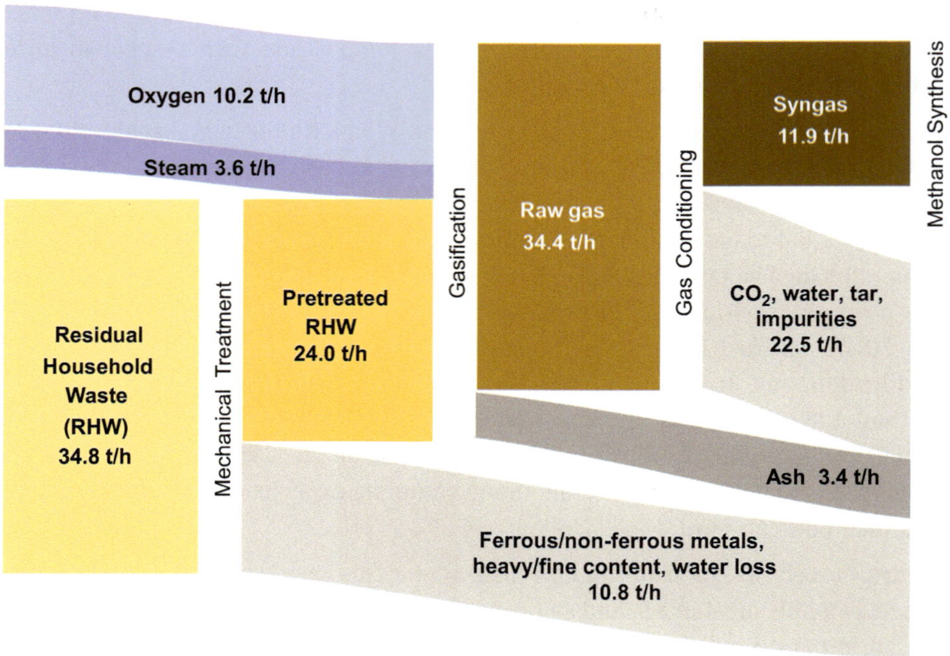

Figure 11.3 Material flows in the process chain for fluidized-bed gasification of residual household waste (RHW) to produce methanol syngas. The heat load of the gasification step is 100 MW

11.3 Pyrolysis for Chemical Recycling as Petrochemical Feedstock

11.3.1 Balancing of a Pyrolysis-Based Waste-to-Chemicals Process Chain

Pyrolysis is a kinetically controlled process. In technical pyrolysis systems, heat and mass transfer control the local reaction conditions within the reactor. Thus, pyrolysis product yields and distributions depend on (i) waste type and composition, (ii) applied pyrolysis and technology, and (iii) the process parameters chosen. Modeling and simulation of these complex multi-phase reacting systems is the scope of research efforts. Pyrolysis process design is thus empirically based and requires experimental data.

The main pyrolysis sub-processes are represented in Figure 11.4. Pretreated waste thermally decomposes in a single- or multi-reactor pyrolysis step, forming permanent gases, condensable vapors, and solid products. A combustion step utilizes the energy of byproducts and destroys undesired carbon-containing substances, whilst the main product is yielded in a condensation step. Phase separation and cleaning steps included are not shown in this figure, and depend on specific pyrolysis technologies, as do removal and utilization of solid byproducts and residues.

In the case of waste mixtures containing biogenic fractions, as well as oxygen-containing plastics which are considered here, aqueous condensates of low calorific value cannot be avoided. The condensable organic components form the product of value that must fulfill the product specification as secondary chemical feedstock. A key issue is the removal of contaminants such as halogens from the process flow in the reactor. Therefore, either sorbents are added to the reacting mixture in the reactor, or pyrolysis gases and vapors are treated before condensation. Also, extensive post-processing of the liquid pyrolysis products (for example, by hydrogenation) can be necessary to adjust the liquid hydrocarbon composition in the condensable pyrolysis products. Typically, permanent gases and aqueous fluids are burnt to enable the autothermal operation of the overall process. Carbon-containing solid products can be burnt for this purpose, too. Residual mineral fractions have to fulfill landfilling requirements. As the permanent gas might contain hydrocarbons (to a large extent depending on feedstock and process), it might be considered for feedstock utilization, which is not the case in this example.

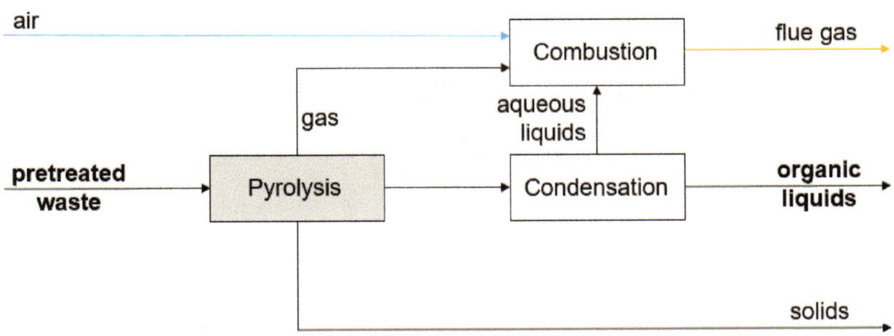

Figure 11.4 Major material flows and process steps in liquefaction of pretreated waste by pyrolysis

Here, we consider non-catalytic pyrolysis of pretreated MSW to provide pyrolysis oil (organic condensate) as a steam cracker feedstock blend. This example case corresponds to the gasification case reported above. Because detailed mass and energy balances that allow process evaluation have not been published so far, the results shown here are based on laboratory-scale experimental data on product yields and distributions for similar waste that have been scaled-up, and have been reported in detail elsewhere [5]. It includes mechanical pretreatment of the MSW collected by applying state-of-the-art SRF production technology, SRF pyrolysis, upgrading of the pyrolysis oil to fulfill the demands of naphtha as steam cracker feedstock, and energetic utilization of the pyrolysis byproducts. Herein, SRF pyrolysis data were taken from [6].

Figure 11.5 shows the material flow diagram for this case. The overall process acts to separate the substances, recovering the metallic value products, separating off mineral products, and recirculating carbon from waste to organic liquids (pyrolysis oil). The carbon mass fraction of these organic liquids as the desired recycled feedstock is significantly higher than the carbon mass fraction of the plastic waste. In the case of pyrolysis, only about one-fifth of the initial waste mass is converted into pyrolysis oil. Based on the carbon used, it is about 45%. All other waste carbon is converted to CO_2 during incineration of the fine mineral residue from mechanical treatment, to achieve the necessary pyrolysis process temperature, and in particular to burn the pyrolysis byproducts. The H/C ratio of the pyrolysis oil is close to that of heavy steam cracker feedstocks, such that consecutive hydrogenation of the pyrolysis oil is applied to remove heteroatoms and to saturate remaining double bonds rather than to provide a higher hydrogen content. In the energy balance (not shown here), the thermochemical process part enables energy-independent operation of the entire process chain due to the supply of high-quality process heat, which also takes into account the electricity requirement. The overall process energy efficiency of the chemical recycling process chain can be up to 90% by making use of both chemical energy of the pyrolysis oil and sensible heat of the combustible byproducts.

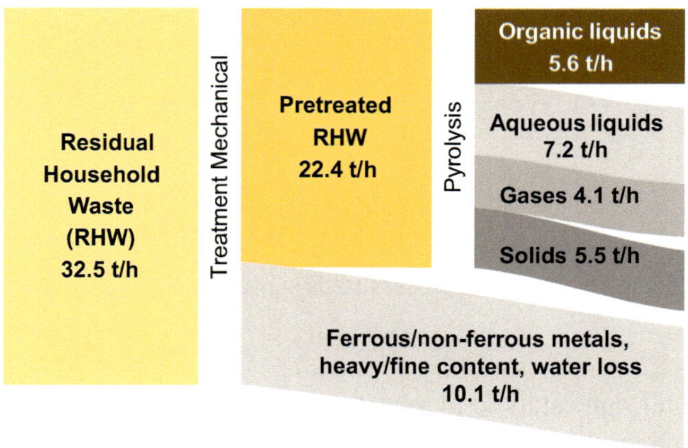

Figure 11.5 Material flows in the process chain for pretreatment and pyrolysis of residual household waste (RHW). The heat load of the pyrolysis step is 100 MW

From this example comparison of chemical recycling of mixed waste by gasification and by pyrolysis, one can see that basically the alternatives perform similarly in terms of indicators related to mass and energy balance. Nevertheless, for pyrolysis this is a theoretical case because no attempts have been made so far to pyrolyze MSW-based SRF with a relatively low plastic content at the relevant scale. Waste streams based on plastic with low biogenic content are the focus of this chapter, as they are the waste streams to be recycled chemically in addition to physical recycling processes.

Little comprehensive data can be found in the literature regarding the reproduction of pyrolysis performance for different types of mixed plastic waste, specifically on a technical scale. With respect to the waste situation and composition in Germany about 25 years ago, most widespread investigations were carried out by Kaminsky's group at Hamburg, applying high-temperature fast pyrolysis in bubbling fluidized-bed reactors on a bench- and pilot-scale to different kinds of waste plastics and mixed waste. A good outline of the chemical recycling performance is available [7].

Zeller et al. [2] recently reported the mass and energy balances of pilot scale pyrolysis of important mixed plastic waste streams shown in Table 11.1. Due to the technology and the moderate process parameters applied, in this case plastics decompose quite slowly and all products were collected separately. Processing occurred non-catalytically in a screw-type reactor without further additives, and in an inert atmosphere. Figure 11.6 shows the product mass yields of these investigations.

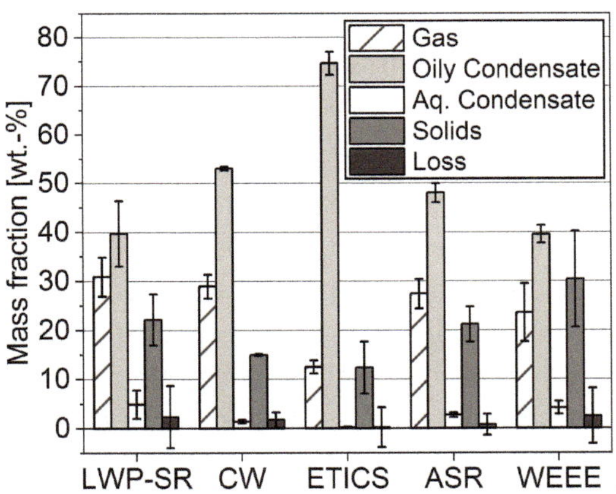

Figure 11.6
Average mass balances for pilot-scale screw-reactor pyrolysis experiments with different mixed plastic wastes at 450 °C and 30 min solids dwell time [2]

The yields of the individual pyrolysis product fractions vary depending on the characteristic composition of the feedstock. Within a specific feedstock, it varies due to inhomogeneities impacting the feed as well as measurement uncertainties. The yield of pyrolysis solids is between 10% and 30%. This product fraction consists of coke and inert materials originating from functionalization of the plastics and contamination of the waste. Thus, the ash content of the feedstock must be considered when interpreting the yield of solids. Between 40% and 75% of the feedstock mass pyrolyzed is found as oily condensate. The thermal insulation system feedstock shows the highest condensate yield, while the oily condensate yield from WEEE and lightweight packaging sorting residue is the lowest. An aqueous phase with up to 5% of the feedstock occurs as a second condensate phase. Since pyrolysis is carried out with dried feedstock, the introduction of chemically unbound water into the reactor as a source for the aqueous pyrolysis fraction is low. The aqueous condensate is thus attributed to oxygen-containing plastics or biomass contamination. The pyrolysis experiments with ASR and WEEE with an expected content of engineering plastics, as well as LWP-SR with biomass contamination, show an increased release of aqueous product and thus confirm this assumption. The permanent gas fraction amounts to 13% to 30% of the total mass.

The carbon yield refers to the amount of carbon that can be recycled in relation to the waste feedstock carbon fraction. For this purpose, the carbon content of the oily condensates was determined and combined with the oil yield. The pyrolysis oil carbon yields are 51.1% for LWP-SR, 60.0% for CW, 74.6% for ETICS, 57.5% for ASR, and 60.5% for WEEE. Depending on the location and detailed design of a large-scale pyrolysis plant, in addition to the carbon in the condensate phase, the carbon contained in the gas phase could also be partially recycled. Hence, the reported carbon yields are to be understood as minimum recycling rates.

At least 50% of the introduced carbon is recovered in the oily phase and is thus in principle available for reintroduction into a value chain. For ETICS, which contains almost solely polystyrene a (a polymer that thermally depolymerizes), about 75% of the feedstock carbon is contained in the oily condensate. The potential for value chain reintegration of the pyrolysis products must be assessed individually for each waste type based on the specific composition.

Since the waste feedstocks investigated are a mixture of plastics and other materials, detailed statements on the influence of the polymers contained in them, in terms of energy demand, cannot be made. Thus, the determined energy balances are of technical nature and relevant only in the context of a given application. The energy demands for heating, melting, pyrolysis, and evaporation determined in the pyrolysis experiments were normalized using the lower feedstock calorific value. They are 5.1% for LWP-SR, 5.2% for CW, 4.9% for ETICS, 5.4% for ASR, and 3.7% for WEEE, and are thus within a narrow range. In a technical process, heat losses and the energy demand of supporting infrastructure would add to these values.

11.3.2 Process and Technologies

Pyrolysis as a recycling process aims to convert solid carbon from (for example) biogenic residues or plastic waste streams to supply the process industry with easily accessible hydrocarbon feedstocks. Pyrolysis was the first dedicated thermochemical process besides combustion, and evolved primarily for the production of coke from wood or coal. Left over from this time before the era of crude oil are two industrial-scale applications:

- The coking plant providing coke from hard coal in large fixed-bed furnaces operated batchwise supplies the pyrometallurgical industries. The solid yield is high due to the feedstock. Byproducts (coal tar and coke oven gas) are mainly used energetically.
- The production of charcoal from wood with a yield typically not higher than 50 wt%. Usually, hard wood is pyrolyzed in batch processes in fixed-bed reactors. As byproduct, vapors are condensed and thermally separated to yield (for example) acetic acid and tar oils. Quite a few such applications worldwide supply the charcoal market.

In addition, there are some industrial-scale petrochemical applications operated continuously, such as those designed for pyrolysis of oil sludge in a rotary kiln. Most large-scale coking processes in refineries are operated with oxygen-carrying gases such as steam, thus not strictly falling under the pyrolysis category defined above.

In recent times, pyrolysis was developed further to have a range of applications to various feedstocks and purposes. Examples are the liquefaction of lignocellulosic biomass for the production of substitute fuels and the pyrolysis of residual household waste for energy utilization. Rotary kilns, auger-type reactors and fluidized-bed reactors are applied in order to continuously operate the processes.

Fast pyrolysis has been investigated extensively in recent decades for the purpose of direct thermochemical liquefaction of sustainable biomass. Resulting bio-crudes and bio-oils are upgraded to bio-fuels or bio-chemical intermediates, for example by means of hydrogenation. So far, only a small number of commercial applications have appeared, mainly in the non-transportation fuel sector. IEA Bioenergy [8] has provided an overview of globally installed thermal liquefaction capacities on the demonstration scale [8].

Recently, pyrolysis has came back into focus as an alternative to polyolefinic waste incineration. In addition to the thermochemical conversion step itself, upgrading or product cleaning steps are necessary to guarantee the qualities of the products for feedstock utilization in process industry. Individual pyrolysis technologies use different reactor types. The most important reactor types applied to continuous plastic waste pyrolysis are:

- **Extruders:** As a standard technology in the polyolefin compounding industry, as well as for processing of polyolefin recyclates, twin-screw extruders are applied to degradative melting and pre-conditioning of mixed polyolefinic waste. Herein, the extruder functions as a first reaction step. Limits are: (i) temperature, which commonly is below the total degradation temperature of most polymers, and (ii) heterogeneity, particle size, and content of non-melting components of the waste feedstock.

- **Stirred-tank reactors/flow reactors:** A stirred tank can be used in continuous operation to pyrolyze meltable plastics. It provides the closest to fully homogeneous reaction conditions of all reactor types. Operation limits can be temperature and size, as well as non-melting ingredients. An auxiliary system is needed to provide homogeneous and molten feedstock and to control flow. This system itself can be used as a flow reactor when it processes a large circular liquid flow to which feedstock is added and from which product is separated.
- **Screw reactors:** In an auger-type reactor that is filled to a small extent only, heterogeneous solid feedstock is conveyed in a well-defined manner during conversion, making this technology very flexible in terms of feedstock composition, morphology, and pyrolysis process parameters. Throughput is limited due to mechanical design constraints, so that scale-up to the t/h of waste scale requires multiple machines.
- **Rotary kilns:** A rotary kiln is the most flexible machine to process any morphology of solid feedstock for thermal decomposition. Furthermore, for purpose of pyrolysis there are no temperature limits and scale-up can lead to high throughput of waste material. Whilst solids mixing and residence time control are well-assured, vapors and gas residence time in the large freeboard are less well-defined. Operational issues can result from sticky materials and from ensuring that the drum sealing of the large rotating machine remains leakproof under pyrolysis conditions.
- **Fluidized-bed reactors:** A bubbling fluidized-bed reactor applied to mixed plastic waste pyrolysis functions as a continuous stirred tank reactor but can be used for non-melting material, too. Furthermore, for the purpose of pyrolysis there are no temperature limits, and scale-up can lead to high throughput of waste material. Operational issues result from the higher demands in terms of feedstock granularity and its homogeneity. Furthermore, the stickiness of meltable polymer waste has to be controlled well to assure operability of the fluidized bed. Whilst reaction conditions can adjusted very well to optimize product yield and composition, the need to fluidize the material typically requires recirculation of post-processed pyrolysis gas, affecting the reaction conditions in the reactor. So far, there have been no circulating fluidized-bed applications for large-scale plastic waste pyrolysis.

Due to the nature of waste, most technologies require pyrolysis product upgrading. In-situ upgrading includes addition and removal of sorbent species to remove heteroatoms, addition and removal of catalysts in polyolefin pyrolysis to adjust chain length of hydrocarbon species (and thus liquid and gaseous product yields and compositions), removal of particulates from condensable species, and control of oily and aqueous condensation to avoid re-polymerization. Ex-situ upgrading in particular includes catalytic hydrogenation of pyrolysis oil at higher pressure to saturate C–C bonds and to allow removal of heteroatoms from the oil. In addition, thermal separation technologies might be applied to recover high-value species from the oil and to adjust the boiling curve. Permanent pyrolysis gases might be treated by adsorption or thermal separation of specific heteroatomic species. Finally, solid pyrolysis residues might be fractionated mechanically to recover high-value products or to segregate streams for disposal.

11.3.3 Large-Scale High-TRL Pyrolysis Applications for Chemical Recycling

Pyrolysis of mixed plastic waste has reached the same high degree of maturity that thermochemical technology already has, with the exception of pyrolysis product upgrading and feedstock utilization.

Mixed plastic waste pyrolysis was operated successfully on an industrial scale for a significant period at the MSW pyrolysis plant at Burgau in Germany (1983–2016). MSW and sewage sludge were pyrolyzed in two rotary kiln lines of ca. 2 t/h capacity each, followed by combustion of the pyrolysis gases and vapors in a steam boiler. According to Quicker et al. [9], the electrical efficiency of the plant did not exceed 22% due to the high amount of char in the solid residues. The plant was finally shut down because of unfavorable economics, due to the low electricity revenues as well as the high disposal cost of the solid residues.

Further successful operational cases for this kind of technology, which use the pyrolysis products energetically in combination with combustion and ash melting, are documented for Japan [4]. In Japan, separation and melting of the slag was the focus of MSW treatment for a long period due to legal requirements. This encouraged application of thermal treatment technologies alternative to incineration on the grate, although the energy efficiency and economic performance of the alternatives was worse.

A mature example of chemical recycling of waste by pyrolysis is the end-of-life tire pyrolysis facility at Hirohata steelworks in Japan. Operating since 2004 with a capacity of ca. 4 t/h in each of the two rotary kilns, shredded waste tires are processed to different valuable products: pyrolysis gas, light and heavy pyrolysis oil, carbon black, metals, and a mineral residue. Whilst the latter is deposited and the pyrolysis gas is energetically used to balance the heat demand of the pyrolysis plant, the other products are used in the steel mill as reducing agents and as carbon feed for metal recycling. Chemical-grade quality of pyrolysis gases and condensable vapors is not the focus of this process. Today, waste-tire-pyrolysis R&D and project activities focus on recovery of the carbon black pigment from the solid pyrolysis residue, in sufficiently high quality that it can be re-reprocessed as recycled carbon black in synthetic rubber.

Most recent activities on chemical recycling of waste by pyrolysis all focus on feedstock utilization of polyolefin plastic waste. Driven by the first EU plastics strategy [10], which sets mandatory recycling quotas for the EU countries, plastics producing industry and waste management companies announced several ambitious projects which will lead to capital investment into plastics liquefaction (see Section 9.1.1). The main purpose is to recover end-of-life plastics as petrochemical liquid feedstock. So far, only two plants (in Almeria and Seville, Spain, operated continuously by Plastic Energy Ltd) fulfill this task at a commercial scale. Both operate on pre-selected polyolefinic feedstocks that, amongst others, are almost free of chlorine, biomass, etc.

As laid out above, the waste streams available at industrially-relevant quantities for chemical recycling are not the pure, sortable plastics that are often amenable to pyrolytic processes but sorting residues that are mixtures of anthropogenic products and biomass. Pyrolysis of these waste streams is more troublesome, and the subject of research and development. Alternatives to rotary-kiln technology are considered in order to homogenize pyrolysis reaction conditions and to achieve high-value product yields adapted to the respective waste feedstock. Current R&D efforts aim at increasing the technology readiness level of pyrolysis processes in order to improve the economics of waste pyrolysis. In the case of mixed plastic waste, the technological challenges are the adjustment of product specifications both in the reactor and by post-treatment, and in managing the plugging or sticking tendency of the intermediates within the reactor and in the downstream gas treatment and condensation steps. As references for waste pyrolysis, the few examples mentioned above are not sufficiently well-documented to allow closure of the mass and energy balances and independent evaluation.

11.4 Outlook

Due to the mixed nature of waste and consequent variation in behavior, achieving the desired recycling quota and promoting the circular economy for carbon in the medium-term requires both waste collection and mechanical sorting systems, and chemical recycling of waste. Chemical recycling complements mechanical sorting by overcoming the downcycling issues of mixed materials and by avoiding final incineration of large amounts of separately collected wastes. Combined mechanical and chemical recycling thus maximizes the recycling rate and minimizes the greenhouse gas emissions. For this purpose, the thermochemical processes of pyrolysis and gasification function as thermal separation processes for mixed wastes including plastics, biomass, and mineral fractions.

Whereas sorting technologies are ready to be implemented in order to increase waste recycling, thermal technologies for chemical recycling still need R&D efforts, specifically concerning the flexibility to process varying complex mixed feedstocks and the scale of processing. Alternative thermal waste treatment technologies applied to waste-to-energy are mostly not competitive with waste incineration due to higher process complexity, and the associated need for specific capital investment. In contrast, pyrolysis and gasification processing costs are of the same order of magnitude as waste incineration gate fees when being applied to complex waste feedstock recycling, because of the recycled product revenues (for example, see López et al. [6]). Accordingly, alternative thermal treatment of waste seems economically feasible, and supports the transition to a circular economy.

The balancing of pyrolysis- and gasification-based entire waste-to-chemicals process chains shows that both thermochemical processes perform similarly. Gasification recycles waste carbon to feedstock for C_1-chemistry. It can be applied centrally on the chemical production site or the petrol refinery site, and it requires high capital investment. Pyrolysis, more directly than gasification, recycles plastics to plastics because it converts a significant part of waste to feedstock for monomer production. This is no value by itself. The pyrolysis plant can be decentralized when the focus is on the organic liquid product only. Pyrolysis requires lower capital investment than gasification. This might explain current the chemical recycling focus on pyrolysis technologies.

Comprehensive life cycle analysis can indicate whether use of a specific recycling technology is advantageous for a specific waste. The highest efficiency of chemical recycling in terms of greenhouse gas emissions is achieved when a high amount of chemical energy is recycled and external energy demand is minimized. Guiding principles for process selection can be:

- Choose a recycled product specification that best fits your waste specification:
  - Maintain the H/C ratio
  - Consider the degradation mechanism of the polymers in the waste
  - Stay flexible with regard to waste specification and its variation
- Choose a recycling technology that maximizes the value of the recycled product:
  - Apply mechanical recycling and solvent-based processes to purely sortable and little-contaminated products
  - Apply thermochemical recycling to hard-to-sort mixtures, to composites, and to polymers that are easy to upcycle thermochemically
  - Consider central large-scale thermochemical conversion to use gaseous products
- Optimize the entire value chain:
  - Adapt pretreatment process to the demands of chemical conversion technology

- Remove as many contaminants and heteroatoms as possible in-situ during conversion
- Try to integrate product upgrading into central structures

References for Chapter 11

[1] Stapf, D., Ciceri, G., Johansson, I.: *Trends and Drivers in Alternative Thermal Conversion of Waste*, IEA Bioenergy, 2020. DOI: *https://doi.org/10.13140/RG.2.2.15439.10409*

[2] Zeller, M., Netsch, M., Richter, F., Leibold, H., Stapf, D.: Chemical recycling of mixed plastic wastes by pyrolysis. *Chem. Ing. Tech.*, 93 (2021) 11, pp. 1–9. DOI: *https://doi.org/10.1002/cite.202100102*

[3] Waldheim, L.: *Gasification of Waste for Energy Carriers – A Review*, IEA Bioenergy, 2018. *http://task33.ieabioenergy.com/content/Task%2033%20Projects*

[4] Vehlow, J.: Overview of the pyrolysis and gasification processes for thermal disposal of waste. In: *Waste Management*, Volume 6. TK Verlag Karl Thomé-Kozmiensky, Neuruppin 2016, pp. 431–444

[5] BKV-study: *Thermal Processes for Feedstock Recycling of Plastics Waste*, A study conducted by KIT and Conversio Market & Strategy GmbH on behalf of BKV GmbH and Plastics Europe AISBL, 2019. *https://www.bkv-gmbh.de/studies/study-thermal-processes-for-feedstock-recycling-of-plastics-waste-kit-conversio.html*

[6] López, A.,de Marco, I., Caballero, B. M., Laresgoiti, M. F., Adrados, A.: Pyrolysis of municipal plastic wastes: Influence of raw material composition. *Waste Management*, 30 (2010) 4, pp. 620–627. DOI: *https://doi.org/10.1016/j.wasman.2009.10.014*

[7] Brandrup, J., Bittner, M., Michaeli, W., Menges, G. (Eds.): *Die Wiederverwertung von Kunststoffen*, Carl Hanser Verlag, Munich/Vienna, 1995

[8] *Direct Thermochemical Liquefaction*, Pyrolysis Demoplant Database, IEA Bioenergy task 34, 2016. *http://task34.ieabioenergy.com/publications/pyrolysis-demoplant-database/*

[9] Quicker, P., Neuerburg, F., Noel, F., Huras, A., Noël, Y., Eyssen, R. G., Seifert, H., Vehlow, J., Thomé-Kozmiensky, K.: *Status of Alternative Techniques for Thermal Waste Treatment*, Expert Report No. 29217 for the Federal Ministry for the Environment, Nature Conservation, Building and Nuclear Safety, 2015. *https://www.umweltbundesamt.de/publikationen/sachstand-zu-den-alternativen-verfahren-fuer-die*

[10] *A European Strategy for Plastics in a Circular Economy*, Communication from the commission to the European Parliament, the Council, the European Economic and Social Committee and the Committee of the Regions, Brussels, 2018. *https://eur-lex.europa.eu/legal-content/EN/TXT/?uri=COM%3A2018%3A28%3AFIN*

12 Recycling of Composites

Mathias Mühlbacher, Volker Altstädt

12.1 Introduction to Composites

Composite materials are now used in all important sectors of the economy, such as shipbuilding, and construction of wind turbines, bridges, automobiles, aircraft, rail vehicles, and trucks. Industry revenue is predicted to be 100 billion US dollars by 2025. Composite materials make a significant contribution to reducing the weight and energy consumption of components during use. However, the life cycle of these components often ends in landfills in many countries of the world. Recycling/reuse is still little understood in this industry. Nevertheless, it can be expected in the future that technological advances in the recycling of composite materials and legal measures such as landfill bans in European countries will lead to components with reduced energy and cost expenditure. A large proportion of composite materials contain reinforcing fibers. Reinforcement fiber recovery by innovative recycling technologies such as pyrolysis and solvolysis can significantly contribute to the recycling of glass, carbon, aramid, and hybrid fibers in thermoset composites.

Composites are reinforced polymer materials with special requirements that cannot be achieved without additional stiffening or reinforcement. They are used in particular for lightweight construction applications with the highest mechanical, thermal, or special requirements. Composites of mixed materials are formed by incorporating particles or reinforcing fibers into a polymer matrix. Due to the reinforcement, which is commonly done with glass fibers and carbon fibers, polymer composites achieve a lightweight construction potential, and so compete with metallic materials (see Figure 12.1). The right combination of a polymer matrix with reinforcing fibers promises better mechanical properties in a composite than any individual material alone.

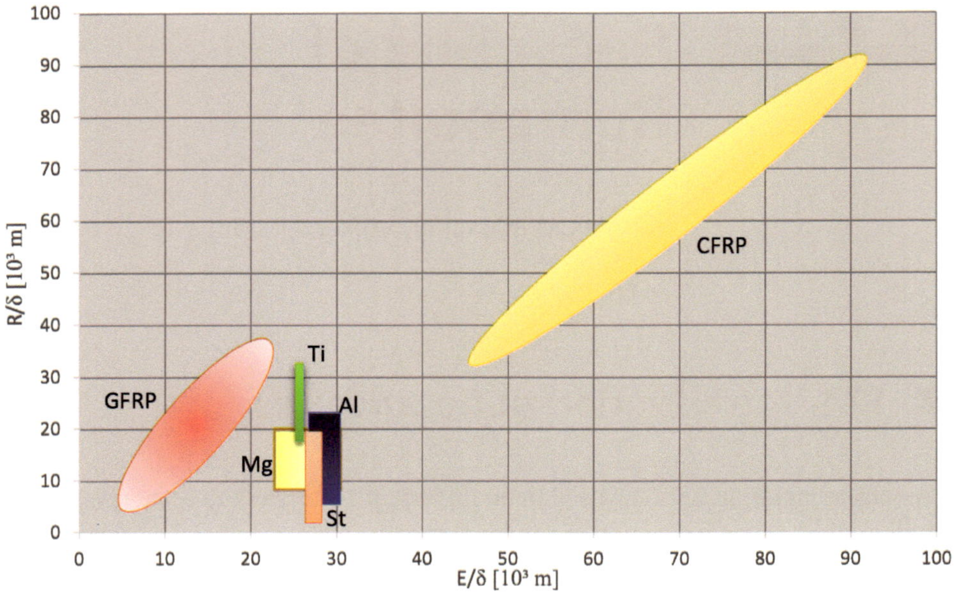

Figure 12.1 Lightweight potential of glass-fiber-reinforced (GFRP) and carbon-fiber-reinforced (CFRP) composites, compared to metals [1]

For lightweight construction, the specific material properties that are related to density of the components are of particular importance. Components made of fiber-reinforced composite materials are in many cases far superior to metallic components in terms of their specific strength and stiffness. Due to the fiber reinforcement, composite materials often have strongly anisotropic properties. Good mechanical component design makes optimum use of this anisotropy to further save weight. In summary, while these materials offer the highest performance, they also have high costs, a questionable environmental footprint in production, and demanding recycling requirements.

Thermoset and thermoplastic polymers are used as matrix materials for embedding the reinforcing fibers. Thermosets are chemically cross-linked. They are non-meltable, insoluble, non-swellable, and have high stiffness and tensile strength. They are also brittle and have good heat resistance. Crosslinking takes place after the shaping of the starting resins in the desired geometry by heating, or by means of chemical additives or catalytic additives. The low processing viscosities allow easy fiber impregnation and a high content of fibers by volume.

Thermoplastics are non-cross-linked polymers that also consist of linear or branched-chain molecules. Under stress and especially at higher temperatures, the chains can slide off each other, so that softening or melting can be achieved. The softening and melting are repeatable. Thermoplastics are also swellable or even soluble in organic solvents. Typical melting temperatures are 130–260 °C. With composite recycling, the fibers and fillers in the polymer melt can be separated from the matrix. Thermoplastics generally have lower stiffness and tensile strength than thermosets, and embrittlement occurs at low temperatures. The ability for repeated melting enables much better and easier recycling routes for thermoplastic composites.

Fibers and fillers in composites are usually used to improve stiffness, strength, heat resistance, and dimensional stability. The fillers used are mainly inorganic powders (chalk, kaolin, and talc). Reinforcing fibers can be made of glass, carbon, polymers (e.g., PP, aromatic polyamides, etc.) or natural fibers (e.g., flax, hemp, sisal, bamboo, etc.), and differ in their fiber length and fiber architecture. Fiber lengths in injection-molded parts range from a few mm (short fibers) to 10–20 mm (long fibers), or are present as continuous fibers in composite laminates. The architecture of continuous fiber composites can be woven fabric, roving, and unidirectional. Fleece and random fiber mats are used as mixed forms.

Complicated material compositions of this kind place ever increasing demands on the recycling process, since the separation of the components is often difficult and therefore requires additional steps that put an additional burden on the recycling balance. In addition, composite components, like rotor blades for wind energy or aircraft parts, can contain other layers such as coatings and paintings. Typical for aircraft parts are also copper layers for lightning strike protection.

A special case in the context of composites are sandwich components for an extremely lightweight design, especially for optimized bending stiffness. In the case of sandwich elements, the cover layers made of composite material are connected to a core material made of foam or aluminum honeycomb (Figure 12.2).

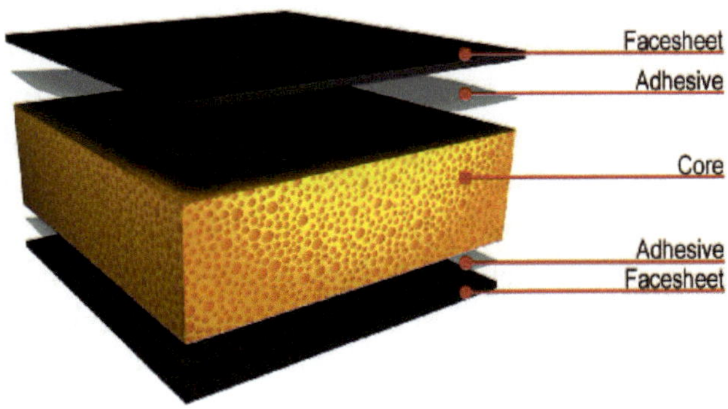

Figure 12.2 Schematic structure of a sandwich element for optimized bending stiffness in an extremely lightweight design with thermoset face sheets, a foam core and additional adhesive layers

In the case of thermoset sandwich components, adhesive layers are often used to bond the face sheets and cores together. Such adhesive layers can subsequently make it impossible to separate the components during recycling. Thermoplastic sandwich materials offer the advantage of homogeneous material composites by connecting cover layers and cores through a thermally cohesive connection. No adhesive layers made of an additional material are needed (Figure 12.3). This means that thermoplastic sandwich structures can be recycled much better than thermoset sandwich structures.

Figure 12.3 Examples of thermoplastic sandwich systems without adhesive layers for optimized recycling [1]

12.2 The Composites Market

The main markets for composites are automotive/transport, electronics, wind energy, pipes, tanks, and aerospace (Figure 12.4). For all of these applications except aerospace, glass-fiber-reinforced polymers are predominantly used. For high-performance applications in the aerospace industry, which need a smaller number of parts, carbon fiber reinforcements are mainly used.

The volume of the global composites market totaled 12.1 million tons in 2021. Compared to 2020, with a volume of 11.2 million tons, growth was around 8%. Overall, Europe's share of the global market is around 25%. The market share for America is similarly high. Asia now accounts for around 50% of the world market. The total European market excluding natural fiber composites was predicted to be 2962 kt in 2021, of which approx. 2% is CFRP, the rest GFRP (glass-fiber reinforced) [2].

If one looks at the application of composite materials, the largest markets for composite materials are transport, the construction industry, and the electrical and electronics industry. In terms of sales, the top four industries that use composites are transport, construction, electrical and electronic goods, and the defense and aerospace industries. The defense and aviation industry, which ranks in final position in terms of volume, with only 0.4% by mass, is among the top three industries in terms of value, representing 13% of the market, indicating the high value-added nature of the composite products employed within this industry.

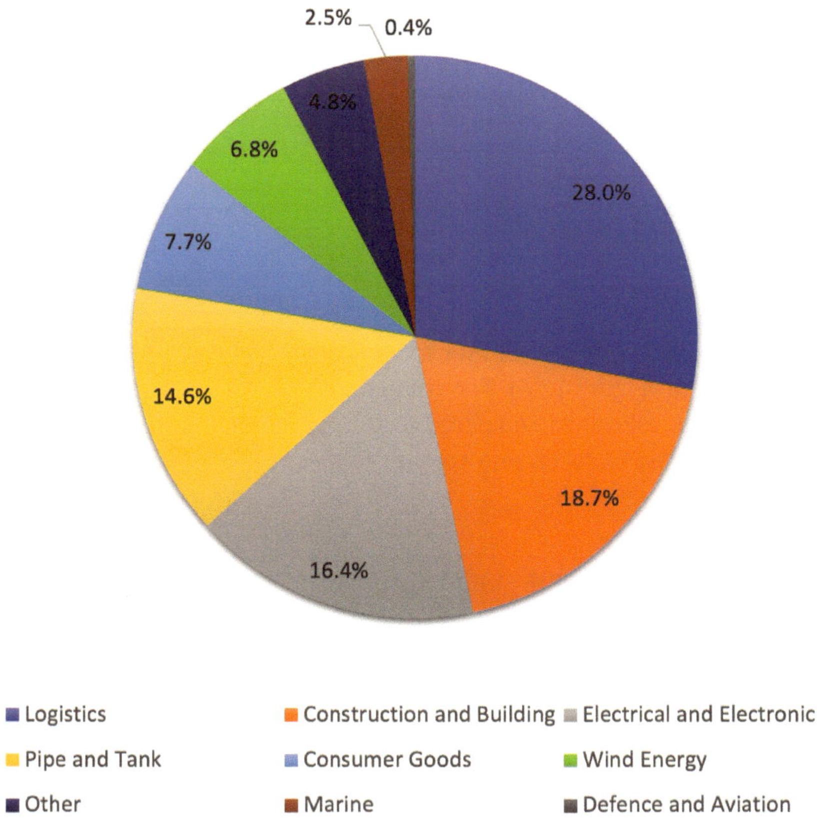

Figure 12.4 Distribution of the composites market by application in terms of share (2016)

From a global perspective, the carbon composites market grew by around 10% in 2020 (Figure 12.5). Even if the market for carbon composites is small in terms of quantity, the energy-intensive production of carbon fibers makes it important to develop sustainability concepts and recycling processes for this market over the next few years. For example, the production of 1 kg of carbon fibers emits 18 kg of CO_2 [3].

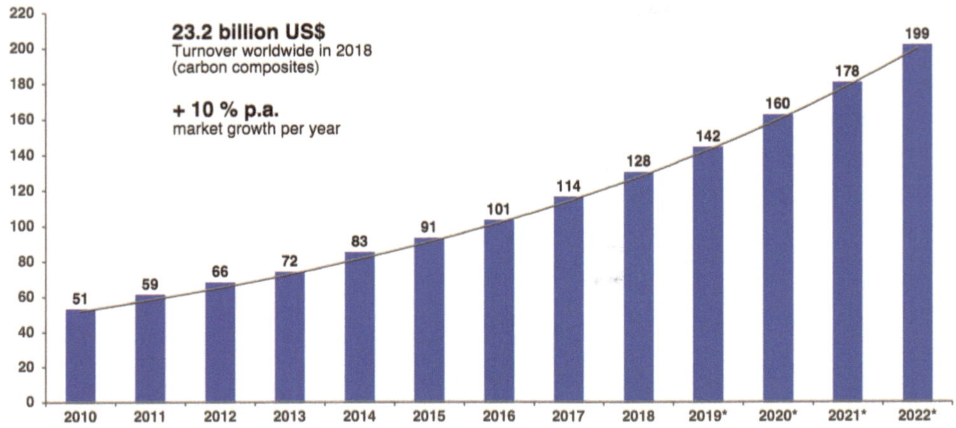

Figure 12.5 Development of the global carbon composites market from 2010–2018, and the forecast for 2019–2022

Current developments in the direction of sustainable composites can already be seen in the rapidly growing market for "bio-composites" with natural fibers as the reinforcement (Figure 12.6). The volume of approximately 500 kt/a in 2020 in Europe shows its relevance compared to the carbon composites market. The increase in price must be taken into account in mass products like furniture or consumer goods.

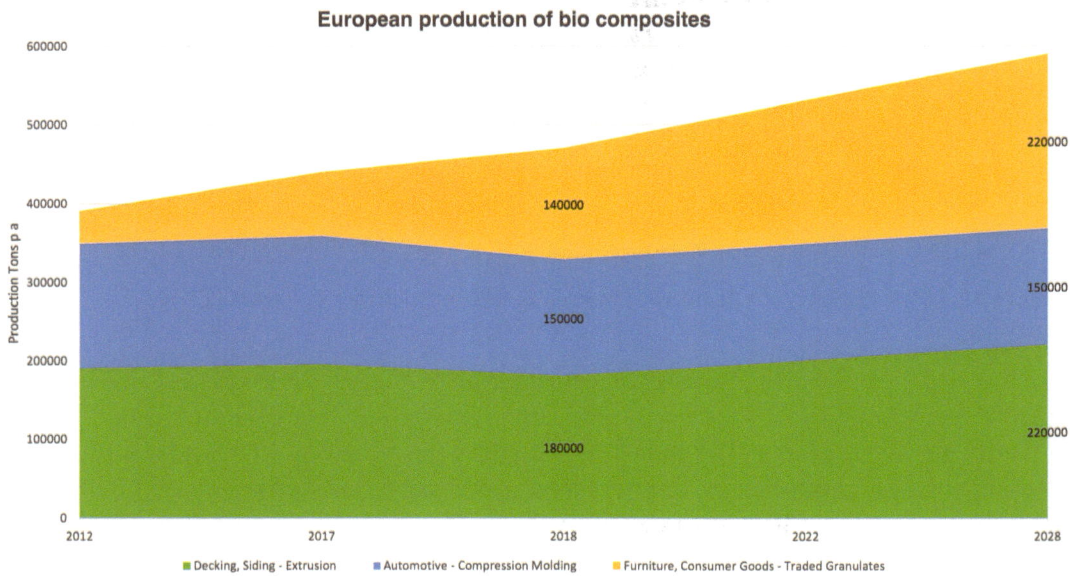

Figure 12.6 European market for bio-composites from 2012–2019, and the forecast for 2020–2028 [4]

12.3 Composite Manufacturing

The first step in composite production is always the manufacture of the individual matrix and fiber components. The production of the carbon fiber is especially energy-intensive and expensive due to the pyrolysis of the PAN precursor. This also explains the high costs of up to 20 €/kg and 18 kg CO_2/kg for some grades. Glass fiber reinforcement is mainly used for rotor blades of wind turbines, boat building, and container construction. Both fibers are used in automotive construction, while the expensive carbon fibers are used in aerospace industry, sports equipment, and occasionally in large rotor blades of wind turbines. Offshore turbines are moving quickly from 3 MW to next-generation turbines rated at 5 MW and larger, on which blade lengths for both on- and offshore systems regularly exceed 45 m.

The next step in production is usually processing of the fiber bundles (rovings), made of up to 12,000 individual fibers, into semi-finished fiber products such as woven fabrics, mats, or unidirectional layers. These semi-finished products can be dry, powdered, or impregnated with a matrix. As a rule, preforms are used as an intermediate step in production. This can lead for the first time to waste in the manufacturing process. This waste volume must be minimized, otherwise the first recycling steps would be necessary during production. Short or long fibers are cut from dry continuous rovings and can also be recycled if necessary. Sections made from thermoset prepregs, on the other hand, are more difficult to recycle, since the additional step of separating the matrix from the fibers has to be carried out here. RTM (resin transfer molding), hand lay-up, infusion, prepreg pressing, and automated tape laying are used in the production of thermoset molded parts. After curing, the parts are often re-contoured by post-processing with water jet, milling, or grinding. This in turn creates waste and dust, which also has to be recycled. To be sustainable, production waste must be avoided as far as possible.

In general, the manufacturing processes for thermoplastic composites can be energy-intensive due to the heat required to melt the polymers. However, the re-meltability also offers easy recycling possibilities. Thermoplastic semi-finished parts can be based on short- or long-fiber-reinforced granules, flat sheets (woven, unidirectional, or mat-reinforced) or unidirectional-fiber-reinforced tapes. Fiber-reinforced thermoplastic components are generally produced by injection molding based on short- and long-fiber reinforcement. A special method is direct compounding (IMC). A twin-screw compounding extruder is combined with the clamping unit of an injection molding machine. In this case, there is no need to use pellets as a semi-finished product, and the long fiber reinforcement can be incorporated directly into the polymer melt in a sustainable and energy-optimized manner. Continuous-fiber-reinforced thermoplastic composites are processed by compression molding. In this process, the semi-finished product must first be heated to deep-drawing temperature in a convection or infrared oven. In this process, the pressed component can also be supplemented with a simultaneous injection molding step using a deep-drawing tool. As a result, a wide variety of functional elements can be added, making subsequent assembly steps superfluous. There is also the option of overmolding the component edges in the tool after pressing in the same system, so molded parts can be produced without trimming. This eliminates the need for a finishing step, which saves costs and, above all, minimizes scrap down to zero.

In a publicly funded research project "MAI", a lightweight component as a thermoplastic mono-material system was researched under special sustainability and recycling aspects, and its feasibility was demonstrated on an industrial scale. A special feature is the significant shortening of the production process chain in connection with minimal production waste and optimal recycling options (Figure 12.7).

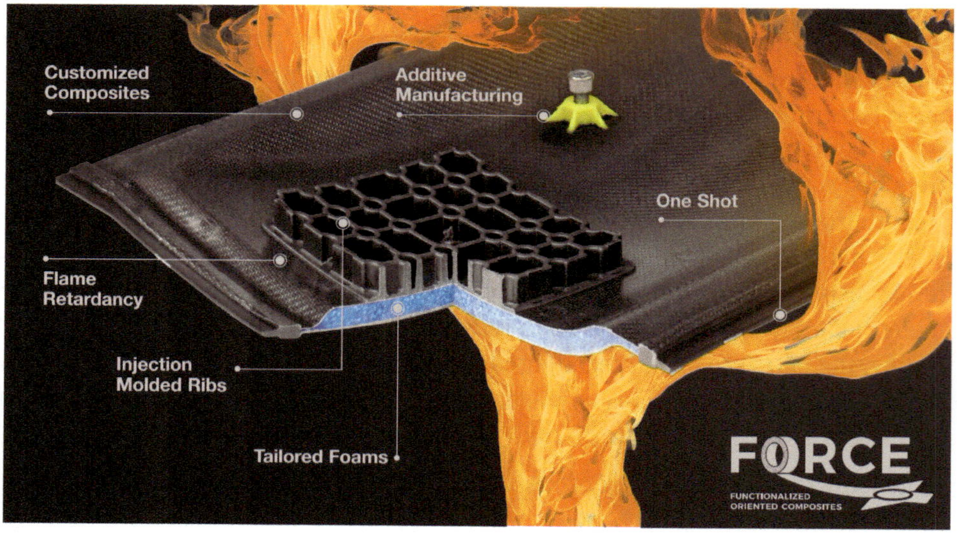

Figure 12.7 For recycling optimized sandwich part in one-shot processing (MAI Sandwich) [1]

A clever combination of processes with just one tool on a vertical injection molding machine made it possible to reduce the cycle time to less than 3 minutes for a finished part. The cover layers were connected to the thermoplastic cores (foams, rebate cores, or lattice struts) in a single process. Curved, complex structures could be created by thermoforming and realized in two injection molding steps. The injection molding step is used to integrate functional elements and eliminate subsequent assembly steps as well as edge refinement, which makes trimming during post-processing superfluous. In general, the component shows the advantages of thermoplastic composites with regard to recycling: the same polymer system can be used for the outer layers and the core materials, so that a purely thermal join is possible without an additional adhesive layer. In addition, this thermoplastic mono-material system (e.g., PP-GF, PA-GF, PESU-CF, etc.) enables the component to be recycled later by granulating and returning the granules to the injection molding process as filler or reinforcing material.

In general, great efforts have been made in recent years to minimize the primary energy consumption of composite materials in production. For example, the German leading-edge cluster MAI-Carbon was able to achieve a significant reduction in primary energy consumption for carbon-fiber-based composites in several projects using demo components (Figure 12.8).

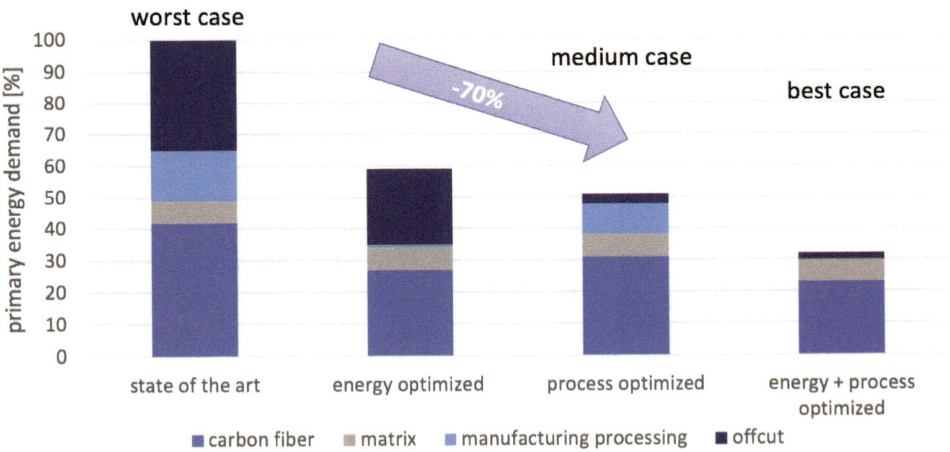

Figure 12.8 Primary energy demand for CFRP components (MAI Enviro)

Material selection, manufacturing process and component design are the essential parameters of a composite component, and must be determined in the early phase of component design with regard to sustainability and recycling. Above all, the manufacturing process determines the energy consumption and the amount of waste in production, while the choice of materials determines the recyclability. Even in the earliest phases of product development, it is important to consider the sustainability of the product, including recycling options. In this context, reference is made to Chapter 17. For lightweight construction in particular, a complete sustainability assessment of a component must be carried out in mobility applications, and thus the footprint savings in use must be shown.

■ 12.4 Challenges and Aims for Composite Recycling

In principle, composite materials can be separated into their starting materials only with great difficulty. For an overview of all the possible recycling methods and to get an idea of future developments, the following articles are recommended:

- Recycling of composite materials: Yang et al., 2012 [5]; A. A.van Oudheusden 2019 [6]
- Recycling technologies for thermoset composite materials - current status: Pickering, 2005 [7]
- Current status of recycling fibre reinforced polymers: Review of technologies, reuse and resulting properties: Oliveux, Dandy, & Leeke, 2015 [8]
- Composites recycling: Where are we now?: Job et al., 2016 [9]
- Technology readiness level assessment of composites recycling technologies: Rybicka et al., 2016 [10]
- Composite Materials Recycling Technology State of the Art and Sustainable Development for the 2020s: Krauklis, Karl, Gagani, & Jørgensen, 2021 [11]

In general, the costs for recycling and reusing are many times higher compared to using new materials. For example, considering glass fiber, the price of virgin glass fiber is so low that no currently available process can provide recycled glass fiber with the same properties as virgin fiber at a competitive price [8].

Recycled materials always show a lower level of properties, since they were already damaged during the first use and their quality is then reduced again by the recycling process [12, 13]. In order to obtain higher mechanical performance over successive recycling loops, it is important to optimize the fiber recovery process and avoid fiber damage [14], both during product use and during the recycling process. Since recycled composite materials have poorer properties, new applications must first be found for the recyclate. For example, recycled fibers are short and fluffy, making them difficult to incorporate into a new material as the bond between the fluffy fibers and the new matrix is reduced [8]. Given these disadvantages, it may not make sense from a sustainable perspective to use and recycle composite materials in the current way. This would indicate that the way the composite materials are currently constructed, used, and/or recycled needs to change. A desirable recycling process would provide near-new quality materials at a lower cost than producing virgin fiber [6]

Individual recycling paths are used depending on the matrix type: with thermoplastic composite materials, no separation of fiber and matrix is necessary for recycling. Thermoplastics can, for example, be converted into a polymer melt in an extruder, granulated from an extruded strand, and fed directly into other processes as granules. However, the shearing of the screws on the cylinder wall in the extruder shortens the fiber length and reduces the length of the macromolecules.

Thermosets are chemically cross-linked plastics and are therefore insoluble and unmeltable, which makes material recycling practically impossible. The process of separating the matrix from the fibers or the fillers is very complex and often incomplete. The following goals can be derived for the recycling of thermoset composites:

- Fiber length distribution sufficient to maintain the mechanical properties
- Recovery without damaging the fiber surface
- High gain of recovery
- Recovery in a CO_2 efficient process

The most common reinforcing fibers are glass and carbon, but also aramid and, above all, increasingly natural fibers (flax, hemp, bamboo, etc.). The reinforcements can be in the form of fibers in a wide variety of arrangements and lengths, as well as fillers with different shapes. In addition, metal inserts can be integrated into composite components, or additional fillers (e.g., flame retardants) can be present in the matrix. In addition, in aerospace components (e.g., the structure of an aircraft), copper grids close to the surface are often found as lightning protection.

Recovery of the reinforcing fibers, in particular the expensive and energy-intensive carbon fibers, is also of great economic interest. The fibers should be recovered in their original length as far as possible in order to minimize any loss of properties regarding the reuse of the recyclate in a new application. Recycling is most difficult in the case of composites with unidirectional fiber reinforcement. In this case, all common recycling technologies lead to a shortening of the fibers and a loss of the targeted alignment. There is currently no recycling technology available to reclaim highly oriented rovings, laminates, or fabrics. These topics are currently being intensively investigated. The greatest challenge is therefore to avoid the downcycling of continuous-fiber-reinforced composites. Current technologies lead to short-fiber-reinforced plastics for processing in injection molding or in a pressing process. A special, complex process is the further processing into nonwovens or yarns.

A sustainability assessment or, even better, a life cycle assessment should already be carried out in the planning phase of a recycling process for composite materials. Initial studies for cradle-to-gate material comparisons show that the reuse of recycled carbon fibers in short-fiber granulates for injection molding can keep up with aluminum die-cast products in terms of primary energy consumption and global warming potential. In the study, primary aluminum with 60 wt% of secondary aluminum was compared with a PA 66 with 60 wt% of recycled PA 66 with proportions of 20, 40, and 60 wt% of recycled carbon fibers. Studies show that compared to recycled aluminum (60% by weight), the primary energy requirement in Europe for a recycled PA 66 rCF compound is around 60% lower. This is associated with a reduction in the greenhouse potential from CO_2 emissions of 55% [15].

In addition, it must be noted that (according to a WHO definition) carbon fibers less than 5 μm in length and with an $L/D > 3$ are carcinogenic. Despite existing recycling plants, there are currently no large-scale applications using recycled fibers. These markets must first be developed in the next few years.

12.5 Routes for Composite Recycling

The main recycling routes for composites can be divided into energy recovery and material recycling. Material recycling saves virgin material and also energy, since production processes using reclaimed material are less energy-intensive than virgin material processes, while energy recovery saves other fuels in different energy systems. In the case of energy recovery, the material is incinerated as part of a thermal recycling process, during which time it releases energy. If a sustainability analysis of a component in the cradle-to-grave framework shows that recycling would be too costly in terms of energy, this option may well make sense if the use of virgin fibers is more favorable overall. Otherwise, mechanical, thermal, and chemical recycling are the main material recovery processes (Figure 12.9).

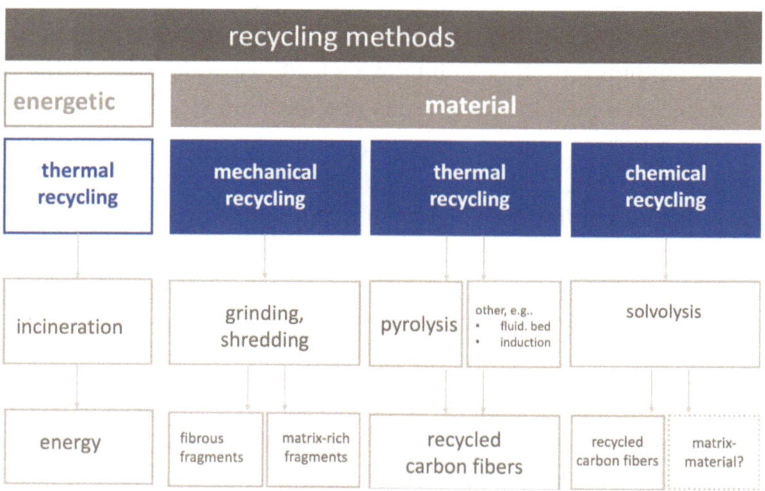

Figure 12.9 Composite utilization and recycling routes [16]

In the case of thermosetting composite materials, reshaping or dissolution after curing are no longer possible because of the crosslinking. For this reason, grinding processes are predominantly used for the production of powders and (more recently) for pyrolysis processes. In mechanical recycling, fibrous and powdery fractions are produced through grinding and shredding. In addition to incineration, there is always the option of comminuting the thermoset and then adding the recycling material as a filler to the new material. For the use as filler, it must be finely ground. With particle recycling, the waste is crushed in a hammer mill while largely protecting the fibers. The resin is obtained as a fine powder and is sieved off, while the short fiber chips remain in the sieve overflow. Both products can be used as fillers and reinforcing materials in new composite materials (especially SMC/BMC).

In the case of GRP, further processing in a cement clinker process can also be considered. In a cross-flow chipper, for example, rotor blades are shredded to a length of 50 mm, the metals are sorted out and the matrix-glass fiber mixture is fed to the cement kiln. The plastic is used to recover energy and the glass is recycled as a cement raw material. However, the demand for clinker production is not so great that larger quantities can be used in the future. In the case of thermoplastic composite materials, the mechanically obtained shredder flakes can be added to the fresh injection molding or extrusion mass. If the shredded material is to be returned to a component as 100% recyclate, then stabilizers must be added to the recyclate. Otherwise, the molecular weight of the matrix is broken down in a thermoplastic recycling cycle by repeated melting.

Pyrolysis is the most common of the thermal recycling processes. The fluidized bed and induction processes are also used. The current focus is on the thermal decomposition or combustion of the thermosetting plastic matrix and thus the recovery of largely preserved long fibers or fabrics. All processes ultimately lead to recycled fibers, but with a shorter length and different surface properties. The pyrolysis process is currently most commonly used for CFRP. In the absence of oxygen, organic substances are thermally decomposed into soot, gas, and oil in the temperature range 400–600 °C. The carbon fibers decompose at temperatures above 3600 °C into carbon dioxide. In comparison, the plastics decompose at temperatures below 600 °C. The solid pyrolysis product is a complex mixture of carbon fibers and soot that can be separated by mechanical processes. In fluidized bed plants with a sand bed, the plastic in CFRP waste can be completely burned off the carbon fiber surface at 550 °C without major fiber damage. At the same time, however, the fiber sizing is also burned off. The sizing serves to mechanically protect the fiber bundles. The carbon fibers only tend to oxidize above 600 °C. Temperatures below 600 °C must be maintained in fluidized bed plants because of the intensive heat treatment and the mass transfer processes. The fibers are entrained with the exhaust gas flow and filtered out of it. Possible metallic components of the CFRP component remain in the sand bed. In the solvolysis processes, the matrix is separated from the fibers by dissolution. Here, too, the recycled fibers can be reused for new fiber composite components. Further research is being conducted into the solvolysis of the plastic component in supercritical fluids. Solvolysis dissolution in water at temperatures of 200–350 °C also enables the resins to be broken down into low-molecular, soluble alcohols. The fibers remain in their original length and shape, including the coating, and can therefore be reused for components of the same shape (Figure 12.10)

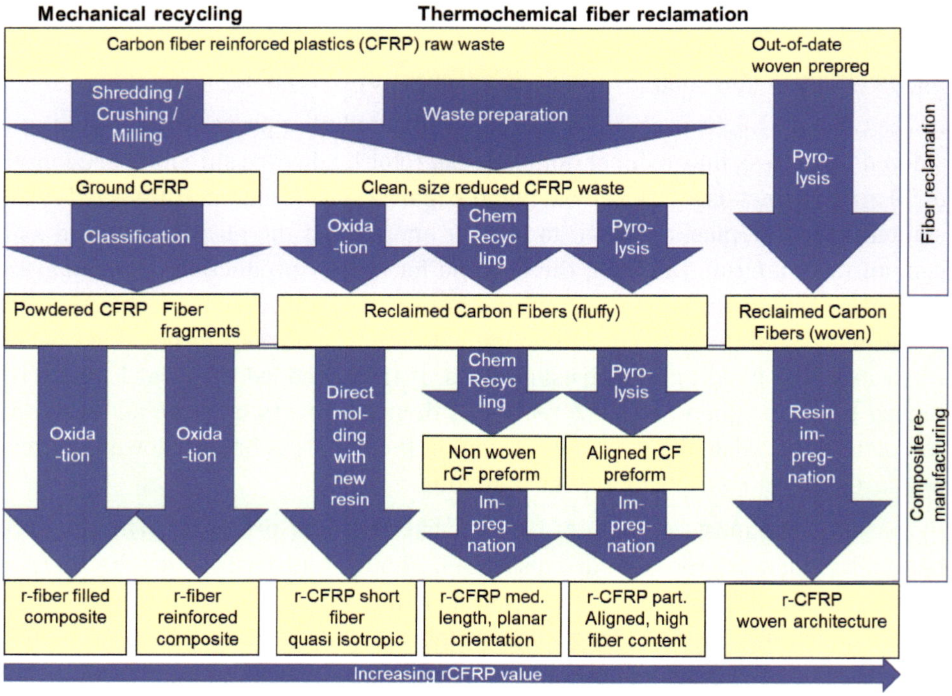

Figure 12.10 Overview carbon fiber recycling [17]

Another process under investigation is the application of pulsed high-voltage technology. This process is based on high-voltage breakdown in the composite material. In tests with CFRP, homogeneous disintegration of the composite components was achieved, whereby the fiber length of the carbon fibers was retained [18].

The company Carbo NXT indicates that the emission (in accordance with GWP100) for a recycled carbon fiber by pyrolysis is approximately 6 kg CO_2/kg, compared to approximately 20 kg CO_2/kg for virgin carbon fibers.

The companies Covestro and Mitsubishi Chemical Advanced Materials GmbH announced that they would cooperate for the recycling of production waste during manufacturing of continuous carbon fiber reinforced polycarbonate (PC-CF). The production waste is reused as recycled PC-CF compound [19].

12.6 Composite Recycling for the Wind Energy Sector

The topic of sustainability is playing an increasingly important role in the wind energy industry. An important question with rotor blades is always: Are there suitable materials that are bio-based or recycled? Balsa wood as the core material in the spars of the rotor blades is a renewable raw material. There are also suppliers of PET foams as the core material, and recycled PET bottles are used as input material [20].

Wind turbines are designed for a service life of 20 to 25 years. Experience so far shows that the actual service life is longer, which increases the profitability of an installation. In the meantime, a market for used wind turbines has developed [21], so that a considerable part of the wind turbines replaced by "repowering" in Germany is overhauled after dismantling, and sold on to developing countries. The old layers of plastic paint are peeled off the rotor blades, and any cracks and holes that have arisen are filled. After the repainting, the systems will be rebuilt and operated at locations in other countries.

In view of the expected amount of >10 kt of rotor blades to be disposed of in 2024 [22], further recycling options have largely been exhausted. Around 85% of a wind turbine can already be recycled today: the reinforced concrete of the tower, the metals of the generator, the electrical cables, and the transformers. With regard to the composite materials in the rotor blades, the situation is as follows:

Landfilling in Germany has been prohibited by law since 2004. Incineration in waste incineration plants is neither attractive nor sustainable due to technical problems with the resulting dust. Therefore, thermal processes were developed that correspond to the current state of the art.

As early as 2010, the company Zajons Disposal Technology described a process in which discarded rotor blades are cut into larger pieces on site with a mobile saw, shredded in a processing plant, and processed into granules. These granulates are to be used in the cement industry and replace fossil fuels and the addition of quartz glass that has been customary up to now (energy recovery). From today's perspective, the company probably came onto the market too early, because at that time the number of rotor blades to be disposed of was small, and the company had to file for bankruptcy. In 2016, the successor company, neocomp GmbH, put a processing plant with a further developed process into operation in Bremen. In this plant, which has an approved capacity of 80 kt/a, the material is processed by a crossflow chopper after coarse pre-shredding. Residual materials from paper production are then mixed in and the mixture prepared for delivery to the cement industry. Zajons was awarded the AVK environmental award in 2011, while neocomp received the GreenTec award in 2017. In the meantime, there are several suppliers in the disposal industry that refurbish rotor blades and separate them into constituent materials.

The increasing proportion of carbon fibers opens up new potential for recycling old rotor blades, even though they are around 10 times more expensive than glass fibers: the carbon fiber strips installed in them are too good for raw material recycling (which, by the way, is not so easy because the conductive carbon fiber dust requires a large technical effort to avoid short-circuits in electrical equipment). Reconditioning is therefore worthwhile, and CFK Valley Stade Recycling GmbH has been operating a multi-stage reconditioning process in a 100 tons per year plant since 2010 [23]. In the first step, the carbon-fiber-reinforced composite components are first sorted and then pyrolyzed. The matrix materials decompose, leaving the pure carbon fiber in the form of the original fiber scrim or woven fabric, which is cut or ground in subsequent steps. The state of the art and a new approach to the production of high-quality recycled carbon fibers from random fibers and/or flakes have been summarized [24]. World market leader "Vestas" published its sustainability strategy for the next few years at the beginning of 2020, which consists of several building blocks [25]. The goal for 2040 is to produce only "zero-residue wind turbines". This goes far beyond the recycling technologies used today and requires the development of new concepts on the part of manufacturers and their suppliers.

The companies Adesso [26] and Connora [27] have described a new approach for splitting epoxy-amine thermosets. This approach uses diamines with acetal or ketal breaking points as hardeners for epoxy resins. The crosslinking points of acetal or ketal structures can be split hydrolytically with aqueous mineral acids. This creates a fusible, linear thermoplastic polymer from which the fibers can be separated again. Connora's technology was acquired by Indian epoxy resin manufacturer Aditya Birla Chemicals in 2019. A comprehensive overview about next generation wind blade recycling is given in Ref [28].

The goal of 100% recycling of wind turbines is a real challenge. In order to achieve this goal, new approaches to a closed-loop economy must be explored, both in the design and in the materials used for rotor blades. On the chemical side, all new building blocks must undergo REACH registration; substances of concern will be subject to corresponding restrictions.

12.7 Composite Recycling for Automotive

The high complexity of vehicle components today poses major challenges for recycling. In particular for carbon-fiber-reinforced plastics - which are increasingly used in structural components in cars, because they can absorb heavy loads - there are still no suitable and economical recycling methods. The disposal of the respective materials must be taken into account as early as the product design stage and particular attention must be paid to recyclability. Vehicle manufacturers should live up to their product responsibilities. Car manufacturers in the EU, for example, are bound by the environmental regulations of the EU. Accordingly, 95 percent of the vehicle weight must be recoverable and 85 percent recyclable. This also applies to vehicles with a carbon-fiber-reinforced (CFRP) structure.

Worldwide, almost 68 million vehicles were produced in 2014, of which 5.6 million were produced in Germany. In 2016 in Germany, around 40 million vehicles were in operation. Of the 3.2 million cars decommissioned in Germany deregistered in 2012, only 0.48 million ended up as end-of-life vehicles for domestic recycling. In some other industrialized countries, the situation is similar (EU, Japan). Of the 3.2 million cars decommissioned in Germany in 2012, only 0.48 million ended up as end-of-life vehicles for domestic recycling; 1.35 million cars were exported as used vehicles and the remaining 1.38 million cars are unaccounted for (theft, non-public use, illegal export). In some other industrialized countries, the situation is similar (EU, Japan) [29, 30].

However, the export of end-of-life vehicles means a considerable loss of secondary raw materials for the industrialized countries, which is often also a global loss, since in the developing world there is often no recycling at all [15].

For conventional combustion-engine-powered vehicles, composites are mainly used as add-on parts and the proportion does not play a significant role, since steel/ferrous materials, light alloys and polymers dominate. For electric cars, the body concepts are often still taken from combustion cars.

The BMW i3 represents one of the first times that composites have been used in a relevant proportion in battery-powered electric vehicles in medium serial production. The chassis (drive-module) consists of a light-alloy frame that supports the running gear, the complete drive and control unit, and the Li-ion batteries. The passenger compartment including the outer skin is called the "life module". This module consists of carbon-fiber-reinforced composite materials. On average, up to 50 percent of the original material can be reused when recycling CFRP."

Due to this design principle, BMW enables the sorting and removal of components and materials for recycling. This means that recycling starts with the dismantling of the life-module from the chassis, followed by the disassembly of the drive-module into its components (Li-ion pack, electric motor, power electronics, chassis, and aluminum frame) [15]. The BMW production network agreed on energy-saving targets and a complete power supply from renewable energies. At the Leipzig site, four wind turbines were erected on the production site of the BMW i3 for this purpose. The materials for the i3 were selected with sustainability in mind. For example, secondary aluminum or aluminum produced with renewable energies is used as far as possible. The carbon fibers for CFRP are manufactured using 100% hydropower, which reduces the greenhouse warming potential by up to 50% compared with conventional production of CFRP components [31].

The functional unit and the reference flow (Figure 12.11) are defined as a passenger car of the BMW i3 BEV series, at SOP (start of production) in 2014, with electric engine in ECE basic equipment over a mileage of 150,000 km in the new European driving cycle. The BMW i3 is a passenger car designed for metropolitan areas, with four seats and an all-electric range of up to 160 km. The drive components include a HV battery (high-voltage lithium-ion battery) and an electric synchronous motor with an output of 125 kW.

Therefore, the GWP as CO_2-equivalent is, over the life cycle of the BMW i3 BEV: approximately 57% production, 40% electricity supply, and 3% recycling [31].

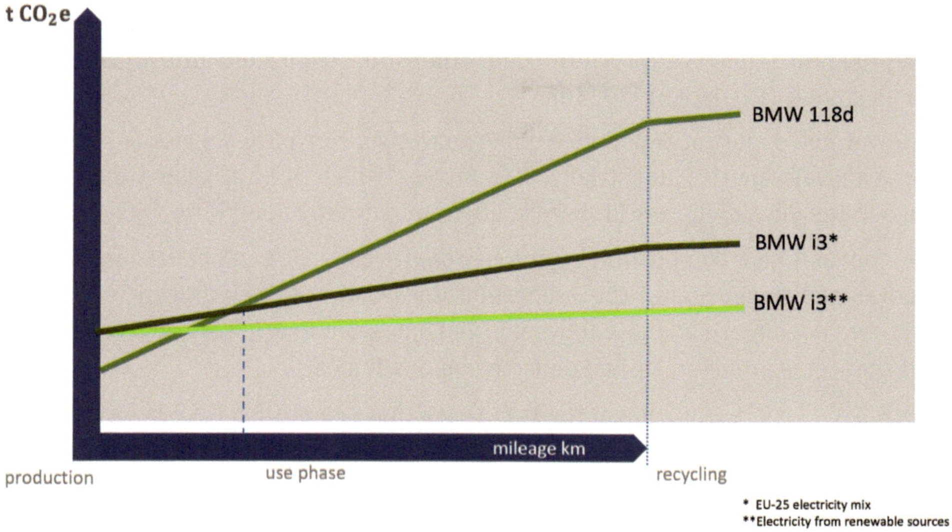

Figure 12.11 Classification of the BMW i3 BEV in relation to a conventional vehicle concept (118d) [31]

References for Chapter 12

[1] Neue Materialien Bayreuth GmbH

[2] Witte, E., Mathes, V.: Der europäische Markt für Faserverstärkte Kunststoffe/Composites 2021, 2022, *https://www.avk-tv.de/files/publications/files/avk_marktbericht_2022_final.pdf*

[3] Zottler, M.: Life-Cycle Analyse von Leichtbaukonzepten für den Automobilbau, Diplomarbeit, TU Graz, 2014

[4] nova-institue.eu, 2019

[5] Yang, Y., Boom, R., Irion, B., van Heerden, D.J., Kuiper, P., de Wit, H.: Recycling of composite materials. *Chemical Engineering and Processing: Process Intensification*, (2012) 51, pp. 53–68

[6] van Oudheusden, A. (TU Delft Industrial Design Engineering): Recycling of composite materials, Delft University of Technology (2019), *http://resolver.tudelft.nl/uuid:0749ed5c-7aeb-4275-abee-0f904a08ea4d*

[7] Pickering, S.J.: Recycling technologies for thermoset composite materials—current status. Advanced Polymer Composites for Structural Applications in Construction: ACIC 2004, (2005) 37, pp. 1206–1215, *https://doi.org/10.1016/B978-1-85573-736-5.50044-3*

[8] Oliveux, G., Dandy, L.O., Leeke, G.A.: Current status of recycling of fibre reinforced polymers: Review of technologies, reuse and resulting properties. *Composite structures*, (2015) 72, pp. 61–99, *https://doi.org/10.1016/j.pmatsci.2015.01.004*

[9] Job, S., Leeke, G., Mativenga, P.T., Oliveux, G., Pickering, S.J., Shuaib, N.A.: Composites recycling: Where are we now?, (2016) 11, *https://compositesuk.co.uk/system/files/documents/Recycling Report 2016_1.pdf*

[10] Rybicka, J., Tiwari, A., Leeke, G.A.: Technology readiness level assessment of composites recycling technologies. *Journal of Cleaner Production*, (2016) 112, pp. 1001–1012, *https://doi.org/10.1016/j.jclepro.2015.08.104*

[11] Krauklis, A.E., Karl, C.W., Gagani, A.I., Jørgensen, J.K.: Composite Material Recycling Technology - State-of-the-Art and Sustainable Development for the2020s. *J. Compos. Sci.*, (2021) 5, p. 28, https://doi.org/10.3390/jcs5010028

[12] Yang, Y., Boom, R., Irion, B., van Heerden, D.J., Kuiper, P., de Wit, H.: Recycling of composite materials. *Chemical Engineering and Processing: Process Intensification*, (2012) 51, pp. 53–68, *https://doi.org/10.1016/j.cep.2011.09.007*

[13] Chawla, K.K.: Composite Materials: Science and Engineering, *https://doi.org/10.1007/978-0-387-74365-3*

[14] Longana, M.L., Ong, N., Yu, H.N., Potter, K.D.: Multiple closed loop recycling of carbon fibre composites with the HiPerDiF (High Performance Discontinuous Fibre) method. *Composite Structures*, (2016) 153, pp. 271–277, *https://doi.org/10.1016/j.compstruct.2016.06.018*

[15] Martens, H., Goldmann, D.: *Recyclingtechnik* (2nd edition), Springer Vieweg, 2016

[16] Müller, W., Schneller, A., Horn, S.: Übersicht der Verfahren zur Trennung von Faser und Matrix, Workshop Carbon Composite Recycling CCeV Strategiekreis Nachhaltigkeit on January 31, 2017

[17] Pimenta, S., Pinho, S.T.: "Recycling of Carbon Fibers" in Handbook of Recycling, Elsevier, 2014, p. 214 ff.

[18] Woidasky, J.: Weiterentwicklung des Recyclings von faserverstärkten Verbunden, in: *Recycling und Rohstoffe*, Vol. 6, Thomé-Kozmiensky, Goldmann, D. (eds), TK Verlag, 2013

[19] Gutbrod, M.: *K-Zeitung*, *https://www.k-zeitung.de/rezyklat-aus-polycarbonat-verbundwerkstoff/*, accessed June 4, 2021

[20] Gurit, Broschüre Shape, (2019) 18, p. 8

[21] *https://www.baywa-re.de/de/wind/weitere-themen/gebrauchte-anlagen/*, accessed February 02, 2021

[22] Weh, A., Monti die Sopra, F.: Zerkleinerung und Aufschluss von Abfallströmen mittels gepulster Hochspannungstechnologie, *Recycling und Rohstoffe*, Vol. 4, ed.: Thomé-Kozmiensky, K., Goldmann, D., TK Verlag, 2011

[23] *https://www.cfk-recycling.de/index.php?id=5*, accessed February 02, 2021

[24] Cherif, C., Hund, R.D., Abdkader, A., Treichel, S. (TU Dresden/BASF): Patent EP3015576A1, 2016

[25] Vestas, press release, January 2020

[26] Liang, B., Pastine, S.J., Qin, B. (Adesso Advanced Materials): Patent WO2012071896A1, 2012

[27] Pastine, S.J. (Connora): Patent US20170114002A1, 2017

[28] Mason, H.: "Moving towards next-generation wind blade recycling", *Composites World*, June 28, 2022

[29] Jahresbericht über Altfahrzeugverwertungsquoten in Deutschland 2012, Umweltbundesamt, 2014

[30] Verbleib von 1,4 Millionen stillgelegten Pkw unklar, Umweltbundesamt, 2014, *https://www.umweltbundesamt.de/themen/verbleib-von-14-millionen-stillgelegten-pkw-unklar*, accessed August, 11, 2021

[31] Umweltbericht (environmental report) BMW i3 BEV, *https://www.bmwgroup.com/content/dam/grpw/websites/bmwgroup_com/responsibility/downloads/de/2016/Umweltbericht-BMW-i3.pdf*, accessed August, 11, 2021

PART C
Value Chain

13 Comparison of Global Recycling Processes

Raphael Kiesel

"For the last 150 years, plastic materials have been key enablers for innovation and have contributed to the development and progress of society." [1]. This quote from PlasticsEurope portrays the importance of plastics in today's industrialized society. Imagining today's world without plastics is virtually impossible. Compared to every other engineered material in the world, plastics have the highest growth rate, which is related to their unique properties. Plastics have become the most important raw material for a variety of products and applications, including automotive components, mobile phones, pipes, foams, and packaging. Their versatility, durability, and high strength-to-weight ratio have led to an exponential 20-fold growth in their production in the past half-century. They constitute about 15% of the weight of a car and 50% of the Boeing Dreamliner [2–6]. Between 2010 and 2019, worldwide plastic production increased by nearly 40% from 270 Mt to 369 Mt [7].

Increasing production consequently results in higher amounts of plastic waste. In 2018, 250 Mt of plastic waste were generated worldwide. Figure 13.1 shows the plastic waste distribution of the different geographic regions.

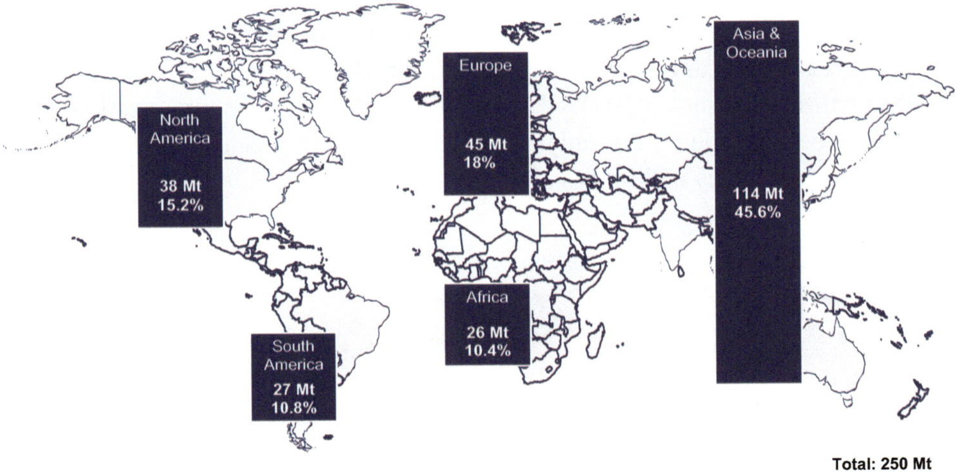

Figure 13.1 Plastic waste production worldwide in 2018 (remark: Europe covers both EU and non-EU countries) [8]

This increased amount of plastic waste as well as a continuously growing environmental awareness - especially due to plastics in the oceans - has put a focus on plastic waste handling and recycling.

In general, there are three waste handling methods - landfilling, incineration with energy recovery (waste-to-energy - WTE), and recycling. Figure 13.2 shows on a high level how these are interrelated. In 2018, 69.2% of plastic waste (173 Mt) was collected. Of this collected waste, 72 Mt were disposed in landfills; 81 Mt were recovered, which is around a third of the total waste. The recovered waste was nearly evenly distributed between recycling and WTE with 40 Mt each.

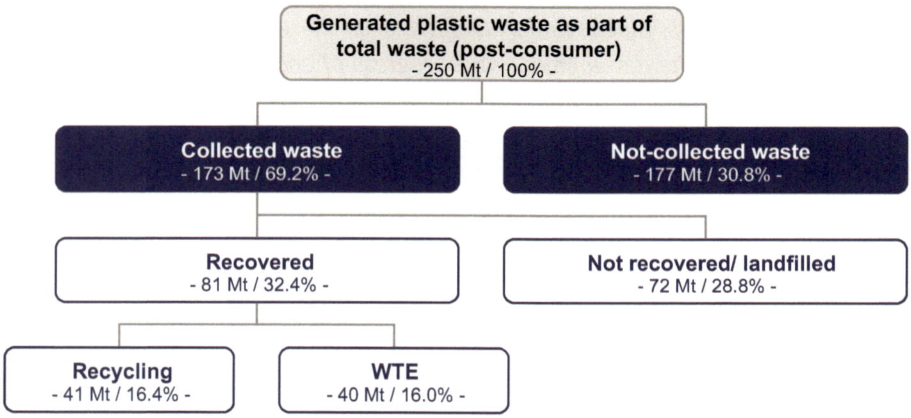

Figure 13.2 General division of plastic waste handling processes and their ratios in 2018 [8]

However, the ratio of landfilling, WTE and recycling varies a lot from region to region. As Table 13.1 shows, the EU28+2 has a plastic recycling rate of 31.3%, whereas the USA recycles 9.3% of its plastic waste and India about 60%. But it is not just the recycling rates that vary: the amount of plastic waste that is not collected varies significantly between regions (from 0% in Japan to 32% in China). As well as political reasons, described in Chapter 5, this also has organizational and process-driven reasons - not only between the different regions and continents, but also the countries within them [8, 9].

Table 13.1 Overview of Plastic Waste Handling in Different Regions in 2018 [8]

Region		Total amount plastic waste [Mt/%]	Recycled plastic waste [Mt/%]	WTE plastic waste [Mt/%]	Landfilled plastic waste [Mt/%]	Not collected plastic waste [Mt/%]
Asia	China	59.1/100	12.6/21.3	19.2/32.5	8.3/14.0	19.0/32.2
	India	9.5/100	5.7/60.0	0.3/3.2	0.9/9.5	2.6/27.4
	Japan	9.0/100	2.7/30.0	5.0/55.6	1.3/14.4	0.0/0.0

Region		Total amount plastic waste [Mt/%]	Recycled plastic waste [Mt/%]	WTE plastic waste [Mt/%]	Landfilled plastic waste [Mt/%]	Not collected plastic waste [Mt/%]
Europe	EU28+2	30.0/100	9.4/31.3	12.4/41.3	7.2/24.1	1.0/3.3
North America	Canada	3.3/100	0.3/9.1	0.1/3.0	2.8/84.8	0.1/3.0
	USA	34.5/100	3.2/9.3	5.1/14.8	25.5/73.9	0.7/2.0

Therefore, in this chapter, the plastic waste handling processes of the different regions are analyzed in more detail. The chapter focusses on the regions and countries in Table 13.1, as these regions contribute 60% of the global plastic waste production and provide a solid database regarding plastic waste handling. As shown in Figure 13.3, this chapter mainly focusses on the **consumed plastics**, **waste collection**, **recycling**, and **recyclates**. For the latter, imports and exports are further considered, to point out what "recycling" really means. For the recycling process itself, mechanical recycling is assumed.

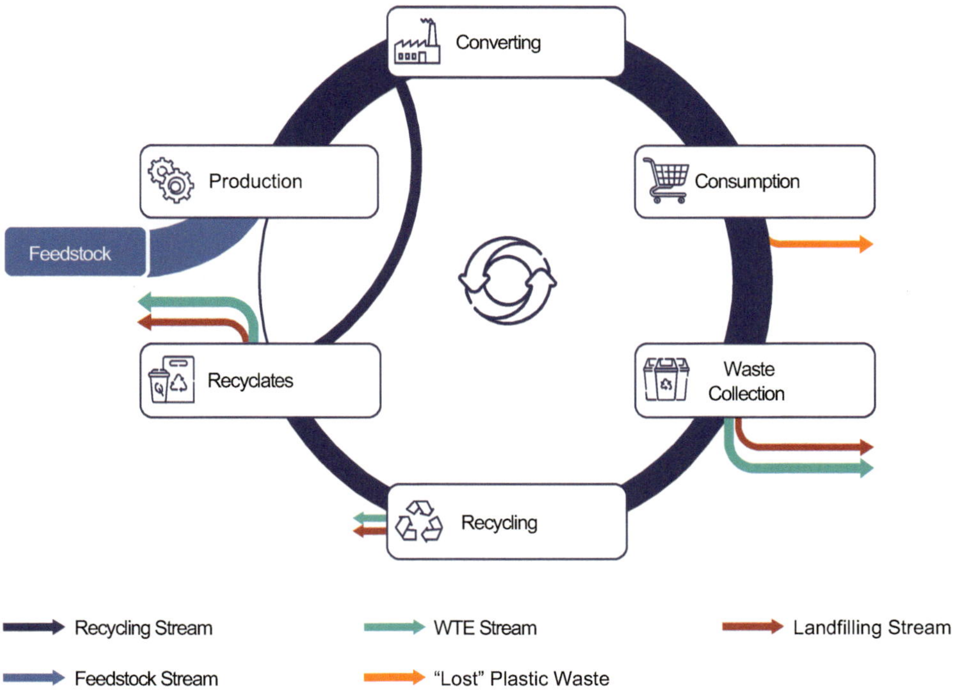

Figure 13.3 Plastic flow from feedstock to recyclate [8]

13.1 Europe

Europe has a very strong plastics industry, which has remained relatively consistent over the last decade. Due to available data, the following data focus on the EU28+2 countries (28 European Union members, plus Norway and Switzerland).

Figure 13.4 shows the **plastic consumption** distribution of the EU28+2 countries in 2018, consuming in total 51.2 Mt of plastic. With 39.9% of the total demand, the packaging sector dominated, and had over twice the demand of the building and construction sector, which with 19.8% ranked second regarding demand [10].

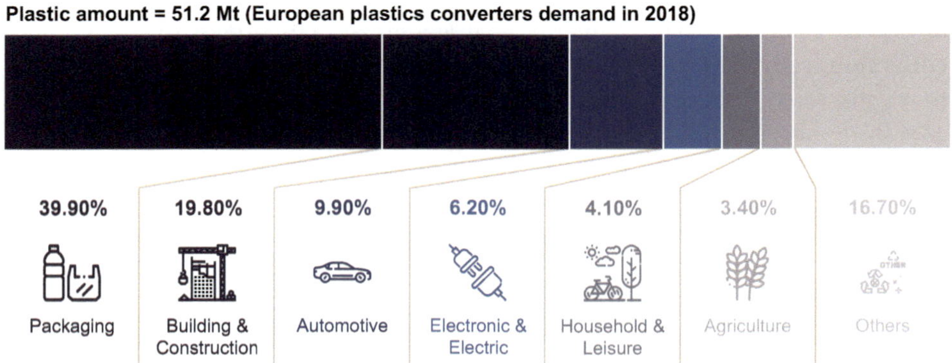

Figure 13.4 Total European plastics converters demand in 2018 [10]

Comparing the data presented for Europe's plastic materials demand (Figure 13.4) with the generated post-consumer waste (Figure 13.5), differences between demand and waste can be noted. Here, packaging is even more dominant, with over 60% of the total plastic waste. This difference arises due to the short service life of plastic packaging, especially single-use plastic (e. g., bottles, plastic bags, and plastic wrap). In contrast, plastics used in construction have a much longer service life, between 30 to 40 years. This explains the discrepancy between the demand for plastic in this sector (19%) and the relatively low contribution (5%) to plastic waste [11].

In Europe, all three presented **waste management treatments** are used: disposal in landfills, WTE, and recycling. In 2016, recycling overtook landfilling for the first time. In 2018, 29.1 Mt of plastic waste were collected through official schemes (out of 30.0 Mt plastic waste in total), from which 42.6% was burned to recover energy, 32.5% was recycled, and 24.9% was landfilled. Figure 13.6 shows the development of waste treatment in Europe between 2006 and 2018 [9].

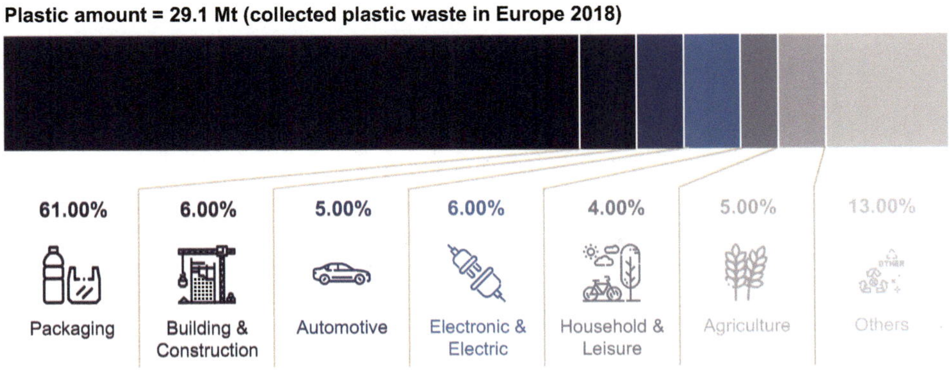

Figure 13.5 European plastic waste generation by sector in 2018 [11]

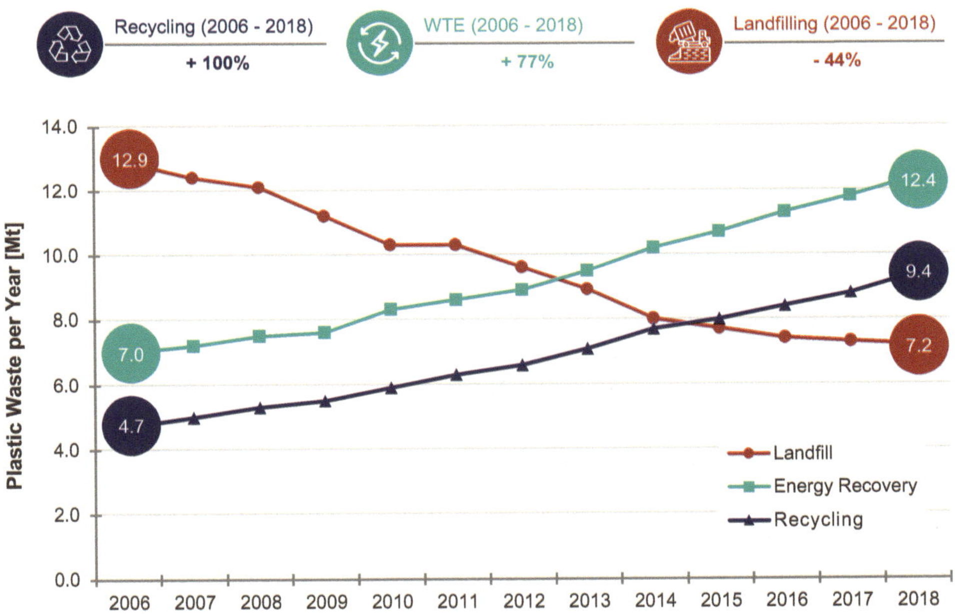

Figure 13.6 Plastic waste treatment in the EU28+2 countries from 2006 to 2018 [9]

However, this development cannot be recorded in every country. Due to a landfill ban in Switzerland, Austria, Netherlands, Germany, Sweden, Luxembourg, Denmark, Belgium, Norway, and Finland, which started partially in 1996, WTE and recycling are dominating in these countries – the proportion of landfilled plastic is nearly zero percent. However, in contrast, Croatia, Bulgaria, Cyprus, Greece, and Malta have landfill ratios for plastic waste of over 60% [9]. Figure 13.7 shows the different plastic waste handling ratios of the EU28+2 region.

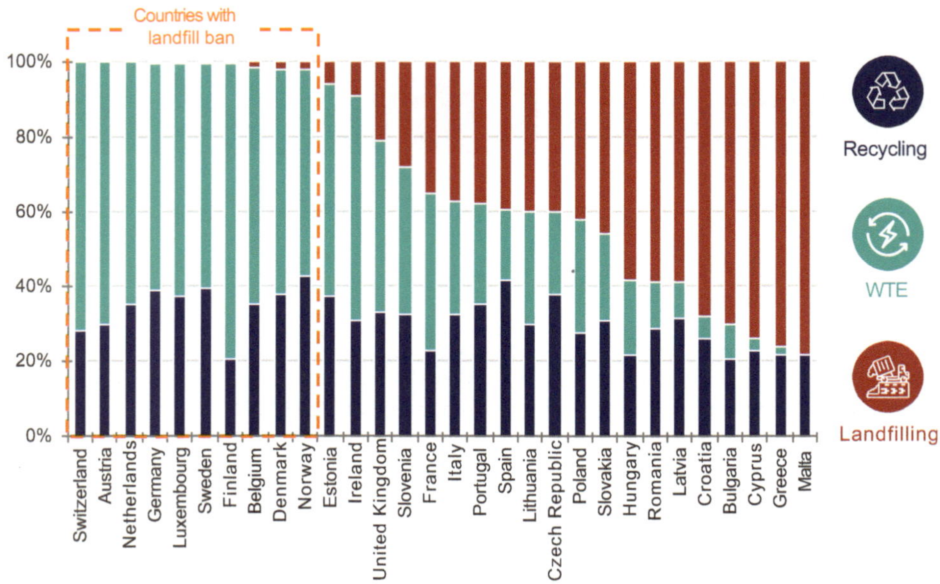

Figure 13.7 Waste handling in the different countries [9]

As mentioned in the introduction of this chapter, laws and political regulations are one reason for this inequality between the countries. However, two other factors play an important role. First, Croatia, Greece, Bulgaria, and Cyprus are the countries with the lowest amount of generated *plastic waste per capita* [12]. This means that the total amount of plastic waste is relatively small, which makes plastic recycling facilities not economic. As described by Rudolph et al., a certain amount of plastic waste is necessary to have a process which is economically profitable or at least economically balanced [5]. Second, all countries are below the EU average of *gross domestic product (GDP) per capita*, with Croatia and Bulgaria being (respectively) third-to-last and last in this list [13]. According to a World Bank report, the lower the income, the lower the amount of recyclates in the waste. This thus exacerbates the first effect.

Furthermore, waste collection rates and thus recycling rates vary widely by income levels. High- and upper-middle income countries typically provide universal waste collection. Low-income countries tend to collect about 48% of waste in cities, but outside of urban areas waste collection coverage is about 26%. In middle-income countries, rural waste collection coverage varies from 33% to 45% [14]. This of course has an impact on the plastic recycling rate.

Different types of **waste collecting systems** are in place. The design of the waste flows (e.g., complete individual flow, complete separation of all recyclable materials, joint collection of materials) varies between countries. Germany, for example, even differentiates on the level of federal state or individual cities. In addition to door-to-door recycling flows, there are also stationary recycling containers (e.g., for paper and glass) or recycling yards (e.g., for electronic scrap). This table shows that 24 out of 28 capitals work with a so-called "dual-stream" or "multi-stream" recycling system, meaning that the end-user separates the waste (e.g., paper from plastics) and that this waste is then collected separately. This will further be analyzed in the discussion on the USA (Section 13.2). However, in three of the 28 capital cities of the EU (Bratislava, Bucharest, and Prague), no waste is collected on the curbside at all, but only collected and recycled through central points. In most countries, mixed forms of recycling are implemented, as shown in Table 13.2. It is therefore clear that countries for which the capitals have no waste collection have high landfilling ratios for plastic waste (Slovakia: 45%, Romania: 58%, Poland: 42%) [15].

Table 13.2 Overview of the Collecting Systems in the 28 EU Capitals [15]

Capital	Complete separation/ multi-stream (curbside)	Partial separation/ single-stream (curbside)	Central private facility	Central public facility
Amsterdam	X		X	X
Athens	X	X	X	
Berlin	X	X	X	X
Bratislava			X	X
Brussels	X	X	X	X
Bucharest			X	
Budapest	X	X	X	
Copenhagen	X		X	X
Dublin	X	X	X	X
Helsinki	X		X	X
Lisbon	X		X	X
Ljubljana	X	X	X	X
London	X	X	X	X
Luxembourg	X	X	X	X
Madrid		X	X	X
Nicosia	X	X	X	X
Paris	X	X	X	X
Prague			X	X
Riga	X		X	X
Rome	X	X	X	
Sofia	X	X	X	X
Stockholm	X		X	X
Tallinn	X	X	X	X
Valletta	X	X	X	X

Table 13.2 Overview of the Collecting Systems in the 28 EU Capitals [15] *(continued)*

Capital	Complete separation/ multi-stream (curbside)	Partial separation/ single-stream (curbside)	Central private facility	Central public facility
Vienna	X	X	X	X
Vilnius	X	X	X	X
Warsaw	X	X		
Zagreb	X	X	X	X

As the recycling system in Germany is quite successful, it is further explained here. Germany's dual system organizes household packaging waste collection (via a yellow bag that is left on the curbside) and ensures presorting by the consumer and professional recycling of packaging in appropriate facilities. The disposal services (collection, sorting, and recycling of packaging) are financed via participation payments of packaging manufacturers and executed by commissioned disposal companies. Participation is mandatory and allows the manufacturers to label their products with the *green dot ("Grüner Punkt").* Since manufacturers are required to pay for their packaging products by weight, material consumption has decreased since the system was introduced due to use of thinner glass, paper, plastic bottles, and films. This further reduces the amount of waste that needs recycling. For plastic bottles a *deposit and return system* exists. The deposit is paid by the consumer upon purchase of a product. It ranges from €0.15 for refillable bottles to €0.25 for disposable bottles. The empty bottle can be returned to stations similar to vending machines that automatically scan and accept or reject the bottle and reimburse the deposit for all refillable bottles. All these improvements are possible due to mechanized collection, efficient recycling, waste treatment facilities, and continued education of the consumer over the last three decades. Last but not least, the economic strength of Germany and other European countries enables the investment in such projects and initiatives [5].

After the collection, plastic is shipped to the mechanical plastic recycling facility to produce recyclates. First, plastic is separated by polymer type. Second, plastics are ground into smaller pieces (shredding). Third, shredded plastics are washed to remove dust and dirt. Fourth, plastics are sorted again and controlled before being extruded. Lastly, plastic flakes are converted into homogenous pellets ready to use in the manufacture of new products. In this way, Europe does not focus on a type of polymer when designing the recycling process – other than the USA and Canada, as Section 13.2 describes.

Not all recycled plastic is converted to **recyclates**. Some parts of the recycled material are used as pellets, while some are exported outside the EU and re-enter other regions of the world's economy. Of the 9.7 Mt of plastic waste in 2018, 5 Mt of recyclates were produced, with 80% of them re-entering the European economy to manufacture new products (Figure 13.8 shows the distribution of the sectors the recyclates re-entered). The remaining recyclates (1 Mt) were exported [9].

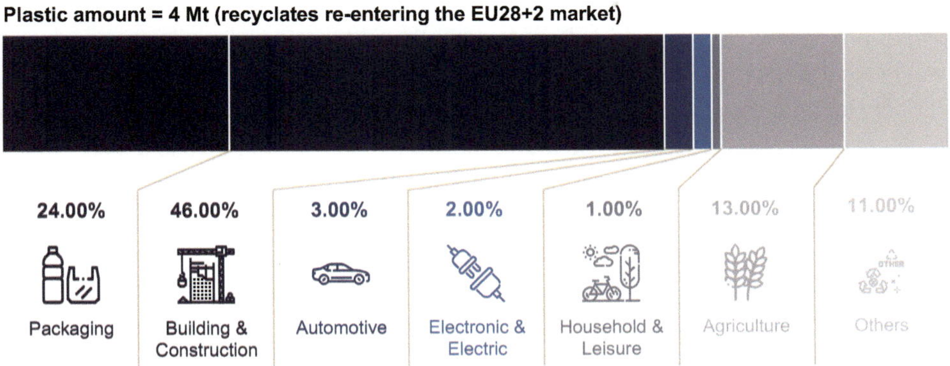

Figure 13.8 Distribution of recyclates re-entering the EU28+2 market in 2018 [9]

In total, 2.3 Mt of plastic waste were exported by the EU in 2018. Compared to 2014, when a maximum of 3.24 Mt was exported, the exported amount decreased by 40% [16]. Germany especially is exporting a high volume of plastic waste. The route of the exported plastics will be further analyzed in Section 13.3.

13.2 North America

The increase in plastic production and the replacement of other materials in many industries is also leading to an increasing amount of plastic waste in North America (or more specifically, the area covered by the North American Free Trade Agreement, NAFTA). Both the demand per sector as well as the usage per capita are similar to that for the EU. In 2015, NAFTA's per capita consumption of plastic material was 139 kg (in Western Europe, it was 136 kg) [17].

The increased consumption results in an increased amount of plastic waste. However, here lies a difference compared to Europe. The treatment of plastic waste differs significantly, as the following sections show for the USA (Section 13.2.1) and Canada (Section 13.2.2).

13.2.1 USA

The total amount of plastic waste in the United States increased by a factor of 80 between 1960 and 2017. In that time frame, the share of plastic waste in municipal waste also grew in parallel, demonstrating the increasing importance of plastic waste treatment in the United States. However, it should be noted that the growth rate of the amount of plastic waste has slightly decreased since 2005 compared to the sharp increase between 1980 and 2005 [6, 18]. However, there is a clear difference compared to Europe in terms of plastic waste treatment. As shown in Section 13.1 (and especially in Figure 13.6), in the EU28+2, landfilling of plastic waste is also decreasing and accounted for the smallest share of plastic waste treatment, at 24.7% in 2018. Recycling accounted for 32.3% in Europe in the same year, and incineration with energy recovery 43.0%. In the USA, the picture is different over the same period. Until 2015, the recycling rate slightly (but constantly) increased. However, since 2015, this trend stagnated and even regressed. In 2018, landfilling dominated waste management of plastic waste at 75.0%. Incineration with energy recovery accounted for 16.3% and recycling for only 8.7%. The evolution of the distribution of the three main waste management measures from 1960 to 2018 is shown in Figure 13.9 [6, 18].

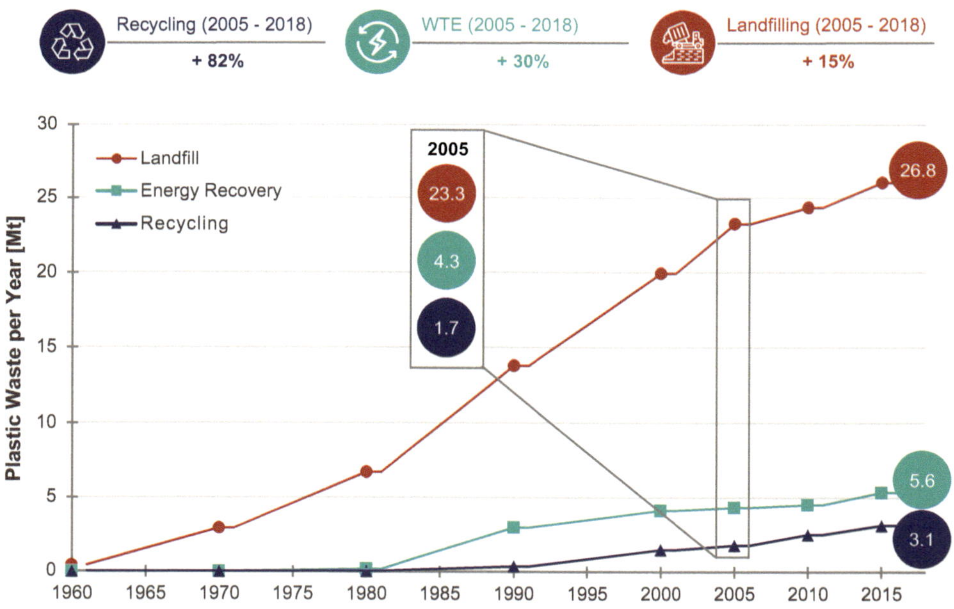

Figure 13.9 Plastic waste treatment in the USA from 1960 to 2018 [6, 18]

In contrast to the European **recycling process**, single-stream recycling is the standard in the U.S. Single-stream recycling works as follows: After collecting the recyclables (e.g., paper, glass, plastics, metal) mainly by curbside collection, they are transported to material recovery facilities (MRFs). A MRF is a processing facility where waste is sorted and prepared for additional processing for the different recyclables. This means that plastic is sorted out from the other recyclables and then sent to a plastic processing facility. Single-stream recycling started in 1989 in Phoenix and Arizona - the first large MRF started in 1993 [19]. Subsequently, many federal states and cities changed from curbside waste collection to single-stream recycling. In 2018, around 80% of the US communities used single-stream recycling. In 650 MRFs in the US, up to 100 kt of plastic waste are handled per day [8].

The growing problem with single-stream recycling is the *contamination* of plastics (around 25%). Contaminated material cannot be recycled and is either incinerated or landfilled. This leads to decreasing economic value of the material for recycling, and often even no economic profitability for the recycling companies. As in Europe, this depends on the state. However, many states have relatively low landfilling fees for mixed plastics, which makes recycling even less attractive [8]. The curve shapes in Figure 13.9 underline the situation described.

A third issue related to organizational processes, besides high contamination and low profitability, is the *focus mainly on PET recycling* in the US. The reason for this is the relatively easy reusability for new products (e.g., plastic bottles and plastic packaging). Besides industrial materials (mainly other resins), PET is the only plastic resin with a double-digit recycling rate. Assuming, that only PET and the "other" resins are attractive for recycling, only 25% of material is left for recycling. These three effects (high contamination, cheap landfills, and low recycling rate) reinforce each other and thus lead to a stagnation or even decrease in plastics recycling.

Table 13.3 Plastic Waste by Polymer in the United States in 2018 [18]

Polymer	Generated		Recycled	
	Amount [kt]	Percentage of plastic waste [%]	Amount [kt]	Percentage of plastic waste [%]
PET	5290	14.8	980	18.5
HDPE	6300	17.7	560	8.9
PP	8150	22.8	50	0.6
LDPE/LLDPE	8590	24.1	370	4.3
PS	2260	6.3	20	0.9
PVC	840	2.4	Negligible	-
Other resins	4160	11.7	1110	26.7
Total	**35,680**	**100.00**	**3090**	**8.7**

Currently, ten states have established beverage container deposit laws (similar to Europe, especially Germany) to increase the recycling rate. However, since 2002, this number did not change [20].

As in the EU28+2 countries, the USA does not reuse all **recyclates**. In fact, large percentages of recyclates were exported. 2015 was the peak of exports with around 2 Mt of exports (63% of recycled plastics). Between 2017 and 2018, a significant shift in exported waste between countries can be observed, as Figure 13.10 shows. Due to the import ban of plastic waste in China, which is discussed in more detail in Section 13.3, US exports to China were reduced by almost 90%. Despite the increase in exports to other countries, particularly Malaysia, Thailand, and Vietnam, about one-third less plastic waste was exported in the first half of 2018 compared to the first half of 2017. This poses a major challenge for the U.S., as about 600 kt of plastic waste was "left over" in 2018. For the recycling industry, this equates to a significant reduction in the selling price, further decreasing the economic viability of recycling in the US. Accordingly, new recycling options for plastic waste need to be developed or plastic waste generally avoided. In other countries, especially in Europe, some measures have already been established in this regard that could also be helpful in the USA [21, 22].

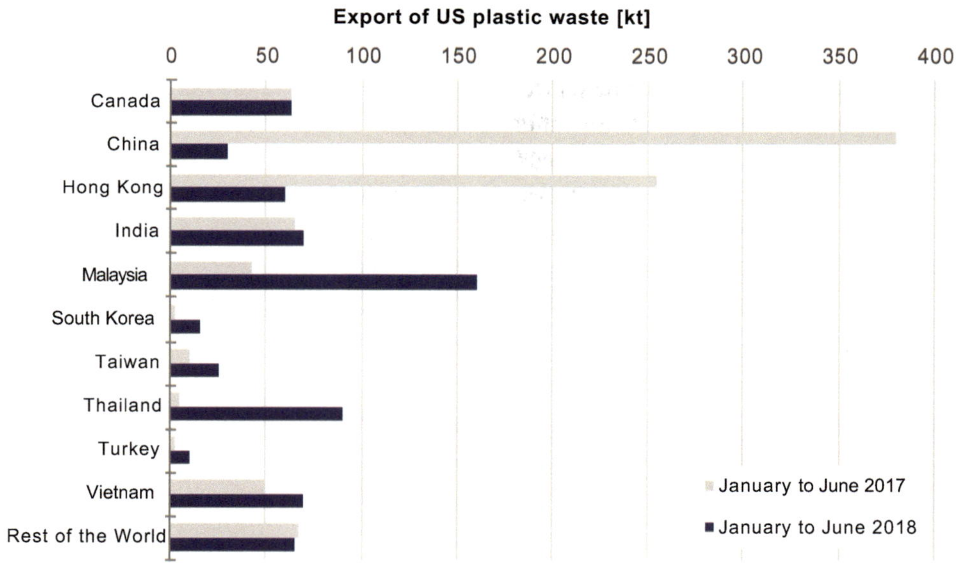

Figure 13.10 Distribution of U.S. plastic waste exports by country [21]

13.2.2 Canada

In 2016, Canada generated around 3.24 Mt of plastic waste, which corresponds to 118 kg of plastic waste per capita. This is slightly higher than the amount per capita in the US (105 kg) [8]. Figure 13.11 illustrates the **consumption** of plastics in Canada. Three sectors account for nearly 70% of plastic use: packaging, construction, and automotive. As in Europe and the US, packaging has the highest amount, but is not as dominant as in Europe and the US [23].

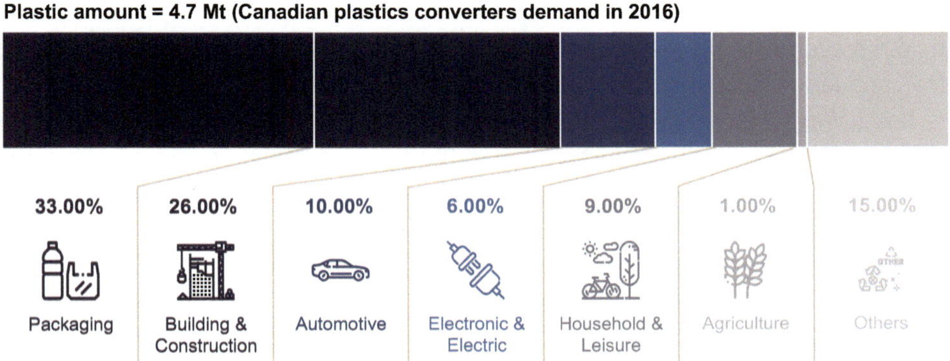

Figure 13.11 Distribution of recyclates re-entering the Canadian market in 2016 [23]

For the waste generation, the same effect as in Europe can be observed: due to the durability of the automotive and building/construction sectors, the waste ratios differ from the demand ratios. For example, while packaging accounted for 33% of plastics introduced to the market in 2016, it accounted for 47% of all plastic waste discarded that same year (Figure 13.12) [23].

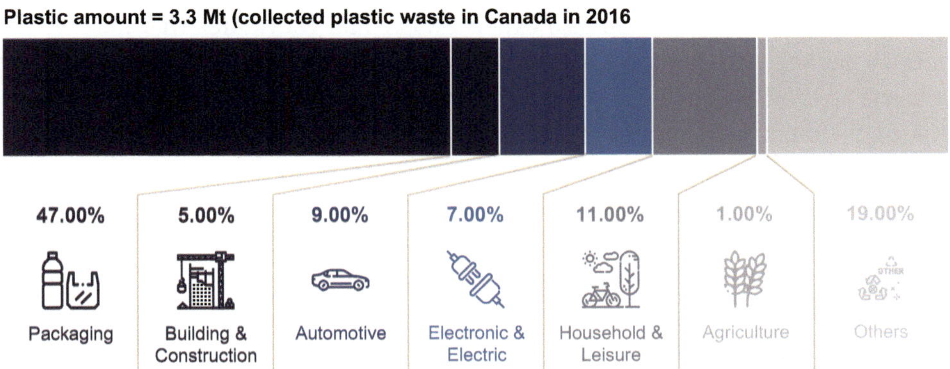

Figure 13.12 Canadian plastic waste generation by sector in 2016 [23]

Of the generated and collected plastic waste in 2016, the majority ended up in landfills (87%). A minority was thermally converted to energy by WTE facilities (4%). The other 9% was recycled, most of it mechanically (8% of total collected plastic waste) and a small parts chemically (1% of total collected plastic waste) [23].

This relatively low recycling rate and especially high landfilling rate has several process- and organization-related causes. This starts with **the collection mechanisms**: Improper sorting at the consumer and collection level - especially due to increasing reliance on single-stream collection systems - results in the contamination of collected plastics. This requires additional sorting and quality at MRFs and additional technologies to remove contaminants. Furthermore, contaminated plastic often cannot be reused as recyclate. Besides improper sorting, it is challenging for Canada to establish a comprehensive and cost-effective collection system, as it is has a *large area* with a distributed waste generation system (e.g., end-of-life agricultural plastics). To decrease the landfilling ratio, many small-scale MRFs or WTEs would be necessary, which would require a certain investment. Currently, only 11 MRFs are running, thus many plastics are landfilled. In addition to the challenge of the infrastructure needed over a large area, the *lack of robust infrastructure of WTEs* limits the potential of routes for hard-to-recycle plastic material which then - again - end up on landfills [23].

In addition to these reasons, the *pricing* of both virgin resins and recyclates has an impact on the recycling rate. Low prices for virgin resin establish the ceiling at which recycled resins can be sold, impacting the amount of end-of-life plastic products that can be cost-effectively diverted for recycling. Currently - due to low oil prices - the virgin resin price is low, which makes recycling less profitable [5]. Added to this the fact that many resins **in general lack a market** for recycled plastics - recycled PET and HDPE are the only resins with a reasonably profitable market in Canada. They account for two-thirds of the recycled plastic waste (PET: 37%; HDPE: 30%) [23].

As in the EU and US, Canada also exports waste and **recyclates** that have no domestic market. Accordingly, Canada has a similar decrease and country-shift for exports as the US (see Figure 13.10). However, the reason for the decrease results not only from the import ban in China, but also the *Basel Convention's plastics waste amendments.* It became effective in Canada on 1st January 2021, and focuses attention on international shipments of plastic waste that may have limited recycling options or be more difficult to recycle. These include soiled or contaminated plastics (e.g., residues remaining in containers or the presence of other household wastes), halogenated plastics, and plastic wastes consisting of more than one polymer. These changes will likely impact all supply chain participants, including collectors and processors who may need to adjust the way materials are collected, sorted, and processed, in order to find markets [24].

13.3 Asia

The region "Asia and Oceania" produces the highest amount of plastic waste, as Figure 13.1 shows. However, this number must be put into perspective with its total population, which is nearly five billion. This means that the plastic waste per capita in this region is 43.9 kg, and thus less than half of the amount per person in the US or Canada. This number varies a lot between the countries, which depends on different factors (e.g., income level and regulations). Therefore, the following subchapters analyze China (Section 13.3.1), India (Section 13.3.2) and Japan (Section 13.3.3) in detail. Section 13.3.4 points out some highlights of other Asian countries.

In general, for more than 25 years by now, Asian countries have played an important role in the whole plastics waste-handling value chain, but unlike Western countries, many of them are developing nations lacking the capacity to manage such waste [25]. Section 13.1 and Section 13.2 have already analyzed this from the exporter perspective. The following subsection will have a closer look at the impact of importing. Special focus lies on the shift of plastic imports after the Chinese import ban in 2018.

It must be noted that the database for the Asian countries is not as reliable as for Europe or North America. Therefore, the analysis is at some points more qualitative than quantitative.

13.3.1 China

As consequence of China's fast economic growth, in 2013 the country bypassed all other countries and regions and became the world's largest producer of plastic products. Its domestic consumption of plastics has also greatly increased [8]. Since then, China has been both the biggest producer and consumer of plastics and plastic products. In 2018, China produced 83.6 Mt of plastic waste and consumed 67.6 Mt [8].

The **plastic consumption** by sector is not as recorded in as much detail as in Europe and North America. The same applies to the recycling, WTE, and landfilling. As Table 13.1 shows, conversion studies estimate that of 40.1 Mt of collected plastics (68.8% of total produced waste) in 2018, 31% was recycled, 48% was incinerated with energy recovery, and 21% was landfilled. However, over 30% of the produced plastic waste was not managed at all in China in 2018.

In-country **collection and recycling** in China is much more difficult than in Europe and the U.S. In many places the infrastructure is not yet well-developed. There is not yet a countrywide, formal recycling system for plastics or indeed other recyclables. The largest share of plastic waste is still collected by informal waste pickers – private individuals collecting plastic bottles and earning a small amount of money for doing so. As a result, many cities turn to private waste management companies that do not yet have adequate recycling strategies [8].

However, individual cities in China are implementing recycling systems. On 1 July 2019, Shanghai was the first city to launch a mandatory waste separation at source. Whereas previously all MSW was disposed of in one bin, Shanghai has now added four bins: one bin for dry residual waste, one bin for wet residual waste, one bin for hazardous materials, and one bin for recyclable material. The bins are emptied every day. In addition, according to media reports, around 30,000 volunteers are being recruited to explain waste separation to Shanghai residents. The first positive effects were quickly noticed, especially with regard to the reduction of waste on the streets. After one year, compared with 2018, the average amount of recyclables collected daily in 2019 increased by 431.8%. The amount of kitchen waste and hazardous waste collected also grew by 88.8% and 504.1% respectively, while the amount of residual waste decreased by 31%. This shows that the initiative is fruitful. The exact numbers for plastic waste have not yet been analyzed, but a positive trend is visible. The success from Shanghai's complex waste segregation regulations has encouraged other Chinese cities to follow suit, including Guangzhou, Kunming, Xi'an and Beijing, where garbage sorting is now mandatory for all households and businesses [26]. This also affects the recycling process as well as the waste handling itself.

Analyzing the Chinese plastic recycling chain, **recyclates** also play an important role. China's economic growth can be traced back to the import of plastic waste in the last century. Due to the low-quality domestic post-consumer plastic waste sources and the high demand for plastics, since the late 1980s China has been the world's largest importer of plastic. Between 1988 and 2016, China (including Hong Kong) imported 170.5 Mt of plastic waste, so taking a 72% share of the total imported plastic waste. China was also the biggest exporter of plastic waste, but with a cumulative weight of 56.1 Mt and a share of 26.1%, the net imports were significantly higher. Table 13.4 summarizes the plastic waste imports and exports for selected countries between 1988 and 2016 [27]. From these countries, China is the only one with more imports than exports. Germany, Japan and the USA also have the lowest net import of plastic waste, and so are the three principal net exporters of plastic waste.

Table 13.4 Plastic Waste Imports and Exports by Country between 1988 and 2016 [27]

Country	Import		Export		Net import [Mt]
	Cumulative amount [Mt]	Global share [%]	Cumulative amount [Mt]	Global share [%]	
Canada	3.83	1.62	3.89	1.81	-0.06
China	170.50	72.4	56.1	26.10	114.4
Germany	5.36	2.27	17.6	8.22	-12.24
Japan	Negligible	Negligible	22.2	10.30	-22.2
USA	8.49	3.60	26.7	12.40	-18.21

Since 2013, however, the import quota of post-consumer plastic waste has been declining – despite the continuing increase in demand. This effect results from the initiative "Operation Green Fence" in 2010, aiming to reduce the amount of poor-quality plastic waste imported into China, and only importing plastics that have already been sorted by type. As a result, a total of 2 Mt of post-consumer plastic waste was imported in the period from January to April 2016, a decrease of 11% compared to the same period in 2015 [27–29].

In this period, China realized that it might be able to completely cover its demand for post-consumer waste through domestic recycling by 2030. Furthermore, the increased pollution in China due to the high levels of importation of low-quality plastics was becoming increasingly visible. Therefore, in 2017, the law was tightened considerably, with a stricter, permanent ban on non-industrial plastics imports being introduced. In the first year of the ban, the dimension of this decision was already becoming apparent: as shown in Figure 13.10, the plastics exports of the U.S. to China recorded a drop from 380 kt (January to June 2017) to 35 kt (January to June 2018). A similar trend is evident in Europe: for example, exports from the United Kingdom fell from around 90 kt (January to April 2017) to around 1.5 kt (January to April 2018). Small amounts of plastic waste were exported to other, primarily Asian, countries (e.g., Malaysia, Taiwan, Vietnam, and Indonesia) [21, 27, 30, 31]. Table 13.5 summarizes the changes in EU exports between 2015 (before the Chinese ban) and 2018 (after the Chinese ban) for chosen Asian countries.

Despite the ban, Japan, Germany, and the USA remained the principal net exporters of plastic waste in 2019. China, on the other hand, was not even in top five of the net-importers: Turkey, Vietnam, and Malaysia were the three countries with the highest amount of net-imports that year [32].

Table 13.5 Import Changes after the China Ban on Plastic Imports from the EU [27]

Country	EU Imports 2015		EU Imports 2018		Change [%]
	Cumulated amount [t]	EU share [%]	Cumulated amount [t]	EU share [%]	
China	1,658,970	54	64,662	3	-96
Hong Kong	775,557	25	211,530	11	-73
India	139,628	5	158,250	8	+13
Indonesia	32,640	1	190,933	10	+485
Malaysia	137,876	4	404,123	21	+193
Taiwan	31,548	1	99,071	5	+214
Thailand	15,414	1	39,676	2	+157
Turkey	19,377	1	270,339	14	+1295
Vietnam	88,760	3	187,378	10	+111
Other	177,742	6	301,135	16	+69
Total	**3,077,512**		**1,927,097**		**-37**

The Chinese regulations, which have already shaken the whole plastics market, will be an even bigger challenge for exporters in the future. Depending on the strictness with which China imposes the import ban, the cumulative displacement of global plastic waste is expected to reach an estimated 110 Mt by 2030, as Figure 13.13 shows [27]. This requires rethinking the correct use of *recyclates*, e.g., in furniture, clothes, building and construction materials, or for packaging material.

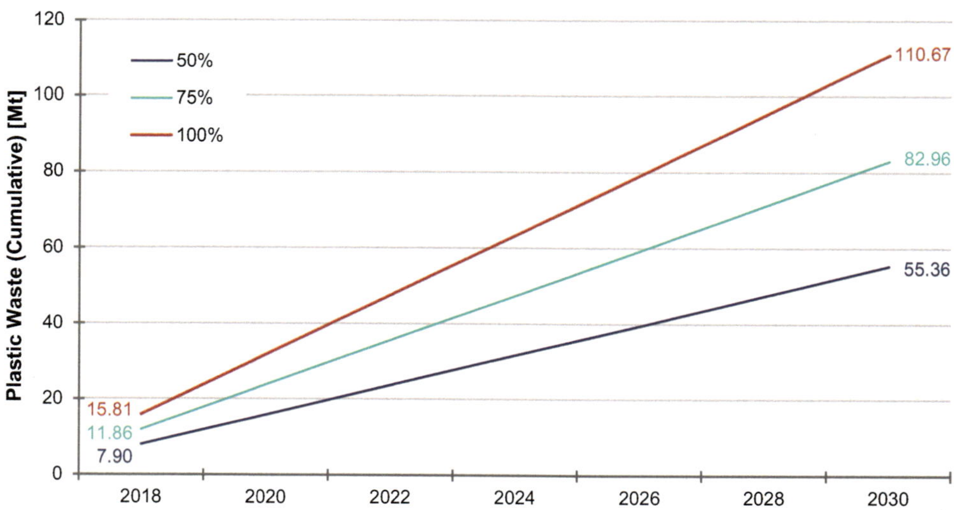

Figure 13.13 Cumulative amount of plastic waste resulting from the Chinese import ban, presented for three possible import ban scenarios (50%, 75%, and 100%) [27]

13.3.2 India

Despite India's population being nearly as high as China's population, its **plastic consumption** was only a fifth that of China. However, due to the development of the country in general and especially the growing construction sector, the demand is higher than the current in-country production in India. This is underlined by the growth of plastic production in India from 8.3 Mt in 2010 to an expected amount of 22 Mt in 2020. Accordingly, India's demand is dominated by plastics with a shorter life time (e.g., packaging, bottles, and plastic bags) [33].

The lower plastic consumption is also reflected in a smaller amount of plastic waste, which Conversio estimates to be 9.5 Mt per year [8]. As mentioned in the introduction of Section 13.3, data quality and availability in Asia is limited, leading to uncertainties in the analysis. Various sources assume plastic waste generation to be between 8 and 15 Mt per year, with the higher of these values being nearly twice the above-mentioned amount. However, the discussion in this chapter relies on the data of Conversio [8].

Of the 9.5 Mt of plastic waste per year, 73% was **collected**, meaning that more than a quarter of plastic waste is dumped somewhere (e.g., oceans, unmanaged landfills). India has not yet implemented systematic collection and recycling systems, but due to low labor costs and existing domestic markets for recycled plastics, the informal collection and recycling industry is relatively large compared to other emerging countries. The informal recycling focusses on PET and HDPE bottles and other rigid plastics. For these materials, India has the highest recycling rate of the emerging countries. In general, the plastics recycling rate of India is high compared to the rest of the world (65% according to Conversio) [8].

The **recyclate** usage ratio is relatively high, as the demand for plastics material exists. Furthermore, India was the fifth biggest net-importer of plastics waste in 2019 (imports surpass exports by 24 kt) [32].

13.3.3 Japan

In 2017, Japan consumed around 10 Mt of plastic. With 41% of this used for packaging, the **plastic consumption** in Japan was similar to the Western countries [8].

In the same year, Japan produced around 9 Mt of plastic waste in, of which all was formally **collected**. The reason for this high collection rate is the multi-stream separation at all households. Plastic, paper, PET bottles, aluminum, and glass are separated by the end-user and separately collected for recycling [8].

Japan also uses the three introduced **plastic waste handling treatments.** From the collected waste, 14% was landfilled, 56% was incinerated with energy recovery

(WTE), and 30% was recycled. It must be noted that some sources quote a recycling rate of over 80%. However, in these sources, WTE is counted as "recycling". Compared to other countries, Japan has a relatively high amount of chemical recycling. Of 3 Mt of recycled plastic waste, 0.4 Mt (13%) was recycled chemically, while 2.6 Mt (87%) was recycled mechanically and transformed into recyclates [8].

A large amount of these **recyclates** are exported. As Japan exported big amounts to China, a decrease in total exports was recorded between 2015 (1.5 Mt exported) and 2019 (0.898 Mt exported) [34]. Despite this decrease, Japan remained one of the principal net exporters of plastic waste in 2019, with 0.5 Mt more exported than imported. With a share of 15.3% of total plastic exports worldwide, Japan is third in this statistic [35].

13.3.4 Other Asian Countries

As Table 13.5 shows, the import ban in China impacts the plastic imports of other countries, with shifts in processing European waste also recorded in other regions.

Since 2018, **Malaysia** has become the new import hub of plastic waste, importing 0.7 Mt of plastic waste. Compared to the plastic waste Malaysia produced itself (1.8 Mt), this import amount was relatively high (40%). PET, LDPE, HDPE, and rigid plastics are thereby recycled in the many operated recycling facilities, leading to a recycling rate of 18%. However, Malaysia does not run any WTE plants. This means that the rest of the waste is landfilled, often in unmanaged facilities. This leads to a share of mismanaged plastics of 39%, which often ends up in the oceans. Exports to Malaysia therefore have to be carefully observed [8].

Indonesia has also stated a massive increase of plastic waste imports (>400%). But even without this additional waste, Indonesia has massive issues in collecting and properly handling waste. In 2018, less than half of the waste was collected (46%) and 30% additionally ended up in landfills. 16% was recycled, but not all of it mechanically. Some of the recycled plastics was also used as alternative fuel [8].

These two examples represent the challenges of the Asian countries importing plastic waste from high-wage countries. Without implementing proper plastic waste handling processes, the import of plastic waste into these countries will just exacerbate current challenges.

13.4 Conclusion & Summary

This chapter has analyzed the plastic waste handling chains, especially the recycling chains, of the biggest plastic waste producers in the world. Accordingly, it has analyzed **consumed plastics**, **waste collection**, **recycling**, and the **recyclates**.

The analysis of the **consumed plastics** showed that the share of plastics with a short lifetime (e.g., packaging material) plays an important role for all waste handling. Packaging material is still difficult to recycle and then either burned or landfilled. Furthermore, if the share of plastic waste in the MSW is too small, plastic recycling is probably not profitable. These countries showed a small percentage of recycling.

Some points relating to the importance of **plastic waste collection** have become clear. In general, it can be said that countries with a multi-stream collection at the curbside seem to be the most successful at achieving a high recycling rate. Despite the "risk" posed by having to educate the customer on how to correctly separate waste, contamination of plastic waste massively decreases when this separation is used. In all regions (EU, North America, and Asia), multi-stream recycling has a higher shares of recycling or WTE than landfilling.

Analysis of the **recycling process** itself has showed that many countries are mainly recycling mechanically. Here, PET, HPDE, LDPE, and rigid materials are of interest. Flexible materials normally cannot be recycled and are incinerated or landfilled.

The analysis of **recyclate** usage shows a big potential. Currently, countries with a relatively high recycling ratio only reuse parts of it. Many of the recyclates are exported to low-wage Asian countries, but it is not clear how they end up in these countries. Most of them have a poor waste handling infrastructure, resulting in plastics being dumped in the oceans. Thus, plastics recycling becomes more of a "green-washing" exercise than an environmentally friendly way of handling waste. The exporting countries, especially Germany and the USA, must therefore rethink their strategy of dealing with these recyclates, with particular emphasis on following Canada's example and tracing the route their exported plastic waste takes.

References for Chapter 13

[1] Plastics – The facts 2015/2016: An analysis of European plastics production, demand and waste data. PlasticsEurope, 2015, pp. 1–40

[2] Kearney, A. T.: Plastics: The future for automakers and chemical companies, 2012

[3] MacArthur, D. E., Waughray, D., Stuchtey, M. R.: The new plastics economy: Rethinking the future of plastic, Ellen MacArthur Foundation, 2016

[4] Osswald, T. A., Baur, E., Rudolph, N.: *Plastics Handbook: The Resource for Plastics Engineers*, Hanser Verlag, 2019

[5] Rudolph, N., Kiesel, R., Aumnate, C.: *Understanding Plastics Recycling: Economic, Ecological, and Technical Aspects of Plastic Waste Handling* (vol. 2), Hanser Verlag, 2020

[6] Advancing sustainable materials management: 2018 Tables and figures assessing trends in material generation, recycling, composting, combustion with energy recovery and landfilling in the United States. US EPA, 2020

[7] Weltweite und europäische Kunststoffproduktion in den Jahren von 1950 bis 2019. Statista, *https://de.statista.com/statistik/daten/studie/167099/umfrage/weltproduktion-von-kunststoff-seit-1950/*, accessed August 02, 2021

[8] Global Plastic Flow 2018. Conversio Market & Strategy GmbH, 2020

[9] Plastics – The facts 2020: An analysis of European plastics production, demand and waste data. PlasticsEurope, 2021

[10] Plastics – The facts 2019: An analysis of European plastics production, demand and waste data. PlasticsEurope, 2020

[11] EU action to tackle the issue of plastic waste. European Court of Auditors, 2020

[12] Generation of plastic packaging waste per capita in the EU 2018 by country. Statista, *https://www.statista.com/statistics/972604/plastic-packaging-waste-generated-per-capita-countries-eu/*, accessed August 02, 2021

[13] Bruttoinlandsprodukt (BIP) pro Kopf in den EU-Ländern 2020. Statista, *https://de.statista.com/statistik/daten/studie/188766/umfrage/bruttoinlandsprodukt-bip-pro-kopf-in-den-eu-laendern/*, accessed August 02, 2021

[14] Kaza, S., Yao, L., Bhada-Tata, P., Van Woerden, F.: What a waste 2.0: A global snapshot of solid waste management to 2050. World Bank Publications, 2018

[15] Seyring, N., Dollhofer, M., Weißenbacher, J., Bakas, I., McKinnon, D.: Assessment of collection schemes for packaging and other recyclable waste in European Union-28 Member States and capital cities. *Waste Management & Research*, (2016) 34, pp. 947–956

[16] Tiseo, I.: Annual plastic waste exports from the EU-27 2005–2020, *https://www.statista.com/statistics/1235915/plastic-waste-exports-european-union/*, accessed August 02, 2021

[17] Per capita consumption of plastic materials worldwide by region 2015. Statista, *https://www.statista.com/statistics/270312/consumption-of-plastic-materials-per-capita-since-1980/*, accessed August 03, 2021

[18] Advancing sustainable materials management: 2018 Fact sheet – Assessing trends in material generation, recycling, composting, combustion with energy recovery and landfilling in the United States. US EPA, 2020

[19] Guttentag, R.: Processing recyclables: What's my line?, *World Wastes*, (1994) 37, pp. 28–35

[20] State beverage container deposit laws. National Conference of State Legislatures (NCSL), 2020, *https://www.ncsl.org/research/environment-and-natural-resources/state-beverage-container-laws.aspx*, accessed August 04, 2021

[21] Clarke, J. S., Howard, E.: US Plastic waste exports to developing countries, causing environmental problems at home and abroad. Unearthed, 2018, *https://unearthed.greenpeace.org/2018/10/05/plastic-waste-china-ban-united-states-america/*, accessed August 04, 2021

[22] Dell, J.: 157,000 Shipping containers of U.S. plastic waste exported to countries with poor waste management in 2018. Plastic Pollution Coalition, 2019, *https://www.plasticpollutioncoalition.org/blog/2019/3/6/157000-shipping-containers-of-us-plastic-waste-exported-to-countries-with-poor-waste-management-in-2018*, accessed August 04, 2021

[23] Economic study of the Canadian plastic industry, markets and waste: Summary report. Environment and Climate Change Canada, 2019

[24] Basel Convention plastic waste amendments: Impacts on Canadian recycling industry. Government of Canada, 2021, *https://www.canada.ca/en/environment-climate-change/services/*

managing-reducing-waste/permit-hazardous-wastes-recyclables/basel-convention-plastic-waste-amendments-impacts-industry.html, accessed August 03, 2021

[25] Parker, L., Elliot, K.: Plastic recycling is broken. Here's how to fix it. National Geographic, 2018, *https://www.nationalgeographic.com/science/graphics/china-plastic-recycling-ban-solutions-science-environment*, accessed August 04, 2021

[26] One year of waste segregation in Shanghai: Success or failure? Green Initiatives, 2020, *https://greeninitiatives.cn/one-year-of-waste-segregation-in-shanghai-success-or-failure/*, accessed August 04, 2021

[27] Brooks, A. L., Wang, S., Jambeck, J. R.: The Chinese import ban and its impact on global plastic waste trade. *Science Advances*, (2018) 4, DOI: 10.1126/sciadv.aat0131

[28] Koty, A. C.: Trash or treasure? Prospects for China's recycling industry. China Briefing, 2016, *https://www.china-briefing.com/news/trash-or-treasure-prospects-for-chinas-recycling-industry/*, accessed August 04, 2021

[29] Plastics market situation report. Waste and Resource Action Programme, 2016, *https://wrap.org.uk/resources/market-situation-reports/plastics-2016*, accessed August 04, 2021

[30] After China's import ban, where to with the world's waste? *Deutsche Welle*, 2019, *https://www.dw.com/en/after-chinas-import-ban-where-to-with-the-worlds-waste/a-48213871*, accessed August 04, 2021

[31] Plastic waste export tide turns to south-east Asia after China ban. Financial Times, 2018, *https://www.ft.com/content/94ee72d0-6f26-11e8-852d-d8b934ff5ffa*, accessed August 04, 2021

[32] Buchholz, K., Which countries export & import plastic waste? Statista, *https://www.statista.com/chart/18229/biggest-exporters-of-plastic-waste-and-scrap/*, accessed August 04, 2021

[33] Padgelwar, S., Nandan, A., Mishra, A. K.: Plastic waste management and current scenario in India: A review. *International Journal of Environmental Analytical Chemistry*, (2019) 101, pp. 1894–1906

[34] Klein, C.: Plastic waste export volume Japan 2016–2020. Statista, *https://www.statista.com/statistics/1193746/japan-plastic-waste-export-volume/*, accessed August 04, 2021

[35] Denyer, S.: Japan wraps everything in plastic. Now it wants to fight against plastic pollution. *Washington Post*, 2019, *https://www.washingtonpost.com/world/asia_pacific/japan-wraps-everything-in-plastic-now-it-wants-to-fight-against-plastic-pollution/2019/06/18/463fa73c-7298-11e9-9331-30bc5836f48e_story.html*, accessed August 04, 2021

14 Value Chain EMEA – Collecting and Sorting

Jürgen Ephan, Arne Köhne

In the following chapter a light is shed on the part of the circular economy value chain where the waste management and recycling industry plays the most important role. Even though in a well-functioning circular economy the waste management industry or rather the recycling industry already becomes involved before products become waste, for instance in the field of design for recycling, this chapter will start at the point where products have already become waste. In this context also the quantities of plastics used on a global level as well as on a European and a German level will be outlined.

Following this, a practically orientated overview is provided in the field of waste collection and sorting as well as in the field of mechanical pretreatment of the waste for mechanical and chemical recycling.

14.1 Waste: Sources and Origins

Plastic waste is divided into two categories:

- Post-industrial plastic waste (originating from production of plastics and from plastic processors before its use as a post-consumer product)
- Post-consumer plastic waste (originating from industry, commercial activities, and households after its use as a post-consumer product)

Material from post-industrial plastics is much cleaner and mainly sorted or single-type. Some plastic processors reuse their own plastic scrap in the way of circular economy in their own production. The material has a high value as it can be ground and pelletized very easily.

While unsorted post-industrial scrap still can be recycled it competes economically with incineration as pre-sorting of the goods is much more expensive. Sometimes it is even uneconomic.

Recycling quotas of above 90% can be expected on a German national level. But even on the European level, recycling quotas of above 90% can be expected for post-industrial plastic waste [1].

Because recycling quotas of above 90% in the field of post-Industrial plastic waste can be expected, we do not see a serious potential for increase. Rather less recycling is conducted in the field of post-consumer plastic waste and this is what we have to challenge. Therefore, we will focus on the post-consumer plastic waste in this chapter.

Post-consumer plastic waste can be divided into two additional subcategories:

- Post-consumer plastic waste that originates after usage in commercial and industrial enterprises. This also includes the following:
  - Commercial packaging waste collected
  - Commercial and industrial packaging waste collected and treated (transport packaging)
  - End-of-life vehicle (ELV), including ASR (Auto-Shredder Residue), plastic waste from dismantling of end-of-life vehicles and of ASR
  - Plastic waste from other national collecting and recycling systems, e.g. pipes, agricultural films, PVC waste products
- Plastic waste from households and recycling depots:
  - Residential waste collected by or on behalf of municipalities (household waste excluding all waste fractions that are collected separately)
  - Bulky household waste collected by or for local authorities
  - Household packaging collected: Used sales packaging collected by "integrated systems" such as (in Germany) "DSD" or "Eco-Emballages" with their own financial system
  - WEEE collection: Waste from electrical and electronic equipment (commonly referred to as WEEE) from households and all kinds of commercial and industrial activities (collected by municipalities, retailers, and private enterprises)

Most of the aforementioned post-consumer plastic waste streams are mainly a mixture of all kind of polymers, laminates, as well as composites and partly other materials like metals or biomass. The latter applies in particular if it is a mixed collection scheme. Therefore, further treatment such as sorting, shredding, and partially washing is usually needed to produce recyclates that can substitute virgin material made from fossil sources.

Normally, only single-sorted polymers fulfill the prerequisites to substitute virgin material. On the other hand, laminates, composites, as well as low-price plastic applications and mixtures of several different polymers tend to be downcycled or to be incinerated in order to produce heat/electricity or to produce secondary fuels for co-incineration plants and cement plants.

The unfulfilled potentials in the field of plastics recycling cannot only be explained by the technological limitations of the currently used recycling processes. Rather, the limitations that come from a lack of economic efficiency have to be taken in consideration as well. The laborious sorting and recycling processes have to be covered by the sales of the recyclates to the plastic distributing and converting companies. If the quality and the available quantities of the recyclates are insufficient for the plastic converting industry (reasons for that are manifold) or the price for virgin material is too low, incineration is normally the only reliable recovery path for the continuously streaming plastic waste quantities.

In Figure 14.1 the mechanism described above is visualized.

The constant and continuous origination of enormous quantities of post-consumer plastic waste has a variety of reasons. One of the major reasons is that plastic applications, especially in packaging, are characterized on the one hand by a huge demand and on the other by a short lifecycle. Lifecycles range as follows:

- Below 1 month for perishable food such as meat or vegetables
- From 0–1 year for food packaging or fast-moving consumer goods packaging
- From 1–5 years for consumer goods like smart phones, TVs, computers, small household goods
- From 5–10 years for large household goods, electronic equipment
- From 10–20 years for automotive parts
- From 20–50 years for applications in building & construction, such as window and door profiles, sewage and fresh water piping

While there are developed recycling concepts in place for "monolithic" applications like agricultural film, HDPE canisters, PET beverage bottles, and PVC-based applications like door/window frames, it is also true that

- such recycling and collection systems are established only within "developed" societies,
- the market price for recycled material determines significantly if recycling of a specific waste is feasible (the market price, on the other hand, depends on the availability of recycled waste and on the conversion cost into flakes or pellets), and
- for applications other than those mentioned above, the heterogeneity of products and ingredients in connection with costs of recycling and achievable sales prices pose a challenge for any recycling route.

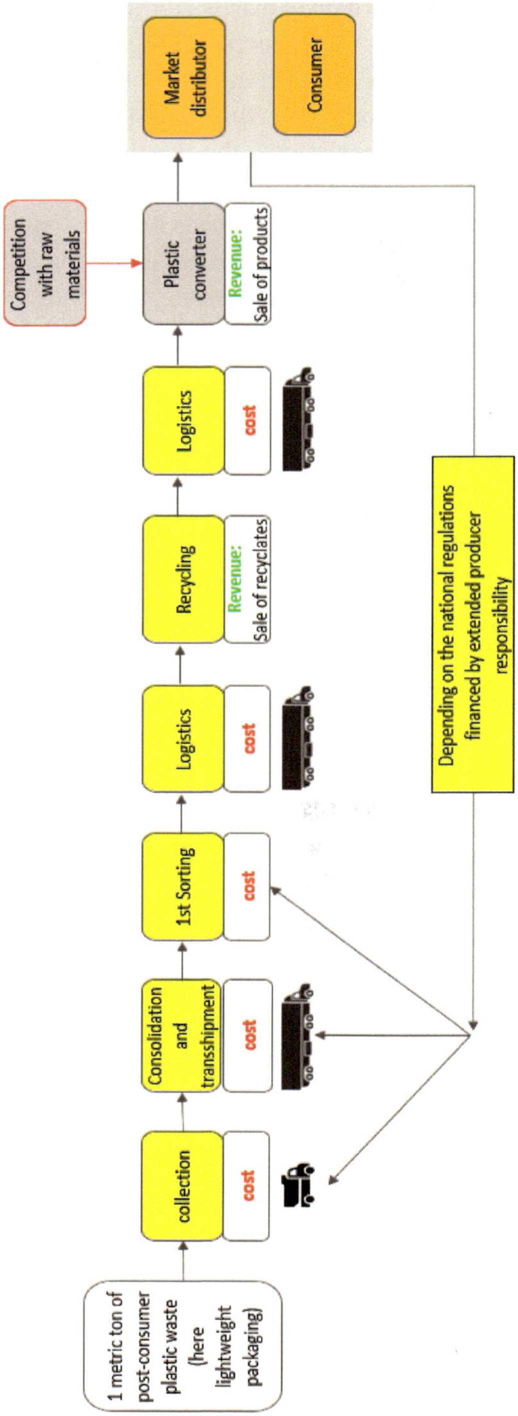

Figure 14.1 Overview of the waste management value chain and the cost-benefit allocation

In fact, the development of a circular economy is lagging behind on a global scale (Figure 14.2).

In summary, approximately 2/3 of global plastic waste ends up in landfill or in the environment. Only approximately 16% is subject to recycling, which is about the same amount that is incinerated, including energy recovery.

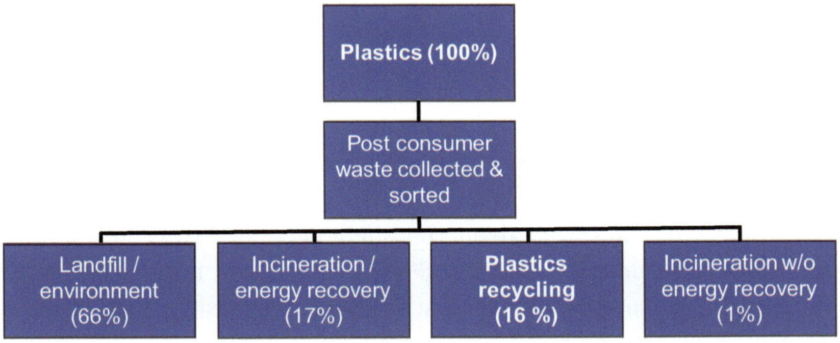

Figure 14.2 Global plastic recycling in 2018 [2] (source: IHS Markit based on data from OECD and World Economic Forum, Ellen Mc Arthur Foundation (IHS Markit 2019)

After having shed a light on the global status of plastics recycling and disposal above, let us dive into the global waste management processes. The first recycling concepts were implemented on a larger scale only 30 years ago. Since then the quantities that end in landfill or in the environment have been significantly reduced. But still approximately half of the total generated waste ends up in landfill or in the environment. The rest is incinerated or recycled. The fact that the current estimated global recycling rate of 16% (in 2018) contains waste exports, e.g. from Western countries to China, Malaysia, and Indonesia, further demonstrates the upside potential (Table 14.1).

Table 14.1 Estimated Global Polymer Recycling Figures in %, Based on Total Waste Generation (source: [3] and authors' estimation)

Year	1980	1990	2000	2010	2020	2030 (forecast	2050 (forecast)
Landfill, discarded	100	93	81	70	50	35	7
Incineration	0	5	10	14	28	35	50
Recycling	0	1	9	16	22	30	43

Driven, for example, by political pressure, the percentage of recycled plastics is predicted to increase significantly. Almost a third of the reported recycling rate for plastic packaging in the EU is due to shipments to third countries.

To address the growing waste problem, the European Commission adopted the Plastics Strategy in 2018, including an update of the Packaging and Packaging

Waste Directive from 1994 and a doubling of the previous recycling target to 50% by 2025, increasing even to 55% by 2030. Achieving these targets would be an important step towards meeting the EU's circular economy goals.

In January 2021, however, a ban came into force that affects the majority of shipments of plastic waste. Combined with the inadequate capacity to deal with this waste within the EU, this creates a further risk for the achievement of the EU recycling objectives. It also harbors the risk of an increase in illegal shipments to third countries and waste crime, against which the EU is insufficiently armed within its current legal framework.

The EU's drive to improve the recycling of plastic waste reflects the environmental challenge. The new approach gives the EU the opportunity to gain a pioneering advantage and strengthen its global leadership position in the field of plastic packaging recycling. However, given the challenges and gaps to be addressed, there is a risk that EU member states will not meet the new targets. Hence, there is an urgent need for a comprehensive concerted effort to nearly double the amount of recycled plastic packaging waste in the EU by 2030 [4].

One of the main reasons why the majority of plastic waste ends up in landfills or unmanaged in the environment is limitations in national economic performance and therefore the lack of a publicly financed system that supports the development of advanced waste management activities and infrastructure. On the other hand, even for foreign investors it is quite challenging to build up capacities of advanced waste management activities because the national regulatory framework often counteracts such endeavors. For instance, landfilling is the least expensive method to get rid of the waste but of course it is quite a problematic option as well, especially in the long-run. If there are no limitations (a ban for specific waste fractions, like used plastics, would be best), from the perspective of economic competitiveness it is almost impossible to implement advanced waste management activities. That is why the EU needs to push the ban for landfilling of certain waste streams proactively.

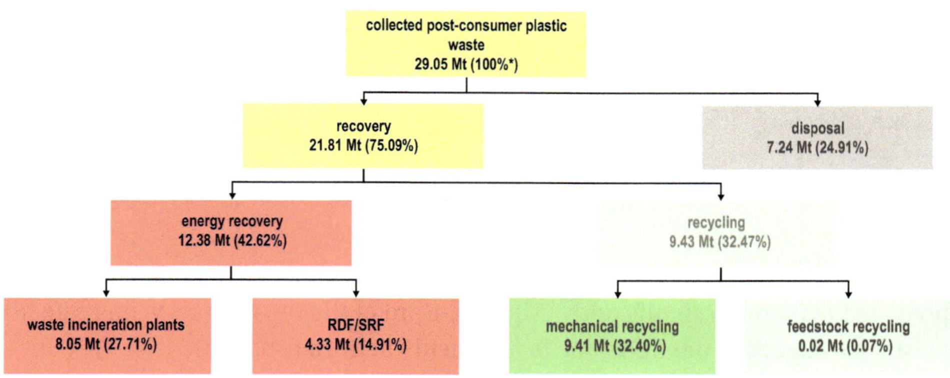

Figure 14.3 Plastics Circular Economy in Europe [5]

The ban for landfilling waste logically forces the buildup of more advanced waste management activities and infrastructure. In 2005 in Germany it became forbidden to landfill untreated residential waste. The strict requirements of the regulations meant that residential waste can only be landfilled after its incineration in the form of slag. The ban of landfilling is one of the major reasons why today more than 99% of German plastic waste is recovered either by energy recovery or from mechanical or chemical recycling. Between energy recovery and recycling exists the same interrelationship as exists between landfilling and advanced waste management activities. If recycling cannot compete economically with energy recovery, the waste tends to be energy recovered. And as already described, the opportunities of recycling are highly determined by the economics of the production of virgin material.

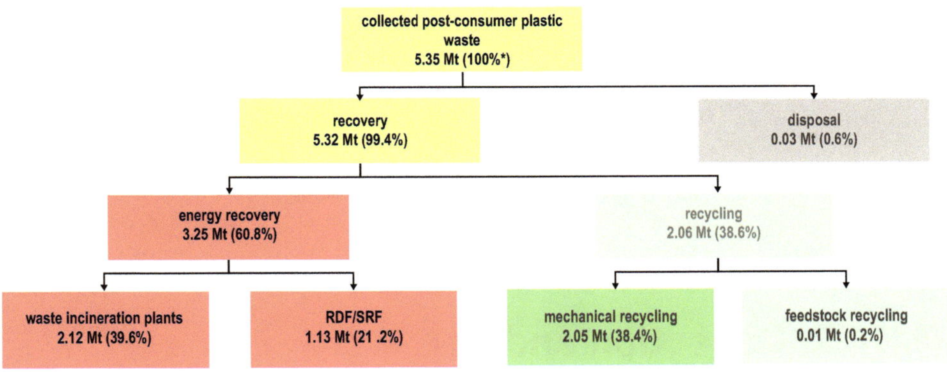

Figure 14.4 Status of plastics recycling and disposal in Germany in 2019 [5]

Even in countries like Germany with highly developed waste management activities and infrastructure, more than half of the post-consumer plastic waste is disposed of via energy recovery. Thus, the potential is still huge to extend the recycling activities. Figure 14.5 shows how mechanical and chemical recycling can complement each other and jointly contribute to strengthen the circular economy.

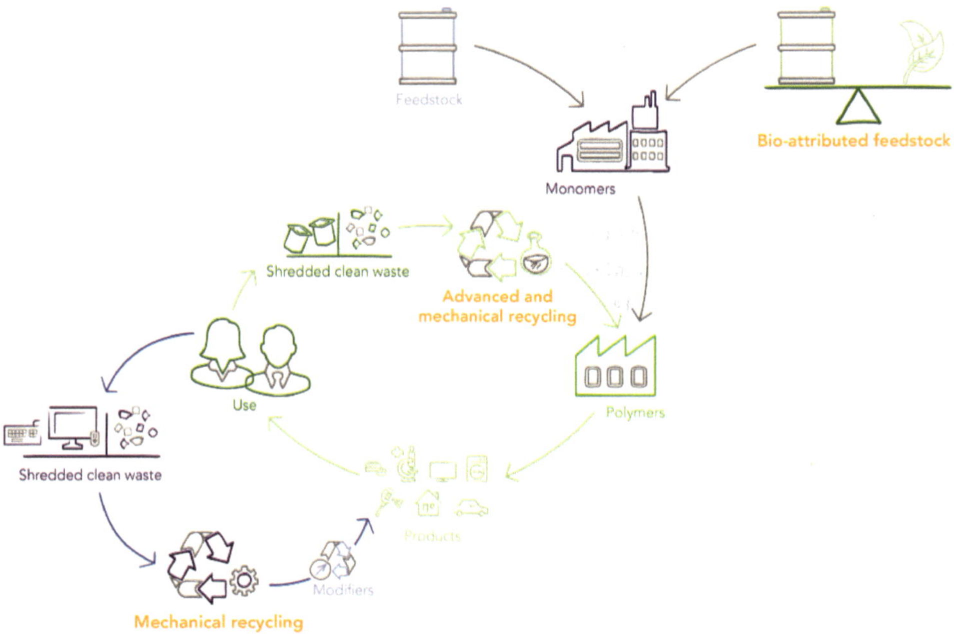

Figure 14.5 A circular economy (Image source: INEOS Styrolution)

■ 14.2 Collection and Sorting

Recyclables originating at waste producer sites are usually kept separate if it is economically advantageous for the waste producer to do so. As previously described, the subsequent recycling processes have to enable a more attractive recovery price for the waste producers.

In order to create a solid basis for an structured and well-functioning collection and disposal/recovery of waste in general and in particular of separate collected recyclables, it is necessary to categorize the waste streams before they are disposed of, collected, and brought to the dedicated recovery plant. Often the categorization is made by the usage of waste code numbers which already provide an indication of from where the waste originates, where and how it has potentially already being treated, and what are the main ingredients. More specific information is usually not included in the logic of the waste code numbers. This has to be included in other standards. Relevant parameters are more precise information about the mix/distribution of the (raw) materials, type and level of contamination, size and mechanical properties of the pieces, etc. In Germany, for instance, this further information is already part of the specifications that were coordinated with the Dual

System. The Dual Systems are in charge of the collection, sorting, and recovery of used post-consumer sales packaging.

It is also possible to collect recyclables in mixed waste fractions. This often means that all kinds of recyclables (metals, paper, plastic, textiles, wood, etc.) are kept and collected together with residual waste or biomass whereby both of the latter ones are generally intended for energy recovery.

A mixed collection scheme of course needs at least a material sorting before the recyclables can be sent to dedicated recycling facilities at all.

If even this opportunity is technically or economically disadvantageous, only energy recovery can ensure a viable recovery.

It is also possible that laws and regulations enforce greater separation of recyclables or sorting of mixed waste fractions before they can be sent to energy recovery at all. For instance, in August 2017 the amendment of the commercial waste ordinance came into force in Germany. In the first step the ordinance requires more effort by the waste producers to keep the recyclables separate. Following up on this it is generally forbidden to keep recyclables mixed unless the separation is economically or technologically unreasonable. These specific circumstances can be given for example if the waste producers only produce small quantities over all recyclable fractions or simply do not have sufficient space to install several containers for the separation of different recyclable fractions (metals, paper, plastic, textiles, wood, etc.).

In the second step, which came into force on January 1, 2019, it is forbidden to send the quantities that were collected mixed straight to energy recovery. The mixed fractions have to go through a mechanical pretreatment cascade/plant with minimum technical requirements in order to sort out recyclables.

However, neither the separate collection nor the subsequent sorting necessarily lead to more recycling to the same extent.

Separate collection of recyclables does not automatically lead to more recycling if the recycled raw material cannot be produced cost-efficiently in comparison to virgin material or it cannot compete with virgin material in terms of quality. Especially in 2020 during the COVID-19 crisis, it was apparent how vulnerable that recycling in general and in particular the recycling of used plastics is to a crash of the oil price.

Keeping and collecting recyclables and residual waste in a mixed state leads necessarily to a contamination of the recyclables. This contamination often makes a recycling more difficult, because it is either technologically or economically challenging.

In addition, the potential applications in which recyclates stemming from mixed recyclables can be used in are heavily restricted. For instance, only PET recyclates stemming from recyclables that were collected mixed can be used in food contact applications. These legal requirements are of course quite important because in many cases it is impossible or at least extremely challenging to ensure that harmful substances do not migrate into the recyclables and stay in the recyclates.

Nevertheless, plastic waste can be collected and transported in the containers/ systems and with the vehicles given in Table 14.2.

Table 14.2 Schematization of the Collection and Transportation Systems

Type of Waste	Containers/Systems	Vehicles[1]
Commercial waste & industrial waste	containers, lattice boxes, big bags, octabins	Hook lifts, dumpsters, rear and front loader
Household waste	container, sack or bin, collection or delivery system	Rear, side, and front loaders

[1] There are many other types of vehicles in the waste management and recycling industry, but the here-mentioned are the most common ones to transport the waste streams that contain most of the post-consumer plastic waste.

A comprehensive description of the different collection schemes is provided in Table 15.1 in Chapter 15. In the aforementioned chapter the different sorting processes are also described quite comprehensively, especially from a technical point of view.

The following explanations will shed a light on the sorting processes from the perspective of the origin of the waste streams.

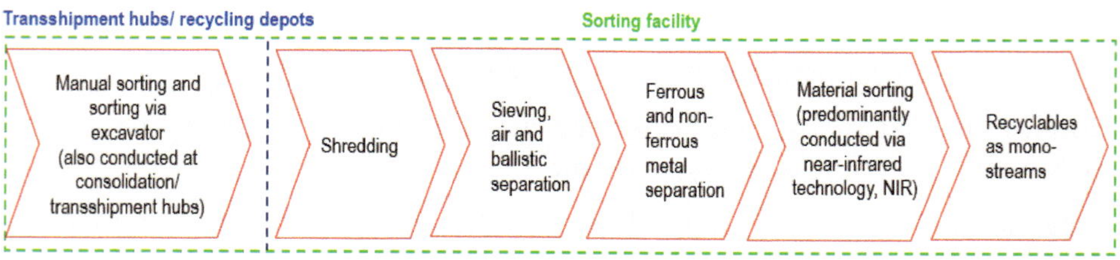

Figure 14.6 Overview mechanical pretreatment of solid waste

When recyclables are collected in a mixed fraction, a precondition is to separate the different materials. That is the point where sorting has to come into play.

Sorting can be conducted in dedicated sorting facilities or preparatively via manual sorting/excavator sorting or in a cascading combination of both options. Furthermore, laws and regulations define what kind of mechanical pretreatment for a specific waste stream is allowed to be conducted at all. For instance in Germany it is forbidden to sort human medical waste even if it can be classified as non-hazardous waste. The quantities have to be transported straight to an energy recovery plant. Therefore, the plastics contained therein cannot be supplied into a classic mechanical recycling system at all.

Especially in a mixed commercial waste stream, which is not collected in a nation- or area-wide harmonized collection scheme, sometimes large recyclable items are still included that can be separated cost-efficiently via manual sorting or via excavator sorting, e. g. wooden transport pallets or intact plastic canisters.

Especially when the majority of the waste stream as it was collected has to be sorted and the material to be sorted has a relatively small particle size (e. g., post-consumer packaging waste) a multistage automated sorting is the only option to process the huge quantities.

On an industrial level, ballistic-separators, all kinds of sieves and sifters (e. g., wind sifters), metal separators, eddy currents, and near-infrared technology (NIR) have proven themselves as the most cost-efficient material sorting technologies and therefore are established nation- and worldwide.

In addition, there are detection technologies like X-ray fluorescence, identification systems (e. g., watermarks), as well as imaging-based and laser-based technologies. In the future, robotics-assisted sorting systems will definitely be utilized as well.

What kind of sorting technologies as well as to what extent they are used as part of sorting facility mainly depends on the input/feedstock quality and the requirements for the output fractions, which are determined by the requirements of the buyers of the sorting fractions. As already described, it is also possible that laws and regulations require specific technical set-ups in order to ensure output recyclables.

■ 14.3 Recycling

More than 50% of all post-consumer plastic waste is collected via different mixed waste collection schemes, in which the share of plastics is between 2 and 8%. A large portion of recycled plastics stems from waste that is collected separately. Experience has shown that, especially for post-consumer plastics stemming from commercial and industrial waste, recycling rates (measured by the actual usage of the recyclates as such) are 10 times higher for separately collected waste, compared to mixed collection schemes.

In 2018, 80% of the 5 million tonnes of plastics recyclates re-entered the European economy in order to manufacture new products. The split by industry is shown in Figure 14.7. The rest was exported outside Europe to other regions.

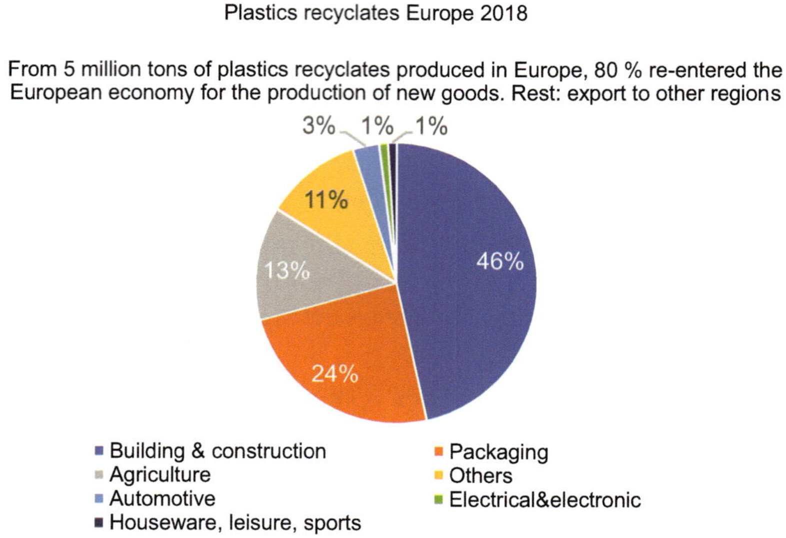

Figure 14.7 Reuse of recycled plastics in the EU [6] (source: Plastics Europe „Plastics – the facts 2020“, Conversio)

The typical mechanical recycling processes were already described in detail in Chapter 8. Therefore, Figure 14.8 shows the typical mechanical recycling processes just in overview.

Figure 14.8 Typical mechanical recycling process for plastics [6]

The layout of a specific mechanical recycling facility is designed to a specific input material:

- Type of raw material (metals, paper, plastics, textiles, wood, etc.)
- Specific varieties of dedicated raw material (in the field of plastics recycling for instance only PE, PP, PS)
- Specific attributes of the raw material or of the input material to be recycled
  - Example: PP or PS derived from WEEE has quite different attributes due to their chemical additives compared to PP or PS derived from packaging material and therefore the potential applications for the recycled materials are quite different

The opportunities and challenges of mechanical recycling of plastic waste thus highly depend on the feedstock and its quality. Accordingly, a mixture of different polymers is more challenging to recycle than clean, sorted post-industrial scrap. Mechanical recycling can be divided into the following main categories:

- Agglomeration
- Grinding
- Extrusion with and without Additives

Additional sorting, for instance via the previously described technologies or via swim-sink-separation, is not mandatory for all the aforementioned recycling methods. Extensive additional sorting or washing come into play if the feedstock quality is relatively far away from the output specifications. Correspondingly in relation to the output quality, it can be distinguished for instance between the requirements coming from the industry producing compressed molded parts for simple applications and the industry producing complex applications with special machines.

In a nutshell, the way in which mechanical recycling is used depends on the requirements of the products as well as of the input feedstock and the tension field between both parameters.

Agglomeration is mainly used to raise the bulk density of 2D material (mainly films) and to make the 3D plastic easier to convey. An agglomerate can only be used in limited ways.

Grinding, as a stand-alone recycling method, is mainly used for clean and presorted post-industrial scrap. The so-called regrind can be used directly in the production stream of, for example, injection molding companies. The color of the regrind will stay the same as in the unground part; the particle size depends in the sieves used. The more complex the requirements of the plastics converter are the more effort the grinding facility has to add to their technical setup. Typical add-ons are metal separators, suction systems for removal of fines, etc. To avoid carryover of recycled materials the mill has to be cleaned directly after the feedstock polymer is changed comprehensively.

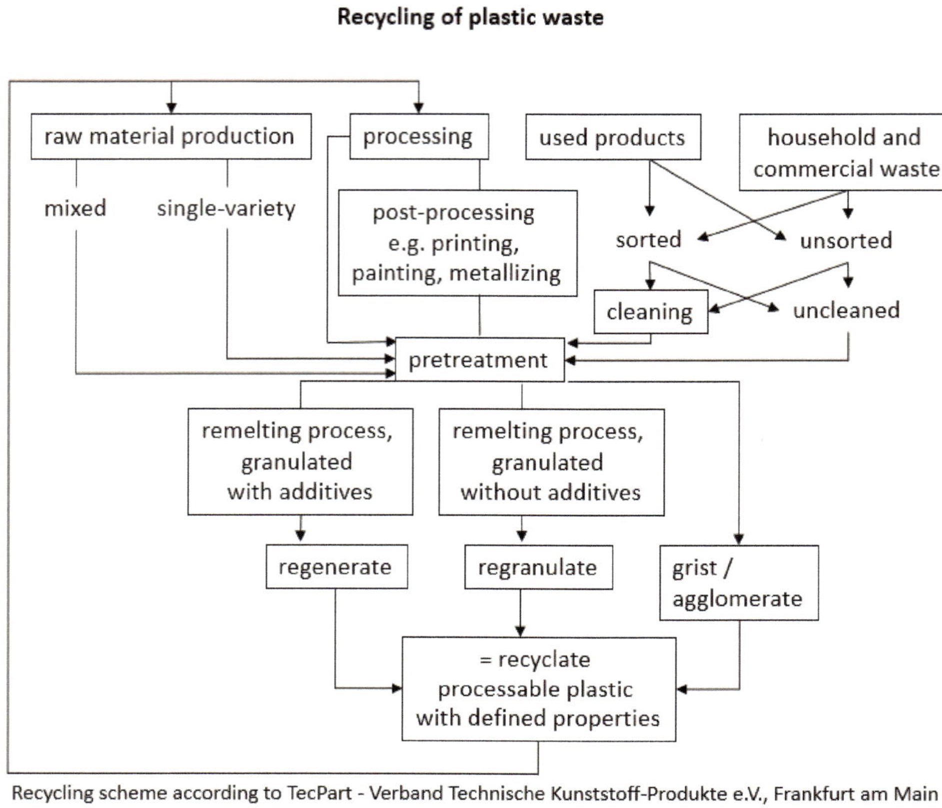

Figure 14.9 Comparison of the processes in mechanical plastics recycling [7] (source: Materials Guide Plastics)

The quality of extruded recycled material is the closest to virgin material in terms of the shape of the recycled products. A typical extrusion set-up consists of a shredder to ensure the right granular size and the extrusion unit itself, often upgraded with a degassing unit and a filter. Additionally, to prevent the extruder undergoing excessive wear and in the worst case serious damage, abrasive materials (foremost metal particles) have to be sorted out prior.

Furthermore, to prevent contamination of the recycled product with other polymer types, it is common to sort the material after shredding using a float-sink separation. For this purpose shredded material is fed into a liquid current. At the end of the swim-sink tank a swim and a sink-fraction is the result. The density of the liquid is the central parameter to separate different polymer types via swim-sink separation. For example, polyethylene and polypropylene without specific additives and/or filler have a density both below 1.0. The use of water as the separation medium is widespread, whereas using of other liquids is not. To a certain and limited extent it is possible to separate polymers that have a different density through their chemical additives (e.g., brominated PP can be separated from non-brominated PP) but in the overall view additives and fillers makes an effective and cost-efficient swim-sink separation often challenging.

The subsequent granulation process itself can be split into dry or underwater granulation. Homogeneous feedstock is heated, melted, and formed into a melt strand which is cut to a specific grainsize. The color of the granule mainly depends on the color of the input material. Mixed-color feedstocks mainly result in a black granules. Flake sorting by NIR can be used to separate colors. To get rid of the odor and to ensure certain chemical properties of the recycled material it is also possible to refresh the material with a thermal follow-up treatment (detailed description of flake sorting can be found in Section 8.4.1.4.3). This is especially needed when recycled material derived from mixed waste streams has to be implemented in packaging material.

Plastics waste that cannot be mechanical recycled (e.g., thermoset fiber composites) is recovered to produce heat and electrical energy. For this purpose, new technologies like chemical recycling and dissolution will provide higher recycling rates also for this type of plastic waste.

The mix of established recycling methods is being joined by new technologies like chemical recycling, dissolution, and high quality (food grade) mechanical recycling (other than PET beverage bottles). These are currently emerging on a (pre) commercial scale and will result in a disruptive change from a rather linear towards a circular economy. This scenario will reshape global waste management as well.

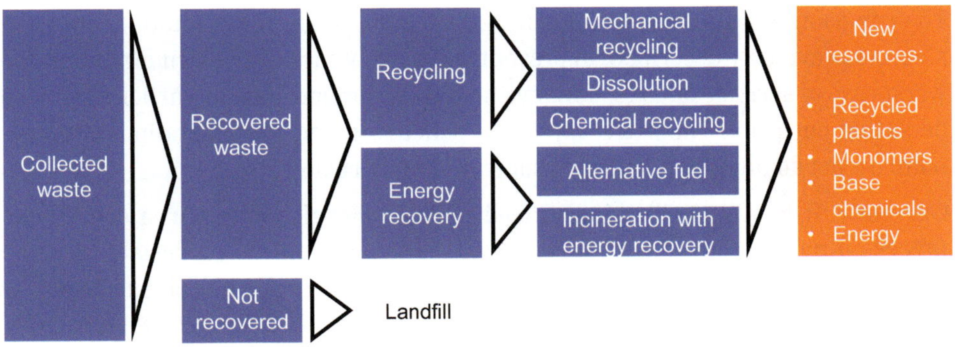

Figure 14.10 New waste treatment technologies will result in a circular economy, providing new resources

In the following the differences between the pretreatment of mechanical and chemical recycling will be shortly comparatively outlined. As already described in this book, chemical recycling on the technological level of meaning is a collective term for several technologies. To make it more concrete the following comparison shows the differences between the pretreatment for a typical mechanical recycling process and for a recycling process via pyrolysis.

Pyrolysis has to be conducted in enclosed spaces in order to ensure the exclusion of oxygen. For this purpose reactors are used. These reactors are of course bounded volumetrically. Based on current knowledge and experience, pyrolysis reactors for the processing of mixed plastic waste should not exceed a capacity around 3 t/h in order to ensure a controllable process environment. This seems to be valid in the short-term but there are several technology-vendors who indicate that even higher throughputs will be possible in the medium term.

To reach an ideal combination of throughput and controllable process environment, a mechanical pretreatment has to be conducted. If the plastic waste stream is too contaminated (here referring predominantly to material that is not plastic) at first an additional sorting might be necessary. But here we assume that the chemical properties of the feedstock are acceptable for the pyrolysis and therefore no additional sorting is needed.

To ensure a sufficient throughput is key for a profitable operation and therefore a throughput-optimized feeding process is also needed. To ensure both is especially challenging when material with a low bulk density and a poor pourability should be processed. In particular, films and laminates from packaging material are characterized by a low bulk density and a pour pourability. In order to fix this issue a compacting process is necessary. But before such a process step can be executed a shredding or granulation step is mandatory because the compacting process can only operate efficiently with a certain grain size.

The compacting process can be conducted via pelletizing, agglomeration, or extrusion. As previously described, to prevent excessive wear of the extruder and in the worst case of serious damage, abrasive materials (foremost metal particles) have to be sorted out prior. The same applies to pelletizing and agglomeration, although both are not so directly vulnerable compared to extrusion.

Currently, it is still challenging to feed pellets or agglomerates in a continuous process directly into the pyrolysis reactor without losing the process heat of the upstream compacting step. In the field of extrusion currently it seems to be easier to ensure the heat transfer between the extrusion unit and the pyrolysis reactors. In direct comparison to the usage of an extruder in the mechanical recycling set-up, it is not mandatory to implement a degassing or a comprehensive filter system to ensure the right quality of the output. Within the pyrolysis process set-up the extruder only has to create a stable melt strand.

For now the pyrolysis of mixed plastic waste working on an industrial scale is still in its infancy and therefore it cannot be assessed finally which kind of compacting process is going to emerge as the preferred one.

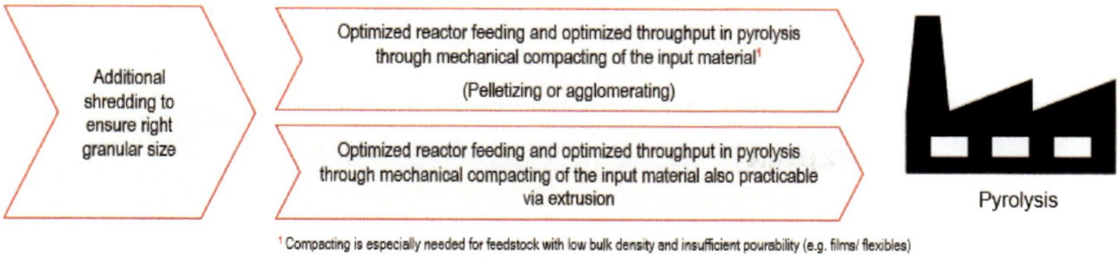

Figure 14.11 Typical mechanical pretreatment process for pyrolysis

To sum up, the pyrolysis of plastic waste does not need such clean and pure (here with the meaning of varietal purity of the polymers) input fractions to create outputs with a value that justifies the upstream effort economically. But there are also certain limits because it may be fundamentally possible to pyrolyze almost everything. However, especially if a complex and challenging feedstock is pyrolyzed, an oil is going to be the output that cannot be sold in this form to anyone and therefore the operation cannot be executed profitably at all. Generally, it is possible to purify the highly contaminated pyrolysis oils subsequently but this of course will cause additional costs. Therefore, in the short term it is going to be key to harmonize the process steps along the value chain in a technologically and economically efficient way.

The strong drive by the EU Commission to avoid waste, and to extend recycling by quotas can most likely not be satisfied by mechanical recycling alone. Therefore, new emerging technologies will complement the progress in conventional recycling and hence will play a pivotal role in future.

Prof. Kaminsky started the "Hamburg pyrolysis process" in the 1980s and still today under lead of Prof. Gerrit Luinstra, this process of cleaving polystyrene back into monomers and a few side products represents the state of the art in PS depolymerization [9]. The first (pre)commercial depolymerization was reported by Agilyx, and there are a few projects in planning using Agilyx technology to thermally cleave polystyrene back to styrene.

The Fraunhofer IVV Institute in Freising, Germany announced the "CreaSolv®" dissolution process in the 2000s (source: website CreaSolv). In the meantime, further parties appeared and offer recycling solutions based on dissolving polymers in solvents and (proprietary) solvent blends, followed by re-precipitation of the purified polymer and work-up of the solvent to be reused again.

All these new recycling pathways will change the way our current value chain functions. The relevance of waste as a precious raw material will increase, the role of detection and sorting technologies will significantly increase as well, and so will the quality of the recyclate materials overall. We do not see a "one size fits all" technology emerging, but rather a whole portfolio of different recycling methods that will partly compete but in the end will complement each other.

A schematic depiction of the options for recycling can be found in Figure 14.12.

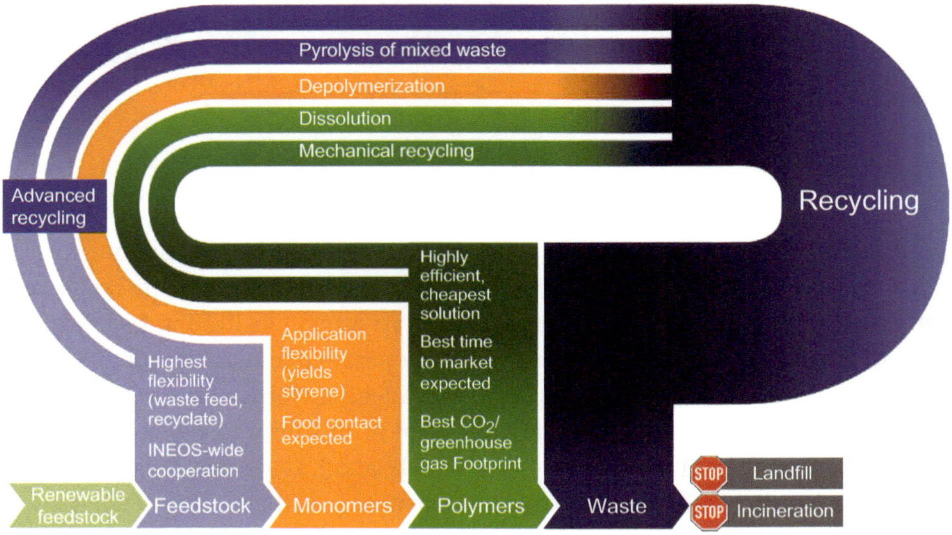

Figure 14.12 Options for plastic recycling (Image source: INEOS Styrolution)

References for Chapter 14

[1] Conversio Study: "Post-consumer Plastic Waste Management in European Countries 2016 – EU 28 + 2", Management Presentation January 30, 2018 to PlasticsEurope, Brussels

[2] IHS Markit Report: "Plastics Recycling", December 20, 2019, p. 10, based on data from OECD and World Economic Forum, Ellen MacArthur Foundation

[3] Geyer, R. in: Letcher, T. M. (Ed.): *Plastic Waste and Recycling*, Elsevier, London, 2020, p. 25

[4] *www.eca.europa.eu*

[5] PlasticsEurope: "Plastics – the Facts 2019"

[6] PlasticsEurope: "Plastics – the Facts 2020"

[7] Baur, E., Harsch, G., Moneke, M.: Werkstoff-Führer Kunststoffe, 11th Ed., Hanser, Munich, 2019, p. 61

[8] Scheirs, J., Kaminsky, W.(Eds.): Feedstock Recycling and Pyrolysis of Waste Plastics, Wiley, 2006, p. 475 ff.

15 Recycling Value Chain and Detection/ Purification Technologies

Jürgen Priesters

15.1 Introduction

Whether addressing food protection, lightweight and high-performance components, or any kind of packaging, in most cases, it is useful to use plastics. Therefore, plastics have become an ever-present material in our everyday lives, and global plastic waste continues to increase at significant rates. Municipalities and waste-management companies have been tasked with new challenges since the dramatic shift of the Chinese National Sword. Meanwhile, our environment, especially our oceans and marine life, face existential threats. Collectively, we need to reduce the environmental impact of single-use plastics and focus on long-term alternatives, recycling whenever possible. Through implementing effective measures in the plastics value chain, we can ensure the long-term health of our economies, communities, and environment.

The complex nature of processing waste streams often requires more than a standard waste management system. The systems in place today used to collect, sort, and recycle plastics are not efficient. We need to rethink, to be open to new solutions and models so that we can move forward in effectively managing waste.

15.2 Recycling Value Chain

It is no wonder why plastic is encountered in practically every part of everyday life: it is inexpensive, versatile in use, lightweight, and resistant. Plastic plays a critical role in the protection and preservation of goods, helps prevents food waste, and, thanks to its light weight, reduces transportation-related emissions. Whether being used in an automobile, electronics, or packaging, plastic helps meet consumers' demands for convenience. What we do with the plastic after use, and how the plastics re-enter the circular economy, remains the crux of the matter. The recovery and reprocessing of plastic depend on the plastics value chain and a shared vision for success without compromising the environment:

Replacing plastics in packaging and consumer products with alternative materials could raise environmental costs nearly fourfold [1].

Even decades after the launch of the first universal recycling symbol, only 5% of plastic material is retained for subsequent use. While some forward-looking governments have introduced restrictions on single-use plastics and continue to aggressively reduce waste in the environment, there is still an urgent need to improve infrastructure, especially in developing countries, in addition to creating value from post-use plastics. Collaborative efforts between industry, governments, and NGOs are focused on advancing new technologies that not only minimize waste but also make recycling and recovering plastics easier [2].

In developed countries, the *real* amount of plastic packaging being recycled is approximately 25%, with the amount of all types of plastic recycled being much less. However, due to inconsistent and supportive measurement methods, the official numbers look much higher. In the past, a large portion of the quoted amount of material being recycled was achieved by shipping millions of tons of plastics per year to China.

The most notable and recent change in plastics recycling came then with the Chinese National Sword Standard, which put an end to the import of over 10 million metric tons of plastics per year. In 2019, India followed suit by also banning the import of plastic waste and other waste importers have introduced new tighter restrictions. This dramatic shift in the industry has created a global ripple effect. Many recycling programs and companies have temporarily reverted to incinerating recyclables, sending plastic waste to landfills, or shipping to alternative but not as well-developed recycling locations (Figure 15.1).

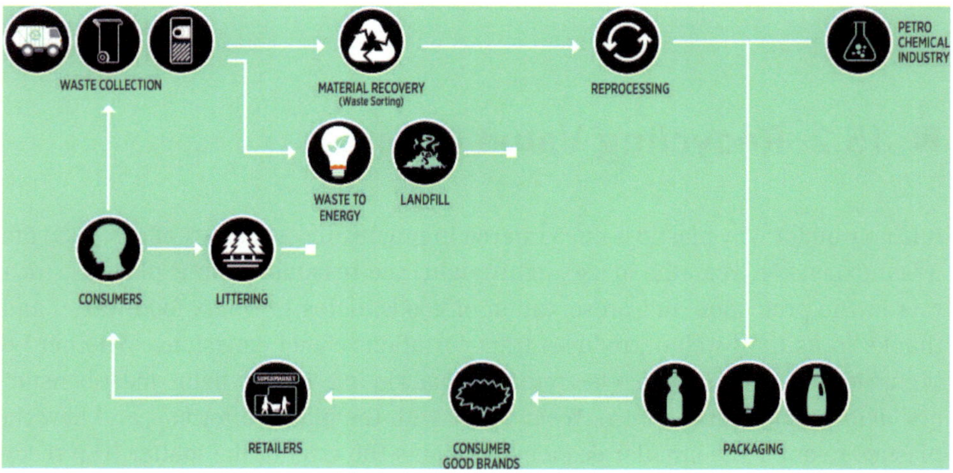

Figure 15.1 The new plastics economy [source: TOMRA e-book]

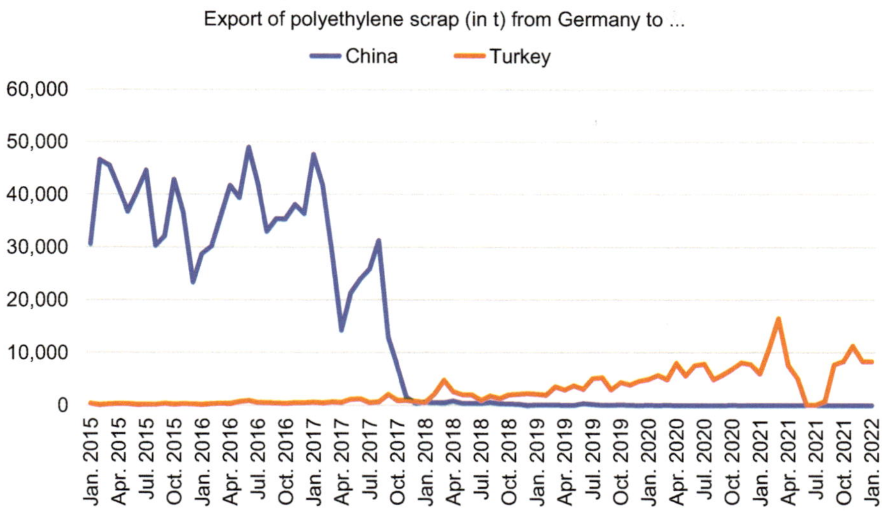

Figure 15.2 Polyethylene scrap exports [source: eurostat]

Today, besides the export of plastic scrap to a small group of foreign countries, in the EU approximately 15–20 million tons of plastic material enters landfills and incinerators as well as cement kilns and coal power plants every year. By doing so, millions of tons of carbon dioxide is generated, and the plastic resource is definitively irrevocably gone.

From an environmental standpoint, it is crucial that we move away from an export mindset as it comes with the innate risk of leakage into the oceans. Improving waste management infrastructures globally, particularly in industrialized countries, would ensure all waste can be locally managed appropriately.

15.3 Marine Litter

With 8 million tons of plastic entering the ocean each year, we urgently need collective action to deliver new solutions to a global challenge of this scale. Innovations and technology-driven waste management have proven successful and also created viable economic opportunities.

Each year, at least 8 million tons of plastics leak into the ocean – which is equivalent to dumping the contents of one garbage truck into the ocean every minute. If no action is taken, this is expected to increase to two per minute by 2030 and four per minute by 2050 [2].

Whereas most see a problem with financing waste extraction from the world's oceans, Paul Manning, Chairman of Ocean Resource Management, sees a unique opportunity. “What the world has missed is the massive wealth of raw materials floating in the oceans,” says Manning. With over 30 years in the waste management industry, Manning is already familiar with optical sorting equipment to recycle just about everything floating in polluted waterways. His plan: to recycle waste at sea on board an ocean processing vessel and clear the oceans while making a profit [3].

15.4 Waste Management Systems and Materials

Dating back to the 19th century, “non-valuable waste” in Europe was collected in bins and bags while the “valuable waste” – such as rags, bones, paper, and metals – was left to the informal sector. Generally, the system was considered to be acceptable at the time.

Much changed, however, with the introduction of industrialization. For example, less manual labor was used to sort out valuables, the waste composition changed, and rags could no longer be used within the paper industry. At that time, one of the top priorities was to collect the waste free of dust due to the ash from the coal furnace heating systems.

Laws and regulations from an early stage supported waste collection. One of the first boosting waste management regulations in Germany was introduced in 1893, in which municipalities were permitted to charge fees for waste collection and sorting.

Later, plastic was developed and entered quickly into waste streams. Coal furnace heating was replaced with central gas or oil heating systems, but with paper and board used to light the ovens, leading to increased generation of waste and disposal. Only a few material streams, mainly paper, glass, and metals, were still collected separately or in the informal sector in order to sell to the market and use for recycling.

If we compare this situation to today in many countries worldwide, we see numerous similarities. People live on landfills or dumps, take out the valuables to sell, and leave the materials without or with only limited value in the landfill or dump. At least waste is collected in these countries. The situation becomes critical in countries where waste is not collected at all: there is no buyer for many materials, and no or very limited infrastructure.

But let us first focus on the developments which developed countries have achieved. Therefore, we go back to 1990 in Europe.

In a number of European countries, paper, metals, and glass were collected separately and a well-functioning refillable system for beverage bottles was installed.

However, the remaining waste was primarily landfilled. Due to the higher costs of incineration facilities when compared to landfill without aftercare costs, only a few of these facilities were in operation.

With time, the mindset of waste management and the political world started to modernize, and landfill would no longer be seen as a sustainable solution for the future. Besides this, the industry was looking for new business opportunities.

As part of this mindset, a new idea was born to collect selected packaging materials separately, and sort and recycle them, thus creating space in landfills and extending their lifetime.

In addition, large amounts of new investment were becoming prevalent in the waste business sector. In the beginning, 1993, approximately 2000 DM/ton (approx. 1000 Euro/ton) was paid from the first German green dot system to the waste management companies in order to collect, sort, and recycle this lightweight packaging material. Today, it is more in the range of 200 Euro/ton of lightweight packaging material.

With the implementation of this new system in Germany, recycling was boosted, technology development heavily increased, automatic sorting technologies were developed, and, finally, the waste sector became an attractive option in comparison to other industries. It is no wonder that the first patent to sort waste automatically using electromagnetic waves was from an energy producer and not from today's suppliers of sorting machines!

Finally, with the closure of landfills for the majority of household waste streams in 2005, sorting and incineration developed key technologies for handling waste streams in Germany.

At that time, a small niche developed in parallel. In 1972, a small Norwegian company developed the first RVM (Reverse Vending Machine). The idea: put a deposit on waste materials to ensure people return these to get their deposit back. Via such a system, a very high collection rate and very pure material stream could be ensured, if only for a very limited amount of packaging waste.

Today deposit systems and RVMs are known worldwide and have a very high acceptance in many countries. Unfortunately, in some countries, such as in Germany, the material collected with very high purities is sometimes mixed after the detection in these machines into one container or bin.

With the goal of reducing waste and increasing reuse and recycling, many different bags, bins, and systems were introduced in the early 1990s, and the idea that recycling is possible only via a separate collection was established. Today, an average of 10 different waste collection systems in the EU exists. Landfills are often closed for household waste streams. Still, the recycling rates are not at all where they should be [4, 5] (Table 15.1).

Table 15.1 List of collection systems in the EU

Waste Material	System min	System max
Cullet glass	Bring system mixed color	Bring system into three colors
Packaging other than glass/ paper	One bag or bin	Two bags or bins, flex and rigid
Paper & board	One bin or bag or bring system	Two bin or bag or bring system
Bulky waste	Pick-up service	Separate wood and rest pick-up
WEEE	Pick-up or bring system	Separated into groups of e-waste
Organic waste	One bin or bag	Garden, household, and food waste separate
Batteries	Bring system	Bring system
Deposit materials	Reuse PET and glass only	Single-use PET and aluminum beverage and reuse beverage bottles
Residual waste	One bin or bag	One bin or bag
Textiles	Into residual waste	Separate shoes, textiles, carpets bring system
Construction & demolition waste	All-in-one container system	Inert, tar, gypsum, and others separate

The idea to collect waste materials separately is an old one, having been established in times when waste was manually sorted in sorting plants and recycling was focused mainly on glass, metals, and paper materials. Though theoretically plausible, with the practicalities of human behavior, even in developed countries, such systems are unfortunately not followed. In each and any of the above-mentioned waste collection systems, a mixture of material is often obtained: wood in textiles; organic waste in packaging; and glass, metals, paper, and plastic in the residual waste bin or in bulky waste. In fact, over 50% of plastic packaging and over 60% of all plastics do not end in the packaging waste bin or bag. Considering that it is only *these* bins or bags that are sorted in many countries, an above 50% recycling figure can only be achieved if all waste streams are sorted.

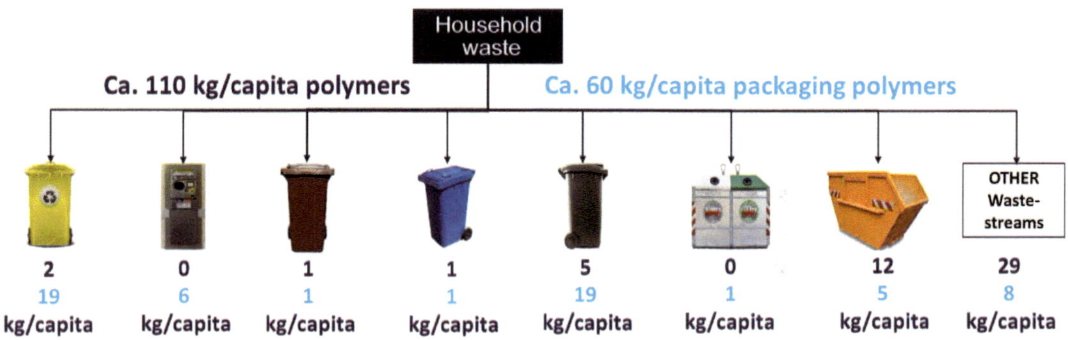

Figure 15.3 Amount of plastics in waste collection systems, average in EU (source: TOMRA Sorting GmbH measurements)

In sorting plants, such as the fully automatic plant ROAF (constructed by Stadler Anlagenbau GmbH, Germany) or I. V. A. R (constructed by Sutco Recycling Technik GmbH, Germany) in Norway, plants in Kozani and Chania in Greece, plants in Gdansk and Warsaw in Poland, and many more, the residual waste is sorted and recyclables that would have ordinarily burned or landfilled are sorted out and sold into the recycling market. By doing so, more than 2 million tons per year of recyclables is sorted out of mixed residual waste, contributing to higher recycling rates and saving millions of tons of CO_2 emissions per year.

As discussed in Section 15.2, the officially reported recycling data have several weaknesses. EU member states do not have a common measurement method. Although all countries include the output of sorting facilities in their figures, in some countries, only the licensed packaging item is counted and not the entire amount of packaging put into the market. This output is usually a bale, 1 m × 1 m × 1 m size, and contains moisture, residual material inside the packaging item, impurities, etc., and non-packaging plastics as well. If the impurities, moisture, and non-packaging items would be subtracted, the real material amount is, in many cases, below 70%. With this taken into consideration, the recycling numbers in the EU would decrease by at least 30% from the reported ones.

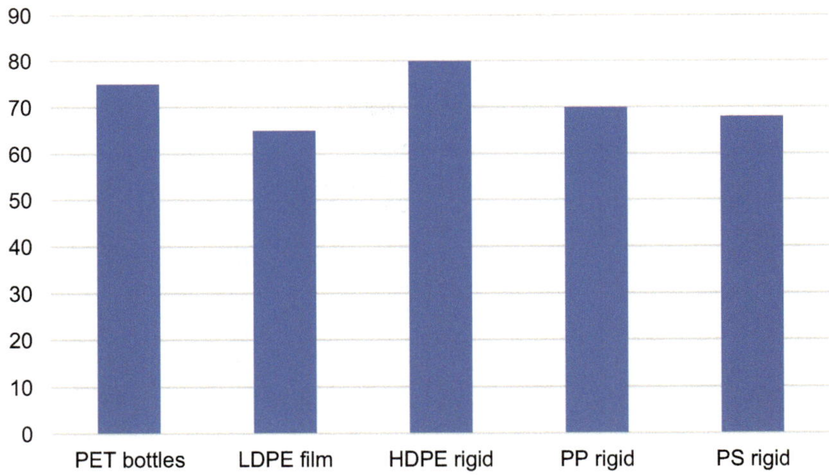

Figure 15.4 Yield (in %) of different packaging materials by being recycled (source: PRE/TOMRA Sorting)

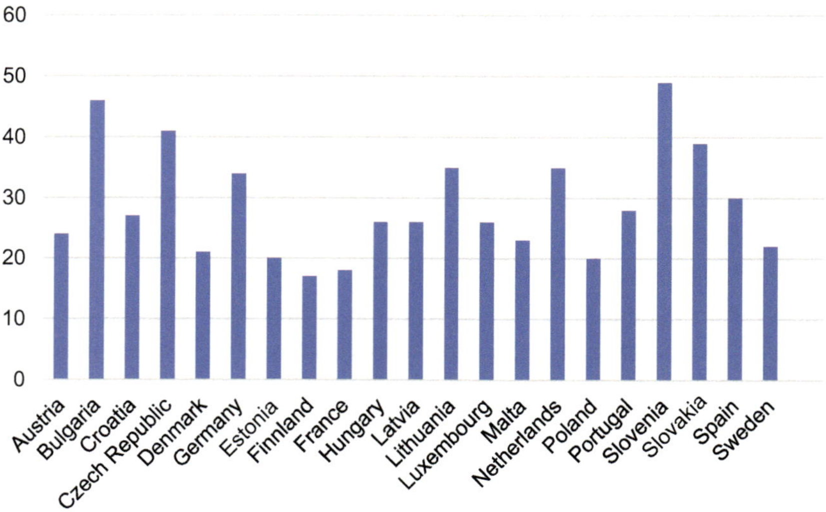

Figure 15.5 Plastic packaging recycling percentages reported to eurostat; yield taken into account (source: TOMRA Sorting)

If each EU member state would simply calculate plastic packaging material introduced into the market and compare this with the final product, flake or pellet, the numbers above shown would look even worse.

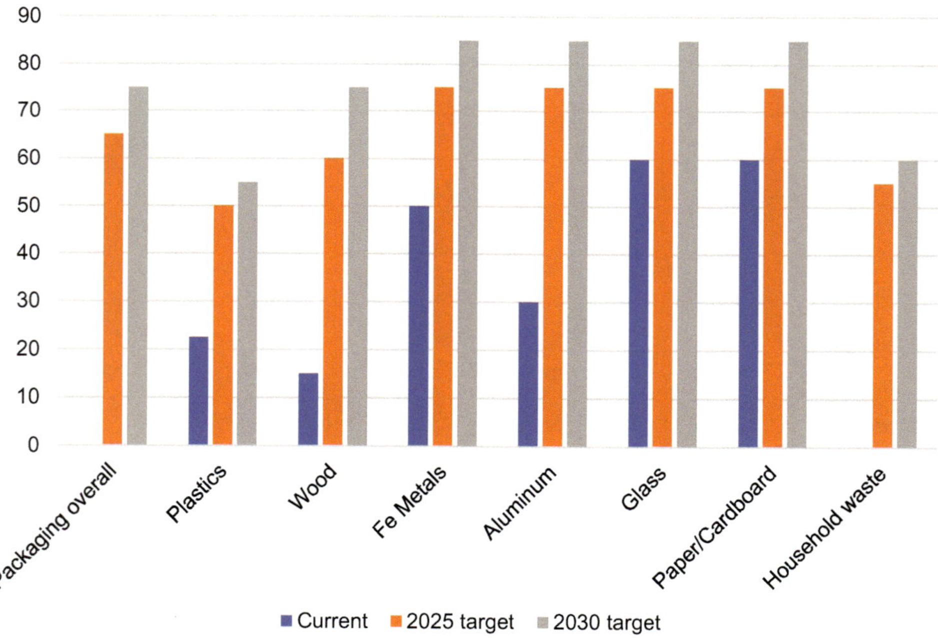

Figure 15.6 EU recycling targets [6]

Some member states have fully understood this along with some EPR systems and municipalities. In these countries, such as Norway, Spain, Italy, Poland, and the Netherlands, plastic packaging materials are collected or sorted out from many waste streams and are counted correctly into the recycling figures for such materials. It is only by applying this system that a well-above 50% recycling rate for packaging materials be achieved. Therefore, a paradigm shift must take place including modern and open thinking on how to treat waste streams and supported by law.

If each ton of plastic that is burned today instead underwent high-quality mechanical recycling, 2–3 tons of CO_2 emissions could be avoided [7].

Beyond the environmental benefits of recycling plastic waste, there are positive economic shifts that are also to be expected. Economies previously built on "use once and discard" are quickly embracing new business opportunities given the advanced sorting technologies that purify and deliver high-quality recyclates. A recent McKinsey report [8] estimates plastics reuse and recycling could generate profit-pool growth of as much as 60 billion USD for the petrochemicals and plastics sector. Plastic-specific recycling goals will soon transform into quotas that will depend heavily on state-of-the-art technology to achieve high recycling rates. Packaging materials of the future will perhaps feature codes that can then be scanned in mechanical recycling plants to increase the quality of recovery. Recycling goals for plastic waste from end-of-life vehicles, large household appliances, and electronic waste will also see a large boost in the near future. Recyclable product design and sensible recycling practices are essential to reducing potential pollutants and unwanted waste.

Sorting solutions are key to the "resource revolution". With the latest intelligent sensor-based recycling solutions, a wide range of technology products specifically engineered for plastics recycling, simultaneous detection and sorting of plastic fractions by color, material type, and object recognition results in much higher quality and higher yields. With such intelligent sorting technologies, valuables are sorted out of many different waste streams and recycled into high quality products. Small plastic property differences can be detected, resulting in consistent and predictable quality.

15.4.1 Bulky Waste

Bulky waste sorting can produce a large number of valuable materials. The accessibility of these materials, however, is challenging due to its cumbersome nature (Figure 15.7).

In order to streamline the separation process, the initial treatment step must involve shredding the input material or at least the oversized part of the input material. Sizing of fractions by screening, i.e. separation of light and heavy components, is required in order to prepare fractions for sorting. With sensor-based sorting systems, the different materials can then be successfully extracted and prepared for reuse.

Figure 15.7 Bulky waste [6]

From bulky waste, usually cardboard, wood, metals, PE film, and rigid plastic types are sorted out. Bulky waste is, at ca. 30–40 kg/capita, a reasonable amount of the household waste stream. Up to 50% by mass of bulky waste can be different grades of wood.

15.4.2 Commercial and Industrial Waste

Commercial and Industrial (C&I) waste sorting separates many elements, such as plastics, paper, cardboard, and film – all of which can be effectively recycled into saleable products.

Figure 15.8 Commercial and industrial waste [6]

Figure 15.9 Commercial and industrial waste [6]

At the other hand, the two examples shown in Figure 15.8 and Figure 15.9, respectively, demonstrate the big variety within C&I waste streams. The composition is primarily dependent on the kind of business of the industry disposing the waste: if a textile manufacturer is close by, a large proportion of such a waste stream will be textiles.

Traditional manual or semi-automated techniques are only able to recover low levels of separate materials from commercial and industrial waste streams. Humans, even when trained, cannot effectively distinguish between different polymer types or cannot sort small items. In contrast, sensor-based sorting and recycling systems convert complex and varied commercial and industrial streams into high quality, separated output. The systems also make manual picking obsolete, thus reducing labor costs and material loss while increasing the efficiency of commercial waste recycling. If a producer changes the material in use, for example from PP to PE, humans may not detect that whereas machines do. NIR (near infrared) based machines used for sorting polymers can distinguish between different as well as slightly altered materials, such as LLDPE from LDPE, or cross-linked PE from "standard PE".

15.4.3 Municipal Waste

The automatic recovery of recyclable fractions such as different plastics, mixed paper and cardboard, metals, and even beverage cartons is becoming increasingly prevalent. One reason, as mentioned above, is that this solution is often cheaper in developing countries where a separated source system often results in improper disposal of 50–70% of plastic recyclables. Another is the refusal of some countries to allow incinerators that burn plastics to claim green energy, whereas another might be CO_2 savings. In the EU alone, approximately 320 MSW sorting plants are equipped with sensor-based sorting technologies and contribute more than 2 million tons per year of different recyclables. Although there are a few poor examples, the majority are working properly and heavily protect the environment.

A typical residual mixed waste sorting plant comprises the following process steps:

- Input: mixed municipal solid waste after bag opening or shredder
- Pretreatment/screening
- Ferrous and non-ferrous metal separation
- NIR sorting into two streams: polymers and others
- Mixed polymer separation of flat and rolling fraction in a ballistic separator
- NIR sorting of paper and board materials
- Rolling fraction sent to sensor-based sorting → separation of either mixed rigid polymers or into different polymer types such as PET bottles, HDPE, PP, and PS
- Flat fraction sent to sensor-based sorting → separation of polyethylene film
- Final NIR cleaning step to get constant purities, allowing the materials to be easily sellable

The technical process is very similar to a process used to sort mixed packaging waste. Such sorting plants could easily be used for sorting separate collected packaging waste and in a separate shift mixed residual waste. This is being done for example in Chania, Katowice, Cordoba, and many more sorting plants [9].

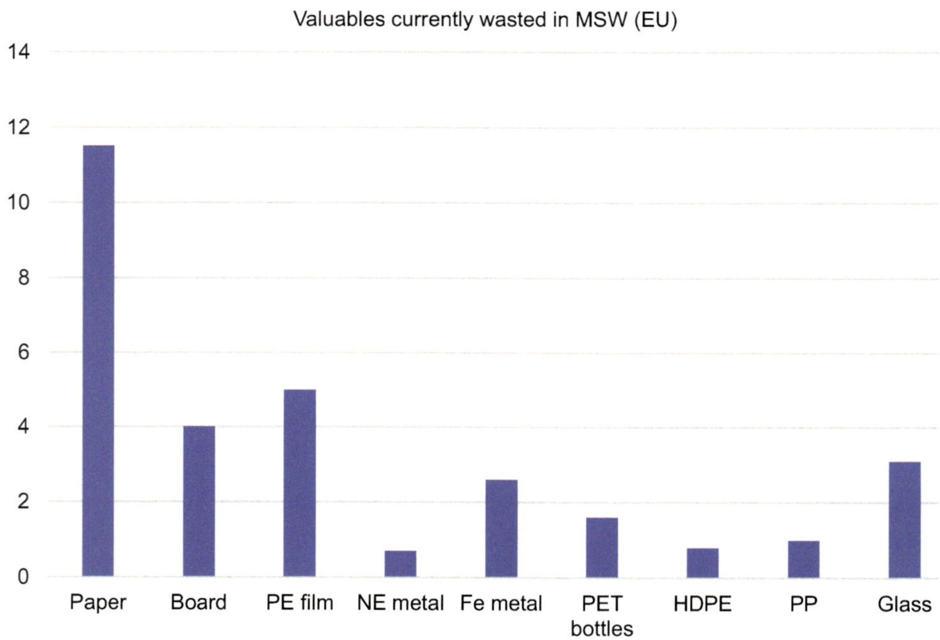

Figure 15.10 Valuables in EU mixed municipal waste [source: TOMRA]

Figure 15.11 Sorted valuables out of residual mixed waste [source: TOMRA]

15.4.4 Organic Waste

The total waste volume contains a high percentage of organic waste material. Depending on the origin country, this fraction is generally not separately collected. Even in cases where organic waste is collected separately, it is often treated in different ways. Within the organic fraction, the amount of impurities is partly significant. In large municipalities, the organic waste bin is more similar to a plastic bin, and the huge amount of inert material, especially glass, can cause significant problems during some of the different treatment processes. Biodegradable plastics very often are collected within the organic waste bin. However, due to inefficient biological treatment or processes, these cannot degrade so they are found as small particles, looking very much similar to standard plastics.

In Germany, there is a standardization for household waste compost in which impurities are described and the maximum amount regulated and checked. Again, sensor-based sorting can play an important role in order to purify the organic or the final compost by sorting out all remaining non-organic materials, such as glass, plastics, textiles, etc.

By purification of the organic waste, before or after composting or biogas production, maintenance can be reduced significantly, and a constant good quality of compost can be achieved.

Figure 15.12 Purification of biowaste [source: TOMRA]

15.4.5 Packaging Waste

The first Green Dot system was founded in Germany in 1992 due to landfill resource savings and as an additional income to the waste management companies. In the following years, the system improved, and massive development progress was achieved in sorting technologies. Unfortunately, the technology used today and the technical sorting plant design as such are still very similar to the ones used 20 years ago, only with plants increasing capacity by having more than one line. Over time, plenty of specified material streams have been developed to be able to fulfill quotas, even though the market price for some of these streams is similar to the price of being incinerated. While regulation has developed over time, and now encourages a green economy, improper disposal of packaging waste in a source-separated collection system and exaggerated recycled yield figures in aluminum waste streams are symptomatic of a system that has yet to be optimized.

A modern packaging waste sorting plant is fully automatic, with only loading, unloading, and maintenance being completed manually. From approximately 2 tons/hour capacity in 1993, current capacity levels of the biggest plants have reached 30–40 tons per hour. In Germany, the amount of sorted plastic materials for mechanical recycling is approximately 20–30% of the collected material. Approximately 40–50% of all collected material is sorted into various plastic streams but often used as RDF, energy recovery, etc., with the remaining 50% being beverage cartons, metals, and residue. As explained above, with a yield of approximately 70% within mechanical recycling this leads to a real recycling level of approximately 15–22%.

In a sorting plant for packaging waste, the following technologies are used:

- Shredder or bag opener
- Screens
- NIR - VIS - AI - X-ray - EM Sensors
- Wind sifter
- Ballistic separators
- Balers
- Eddy current units
- Magnets

The units above are used to prepare the material stream for the metal sorting units and sensor-based machines that sort out final products. In a sensor-based unit, several different sensor technologies are implemented, depending heavily on the target to be achieved. In general, the principle is as shown in Figure 15.13. One or more sensors, located in a scanner box, detect all materials, colors, shape, form, brand names, molecular weight, or electromagnetic properties. The desired material or item is then sorted out using compressed air and valves. Usually, a material stream is divided only into two streams: product and remaining material. The product can be just one particular material or a combination of several. The throughput into such units varies from approximately 700 kg/h up to 10 t/h, depending on the infeed material and the width of the sensor-based unit.

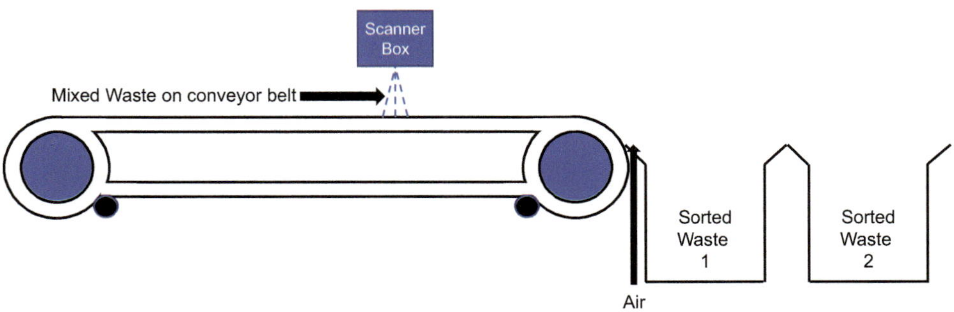

Figure 15.13 Technology of a sorting machine [6]

15.4.6 Paper Recovery

Early on, the paper industry understood the need to put voluntary recycling targets in place and invest to achieve those targets. Sensor-based sorting technology is mostly used to separate mixed collected paper into different grades. In the early 1990s, all this was done manually, at 10 tons/hour infeed with up to 12 people per shift doing an acceptable level of sorting into the different paper grades. Today, modern sorting plants treat up to 50 tons/hour and separate mixed paper into at least three grades: newspaper and magazines, brown board, and mixed paper. In the near future, paper will be sorted into more grades: wood-free white, paper containing wood, gray board, special papers, and more.

One example is the state-of-the art Regional Materials Recovery Facility in Ballymount, Dublin, which processes household dry recycling material from approximately 400,000 homes in Dublin and the surrounding areas. 70% of the infeed material is privately collected by Panda Recycling and Greenstar, with the remainder collected by other operators.

The facility is the largest of its kind in Ireland, employs 105 staff, is licensed to process 100,000 tons per annum, and is currently running at 90% capacity and operating 20 hours a day (two shifts), five days a week.

Every hour, 20 tons of household dry recycling is separated into its constituent parts: two paper grades, cardboard, aluminum, steel, and five grades of plastic. Recovered materials are formed into bales, which Irish Packaging Recycling (IPR) sells to markets around the world. Such trends are visible in all material streams. For example, today, scrap aluminum, which contains up to 10 different aluminum alloys, are split into such alloys, plastics are sorted by color and mechanical properties such as MFI (melt flow Index), magnetic ferrous metals are sorted into different grades, and so on. Only by applying these measures can high-quality recycling be achieved, and the market be big enough to take the recycled materials.

By introducing TOMRA equipment, the plant is now able to run at 20 tons per hour (an increase from 16.5 tons per hour) and achieve 98.3% purity on paper grade before the material is manually quality controlled in one of the three new paper picking cabins. The operator now benefits from greater efficiencies, reduced labor costs, and paper that is far cleaner than before.

15.4.7 Polymer Recovery – Upgrading and Recycling

The demand of virgin plastics by industry segments strongly depend on the applications (Figure 15.14). A major portion of plastics is used for packaging purposes, and here especially for food packaging. A second important sector is construction, with plastics used in door/window frames, and in piping/tubes. Automotive is the next important industry sector, with plastics used in manifold applications, ranging from interior dashboard/console parts over mirror housings and grilles, up to high-temperature applications "under the hood". Another important sector is "electrical & electronics" applications, such as housings for computer, printer, or household appliances. Substitution of virgin plastics by recycling grades in such segments is possible; however, several boundary conditions have to be observed:

- **Packaging:** Approx. 40% is used in food packaging; therefore, the highest possible requirements and official approvals are required for most cases. These issues are mainly resolved for r-PET; r-HIPS for food is on the way (see Section 8.4.1); recycled polyolefin projects for food applications have been initiated, but are challenging due to intrinsically low barrier properties of PO versus migration of impurities (due to the lower glass transition temperature of PE and PP vs PET or HIPS).
- **Construction:** Recycled grades have to meet ambitious technical specifications; solvable
- **Automotive:** Recycled grades have to fulfill highest demand on low odor and low volatile organic compounds; this, however, seems solvable due to technical degassing steps during melt processing of recycled grades.
- **Electrical & electronics**: Ambitious technical specifications for recycled grades; solvable

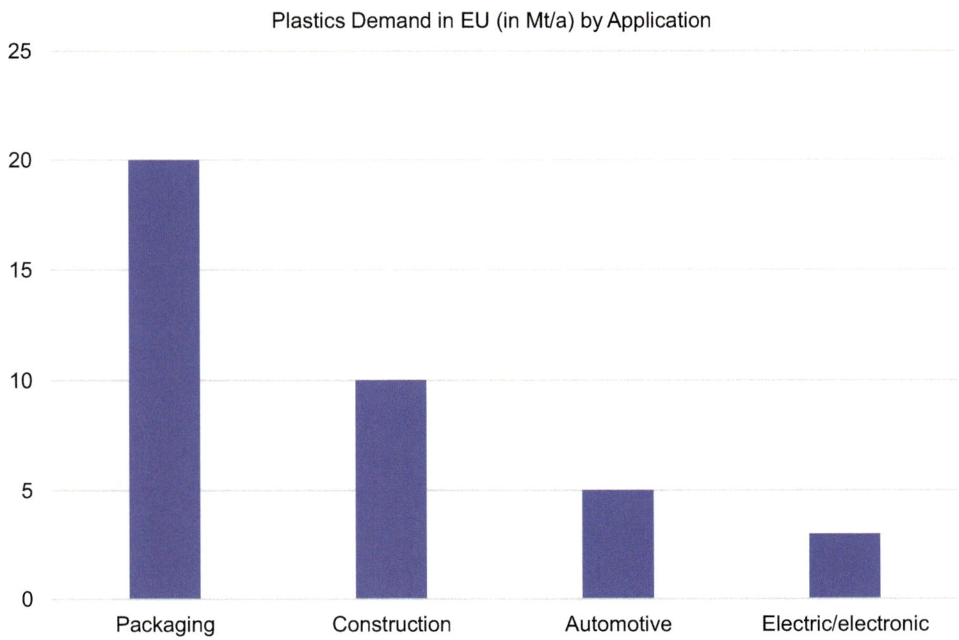

Figure 15.14 Virgin plastic demand per industry [source: TOMRA]

The sections above describe what is done - or needs to be done - to collect polymers. Collection can be defined as separate collection in a bin/bag or RVM (Reverse Vending Machine), collected from an informal sector but also from sorting machines out of waste streams prior to incineration or landfill, for example. The question remains as to what to do with the collected plastic. The recycling industry is widely known to produce materials of inconsistent quality, with odor and purity levels in a 95% range (Figure 15.15). One exception is PET beverage bottles, which account for only approximately 5% of the entire virgin plastic market. As already explained, the remaining 95% is either not collected or, compared to PET beverage bottles, recycled into low-quality products.

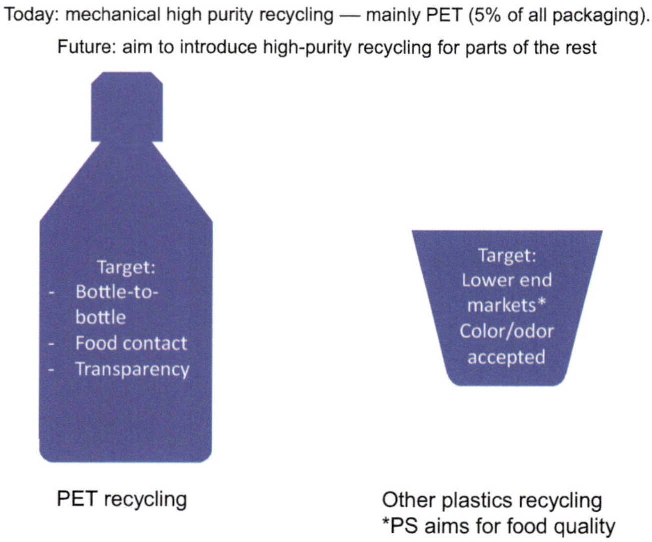

Figure 15.15 Comparison between material purities after recycling processes [source: TOMRA]

It is true that it is easier to reach 99% purity levels with PET beverage bottles when compared to PS, PE, and PP, and even more so than with flexible materials. However, to aim for the same high purities in the non-PET bottle streams is possible and makes financial sense. To get there, the PET beverage industry, which has approximately 30% closed loop recycling (bottle-to-bottle), shows what needs to be done. A combination of top-quality sorting and top-quality washing is needed to achieve purities that will be accepted by brand owners and thus can be used in their standard products. If this is done to scale, the market, which is currently skeptical of using recycled materials outside of PET in branded products, will accept and agree to use recycled materials of multiple polymer streams. Once pre-sorting, hot washing, flake sorting, and super clean plants are in operation, there is no difference in the final product properties, whether the material was collected separately or collected by sorting out from mixed waste streams. Taking into consideration that by having sorted all waste streams, the collection rate is double, meaning the overall yield calculated on polymers put into the market is nearly double as well (Figure 15.16).

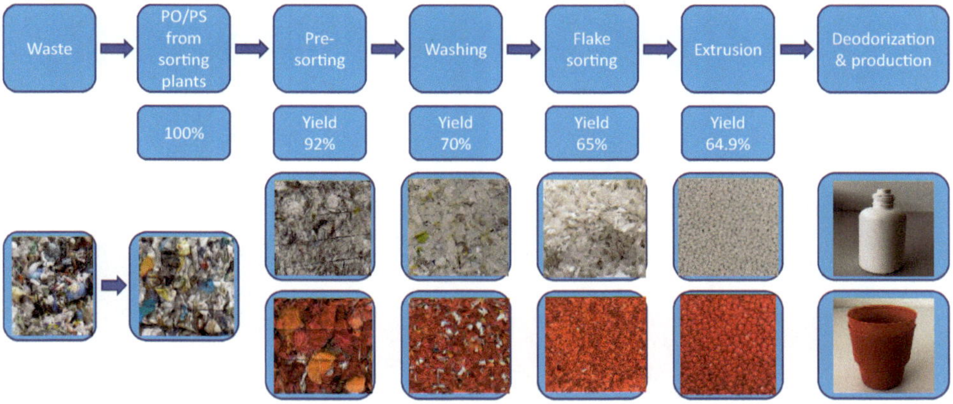

Figure 15.16 Future of recycling polymers [source: TOMRA]

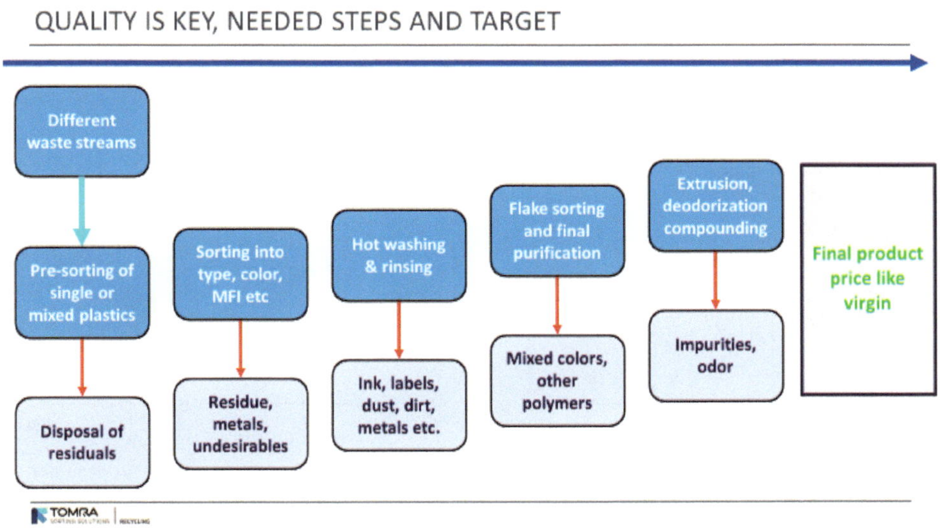

Figure 15.17 Needed steps in polymer recycling [source: TOMRA]

At present two successful plants are in operation, one in the Netherlands and one in Germany. Coupled with a modern washing plant, all available sorting technologies are in place to allow production of high-quality recyclates. With perpetual support from several laboratories, the quality achievements are permanently documented (Figure 15.19 and Figure 15.20).

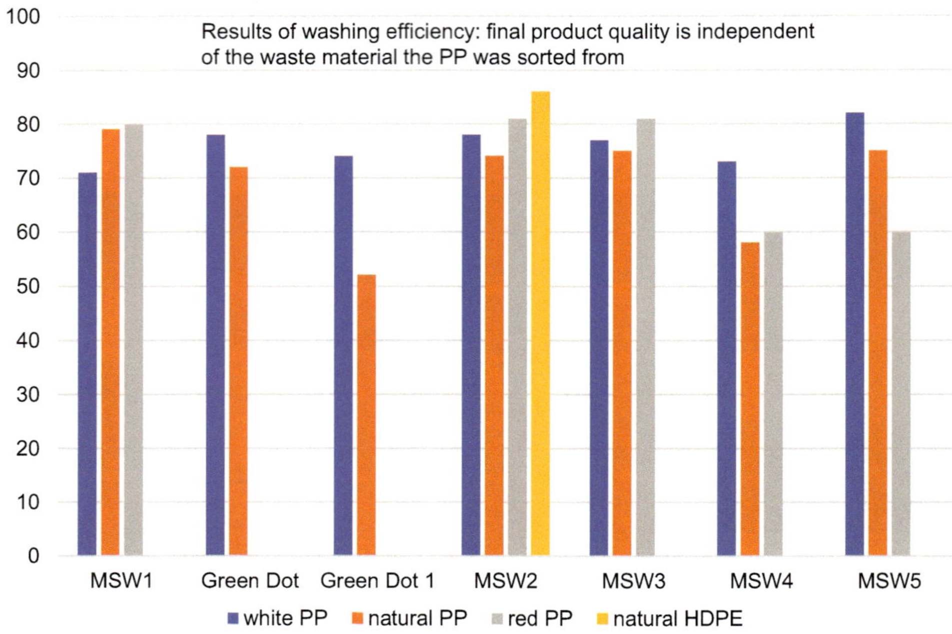

Figure 15.18 Yield by washing/flake-sorting various materials pre-sorted out of different waste streams [source: TOMRA]

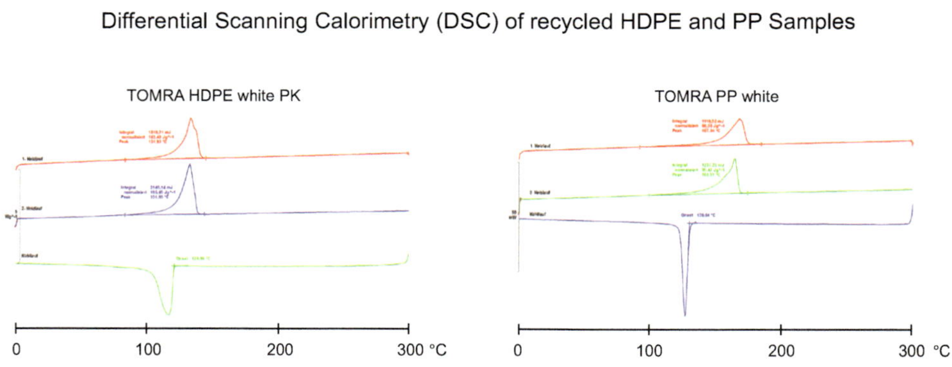

Figure 15.19 Purity of various recycled materials from different waste streams: the recycled materials have a clean meting peak in DSC, and hence no unwanted (crystalline) by-products are apparent, which would appear as additional peaks in the DSC [source: TOMRA]

Substances	Measured residuals
Organic acids	All results << limits
Parabene	
Further preservatives	
Azo (22 amine + xylidine)	
Biocides	
Formaldehyde	
Bisphenol-A migration	
Chloroparaffins	
Dimethylfumarate	
Organic bromine, bisphenyls, bisphenyleters	
Flame retardants, brominated phenols	
Flame retardants TEPA	
Metals	
Heavy metals	
Organotin compounds	
Plasticizer	
Polycyclic aromatic hydrocarbons	
Chromium	
Triclosane	

Figure 15.20 Measured chemicals from PP sorted out of mixed residual waste streams [source: TOMRA]

15.5 The Technologies

TOMRA's AUTOSORT combines leading-edge features and technologies in one machine (Figure 15.21). Compact and flexible in construction, AUTOSORT allows for an uncomplicated integration into existing and new plants. Featuring TOMRA's FLYING BEAM® technology, an excellent homogeneous light distribution for better detection and monitoring across the entire belt width is guaranteed, and results in heightened performance and operational efficiency. An optional integration of TOMRA's DEEP LAISER technology moreover enables the detection of previously undetectable materials and drives the levels of sorting purity even further. Object recognition is as an add-on technology available to distinguish theoretically everything that could be detected and distinguished by humans.

Figure 15.21 TOMRA AUTOSORT [6]

In general, there is a split between sorting on an item level and purification once the pre-sorted material is ground, washed, and dried. This can be easily understood when considering that sorting one plastic bottle is easier and more effective than sorting the same bottle ground into 1000 pieces. The pre-sorting is preferably used to pre-concentrate materials or colors, with high throughput, high yield, low operational cost, and very low losses.

Figure 15.22
TOMRA AUTOSORT [6]

After such precise pre-sorting, a washing process must follow. Until recently, most of such washing plants used cold washing, which is not able to wash off ink, glues, and, especially, odor. Hot washing is essential if high quality is to be achieved. Thereafter, another sorting step is needed. For example, when considering PET beverage bottles, the lids and caps are made from PP and PE. In a simple density separation, they swim (float) and PET sinks, meaning a separation of these materials is very easy. Within PE and PP, the density is similar, so a PP lid on a PE item cannot be separated with a density separation. A sorting step with sensor-based technology is needed. Such machines are able to sort materials, colors, and metals down to 2 mm size and are widely used for the last, remaining material purification step. After this final sorting step, a deodorization or super clean process and extrusion or compounding follows. The result is odorless pellets, of a specified constant quality, property, and color, to be used in standard converter processes and standard products. Neither an injection molding machine nor a blow molding machine need to be used to produce items from the recycled materials if the described process is used [10].

Figure 15.23 Flake and pellets produced from different polymers out of waste [6]

■ 15.6 Example: Ultimate Pure Sorting of Polystyrene

Under the umbrella of Styrenics Circular Solutions (SCS), TOMRA has recently completed tests revealing that polystyrene is a very good plastics for both recycling and sorting.

With TOMRA's near-infrared (NIR) sensor technology, post-consumer plastic waste was sorted in a multistep process, including initial sorting from post-consumer waste, sorting a second time into white dairy HIPS products and remaining mixed color HIPS, grinding into smaller flakes, hot washing, drying, and flake sorting. The resulting purity of PS turned out to be higher than 99.9%. Styrenic compounds have a unique signal that enables easy and very precise sorting, an advantage which some of the other polymers do not have. Additionally, a large portion of white HIPS is used in the dairy industry: packaging for food contact purposes (Figure 15.24). Within PS materials, small differences, for example the amount of PS within HIPS and other styrene polymers, can be detected and the different fractions separated from each other and of course from other plastics and materials as well.

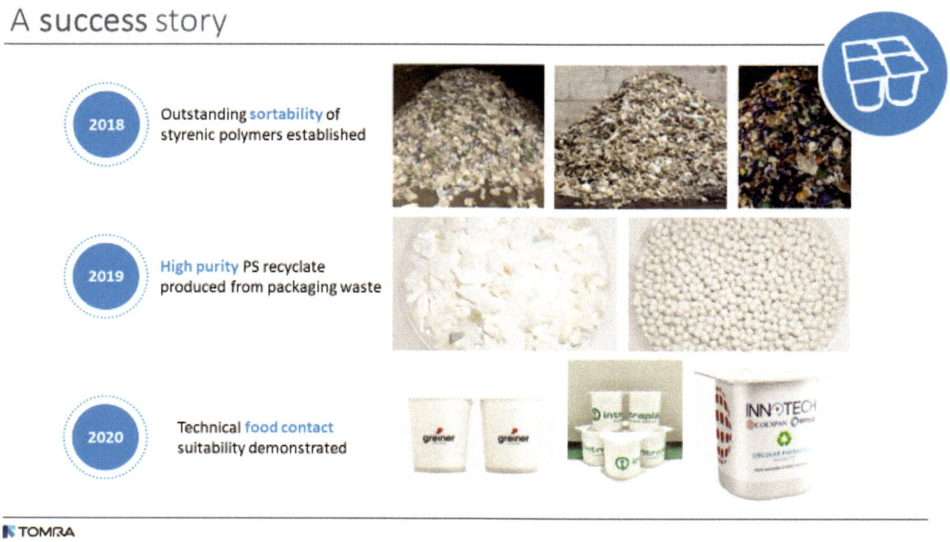

Figure 15.24 Process flow, from mixed PS via HIPS food containers to new yogurt pots [6]

TOMRA's findings prove that today's sorting technologies achieve a purity level beyond what is required to successfully recycle polystyrene through both mechanical and advanced recycling processes. Along with INEOS Styrolution's investments in multiple recycling projects, this takes the company another step closer to developing closed-loop solutions for styrenics.

References for Chapter 15

[1] Alliance to End Plastic Waste – *www.endplasticwaste.org*
[2] Ellen MacArthur Foundation – The New Plastics Economy
[3] *www.oceanresourcemanagement.co.uk*
[4] TOMRA measurements
[5] EU recycling targets for packaging materials
[6] TOMRA.com
[7] EUNOMIA study for TOMRA
[8] *https://www.mckinsey.com/industries/chemicals/our-insights/how-plastics-waste-recycling-could-transform-the-chemical-industry*
[9] TOMRA, internal information
[10] Greiner Packaging, TMP, Dressel GmbH

16 Reuse

Josefa Nüßlein, Thomas Neumeyer

16.1 Introduction

The increasing awareness for sustainability aspects as well as the intention for high quality products are driving manufacturers and consumers towards sustainable solutions for a product's life cycle and waste management. In this book, a comprehensive overview of recycling possibilities for various plastics has been provided. Nonetheless, before considering recycling, reuse of products and materials should be considered, according to the waste pyramid shown in Figure 16.1 and the European Green Deal. In the "Circular Economy Action Plan" [1], which includes a guideline for sustainable products, reuse is promoted with a strong emphasis on electronics, textiles, vehicles, batteries, and packaging.

For a clear understanding of the definition and the types of reuse, firstly the relevant terms will be explained. For each type of reuse, several examples will be given to illustrate the approach of the different reuse cases. The requirements and challenges that come along with reuse options regarding polymeric materials will also be considered. Since an important consideration for a decision towards reuse is sustainability, a systematic view on the effects of reuse models on the environmental benefit is provided. Reuse approaches can save the energy that is required for producing a product, but enabling reuse also has an impact on the environmental footprint. Thus, evaluating the carbon footprint for a reuse product allows judgement on the justifiability of reuse. Also, consumer behavior and the decision factors that have impact on the behavior, such as price, design, and familiarity, will be considered.

16.1.1 Definition of Reuse

According to the European Union's guidelines on waste management released in 2008, reuse is incorporated into the waste management hierarchy. The waste management hierarchy described in this guideline is summarized in Figure 16.1 [2].

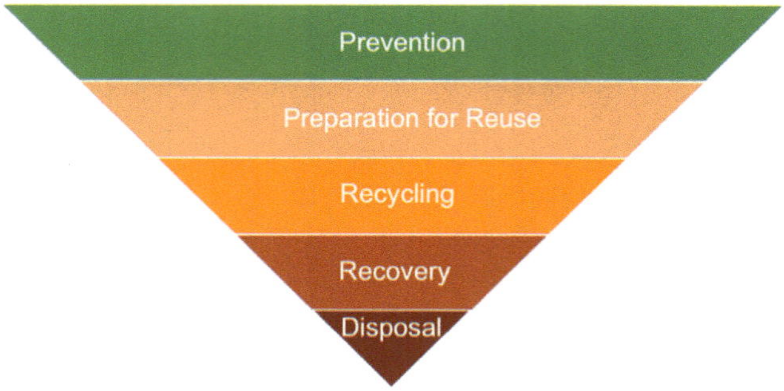

Figure 16.1
Waste management hierarchy

Here, the term "reuse" is relevant in two different stages of the pyramid. The definition of waste "prevention" as the basis of the pyramid states that reuse leads to the extension of the life phase of the product and to the avoidance of the premature disposal of products by development of products that can be used multiple times. The second level, which is labeled "preparation for reuse", includes checking and cleaning, as well as repairing goods to allow them to be reused for their original purpose. The definition of reuse itself that is used in the guideline is the following:

"'re-use' means any operation by which products or components that are not waste are used again for the same purpose for which they were conceived"

Accordingly, the term "reuse" by this definition excludes using a product for a different purpose after its first life cycle [2]. However, this contradicts a large number of scientific publications that define the term reuse within a broader spectrum. The *Material Reuse Good Practice Guide* [3] describes reuse as:

"a nondestructive process that finds a second or further use for end-of-first-life solid materials (products or components) without a change of state, excluding melting for metals, plastics and glasses, and pulping for paper. The further use of a product life extension overlaps with the activity of product resale."

This definition includes the activities and predictions described in the literature and is very close to other sources. Also Allwood et al. use a similar approach and include life extension, repair, resale, and also some remanufacturing processes as well as component reuse as processes that are referred to as reuse [4]. The underlying definition in the works of Kerr et al. as well as Hunka et al. are also close to the one used by Cooper et al. [5, 6].

Accordingly, in this chapter, reuse is defined as the practice of using a good for more than one life cycle or prolonging the first life cycle by relocating or reselling. Also remanufacturing and component reuse approaches, including refill, repair, and adaptive reuse will be considered. From a material perspective, by reusing, the energy required in production and manufacturing of goods is preserved. Therefore, changes of state such as melting for polymers or metals are not referred to as reuse.

Since, in this book, the focus is on sustainable options for plastic products, also in this chapter, reuse of plastic products will be emphasized. In producing plastic materials, from the synthesis of the raw polymer to compounding to the final production step, there is a significant consumption of chemicals, electricity and/or other energy sources, and water. Any process that allows the further use of plastic materials after the first use phase with only minimal remanufacturing steps therefore increases the efficiency and value. In recycling processes such as mechanical or chemical recycling or dissolution, a greater consumption is required for recovering raw materials compared to reuse, which can be seen as the ecological benefit of reuse over recycling.

If a product is designed to be used multiple times by the same user, it is also often referred to as reuse. Since the actual purpose of the product is to be used often and not just once, the end of first life is here defined as occurring after the intended numbers of uses. Therefore, in this definition, these cases are not referred to as reuse, but as life cycle extension. For a complete picture, some examples of products designed for extending the use phase are given in Chapter 17.

16.1.2 Types of Reuse

There are several reuse options, which are differentiated in terms of more or fewer manufacturing steps and they are taking place at different life stages of the product. These approaches then lead to a more circular product manufacturing environment. In Figure 16.2, the types of reuse – displayed in orange boxes – are assigned to the relevant life phases of a product.

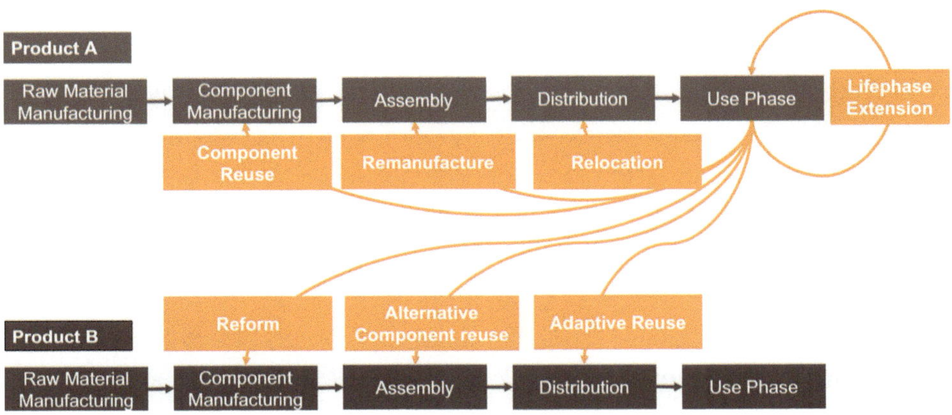

Figure 16.2 Types of reuse

The concept of reuse includes a variety of processes that play a role at different stages of the life cycle. Reuse comprises using a product either in the same or in a different application.

In the case of reusing for the same product (Figure 16.2, upper half), four different types can be classified. First, there is a trend for **life cycle extension** of products by designing them to be used repeatedly by the same group of users without an intervening external distribution. In **relocation**, products are sent for repeated distribution without an intermediate processing. After that, another user reuses the item. Unlike relocation, **remanufacturing** involves an intermediate processing step that serves either to repair or refill the product. Alternatively, in **component reuse** a component is upgraded or modified, after which it is reinserted into a modular assembly [3, 4].

If a divergence of the purpose between the first and the second life cycle of a product occurs (Figure 16.2, lower half), the term reuse can still be applied following the definition introduced before. There are three types to be distinguished, which are adaptive reuse, alternative component reuse, and reform.

- **Adaptive reuse** is understood as the change of purpose of a good without converting the form and function. This can be carried out either by distributors or by the users themselves.
- **Alternative component reuse** is another option, in which one component or part of a product is used in a different application, whereby a certain reconstruction or modification must be made.
- **Reforming** is the most destructive process of reuse. In this process the material is isolated and shredded or cut to a certain point to allow it to be reshaped. Here it is crucial that the physical state is not changed, meaning that there is no melting of the reused material, as this is process would be defined as recycling and not any more as reuse [4, 5].

16.2 Examples

Corresponding to the different types of reuse that have been described and delineated, examples for every reuse approach will be introduced here.

16.2.1 Life Cycle Extension

The extension of the life cycle is becoming increasingly popular. As the European Union decided to ban many single-use plastic items such as plates, balloons, food containers, beverage containers, plastic bags, and packets from 2025, alternatives made of either other materials or multiple-use plastics are being implemented [6, 7].

The company Recup (Munich, Germany) developed a reusable mug for beverages such as coffee or tea. The mug is made of polypropylene and therefore can be rinsed up to 1000 times. In a system that has been implemented in cafés, bars, and to-go stores, beverages are served in these reuse mugs, which can then be brought back to any other shop that uses them. To promote the reuse in high numbers, a deposit is applied on the Recup. A reusable lid can be bought additionally that is then owned and cleaned by the customer him-/herself. An environmental benefit of the reusable mugs is the reduction of waste from some of many single-use plastic mugs that are produced currently. Thus, in Germany 40 000 tons of waste are produced per year due to single-use cups, which Recup wants to replace with their product [8, 9].

For several detergents, such as laundry detergent, dish soap, and handwash, companies like BLUELAND (New York City, USA), everdrop (Munich, Germany), and klaeny (part of Green Home Living GmbH, Berlin, Germany) developed detergent tabs intended to replace ready-made liquid detergents that are sold in single-use plastic containers. In local grocery stores and online shops, customers can buy multiple-use plastic containers that can contain just the right amount of water to dissolve the tab to obtain an optimal cleaning solution. For this system, everdrop won the German sustainable design award and, as well as the other companies, gained a lot of attention in social media and the press. Since then, also other brands, such as Blaue Helden (Bad Homburg, Germany) and Biobaula (Bruckmühl, Germany), followed the trend with their own product lines [10–13].

16.2.2 Relocation

Reuse by reselling has been practiced for a long time in a direct and personal approach. Nowadays, resale takes place, in addition to in stationary secondhand stores and markets, especially on the Internet. There are various websites where private sellers and buyers offer their goods for a predetermined price or a price to be offered. One of the well-known providers is eBay, which was one of the first portals of its kind.

A market survey commissioned by eBay (Kleinmachnow, Germany) examined user behavior and intention [14]. It was apparent therefrom that a key aspect for the customer's choice of buying or selling a secondhand product is extending the life cycle of products. The results indicated that 95% of the surveyed online shoppers consider it important to very important to use products for as long as possible. Furthermore, consumers take sustainably after their own use into account, as reselling them was, at 35%, the most frequent answer to the question of what happens to products that are no longer needed. The options of giving them away freely, disposing and donating them were named in the given order. For reselling, over 80% of those surveyed chose online markets over local stores for selling and buying secondhand goods [14, 15]. By contrast, a study by Clausen et al. revealed that the motives of buying or selling goods on eBay is most dominantly a practical or financial decision [16].

There are several goods being purchased and reused. The graph in Figure 16.3 shows which goods those surveyed had bought before and which ones they prefer to buy used rather than new.

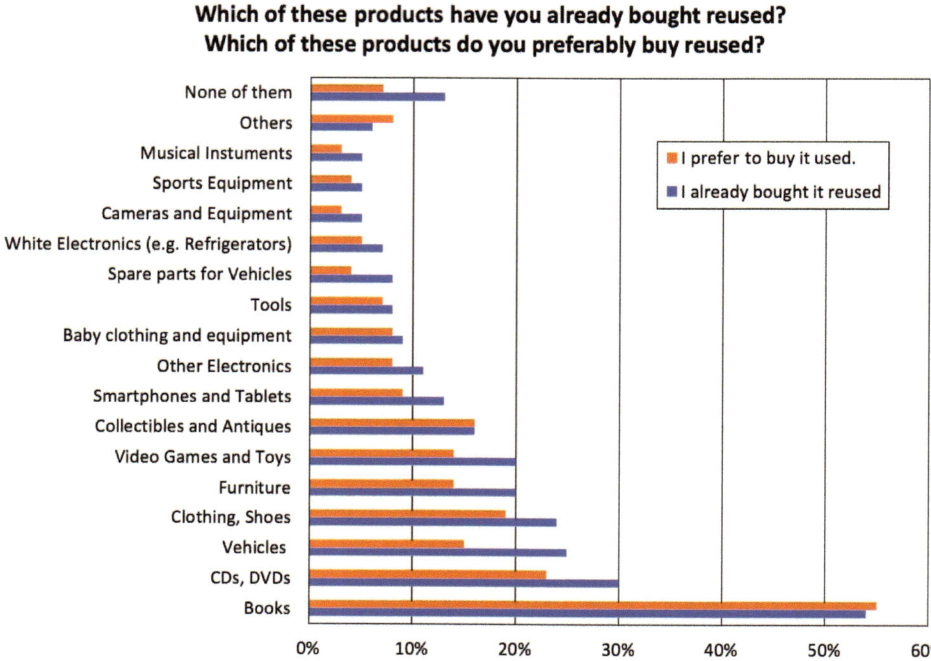

Figure 16.3 Survey on customer choice of reuse by relocation via an online platform [14]

The results show that books are the goods that were bought by most of the people in the study, followed by CDs, DVDs, and Blu-rays, as well as cars and clothing [15]. In another study an environmental perspective was taken into account. Accordingly, the greatest savings in the carbon footprint are achievable in furniture, household and kitchen devices, as well as electronics such as computers, laptops, or cell phones [17]. Clausen et al. also investigated the environmental aspect of reselling. They structured the environmental impact in three categories. First, by using the online platform, there are emissions caused by the generation of electricity for servers and computers. Second, reselling a good replaces the purchasing in stationary or other online markets, as well as the fact that the product that is purchased is not brought to other end-of-life options. Finally, they also include factors such as employment and number of trades. In summary, they stated that trading a reuse good can mostly be considered sustainable, as long as the initial use phase is long enough [16].

The number of reused goods purchased is still very low compared to new ones. The German Federal Ministry for the Environment stated in a report in 2018 that the percentage of reused electronic goods of those that were brought to disposal was only below 1%. The main market for reused goods is still the private sector. To increase the overall amount of relocation and resale, broad infrastructure has to be installed [18].

For the product group of textiles, Sandin et al. showed that there are various publications on reuse and recycling approaches. Most of the studies investigated reuse of textiles, followed by fiber recycling and fabric recycling. The materials in these cases were cotton, polyester fiber, viscose, wool, but also other polymers such as PET, polyamide, polyacrylic, and polypropylene. An environmental benefit from recycling and reuse options compared to incineration or landfill was reported in all the reviewed studies but one. Between reuse and recycling, reuse was promoted as the preferable option [19]. In another life cycle assessment, the electrical energy footprint of textiles made of cotton or polyester was carried out. Thereby, it was shown that the electrical energy saving of reuse clothing by avoiding virgin textiles is approximately 90 kWh per kg polyester fabric and 65 kWh per kg cotton. Reuse was therefore promoted to replace the purchase of newly produced virgin materials [20].

16.2.3 Remanufacturing

Reuse via remanufacturing encompasses several processes, such as refill, refurbish, and repair.

Repair and Refurbish

Since repairing a good is a process that is mostly performed in the private sector or by services, there are few statistics or published investigations. Starting in Amsterdam, events called Repair Cafés started to grow within Europe that allow self-performed repairs of e. g. electronic devices with the help of professionals and tools provided by the organizers. One of the goals of these events is to promote repairing options by increasing attention to that topic [21].

Repairing and refurbishing of WEEE goods should be the most practical, as well as the most environmentally friendly, end-of-life option since it is less energy- and material-consuming [22]. However, most of the waste is given to disposal, which leads to landfill in most of cases. One reason for that is the consumer's behavior, which is partly driven by advertisements to promote the decision to buy newly designed and optimized products. Other reasons are a missing structure that makes repairing easy and the equipment needed accessible, furthermore damaging products during waste collection, and missing knowledge for the repairing process itself. In providing regulations for repair activities, the European Union is granting consumers a "right to repair" [22–24]. Repairing products should become more attractive and be carried out systematically. This is to be helped by a longer guarantee period, guaranteed availability of replacement parts, and better access to information on repair and maintenance [25]. In fact, the possibility of recycling of WEEE goods, which would be the next option in the waste pyramid, is limited by the material compositions of components. The bodies are often made of polypropylene, polystyrene, or ABS, which can be mechanically recycled, whereas for the electronic parts, which include also metal components, or sealings made of rubber, recycling options are very limited. Reuse by repair is therefore even more attractive since disposal is to be avoided.

A sustainable development of consumer electronics is performed by the companies Fairphone (Amsterdam, Netherlands) and Shiftphone (Falkenberg, Germany). In their reuse and recycle concept, a smartphone with a modular structure, made of recycled components, is designed to be used for as long as possible and to be easily repairable. Not only a repair but also a refurbishment can be performed as, for example, cameras or batteries can be upgraded when available [26].

Refill

For the product category of containers for packaging or food and beverages, the predominant reuse option is refill. Well-established return and refill systems for beverage containers have already existed in Germany since the 1990s, when the Coca-Cola Company started to sell reusable PET bottles. Nowadays, the market volume of reusable PET bottles compared to other options for beverage containers such as cans, single-use PET or glass bottles in Germany is at 13% [27]. In this use case, a plastic bottle is usually filled and delivered to retailers up to 25 times before it is discarded. To remain in a food contact quality, returned bottles must be closely inspected. Contaminated or mechanically damaged bottles are sorted out and sent for recycling. All intact containers are stripped of their old labels and hygienically cleaned in several cleaning processes at more than 60 °C. Today, a special computer nose, a so-called "sniffer", is used in a bottle washing line to inspect and sort out contaminated bottles by detecting odors [28]. For evaluating the environmental impact of several beverage bottles, the German central environmental authority carried out a life cycle assessment. It showed that the reusable bottles made of PET as well as glass are in most cases preferable compared to single-use options. For example, a Cradle-to-Grave analysis showed that a 0.5 l single-use bottle made of PET results in almost double the CO_2 footprint compared to a reusable bottle made of the same polymer [29].

Especially for food and beverage containers, there are several companies working on innovative multiple-use and sustainable solutions. The company OZZI (Bogota, Colombia) developed reusable food containers made of polypropylene that can be used in various take-away services such as restaurants, hotels, or cafeterias. The company uses a patented return container that is meant to be accessible at any time. In this return system, the food containers are placed into a case that can be used in an industrial dishwasher system with no intermediate manual step. As the product requires several washing steps in industrial washing machines, the plastic material has to be resistant to temperature and detergents without changing color, cracking, or deformation. This product, like the Recup coffee mugs that were mentioned in Section 16.2.1, is made of polypropylene, which indicates that this polymer is suitable for such applications [30–32].

In an award-winning refill concept installed in three cities so far, the company Algramo (Santiago, Chile) sells and promotes reusable packaging systems. In a collaboration with big brands for cosmetics, detergents, but also rice and other long-life food as well as pet food are provided at a refill station, mounted on a tricycle. Besides, they installed refill machines in grocery stores for avoiding the use of single-use plastic packaging. Furthermore, the Algramo refill system allows their customers to book an appointment for a refill cargo bike that comes to their location and can be paid for via an app. Besides the ecological factor, they also managed to reduce the price of the purchased groceries [33–35].

With a similar concept, the platform "loop" sells goods from several leading brands in containers that can be refunded when brought back. Therefore, customers can bring containers back to a delivery station or order a pickup from home. The packaging for delivery is also reused, for omitting any packaging waste. After the containers are brought back, they are cleaned and refilled for the next customer to be used [36, 37].

Besides the previously mentioned grocery packaging solutions, also delivery packaging for reuse is becoming more prominent. The business-to-consumer (B2C) delivery market is constantly growing, requiring innovative packaging solutions to avoid the use of single-use packaging made of plastics or paper [38]. However, also business-to-business (B2B) deliveries are very suitable for reuse options since they are easy to track and return and a larger number of the same goods can be transported with the same packaging solution. This makes a reuse of even a very specialized packaging solution for a certain product possible. Packaging solutions made of bead foam are often used in packaging for sensitive products. For these materials, the choice for the type of polymer strongly depends on the number of uses. Polystyrene is most commonly used in single-use packaging, whereas polypropylene is predominant for multiple-use boxes, due to its excellent ductility. In absorbing the energy of a shock, polystyrene becomes cracked and is therefore only able to protect loads once until the material becomes damaged. Hence, polystyrene is used in single-use bead foam packaging. In contrast, polypropylene only deformed due to its higher ductility and can be used more often, without losing the ability to protect.

In the B2C market, the company LivingPackets (Pully, Switzerland) developed a not only reusable but also smart packaging system called "The Box" with 4G connectivity that includes also a temperature, shock, and humidity sensor. The box is shown in Figure 16.4. It allows the quality control of both the product and the delivery itself. The box is made of recycled EPP, which makes it more sustainable, both by a low weight due to the bead foam material and by the avoidance of virgin plastics. The box also detects unauthorized opening and the exact location and sends all information to the receiver. For shipping labels, it has a display to avoid paper and ink [39, 40]. With a similar approach, the company called Returnity provides customable and reusable packaging solutions that are mostly used in renting applications, e.g. for clothing or camping equipment [41].

Figure 16.4
Smart transport box from LivingPackets made of recycling EPP [40]

For the B2B market, there are many applications for EPP carriers. For example, parts for car manufacturing that are transported are often of high value, so the shape and quality of the packaging solution are crucial. Bead foam can be formed into customer-specific shapes. As a result, a reusable packaging is created that also has many other advantages such as high resistance against detergents as well as weather in addition to its high durability. The EPP can be varied in density and elasticity according to individual product requirements. Due to the very low density of between 60 and 80 kg/m^3, the weight of the carrier box is very low compared to the weight of its load, so this packaging option is also environmentally efficient in transport. A picture of a carrier box for windshield wipers is shown exemplarily in Figure 16.5 [42, 43].

Another reuse approach by refill is the refilling of inkjet or print cartridges. For reuse of these cartridges, they have to be given back to a refill station or shop, where they are cleaned and refilled with ink. An LCA study carried out by Farouk compared three different end-of-life scenarios, one being disposal of the cartridges after use, the second being the partial reuse of some components, and the third being refill of the cartridges. In all impact categories, the refill approach showed the best results and can therefore be considered the most sustainable option [44]. The environmental benefit of cartridge refill compared to other end-of-life options was also shown in a study by Krystofik et al. [45].

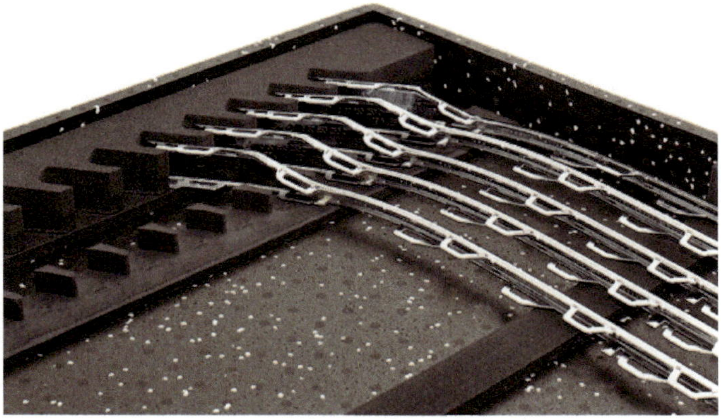

Figure 16.5 Reuse carrier box made for windshield wipers [43]

16.2.4 Component Reuse

When a product's first life cycle has ended but there are still undamaged functional components, they can be removed and brought back to assembly for use in new or secondhand products.

Especially in car applications, there is a large market for reused components that are brought back to a second life phase. In an end-of-life vehicle that is either sold to a dismantling company or brought to a collection point there are several components such as engine parts, tires, or batteries that can be reused. Compared to the replacement of broken components with newly manufactured spare parts, the assembly using second-life parts is both cost efficient and sustainable [46, 47]. Because of that, the average end-of-life vehicle reuse and recycling rate by weight has already reached 81% in Germany [48]. A comparative LCA study carried out by Gabhane and Kaddoura revealed an environmental benefit of using a remanufactured gearbox as a spare part, with a decrease in the carbon dioxide emissions of 36% compared to a newly manufactured one. Nevertheless, the environmental hotspot of the car in general is identified as the use phase, which shows that another focus should be using spare parts that have a potential to increase the efficiency.

Another example for a good practice component reuse is from the company Xerox (Norwalk, CT, USA), which produces photocopiers. Their approach of taking back end-of-life products and remanufacturing them as spare components or in new products, combined with a repair service, includes several types of reuse in one system. With those techniques, a study carried out by Kerr et al. showed that over one photocopier life cycle, material savings of up to 49% and energy savings of up to 68% can be achieved, compared to a non-reuse case study of the same type of photocopier [49]. This circular life cycle including several reuse and recycling approaches is shown in Figure 16.6.

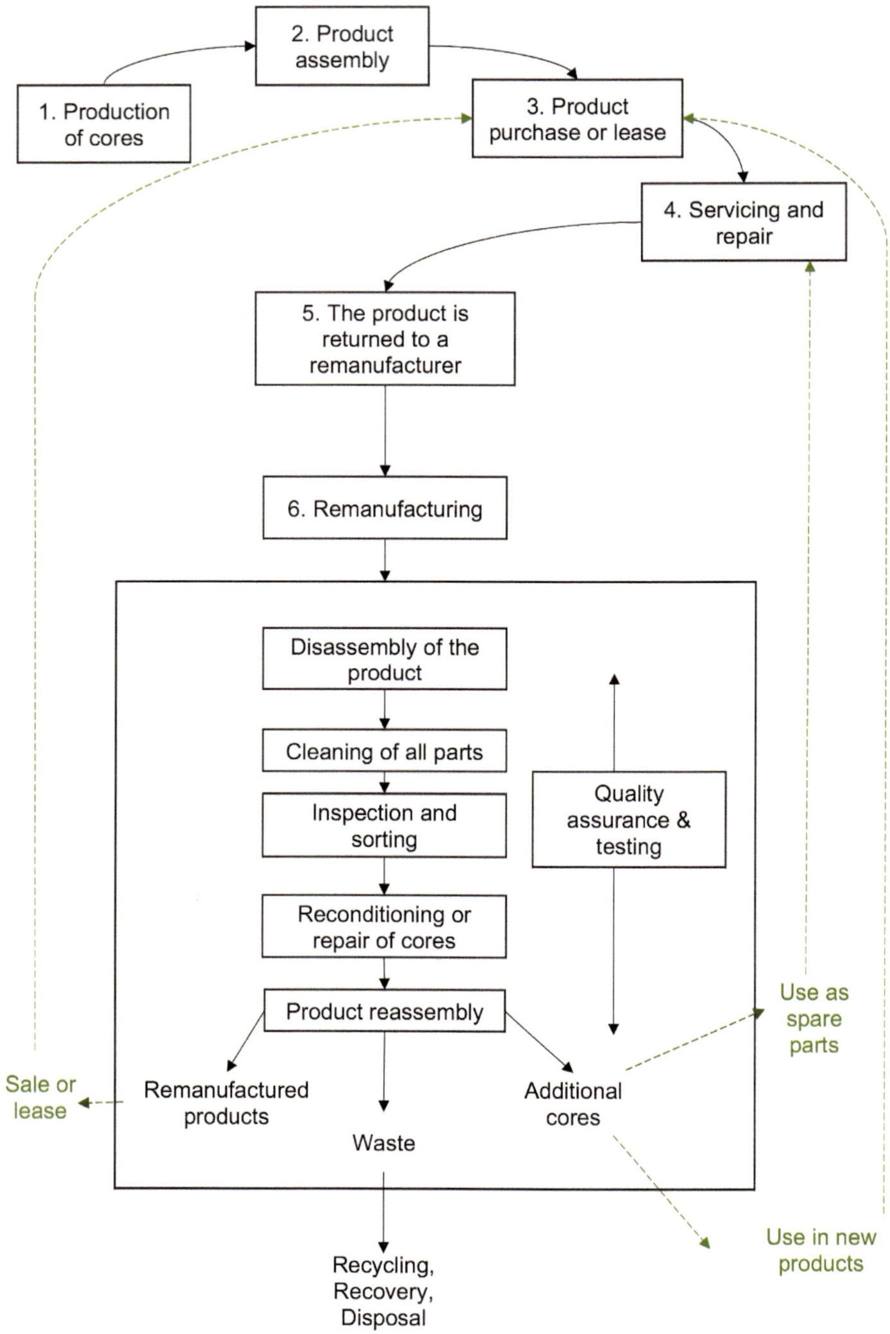

Figure 16.6 Schematics of Xerox reuse approach [49]

16.2.5 Alternative Component Reuse

An end-of-life component can be used in several ways. Especially for products that have a broad range of design possibilities, reused components can easily be included.

In clothing and accessories, alternative component reuse is becoming a trend, since customers' awareness for sustainability is rising [50]. The brand AIRPAQ (Cologne, Germany) is using reused airbags as a textile material to produce backpacks, belly bags, and bow ties. The locks of these backpacks are reused seat belts. Both components are made of components from end-of-life cars that were brought to disposal. So far, with this alternative reuse, 43,293 airbags and 48,705 seatbelts have been given a second life, meaning that airbags and seat belts of around 10,000 cars were reused [51, 52]. End-of-life wheels have been given a second life by the company Manguetown (Lausanne, Switzerland), which are also using teabags and fire hoses [53]. The French company PlastiKetic designs and produces rocking chairs and flower pots made from PET plastic bottles [54]. The company BAG TO LIFE (Lichtenfels, Germany) uses spare parts from airplanes to produce lifestyle goods. Accessories such as cosmetic and laptop bags have been produced from life vests, parachutes, and blankets. With these products, they have won several sustainability and design awards [55]. Two examples of these products, a bag from AIRPAQ and a part of the backpack collection of BAG TO LIFE, are shown in Figure 16.7.

Figure 16.7 Packpacks from the companies AIRPAQ (left) [52] and BAG TO LIFE (right)

Lufthansa's upcycling collection uses different end-of-life materials to produce several lifestyle products. An Airbus that was used for 30 years, called "Papa Charlie", has been used to produce an interior box to organize magazines, toys, or other private goods. From other end-of-life airplane parts, mostly from doors or the outside wall, Lufthansa produced coffee or couch tables, lamps, and clocks and wardrobes [56, 57].

16.2.6 Reform

The process of reforming is very close to an actual recycling process, as the form of the reused item is changed generally. The difference is that there is no change of state, such as melting, but only a cutting or shredding process included in the reforming routine.

The company ecoBirdy® (Antwerp, Belgium) produces furniture such as chairs, tables, and lamps for children that are made of the material ecothylene®, which is composed of polyolefins. Figure 16.8 shows a chair made of this plastic.

Figure 16.8 ecoBirdy® Charlie chair made of ecothylene® [58]

In that material, post-consumer plastic waste, mostly from plastic toys, is manually checked, sorted by color and material type, and then bound together by a polyolefin matrix. This process was developed by ecoBirdy® and is patented worldwide [58, 59].

For improving relevant material properties, PET waste can be used in asphalt mixtures up to 20% in volume. This can help the problem of landfill, but from a circular economy perspective, does not allow further recycling cycles. In publications, the term reuse is applied to these cases, but since the actual purpose of promoting reuse is increasing circularity in product life cycles, this can be questioned [60]. Since the thermoset plastics cannot be recycled by melting processes, this plastic group is preferably used as a filler for lightweight concrete as it can also improve concrete properties in well-formulated mixtures [61].

16.3 Environmental Impact of Reuse

In order to quantify the environmental impact of reuse, different factors must be taken into account. The distribution of the environmental impact of different life cycle stages influences the efficiency of reuse cases immensely. For exemplary products, this distribution is shown in Figure 16.9.

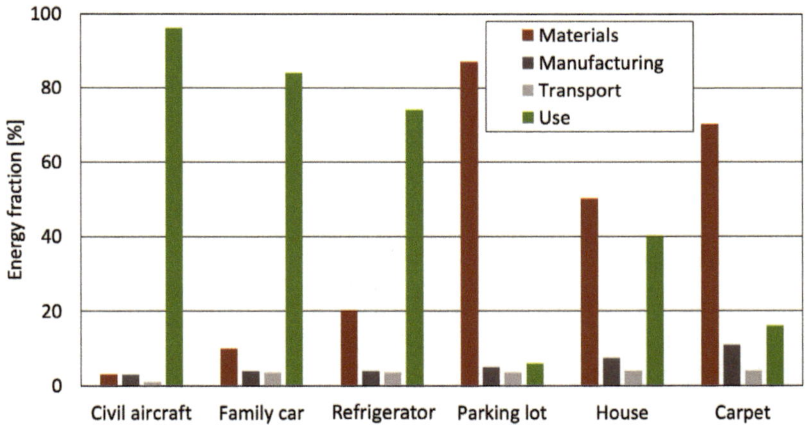

Figure 16.9 Life cycle contribution of different product groups

Depending on the good itself, the environmental hotspot can be identified either in the materials, the manufacturing step, the use phase, the end of life, or, in some cases, the transport. If the use phase requires energy or fuels, for example electricity for white goods such as refrigerators or fuel for cars, it is to be expected the environmental impact of the use phase is significantly higher than in goods that do not require resources in using them. For example in housing and furniture, the environmental hotspot can be found in material and manufacturing [4].

In a scenario where a decision between bringing a reuse item or a newly manufactured item into application must be made, the environmental benefit between each of those two options should be compared. A systematic approach of evaluating this benefit is shown in Figure 16.10.

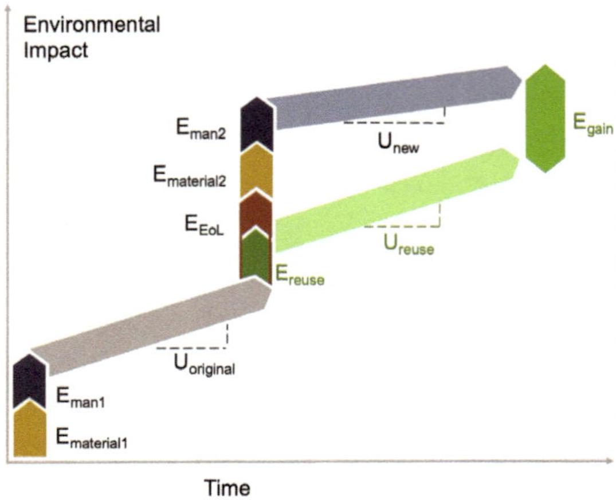

$E_{material1/2}$	Impact of raw materials for original product 1 or new product 2
$E_{man1/2}$	Impact of manufacturing of the original product 1 or new product 2
$U_{original}$	Impact during the use phase of the original product
E_{EoL}	Impact for end-of-life option for first product
E_{reuse}	Impact of manufacturing step to prepare the reuse
U_{reuse}	Impact during use phase of the reuse product
U_{new}	Impact of use phase of a newly manufactured product
E_{gain}	Gained environmental impact

Figure 16.10 Environmental impact of reuse and new products [62]

For the systematic evaluation of the environmental impact of reuse products, firstly the material manufacturing process for the initial product must be assessed. The impact that was required for the first use phase of a product must also be considered. Afterwards, two different scenarios are shown:

- Reuse scenario

  The reuse scenario includes the impact of the process that prepares a good for reuse and the impact during the use phase of the reuse product.

- Non-reuse scenario

  The non-reuse scenario includes the environmental impacts of the end-of-life scenario of the initial product, manufacturing and materials of the newly manufactured product, and its use phase.

The environmental benefit is then defined as the difference in the environmental impact between those two scenarios. It is therefore influenced by differences in material and impact of production of the first good compared to reuse, which depends on the type of reuse. Additionally, an efficiency gap between the reused and new product and the end-of-life option of the initial good can be crucial to the benefit. The impact of the materials is defined by the global warming potential (GWP) of the material type and the mass that is used in a product. For polymers with high GWP like polyamide, the possible GWP saving by reuse is higher compared to lower impact polymers, such as polypropylene or polystyrene [63].

The energy that is required to allow the reuse of a product is strongly dependent on the type of reuse. In relocation, where only a transportation step is needed, there is a comparatively low energy demand. However, the remanufacturing steps of repairing, refurbishing, or refill can cause emissions that lower the energy saved in reuse. In component reuse, for producing a functional good, the overall structure with the exception of the reuse component has to be produced; hence, the expected savings from reduced manufacturing impact will be lower.

A further factor relevant to the environmental benefit of reuse is efficiency differences during use phase. Boustani et al. [64] showed that, over the years, there have been large technological improvements resulting energy savings during the use of various product types such as refrigerators, dish washing machines, and washing machines for clothing. The percentage energy savings from 1981 to 2008 are as follows:

Refrigerator	62%
Dish washing machine	45%
Clothing washing machine	70%

In these cases, the environmental impact of the second life use phase will be higher compared to a use phase of a newly manufactured one [55]. Similar effects can be also seen in the decreasing fuel consumption of new cars over the years. For other goods, which do not consume energy or fuel during the use phase, this gap of efficiency between reuse case and new product can be neglected [2].

An end-of-life option for the initial product has to be applied for a non-reuse scenario, where the initial product is disposed of and a new one is manufactured. The end-of-life option can be either recycling or incineration. Landfill should not be considered anymore today. The environmental impacts of these options vary significantly, so that the decision for the end-of-life option also has a strong influence on the saved energy for a reuse case.

In packaging applications, the environmental impact is often allocated to the product the packaging is supposed to carry. Therefore, it can be seen as included in the material and manufacturing for the first product. The efficiency of transportation is assumed to be the same for every time that the packaging is refilled and brought back. This leads to the conclusion that the saved environmental impact of refilling a packaging system is the saving compared to the end-of-life treatment option that would be applied to the first packaging and the manufacturing and materials for the second packaging in addition. This can be continued with every following refilling step. The underlying chart is shown in Figure 16.11.

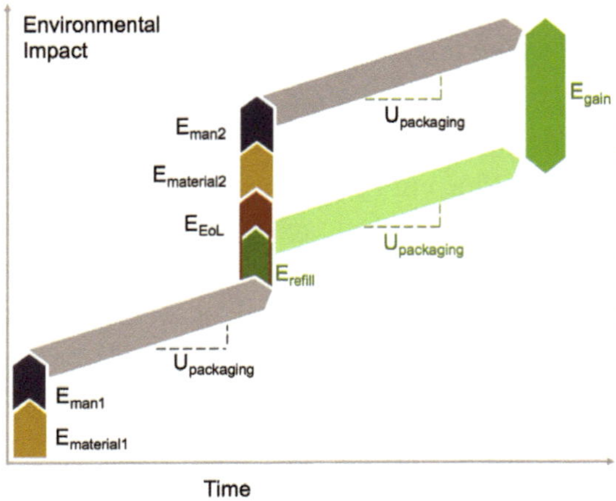

$E_{material1/2}$	Impact of raw materials for original product 1 or new product 2
$E_{man1/2}$	Impact of manufacturing of the original product 1 or new product 2
$U_{packaging}$	Impact during the use phase of the packaging, e.g. transport
E_{EoL}	Impact for end-of-life option for first product
E_{refill}	Impact of manufacturing step to prepare the reuse, here: refill
E_{gain}	Gained environmental impact $E_{gain} = E_{man2}+E_{material2}+E_{EoL}-E_{refill}$

Figure 16.11 Systematic approach to evaluating the environmental impact of refill packaging

16.4 Consumer Behavior

To benefit from reuse, it is important that consumers decide to consume reuse goods instead of newly manufactured goods. There are various factors that influence such choices.

Hunka et al. investigated the decision-making process for reuse goods based on different factors. In their study, the price difference between the reuse and new goods, repairability, appearance, customer service, and shop type were investigated. Their results reveal that there is not only an interest in low-price products, but also high-quality refurbished or partly reused products with an equivalent price compared to new products. For mobile phones they reported a significance of functionality, appearance, and battery life; in the case of buying a vacuum cleaner, the possibility of a fully self-performed repair option was considered important. Modularity was reported beneficial for buying a reuse good as the results for 50% circular products were the most positive [65].

Kunamaneni et al. reported that the design of a reuse product has a strong influence on the decision of a potential consumer towards reuse products. In general, it was stated that the cognitive effort in the decision-making process should be minimized, so that reuse is promoted. A familiar design can help increase the ease of a choice. If this is not possible or not intended, a functional benefit of the product should be implemented compared to the single-use threshold. Also, aesthetic aspects should be considered intensively to promote the intended reuse behavior. A familiar brand can influence the consumer's choice compared to a new or unknown brand. The materials in the product may also be important as it was reported that recycled or recyclable materials in reuse products are more likely to be considered sustainable compared to virgin materials [66].

Finally, it can be a question of which product type is consumed. In a survey of users of the secondhand website eBay in the year 2016, shown in Figure 16.12, people were asked whether they would generally consider buying a reuse good or not.

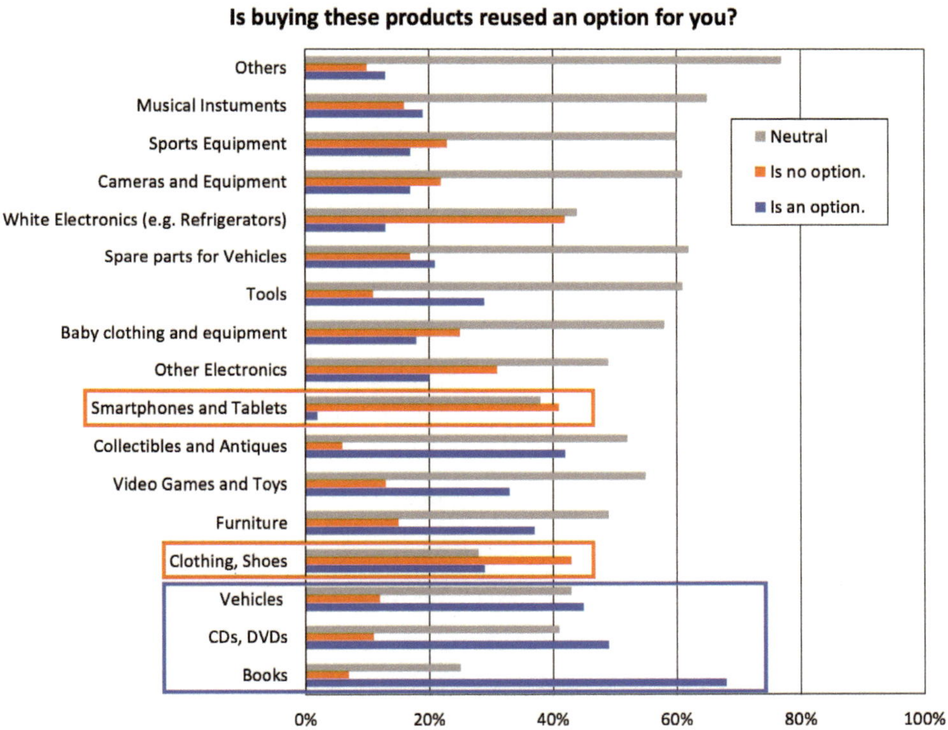

Figure 16.12 Survey of consumer behavior on product types purchased in reuse by relocation [14]

The results show that vehicles; CDs, DVDs, and Blu-rays; books; and collectibles and antiques are most preferred to be bought reused. On the other hand, the surveyed stated that clothing and accessories, smartphones and tablets, and white goods such as refrigerators are most preferably bought new. Other goods such as sports equipment, cameras, and musical instruments are specified as neutral.

This survey was about reuse by relocation, which implies that there is no repair or washing step between first life and reuse. The results of this study can be strongly influenced by the type of reuse, since over 80% of the consumers stated that sustainability is the an important factor to promote secondhand products [15].

■ 16.5 Conclusion

This review of different aspects of reuse reveals that there is a trend towards reuse as a more sustainable and circular approach through several product groups. The type of goods that are reused strongly depends on the type of reuse, corresponding to differences in suitability for different goods types. A summary of types and examples as well as type of plastic is shown in Figure 16.13.

Type	**Lifecycle Extension**	**Relocation**	**Remanufacturing**	**(Adaptive) Component Reuse**	**Reform**
Examples	Food packaging, beverage cups, containers for detergents	Several consumer goods e.g. books, vehicles, CD´s, electronics	**Repair:** WEEE goods, smartphones **Refill:** Bottles, food containers, packaging	Vehicles, photocopies, Adaptive: Clothing, accessories, furniture	Furniture
Plastic Type	Polyolefins	Several	PC, ABS, PP, EPP	PA6, PP, ABS, PC Adaptive: PA6, PA6.6, Polyester	Polyolefins

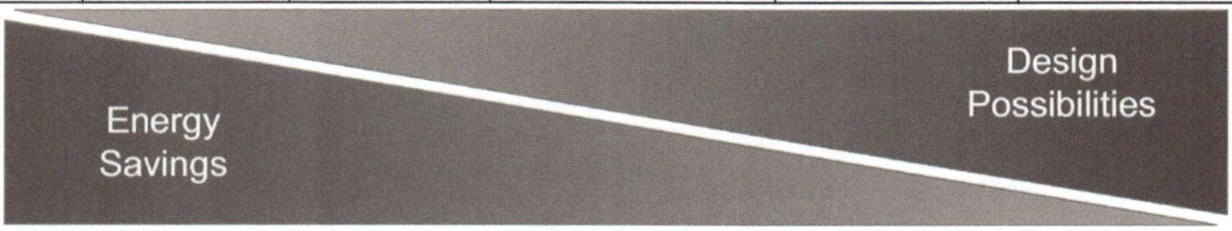

Figure 16.13 Summary of characteristics and examples of different reuse types

Generally speaking, environmental and energy savings can be achieved by reusing. The environmental benefit has to be evaluated via life cycle assessment considering materials, initial manufacturing, manufacturing for reuse, the initial product's end-of-life option, as well as the gap between the efficiency of a reuse good and a state-of-the-art product. Through the different types of reuse, it can be seen that from life cycle extension, which is by definition not considered reuse, relocation, remanufacturing, component reuse to reform, the extent of manufacturing steps increases, which can result in a lower environmental benefit. Instead, the design possibilities for a reuse good increase. In relocation, because the same product is reused by another user, there are no changes in design or functionality to be implemented, whereas in remanufacturing, through refill or repair and refurbish, small changes can be made. Finally, in component reuse or reform, an entirely new product is designed and manufactured.

In reuse of products, the requirements on materials differ from single-use products. For example, containers or bottles for food, beverages, or detergents must resist several washing and refilling steps, where high temperatures and washing detergents are applied. These factors have to be considered in the choice of material. Also, highly stable and strong fabrics such as from airbags, seat belts, and life vests are reused due to their outstanding properties. In reused bead foam packaging, EPP is preferable compared to EPS due to higher ductility. For products or components that include more than one material type, such as electronic parts, other sustainable end-of-life options are difficult to implement. In these cases, reuse is a very promising approach to increase circularity in a product's life cycle.

References for Chapter 16

[1] Publications Office of the European Union: *Circular economy action plan*, Luxembourg, 2020

[2] The European Parliament and the Council of the European Union: *Directive 2008/98/EC on waste and repealing certain Directives*, 2008

[3] Miller, S., McGoughlin, J., Gaillot, O., Connolly, L.: *Material Reuse Good Practice Guide*, 2017

[4] Cooper, D.R., Gutowski, T.G.: *Journal of Industrial Ecology*, (2017) 1, pp. 38–56

[5] Allwood, J.M., Cullen, J.M., Milford, R.L.: *Environmental science & technology*, (2010) 6, pp. 1888–1894

[6] European commission: *Circular Economy: new rules on single-use plastics*, 2019

[7] The European Parliament and the Council of the European Union: *Directive (EU) 2019/904 on the reduction of the impact of certain plastic products on the environment*, 2021

[8] Clarke, C.: *Die Coffee-to-go-Revolution: Startup RECUP führt ein deutschlandweites Pfandsystem für Kaffeebecher ein, https://www.nachhaltigejobs.de/recup-coffee-to-go-pfandsystem/m*

[9] *Nachhaltigkeit - RECUP, https://recup.de/nachhaltigkeit/*

[10] *Eco-Friendly Cleaning Products | Blueland, https://www.blueland.com/*

[11] *Everdrop, Blaue Helden, BIOBAULA, klaeny und denkmit im Test: Was können die nachhaltigen Putzmitteltabs?, https://www.trendyone.de/news/bye-bye-plastik-nachhaltig-putzmitteltabs-im-test*

[12] *Unsere Auszeichnungen und Zertifizierungen, https://mission.everdrop.de/zertifizierung-auszeichnungen*

[13] *Zertifikate, https://klaeny.de/pages/zertifikate*

[14] *eBay-Umfrage zu Nachhaltigkeit im Onlinehandel - Kauf und Verkauf von gebrauchten Produkten erfolgt vorrangig im Netz, https://www.ebayinc.com/stories/press-room/de/ebay-umfrage-zu-nachhaltigkeit-im-onlinehandel-kauf-und-verkauf-von-gebrauchten-produkten-erfolgt-vorrangig-im-netz/*

[15] Wilts, H., Fecke, M.: *ReUse und Secondhand in Deutschland*, Wuppertal Institut, Wuppertal, 2020

[16] Clausen, J., Blättel-Mink, B., Erdmann, L., Henseling, C., *Sustainability*, (2010) 6, pp. 1810–1830

[17] Reiter, T., *Österreichweite willhaben-Nutzung schafft Öko-Mehrwert von rund 15 Millionen Bäumen, https://www.ots.at/presseaussendung/OTS_20170731_OTS0030/oesterreichweite-willhaben-nutzung-schafft-oeko-mehrwert-von-rund-15-millionen-baeumen*

[18] Umweltbundesamt: *Elektro- und Elektronikaltgeräte in Deutschland: Daten 2018 Sammlung und Behandlung (KOM), https://www.umweltbundesamt.de/daten/ressourcen-abfall/verwertung-entsorgung-ausgewaehlter-abfallarten/elektro-elektronikaltgeraete#wo-steht-deutschland*

[19] Sandin, G., Peters, G.M.: *Journal of Cleaner Production*, (2018), pp. 353–365

[20] Woolridge, A.C., Ward, G.D., Phillips, P.S., Collins, M., Gandy, S.: *Resources Conservation and Recycling*, (2006) 1, pp. 94–103

[21] *Über Repair Café (DE), https://www.repaircafe.org/de/uber/*

[22] Cole, C., Cooper, T., Gnanapragasam, A.: *Extending product lifetimes through WEEE reuse and repair: Opportunities and challenges in the UK*, in: *Electronics Goes Green*, 2016+ (EGG), IEEE 2016, pp. 1–9

[23] King, A. M., Burgess, S. C., Ijomah, W., McMahon, C. A.: *Sust. Dev.*, (2006) 4, pp. 257–267

[24] Ongondo, F. O., Williams, I. D., Cherrett, T. J.: *Waste management*, (2011), 4, pp. 714–730

[25] *Parlament will Verbrauchern in der EU "Recht auf Reparatur" einräumen, https://www.europarl.europa.eu/news/de/press-room/20201120IPR92118/parlament-will-verbrauchern-in-der-eu-recht-auf-reparatur-einraumen*

[26] Wolf, F.: *wmn*, (2020), *https://www.wmn.de/lifestyle/diy-living/nachhaltige-smartphones-fairphone-vs-shiftphone-id37343,*

[27] Umweltbundesamt: *Neuer Tiefststand: Immer weniger Mehrwegflaschen, https://www.umweltbundesamt.de/themen/neuer-tiefststand-immer-weniger-mehrwegflaschen*

[28] Heimrich, M., Hermannsdörfer-Tröltzsch, G.: *Wieder verwendbare Getränkeflaschen aus PET, https://www.vis.bayern.de/produktsicherheit/produktgruppen/haushaltswaren/pet_flasche.htm*

[29] Schonert, M., Motz, G., Meckel, H., Detzel, A., Giegrich, J., Ostermayer, A., Schrob, A., et al.: *Ökobilanz für Getränkeverpackungen II Phase 2*, Berlin, 2002

[30] Gowhari, M.: *A deep dive into reusable packaging solutions - tebpalstic holding, https://en.tebplastic.com/a-deep-dive-into-reusable-packaging-solutions/*

[31] Koelsch Sand, C.: *Orchestrating More Sustainable Reusable Food Packaging, https://www.ift.org/news-and-publications/food-technology-magazine/issues/2020/december/columns/packaging-orchestrating-more-sustainable-reusable-food-packaging*

[32] *OZZI – Solutions, https://www.planetozzi.com/solutions.php#ozzispec*

[33] *Algramo Website, https://algramo.com/en/home/*

[34] *Algramo - New Plastics Economy, https://www.newplasticseconomy.org/innovation-prize/winners/algramo*

[35] *Algramo Introduces State-of-the-Art Refill Model to Deliver Affordable Cleaning Product Without Waste in New York City - Closed Loop Partners, https://www.closedlooppartners.com/algramo-introduces-state-of-the-art-refill-model-to-deliver-affordable-cleaning-product-without-waste-in-new-york-city/*

[36] *Loop US - How It Works, https://loopstore.com/how-it-works*

[37] Makower, J.: *Loop's launch brings reusable packaging to the world's biggest brands, https://www.greenbiz.com/article/loops-launch-brings-reusable-packaging-worlds-biggest-brands*

[38] Handelsverband Deutschland (HDE): *Online Monitor 2020*, Berlin, 2020

[39] *LivingPackets | Intelligent, Reusable Packaging, https://livingpackets.com/announcement*

[40] KunststoffWeb (2020), https://www.kunststoffweb.de/branchen-news/living_packets_sprechende_e-commerce-verpackung_t245479

[41] *How to Build Circularity – Returnity, https://www.returnity.co/buildingcircularity*

[42] Schmider, V.: *EPP-Ladungsträger für Industrie und Gewerbe, https://www.de-pack.de/prd/epp-ladungstraeger-fuer-industrie-und-gewerbe*

[43] *EPP-Verpackung, https://www.de-pack.de/coll/epp-verpackung?gclid=EAIaIQobChMIw93wweGC8AIVU-R3Ch2v9gNLEAAYASAAEgJfu_D_BwE*

[44] Farouk, O.: *Life Cycle Assessment of a Printing Toner Cartridge: Comparative Study for Different End of Life (EoL) Scenarios in AIT Campus*, Asian Institute of Engineering and Technology, 2016

[45] Krystofik, M., Babbitt, C. W., Gaustad, G.: *Int J Life Cycle Assess*, (2014) 5, pp. 1129–1145

[46] Dalmijn, W. L., Jong, T. P. R. de: *JOM*, (2007) 11, pp. 52–56

[47] Kanari, N., Pineau, J.-L., Shallari, S.: *JOM*, (2003) 8, pp. 15–19

[48] McKenna, R., Reith, S., Cail, S., Kessler, A., Fichtner, W.: *Journal of Cleaner Production*, (2013), pp. 103–112

[49] Kerr, W., Ryan, C.: *Journal of Cleaner Production*, (2001) 1, pp. 75–81

[50] Smend, M.: *Vogue Germany*, (2018), *https://www.vogue.de/mode/artikel/nachhaltige-mode-berlin-fashion-week*

[51] *Upcycling-Startup Airpaq vervierfacht Produktion: Steigende Händlerzahlen und neue Produkte, Innatex und ILM im Fokus, https://deutsch.airpaq.de/a/press-kit/press-release/szjtqbuc/upcycling-startup-airpaq-vervierfacht-produktion-steigende-handlerzahlen-und-neue-produkte-innatex-und-ilm-im-fokus*

[52] *Airpaq - Unikate aus recyceltem Autoschrott, https://deutsch.airpaq.de/*

[53] Galbiati, S.: *Nachhaltigleben.ch*, (2018), *https://www.nachhaltigleben.ch/mode/upcycling-taschen-diese-11-schweizer-labels-musst-du-kennen-4549*

[54] *PET Bottle furniture by PlastiKetic, https://www.upcycledzine.com/pet-bottle-furniture-by-plastiketic/*

[55] *B2L - Bag to Life, https://shop.b2l.group/B2L/*

[56] *Lufthansa Upcycling Collection, https://www.worldshop.eu/en/category/Lufthansa-Upcycling-Collection/WorldShop_Sortiment_Lufthansa_And_Aviation_Lufthansa_Upcycling_Collection_?p=Rummabm8CUQ&etcc_cmp_onsite=upcycling&etcc_med_onsite=alle_marken_logo&etcc_cu=onsite*

[57] Reese, J.: *Frankfurter Allgemeine Zeitung*, (2020), *https://www.faz.net/aktuell/stil/mode-design/lufthansa-upcycling-collection-aus-alten-airbus-teilen-16537014.html*

[58] *Ecobirdy - Collection, https://www.ecobirdy.com/collections*

[59] Doveil, F.: *LifeGate*, (2019), *https://www.lifegate.com/plastics-reuse-recycling-design*

[60] Hassani, A., Ganjidoust, H., Maghanaki, A. A.: *Waste management & research the journal of the International Solid Wastes and Public Cleansing Association, ISWA*, (2005) 4, pp. 322–327

[61] Panyakapo, P., Panyakapo, M.: *Waste management*, (2008), 9, pp. 1581–1588

[62] Ashby, M. F.: Materials and the environment, Elsevier Butterworth-Heinemann, Amsterdam, 2013

[63] PlasticsEurope: *Eco-profiles*, *https://plasticseurope.org/sustainability/circularity/life-cycle-thinking/eco-profiles-set/*

[64] Boustani, A., Sahni, S., Graves, S. C., Gutowski, T. G.: *Proceedings of the 2010 IEEE International Symposium on Sustainable Systems and Technology*, (2010), pp. 1–6

[65] Hunka, A. D., Linder, M., Habibi, S.: *Bus Strat Env*, (2021) 1, pp. 535–550

[66] Kunamaneni, S., Jassi, S., Hoang, D.: *Sustainable Production and Consumption*, (2019), pp. 253–272

PART D
Design for Recycling

17

Design for Recycling

Robin Fachtan, Volker Altstädt

17.1 Introduction

In 2019 the global annual primary plastics production was reported to be as high as 368 Mt [1] with the global demand for plastics almost constantly rising each year (see also Chapter 1). The latest available data on the plastic demand in different industrial sectors for Europe are for 2019, but worldwide are only for 2015. As shown in Figure 17.1, the highest usage of plastics in 2015 worldwide is found in the packaging market, followed by the building & construction sector, textiles and consumer/institutional products. The transportation and electronics sectors are responsible for less than 10% of the annual usage each [2]. The medical market has a negligible contribution. More recent statistics from 2019 concerning the annual demand for plastics (but limited to the European market) can be found in the annual review of Plastics Europe, suggesting that the spread of annual demand, divided into the above-mentioned markets, is largely similar, but with two slight differences: textile production from resins is only a small industry in the European market, and is therefore not mentioned as a special category, and the use of plastics in the building sector is more pronounced [1].

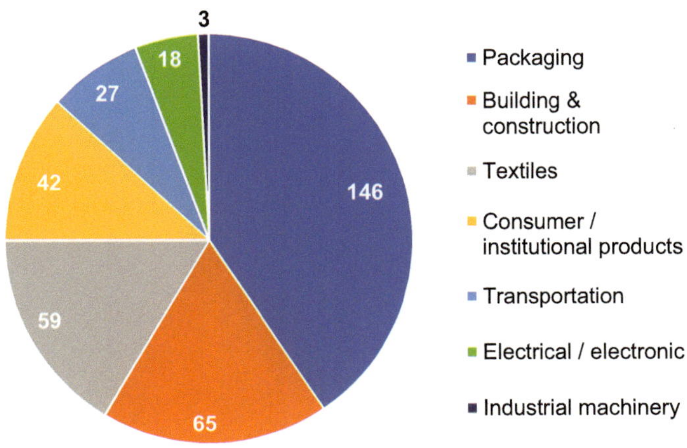

Figure 17.1 Global annual primary plastics production of 2015 in Mt [2]

When it comes to the topic of recycling and circular economy, it is important to look not only at the annual production rates of virgin plastics but especially at the amount of plastic waste that is generated each year. Figure 17.2 shows the global primary plastic waste generation in 2015, which can therefore be directly compared to the global primary plastics production shown in Figure 17.1.

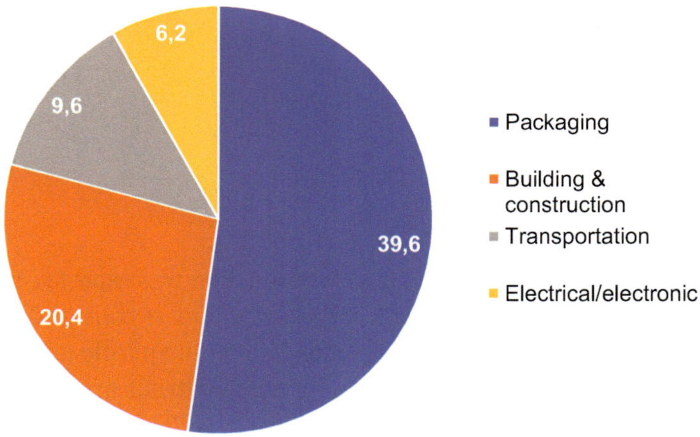

Figure 17.2 Global primary plastics waste generation of 2015 in Mt [2]

It is evident that roughly 47% of the annual plastics waste is generated in the packaging sector. If one compares the specific quantities for this sector, it is also noticeable that the newly produced quantity almost corresponds to the quantity of waste. The temporal offset for the return in this branch is therefore only about one year. In the sector "Consumer & Institutional Products" a comparatively small temporal offset (production: 42 Mt; waste: 37 Mt) can be seen. All other industries show a much more pronounced temporal offset due to higher product lifetimes (see also Figure 17.3), with the highest offset in the sector "Building and Construction" (production: 65 Mt; waste: 13 Mt) due to estimated product lifetimes of up to 100 years [2, 3].

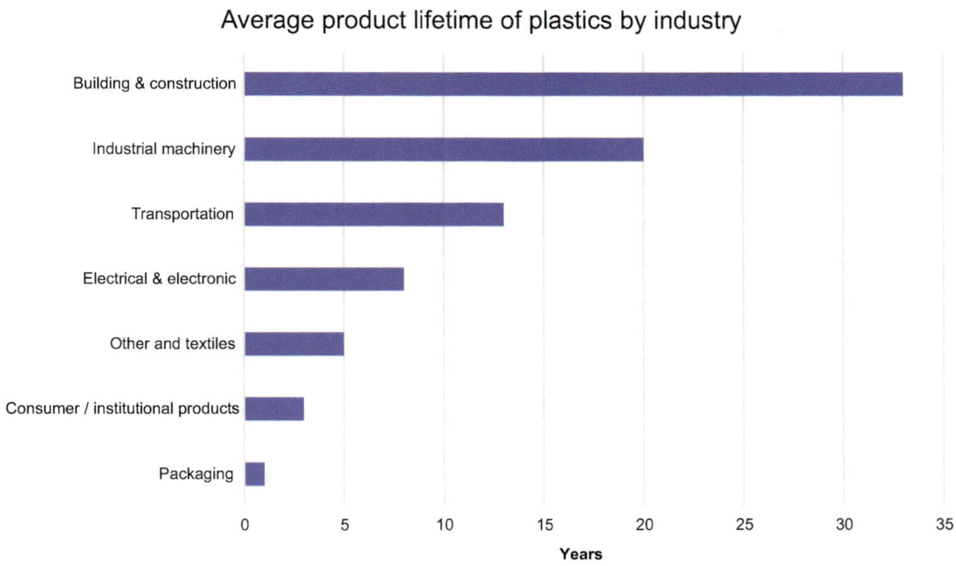

Figure 17.3 Average product lifetime of plastics by industry for the eight industrial use sectors [2]

The current recycling rates for plastics are alarming. The Veolia Institute estimates recycling rates of around 30% in developed economies with regulations encouraging recycling, whereas they see the recycling rates in economies with a minimal industrial base to be close to 0% [4]. The OECD (Organisation for Economic Co-operation and Development) estimates the global recycling rates to be only around 18%, resulting in a global market share of recycled plastics to be a maximum of 12%, as shown in Figure 17.4 [5]. In a comprehensive study by Geyer et al. it was estimated that of the 8300 Mt of primary plastics produced between 1950 and 2015, only 500 Mt were recycled through raw material recycling and 700 Mt through incineration. An incredible 4900 Mt of the plastics produced were simply thrown away and ended up as trash in regulated or unregulated locations, the latter being responsible for polluting natural environments such as beaches, estuaries, and oceans [2].

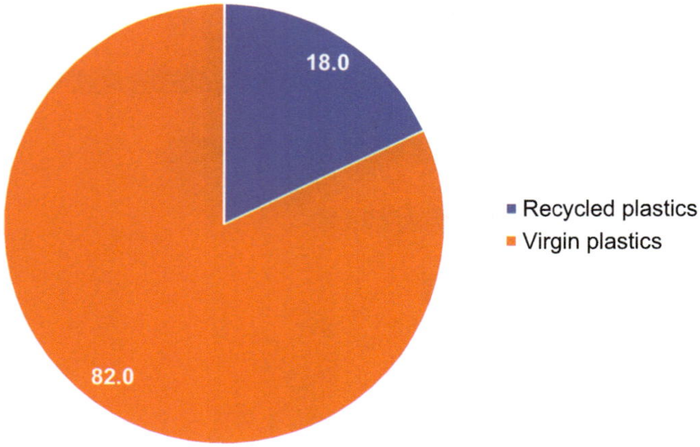

Figure 17.4 Estimated global market share of virgin and recycled plastics [2, 5]

Even developed economies like the USA have poor recycling rates of around 10%, as the OECD found for 2014 [5], because there is a lack of regulatory incentives.

In effect, this means that 82% of the annual plastic waste is not recycled but incinerated (24%) or discarded in other ways (58–62%) [2, 5]. It should be said the incineration of plastics can be an ecofriendly approach as long as it is used to generate energy, and thus displacing fossil fuels like coal, oil, or gas. Unfortunately, construction and operation of those incineration plants is costly, often leading to plastic waste being instead treated as landfill [6].

In developed countries like the USA the traditional way to discard plastic waste that is neither recycled nor incinerated would be as landfill. In less developed economies the waste management infrastructure is usually non-existent or poorly developed with no systematic waste collection systems, which results in industrial and household waste being dumped at unofficial and unregulated sites. The waste dumped at those unmonitored sites has a high risk of ending up in rivers and oceans [4]. Figure 17.5 shows the 20 countries with the worst plastic waste management as found by the Ocean Atlas of 2017 [7].

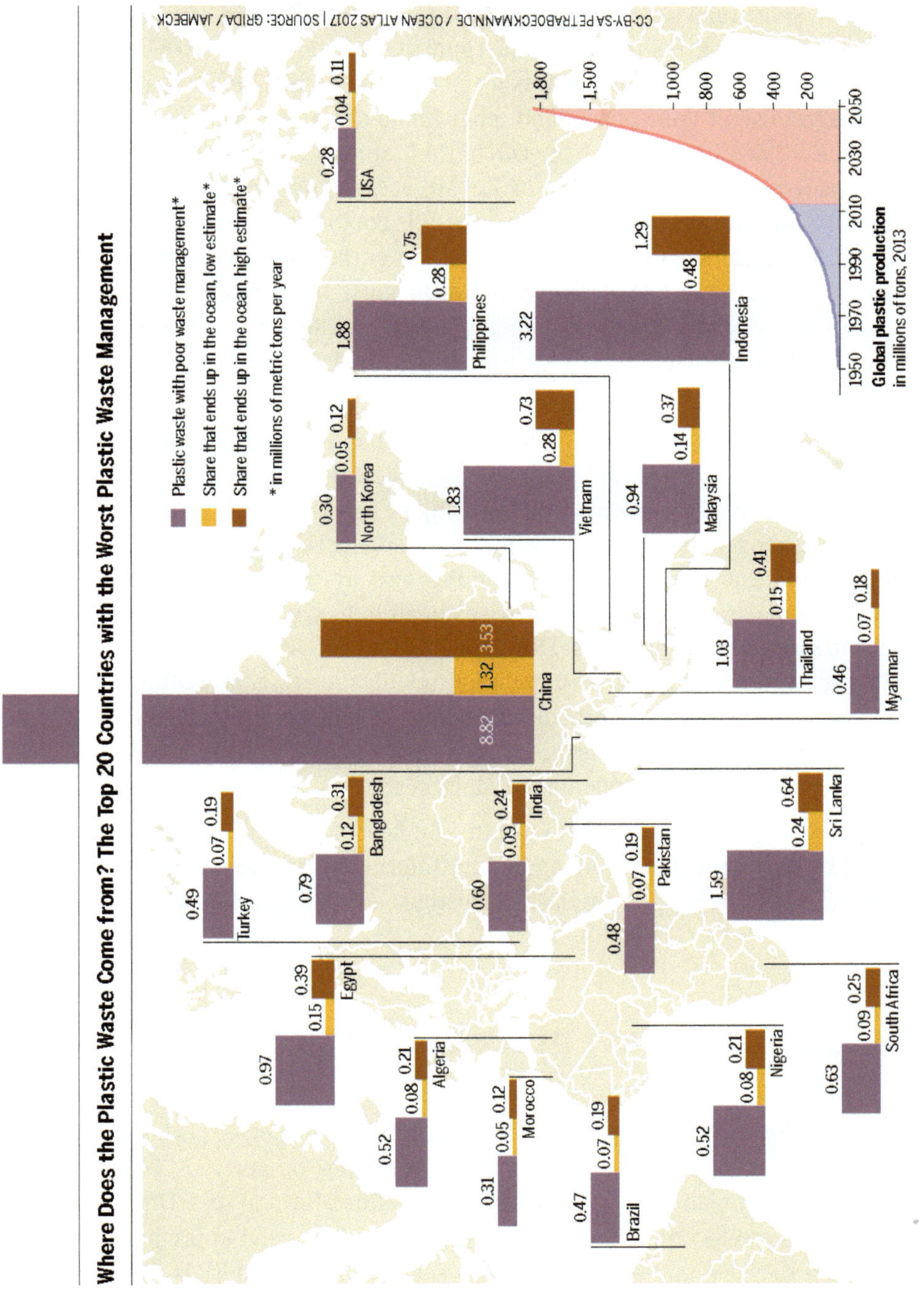

Figure 17.5 Top 20 countries with the worst plastic waste management [7] (estimates based on available comparative waste management data from 2010 [8])

However, since the year 2010 (on which those statistics are based), there has been already a lot of progress in some of the countries mentioned. In 2018 China stopped the import of most of the plastic waste and other disposals. Before that, many countries used to send their plastic and other kinds of waste to China, with countries from the EU sending as much as 87% of their annual plastic waste there. In 2016 China saw plastic waste imports of 7.3 Mt, accounting for more than half of the global plastic waste imports in that year [9]. Furthermore, China started a pilot project for waste separation in 2019 (though currently limited to Shanghai), requiring that private and commercial waste must be separated into four waste streams: recyclables (e.g. glass, paper, plastics, and textiles), wet wastes like compostable kitchen waste, hazardous wastes (e.g. batteries and medicines), and dry waste [10].

Keeping those statistics in mind, the following chapter will give some general strategies suitable for Design for Recycling to make the recycling of plastic products easier and more efficient. This is followed by a deeper look into the sectors that are generating most plastic waste in order to investigate the individual challenges and strategies and to provide design examples already established in those sectors. One of the major challenges which arises in all industry sectors is the lack of varietal purity in plastic products. Only few applications allow for the use of a pure polymer. Most applications require either modification (e.g. by additives like stabilizers or flame retardants) of the used polymer, reinforcement by fillers (e.g. fibers), or blending with another polymer type. Furthermore, polymers are often filled with cheap materials like talcum to reduce costs.

To ensure the timeliness of this keen insight based on published findings, this report will be supplemented by additional information gained through interviews with experts from big players in the investigated industry branches.

■ 17.2 Overview of Design for Recycling

Design for Recycling (DfR) is one of many eco-design strategies [11]. It targets similar challenges as Design for Disassembly, since for proper recycling it is often necessary to provide a way to disassemble a product. Beyond that, DfR additionally addresses the selection of materials and highlights the need to produce new products from recycled materials, meaning a second life for materials [12]. The main target of this design strategy is to ensure high-quality waste streams, which are then recycled and reused. This mainly addresses post-consumer-waste (PCW) from End-of-Life (EoL) products but can also address post-industrial waste (PIW) streams. DfR therefore has influence on post-consumer-recycling (PCR) and post-industrial recycling (PIR). The association "Verein Deutscher Ingenieure" (VDI) distinguishes recycling into three stages [13]:

1. Recycling during production
2. Recycling during use
3. Recycling after use

DfR targets the first and the third stage and thus greatly differs from Design for Reuse, which addresses the second stage and concentrates on design for maintenance, repairability, upgradability, and remanufacturing (see also Chapter 16) [14].

Post-consumer waste (PCW)

Waste from disposed products after end-of-life. Usually those waste streams consist of a material mix (different plastics, metal, paper, or product residues) which complicates recycling processes and/or results in recyclate grades with lower quality.

Post-industrial waste (PIW)

Waste occurring during production (e.g. scrap). Those waste streams are usually good to recycle, since the company can easily separate and sort this waste and knows the detailed material composition. Even if the recycled material cannot be used to produce the same product as previously intended (e.g. due to minor downcycling), companies can often use them for products with different requirements.

All the above-mentioned eco-design strategies contribute to a circular economy and help to close the loop that is still open. Taking a closer look at the schematic for a circular economy in Figure 17.6, it becomes apparent that "Design for Recycling" has the highest impact on the "Recycling" stage but needs to start in the "Design" stage, must be implemented in the "Production" stage, and can even effect the "Collection" stage of that circle. With product lifetimes of up to 100 years as mentioned in the introduction, it is inevitable to already consider the DfR during product development.

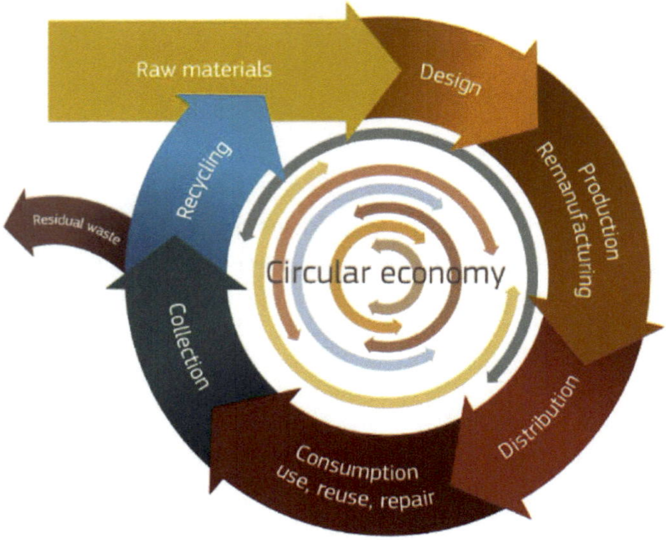

Figure 17.6 Schematic of a circular economy as described by the European Commission [14]

■ 17.3 General Design Strategies

Concerning plastic products, the design or product development step has a major impact on the later recyclability at the end-of-life of the product. This is especially important for products with long lifetimes like in the transportation sector (mean lifetime: 13 years) or building sector (mean lifetime: 35 years). Unfortunately, there are no global rules for the DfR established yet. But the following guidelines can help designers to improve the recyclability of their products [2, 14–17].

- **Reduce or avoid hazardous materials.**
- **Design the product so that harmful/hazardous or valuable components are easily accessible.**
- **Reduce or avoid materials or components that are difficult to recycle with common recycling processes.** Prefer materials that can be easily recycled (e.g. PE, PP, PET, PVC) [18].
- **Reduce or minimize the variety of materials during the development phase. Aim for a monomaterial design if possible.**
- **Aim for "Design for Disassembly" if a monomaterial design, or a design which allows automated sorting of shredder residues, is not possible:**
  - Use joints that allow easy and preferably tool-less dismantling of subassemblies.
  - Make detachable connections easily accessible.

  - Minimize the number and variety of the used connection elements. Standardize if possible.
- **Label the product correctly to ensure the polymer type can be identified.**
  - Packaging products are usually already labelled according to ASTM D7611 (also known as SPI standard), though only six resin types can be unmistakably identified (codes 1–6) with the seventh code being used for all other resin types [19, 20]. Usually only the polymer name is given without explicitly naming additives.
  - All carmakers in the EU need to label plastic parts with a weight above 100 g according to ISO 1043-1, ISO 1043-2, and ISO 11469, as required by directive 2000/53/EC (see Figure 17.7) [17, 21].
  - Machine-readable codes could help to automatically identify polymer types, but their use is not yet established in the economy.
- **Avoid additives or fillers if possible:**
  - *Color pigments* are hard to remove during recycling, and in consequence most recycling grades are sold in black or gray due to otherwise ugly color mixes. Infrared-absorbing colors can make an automated sorting process impossible.
  - *Additives* like flame retardants or plasticizers can make reuse after recycling impossible in some sectors (e.g. food contact, cosmetics) if products containing those additives are not removed before the recycling process. This requires correct and detailed labeling (see point 6).
  - *Fillers*, which can be inert (e.g. talcum) or reinforcing (e.g. glass fibers), can complicate recycling if they have to be removed during the recycling step.
- **Provide information:**
  - Information relevant to recycling: e.g. disassembly procedures, available recycling processes, and material contents.
  - Motivate the end consumer to initiate the recycling process, and provide information on how to start it.

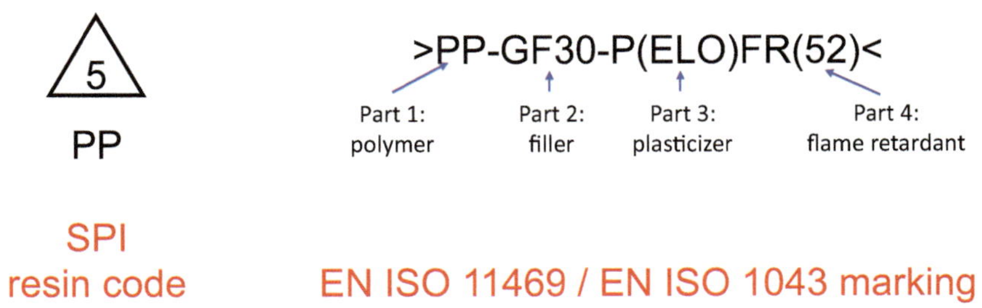

Figure 17.7 Examples of the marking that would be required for a polypropylene with three additives: (1) 30 weight% glass fibers; (2) epoxidized linseed oil plasticizer; (3) red phosphorus flame retardant, according to the SPI standard (left) and the ISO 11469/ISO 1043 standards (right) [21]

With all of these newly introduced methods, it has to be mentioned that DfR is just one of many design goals that must be met by a designer. For example, better better DfR can never take priority over the appearance or the safety of the product. And if a less recyclable design is lighter, then it may nevertheless be preferred because it would be a cheap way of reducing the carbon footprint, in a situation when most of the emissions are generated during product use.

Nevertheless, those thoughts should always be considered at the design stage. In this design step, various strategies such as DfR, dismantling, lightweight construction, or cost efficiency should be thought-through. The well-known "rule of ten" for the costs of errors in the value chain says that the costs of an undetected error or a design that is wrong from the start increase step by step within the value chain by a factor of 10. The sooner a flaw in the design is discovered and eliminated, the lower the cost of the product and the process for the company [22]. This also applies when it comes to recycling at the end of its life.

17.4 Design for Recycling by Industrial Sector – Strategies and Recent Developments

17.4.1 Packaging Industry

The packaging industry is the industry with the shortest product lifetimes concerning plastic products, with a mean lifetime of as low as 0.5 years [2]. Those waste streams were already identified by the industry and legislators as a valuable resource and as a driver for pollution if the waste management is poorly developed. Looking at the numbers provided by the European Commission [23], there has been a mostly steady increase in the recycling rates for plastic waste in the EU since 2006 (see Figure 17.8).

A very good example for an almost perfectly closable loop is the sector of PET bottles. In Germany there is a deposit of 25 cents on most PET bottles, resulting in a recycling rate of deposit PET bottles of 97.2%. PET bottles without a deposit are also recycled at a rate of 93.6% due to a well-established plastic waste collection system and automated sorting. 30% of the recovered PET is recast into new bottles with another 30% each used to produce fibers or films. An essential part of this success are the consumers, who return 96% of the plastic bottles [24].

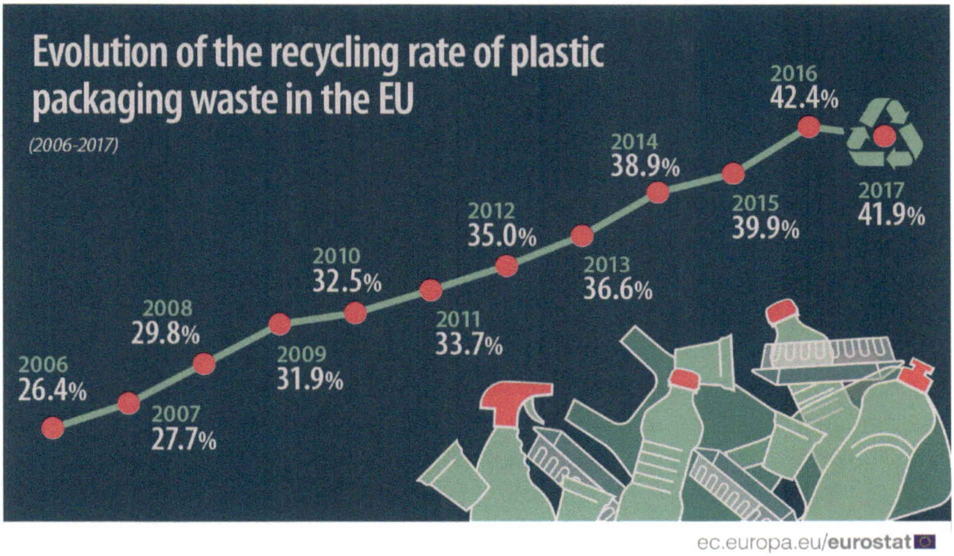

Figure 17.8 Evolution of the recycling rate of plastic packaging waste in the EU (2006–2017) [23]

Nevertheless, there is still a big gap between packaging production and overall recycling (see Figure 17.8), with 39.5% being discharged for incineration and another 18.5% of post-consumer waste ending up as landfill, according to figures for 2017 [1]. One way to improve those rates is DfR.

For the packaging industry, a monomaterial and monolayered design must be the main objective, with commodity thermoplastics like PP, PE, and PET being favored due to established recycling industries for those resins [1, 16, 25]. It should also be noted that an application for EU authorization of mechanically recycled polystyrene as a food contact material has recently been filed [26].

If a multi-material design is needed, inseparable or incompatible combinations of polymers, barrier layers, etc. should be avoided. For example, polyamides and polyolefins are incompatible with each other, whereas PE or PP products with EVOH barriers are recyclable, and PE and PP are considered to be compatible with each other. In addition, inseparable combinations of plastics with other types of material (e.g. paper and metal) should be avoided, although metal flakes with plastic residues can be separated and sent for metal recycling [16, 25]. Vacuum bags that previously were made of an PA/PE multilayer can be replaced by fully recyclable vacuum bags made from 95% PP exhibiting the same barrier properties [27]. Similar progress can be seen for toothpaste tubes, which are usually multi-layered and hard to recycle. A new development of a recyclable toothpaste tube was recently published with the bottle caps changed to PE-HD [28]. At the K-fair 2019 in Düsseldorf an enhanced IML-T (In-Mold-Labeling-Thermoforming) process was presented, which allows the production of plastic/cardboard composite components where the plastic inlays can easily be separated from the cardboard and only the cardboard is decorated. The plastic inlays remain in their natural color [29].

Tetra Pak® is a multi-layered design for beverage containers that is publicly recognized as environmentally friendly (see Figure 17.9). The Tetra Pak group states that the paper fraction (70% of the carton) can be recycled into other paper products several times and that the polymer films can be blended with other plastics and possibly recycled into (for example) roofing tiles [30]. However, Deutsche Umwelthilfe e. V. (Environmental Action Germany) calculated a recycling rate of only 29.9% and furthermore states that 38% of the Tetra Pak cartons are not disposed correctly [31, 32]. Nevertheless, a recent study, albeit commissioned by Tetra Pak International SA, came to conclusion that the Tetra Pak beverage cartons show better life cycle assessment results than single-use plastic bottles or glass bottles [33].

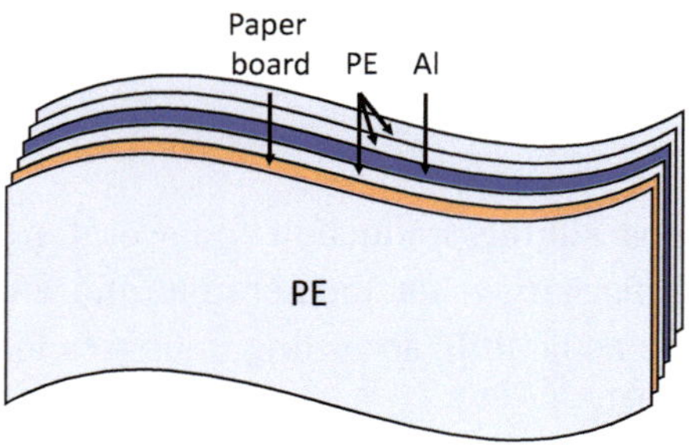

Figure 17.9
Typical composition of a Tetra Pak® carton [34]

Concerning the use of colors, pigments should be minimized wherever possible since these are hard or costly to remove (e.g. by a washing step). Furthermore, carbon black should be avoided since it is infrared-absorbing and thus prevents automated sorting, even with state-of-the-art technologies [25]. Some companies are already providing carbon-black-free, near-infrared detectable black masterbatches (e.g. Berry M&H, Ampacet, and Gabriel Chemie Group). Unfortunately, these masterbatches are not only far more expensive, but also often lead to rather dull black product surfaces. As interviews with representatives of the plastic processing industry showed, the duller surface appearance is problematic for products where the customer expects a piano-black, glossy look. The dull appearance leaves the impression on the buyer that the product is of lower quality, meaning that it is not purchased.

There also is progress being made to develop colors and recycling processes that allow separation of pigments and other additives from the polymer (e.g. PA, PE) by a solvolysis step [35]. Those recycling grades have 47–88% less carbon emission than virgin grades, though the avoidance of colors is still the ideal design way.

In addition, labels should not be printed directly on the packaging, but rather applied to carriers that can be separated from the main part of the packaging according to use [16, 25]. The labels should be designed as wash-off labels that can be automatically removed in a washing step. The Avery Dennison Corporation, Herma GmbH, and etifix GmbH are just three of many companies providing wash-off labels. Ideally, the material of the label should be from the same resin as the packaging or made from paper. Paper labels should contain less than 5% polymer (e.g. glue) to be recyclable in a standard paper recycling processes, as stated in interviews with industry representatives. There are also labels available that are fully compostable.

A notable eco-design for pouch packaging was presented by the Werner & Mertz GmbH for their Frosch® product line and developed together with the Mondi Consumer Packaging Technologies GmbH. The printed labels are easily detachable and fully recyclable. The inner pouch is made up from up to 100% recycled material and not colored, while the cap is from the same material and also not colored [36].

© Werner & Mertz GmbH

Figure 17.10 Recyclable detergent pouch with easily detachable, recyclable labels [36]

Improvements are needed in terms of identifying the types of plastics used in packaging. Labeling according to ASTM D7611 (see Figure 17.11) only gives information about the main resin type used but no information about the additives used, the multi-material design, etc. Although these codes help to identify the most important plastic components, they do not provide any information about whether the product is recyclable in the first place, or which recycling processes should be used [19].

Resin	Resin Identification Code – Option A	Resin Identification Code – Option B
Polyethylene terephthalate	1 PETE	01 PETE
High density polyethylene	2 HDPE	02 PE-HD
Polyvinyl chloride	3 V	03 PVC
Low density polyethylene	4 LDPE	04 PE-LD
Polypropylene	5 PP	05 PP
Polystyrene	6 PS	06 PS
Other resins	7 OTHER	01 O

Figure 17.11 Resin identification codes according to ASTM D7611 [20, 37]

To tackle this problem, AIM (the European Brands Association) and more than 160 companies and organizations across the complete packaging value chain have formed an initiative called the Digital Watermarks Initiative or "Holygrail 2.0", which is assessing the technical and economic feasibility of imperceptible digital watermarks at large scale. Those postage-stamp sized codes can be applied on the surface of plastic packaging and will be automatically detected by cameras in the sorting line [38, 39]. Those digital codes "can carry a wide range of attributes such as manufacturer, SKU, type of plastics used and composition for multilayer objects, food vs. non-food usage, etc.", as one of the participating companies stated [40].

Figure 17.12 Digital watermarks for automated and precise sorting of packaging waste as developed by the Digital Watermarks Initiative HolyGrail 2.0 – Driven by AIM – European Brands Association and Powered by the Alliance to End Plastic Waste [39]

This technology could be very beneficial for the amount of recycled plastics actually being reused for packaging applications. Discussions with representatives of the packaging industry show that the use of PCR grades (see Section 17.2) from mechanical recycling (mPCR grades) often has enormous disadvantages for the product made from it due to downcycling:

- Strange odors
- Only available in black or gray due to pigment residues otherwise resulting in colored streaks on the product
- Varying quality (fluctuations in MFI, viscosity, melt strength, etc.)
- May contain metal residues which can damage production facilities or stop production if an alarm is triggered by metal detectors
- May contain chain extenders, oxidizing agents, plasticizers, or heavy metals and is therefore not approved by the FDA

One design flaw that can be critical especially for proper mechanical recycling is that packaging is often not completely emptied by the consumer. Those residues can result in contamination during the recycling process leading to worse quality (odor, color) of the product. Furthermore, the definition of an “ideal” packaging size should be targeted. Oversized packaging creates additional waste, while small items in the packaging waste (<2 cm) usually have to be sorted out before recycling

[16, 25]. It is expected that the size limit will decrease within the next years as discussions with the packaging industry indicate.

Blending of mPCR grades with virgin plastics often results in better quality and less fluctuation of properties, but those blends are not FDA-approved.

Chemical or pyrolytic recycling could be an opportunity for mixed waste streams, since the plastics are broken down into their monomeric units, cleaned, and then converted into new polymers. Unfortunately, those technologies are yet far more expensive than mechanical recycling. Another challenge is marketing issues, as a recently published interview indicates [41]. In the chemical recycling process scrap and virgin feedstocks need to be mixed and the amounts often vary depending on the quality of the feedstock. Actual numbers for the recycled content in a final product can therefore not be given (or marketed on labels), but only a mean value calculated by mass balance accounting over a defined time period. Education of consumers is needed to fix this issue.

For more information on recycling technologies (e.g. mechanical, chemical, pyrolytic), see Chapters 7–12.

17.4.2 Transportation

17.4.2.1 Automotive Industry

Conventional Cars

As shown in Figure 17.13, most of the energy and, directly related to it, the CO_2 emissions are consumed during the use phase of a car. This is shown by a study of the American car market in 2002 [42]. More recent life cycle assessments for the Bulgarian market from 2018 onwards show similar results, in which it was found that 86.5% of the total energy consumption for the life cycle of a car is used for locomotion, and only 13.5% is accounted for by the production and recycling of the vehicle [43].

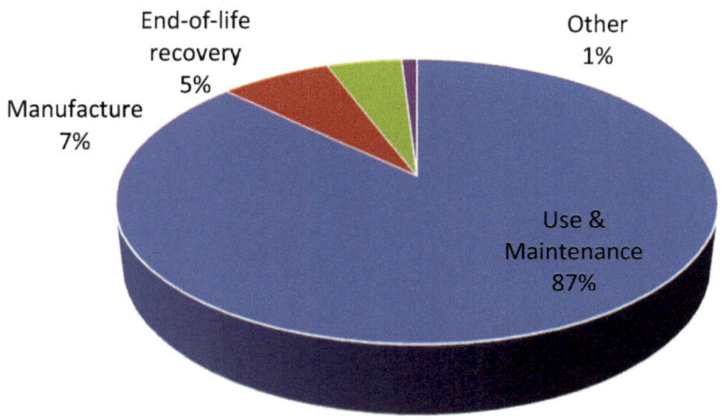

Figure 17.13 Energy consumption in the life cycle of a car (American market, 2002) [42]

Requirements for higher fuel efficiency and reduced CO_2 emissions have already resulted in a trend for lightweight design [42, 44]. Soo et al. performed a life cycle impact assessment for 1980, 1995, and 2010 (see Figure 17.14), and came to the conclusion that the global warming potential (GWP) still has the highest environmental impact due to the CO_2 emission during the use phase, although they recognized a decrease over the years caused by increasingly lightweight designs. However, they found that steel has increasingly been replaced with aluminum or magnesium and plastics, resulting in increasing waste production: they cited the lower recovery rate of aluminum (25%) compared to steel (96%), and the treatment of composite materials as landfill [45].

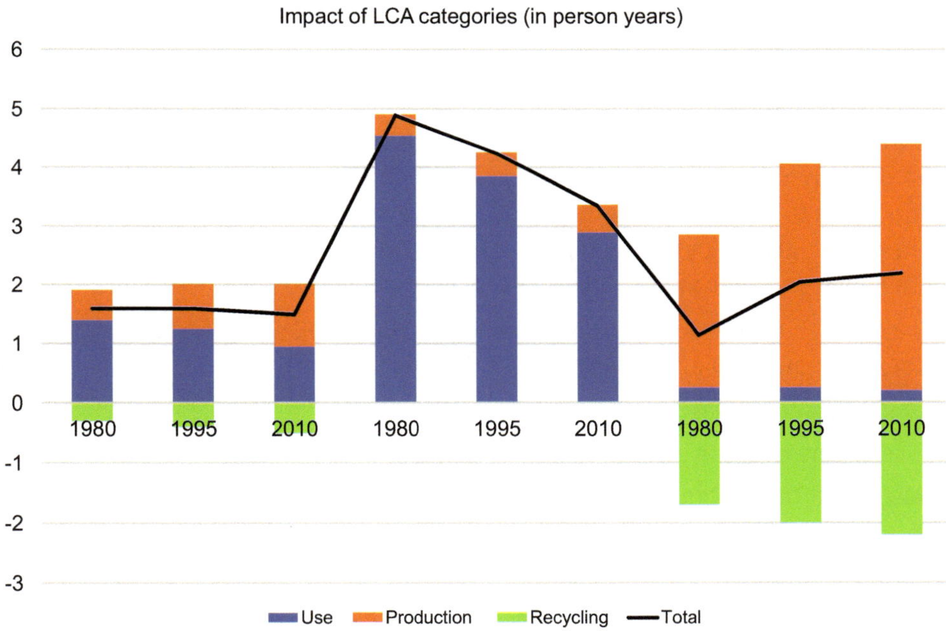

Figure 17.14 Normalized results of a life cycle impact assessment of cars for 1980, 1995, and 2010; left bars: resources, middle bars: global warming potential, right bars: waste [45]

This shows that apart from the usual design criteria of safety, performance, and appearance, a lightweight design is very important in the automotive sector especially if the environmental impact is to be reduced. Alongside the lightweight design, a design optimized in terms of air resistance is of highest interest, because it increases exponentially with increasing speed. As a result, above a vehicle-specific "limit speed", air resistance predominates and becomes the dominant influence on fuel consumption. Gaensicke et al. showed that even for a super lightweight vehicle (VW XL1, Volkswagen AG) the vehicle-specific limit speed is as low as 78 km/h [46]. For other less-well-optimized models (SUV, pick-up) with a larger air-attack surface, this means that when the vehicle is being driven, the air resistance has a decisive influence on fuel consumption at much lower speeds (approx. 50 km/h).

Nevertheless, DfR is also getting more relevant in this sector since the European Parliament approved the directive 2000/53/EC, which deals with end-of-life of vehicles [47, 48].

Table 17.1 EU Recovery Requirements by Law for Vehicles Produced After 1980 by Weight According to EU Directive 2000/53/EC [47, 48]

	2006	2015
Re-use and recovery	85%	95%
Re-use and recycling	80%	85%
Implied allowable energy recovery	5%	10%

As Table 17.1 shows, since 2015 the required re-use and recovery rates for materials from EoL cars need to be as high as 95% within the EU. Before 2006, the reuse and recovery rates for cars were around 75–80%. Therefore, the required 10–15% extra needs to be gained from the 20–25% of previously non-recycled content (polymer, rubber, glass, or electronic parts). Since plastics make up the highest amount of the non-recycled content, automotive plastics recycling needs to increase [47].

To achieve this, DfR can be an opportunity, with the following design rules being most important (and often going hand in hand):

- Reduction of polymer classes
- Use of single family plastics in subassemblies
- Monomaterial design
- Design for disassembly
- Meet the needs
- Provide information relevant for recycling (see Section 17.3)

In modern cars, a mix of different types of plastics (see Figure 17.15) are used, which are often incompatible to each other and therefore hard to recycle [49, 50]. Minimizing the material diversity would be a feasible way to enhance recyclability, as many studies suggest. Furthermore within the individual plastic types, there are various additional chemicals used – for example, flame retardants, plasticizers, or stabilizers.

One idea to reduce this diversity could be to tailor polymers and integrate certain functions directly into the backbone of the polymer instead of using additives. Another way would be to consolidate parts in one polymer family as a so called "monomaterial design". Those parts could be disassembled at end-of-life, gathered, and recycled together [42, 51]. Monomaterial design is often discussed in the automotive industry. For example, metallocene polyolefins could be used for door or instrument panels since they can be engineered to certain design properties [51].

From dialogs with automotive suppliers, it was learned that polypropylene is often targeted for monomaterial designs, since it can be supplied with a wide variety of properties and furthermore tailored in its mechanical strength when fibers are added. Unfortunately, PP is not applicable to all car components. Car ceilings, for example, cannot be produced from PP since high fit accuracy and low shrink are needed. Furthermore, car ceilings are often varnished and have integrated sensors, which is not compatible with a monomaterial design. Therefore design and development need to work more closely together in the future, and already consider the cradle-to-cradle approach in the design stage. Additionally even monomaterial waste streams of polypropylene could be problematic to treat if the molecular weights of the fractions are too diverse, as this would cause phase separation in melt state.

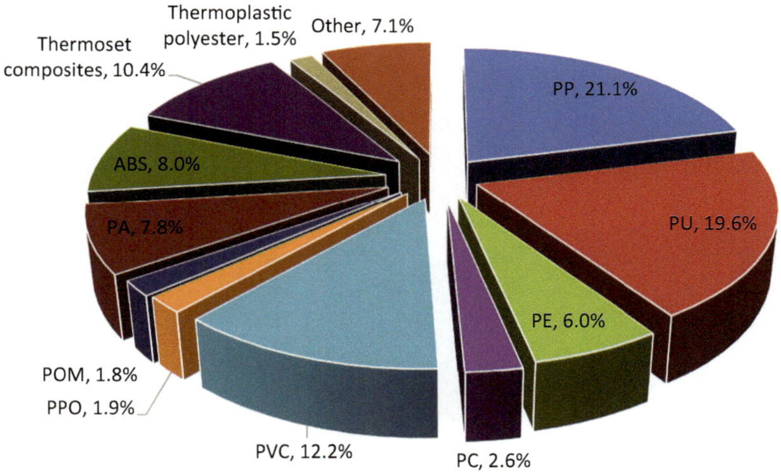

Figure 17.15 Varieties and amounts of common plastics used in automobiles [49]

Since a monomaterial approach, as already mentioned, is not always possible, a reduction in the number of polymer classes and, in particular, optimization of the design of subassemblies with a focus on just one plastic family or on easily separable families can be an alternative approach. This would be applicable to dashboards, for example.

Tian et al. did a case study on car dashboards and found that the skin is mainly from PVC, followed by polyurethane foam, and a rigid framework consisting of PC, ABS, and PP. At end-of-life this inseparable material mix is discarded and crushed. The PU foam can then be separated by flotation and recycled, but the residue will be discarded as landfill [52]. This clearly shows that there is still a need for improvement.

In a publicly funded research project (funding code: BMBF 03X3033E) the research institute “Neue Materialien Bayreuth GmbH” together with industrial partners (BASF SE, BMW AG, Schraml Metallverarbeitung GmbH, and Teubert Maschinenbau GmbH) pursued a new approach to produce a mechanically stable yet light interior component based on a single polymer family (polyolefins). The dashboard developed (see Figure 17.16) consisted of a rigid, injection-molded PP-GF carrier, a center layer from EPP (expanded polypropylene) and a thin decorative film from TPO (thermoplastic polyolefin). The carrier together with the EPP middle layer granted sufficient stability, and the use of the EPP combined with the TPO surface provided a soft and appealing touch. Furthermore they could achieve a weight reduction of 25% and a predicted cost reduction of 6% compared to state-of-the-art dashboards based on PUR systems. Due to a material combination from one polymer family granting varietal purity, it also has better recycling properties than PUR-based instrument panels.

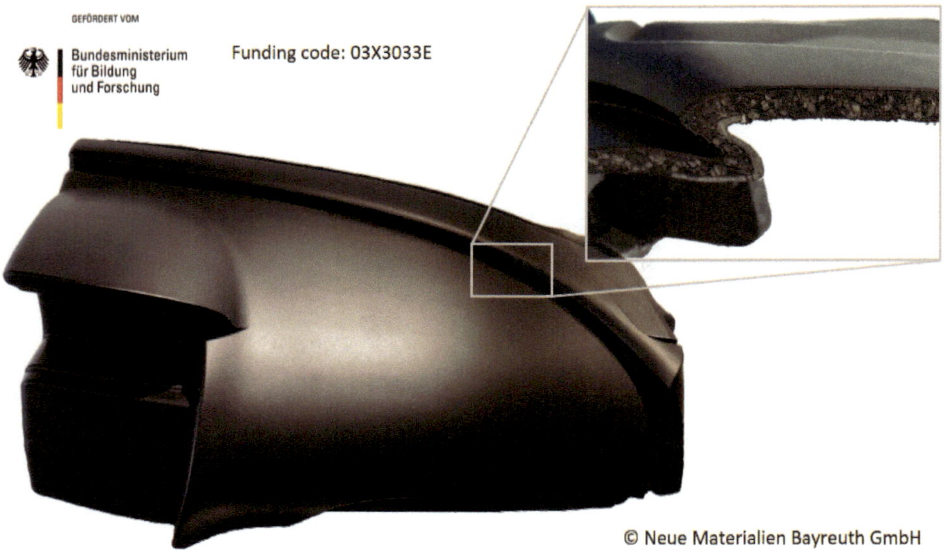

Figure 17.16 Dashboard made from a single polymer family (polyolefins) using PP-GF, EPP, and TPU

Grimmer invented a method to separate the fractions of a dashboard composed of polyvinyl skin, polyurethane foam and a polystyrenic carrier. He was able to remove the PU foam by air separation after granulation of the dashboard. Afterwards they applied pressure to the residue, resulting in the polystyrene particles increasing in size while the polyvinyl particles maintained their shape [53].

Gerrard found in his study that many automotive OEMs are already adapting to the new regulations by encouraging monomaterial design, developing subassemblies with resins only from one family, and encouraging the use of recycled grades. Furthermore, the use of plastics or additives (e.g. flame retardants) which are hard to recycle is reduced or replaced. Gerrard also states that there are worries that the EU Directive could induce a step backwards to heavier but more easily recyclable materials like metals (in the worst case) [47].

Coulter et al. suggest in their 1998 study that the manual dismantling of subassemblies should be enabled. The correct selection of the fastening elements is important in order to shorten the dismantling time. The number of fasteners should be reduced and the fasteners themselves should be standardized. Furthermore, snap fits should be used and non-removable fasteners avoided. When it comes to the fastener design, they should preferably made from plastics compatible with the plastics of the subassemblies to improve recyclability in possible later mechanical recycling steps. If this is not possible, the fasteners should be made out of metal to magnetically separate them from the shredder residue during mechanical recycling [17, 50, 54].

Christensen et al. investigated poly(diketoenamine)s, which are connected via reversible covalent bonds. When mixed with a strong aqueous acid, the bonds open and the monomer units can be recovered and prepared for the next polymerization step. This allows other plastics, flame retardants or fibers to be removed in a solvolytic process within less than 12 hours. Furthermore, this process is not as energy-intensive as other chemical recycling processes [55].

Discussions with automotive suppliers show that automotive manufacturers are still reluctant to use recycled materials, because they require them to have the same properties and meet the same requirements as the virgin material currently used. As Anastas et al. already propose, it is always necessary to design for the use-case ("Meet the needs"). Overdesigning for worst-case scenarios or optimizing for unrealistic conditions is always to be avoided, because it results in high material and energy costs [51]. Recycled grades should therefore be taken more into account.

While the EU Directive 2000/53/EC requests high reuse and recovery rates of 95%, those rates can naturally only be met if the treatment of EoL vehicles is controllable. However, this is contradicted by a large number of used vehicles being exported to countries in Africa or Asia and nations like Russia after a certain lifetime. In Germany, for example, the majority of the used vehicles are exported after a mean operating life of 10–12 years. To put this in numbers, only 560,000 of the approximately 3 million vehicles that are annually put out of service and deleted from the vehicle register are treated in German waste utilization plants. This results in an annual export of 0.33 Mt of automotive plastics opposed to an annual plastics demand of 1.1 Mt in the German automotive industry [56].

e-Mobility

The switch to electric vehicles poses new challenges, with special attention needing to be paid to the recycling of batteries and the electric drive (for the chassis etc., similar DfR principles as for combustion engine cars can be applied). In Figure 17.17 a schematic assembly of a battery pack is shown.

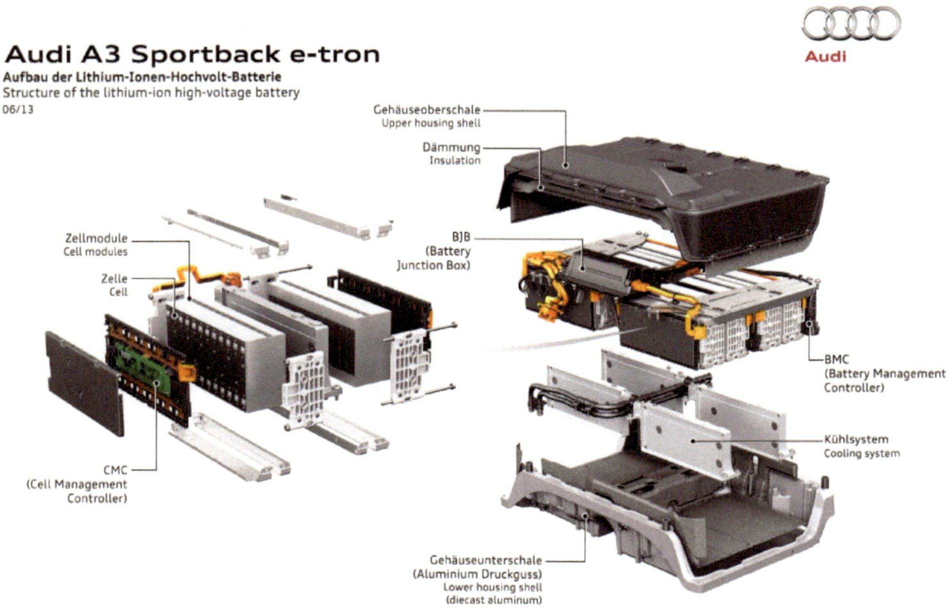

Figure 17.17 Assembly of an electric vehicle battery using the example of an Audi A3 Sportback e-tron [57]

Many parts of the battery housing are made from aluminum (e. g. the lower housing shell), but some parts (e. g. housing covers) are made from technical thermoplastics (see Figure 17.17). As learned from discussions with a car battery recycler, up to 30–40 weight% of the materials used in a battery pack do not primarily contribute to energy storage. Furthermore, those battery packs are located in the vehicle underbody and weigh up to 700 kg, which makes direct recycling of the whole part difficult. Recyclers would therefore prefer a modular design where the single battery modules can be easily disassembled from the pack. The recycling of batteries will always focus on the recovery of the valuable battery metals (rare earth elements), but with a modular design the recovery rates of plastics and other metals could be enhanced. “Design for Recycling” and “Design for Disassembly” are therefore very important.

As learned from discussions with recyclers, but also from the literature and recent talks [58, 59], the following design principles can contribute to an ecofriendly battery design:

- Design for Disassembly
- Standardization of battery materials and designs
- Labeling
- Avoidance of toxic materials

Standardization of battery materials and the casing design that allows disassembly would not only benefit the recyclability but would also allow a second-life use. To achieve this, battery packs should be dismountable to the cell module level. In order to avoid safety problems, the cell modules themselves do not necessarily have to be further dismountable. Removable cell modules, however, lead to a higher recycling quality of the rare earths due to less contamination. On the other hand, this would allow a second life of the battery. Used car battery cells could be systematically separated and given a second life in commercial vehicles (e-bus) or in stationary energy storage devices. In those applications the performance losses in the used cell modules will not come into effect that much, since the battery packs can be larger in dimension and weight. To allow for this, the battery modules should be standardized, ideally worldwide but at least within the different major automotive groups. To optimize for disassembling, reuse, and recycling, the design should be developed together with recyclers and experts from the stationary energy storage business.

An ecofriendly design would moreover benefit from a reduction of material types (e.g. as few different plastic grades as possible, and only one polymer type per subassembly) and reversible joining. This means that adhesives or welding should be avoided in favor of reversible joints like nuts and bolts. This would enable an additional recycling or reuse of (for example) housings, control electronics, or cooling systems [59]. Naturally, those design principles must never result in safety issues. In addition, they should not adversely affect the performance of the battery pack or an economical, automatable manufacturing process.

Last but not least, thorough labeling of the battery modules is essential. The Li-ion batteries need to be distinguishable from other battery types like lead-acid batteries or nickel-metal hydride batteries. The ILA (International Lead Association) reported that Li-ion batteries mistaken for lead-acid batteries and therefore being sent to recycling in secondary lead smelters pose a security risk, and in some cases resulting in fires or explosions [60]. Proper labeling can provide information about the contents, reuse possibilities, and recycling paths, and therefore aid an ecofriendly design.

17.4.2.2 Aviation Industry

When it comes to aviation, DfR is rightfully of little importance in the development phase. The energy consumption and therefore greenhouse gas emission (see Figure 17.18) during use is much higher than during the manufacturing of the aircraft [61]. With a useful life of 20 to 30 years, the focus in the development of an aircraft is rather on fuel efficiency, for example through a lightweight design than on the recycling after end-of-life.

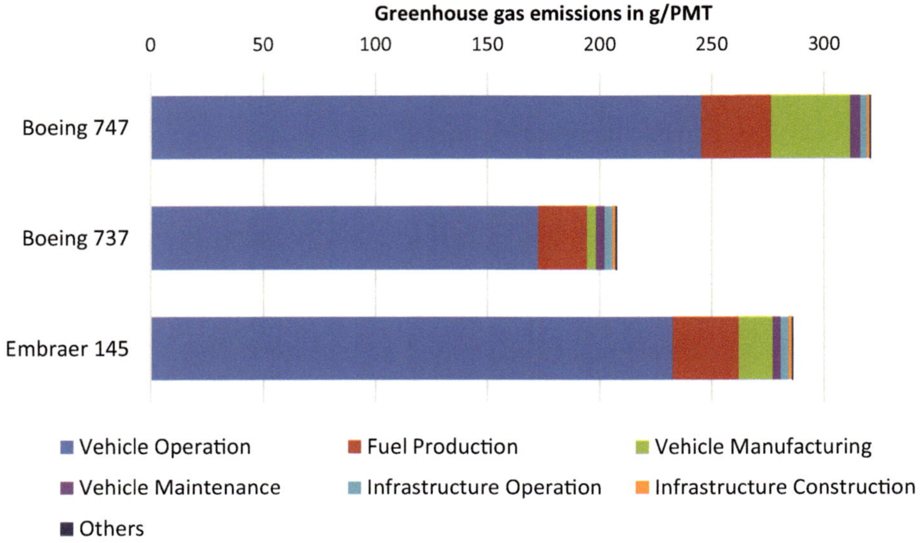

Figure 17.18 Life-cycle assessment of passenger transportation; PMT = Passenger miles traveled (based on and modified from [61])

As result, many structural parts formerly consisting of aluminum are now replaced by carbon-fiber-reinforced composite parts. The Boeing 787 Dreamliner or the Airbus A350, for example, consist of more than 50% composites by weight, with the manufacturers claiming a 20–25% reduction of fuel consumption and therefore reduced greenhouse gas emissions [62, 63].

From discussions with an aircraft manufacturer, it was found that the higher carbon footprint during the production of carbon fiber composites (CFC) compared to primary or secondary aluminum (see Figure 17.19) is outweighed by 50–100 times due to the enhanced lightweight design during service life.

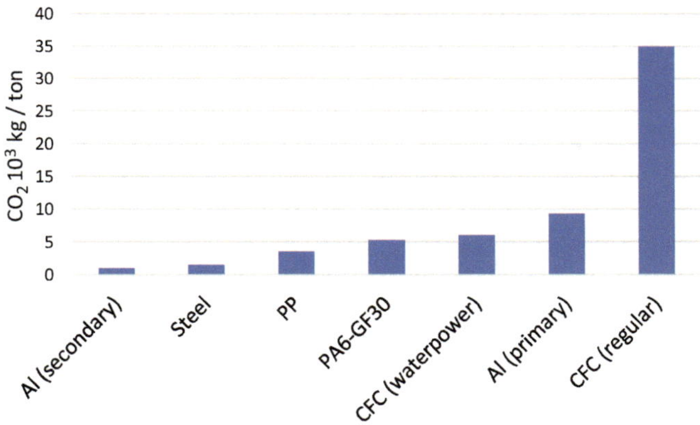

Figure 17.19 CO_2 emission during production of certain materials (according to ProBas [64])

The highest amounts of composites are usually used in the wings (CFC), interior trim (glass fiber reinforced), flooring (CFC) and fuselage, with the fuselage made from CFC or aluminum depending on the aircraft type. The use of unidirectional CF-composites for the wings is especially beneficial, since the wings are long beams and therefore the highly oriented fibers up to 30 m long can work to their full potential.

The high safety standards in aviation also counteract a DfR. Discussions with an airplane manufacturer have shown that structural parts can mostly not be fully disassembled. The fuselage, for example, is additionally riveted. Furthermore, even the use of recycled materials in aviation is challenging. The approval of materials for aircraft is very cost-intensive, and the suppliers are required to keep thorough documentation. For recycled plastics, the supplier would, for example, have to prove the varietal purity. For recycled fibers, the downgrading of the average fiber length, even during state-of-the-art recycling processes, presents an exclusion criterion for wings or fuselage (for example). Nevertheless, structures made of impregnated nonwovens, which consist of recycled but shorter fibers and have a similar fiber area weight, could be used more frequently in the future. The prerequisite for this, however, is that the mechanical properties are sufficient for the designated application, and that the cost savings from the recycled fibers outweigh the costs for approval and documentation.

This raises the question of how aircraft are currently recycled. Decommissioned aircraft are sent to so-called "airplane graveyards" until the airline decides whether to sell them, dismantle them, or put them into service again. These graveyards usually are located in deserts where the dry climate helps to preserve the airplane [65]. Upon arrival, fluids are drained and all inlets are covered. When the decision to dismantle the aircraft is made, valuable parts that can be used as spare parts (e.g. engines, seats, and electronics) are extracted and sold. Some parts like the seats are even sold as furniture to interested customers. The aluminum structures are usually crushed and sent to smelteries, though the large proportion of alloying metals used in airplane aluminum makes recycling more difficult and costlier (and therefore economically less attractive) than other secondary aluminum sources. Finally, the carbon fiber composites could be removed and the carbon fibers recovered by solvolytic or pyrolytic processes [66, 67]. The recovered but shorter carbon fibers could then be used in different applications like cars or the sport industry [68]. Unfortunately, this recycling industry is not yet fully developed.

For more information on solvolytic or pyrolytic recycling technologies, see Chapters 9 and 11, respectively

Although design is less of a concern for recycling in the aviation industry, there are a few strategies that could improve the carbon footprint. For example, in-flight duty-free shopping should be abolished, since it results in additional weight and can easily be compensated by the already existing duty-free shops at airports. Furthermore, as learned from discussions, the largest amount of plastic waste actually originates from the onboard service (e.g. cups, dishes, cutlery), and over the lifetime of an airplane easily outweighs the amount of plastics installed in an aircraft. To close this loop, design principles from Section 17.4.1 can be utilized, or the plastics can be replaced by biodegradable alternatives to which some airlines have already committed [69].

17.4.3 Electronics

When it comes to recycling of electronic components, the focus of recyclers is driven economically by recovery of the contained metals [70, 71].

The following design principles can help in the electronics industry to also ensure plastic waste streams of high quality from electronics:

- Reduction of polymer types
- Avoidance of paint coatings
- Design for Disassembly
- Thorough labelling

Masanet et al. investigated in their study the market share of plastics used in computer products and compared the amounts to recent recycling studies (see Table 17.2). They found that high-impact polystyrene (HIPS), acrylonitrile-butadiene-styrene (ABS), and polycarbonate (PC) have the highest market share. Based on various recycling studies, they found that the composition of waste streams in the electronics industry can vary significantly [70]. A reduction of polymer types would therefore be beneficial for recycling.

Table 17.2 Market Share of Computer Plastics Found in Recent Recycling Studies [70]

Polymer type	Market share	Study 1	Study 2	Study 3
HIPS	22.7%	25%	10%	5%
ABS	20.5%	39%	34%	57%
PC	19.2%	4%	5%	–
PC/ABS	11.7%	6%	29%	2%
PPO	6.5%	17%	12%	36%
Other	19.4%	9%	10%	>1%

Masanet et al. also assessed labelling, and came to the conclusion that thorough labelling (in accordance to Figure 17.7) helps to improve the quality of waste streams that are manually sorted, provided that the labeling is correct for each part. For the investigated automated sorting processes (flotation and density separation) the used labels had no influence. They also found that molded-in or glued-on metal parts usually impose no problem for mechanical recycling since metal residues can be removed, and that painted coatings should be avoided since the paints remain as inclusions and affect the mechanical and aesthetic properties of later products [70].

In their study, Peeters et al. analyzed monitors and TVs from different suppliers. They found that the housings of flat-panel display (FPD) monitors and cathode-ray-tube (CRT) televisions are largely made from HIPS or ABS and mainly contain no flame retardants. They state that those housings are therefore readily recyclable in density-based recycling processes. In contrast, they found that end-of-life FPD televisions and CRT monitors are made from various plastics, mostly containing flame retardants. Since density-based processes cannot separate a given plastic with flame retardants from one without, they forecast that the separable amount will decrease from 43% in 2005 to 28% in 2025. On the other hand, they predict that the share of plastics that is separated by a disassembly-based EoL treatment and innovative plastic identification techniques for later recycling will rise from 70% in 2005 to 99% in 2025. Therefore, they suggest that recycling companies should adopt systematic disassembly techniques and identification processes to assure an economically attractive recovery of plastics [72]. To reduce dismantling time and costs, the use of elastomer-based or foamed fasteners can be attractive and should be considered in the product design step. Elastomer-based fasteners can be easily disconnected by applying enough force. Foamed fasteners could be automatically but non-reversibly disconnected by applying negative pressure leading to a collapse of the foam cells and therefore a shrinkage of the joint [71, 73].

Discussions with a recycling company showed that if the above-mentioned dismantling technologies are combined with machine-readable labeling providing information about ingredients like plasticizers or flame retardants, the recycling quality could further be enhanced.

Another study by Peeters et al. investigated the economic benefits of active disassembly based on the material composition of modems, digicorders, tablets, laptops, TVs, and other electronics. They found that the economic and environmental benefits are strongly product-dependent. The benefit is the highest for products with a:

- High collection rate of EoL products
- High product complexity and little product knowledge at EoL, resulting in a high disassembly time
- Low cost for the implementation of design for demanufacturing
- High concentration of precious metals
- High concentration of plastics that cannot be separated in a size-reduction based EoL treatment, such as flame-retardant-containing plastics
- High failure rate and/or short average lifetime

They also stated that incentives have to be provided for OEMs to implement a design for disassembly, since EoL products are usually collected by collection schemes mostly charging a fixed fee per product [71].

Note from the authors:

At this point it must be emphasized again that plastics waste collection rates do *not* equal plastics waste recycling rates. Currently only about 40% of the collected plastics waste ends up being recycled via material recycling (feedstock recycling) or used for energy recovery by waste incineration (thermal recycling), with the rest ending up as landfill or worse (see Section 17.1). Nevertheless, higher collection rates will most likely benefit recycling rates.

Regarding the design of printed circuit boards (PCBs), the Department for Polymer Engineering of the University of Bayreuth, together with partners from the industry, investigated a more recycling-friendly approach in various publicly funded projects (BFS AZ-858-09 and BFS AZ-911-11) by employing thermoplastics instead of thermosets for the circuit board carrier. According to the state of the art, printed circuit board substrates are usually manufactured in labor-intensive processes from glass-fiber-reinforced epoxy resin or in rare cases from PTFE or ceramic. In order to meet flame retardancy requirements such as the FR4 standard, these matrices must also be equipped with potentially toxic flame retardants such as tetrabromobisphenol. By using high-performance thermoplastics (e.g. PEI, PES, PPS, and PEEK) as the matrix polymer they developed "green" circuit board substrates that are intrinsically flame-retardant and therefore halogen-free, and that can be produced in a continuous extrusion process. The substrates are also characterized by their 3D formability and their excellent recyclability. Through either a foam extrusion process (weight and material reduction) or the use of fillers they could further tailor the properties for the use as either standard PCBs or high-performance PCBs (see Figure 17.20) [74, 75].

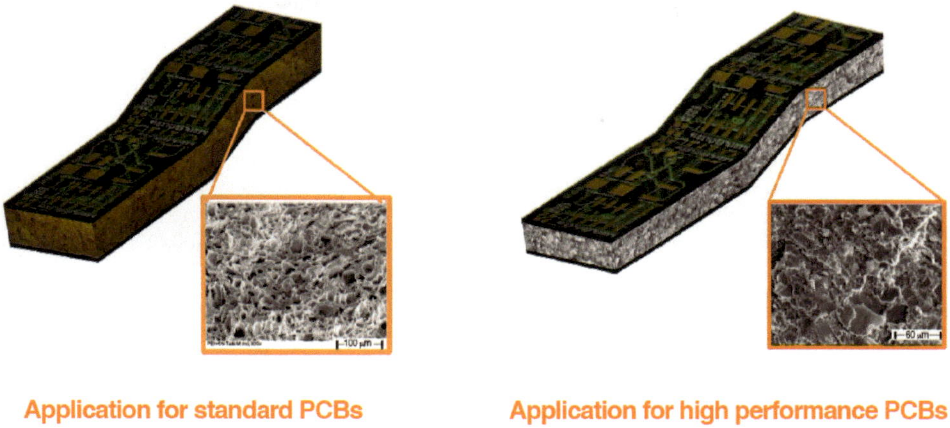

Figure 17.20 PCBs from high-performance thermoplastics. Left: foamed substrate; right: highly filled substrate (source: Polymer Engineering Department of the University of Bayreuth)

As learned from discussions with recyclers, smartphone recycling usually targets precious metals because of their highly integrated designs. Due to the lack of a design for disassembly, economical dismantling of smartphones is not possible, resulting in the loss of metals like cobalt and nickel and the contained plastics. Even a design for disassembly that simply allows automated removal of the battery would result in a higher recovery ratio.

Unfortunately this is often not the case at the moment. As the economic journal *Bloomberg* found, one of the major manufacturers of smartphones does not even allow commercial reuse of spare parts from EoL smartphones, but demands mechanical recycling of the whole phone with the extracted gold, copper, aluminum, and glass being recast in other products. The company justifies this procedure with the concern that otherwise product counterfeits could appear on secondary markets. According to the chief strategy officers of one of these recycling companies, recycling the phones mechanically even requires more energy than reusing components. Suitable for a reuse in the same or other products would be (for example) the camera modules, screens, or chips [76].

Furthermore, there is a strong need to improve the collection rates of electronics. As Hagelueken showed in his talk, the recycling rates for electronics are estimated at only around 25% due to an open loop mostly caused by the business-to-consumer relationship [58]. This was confirmed in a discussion with a recycler. A large part of the electronics is incorrectly disposed of in the residual waste. Therefore a better education of the consumer and improved collection systems within easy reach of the consumer are needed. Since electronic components with a self-sufficient power supply represent a safety risk (e.g. fire or explosion), disposal in unsupervised waste containers is not possible. Therefore, legislators should require retailers of a certain size (like the bigger electronics stores) to take back electronic devices containing batteries. Electronic waste not containing autarkic power supplies could also be disposed in public electronic waste containers, provided that consumers are thoroughly educated.

Additional information: The 9-volt fire risk

9-volt batteries have a fire risk due to the design flaw that the positive and negative terminals are very close to each other. If disposed with the residual waste, these batteries can start fires on landfill dumps or in waste sorting plants if the terminals get into contact with metals. Apart from that, those batteries should always be stored safely apart from other batteries or metals and in an upright position. Always check the regional rules for battery disposal and ideally cover the terminals with duct or electric tape before disposal.

17.4.4 Building and Construction

Due to the average lifespan of 35 years for plastic products in the construction sector, there is a large discrepancy between the annual plastic demand and the annual waste return rates in this sector. In 2015, the global annual consumption of plastic in the construction sector of 65 Mt was offset by an annual waste return of only 13 Mt. This corresponds to an annual waste return of only 20% [2].

The plastic demand in the building sector is dominated by four resin types (see Figure 17.21). The highest proportion of plastics used is poly vinyl chloride (PVC) for products such as cable insulation, window/door profiles or floorings. In 2015, PVC was followed by a high share of polyolefins (HDPE: 18%; LDPE/LLDPE: 6%; PP: 6%) which are used for pipes, fittings, and foils. Polyurethane (PUR) with 13% (e.g. for building foam, pipe insulations) and polystyrene (PS) with 12% (e.g. for wall insulation) complete the list [2].

Discussions with manufacturers of plastic products and semi-finished products for the construction sector show that the largest quantities of plastic in buildings are used in cable insulation, pipes, door and window profiles, wall insulation, and facade elements. Though the building and construction industry has the second highest annual demand for plastics [2], plastics make up only a small proportion of the diverse building materials. Plastics only account for 1–3% in demolition waste, as reported by various studies [77–79].

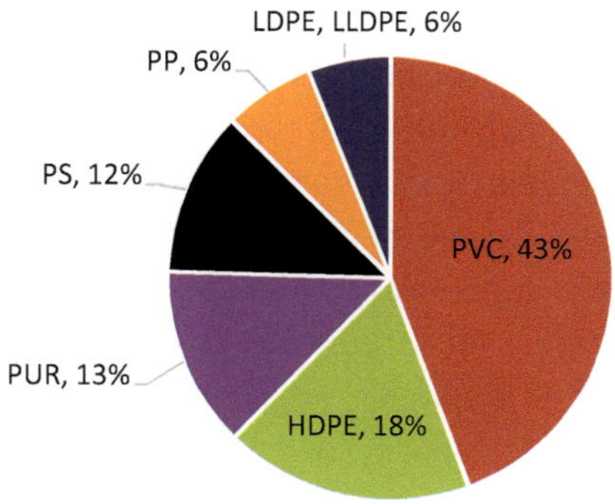

Figure 17.21
Global annual plastics demand in the building and construction sector by polymer type in 2015 [2]

Unfortunately, this low amount of plastics in the demolition waste, as reported by studies and also confirmed in talks with manufacturers, makes material recycling economically unviable. The plastics fraction is usually discarded as landfill or incinerated, though incineration for energy generation (and so replacing coal or oil) is an environmentally commendable treatment.

A particular challenge is the often difficult separation and recovery of the plastics used. For example, pipes and cables are installed inside the walls or ceilings, making "clean" recovery impossible or uneconomical. Therefore, after the demolition of the building, the waste is crushed and either discarded as landfill or the inorganic fractions are separated (for example, by flotation) and then thermally recycled. Metal fractions can also be recovered (for example, by magnetic separation). Underfloor heating panels would not be recyclable at all even if properly recovered, because they are made of chemically cross-linked polyethylene (uncured polyethylene would pose a leakage risk).

For EPS (expanded polystyrene) panels used as wall insulation, "clean" recovery is also difficult and costly since they are usually glued to the wall with a PUR glue and covered by two layers of plaster. Nevertheless in the EU and some other countries the disassembly and collection of EPS or XPS (expanded polystyrene) waste from building sites is required by law. The reason for this is hexabromocyclododecane (HBCD), which at the time was commonly used as a flame retardant in these products, and which was listed as a persistent organic pollutant under the Stockholm Convention in 2013. In order to reduce the amount of HBCD in circulation, EU Directive 2016/460/EU stipulates that materials with more than 1000 mg (HPCD) per kg must be treated as hazardous waste (incineration/heat treatment required). Since this Directive came into force, many companies specializing in the collection and disposal of EPS and XPS construction waste have been founded. These companies can also separately collect and recycle newer EPS/XPS materials that do not contain HBCD [80, 81].

Finally, the necessary use of flame retardants in (for example) cable insulations or wall insulation also hinders material recycling and reuse in other products.

Despite high hurdles, some examples for material recovery with subsequent recycling for reuse in the building sector can be given. In Germany and many other European countries, doors and windows with PVC profiles are removed before demolition and collected (as learned from discussions with a manufacturer). If possible and profitable they are directly reused, but if not they are milled for subsequent recycling. Metal and glass residues are separated and other residues like sealing rubbers are separated by a PVC-melt filter. All scrap from production or assembly is also collected for recycling. Since PVC is readily recyclable, new window or door frames can be made from the recycling grades.

Other PVC building products like floorings, facade elements, pipes or roofing and sealing membranes can also be recycled if they can be recovered in an acceptable, usable purity. Major resin manufacturers have already reacted and are using mass-based chemical recycling to offer PVC recycling grades [82]. Thorough labeling of all plastic types and their additives as well as incentives from legislators could help to improve the recovering quota.

Janajreh et al. have examined the mechanical recycling of PVC waste from the cable industry, and have found that even after multistage remolding and aging, the PVC showed good stability during reprocessing and mechanical properties close to virgin PVC. Through addition of stabilizers or other additives the properties could be further improved and tailored [83].

Experts point out that the areas of application for recyclates from cable insulation are very limited. The metal particles that may be present in the recyclate can trigger metal detectors in the production lines and therefore lead to emergency stops; a high flame retardant content could be also critical. This kind of PVC recycling could also be applied for cable insulation used in electronics (see also Section 17.4.3).

Another concept that is considered as a sustainability strategy by the industry, as confirmed by a manufacturer, is a coextrusion of pipes with an inner liner made from a virgin resin and the outer liner made from a recycled grade of the same resin (e. g. PP and recycled PP). Due to the monomaterial design, such pipes could also easily be recycled at EoL. There are some concerns regarding a possible migration of pollutants from the recycled liner through the virgin liner into the transported liquids during the long pipe lifetimes, which is why those kind of tubes are not yet certified for use. But normative restrictions regarding the use of recycled grades are already changing, as learned from an expert from the industry.

■ 17.5 Conclusion

With global warming and exploitation of natural raw materials continuing, recycling is and will continue to be a major topic. The incorporation of DfR at the product development stage will become more important and (as confirmed by representatives from a wide variety of industries) this concept is already getting more exposure. Increasing incentives by legislators, as well as rising pressure from consumers on the industry regarding socio-ecological responsibility, will strengthen this development and be a driving force.

Closing this chapter, there is one thing that needs to be mentioned and kept in mind when talking about plastics recycling and the impact of plastics on carbon emissions. When taking a closer look at the annual global usage of oil and gas, only 4% is actually used for the production of plastics (although some other sources cite annual consumption at 4–8% [84]). 87% percent of fossil fuels is used for transportation or heating and energy generation, and thus are the bigger driver of carbon emissions. The use of green energy, implementation of better wall insulation or alternative transportation concepts such as e-mobility, car sharing, or public transportation will have a greater environmental impact at the moment. However, if the demand for plastics continues growing as expected, the World Economic Forum forecasts the plastics industry to account for 20% of the total oil consumption in 2050, making plastics recycling even more important [84–86].

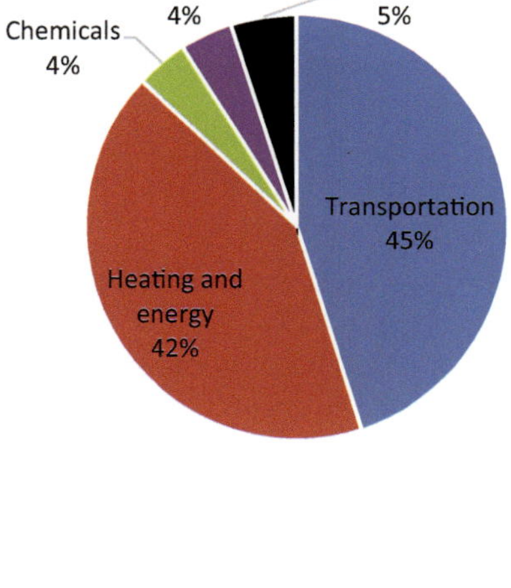

Figure 17.22
Annual global oil and gas consumption by sector [85, 86]

References for Chapter 17

[1] Plastics Europe e.V.: *Plastics – the Facts 2020* [online]. An Analysis of European Plastics Production, Demand and Waste Data, *https://www.plasticseurope.org/download_file/force/4261/181*, accessed March 25, 2021

[2] Geyer, R., Jambeck, J.R., Law, K.L.: Production, use, and fate of all plastics ever made [online], *Science Advances*, 3 (2017) 7, e1700782, 10.1126/sciadv.1700782

[3] Institut Bruxellois pour la Gestion de l'Environnemen: *Assessing the Potential of Plastics Recycling in the Construction and Demolition Activities* [online]*, Guide: Towards sustainable plastic construction and demolition wastemanagement in Europe*, Institut Bruxellois pour la Gestion de l'Environnemen, *https://www.acrplus.org/images/pdf/document142.pdf*, accessed March 30, 2021

[4] d'Ambrières, W.: Plastics recycling worldwide: Current overview and desirable changes [online], *Field Actions Science Reports – The Journal of Field Actions*, (2019) 19, *https://journals.openedition.org/factsreports/5102-7Ki6*

[5] Improving Plastics Management: Trends, policy responses, and the role of international co-operation and trade [online], *OECD Environment Policy Paper*, (2018) 12, *https://www.oecd.org/environment/waste/policy-highlights-improving-plastics-management.pdf*

[6] Royte, E.: *Energie durch Plastikverbrennung – (K)eine gute Idee?*, March 14, 2019, *https://www.nationalgeographic.de/planet-or-plastic/2019/03/energie-durch-plastikverbrennung-keine-gute-idee*, accessed April 18, 2021

[7] Pösel, N., Bähr, U., Kronfeld-Goharani, U.: *Ocean Atlas 2017* [online], *Facts and Figures on the Threats to our Marine Ecosystems*, Heinrich Böll Foundation, *https://www.boell.de/sites/default/files/web_170607_ocean_atlas_vektor_us_v102.pdf?dimension1=ds_meeresatlas*, accessed March 30, 2021

[8] Jambeck, J.R., et al.: Marine pollution. Plastic waste inputs from land into the ocean [online], *Science*, 347 (2015) 6223, pp. 768–771, 10.1126/science.1260352

[9] ntv Nachrichtenfernsehen GmbH: *China verhängt Import-Stopp: Deutschland sucht eine neue Müllkippe*, January 9, 2018, *https://www.n-tv.de/wirtschaft/Deutschland-sucht-eine-neue-Muellkippe-article20222136.html*, accessed April 18, 2021

[10] Schmitt, S.: *China geht in Shanghai neue Wege der Mülltrennung*, October 3, 2019, *https://www.gtai.de/gtai-de/trade/branchen/branchenbericht/china/china-geht-in-shanghai-neue-wege-der-muelltrennung-163298*, accessed April 18, 2021

[11] Maris, E., et al.: From Recycling to Eco-design, in: Worrell, E., Reuter, M.A. (eds.): Handbook of Recycling. State-of-the-art for practitioners, analysts, and scientists, Elsevier, 2014, pp. 421–427

[12] Kelly, A. (ed.): Comprehensive composite materials, Elsevier, 2006, 9780080429939

[13] Leterrier, Y.: Life cycle engineering of composites, in: Kelly, A. (ed.): Comprehensive Composite Materials, Amsterdam, Elsevier, 2006, pp. 1073–1102

[14] European Commission: *Towards a Circular Economy* [online], *A Zero Waste Programme for Europe. European Commission*, *https://ec.europa.eu/environment/circular-economy/pdf/circular-economy-communication.pdf*, accessed March 30, 2021

[15] Kriwet, A., Zussman, E., Seliger, G.: Systematic integration of design-for-recycling into product design [online], *International Journal of Production Economics*, 38 (1995) 1, pp. 15–22, 10.1016/0925-5273(95)99062-A

[16] Scriba, M.O.: *Ten Common Rules of Design for Recyclability (DFR) for Plastic Packaging* [online], MTM Plastics GmbH, ed., *https://www.mtm-plastics.eu/fileadmin/user_upload/TEN_DFR_fin_2018-08-21.pdf*, accessed March 30, 2021

[17] Zhao, Q., Chen, M.: Automotive plastic parts design, recycling, research, and development in China [online], *Journal of Thermoplastic Composite Materials*, 28 (2015) 1, pp. 142–157, 10.1177/0892705713519810

[18] Schubert, N., Steilemann, M., Schiller, C.: *Plastik: In Zukunft ohne?*, March 16, 2021, *https://www.ihre-chemie.de/oeffentliche-veranstaltungen/klimaschutz/*, accessed March 30, 2021

[19] Moss, E., Harris, R.: *"Can I Recycle This?"* [online], *A Global Mapping and Assessment of Standards, Labels and Claims on Plastic Packaging*, United Nations Environment Programme & Consumers International, *https://consumersinternational.org/media/352255/canirecyclethis-finalreport.pdf*, accessed March 30, 2021

[20] ASTM International; D20 Committee: ASTM D7611, *Standard Practice for Coding Plastic Manufactured Articles for Resin Identification // Practice for Coding Plastic Manufactured Articles for Resin Identification*, ASTM International

[21] Donatello, S., Gama Caldas, M., Wolf, O.: *Revision of the EU Green Public Procurement (GPP) Criteria for Furniture, Technical Report: Final Version*, Publications Office of the European Union, 2017 (EUR Scientific and technical research series 28729), 978-92-79-71863-2

[22] Barth, M., et al.: Object-oriented engineering data exchange as a base for automatic generation of simulation models, IECON 2009, Porto, Portugal, November 3–5, 2009, *IEEE*, (2009), pp. 2465–2470

[23] European Commission: *How Much Plastic Packaging Waste do we Recycle?* [online], November 5, 2019, *https://ec.europa.eu/eurostat/web/products-eurostat-news/-/DDN-20191105-2*, accessed March 30, 2021

[24] Žmak, I., Hartmann, C.: Current state of the plastic waste recycling system in the European Union and in Germany [online], *Technical Journal*, 11 (2017) 3, pp. 138–142, 10.31803/tg

[25] IK Industrievereinigung Kunststoffverpackungen e. V.: *Eco Design of Plastic Packaging [online], Instruction – Design for Recycling*, IK Industrievereinigung Kunststoffverpackungen E. V., *https://ecodesign-packaging.org/wp-content/uploads/2019/09/02-Strategieelement-04a-Df-Recycling-Factsheet-final-EN.pdf*, accessed March 30, 2021

[26] Styrenics Circular Solutions: *Styrenics Circular Solutions Seeks EFSA Opinion on Mechanically Recycled Polystyrene as Food Contact Material*, March 9, 2021, *http://styrenics-circular-solutions.com/wp-content/uploads/2021/03/20210309-SCS-seeks-EFSA-opinion-on-rPS_Press-Release.pdf*, accessed March 30, 2021

[27] Carl Hanser Verlag: *Vakuumbeutel aus Monomaterial ersetzt PA/PE-Verbund* [online], Carl Hanser Verlag, *https://www.kunststoffe.de/a/article-313314?etcc_cmp=Newsletter+KU-News&etcc_med=Newsletter*, accessed March 30, 2021

[28] *Vollständig recycelbare Zahnpastatuben* [online], Carl Hanser Verlag, *https://www.kunststoffe.de/a/produktmeldung/vollstaendig-recycelbare-zahnpastatuben-313307*, accessed March 30, 2021

[29] *Nachhaltige Verpackungslösungen für die Kreislaufwirtschaft* [online], Carl Hanser Verlag, *https://www.kunststoffe.de/a/produktmeldung/nachhaltige-verpackungsloesungen-fuer-di-252589*, accessed March 30, 2021

[30] *Recycling & Recyclability* [online], Tetra Pak Group, *https://www.tetrapak.com/en-gb/sustainability/planet/recycling*, accessed March 30, 2021

[31] *Das Märchen vom umweltfreundlichen Getränkekarton* [online], Deutsche Umwelthilfe e.V, *https://www.duh.de/fileadmin/user_upload/download/Projektinformation/Kreislaufwirtschaft/210114_Mythenpapier_Getr%C3%A4nkekarton.pdf*

[32] *Recycling von Getränkekartons in Deutschland 2018* [online], Deutsche Umwelthilfe e. V., *https://www.duh.de/fileadmin/user_upload/download/Projektinformation/Kreislaufwirtschaft/210108_DUH_Recyclingquote_Getr%C3%A4nkekartons_2018.pdf*, accessed March 30, 2021

[33] Markwardt, S., et al.: *Comparative Life Cycle Assessment of Tetra Pak® Carton Packages and Alternative Packaging Systems for Liquid Food on the Nordic Market* [online], Institut für Energie und Umweltforschung Heidelberg (ifeu) (ed.), *https://www.ifeu.de/fileadmin/uploads/oekobilanzen/pdf/LCA_Nordic_final_report_incl_Critical_Review.pdf*, accessed March 30, 2021

[34] Georgiopoulou, I., et al.: Recycling of post-consumer multilayer Tetra Pak® packaging with the selective dissolution-precipitation process [online], *Resources, Conservation and Recycling,* (2021) 165, 10.1016/j.resconrec.2020.105268

[35] *Siegwerk und APK AG machen die Entfärbung beim Recycling von Kunststoff-Folien möglich* [online], *Recyclingfähigkeit von Verpackungsanwendungen*, Carl Hanser Verlag, *https://www.kunststoffe.de/a/produktmeldung/nachhaltige-verpackungsloesungen-fuer-di-252589*, accessed March 30, 2021

[36] Kälin, A.: *Cradle to Cradle Design Innovations. Rethink the way we make things*, CU Online – CU prepared for future, *https://www.youtube.com/watch?v=CSa0zDfH204*, accessed March 30, 2021

[37] ASTM International: ASTM D7611 [online], *Standard Practice for Coding Plastic Manufactured Articles for Resin Identification, https://www.astm.org/COMMIT/d7611.pdf*, accessed March 31, 2021

[38] *Pioneering Digital Watermarks for Smart Packaging Recycling in the EU. AIM, the European Brands Association, launches cross-value chain initiative to drive circular economy goals*, European Brands Association, September 8, 2020, *https://www.aim.be/news/pioneering-digital-watermarks-for-smart-packaging-recycling-in-the-eu/*

[39] *Digital Watermarks Initiative HolyGrail 2.0 – Driven by AIM – European Brands Association and Powered by the Alliance to End Plastic Waste*, European Brands Association, 2022, *https://www.digitalwatermarks.eu/*

[40] *P&G Continues Support of HolyGrail with AIM Test Market*, Procter & Gamble, October 9, 2020, *https://us.pg.com/blogs/HolyGrail/*

[41] Paben, J.: *The Big Issues in Chemical Recycling? They'll Sound Familiar*, Takeaways from a chemical recycling workshop at the Plastics Recycling Conference and Trade Show (Nashville 2020), February 2020, *https://resource-recycling.com/plastics/2020/02/26/the-big-issues-in-chemical-recycling-theyll-sound-familiar/*, accessed March 30, 2021

[42] McAuley, J. W.: Global sustainability and key needs in future automotive design [online], *Environmental Science & Technology*, 37 (2003) 23, pp. 5414–5416, 10.1021/es030521x

[43] Evtimov, I., et al.: Life cycle assessment of electric and conventional cars energy consumption and CO 2 emissions [online], *MATEC Web of Conferences*, (2018) 234, 2007, 10.1051/matecconf/201823402007

[44] Duflou, J. R., et al.: Environmental impact analysis of composite use in car manufacturing [online], *CIRP Annals*, 58 (2009) 1, pp. 9–12, 10.1016/j.cirp.2009.03.077

[45] Soo, V. K., Compston, P., Doolan, M.: Interaction between new car design and recycling impact on life cycle assessment [online], *Procedia CIRP*, (2015) 29, pp. 426–431, 10.1016/j.procir.2015.02.055

[46] Friedrich, H. E. (ed.): Leichtbau in der Fahrzeugtechnik, Springer Fachmedien, Wiesbaden, 2013. 978-3-8348-1467-8

[47] Gerrard, J., Kandlikar, M.: Is European end-of-life vehicle legislation living up to expectations? Assessing the impact of the ELV Directive on 'green' innovation and vehicle recovery [online], *Journal of Cleaner Production*, 15 (2007) 1, pp. 17–27, 10.1016/j.jclepro.2005.06.004

[48] Ferrão, P., Amaral, J.: Assessing the economics of auto recycling activities in relation to European Union Directive on end of life vehicles [online], *Technological Forecasting and Social Change*, 73 (2006) 3, pp. 277–289, 10.1016/j.techfore.2004.03.010

[49] Zhang, H., Chen, M.: Current recycling regulations and technologies for the typical plastic components of end-of-life passenger vehicles: a meaningful lesson for China [online], *Journal of Material Cycles and Waste Management*, 16 (2014) 2, pp. 187–200, 10.1007/s10163-013-0180-3

[50] Froelich, D., et al.: State of the art of plastic sorting and recycling: Feedback to vehicle design [online], *Minerals Engineering*, 20 (2007) 9, pp. 902–912, 10.1016/j.mineng.2007.04.020

[51] Anastas, P. T., Zimmerman, J. B.: Design through the 12 principles of green engineering [online], *Environmental science & technology*, 37 (2003) 5, pp. 94A–101A, 10.1021/es032373g

[52] Tian, J., Chen, M.: Sustainable design for automotive products: dismantling and recycling of end-of-life vehicles [online], *Waste management*, 34 (2014) 2, pp. 458–467, 10.1016/j.wasman.2013.11.005

[53] DAVIDSON TEXTRON INC [US]: Method for recovering PVC and urethane foam from automotive interior trim waste, Inventor: Grimmer, R. A., appl: June 15, 1991, EP. EP0469270 (A2)

[54] Coulter, S., et al.: Designing for Material Separation: Lessons From Automotive Recycling [online], *Journal of Mechanical Design*, 120 (1998) 3, pp. 501–509, 10.1115/1.2829179

[55] Christensen, P.R., et al.: Closed-loop recycling of plastics enabled by dynamic covalent diketoenamine bonds [online], *Nature chemistry*, 11 (2019) 5, pp. 442–448, 10.1038/s41557-019-0249-2

[56] Plastics Europe e.V.: *Stoffstrombild Kunststoffe in Deutschland 2019 - Kurzfassung* [online], *https://www.plasticseurope.org/download_file/force/4020/319*, accessed April 6, 2021

[57] Audi Technology Portal: *Audi Technology Portal - Audi A3 e-tron* [online], AUDI AG (ed.), May 5, 2022, *https://www.audi-technology-portal.de/en/drivetrain/hybrid-technologies/audi-a3-e-tron_en*, accessed May 5, 2022

[58] Hagelueken, C.: *Recycling von Lithium-Ionen Batterien*, November 28, 2019, Themennetzwerk Materialwissenschaft und Werkstofftechnik

[59] Gaines, L.: The future of automotive lithium-ion battery recycling: Charting a sustainable course [online], *Sustainable Materials and Technologies*, (2014) 1–2, pp. 2–7, 10.1016/j.susmat.2014.10.001

[60] Binks, S.: Lead, lithium recycling mix: a clear and present danger, *Batteries International*, (2015) 95, pp. 12–13

[61] Horvath, A., Chester, M.: Environmental Life-cycle Assessment of Passenger Transportation An Energy, Greenhouse Gas, and Criteria Pollutant Inventory of Rail and Air Transportation [online], University of California Transportation Center, Working Papers, 2008, *https://escholarship.org/uc/item/6m5865v5*

[62] Gondaliya, R., Sypeck, D., Feng, Z.: Improving Damage Tolerance of Composite Sandwich Structure Subjected to Low Velocity Impact Loading: Experimental Analysis, 31st American Society for Composites

[63] Timmis, A.J., et al.: Environmental impact assessment of aviation emission reduction through the implementation of composite materials [online], *The International Journal of Life Cycle Assessment*, 20 (2015) 2, pp. 233–243, 10.1007/s11367-014-0824-0

[64] Umweltbundesamt Deutschland: *ProBas - Prozessorientierte Basisdaten für Uweltmanagementsysteme* [online], April 9, 2021, *https://www.probas.umweltbundesamt.de/php/index.php*, accessed April 9, 2021

[65] Asquith, J.: Where Do Planes Go To Die? [online], *Forbes*, October 13, 2019, *https://www.forbes.com/sites/jamesasquith/2019/10/13/where-do-planes-go-to-die/?sh=4204647530fa*, accessed April 9, 2021

[66] Rademacker, T.: *CFK-Recycling – durch Nachhaltigkeit zum Mehrwert*, Virtual Composites Show, *https://www.youtube.com/watch?v=xxbU1A8ioWU*, accessed March 30, 2021

[67] Brühl-Saager, M.: *Reststofffreies, nachhaltiges CF Recycling*, CU Online - CU prepared for future, *https://www.youtube.com/watch?v=L_7T7Uv9OKA*, accessed March 30, 2021

[68] Asmatulu, E., Overcash, M., Twomey, J.: Recycling of Aircraft: State of the Art in 2011 [online], *Journal of Industrial Engineering*, (2013) 1–8, 10.1155/2013/960581

[69] Garcia, M.: Recycling aircraft interiors [online], *Aircraft Interiors International*, August 14, 2019, *https://www.aircraftinteriorsinternational.com/features/recycling-aircraft-interiors.html*, accessed April 9, 2021

[70] Masanet, E., et al.: An assessment and prioritization of "design for recycling" guidelines for plastic components, Conference Record 2002 IEEE International Symposium on Electronics and the Environment (Cat. No.02CH37273), *IEEE*, (2002), pp. 5–10

[71] Peeters, J.R., et al.: Economic and environmental evaluation of design for active disassembly [online], *Journal of Cleaner Production*, (2017) 140, pp. 1182–1193, 10.1016/j.jclepro.2016.10.043

[72] Peeters, J .R., et al.: Forecasting waste compositions: A case study on plastic waste of electronic display housings [online], *Waste management*, (2015) 46, pp. 28-39, 10.1016/j.wasman.2015.09.019

[73] Peeters, J. R., et al.: Elastomer-based fastener development to facilitate rapid disassembly for consumer products [online], *Journal of Cleaner Production*, (2015) 94, pp. 177–186, 10.1016/j.jclepro.2015.01.081

[74] Apeldorn, T., Altstädt, V., Glöde, S.: *Thermoplastik als Leiterplatten-Basismaterial* [online], May 5, 2021, *https://www.elektroniknet.de/elektronikfertigung/leiterplatten/thermoplastik-als-leiterplatten-basismaterial.1555.html*,accessed May 5, 2021

[75] Apeldorn, T.: *Untersuchung Leiterplattensubstrat-spezifischer Eigenschaften von ausdehnungsarmen Compounds auf Basis von Thermoplasten*, Zugl.: Bayreuth, Univ., Diss., 2013, Hamburg: TuTech Innovation, 2013, 978-3-941492-61-5

[76] Culpan, T.: *Where Your iPhone Goes to Die (and Be Reborn)*, February 16, 2016, *https://www.bloomberg.com/news/features/2016-02-16/where-your-iphone-goes-to-die*, accessed April 25, 2021

[77] Jeffrey, C.: *Construction and Demolition Waste Recycling. A Literature Review*, 2011, *https://cdn.dal.ca/content/dam/dalhousie/pdf/dept/sustainability/Final%20C%26D%20literature%20review.pdf*, accessed April 12, 2021

[78] Islam, R., et al.: An empirical study of construction and demolition waste generation and implication of recycling [online], *Waste management*, (2019) 95, pp. 10–21, 10.1016/j.wasman.2019.05.049

[79] Yeheyis, M., et al.: An overview of construction and demolition waste management in Canada: a lifecycle analysis approach to sustainability [online], *Clean Technologies and Environmental Policy*, 15 (2013) 1, pp. 81–91, 10.1007/s10098-012-0481-6

[80] Wurbs, J., et al.: Hexabromcyclododecan (HBCD). Antworten auf häufig gestellte Fragen, 7th ed., 2017, Chemikalien, *https://www.umweltbundesamt.de/sites/default/files/medien/421/publikationen/faq_hbcd_de_17.pdf*, accessed April 26, 2021

[81] Weiss, T.: *Styropor richtig entsorgen, Tipps, Kosten und Risiken - Ratgeber 2021* [online], April 26, 2021, *https://www.entsorgo.de/entsorgung-styropor/*, accessed April 26, 2021

[82] PVCH - Arbeitsgemeinschaft der Schweizer PVC-Industrie. Starke Seiten [online]. Mensch, Welt und PVC. PVCH - ARBEITSGEMEINSCHAFT DER SCHWEIZER PVC-INDUSTRIE, ed. Fruehjahr 2021, *https://pvch.ch/Bilder/Aktuell/Publikationen/Starke-Seiten/Starke-Seiten-Fr%C3%BChjahr-2021.pdf*, accessed April 12, 2021

[83] Janajreh, I., Alshrah, M., Zamzam, S.: Mechanical recycling of PVC plastic waste streams from cable industry: A case study [online], *Sustainable Cities and Society*, (2015) 18, pp. 13–20, 10.1016/j.scs.2015.05.003

[84] World Economic Forum, Ellen MacArthur Foundation, and McKinsey & Company: *The new plastics economy: rethinking the future of plastics*, January 19, 2016, *https://www.ellenmacarthurfoundation.org/publications/the-new-plastics-economy-rethinking-the-future-of-plastics*, accessed April 16, 2021

[85] Beckman, E.: *The world of plastics, in numbers* [online]. THE CONVERSATION TRUST (UK) LIMITED (ed.), July 8, 2019, *https://theconversation.com/the-world-of-plastics-in-numbers-100291*, accessed April 18, 2021

[86] British Plastics Federation: *Oil Consumption* [online], April 18, 2021, *https://www.bpf.co.uk/Press/Oil_Consumption.aspx*, accessed April 18, 2021

PART E

Future Trends and Developments

18 Future Trends

Holger Ruckdäschel, Tobias Standau

The plastics industry is facing a change of paradigm. Sustainability has emerged to a major driver in product and application development, and more importantly, is evolving increasingly to be part of the license to operate for many companies. Consumers are asking for sustainable solutions and thus are igniting the change. Therefore, the industry is strongly committing to sustainability goals and is driving the change to a circular economy of plastics [1]. Adequate recycling strategies are playing a key role at each stage of the value chain. New business and operating models are currently evolving, and the trend will continue.

Reusing plastics waste could also become an important driver for profitability of chemical companies [2]. Today, most of the feedstock is still based on fossil and, to a minor extent, renewable resources (Figure 18.1). Additionally, only a minor proportion has been mechanically recycled, based on the data from 2016.

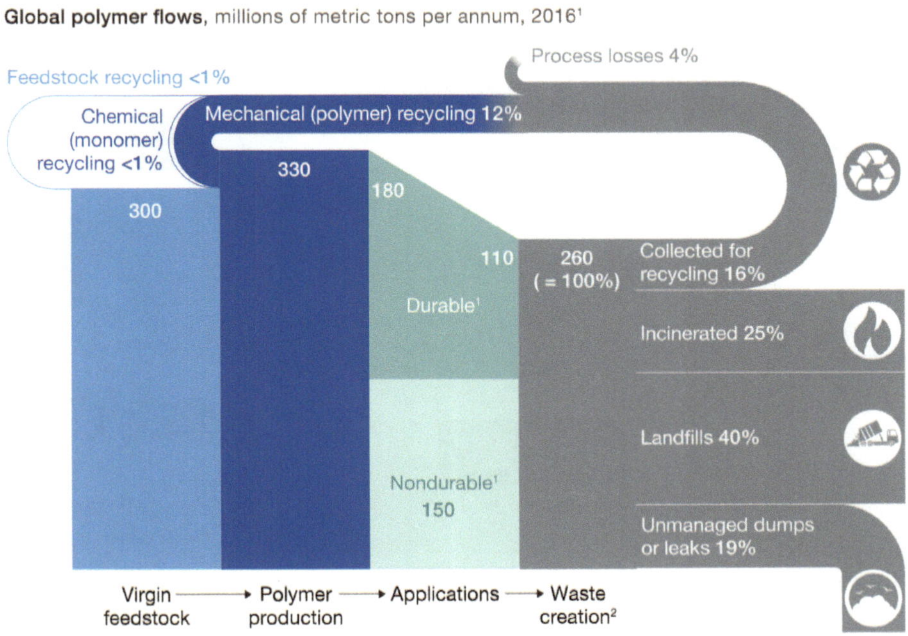

Figure 18.1 Global polymer flows in 2016 [2]

18.1 Thinking in Closed Loops at Every Stage

The circular economy is a model of production and consumption that involves sharing, leasing, reusing, repairing, refurbishing, and recycling existing materials and products as long as possible. Moving from a linear to a circular economy, or, in simple words, from a take-make-use-dispose to a take-make-use-recycle industry poses significant challenges. Four strategic pillars are seen as important elements to support this process: redesign, reduce, reuse, and recycle [3]. For plastics and their applications, all four elements are of importance and interact with each other. Besides purely ecological elements, economic and social impact should not be ignored to ensure a sustainable development [4].

Following the concept of circular economy of plastics, the life cycle of products is extended. In practice, it implies reducing waste to a minimum. Recycling plays a key role here. However, the circular economy should not only be considered at the end of use of the material, but as an integral part of each step from first product idea to recycling to renewed use (Cradle-to-Cradle concept).

Particularly in the early steps of material selection and product design, decisions are taken that impact the overall life cycle. Recycling should therefore not be regarded as isolated but as an integrated approach, targeting circularity of all processes. The next sections will therefore cover the stages of design and material selection, usage, end-of-life, and recycling (Figure 18.2).

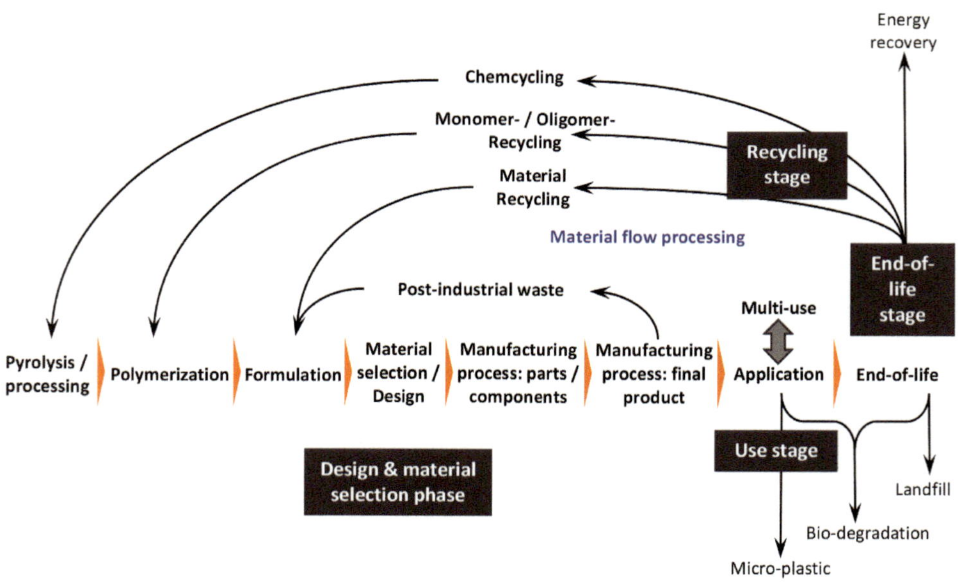

Figure 18.2 Circular economy of plastics and their different stages

■ 18.2 Design and Material Selection Stage

Already at the design and material selection stage, important decisions are taken that further influence life cycle and recyclability of plastic goods. During the last decades, multi-material concepts and integrated designs have dominated the development of new applications, mainly driven by higher performance and lower costs. Examples are manifold: multi-layer food packaging, polymer-coated paper cups, fiber-reinforced resins, powder-coated plastic bumpers, or glued multi-plastic sports shoes, to name just a few. In any of these cases severe challenges are obviously faced at the recycling stage. Particularly for consumer goods and packaging, a trend toward more sustainable design is therefore required from an ecological point of view. Moreover, new regulations further push suppliers, manufacturers, and brands to rethink their current solutions.

In the packaging industry, mono-material solutions are becoming increasingly prominent [5]. One example is toothpaste packaging. Today, most toothpaste tubes are still made of multiple layers of different plastics and a thin aluminum layer, hindering adequate recycling routes. Recently, the company Albéa introduced its Greenleaf technology with all-plastic laminate tubes for oral care and personal care products [6] (Figure 18.3). Such tubes are compatible with HPDE bottle recycling streams and thus enable efficient recycling by entering established recycling routes. The technology has been approved for recycling readiness by North American and European certification bodies such as the Association of Plastic Recyclers (APR). It is claimed that the solution offers a lower environmental impact and thus contributes to sustainability. Brands owners such as Procter & Gamble and Henkel are already using these tubes.

Figure 18.3 Greenleaf technology for mono-material solutions in cosmetics packaging [6]

In another example of the application of recycled materials in the highly demanding cosmetics and body care industry, the company Gerresheimer has demonstrated the use of recycled PET (rPET) from post-industrial and post-consumer waste in refillable plastic cosmetic bottles [7]. Sustainability is not only approached by material selection, but also the shape-optimizing transport. The bottle has straight edges like a cube, offering four sides that are covered with transparent labels. An indentation in the base makes it easy for the angular containers to be stacked on top of each other, saving space and thus reducing CO_2 emissions per shipment.

Mondi, a company specialized in packaging solutions, is cooperating with multiple OEMs to develop sustainable food packaging. For instance, they are working with the retailer REWE and the food manufacturer Hazelager Kaas on cheese packaging, and with Unilever on instant soup packaging [8]. The new solutions are designed to show similar shelf life and can be used as drop-in packaging machines. For the cheese packaging, multilayer films have been replaced by a lighter, fully recyclable polypropylene mono-material packaging solution. Like the previous cases, certification plays a key role to prove the recyclability, and has been passed at the Cyclos-HTP institute.

Printing inks, labels, and closures often are challenging when it comes to recycling. In a five-year joint project, the manufacturer of detergents Werner & Mertz developed with Mondi a completely recyclable pouch which is made of a mono-material (polyethylene) with removable wrap-around label [9]. The patented pouch design solves the problem of recycling printed plastic; 85% of all the packaging material is not printed and can therefore be recycled without any significant loss of quality. The printed label, the remaining 15%, is free of adhesives and bonding agents. It can thus be recycled in a completely separate stream. Recyclability was again confirmed by the Institute Interseroh and Cyclos-HTP.

The previous examples demonstrate that standardization and mono-material solutions can ease recycling in the consumer and food packaging industry. The most prominent and well-established role model is PET bottle recycling. The overall collection rate in Europe was 64% in 2019, outperforming all other packaging materials [10]. Major reasons besides the increased ecological awareness are the excellent standardization, the use of a deposit, as well as extensive return systems in many European countries. rPET shows another difference compared to all previous examples: it is the only material that can be reused in food packaging applications, at least up to contents of 25 wt%. In the case of all other polymers, hurdles in regulation are still ahead, such as allowing higher content of recycled materials and standardized rules per region. Therefore, rPET is also often used for fibers in textile applications.

It is not only packaging industry that is adopting one-material concepts, but also the consumable goods industry. Sports shoes are a prominent example for hard-to-recycle, high-volume applications. Typically, several materials such as polyester yarns, adhesives, thermoplastic foams and others are used to assemble each shoe. Adidas rethought the concept and developed a one-material shoe [11].

Under the name futurecraft.loop, they introduced the first running shoe that is made to be remade. The idea is to close the loop and build new high-performance running shoes based on used ones. In a circular manufacturing model, the raw materials can thus be repurposed again and again. The complete shoe is made from 100% reusable TPU, which is spun to yarn, knitted, molded, and clean-fused to a midsole.

It will be interesting to see how the complexity of plastic goods and the number of polymer compounds will develop in the future. Since the invention and first industrial production of plastics, the diversity of grades has continuously increased, always meeting the specific demand of the respective application. This successful history contrasts with current trends, asking for large-volume material streams that easy to classify and recycle. It is anticipated that three concurrent pathways will exist: specific solutions for demanding applications, easy-to-disassemble multi-material solutions, and standardized or even one-material concepts for mass applications. In summary, the number of polymer grades is expected to decrease over time.

Another pathway to reduce plastic waste in the environment, and to "recycle sustainably", is the use of biopolymers. It should be noted that biodegradable polymers are not the same as bio-based polymers, although both can be classified as biopolymers. Bio-based products such as bio-based polyethylene (bio-PE) derive from renewable resources but can be very stable in natural environments. Here, the target is to efficiently capture carbon dioxide via biological systems and to minimize additional resource-demanding steps in production, thus being carbon neutral or at least being close. At the end of the product lifetime, standard recycling routes must be used due to the identical chemical nature of the materials to their fossil-based analogs.

In contrast, biodegradable plastics are natural or chemically derived materials made from renewable or fossil-based resources targeting complete biodegradability or compostability [12]. The market size of biodegradable plastics was about 2.3 million tons in 2020, and thus far below the total plastics production volume of 368 million tons. Among the bio-based, biodegradable materials, cellulose derivatives, thermoplastic starch, and their blends have the highest production capacity, replacing standard polymers in flexible film packaging mainly [13]. Recently, most attention has been placed on polylactic acid (PLA) due to its properties, which are close to those of polystyrene, its excellent barrier performance, and its low carbon footprint [14]. Moreover, PLA can be blended with a number of different polymers such as poly(butylene adipate-co-terephthalate) (PBAT) to improve its processing characteristics and ductility. All these materials have in common that they can be biodegraded under certain conditions [15] (Figure 18.4). If a suitable biodegradable polymer is used for the respective application, sustainable disposal can simply mean (controlled) biodegradation as an end-of-life option to prevent plastic pollution.

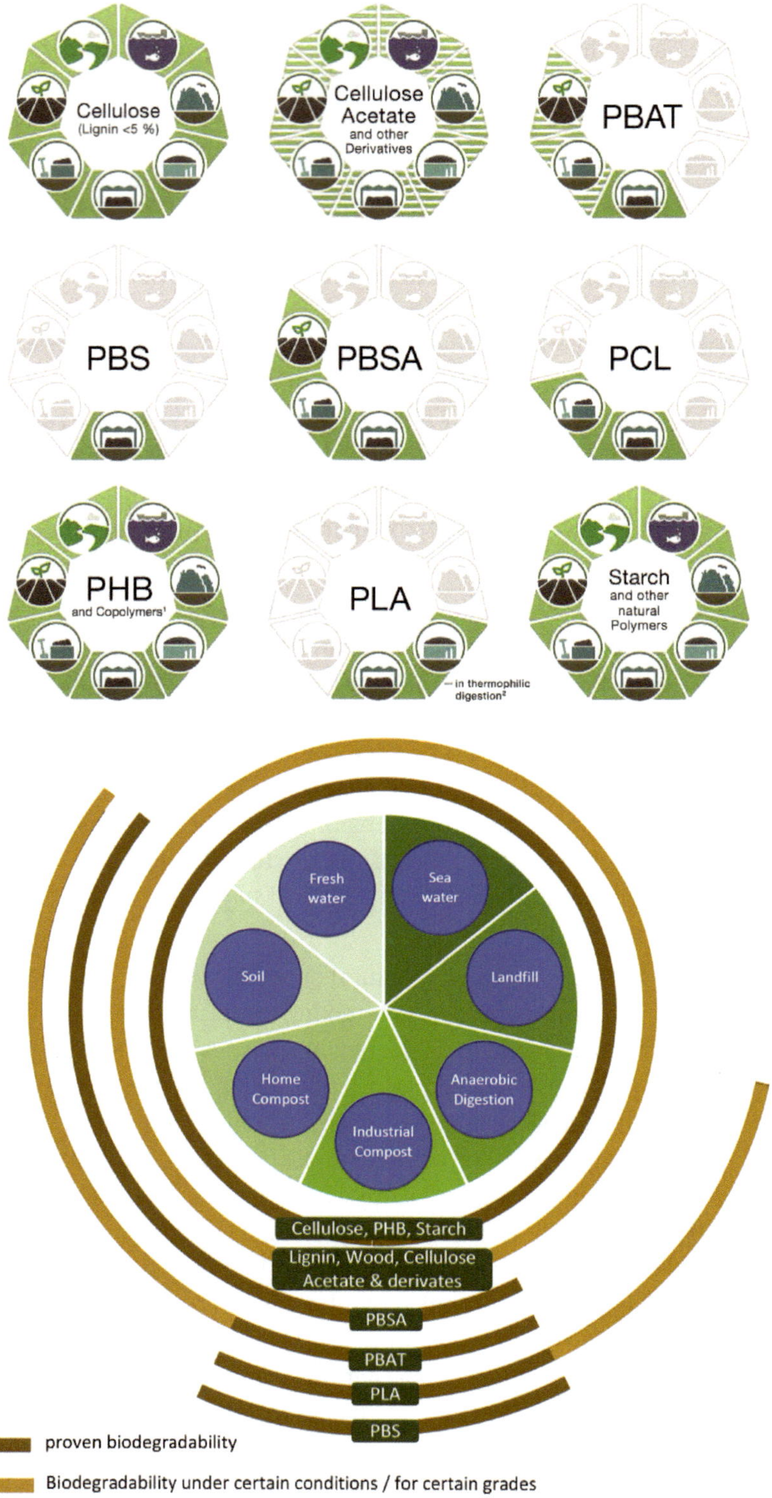

Figure 18.4 Degradation of biopolymers under various conditions [15]

In any of the examples, the complete life cycle of the material should be analyzed to ensure sustainability. Material selection and design for recycling comprise one important element but should not be considered in isolation. Life Cycle Analysis (LCA) provides an appropriate framework to estimate and compare the impact of different solutions. For instance, a material might have a small carbon footprint and be perfect in mechanical recycling but necessitate an enormous logistic setup to return parts after use and require a multiple of the material amount used in application. Like sustainability, recyclability and biodegradability of different material solutions are often understood differently. Consumers, brands, and OEMs demand transparent and simple assessments. Thus, the role of independent institutes, certifying bodies, and objective, data-based tools will gain weight.

18.3 Usage Stage

At first glance, the usage stage has only little influence on the recycling of plastics. However, the impact can be quite severe, as will be shown by several examples.

Following the principle of Reduce-Reuse-Recycle, the multi-use of plastic parts contributes strongly to sustainability. However, reuse has several consequences regarding recycling, notably the following: (i) The volume of plastics to be produced and to be recycled is reduced, at least temporarily. Thus, the production and recycling facilities need to be dimensioned accordingly. (ii) The material stays longer in use. Thus, the material might degrade differently and require a distinct type of recycling.

In several segments, the multi-use of plastic goods is already state-of-the-art. Plastic boxes and pallets used for transporting food or technical goods are widely used in industry. Producers such as Schöller Allibert and Crämer are able to manufacture high-performance boxes and pallets, which withstand harsh conditions and manifold cycles during the lifetime of several years to decades. And the companies using the plastic carriers have established internal logistics systems or follow pooling solutions [16, 17]. Efficiently controlling the flow of the carriers has recently gained enormous attention. Implementation of tags and sensors to observe the flow in real time to efficiently steer the supply chain can further minimize the use of plastics.

The principle of tracking and tracing materials and plastic parts is another interface to recycling. Tracking refers to following the course in real time of, for example, the temperature or position. In contrast, tracing is analyzing the past track that was left behind at any time. Digital tools, sensors, and coding systems (e.g., QR codes) enable following of the product along its complete lifetime, which provides an excellent starting position for subsequent recycling. For instance, tracking can be used in plastic applications to monitor their durability or to compare the use of virgin vs. recycled material.

Tracing can be seen as an enabling technology to select the best recycling method for the material after elapse of the product lifetime, knowing its origin, composition, processing, and use phase. For instance, it might be important to know which additives have been used or whether the material has been recycled once or several times already. Costs and device size of track-and-trace are still prohibitive for its widespread use. But as technology advances, the application shifts from large and expensive to decreasingly small parts. A broader use can thus be expected.

Chemically marking materials with "material-embedded" markers is regarded as an alternative to externally added devices and codes. Several companies are active in this field, as it would enable a simple and straightforward identification on the material level. For instance, BASF is partnering with Security Matters to enable physical and digital tracking of closed loop recycling [18]. The track and trace solution of Security Matters marks physical objects with a unique and unalterable chemical-based barcode and connects them to a digital twin. It is stated that the barcode withstands manufacturing and recycling processes, without altering the appearance or performance of the object. Using proprietary technology, the barcode captures a wide variety of information embedded in the plastic and can be used to authenticate sustainability claims and enhance sorting accuracy.

During the usage stage, another important aspect plays a role for ecological considerations: microplastics. Microplastics are fragments of any type of plastic smaller than 5 mm in length and larger than 100–1000 nm, depending on the definition used [17, 19]. These fragments enter natural ecosystems and thus cause pollution. Microplastics are classified as primary and secondary microplastics. Primary microplastics are already produced smaller than 5 mm, as this size is required in the respective application. In contrast, secondary microplastics are formed by degradation processes of larger plastic products through wear or weathering during or after application. Major contributors to secondary microplastics are tires, textiles, packaging materials, and industrial processes.

As polymers are often designed for longevity and stability under extreme conditions, microplastics accumulate in ecological systems, and are likely ingested as well as incorporated into bodies and tissues of organisms. Several routes are proposed to minimize the impact of microplastics in the environment which also affect recycling strategies of plastics. Besides reactive measures such as filtering, proactive education for collecting and recycling of plastics plays an important role. Thus, the disposal of plastics is reduced. In addition, controlled microbial degradation of microplastics is receiving increasing attention [20–22]. Rates of biodegradation can strongly differ based on the polymer, its pretreatment, and the biotic environment [21].

Most research on enzymatic degradation of synthetic polymers has been performed on poly(ethylene terephthalate) (PET) [23–25] (Figure 18.5). For instance, PET esterase catalyzes the breaking of the polymer chains in PET. Current research targets the development of more efficient enzymes and the transfer to other polymeric systems. Successful transfer of concepts has been shown for polyurethanes, enabled by the similar chemical backbone to PET. In most other cases, including for polystyrene and polyamide, no defined enzymes have been at the level of amino acid or DNA sequences yet. Only recently, it was reported that invertebrates can degrade different plastics [DAN_2019]. These insect perform a kind of mechanical grinding, increasing the surface area, and thus promote further degradation. Similarly, it was shown that mealworms digest polystyrene-based foams, with conversion of 50% of the polystyrene into carbon dioxide within 16 days [26].

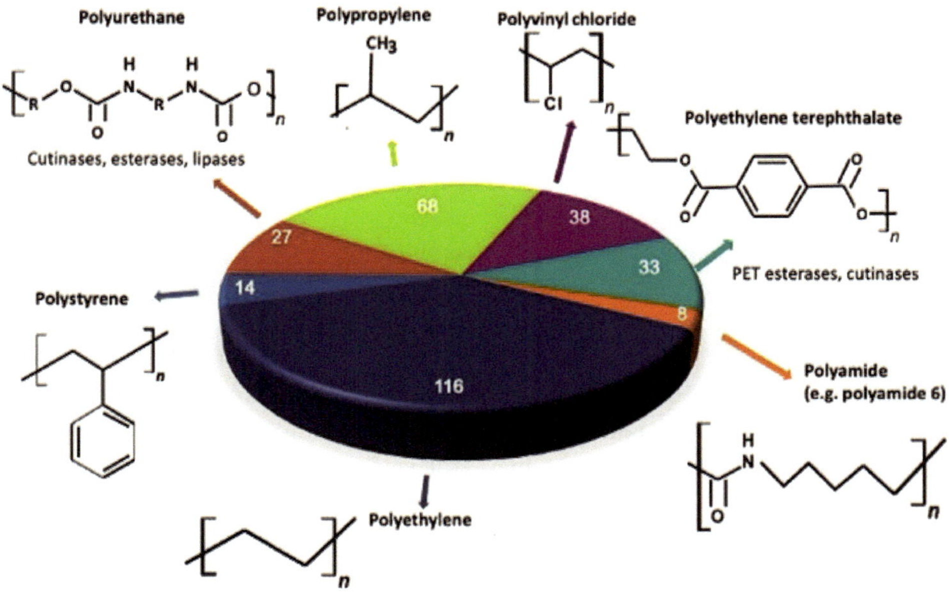

Figure 18.5 Main synthetic polymers: numbers in the pie chart are global annual production (2016) in millions of tons. For PA, PP, PE, PS and PVC, no defined enzymes that act on the polymer have been identified; PET is degraded by PET esterases and cutinases (from *Thermobifida, Ideonella, Bacillus, Pseudomonas,* non-cultivated and others); PUR is degraded by cutinases, esterases, and lipases (from *Comamonas, Pseudomonas, Thermobifida, Bacillus, Fusarium, Candida,* and others) [26]

Microbial degradation can be seen as a potential solution to microplastic pollution. Adding suitable microbes into natural environments could reduce the amount of microplastics therein. However, these enzymes will not only attack microplastics, but also plastic parts in use, and deteriorate their physical-mechanical properties. Premature failure and reduced lifetime of plastics in various applications may result. Therefore, care should be taken when releasing enzymes into non-controlled environments.

As can be seen, the usage stage is strongly linked to the previous material selection and to the subsequent end-of-life stage. Different strategies exist to optimize controlled "recycling" of plastics, starting from track-and-trace technologies to biodegradation of microplastics following uncontrolled release.

18.4 End-of-Life Stage

After use of the plastic article, the end-of-life stage and complexity starts. Relatively few polymers are typically synthesized, converted, and distributed via few production and distribution sites globally. But they are used at millions of places under different environments and for various lifetimes. Efficient sorting, collecting, and standardization measures are logistically extremely challenging. Tolling and collecting systems are a first step, while limited to certain regions and applications.

To promote simplified access to post-consumer and post-industrial plastic waste, several trends can be identified. Besides the already highlighted track-and-trace technologies, digitalization of material information and logistics is further evolving currently. Trading of plastic waste via digital marketplaces is an excellent example of how modern digital technologies support simplification and contribute to sustainability. KraussMaffei, traditionally know as plastic processing equipment manufacturer, has recently launched its marketplace for plastic waste, recyclates, and recompounds [27]. Like other marketplaces, Polymore connects suppliers and customers for plastic waste on their platform. Flow of information and material is simplified, and particularly smaller to medium-sized companies can find adequate materials. Continuous digitalization will enable optimized recycling routes and material flows in future. It is expected that other companies will enter this area and contribute to a connected value and recycling chain. Quality management systems will be established to provide a faster access to reliable plastic waste information, well beyond today's minimum data on size, shape, and major polymer type.

Besides more or less controlled return logistics of plastic waste, the uncontrolled release into the environment is a major issue. Marine plastic litter is being deposited in increasing amounts onto the world's beaches and into the oceans [28]. It originates from land-based and sea-based sources. Depending on composition, density, size, ocean dynamics, and other factors, marine litter can accumulate at different ocean depths and in different regions. Ideally, marine litter is avoided by prevention. While cleanups and collection from the oceans are time-consuming and costly, they are the only way to reduce the amount of plastics already disposed of. Several alliances such as Waste Free Oceans and One Earth One Ocean have been formed to develop technologies and to mobilize fisheries, recyclers, manufacturers, as well as policy makers [29]. Well-known brands have meanwhile developed products made of plastic marine waste, including hats, chairs, backpacks, shoes, and other items [16]. Interestingly, consumers have demonstrated a strong interest and are willing to pay a premium on such products [30].

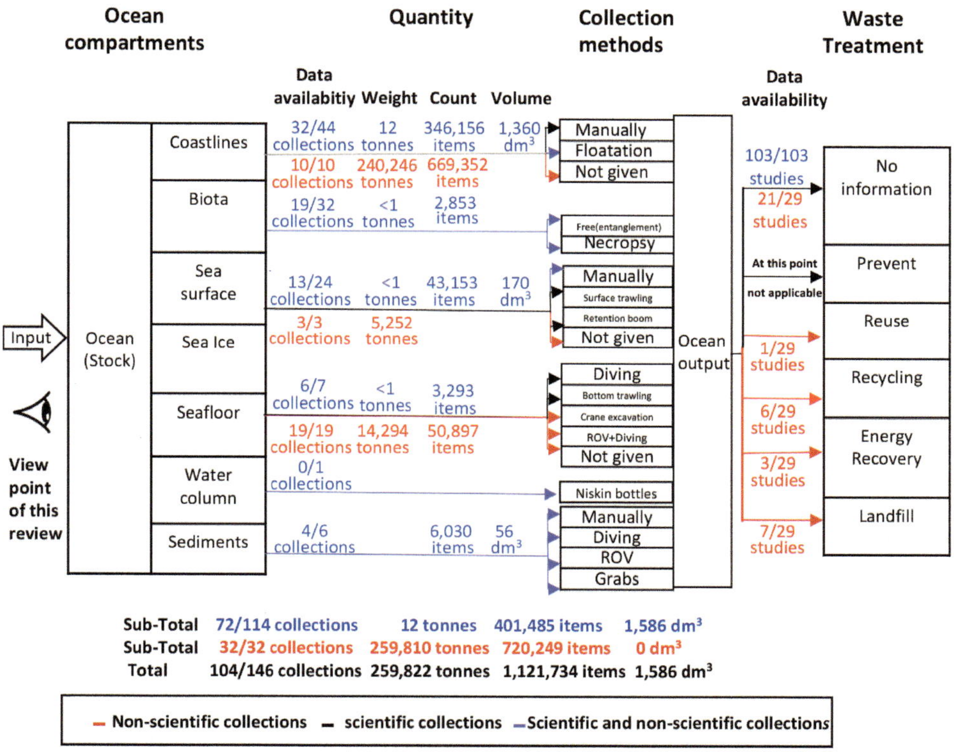

Figure 18.6 Collection of marine litter from the ocean to waste treatment

Collection of marine litter is notably growing but is still at a very low level (Figure 18.6). The overall amount of plastic waste in oceans is still growing. Reliable data on a global level are not available, but first attempts summarizing some activities are available [31]. Only comparing the carbon footprint of products might not be sufficient to judge whether to collect ocean plastics or not. Additional factors such as the impact on marine life and aquatic ecosystems need to be included to provide a conclusive picture [32]. First steps are currently being made to enrich the LCA methodology with littering indicators, and thus quantitatively integrate the impact of ocean plastics [33, 34].

In summary, efficiently collecting plastic waste and adding information on the respective plastic material remains a future challenge. Digital technologies will play a vital role in both areas. From track-and-trace to orchestrating complex return logistics to waste platforms, they will play a key role in future recycling strategies. In addition, a holistic view on the complete life cycle of materials, including, for instance, their risk and impact of ending up as marine litter, will be established.

18.5 Recycling Stage

Various recycling strategies have been discussed in detail in the previous chapters. Here, the main technologies will only be briefly summarized. Perspective will be added to link the technologies to the circular economy aspects highlighted in this chapter. The technology selected and feasibility strongly depend on the plastic. Three major routes are established today: mechanical recycling, monomer or oligomer recycling, and pyrolysis [35].

Mechanical recycling has been used in the plastics industry for decades [36]. Post-consumer or post-industrial waste is re-fed into existing melt-processing routes of polymers such as compounding to generate plastics for new applications. The process is applied to many polymers including the commodities high-density polyethylene (HDPE), polypropylene (PP), and poly(ethylene terephthalate) (PET). The plastic waste starting material might be degraded after its previous use. Therefore, stabilizers or reactive substances are often used to recover the previous performance level.

For instance, the plastic additives division of BASF is promoting stabilizer packages for different polymer grades [37]. The processing and light stabilizers ensure a sufficient durability in manufacture and use. Another more complex modification is performed via reactive extrusion. Here reactive modifiers are added to increase molecular weight by chain extension or branching. The most prominent example concerns PET. Due to the chemical nature of polyester, hydrolysis can readily take place during use and processing. Chain extenders help to counterbalance this effect and recover similar performance levels to virgin material [38].

Mechanical recycling is not feasible for all plastics and may lead to minor downcycling. Nevertheless, it is an already profitable and sustainable route, as the starting material is a polymer already. Thus, extensive investments in monomer as well as polymer synthesis, and required energy and resources, can be saved. It is therefore often regarded as first choice and will find an even more prominent place in future. Mono-material concepts will further contribute to its standing.

Monomer or oligomer recycling is the second route. It is inherently restricted to polymers that allow controlled degradation to their building blocks. Monomer recycling has been proven for condensation-type polymers such as PET and polyamide, and polymers that can be thermally depolymerized such as polystyrene. Oligomeric systems have been observed when cracking polyurethanes to smaller parts [39]. The oligomers are subsequently purified, modified, and reused in polyurethane systems. Like mechanical recycling, the energy- and capital-intense step of monomer synthesis and purification is omitted, making this route attractive from an ecological and economic point of view.

Finally, pyrolysis is often used for reconverting plastic waste into feedstocks for steam cracking. This route is inevitable for all polymers that cannot be recycled via the other routes. The competitiveness of this process depends on oil prices [2]. Large chemical companies are driving this technology, planning and investing in large-scale plants.

In summary, shorter loops are preferred from ecological and economic perspectives. However, not all recycling loops are technically and performance-wise feasible. Therefore, all three technologies will grow and co-exist to ensure the circular economy of plastics.

■ 18.6 Circular Economy and New Business Models

Traditional linear economy models ("take-make-use-dispose") are from an economic, social, and environmental perspective no longer sustainable today. A shift to a circular economy is required but not widely practiced yet. Using an optimistic scenario, 50% of all plastic materials will be collected for recycling and most can be recycled [2] (Figure 18.7). Most of the rest will be either thermally utilized by incineration (31%) or ends up in landfill (18%). It is thus clear that the aspects of recycling and the circular economy are interconnected. Such massive changes in the economy will also result in new business models [40]. Already today, disruptive models are being introduced into the market. For instance, materials and plastic goods, instead of being sold by the producers and OEMs, can be rented out to their customers and consumers, comparable to a subscription model material-as-a-service or part-as-a-service. After the lifetime, the product is returned to its producer and recycled. Concepts are currently being piloted, for example for sports shoes.

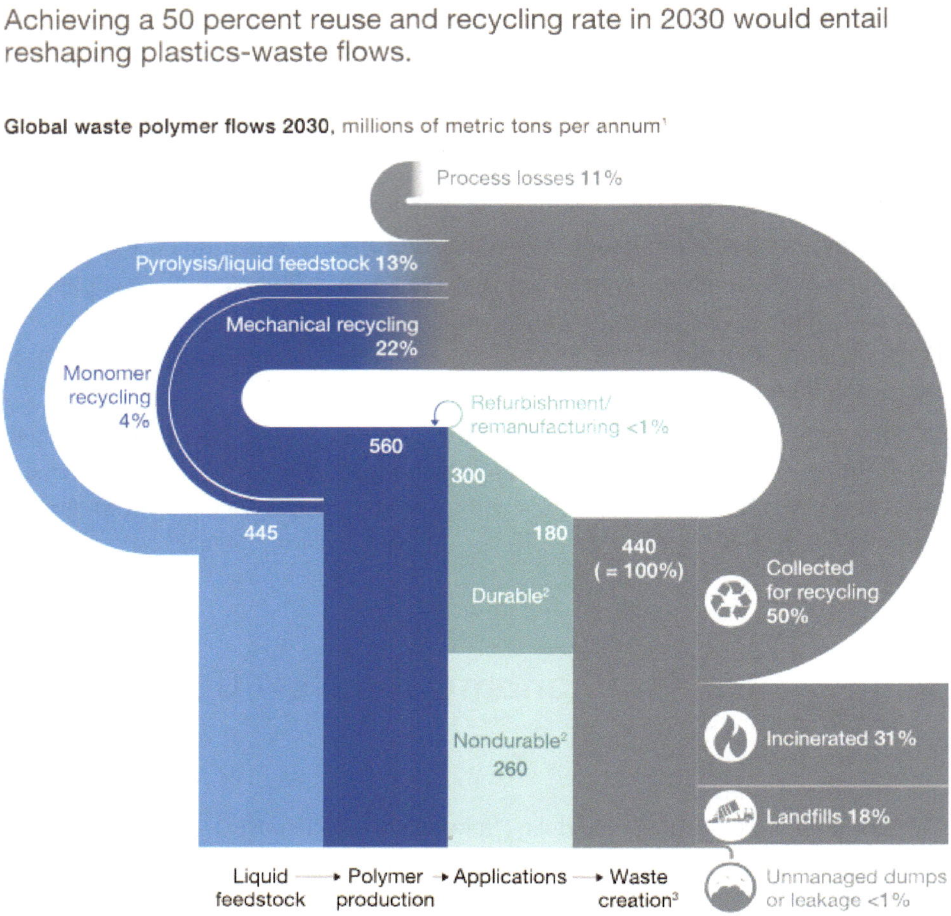

Figure 18.7 Potential re-shaped plastic-waste flows and circular economy of plastics [2]

In order to achieve the goal of a sustainable circular economy of plastics, efforts need to be bundled:

- Collaboration between market players is required, as proven by game theories.
- Designers and engineers must rethink products.
- Education at all levels must be ensured – for experts developing new technologies and solutions, and for the broad public to act sustainably.
- New business models need to be developed to harvest the advantages of circular operating models.
- Investments in technology, research, and assets are required to operate a globally connected supply chain network.

First results are encouraging and will hopefully pave the way into a bright, circular future of plastics.

References for Chapter 18

[1] Bühler, I., in: *ChemManager*, 18 (2021), pp. 1

[2] Hundertmark, T., Mayer, M., McNally, C., Simons, T. J., Witte, C., in: McKinsey: "How plastics waste recycling could transform the chemical industry", 2018

[3] Elisha, O. D.: Moving beyond take-make-dispose to take-make-use for sustainable economy, *International Journal of Scientific Research and Education*, 13 (2020) 3, pp. 497–516

[4] Murray, A., Skene, K., Haynes, K.: The circular economy: An interdisciplinary exploration of the concept and application in a global context, *Journal of Business Ethics*, (2017) 140, pp. 369–380

[5] Middendorf, E., in: *Plastverarbeiter*, (2021), pp. 30–33

[6] *www.albea-group.com*

[7] *www.gerresheimer.com*

[8] *www.mondigroup.com*

[9] *www.werner-mertz.de*

[10] *www.icis.com*

[11] *www.adidas.com*

[12] Wierckx, N., Narancic, T., Eberlein, C., et al.: Plastic Biodegradation: Challenges and Opportunities. In: Consequences of Mircobial Interactions with Hydrocarbons, Oils, and Lipids: Biodegradation and Bioremediation, Springer, Cham, 2018, pp. 1–29

[13] Khan, B., Bilal Khan Niazi, M., Samin, G., Jahan, Z.: Thermplastic Starch: A possible biodegradable food packaging material – A review, *Journal of Food Process Engineering*, (2017)

[14] Vink, E. T. H., Davies, S.: Life Cycle Inventory and Impact Assessment Data for 2014 Ingeo® Polylactide Production, *Industrial Biotechnology*, 11 (2015) 3, pp. 167–180

[15] *www.renewablecarbon.eu*

[16] Schildhouse, J.: *www.rd.com*

[17] Blair Crawford, C., Quinn, B.: Microplastic Pollutants (1st ed.), Elsevier Science, 2016

[18] *https://www.basf.com/cn/en/media/news-releases/global/2020/04/p-20-183.html*

[19] Pinto da Costa, J.: "Nanoplastics in the Environment", in: Harrison, R. M., Hester, R. E. (eds.): Plastics and the Environment, *Issues in Environmental Science and Technology*, London, Royal Society of Chemistry, (2018) 47, p. 85.

[20] Jaiswal, S., Sharma, B., Shukla, P.: Integrated approaches in microbial degradation of plastics, *Environmental Technology and Innovation*, (2020) 17, pp. 100567

[21] Wei, R., Wierckx, N.: Microbial Degradation of Plastics, *Frontiers in Microbiology*, (2021) 12, pp. 1–3

[22] Kale, S. K., et al: Microbial degradation of plastic: a review, *Journal of Biochemical Technology*, 6 (2015) 2, pp. 952–961

[23] Austin, H., et al: Characterization and engineering of a plastic degrading aromatic polyesterase, *Proceedings of the National Academy of Sciences*, 115 (2018) 19, pp. 4350–4357

[24] Yoshida, S.: A bacterium that degrades and assimilates poly(ethylene terephthalate), *Science*, 251 (2016) 6278, pp. 1196–1199

[25] Tournier, V., et al.: An Engineered PET depolymerase to break down and recycle plastic bottles, *Nature*, (2020) 580, pp. 216–219

[26] Danso, D., Chow, J., Streit, W. R.: Plastics: Environmental and biotechnological perspectives on microbial degradation, *Applied and environmental microbiology*, 85 (2019) 19, pp. 1–14

[27] *www.polymore.com*

[28] Williams, A. T., Rangel-Buitrago, N.: Marine litter: solutions for a major environmental problem, *Journal of Coastal Research*, (2019)

[29] *https://www.wastefreeoceans.org*

[30] Magnier, L., Mugge, R., Schoormans, J.: Turning ocean garbage into products – Consumers' evaluations of products made of recycled ocean plastics, *Journal of Cleaner Production*, (2019) 215, pp. 84–98

[31] Schneider, F., Parsons, S., Clift, S., Stolte, A., McManus, M. C.: Collected marine litter – A growing waste challenge, *Marine Pollution Bulletin*, (2019)

[32] Trotter, B., Ramsperger, A. F. R. M., Raab, P., Haberstroh, J., Laforsch, C.: Plastic waste interferes with chemical communication in aquatic ecosystems, *Nature*, (2019)

[33] Civancik-Uslu, D., Puig, R., Hauschild, M., Fullana-i-Palmer, P.: Life cycle assessment of carrier bags and development of a littering indicator, *Science of the Total Environment*, (2019) 685, pp. 621–630

[34] De Sadeleer, I., Askhma, C., Baxter, J., Stensgard, A.: Integration of plastic littering in LCA methodology and eco-design tips for the avoidance of littering, Norwegian Institute for Sustainability Research, 2021

[35] Ragaert, K., Delva, L., Van Geem, K.: Mechanical and chemical recycling of solid plastic waste, *Waste Management*, (2017) 69, pp. 24–58

[36] Schyns, Z. O. G., Shaver, M. P.: Mechanical recycling of packaging plastics: A review, *Macromolecular rapid communications*, (2021) 42

[37] *www.basf.com*

[38] Standau, T., et al.: A review on multifunctional epoxy-based Joncryl chain extended thermoplastics, *Polymer Reviews*, (2021) 62, pp. 296-350

[39] Nikje, M. M. A., Garmarudi, A. B., Idris A. B. G.: Polyurethane Waste Reduction and Recycling: From Bench to Pilot Scales, *Designed Monomers and Polymers*, (2011) 14, pp. 395–421

[40] Fogarassy, C., Finger, D.: Theoretical and practical approaches of circular economy for business models and technological solutions, Resources, 9 (2020) 76, pp. 1–9

19 Future Developments

Achim Schmiemann, Eric Homey

In this chapter, the importance of plastics recycling for the future, innovations along the thermoplastic materials cycle, and an outlook are provided.

A look at the past shows the enormous task that will have to be solved by recycling plastics in the future. From the Figure 19.1, it can be seen that of the approximately 8.3 billion tons of plastic that have been produced since the beginning of the plastics era, about 60% is in landfills and in the environment [1].

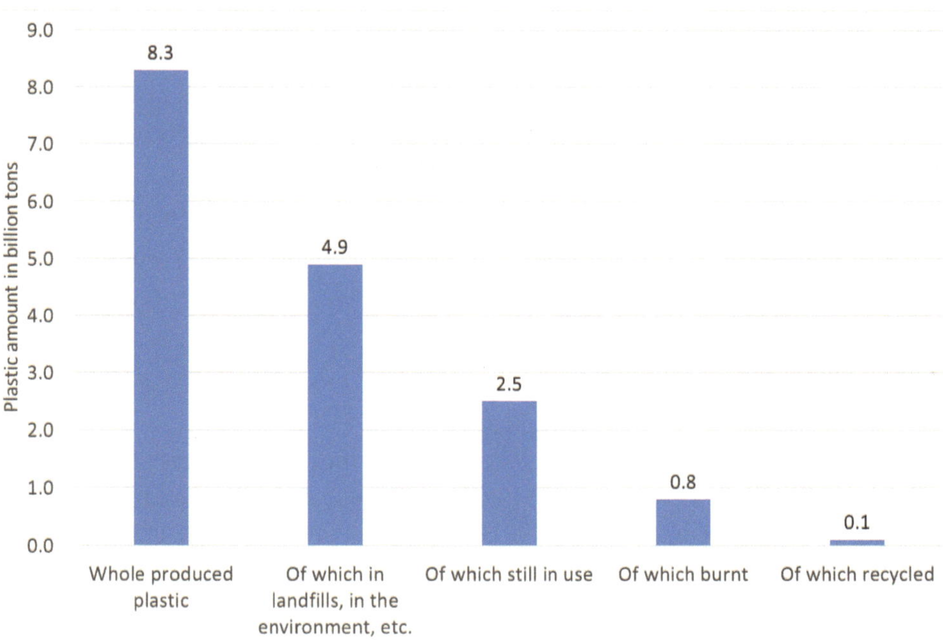

Figure 19.1 Recycling of plastics produced worldwide between 1950 and 2015

This is more than 13 times the global plastic production in 2019 of 368 million tons or, in other words, almost 1900 times the volume of the Cheops pyramid (assuming a density of 1 t/m^3 plastic) [1, 2]. This fact and the further increasing amount of plastic waste result on the one hand in a great opportunity for the recycling industry and, on the other, in a high degree of environmental pollution. In order to take advantage of this opportunity for the recycling industry, the following demands are formulated by the German Federal Association for Secondary Raw Materials and Waste Management (bvse) [3]:

- improvement of collection through consistent separation
- reorganization of plastics sorting
- expansion and promotion of existing plastics recycling
- consistent enforcement of recycling-friendly product design
- promotion of plastic recycling
- priority of waste law over substance law
- information to stakeholders about plastics recycling

Implementing the above requirements would improve enormously the handling of future plastics and those currently in use, but would not overcome the challenge with the plastic waste of the past. Expanding and promoting existing plastics recycling will not be enough. A new industry to break down waste plastic should emerge. Numerous projects for recycling plastics and cleaning up the world's oceans and inland waters are leading the way. Energy generation through thermal recycling must and will also play a major role in this process.

For thermoplastics, there are a variety of opportunities for sustainable value creation in new products (Figure 19.2). Overall, recycling of plastics can be divided into three broad areas:

- the product shape is retained,
- the polymer is preserved, or
- the polymer is depolymerized.

The larger the "circle" shown in Figure 19.2, the more the polymer chemistry is affected. If the first two areas are also known as reuse or further use and mechanical or material recycling, the third area means depolymerization of the polymer chains. Produced therefrom are either oligomers or monomers, or pyrolysis oil, with components such as carbon monoxide, carbon dioxide, or low-molecular hydrocarbons.

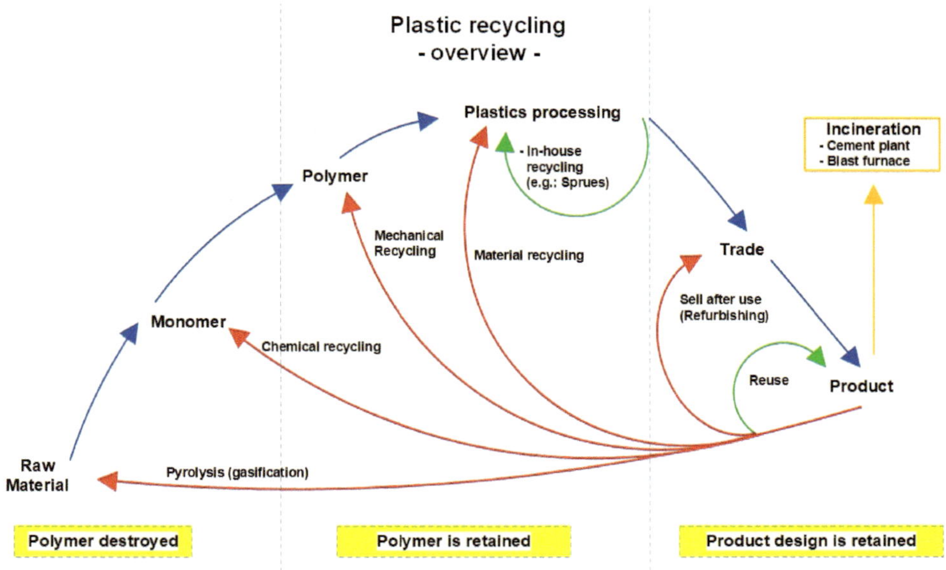

Figure 19.2 Recyclable materials cycle of thermoplastics

In the wake of growing environmental awareness, there is an increasing demand for packaging made of recycled plastic, for example. Political framework conditions also demand high recycling rates. The price of virgin materials is linked to the oil price. As a result, the price of recycled materials may well be higher than the price of virgin materials in the future if the price of oil is low.

In order to achieve high quotas more easily, some conditions are required, such as a better understanding of the influence of recyclates on the quality of the end products.

19.1 The Thinning Effect

Resource-conserving handling of recyclable materials begins with synthesis, preparation, and forming into semi-finished products and articles for daily use. During injection molding, for example, waste can be produced in the form of sprues, especially in small and medium-sized series. This is because cost-intensive hot runner molds are often not worthwhile. The sprues can be shredded and then returned to the production process. In addition, high-priced thermoplastics are often processed, so that production waste recycling would be worthwhile for economic reasons alone.

If relatively small components are produced in the process, the proportion of production waste thus generated is high in relation to the mass of saleable products. There are always reservations about recycling. However, reservations based on a presumed deterioration of the material at 25 or more mass % recycled material are scientifically unfounded in many cases. In an extreme example, 90% production waste is taken to occur, i. e. 90% is recycled at each production cycle. In other words, only 10% virgin material is used each time. In the second cycle, 0.9 × 90% = 81% has already passed through the melting process twice. In the third cycle, 72% of the material has already been melted three times. The question arises as to how often a polymer molecule passes through the melting process on average in "24/7" production. This certainly requires a limit value consideration. To determine an average value, a sum is formed from the reflow rate *p* multiplied by the number of cycles, and divided by accumulated rate *p*. A limit value consideration follows:

$$\frac{\sum_{i=1}^{n}\left(p^{i}\cdot n\right)}{\sum_{i=1}^{n}p^{i}}=\frac{n\cdot p^{n}}{p^{n}-1}-\frac{1}{p-1}$$

$$\lim_{n\to\infty}\left(\frac{n\cdot p^{n}}{p^{n}-1}-\frac{1}{p-1}\right)=\frac{1}{1-p}\quad\text{for } 0<p<1$$

$$\text{with }\lim_{n\to\infty}\left(\frac{n\cdot p^{n}}{p^{n}-1}\right)=1\text{ because }\left(n\cdot p^{n}\right)\to 0\text{ and }\left(p^{n}-1\right)\to -1\text{ for } 0<p<1$$

The result is 1/(1 – *p*). For the thought example with *p* = 0.9, this results in a limit value of 10. In this extreme example, the "average" polymer molecule has gone through the melt cycle 10 times. This result can also be interpreted to mean that at this high rate of return, the polymer must be stabilized enough to go through 10 melt cycles without noticeable damage. In the case of a return rate or recycling rate of 50%, the result is a limit value of 1/(1 – 0.5) = 2. In this case, a plastics processor can therefore assume that his/her product behaves like a component that has been completely recycled once. Of course, it must be assumed that no further influences, such as impurities introduced from outside, are introduced into the process. With the help of these outlined considerations, the use of production waste can become more controllable now and in the future. Even in the case of small components with a disproportionately high share of sprues or production aids that can only be used once, it is possible to estimate how dilution with production waste will affect the end product.

19.2 Prediction of Properties

Following an idea of Boydell [4], a measurement of the properties of thermoplastics in the as-received condition as well as after one, two, and three processing steps enables a prediction of the properties in the end product. The focus here is also on production waste.

With a realistic recycling rate of p = 25%, the proportions of the respective batches after 0 to 10 production cycles are given in Figure 19.3.

Number of cycles	0	1	2	3	4	5	6	7	8	9	10
0	1	0.75	0.75	0.75	0.75	0.75	0.75	0.75	0.75	0.75	0.75
1		0.25	0.188	0.188	0.188	0.188	0.188	0.188	0.188	0.188	0.188
2			0.063	0.047	0.047	0.047	0.047	0.047	0.047	0.047	0.047
3				0.016	0.012	0.012	0.012	0.012	0.012	0.012	0.012
4					3.91E-03	2.93E-03	2.93E-03	2.93E-03	2.93E-03	2.93E-03	2.93E-03
5						9.77E-04	7.32E-04	7.32E-04	7.32E-04	7.32E-04	7.32E-04
6							2.44E-04	1.83E-04	1.83E-04	1.83E-04	1.83E-04
7								6.10E-05	4.58E-05	4.58E-05	4.58E-05
8									1.53E-05	1.14E-05	1.14E-05
9										3.81E-06	2.86E-06
10											9.54E-07
	1	1	1	1	1	1	1	1	1	1	1

Figure 19.3 Composition of recycling stages 1 to 10 with a recycling rate p = 25%

For simplification, it is assumed that low mass fractions of <0.4% do not really adversely affect the molding material. For the further considerations, it is therefore assumed that the (circled) composition in Figure 19.3 represents the properties of the component with recycled production residues. The following formula can then be used to determine all possible properties as:

$$a = (1-p)a_0 + p(1-p)a_1 + p^2(1-p)a_2 + p^3(1-p)a_3$$

The individual terms can be easily explained; it is a geometric series with four terms according to Table 19.1.

Table 19.1 Calculation of the Proportions for Each Recycling Step

$(1-p)$	Percentage of material processed once
$p(1-p)$	Percentage of material processed twice
$p^2(1-p)$	Percentage of material processed three times
$p^3(1-p)$	Percentage of material processed four times

Figure 19.4 shows results on glass fiber reinforced polyamide, which can confirm the applicability of this simple calculation

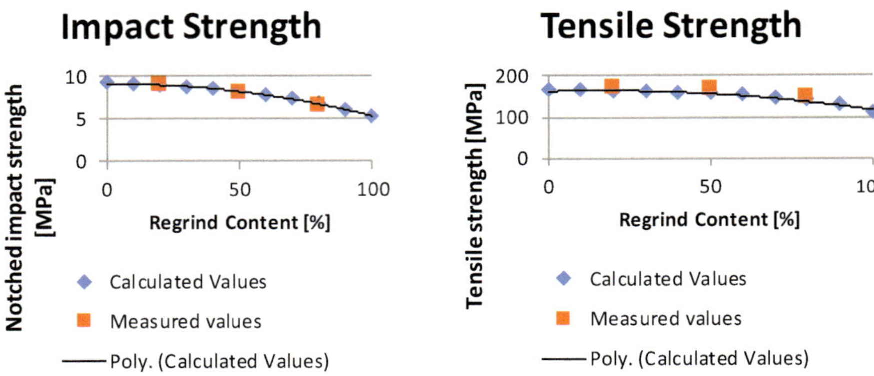

Figure 19.4 Measured and calculated property values of recycled PA-66 GF30 (dry) with different regrind contents

Further practical tests with flame-retardant polyamide for fuse carriers (PA-66 GF25 FR) have confirmed the practical applicability of this approach. In addition to the mechanical properties, it was also possible to demonstrate that the flame retardancy is not impaired by the red phosphorus used in the process and that an addition of up to 50% by mass consisting of faulty batches and sprues to the virgin material does not impair the fuse carriers in the required quality [5].

19.3 Digitization in Processing Technology

The largest market for plastics is packaging. In most cases, packaging is a short-lived product. This means that plastics from packaging, from so-called lightweight packaging (LWP), become waste comparatively quickly and end up in the waste stream. In order to dispose of these waste streams in an environmentally sound manner, it is important to recycle this light packaging as much as possible. In this context, the sorting of LWP mixtures, in addition to the separation of films (mixture of PP and PE), is focused on the plastic types PE, PP, and PS as standard. The PET fractions in a LWP mixture are not qualitatively comparable with the PET fraction from deposit bottle systems. So-called mixed PET is only sorted out on an as-needed basis if there is a sales market for highly mixed PET fractions. Packaging made of expanded polystyrene (PS-E) is also not separated as standard. In principle, PS-E is well suited for mechanical recycling, but the material from an LWP collection mixture has a high degree of contamination.

The product specification - white and uncontaminated (purity of >97%) - cannot usually be met.

The standard treatment processes for such material streams are shown in Figure 19.5.

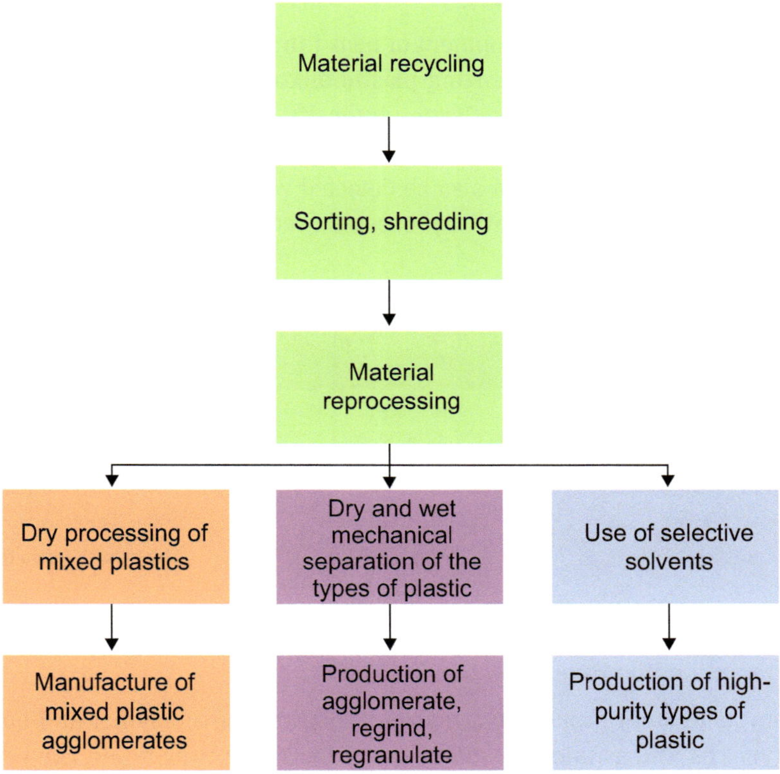

Figure 19.5 Typical processes for the material recycling processing of plastics [6]

As a rule, the middle path (highlighted in purple) is followed in practice. In the processing of plastics from the LWP sorting into agglomerates or regranulates that can be used in high quality, wet mechanical processing is carried out as standard. After dry pre-shredding, metal removal and, if necessary, subsequent wet shredding and washing (separation of labels, sand, and other impurities), plastics of different densities are separated from each other in the wet medium. Centrifuges, hydrocyclones, and floating-sink tanks are mainly used for this purpose. Density separation in the wet medium has proved particularly successful for bulk plastics (PP, PE, PS, PVC) and has been used for more than 30 years [7]. However, due to overlapping density ranges, polyolefins cannot be separated from each other.

Digitization essentially means a yes-no decision. In order to be able to use digitization in recycling or in reprocessing processes for used plastics, it is necessary to separate and detect the separated parts or pieces. With suitable sensors, a decision can then be made as to whether the particle is to be processed or discarded. Further developments then represent an automation of the picking process (combination of “detection” and “removal”).

In particular, the further development of near infrared (NIR) sorting brings significant progress in the field of sorting plastic mixtures. The NIR technique makes use of the material-dependent reflection intensity at different wavelengths. In this way, different types of plastic can be separated from a mixture, depending on the objective. However, plastics can also be detected with visible light. Table 19.2 provides an overview of the separation processes used in plastics recycling.

Table 19.2 Different Separation Technologies

Separation Technology	Wavelength [nm]	Material Property
NIR spectroscopy	780–3000	Absorption bands
VIS spectroscopy	400–780	Color
CCD color camera	400–1000	Color, contour
Laser & camera	200–1000	Color, fluorescence
XRF spectroscopy	0.0001–1	Chemical composition

Despite the many possibilities of plastic detection, there are other challenges in plastic recycling that have not yet been solved:

- PP recyclates from packaging are always mixtures of copolymer and homopolymer types as well as injection molding and thermoforming types. Even PE-HD or HDPE, which is used in the packaging sector almost exclusively for extrusion blow-molded bottles, can be divided into different types. These different types differ in their molecular weight distributions. These differences characterize materials that are easily miscible and can be used together. However, a separation would expand potential applications and sales markets for the corresponding recyclates and mean an increase in value creation.
- There are cycle-restricting features that have a negative impact on the properties of the recyclates produced from it, such as:
  - composite film components that are not or not well miscible (compatible) with the main material
  - firmly adhering adhesives and labels
  - compounded-in secondary ingredients (to create special functions such as barrier properties)
  - blends of black plastics that cannot be separated in density separation

- There may also be an interest in being able to separate plastics according to their type of use or application, e.g. food/non-food, silicone cartridges.

19.3.1 Tracer-Based Sorting

Mechanical recycling of packaging plastics could be increased by a sorting technology that can efficiently separate plastics with the smallest differences or the same plastics for different applications (food and non-food). Tracer-based sorting (TBS) is a possible approach to this. It is based on special substances that fluoresce under certain conditions and are used as markers (in very small quantities) to identify plastic packaging. Another advantage of this approach is that in everyday use neither the sun nor other light sources are intense enough to produce the visible fluorescence.

For tracer-based sorting, the relevant range of fluorescence is that which can be excited cheaply and quickly with electromagnetic radiation (for example, with laser diodes, LEDs) and detected non-destructively with simple detectors (such as camera chips). Excitation and detection can take place non-destructively in milliseconds. The marking is invisible in everyday life and only fluoresces when specifically excited (by light in the non-visible range). The phosphors used are materials with a high efficiency between excitation and emission (quantum yield), so that even traces (1–100 ppm) are sufficient to detect characteristic fluorescence.

In addition to the "classical" Stokes fluorescence (excitation wavelength is smaller than emission wavelength) with quantum yields in the two-digit percentage range, there is also the so-called anti-Stokes fluorescence (upconversion fluorescence). Here, two or more photons are collected before a higher-energy photon is emitted. This makes it possible, for example, to excite in the IR range and emit and detect in the visible range. Although only quantum yields in the single-digit percentage range are achieved, the signal-to-noise ratio is generally much better than with Stokes fluorescence. The reason is that the excitation radiation can only produce visible radiation via the anti-Stokes effect. Otherwise, there is no possible physical process that can produce visible light. Consequently, there is no physically induced background noise to the upconversion fluorescence and thus there is a very good signal-to-noise ratio. In the past decades, numerous inorganic phosphors have been created, usually consisting of a crystalline host lattice, the dopants (activators, sensitizers), as well as defects and impurities, and produced via complex syntheses. Fluorescence is due to the activators, which dissipate energy through electronic transitions. Sensitizers support the energy absorption process through suitable energy transitions. In order to achieve high quantum yields, defects and impurities must also be kept to a minimum. Specific, comparatively wavelength-sharp emissions can be generated via the dopants and the formulation. With a portfolio of five different emissions, presumably at the same excitation wavelength, 31 ($2^5 - 1$) distinguishable codes can already be generated combinatorially.

19.3.2 Black-Black Sorting

A decisive breakthrough in separation technology came with the introduction of sensor-based sorting, which is characterized by contactless detection of material properties. This processing method was able to benefit from the rapid development in the field of sensor technology and digital electronic data processing.

Carbon black (and also dirt) give plastics their dark color. This absorbs a large part of the signal and so the particles can be poorly seen by the optical system. To close this gap in the recycling loop, work is being done on sensor-based sorting machines that also open up new separation features. Here, analysis methods such as X-ray transmission measurement or intelligent image processing and evaluation are successfully applied, see e.g. [8]. The Fraunhofer FHR and Fraunhofer IOSB institutes in Germany are working together on a solution with a concept consisting of a combination of optical sensor technology and a terahertz camera. The aim here is to develop a system that can be used in the sorting of plastic particles with a high separation quality and capacity, with sorting of >98% purity [9]. STEINERT presented the "BlackEye" in 2016. This development is based on a combination of mid-infrared technology and hyperspectral imaging [10]. A process variant with laser excitation (laser spectroscopy) was developed by UNISENSOR and used on a trial basis in the sorting of shredder residues [11].

For the quality of the detection and sorting, it is ultimately decisive whether it is possible to separate the feed material before it passes through the measuring range of the sensor system, for example on a belt conveyor running at a speed of approx. 3 m/s, in such a way that a monolayer is created. Individual particles must not lie on top of each other and should not overlap. This is usually achieved by a cascade of belt conveyors and/or inclined chutes that ensure an increasing transport speed. Positive as well as negative sorting can be carried out with automated individual grain separation. Positive sorting means that valuable substances are enriched in the discharged product, whereas in negative sorting, unwanted components are selectively depleted [12].

19.3.3 MFI Sorting

A simple method for characterizing the melt or processing viscosity is the determination of the "melt flow resistance" or "melt volume resistance", for example, according to DIN EN ISO 1133. In this method, molten polymer is forced at a defined pressure through a standardized nozzle at a specific temperature and the flow time is measured. Sorting e. g. polyethylenes from household or commercial waste into two fractions, namely an easy-flow fraction with an MFI value of e. g. >0.6 g/10 min and a second fraction with a flow value of clearly <0.6 g/10 min would be of considerable advantage for the subsequent marketing of the recyclate. If such fractionation is successful, it becomes possible to split grade-cleaned fractions into subgroups such as injection molded and extruded goods. The added value is then much greater.

Due to constant further developments in spectroscopic sorting methods and the associated software architectures (keyword "artificial intelligence"), it is conceivable that sorting into (at least) two MFR classes can succeed. NIR spectra of PE with different melt viscosities prove this (Figure 19.6).

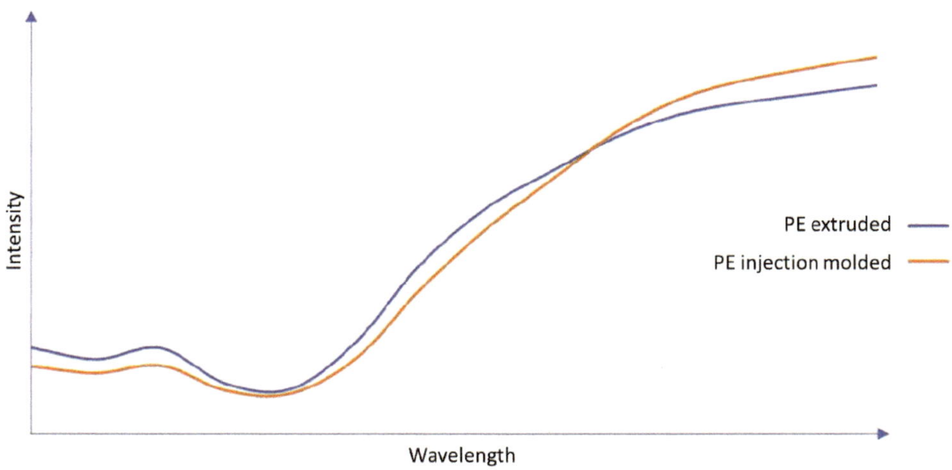

Figure 19.6 NIR spectra of extruded (blue) and injection molded (orange) products

After learning phases, sorting could succeed with sufficient yield. This technique should be supported by optimized input control. This means that with the help of a rough pre-sorting, an enrichment in one or the other direction should already be prepared.

■ 19.4 3D Printing

Plastic recycling at the highest level is the endeavor to consistently close material cycles. This has been achieved, for example, with the "bottle-to-bottle" recycling of PET bottles. But it is not always possible to put such high-value recycling into practice. That is why innovative ideas are needed now and in the future to find recyclate applications with the greatest possible potential for added value. In the case of new product ideas, it is in any case advisable to proceed in a resource-conserving manner and to at least consider the use of recycled materials. A new and promising technology is additive manufacturing, commonly referred to as 3D printing. Here, a new industrial sector is growing - alongside the classic manufacturing processes - at the same time as an increasing awareness of the use of recyclates.

Due to the large amount of existing waste plastics, it seems to make sense to generate products for which large quantities of plastic are needed and which enable long-term sustainable use. In contrast to classic plastics processing, with many size and geometry limitations of the components, 3D printing offer distinctive possibilities. In 3D printing, the size of the resulting products is limited by the installation space of the printer. The world's largest injection molding machine, according to the manufacturer, can handle shot weights of up to 140 kg, while 3D printing technology could be used to print entire houses. Transferring the technology from the construction sector to the plastics industry will lead to the printing of building insulation, truck transport containers, or blades for wind turbines in the future, for example.

3D printed structures with overhang need a temporary support structure during the printing process. This can best be done with materials that can be recirculated. This makes it possible to crush and reuse support material once it has been used.

The most important requirement for building insulation is a high insulation value. Plastics provide this anyway. By foaming thermoplastics, this can be increased considerably, which has led to the triumph of expanded polystyrene as a sheet material in this application. Now it seems possible to scan a building front and use this data set to print a building insulation with a precise fit using "plastic foam". In this way, a sink for plastic waste and its sorted storage can be created with a simultaneous utilization effect. With the various techniques of 3D printing, products can be made from different materials using different processes. In the fused layer process (FFF for Fused Filament Fabrication), plasticizable filaments are melted in a nozzle (Figure 19.7). The emerging melt can be deposited in three spatial directions, corresponding to the x, y, and z axes.

Figure 19.7
Different colored filaments for FFF 3D printing

Instead of filaments, granulates can also be plasticized, but then a set-up with a conveying mechanism, e. g. a screw inside the nozzle, is necessary. Continuous fiber reinforcement is also possible with this process and there is a wide range of printers, including larger machines with installation spaces of several cubic metres.

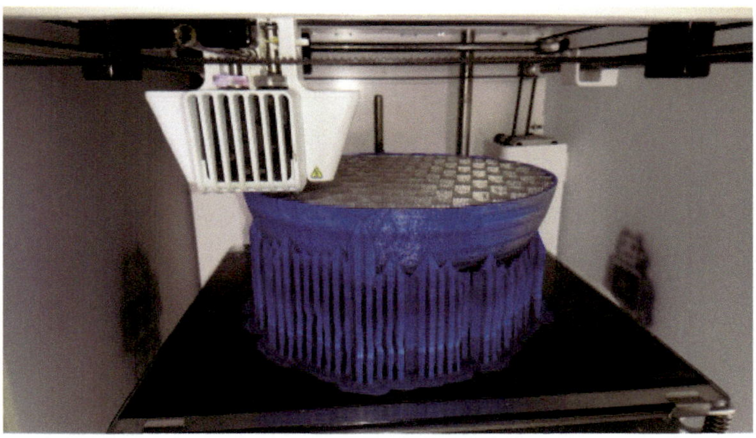

Figure 19.8 Printing a "moon model"; view into the printing room

In addition to the FFF process, resin-based (stereolithography and PolyJet processes) or powder-based (selective laser sintering) technologies have become established for plastics. These processes are very limited in terms of material selection and cycle management. The resin processes use mixtures of monomers and photoinitiators that enable local polymerization - stimulated by specific wavelengths. The products are cross-linked plastics that cannot be repeatedly plasticized and thus do not permit remelting. In the laser sintering process, the plastic powder is heated to just below the melt state in the entire construction space and the required energy for sintering is introduced with a laser. The energy input is high and the respirable powders pose a health hazard. In addition to PA, PP and TPU can now also be processed with this method.

In an initial screening, various thermoplastic production wastes were collected and processed according to type. It quickly became apparent that it is possible to use a number of different, unmixed thermoplastic wastes as 3D printing material after a few process steps. In a second step, compounds were printed. In this process, selected fractions from construction waste were introduced into plastics as fillers, processed into filaments, and then printed. In a third step, post-consumer recyclates were also used as matrix materials.

In fact, various printable recycled filaments are available on the market. For example, recycled filaments are available from production waste and PET (bottles) and from PLA, PS, or ABS. In the case of PET, one roll of filament (750 g) is produced from about 22 bottles. The process chain from waste to print is shown in Figure 19.9 using ABS as an example. It includes the following steps: shredding, extrusion, regranulation, drying, filament extrusion, and 3D printing.

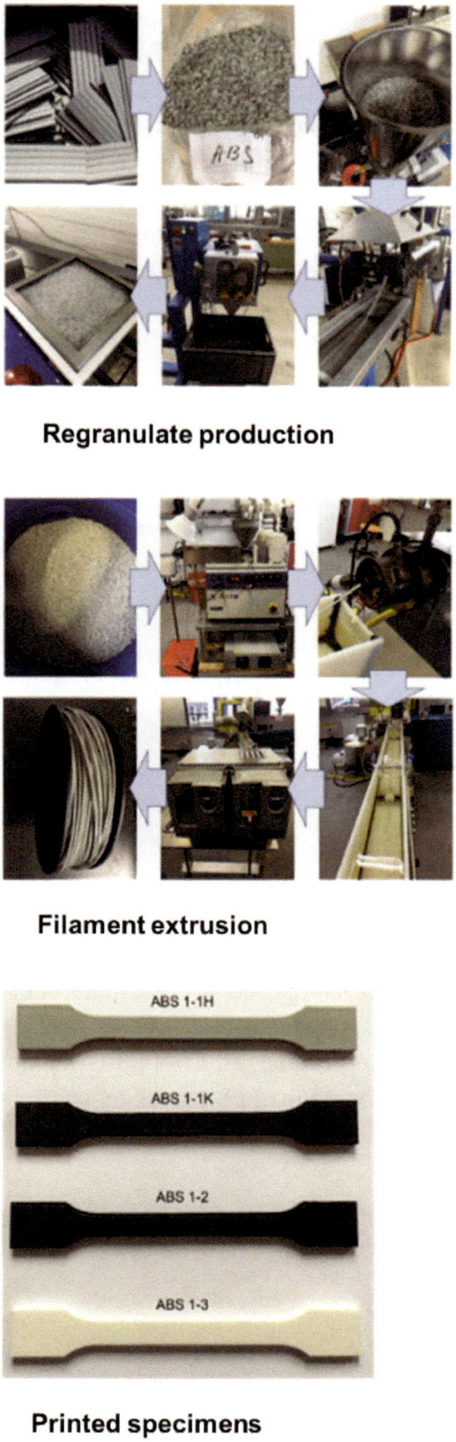

Figure 19.9
Process chain from ABS profiles to 3D printed test specimens

Following this process chain, a variety of recycled filaments were produced and printed, including PLA, ABS, ASA, PEEK, PEI, TPU, PS, PC, and PET. Three processed plastics will be examined in more detail here: ABS, PS, and PET. The characteristic values of the filaments and printed test specimens from the tear test provide information about the areas of application of the printed structures. To simplify matters, the characteristic values of filaments are considered here that were also printed afterwards. The tests were carried out according to DIN EN ISO 527-1 with a clamping length of 100 mm and a test speed of 50 mm/min, see Figure 19.10.

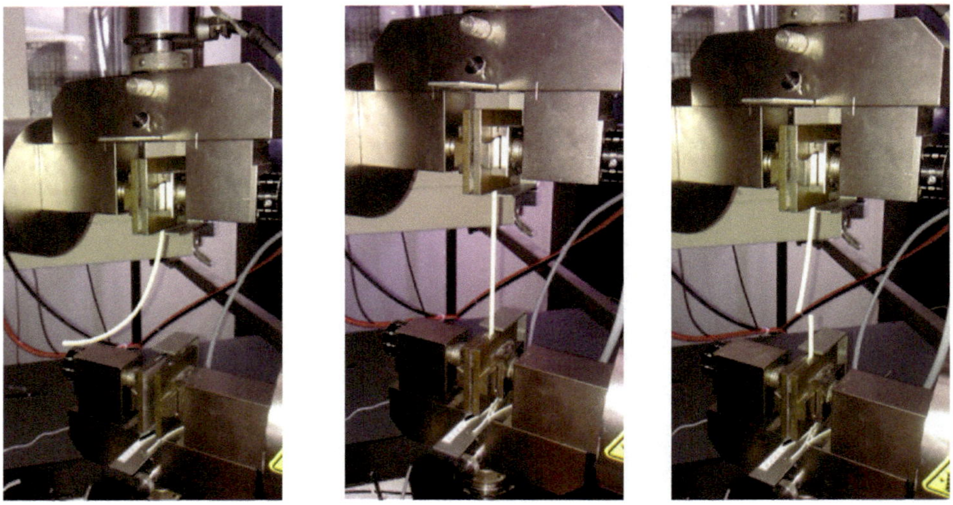

Figure 19.10 Tensile test with filament, image sequence: clamping, test, breakage

In the Figure 19.11(a)–(f), recycled filaments are compared with commercially available new products. The recyclate for the ABS filament comes from profiles from the construction industry, the PET is a bottle product, and the PS is a plant tray. The plant trays are a product that was already made of PS recyclate, so there is multiple recycling.

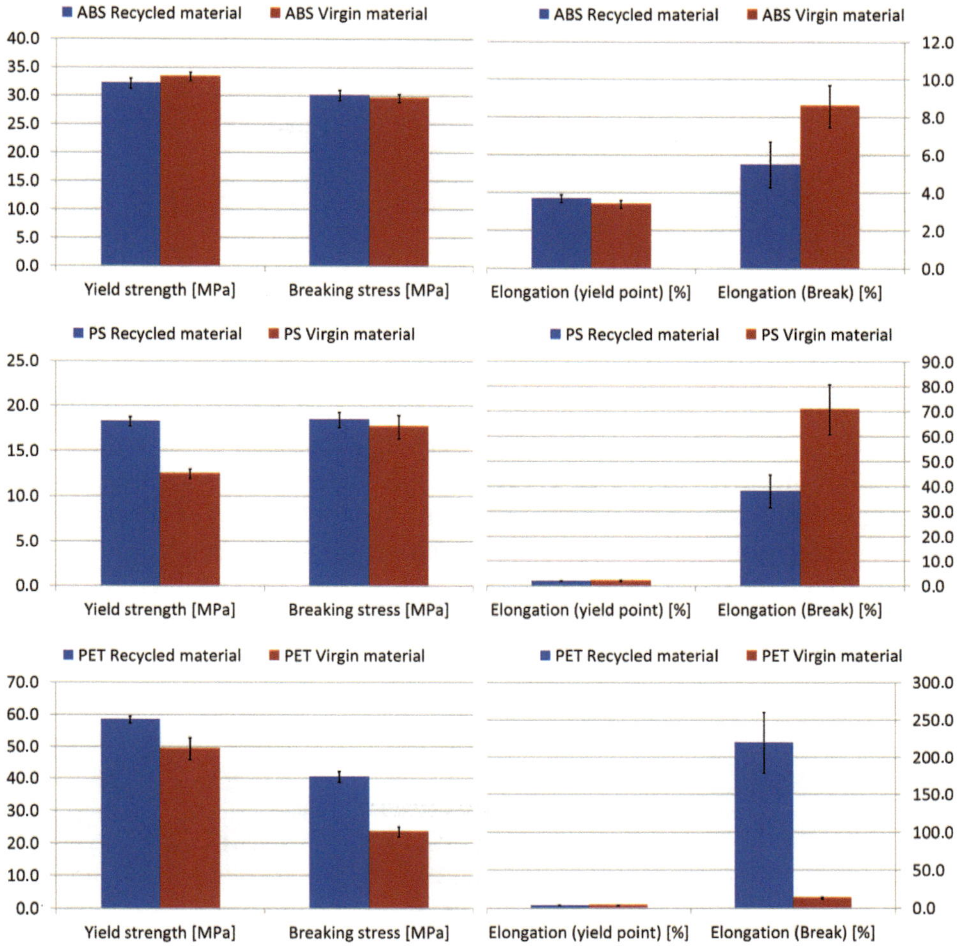

Figure 19.11 Tensile tests with filaments, results in comparison

The recycled filaments made of ABS, PET, and PS have similar tensile strengths and breaking stresses as purchased filaments made of virgin material. In the case of elongation at break, the styrene-containing polymers have higher characteristic values. In the case of the PET types examined, the recycled material has a much higher value. As is usual with 3D printing filaments made of PET, the virgin material is PET-G, a glycol-modified PET with an amorphous structure that solidifies transparently even when cooled slowly. PET-A from bottle production does not contain any additional glycol and must already exhibit extreme stretch stability for blow molding during bottle production. In this case, transparency is achieved by freezing in the amorphous (-A) state through rapid cooling. Printed PET recyclate appears bluish-opaque crystalline (PET-C), see Figure 19.12.

Figure 19.12 PET test bars in the shoulder area, from left to right: PET-G, PET CB5, PET-C

Overall, with regard to the values determined and the qualities achieved, it can be stated that recyclates are fundamentally suitable starting materials for manufacturing high-quality products using the FFF process. The large number of successfully used materials attests to the high degree of robustness of the process. The recycled filaments have been used to print various items such as tools, containers, etc. Figure 19.13 shows an example of a brush holder for a cordless screwdriver. The thermally damaged component was replaced by a printed component made of heat-resistant PEEK. The starting material for the printed filament was sprue material from an injection molding company.

Figure 19.13 Substitution of a PA 6 brush holder in a burnt-out cordless screwdriver by 3D printed PEEK recyclate (brown) with a much higher flash point

The manufacturability of filament and printability with FFF 3D printers (Ultimaker 3 and Tractus P650T) were investigated. The filaments were produced with a laboratory extruder (Brabender compact extruder E 19/25 D). In addition to the warpage of the components, the adhesion to various substrates (print bed) was also taken into account for printability. The reason why some polymers can be printed better than others is due to different material properties. It was recognized that melt viscosity and expansion behavior play a special role. PLA has proven to be the most printable. Therefore, compounds with stone powder and PLA were used to produce samples for measuring the expansion behavior. The expansion behavior was investigated using a dilatometer (Linseis L75 Platinum Series) on printed round samples of a length of 30 mm with a diameter of 5 mm at temperatures of 40 to 150 °C. Above 150 °C, the PLA begins to expand. Above 150 °C the PLA starts to flow and the measurement cannot be continued with a solid dilatometer. The linear expansion of 1.1% could be reduced to 0.9% by adding 30 mass % stone meal. With an increasing content of stone powder, the heat-related length change decreases and thus also the distortion during 3D printing, the so-called "warping". This means that materials with high shrinkage properties can be adapted to 3D printing by compounding them with materials with a low or negative coefficient of expansion (C-fiber).

Currently used plastics in mass production (packaging, construction, automotive, etc.) are not designed for 3D printing. In this context, new combinations of materials, including construction waste, can lead to new products. To illustrate this, experiments were carried out with Poroton bricks containing a filling of mineral wool and PLA as a matrix plastic. In order to incorporate the fillers into the matrix, they were first separated from each other and crushed, see Figure 19.14.

Figure 19.14 Poroton brick filled with mineral wool

In the case of Poroton, further comminution was carried out with the help of an impact crusher and a knife mill was used for the mineral wool. The mineral wool formed agglomerates with diameters <1 cm and a fraction with grain sizes <250 μm was sieved out of the ground Poroton and used further. Analogous to the filament production, see Figure 19.15, filament was drawn, with the difference that in the regranulate production the respective building residue was added with 20 mass % by means of gravimetric dosing. The compounds were then printed into test specimens and tested.

Figure 19.15
Tensile test with printed test specimens made of PLA and mineral wool, time sequence from left to right

First mechanical tests with compounds made of PLA and Poroton or mineral wool led to positive results, see Table 19.3.

Table 19.3 Results of Tensile Tests with 3D Printed Test Specimens

Material	Modulus of Elasticity [MPa]	Yield Strength [MPa]	Elongation (Yield Point) [%]	Breaking Stress [MPa]	Elongation (Break) [%]
PLA	2939.1	44.0	1.6	44.0	1.6
PLA + 20% Poroton	3360.4	44.0	2.0	39.7	3.6
PLA + 20% mineral fiber	4078.6	46.3	1.8	43.1	2.7

While Poroton allows the material higher elongations with the same strength, the mineral wool increases the strength characteristics and allows higher mechanical loads. Poroton and mineral wool can thus be considered active fillers due to their property-improving effect.

References for Chapter 19

[1] DER SPIEGEL No. 38 / 14.09.2019, Page 67, Secondary source: *https://de.statista.com/statistik/daten/studie/1078558/umfrage/verwertung-des-weltweit-produzierten-kunststoffes/*, accessed January 05, 2021

[2] Die Cheops Pyramide (Khufu Pyramide) - Informationen und Zahlenmaterial zur Grossen Pyramide und dem Giza Plateau, *https://www.cheops-pyramide.ch/grosse-pyramide.html*, accessed January 05, 2021

[3] bvse-Bundesverband Sekundärrohstoffe und Entsorgung e. V., Bonn/Berlin/Brussels/Munich: 7 FORDERUNGEN FÜR DIE ZUKUNFT DES KUNSTSTOFF-RECYCLINGS IN DEUTSCHLAND, *https://www.bvse.de/dateien2020/2-PDF/06-Publikationen/04-Broschueren/1706_Forderungskatalog_A4_Kunststoffrecycling_screen.pdf*, accessed January 05, 2021

[4] Brandrup, J. (Ed.): Die Wiederverwertung von Kunststoffen. Hanser 1995

[5] Schmiemann, A., Böger, J.: Recycling flammgeschützter Produktionsabfälle. Kunststoffe, (2020) 11, pp. 60–62

[6] Krüger, F. (Ed.): Entwicklung von Instrumenten und Maßnahmen zur Steigerung des Einsatzes von Sekundärrohstoffen-mit Schwerpunkt Sekundärkunststoffe. UBA, Dessau-Roßlau, 2016

[7] Bahr, R., et. al.: Forschungsprogramm Wiederverwertung von Kunststoffabfällen. Schlussbericht zum Teilprojekt 4: Sortieren von Kunststoff-fabfällen, Institut für Aufbereitung der TU Clausthal, Clausthal, 1980

[8] Uepping, R.: Sensorgestützte Sortiertechnik. In: Recycling und Rohstoffe – Vol. 7, TK-Verlag, Neuruppin 2014

[9] *https://www.kunststoffe.de/fachinformationen/technik-trends/artikel/schwarze-kunststoffesortenrein-trennen-1422484.html*, accessed September 09, 2020

[10] *https://steinertglobal.com/de/magnete-sensorsortierer/sensorsortierung/nir-sortiersysteme/unisort-blackeye/* accessed September 09, 2020

[11] Meyer, J., Fey, D., Krieg, S.: Kunststoff-Sortierung aus Schredderrückständen mit Hochleistungs-Laserspektroskopie. In: Thomé-Kozmiensky, K.-J., Goldmann, D. (Eds.): Recycling und Rohstoffe – Vol. 10, TK-Verlag, Neuruppin, 2017

[12] Pretz, T., Julius, J.: Stand der Technik und Entwicklung bei der berührungslosen Sortierung von Abfällen. *Österreichische Wasser- und Abfallwirtschaft,* (2008) 7–8

Index

S

T